Methods in Cell Biology

VOLUME 66

Apoptosis

Series Editors

Leslie Wilson
Department of Biological Sciences
University of California, Santa Barbara
Santa Barbara, California

Paul Matsudaira
Whitehead Institute for Biomedical Research and
Department of Biology
Massachusetts Institute of Technology
Cambridge, Massachusetts

Methods in Cell Biology

Prepared under the Auspices of the American Society for Cell Biology

VOLUME 66

Apoptosis

Edited by

Lawrence M. Schwartz
Department of Biology
University of Massachusetts
Amherst, Massachusetts

Jonathan D. Ashwell
Laboratory of Immune Cell Biology
National Cancer Institute
National Institutes of Health
Bethesda, Maryland

San Diego San Francisco New York Boston London Sydney Tokyo

This book is printed on acid-free paper. ♾

Academic Press
A Harcourt Science and Technology Company
525 B Street, Suite 1900, San Diego, California 92101-4495, USA
http://www.academicpress.com

Academic Press
Harcourt Place, 32 Jamestown Road, London NW1 7BY, UK
http://www.academicpress.com

International Standard Book Number: 0-12-544165-7 (casebound)
International Standard Book Number: 0-12-632447-6 (paperback)

PRINTED IN THE UNITED STATES OF AMERICA
01 02 03 04 05 06 EB 9 8 7 6 5 4 3 2 1

CONTENTS

CONTRIBUTORS

Numbers in parentheses indicate the pages on which authors' contributions begin.

Myriam Adam (469), Centre d'Immunologie INSERM-CNRS de Marseille-Luminy, 13288 Marseille Cedex 9, France

Manzoor Ahmad (1), Center for Apoptosis Research and the Department of Microbiology and Immunology, Kimmel Cancer Institute, Thomas Jefferson University, Philadelphia, Pennsylvania 19107

Emad S. Alnemri (1), Center for Apoptosis Research and the Department of Microbiology and Immunology, Kimmel Cancer Institute, Thomas Jefferson University, Philadelphia, Pennsylvania 19107

Jonathan D. Ashwell (417), Laboratory of Immune Cell Biology, National Cancer Institute, National Institutes of Health, Bethesda, Maryland 20892

Peter Bangs (321), Cutaneous Biology Research Center, Massachusetts General Hospital, Harvard Medical School, Charlestown, Massachusetts 02129

Elzbieta Bedner (69), Brander Cancer Research Institute, New York Medical College, Hawthorne, New York 10532, and Pomeranian School of Medicine, Szczecin, Poland

Chris A. Benedict (499), La Jolla Institute for Allergy and Immunology, San Diego, California 92121

Lawrence H. Boise (29), Department of Microbiology and Immunology, University of Miami School of Medicine, Miami, Florida 33101

Carl D. Bortner (49), The Laboratory of Signal Transduction, National Institute of Environmental Health Sciences, National Institutes of Health, Research Triangle Park, North Carolina 27709

Claudia Boucher (289), Health and Environment Unit, Laval University Medical Research Center, CHUQ and Faculty of Medicine, Laval University, Quebec, Canada G1V 4G2

Enrique Cepero (29), Department of Microbiology and Immunology, University of Miami School of Medicine, Miami, Florida 33101

Salem Chouaib (417), Cytokines et Immunite Antitumorale, Institut Gustave Roussy, INSERM, 94805 Villejuif Cedex, France

John A. Cidlowski (49), The Laboratory of Signal Transduction, National Institute of Environmental Health Sciences, National Institutes of Health, Research Triangle Park, North Carolina 27709

Silvia Coimbra (437), Instituto de Biologia Molecular e Celular, Universidade do Porto, 4150-180 Porto, Portugal

Amy Cook (187), Department of Cell and Cancer Biology, Medicine Branch, Division of Clinical Sciences, National Cancer Institute, National Institutes of Health, Bethesda, Maryland 20892

Sophie Cornillon (469), Centre d'Immunologie INSERM-CNRS de Marseille-Luminy, 13288 Marseille Cedex 9, France

Zbigniew Darzynkiewicz (69), Brander Cancer Research Institute, New York Medical College, Hawthorne, New York 10532

Rick T. Dobrowsky (135), Department of Pharmacology and Toxicology, University of Kansas, Lawrence, Kansas 66045

William C. Earnshaw (289), Institute of Cell & Molecular Biology, University of Edinburgh, Scotland, United Kingdom EH9 3JR

Stefan van den Eijnde (339), Cardiovascular Research Institute Maastricht, Department of Biochemistry and Molecular Cell Biology and Genetics, University of Maastricht, 6200 MD Maastricht, Netherlands

Angelika Fath (437), Department of Plant and Microbial Biology, University of California, Berkeley, California 94720

Teresa Fernandes-Alnemri (1), Center for Apoptosis Research and the Department of Microbiology and Immunology, Kimmel Cancer Institute, Thomas Jefferson University, Philadelphia, Pennsylvania 19107

Howard O. Fearnhead (167), Apoptosis Section, Regulation of Cell Growth Laboratory, NCI-FCRDC Frederick, Maryland 21702

László Fésüs (111), Department of Biochemistry and Molecular Biology, University of Debrecen Medical and Health Sciences Center, Debrecen, Hungary H-4012

Nathalie Franc (321), Cutaneous Biology Research Center, Massachusetts General Hospital, Harvard Medical School, Charlestown, Massachusetts 02129

Stéphane Gobeil (289), Health and Environment Unit, Laval University Medical Research Center, CHUQ and Faculty of Medicine, Laval University, Quebec, Canada G1V 4G2

Joshua C. Goldstein (365), Division of Cellular Immunology, La Jolla Institute for Allergy and Immunology, La Jolla, California 92121

Pierre Golstein (469), Centre d'Immunologie INSERM-CNRS de Marseille-Luminy, 13288 Marseille Cedex 9, France

Douglas R. Green (365), Division of Cellular Immunology, La Jolla Institute for Allergy and Immunology, La Jolla, California 92121

Lloyd Greene (417), Department of Pathology, Taub Center for Alzheimer's Disease Research, College of Physicians and Surgeons, Columbia University, New York, New York 10003

Yanhui Hu (417), Program in Molecular and Cellular Biology, University of Massachusetts, Amherst, Massachusetts 01003

Bryan W. Johnson (29), Department of Microbiology and Immunology, University of Miami School of Medicine, Miami, Florida 33101

Alan M. Jones (437), Department of Biology, The University of North Carolina at Chapel Hill, Chapel Hill, North Carolina 27599

Ruth M. Kluck (365), Division of Cellular Immunology, La Jolla Institute for Allergy and Immunology, La Jolla, California 92121

Richard N. Kolesnick (135), Laboratory of Signal Transduction, Memorial Sloan-Kettering Cancer Center, New York, New York 10021

Phani Kurada (321), Cutaneous Biology Research Center, Massachusetts General Hospital, Harvard Medical School, Charlestown, Massachusetts 02129

Jean-Pierre Levraud (469), Centre d'Immunologie INSERM-CNRS de Marseille-Luminy, 13288 Marseille Cedex 9, France

Joseph Lewis (187), Department of Cell and Cancer Biology, Medicine Branch, Division of Clinical Sciences, National Cancer Institute, National Institutes of Health, Bethesda, Maryland 20892

Xun Li (69), Brander Cancer Research Institute, New York Medical College, Hawthorne, New York 10532

Simonetta Lisi (321), Cutaneous Biology Research Center, Massachusetts General Hospital, Harvard Medical School, Charlestown, Massachusetts 02129

Zheng-gang Liu (187), Department of Cell and Cancer Biology, Medicine Branch, Division of Clinical Sciences, National Cancer Institute, National Institutes of Health, Bethesda, Maryland 20892

Scott W. Lowe (197), Cold Spring Harbor Laboratory, Cold Spring Harbor, New York 11724

András Mádi (111), Department of Biochemistry and Molecular Biology, University of Debrecen Medical and Health Sciences Center, Debrecen, Hungary H-4012

Lyuben N. Marekov (111), Laboratory of Skin Biology, NIAMS, National Institutes of Health, Bethesda, Maryland 20892

David J. McConkey (229), Department of Cancer Biology, U.T.M.D. Anderson Cancer Center, Houston, Texas 77030

Mila E. McCurrach (197), Cold Spring Harbor Laboratory, Cold Spring Harbor, New York 11724

Donald L. Mykles (247), Department of Biology, Cell and Molecular Biology Program and Molecular, Cellular, and Integration Neurosciences Program, Colorado State University, Fort Collins, Colorado 80523

Stephen Naber (393), Department of Pathology, BayState Medical Center, Springfield, Massachusetts 01199

Zoltán Nemes (111), Departments of Psychiatry and Biochemistry and Molecular Biology, University of Debrecen Medical and Health Sciences Center, Debrecen, Hungary H-4012

Don D. Newmeyer (365), Division of Cellular Immunology, La Jolla Institute for Allergy and Immunology, La Jolla, California 92121

Paula S. Norris (499), La Jolla Institute for Allergy and Immunology, San Diego, California 92121

Leta K. Nutt (229), Department of Cancer Biology, U.T.M.D. Anderson Cancer Center, Houston, Texas 77030

Barbara A. Osborne (417), Program in Molecular and Cellular Biology, and Department of Veterinary and Animal Sciences, University of Massachusetts, Amherst, Massachusetts 01003

Mauro Piacentini (111), Department of Biology, University of Rome "Tor Vergata" and Laboratory of Cell Biology and E. M., IRCCS "L. Spallanzani," Rome, Italy I-00133

Guy G. Poirier (289), Health and Environment Unit, Laval University Medical Research Center, CHUQ and Faculty of Medicine, Laval University, Quebec, Canada G1V 4G2

John C. Reed (453), The Burnham Institute, La Jolla, California 92037

Isabelle A. Rooney (499), La Jolla Institute for Allergy and Immunology, San Diego, California 92121

Ayman Saleh (1), Center for Apoptosis Research and the Department of Microbiology and Immunology, Kimmel Cancer Institute, Thomas Jefferson University, Philadelphia, Pennsylvania 19107

Kumiko Samejima (289), Institute of Cell & Molecular Biology, University of Edinburgh, Scotland, United Kingdom EH9 3JR

Robert A. Schlegel (339), Department of Biochemistry and Molecular Biology, Pennsylvania State University, University Park, Pennsylvania 16802

Lawrence M. Schwartz (393, 417), Department of Biology, and Program in Molecular and Cellular Biology, Morrill Science Center, University of Massachusetts, Amherst, Massachusetts 01003

Mariana Sottomayor (437), Instituto de Biologia Molecular e Celular, Universidade do Porto, 4150-180 Porto, Portugal

Srinivasa M. Srinivasula (1), Center for Apoptosis Research and the Department of Microbiology and Immunology, Kimmel Cancer Institute, Thomas Jefferson University, Philadelphia, Pennsylvania 19107

Ivan Stamenkovic (307), Department of Pathology, Harvard Medical School and Molecular Pathology Unit, Department of Pathology and MGH Cancer Center, Massachusetts General Hospital, Charlestown Navy Yard, Boston, Massachusetts 02129

Peter M. Steinert (111), Laboratory of Skin Biology, NIAMS, National Institutes of Health, Bethesda, Maryland 20892

Howard Thomas (437), Cell Biology Department, Institute of Grassland and Environmental Research, Aberystwyth, Wales SY23 3EB

Christos Valavanis (393, 417), Department of Biology, Morrill Science Center, University of Massachusetts, Amherst, Massachusetts 01003

Tzu-Hao Wang (187), Department of Obstetrics and Gynecology, Chang-Gung Memorial Hospital, Lin-Kou Medical Center, Taoyuan, Taiwan

Carl F. Ware (499), La Jolla Institute for Allergy and Immunology, San Diego, California 92121

Nigel J. Waterhouse (365), Division of Cellular Immunology, La Jolla Institute for Allergy and Immunology, La Jolla, California 92121

Kristin White (321), Cutaneous Biology Research Center, Massachusetts General Hospital, Harvard Medical School, Charlestown, Massachusetts 02129

Patrick Williamson (339), Department of Biology, Amherst College, Amherst, Massachusetts 01002

Yili Yang (417), Laboratory of Immune Cell Biology, National Cancer Institute, National Institutes of Health, Bethesda, Maryland 20892

Hong Zhang (453), The Burnham Institute, La Jolla, California 92037

PREFACE

While the study of cell death has a rich tradition that spans more than 100 years, it has only been during the past decade that this topic has attracted widespread attention. Before this time the field was relatively obscure and was populated primarily by immunologists and developmental biologists, neither of whom would (or could) speak the same language. In fact, in an attempt to create a fundamental understanding, or at least a common set of terms for the field, a noble experiment was attempted in 1990. INSERM sponsored a meeting with the express purpose of sequestering immunologists and neurobiologists together in a monastery in France. While this meeting facilitated the formation of personal friendships between members of the two "teams" (in fact it was at this meeting that the two editors of this volume first met), it did not quite succeed in creating a common foundation for cell death researchers.

One important step in filling this void was the demonstration that almost all cell deaths, independent of cellular lineage, resulted in a stereotypic pattern of morphological changes termed apoptosis (Kerr *et al.,* 1972; Wyllie *et al.,* 1980). Typically, apoptotic cells display shrinkage, membrane blebbing, genomic DNA fragmentation, and the deposition of electron-dense chromatin along the inner aspect of the nuclear envelope. Some or all of these morphological changes can be detected in dying cells from such diverse organisms as nematodes and humans. In fact, components of apoptotic and nonapoptotic cell death morphology have been observed in some prokaryotic and lower eukaryotic organisms (Cornillon *et al.,* 1994; Ameisen, 1996).

The next significant event in the field was the demonstration by Robert Horvitz and students that apoptosis is not a random process, but instead is regulated by a phylogenetically conserved genetic cascade (reviewed in Horvitz, 1999). Subsequent studies by other investigators have demonstrated that misregulation of this genetic machinery serves as the basis for a variety of clinical disorders, including cancer, autoimmunity, and neurodegeneration.

In 1995, Academic Press published Volume 46 of the "Methods in Cell Biology" series, entitled *Cell Death.* The purpose of this volume was to provide the cell biology community with detailed methodology that would allow new investigators to study cell death. The 17 chapters covered a wide range of topics (electron microscopy to differential gene expression) and model systems (nematodes to mice). This volume served its purpose well and proved to be one of the most popular volumes in the "Methods in Cell Biology" series.

While the earlier volume retains its value as a source of protocols and models, progress in the field has rendered it incomplete. Some of the most intensively investigated topics in the field of cell death had not yet been discovered at the time of its publishing, such as the roles of cytochrome *c* and caspases. Other topics were in their infancy, such as our understanding of the Bcl-2 family of apoptosis-modulating proteins and the use of cell-free systems. The current volume has 22 chapters that cover a wide range of molecules, models, and methods for the study of cell death. To distinguish this volume from its predecessor in the "Methods in Cell Biology" series, we have named it *Apoptosis. Cell Death* and *Apoptosis* are intended to be complimentary volumes. They share virtually no overlap and both are intended to be valuable resources for both new investigators to the field and long standing deathophiles.

This is a very exciting time in the field of cell death. Ten years ago we were asking such basic questions as "is the nucleus essential for apoptosis" or "are there such things as "deathases." Now the field has evolved to the stage where we are determining where each of the myriad of caspases acts in the proteolytic cascade that results in specific substrate cleavage during apoptosis. Soon we will be able to manipulate individual components of the apoptotic machinery to treat or prevent major illnesses. It is hoped that this volume will play some small role in facilitating these advances.

Lawrence M. Schwartz and Jonathan D. Ashwell

References

Ameisen, J. C. (1996). The origin of programmed cell death. *Science* **272,** 1278–1279.

Cornillon, S., Foa, C., Davoust, J., Buonavista, N., Gross, J. D., and Golstein, P. (1994). Programmed cell death in *Dictyostelium. J. Cell Sci.* **107,** 2691–2704.

Horvitz, H. R. (1999). Genetic control of programmed cell death in the nematode. *Caenorhabditis elegans. Cancer Res.* **59,** 1701s–1706s.

Kerr, J. F. R., Wyllie, A. H., and Currie, A. R. (1972). Apoptosis: A basic biological phenomenon with wide ranging implications in tissue kinetics. *Br. J. Cancer* **26,** 239–257.

Wyllie, A. H., Kerr, J. F. R., and Currie, A. R. (1980). Cell death: The significance of apoptosis. *Int. Rev. Cytol.* **68,** 251–306.

CHAPTER 1

Isolation and Assay of Caspases

Srinivasa M. Srinivasula, Ayman Saleh, Manzoor Ahmad, Teresa Fernandes-Alnemri, and Emad S. Alnemri

Center for Apoptosis Research and the Department of Microbiology and Immunology
Kimmel Cancer Institute, Thomas Jefferson University
Philadelphia, Pennsylvania 19107

I. Introduction

Significant progress toward understanding the molecular control of apoptosis has been made in recent years, due largely to the discovery of a family of cysteine proteases named caspases (Alnemri, 1997; Cohen, 1997; Los *et al.*, 1999; Salvesen and Dixit, 1997; Thornberry and Lazebnik, 1998; Wolf and Green, 1999). Caspases are so named for their characteristic ability to cleave their substrates and their proenzymes at defined aspartic acid residues. Caspase designation for 11 human members of this family, their structural organization, deduced biological function, and phylogenic relationship are depicted in Fig. 1 (see Color Plate). While the intracellular targets of caspases and the importance

METHODS IN CELL BIOLOGY, VOL. 66

0091-679X/01 $35.00

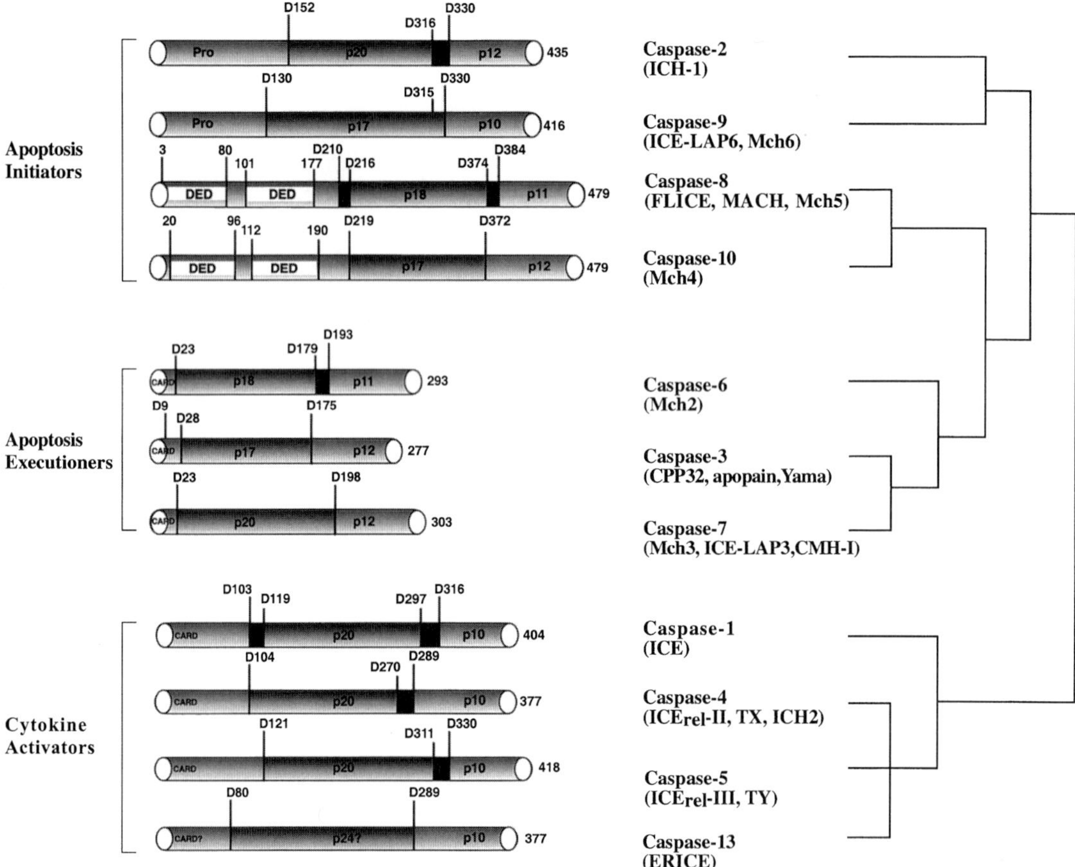

Fig. 1 Structure and organization of human caspases, as well as their functional and phylogenic relationships. Prodomains and large and small subunits are each shown in different colors with markings to indicate the position of the cleavage sites. The small interdomain link region between subunits is also marked. (Left) Deduced biological functions of these caspases. (Right) The phylogenic relationship among caspases. (See Color Plate.)

of caspase cascade in immune functions, development, and cell death have been well recognized, the physiological function(s) of individual caspases and the regulation of their activation are not fully studied. The following sections provide a brief overview of the structure and organization and the mechanism of activation of caspases.

A. Structure and Organization of Caspases

The human caspase family consists of 11 members and is divided into three subfamilies based on phylogenetic analysis (Alnemri *et al.,* 1996) (Fig. 1). The percentage of

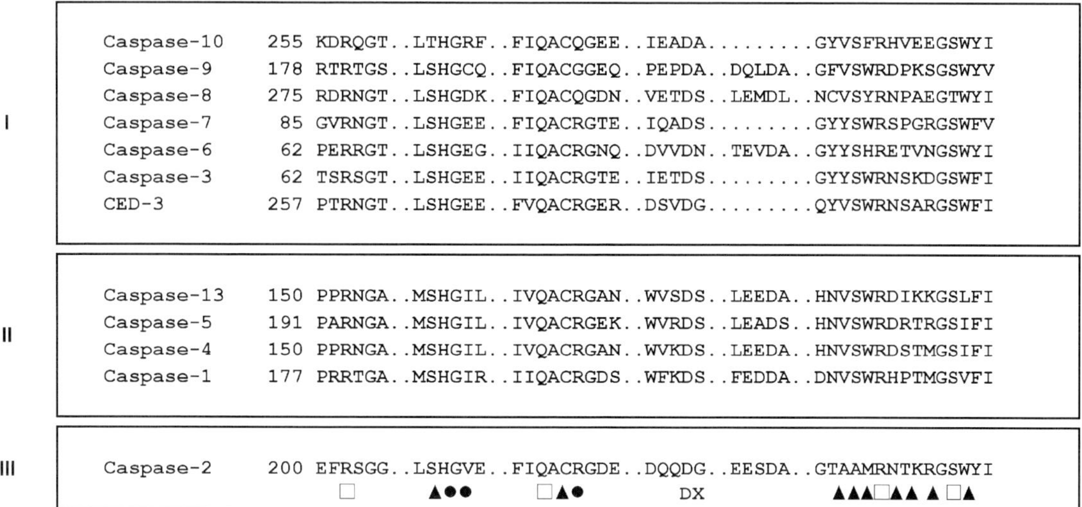

Fig. 2 Multiple sequence alignment of all known human caspases and nematode Ced-3. Based on crystal structure analysis of caspase 1 (ICE), the symbols below the sequences indicate residues that are involved in catalysis (●), binding the substrate-carboxylate of P1 aspartate (□), or adjacent to the substrate P2–P4 amino acids (▲). D/X indicates known and potential processing sites between small and large subunits. The roman numbers on the left indicate the three caspase subfamilies: the Ced-like subfamily (I), the ICE-like subfamily (II), and the Nedd2/ICH-1 subfamily (III).

sequence identity among these caspases ranges between 20 and 77%. X-ray crystallography and sequence analysis suggest that all caspases share a common structure (Wolf and Green, 1999). They are synthesized as inactive zymogens, with molecular sizes ranging from 30 to 55 kDa. Each zymogen contains an N-terminal prodomain, followed by a large subunit and a C-terminal small subunit. Under normal conditions these proteins exist as inactive, single polypeptide chains known as procaspases. Activation requires cleavage at specific aspartic acid residues to generate the large and small subunits and subsequent conformational rearrangement. As shown in Fig. 2, caspases-3 and -7 have a single cleavage site between the two subunits, whereas caspases-1, -2, -4, -5, -6, -8, and -10 contain two cleavage sites. The physiological significance of the second cleavage site is not clear and may be used as a means of regulation in some cell types. Activation accompanies proteolysis of this interdomain linker. In the case of procaspase-9, however, proteolytic cleavage between the subunits is not essential for its activity, and the uncleaved molecule is capable of forming an active holoenzyme complex with Apaf-1 and cytochrome *c* (Rodriguez and Lazebnik, 1999; Stennicke *et al.*, 1999).

Within the large subunit (17–21 kDa) of all caspases is a conserved pentapeptide motif "QACXG" containing the active site cysteine residue, and within the small subunit (9–12 kDa) are several conserved residues that determine substrate binding and specificity

(Fig. 2). The active enzyme is a tetramer consisting of two large and two small subunits. It has been shown that subunits of closely related caspases, such as caspases-3 and -7, can form functional intermolecular heterocomplexes, suggesting perhaps some degree of redundancy (Fernandes-Alnemri *et al.,* 1995).

Based on functional analysis of caspases, they are broadly classified into three groups, namely cytokine activators, upstream apoptosis initiator caspases, and downstream apoptosis executioner caspases as outlined in Fig. 1.

B. Mechanism of Caspase Activation

Caspases exist as latent zymogens in normal cells with extremely low intrinsic protease activity (Muzio *et al.,* 1998; Yamin *et al.,* 1996). An important distinction between caspases is the size of their prodomains. Initiator caspases such as caspases-8, -9, and -10 possess relatively large prodomains compared to the downstream executioner caspases such as caspases-3, -6, and -7. Until quite recently, the significance of this distinct difference in prodomain length was not clear. It is now known that initiator procaspases require their prodomain for recruitment to the death complex or apoptosome via interaction between the prodomain, also known as caspase recruitment domain (CARD), and the death complex. This is a crucial step and is believed to cluster or sequester the proenzyme, enabling its autoactivation (Ahmad *et al.,* 1997; Ashkenazi and Dixit, 1998; Boldin *et al.,* 1996; Duan and Dixit, 1997; Muzio *et al.,* 1996). Such clustering/forced oligomerization of caspases in mammalian cells is believed to increase their local enzyme concentration, facilitating autoactivation and apoptosis (Alnemri *et al.,* 1992; Martin *et al.,* 1998; Muzio *et al.,* 1998; Srinivasula *et al.,* 1998; Yang *et al.,* 1998). Autoactivation of procaspase-9, for example, is accomplished by its recruitment and subsequent oligomerization by the Apaf-1–cytochrome *c* oligomeric complex (Saleh *et al.,* 1999; Srinivasula *et al.,* 1998; Zou *et al.,* 1999). Interestingly, overexpression of most recombinant procaspases in *Escherichia coli* cells induces autoactivation and is believed to be the result of overexpression-induced aggregation of the recombinant proteins.

In caspases with short prodomains (caspases-3, -6, and -7), no significant autoactivation is observed in mammalian cells. These caspases are known as downstream caspases (executioner caspases) as they depend on active initiator caspases for their activation by proteolytic cleavage. Once cleaved, these downstream caspases can cleave several cellular substrates and activate both upstream and downstream procaspases, thus initiating both an amplification cascade and positive feedback. For example, active caspases-3 and -9 can efficiently cleave each other's proenzyme *in vitro,* and also other procaspases, such as procaspases-6 and -7 (Srinivasula *et al.,* 1996, 1998). *In vivo,* however, such amplification of the caspase cascade depends on both relative concentrations of each caspase and the efficiency of transactivation reaction under physiological conditions.

Another mechanism of caspase activation is through aspartate-specific serine proteases. The best known example is granzyme B, found in cytolytic granules of activated cytotoxic T lymphocytes and natural killer cells. Granzyme B is a potent inducer of apoptosis. It efficiently activates several caspases, including procaspase-3 and procaspase-7, by cleaving their proenzymes at critical Asp residues in the interdomain linker loop between their large and small subunits (Darmon *et al.*, 1995; Shresta *et al.*, 1995; Simon *et al.*, 1997; Stennicke *et al.*, 1998; Zhou and Salvesen, 1997). The interdomain linker loop between large and small subunits of caspases is also susceptible to proteolysis by a number of serine proteases, which have no apparent specificity for the conserved aspartate residue favored by cysteine proteases. These include cathepsin G and subtilisin Carlsberg (Zhou and Salvesen, 1997). Several other proteases implicated in apoptosis, such as the proteasome, apoptotic serine proteinase p24, cathepsin D, and calpain have been described; however, their role in the activation of procaspases is unknown (Deiss *et al.*, 1996; Hirsch *et al.*, 1998; Squier *et al.*, 1994; Wright *et al.*, 1994).

Understanding the regulation of caspases and their substrate specificities is important because of their potentially valuable therapeutic applications. Dysregulated apoptosis is observed in many human degenerative diseases, autoimmune disorders, and several forms of cancer. Direct activation of caspases in cancer cells may be an effective strategy to kill cancer cells, whereas inhibition of caspases could prevent the excessive cell death that characterizes cerebral ischemia and degenerative diseases such as Alzheimer's. During the past decade, a tremendous amount of evidence has emerged demonstrating a prominent role for caspases. It almost seems like caspases have a niche in every area of basic science and therapeutic research on cell survival and disease. Obviously, a prerequisite to understanding caspase functioning and defining their catalytic activities is the ability to generate highly purified and functionally optimal caspases. To this end, we provide detailed protocols that we have standardized for the isolation and assay of recombinant and cellular caspases *in vitro*. Most of the protocols presented in this chapter are those used routinely in our laboratory. Select protocols published from other laboratories are included and cited.

II. Materials Required

In addition to the basic equipment in molecular biology laboratories, the following equipment is essential for the protocols described in this chapter.

Luminescence spectrometer LS50B (Perkin-Elmer) or an equivalent spectrometer
Branson Sonifier 450 with microtip attachment
FPLC system (Amersham-Pharmacia)
Refrigerated high-speed centrifuge
Ultracentrifuge (Beckman)

Buffers

Lysis buffer (pH 7.5): 20 m*M* Hepes, 0.1% CHAPS, 0.1 m*M* phenylmethylsulfonyl fluoride (PMSF), and other protease inhibitors (optional) (Roche Molecular Biochemicals).

Denaturing buffer (pH 7.5): 25 m*M* Tris, 5 m*M* EDTA, 100 m*M* dithiothreitol (DTT), and 6.5 *M* guanidine–HCl

Cell extraction buffer (pH 8.0): 10 m*M* Hepes, 150 m*M* NaCl, 500 m*M* sucrose, 1 m*M* EDTA, and 1% NP-40

Chromatography buffer (pH 7.5): 100 m*M* Hepes, 10% sucrose, 4 m*M* DTT, 0.1% CHAPS (for caspases-3, -7, -8, and -9), or 0.1% Triton X-100 (for caspases-4 and -5)

Refolding buffer 1 (for caspases-1, -4, and -5): 100 m*M* Hepes (pH 7.5), 10% sucrose, 1% Triton X-100, 10 m*M* DTT, and 10 μM Ac-YVAD-CHO

Refolding buffer 2 (for caspase-2): 100 m*M* Hepes (pH 7.5), 20% sucrose, and 10 m*M* DTT

Refolding buffer 3 (for caspase-3): 100 m*M* Hepes (pH 8.0), 10% sucrose, 10 m*M* DTT, 0.1% CHAPS, and 150 m*M* NaCl

Refolding buffer 4 (for caspases-7, -8, and -9): 100 m*M* Hepes (pH 7.5), 10% sucrose, 10 m*M* DTT, and 0.1% CHAPS

Buffer A (pH 7.5): 25 m*M* Hepes, 10 m*M* KCl, 1.5 m*M* $MgCl_2$, 1 m*M* EDTA, 1 m*M* EGTA, 1 m*M* DTT, and protease inhibitors

Buffer B (pH 7.5): 25 m*M* Hepes, 300 m*M* NaCl, 10 m*M* KCl, 1.5 m*M* $MgCl_2$, 5% glycerol, 10 m*M* imidazole (Sigma), and protease inhibitors

Buffer C (pH 7.5): 25 m*M* Hepes, 20 m*M* NaCl, 5 m*M* KCl, 2 m*M* $MgCl_2$, 5% glycerol, and protease inhibitors

Buffer D (pH 7.5): 25 m*M* Hepes, 50 m*M* NaCl, 10 m*M* KCl, 1.5 m*M* $MgCl_2$, 5% glycerol, bovine serum albumin (BSA, 200 μg/ml) (Sigma), 1 m*M* DTT, and protease inhibitors

Chromatography columns (Amerham-Pharmacia): Hi-Trap SP column, Hi-Trap Q column, Mono Q column, Superose 12 column, and

Talon metal affinity resin (cobalt-based) (ClonTech)

Cloning vectors: pET vectors (Novagen) and baculovirus recombinant expression system

Host strains required: DH5α (for plasmid storage and propagation) and BL21 (DE3) or BL21 (DE3)pLysS (Novagen) (for induction of proteins) *Spodoptera frugiperda* Sf 9 insect cells

III. Methods

A. Cloning of Caspases

Caspase cDNAs and their expression constructs can be obtained from the original laboratories that cloned them. Otherwise, full-length cDNA can be obtained using caspase-specific polymerase chain reaction (PCR) primers for reverse transcriptase PCR (RT-PCR) or for PCR screening of cDNA libraries using standard PCR amplification techniques. All sequences, especially those obtained by PCR, must be confirmed by sequence analysis. Furthermore, the open reading frame must be checked for common PCR-related errors that could result in frameshifts or base pair substitutions that could have serious effects on the expression and activity of the protein. [We recommend the use of recombinant Pfu polymerase (Stratagene) for PCR amplifications.] Any modifications to the cDNA, such as in-frame tags or base pair substitutions, must also be confirmed by sequencing.

B. Expression of Recombinant Caspases

Bacterial overexpression is the preferred method for the large-scale production of recombinant caspases. It is an easy, fast, and relatively inexpensive method for obtaining pure and functionally active recombinant caspases. Unlike the baculovirus insect expression system, bacteria have no endogenous caspase-like activity. Also, caspases do not require posttranslational modifications for their functional activity. Posttranslational modifications such as phosphorylation and nitrosylation of certain caspases have been reported to occur *in vivo* in mammalian cells but are believed to be associated with the regulation of caspase activation rather than activity (Martins *et al.,* 1998).

1. Selection of Vectors for Caspase Expression

From among a number of available bacterial expression systems we have had the most success using the pET expression vector system from Novagen for caspase expression. For purification of active caspases, a C-terminal (His) 6 tag is recommended, as many caspases have cleavage sites within their N-terminal region, resulting in its removal on overexpression. Most of the available glutathione *S*-transferase (GST) fusion vectors are therefore not suitable for caspase purification.

2. Cloning of Caspases in pET Vectors

Using standard PCR subcloning techniques, amplify the caspase open reading frame with appropriate primers [an example using caspase-3 in pET21(b) is described later]. PCR primers should be designed with overhanging restriction enzyme sites that are com patible with the cloning site(s) of the vector. Purify the PCR products using a QIAquick PCR purification kit (Qiagen) and digest with the appropriate restriction enzyme(s). Allow at least 4–6 h for complete restriction enzyme digestion. However, incubation times will vary depending on the enzyme used and the length of the PCR primer 5′ to the restriction enzyme site. (An additional four to six nucleotides 5′ to the enzyme recognition site are generally sufficient.)

Run the digested sample on a 1% low melting point agarose gel containing ethidium bromide. Excise the DNA band and purify using a Qiagen gel extraction kit. Meanwhile, digest the pET vector with the selected restriction enzymes for 4–6 h, followed by dephosphorylation (to reduce self-ligation of vector). Clone the digested product into the pET vector in-frame with the N-terminal and C-terminal tags. It is essential to maintain the reading frame because any shift in reading frame with N-terminal T7 tag will not give the desired protein, whereas a shift in the C-terminal (His)6 tag will result in the loss of the tag and in an inability to purify the recombinant protein. As discussed earlier, although overexpression of certain caspases may result in cleavage of the N-terminal T7 tag, it is not necessary to remove the tags before cloning, as these sequences provide an optimal initiation of the protein expression. In some experiments the T7 tag can also be used to probe the translational products using the T7-HRP antibody (Novagen).

Example: Cloning of caspase-3 in pET 21(b):

a. Digest 10 μg pET21(b) with *Bam*HI and *Xho*I for 6 h at 37°C in a 50-μl reaction volume.

b. Incubate at 65°C for 10 min to inactivate restriction enzymes.

c. Add 1 unit alkaline phosphatase (AP) + 5.6 μl AP buffer; incubate at 37°C for 30 min.

d. Fractionate on 1% ethidium bromide-stained agarose gel; excise the 5.4-kb vector DNA band visualized under ultraviolet light and purify using a Qiagen gel extraction kit. Elute sample in 50 μl water.

e. Design PCR primers:

Caspase-3 5′ *Bam*HI start primer: (30-mer)
5′-CTG-GCG-GAT-CCG-ATG-GAG-AAC-ACT-GAA-AAC-3′

Caspase-3 3′ *Xho*I end primer: (30-mer)
5′-CCG-GAC-CTC-GAG-GTG-ATA-AAA-ATA-GAG-TCC-3′

f. Set up a 200-μl PCR reaction as follows: 30 ng caspase-3 cDNA, 16 μl dNTP mix (2.5 m*M* each), 20 μl 10 × PCR reaction buffer, 100 ng caspase-3 5′ *Bam*HI start primer, 100 ng caspase-3 3′ *Xho*I end primer, and 2.5 μl Pfu DNA polymerase (Stratagene) to 200 μl with PCR grade water.

g. Vortex gently to mix, spin briefly, and transfer 100 μl each to two 0.2-ml thin-walled PCR tubes. A GenAmp 9600 or equivalent thermocycler is recommended. Perform a 30-cycle reaction (95°C/45 s, 57°C/45 s, 75°C/2min) each cycle.

h. Purify the PCR product using a Qiagen PCR purification kit or a standard phenol–chloroform extraction method and resuspend the DNA with water to a final volume of 40 μl.

i. Digest with *Bam*HI and *Xho*I restriction enzymes (10 U of each enzyme) for 6–8 h at 37°C in a 50-μl final reaction volume.

j. Fractionate on a 1% agarose gel as described earlier. Excise the 0.9-kb insert DNA band and purify.

k. Standard procedures for ligation of the vector and insert and subsequent transformation into DH5α should be used.

l. Transformants selected on LB-ampicillin plates can be screened by standard PCR-screening techniques. Grow a positive transformant in 10 ml LB-ampicillin overnight, isolate plasmid DNA, and sequence using the T7 primer to confirm authenticity. Glycerol stocks of positive transformants must be stored at −80°C.

3. Induction of Caspases

a. Transform the expression host BL21 (DE3) or BL21 (DE3)pLysS (Novagen) with the pET expression vector containing the desired caspase cDNA using standard heat-shock transformation methods.

b. Grow the transformants on either ampicillin or kanamycin selection plates, depending on the type of plasmid used (detailed description of pET vectors is provided in the Novagen catalog).

c. Select a single well-isolated colony from the plate and transfer to 10 ml LB medium plus antibiotic and grow overnight at 37°C with vigorous shaking.

d. The next day, innoculate 1 liter LB medium (with appropriate antibiotic) with the overnight culture so that the initial OD_{600} is nearly 0.1 and grow at 37°C with vigorous shaking until the culture reaches an OD_{600} of approximately 0.5–0.8.

e. Add IPTG to a final concentration of 1 m*M* and reduce the incubation temperature to 25°C. (IPTG induces the expression of the recombinant protein, and lowering of induction temperature allows for a more soluble protein.) Shake the culture vigorously for 3 h.

f. Harvest the bacteria by centrifugation at 6000*g* for 15 min at 4°C. Discard the supernatant and wash the cell pellet twice thoroughly with cold phosphate-buffered saline, pH 7.6 (PBS). (At this stage the cell pellet can be stored at −80°C.)

C. Purification of Caspases

For assay and storage of caspases we use the ICE assay buffer described initially for caspase-1 by Thornberry *et al.* (1992) with some modifications (see Tables I and II). This buffer is generally good for most of the caspases. However, for optimal buffer conditions with individual caspases-3, -6, -7, and -8, researchers may also refer to other publications (Garcia-Calvo *et al.,* 1999; Stennicke and Salvesen, 1997). All purification steps must be performed at 4°C unless otherwise stated.

Table I
Optimal Caspase Cleavage Sequences and Buffers

Caspase	Optimal sequence	Optimal buffer
1	WEHD	1
2	DEHD	2
3	DEVD	3
4	WEHD	3
5	WEHD	4
6	VEHD	1
7	DEVD	5
8	LETD	3
9	LEHD	2
10	LEXD	3

Table II
Composition of Optimal Buffers for Different Caspases

Buffer	Buffer 1	Buffer 2	Buffer 3	Buffer 4	Buffer 5
0.1 *M* Hepes pH	7.5	–	7.0	7.5	7.0
0.1 *M* MES pH	–	6.5	–	–	–
10% sucrose	+	–	–	+	–
10% PEG	–	+	+	–	+
0.1% CHAPS	+	+	+	+	+
10 m*M* DTT	+	+	+	+	+
0.2 *M* NaCl	–	–	–	+	–
5 m*M* $CaCl_2$	–	–	–	–	+

1. For Soluble Proteins

a. Resuspend the cell pellet (from 1 liter induced culture) thoroughly in 15–20 ml of ice-cold lysis buffer (Section II).

b. Sonicate the cell suspension, on ice, using a Branson sonifier 450 with microtip attachment. Sonication of cell suspension in small volumes of 1 ml in 1.5-ml microfuge tubes gives better lysis. (Settings recommended for 1-ml volumes using a microtip with the machine just described are 2 × 20 strokes at output control 4 and duty cycle 20.) Mix the cell suspension intermittently during sonication by gentle tapping.

c. Centrifuge the cell homogenates at 14,000*g* for 15 min at 4°C. Transfer the supernatants to fresh prechilled tubes on ice. The insoluble fraction (pellet) may be saved for inclusion body isolation and refolding, although usually it is not necessary for native caspase purification. If the supernatant is not clear, repeat the centrifugation step in order to obtain a clear lysate. For cultures of 1 liter or less, the removal of nucleic acids from soluble cell extracts prior to purification is not necessary.

d. Use a batch/gravity-flow column purification method using Talon metal affinity resin from Clontech for purification of (His)6-tagged caspases. Refer to Clontech user manual PT1320-1 for detailed information about this resin. Transfer 1-ml of thoroughly suspended Talon resin to a sterile 25- to 30-ml screw cap tube. Centrifuge at 700*g* for 2–3 min to pellet the resin. Aspirate and discard the supernatant, and add 5 ml of lysis buffer (Section II) to equilibrate the resin. Centrifuge at 700*g* for 2 min to pellet the resin. Discard the supernatant.

e. Add the cleared cell lysates to the resin and rotate the tube gently for an hour on a rotator at 4°C.

f. Centrifuge bound resin at 700*g* for 2 min at 4°C. Carefully aspirate the supernatant without loosening the pelleted resin. Wash the bound resin twice with 10 ml of lysis buffer, pH 8.0.

g. Resuspend the washed resin in a small volume (1 ml) of lysis buffer, and transfer the suspension to a clean 1.5 × 12-cm gravity-flow column (Bio-Rad) with its bottom cap in place. Allow the resin to settle out of suspension. Snap off the bottom cap and

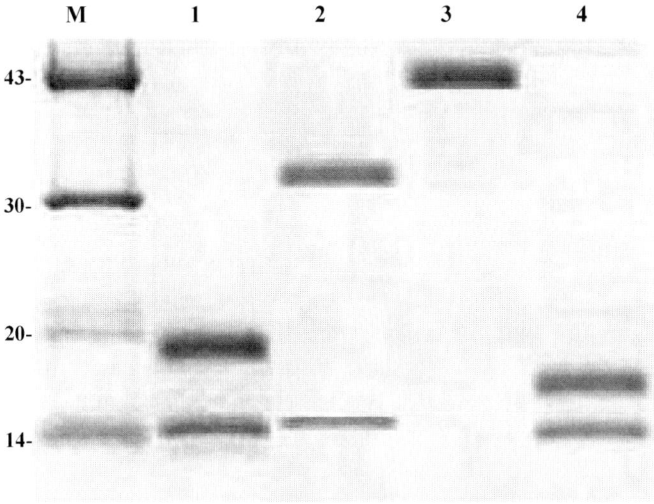

Fig. 3 Processing of procaspases. Procaspases cloned in pET vectors were expressed in *E. coli* and purified to homogeneity as described in the text. The processed caspases were analyzed by 10% SDS–PAGE and Coomassie staining. Molecular weight marker (lane M), purified truncated (amino acids 213–496) wt-caspase-8 (lane 1), purified wt-caspase-9 (lane 2), purified mutated (C287A) caspase-9 (lane 3), and purified truncated (amino acids 194–479) wt-caspase-10 (lane 4), respectively.

allow the buffer to drain (care should be taken to make sure no air bubbles are trapped in the resin).

h. Wash the resin three times with 5 ml of lysis buffer, pH 8.0.

i. Wash once with lysis buffer, pH 7.5, plus 10 m*M* imidazole.

j. Elute the protein with 4 ml of lysis buffer with pH 8.0 plus 100 m*M* imidazole.

k. Collect the elution buffer in 500-μl fractions by gravity flow.

l. Check each eluted fraction by SDS–PAGE and Coomassie blue staining. (Run 20 μl of each fraction on a 12.5% polyacrylamide gel.)

m. Figure 3 shows the processed pure caspases-8, -9, and -10 and also the unprocessed caspase-9 C287A mutants. Most of the caspases show similar patterns on the acrylamide gel.

n. Remove imidazole by dialyzing the fractions against the corresponding ICE buffer overnight at 4°C. If necessary, concentrate the fractions using a Millipore Centricon 30 filter. In this case imidazole can be removed by diluting the concentrated fractions two to three times with the corresponding caspase buffer without sucrose, followed by reconcentrating. Repeating this step should reduce the concentration of imidazole in the sample to less than 1 m*M*.

o. Assay the caspase activity using a peptide substrate (described in Section III,F,1) and store as aliquots at −80°C.

Note: For elution of bound proteins (step j), 100 m*M* EDTA can be used instead of 100 m*M* imidazole, which will basically elute all proteins containing six or more histidine residues. However, this will also strip the metal cobalt off the resin. Cobalt, however, does not inhibit the activity of caspases significantly.

Do not use buffers containing reducing agents such as DTE or DTT or chelators such as EDTA or EGTA in any of the steps prior to elution.

D. Purification of Caspases from Inclusion Bodies and Refolding

In general, this procedure is not necessary for caspase purification as all the caspases expressed in bacteria produce sufficient amounts of soluble, active protein. In case of procaspases such as procaspases-8 and -10 with long prodomains, removal of the prodomain facilitates solubility without apparent change in its *in vitro* activity (Srinivasula *et al.*, 1996). However, insoluble proteins from inclusion bodies can be purified under denaturing conditions.

1. Method 1

a. Lyse the induced bacterial cells by sonication on ice as described earlier. Centrifuge the cell extract at 10,000*g* for 15 min.

b. Discard the supernatant and resuspend each pellet in 1 ml of cold lysis buffer plus 5 m*M* calcium chloride.

c. Add 1 μl of Nuclease-S7 (1 μg/μl, Roche) to each tube, mix well, and incubate at 37°C for 30 min.

d. Centrifuge at 10,000*g* for 15 min. Aspirate the supernatant and resuspend each pellet in 1 ml of lysis buffer.

e. Centrifuge at 10,000*g* for 15 min at room temperature. Carefully aspirate all the supernatant.

f. Resuspend the pellet in lysis buffer plus 6 *M* guanidine–HCl. Incubate the suspension at 37°C for 30 min.

g. Centrifuge at 10,000*g* for 15 min at room temperature.

h. Carefully transfer the clear supernatant to the buffer-equilibrated Talon affinity resin (see Section III,C,d and e). Rotate the tube gently for an hour at room temperature.

i. Centrifuge at 700*g* for 2 min.

j. Carefully aspirate the supernatant without disturbing the resin.

k. Wash the resin twice with 10 ml of lysis buffer, pH 8.0, containing 6 *M* guanidine–HCl.

l. Suspend the resin in a small volume (~1 ml) of lysis buffer and transfer the suspension completely to a 2- to 5-ml column with the bottom cap in place. Allow the resin to settle out of suspension (3–5 min) before removing bottom cap.

m. Wash the resin three times with 5 ml of lysis buffer, pH 8.0, containing 6 *M* guanidine–HCl.

n. Wash once with lysis buffer, pH 7.5, containing 6 *M* guanidine–HCl plus 10 m*M* imidazole.

o. Elute the bound protein with 4 ml of lysis buffer, pH 8.0, containing 6 *M* guanidine–HCl plus 100 m*M* imidazole. Collect the eluant in 500-μl fractions.

p. Several methods are available for refolding of insoluble proteins (Kurucz *et al.*, 1995; Mukhopadhyay, 1997; Rudolph and Lilie, 1996). Otherwise, follow Method 2, steps j–q.

2. Method 2

Recombinant caspases can also be produced by methods involving the folding of active enzymes from their constituent subunits that are expressed separately in *E. coli,* followed by ion-exchange chromatography. The following method was described by Garcia-Calvo *et al.* (1999) for the purification of recombinant caspases.

a. Subclone the large and small caspase subunits independently in an expression vector (pET28/Novagen) in-frame with the tag sequence of the vector.

b. Transform *E. coli* BL21 (DE3)pLysS cells with the respective subunit-expression plasmids. Grow single transformants in selection media overnight.

c. Induce exponentially growing cells ($OD_{600} \sim 0.5$) in M9 media containing 1 m*M* IPTG 37°C overnight. (Under these conditions, most of the expressed subunit protein is precipitated in the insoluble fraction.)

d. Collect the cells by centrifugation, and wash the pellets with cold PBS.

e. Resuspend the cells in cold lysis buffer and lyse the cells using a French pressure cell at 15,000 psi.

f. Pellet the inclusion bodies by centrifugation at 27,000*g* for 15 min.

g. Wash the pellets four times by alternating between the 25 m*M* Hepes lysis buffer containing 1% CHAPS, 1 m*M* EDTA, and a 25 m*M* Hepes (pH 7.5) plus 1 *M* urea buffer. Pellets must be resuspended in wash buffer each time.

h. Resuspend the inclusion bodies in a small volume of denaturing buffer to solubilize the proteins, on a rotator, for 1 h at 4°C.

i. Centrifuge the samples at 27,000*g* for 15 min. Supernatants now contain soluble, denatured caspase subunit. (At this stage, samples can be stored at −80°C.)

j. To enable refolding of subunits into active caspases, first mix the two subunits of each caspase (from step i) in refolding buffer to a final concentration of 100 μg/ml each subunit and incubate the reaction at room temperature overnight. The refolding buffers for various caspases used by Garcia-Calvo *et al.* (1999) are provided in Section II.

k. Centrifuge the mix at 27,000*g* for 10 min to remove any precipitated protein.

l. Concentrate the sample using a Centricon-30 (Amicon) filter and adjust the final concentration of salt to 30 m*M*. (Remove any precipitate in the sample by passing through a 0.2-μm pore size filter.)

m. Ion-exchange chromatography: Use a 1-ml Hi-Trap SP column (Amersham-Pharmacia) for caspase-5 purification and a 1-ml Hi-Trap Q column for other caspases. Equilibrate the column with 10 ml of chromatography buffer, pH 7.5 (as described in Section II).

n. Load the clear sample obtained in step 1 onto the equilibrated column at a flow rate of 0.5 ml/min.

o. Wash the column with 20 ml of buffer at a flow rate of 1.0 ml/min.

p. Elute the bound protein as 0.5-ml fractions at a rate of 10 m*M*/min using a linear gradient of 0–500 m*M* NaCl in chromatography buffer. (Most of the protein elutes in a range of 100–300 m*M* NaCl.)

q. Assay each fraction for caspase activity as described in Section III,F,1. Pool the active fractions, concentrate, and store at −80°C.

E. Preparation of S-100 Extracts from Normal and Apoptotic Cells for Caspase Assay

Mammalian cells express procaspases as inactive zymogens. In most cells, these inactive caspase precursors are present within the cytosol. Therefore, cytosolic S-100 extracts from normal and apoptotic cells can be used for studies on caspase activation.

a. Collect the normal or apoptotic cells by centrifugation at 600*g* for 3 min.

b. Wash the cell pellet gently with cold PBS.

c. Centrifuge at 600*g* for 3 min. Aspirate the supernatant carefully (at this stage cells can be stored at −80°C). Frozen cells must be thawed on ice and used immediately.

d. Resuspend the cell pellet in 4 volumes of cold buffer A with protease inhibitors (see Section II) and lyse the cells by passing 20 times through a syringe fitted with a 21-gauge needle. Keep the extract on ice for 10 min.

e. Centrifuge the extract at 10,000*g* for 15 min at 4°C.

f. Collect the supernatant carefully into a fresh tube on ice and recentrifuge the supernatant at 100,000*g* at 4°C for 60 min.

g. Collect the supernatant. This is the S-100 extract. Estimate the protein concentration using standard protein assay methods (Bradford's colorimetric assay).

h. Aliquots of the S-100 extract must be stored at −80°C.

F. Assay of Caspases

Caspases play a central role in mammalian apoptosis by proteolytic cleavage after defined aspartic acid residues in a variety of cellular substrates. All caspases hydrolyze peptide bonds on the carboxyl side of an aspartate residue (termed the P1 residue). Analysis of caspase-1 and caspase-3 active sites reveals that this cleavage most likely involves deprotonation of the sulfhydryl within the active site cysteine (Cys-285 in caspase 1) by the imidazole ring of histidine (His-237 in caspase 1), thereby resulting in hydrogen bonding to the amide nitrogen of glycine (Gly-238 in caspase-1) through a tetrahedral intermediate formation (Wilson *et al.,* 1994). These catalytic residues, Cys-285, His-237, and Gly-238, are conserved in all 11 human caspases (Fig. 2). The two amino acid residues N-terminal to the P1 aspartate (termed P2 and P3) have a limited effect on substrate cleavage (Thornberry *et al.,* 1992; Wilson *et al.,* 1994). In contrast, the amino acid residue (P4) three amino acids N-terminal to P1 aspartate confers distinct substrate specificity to each caspase. The observation that caspases are active against tetrapeptides blocked at their N and C termini permitted the development of both fluorogenic substrates and effective inhibitors (Howard *et al.,* 1991; Sleath *et al.,* 1990; Thornberry *et al.,* 1992). Thornberry *et al.* (1997) used a novel positional scanning substrate combinatorial library method to define the substrates of human caspases 1 to 10.

1. Fluorometric Assays

Caspase activity can be assayed using either a fluorophore (AFC, AMC) or a chromophore (pNA) attached to a substrate sequence. Fluorogenic substrates offer greater sensitivity of detection compared to colorometric substrates. Several companies (Alexis Biochemicals, Calbiochem, Enzyme Systems Inc., and Pharmingen) offer these peptide substrates for assaying the activity of different caspases. The optimal tetrapeptide sequences preferred by each caspase for cleavage are provided in Table I. The optimal buffer conditions for each caspase for cleavage of peptide substrates are provided in Table II (Garcia-Calvo *et al.,* 1999).

a. Set up the following reaction:

Pure caspase	10–100 n*M* or
S-100 cell extract	50–100 μg
2 *x* buffer X	50 μl
1 m*M* substrate	10 μl
Water	to 100 μl

b. Incubate at 37°C.

c. At different time points (0–60 min) take 20 μl of reaction mix and dilute it to 120 μl with reaction buffer.

d. Read the value immediately in a fluorometer using a microcuvette: 380 nm excitation and 460 nm emission for AMC substrates or 400 nm excitation and 505 nm emission for AFC substrates. For pNA substrates the reaction samples should be read at a 400- to 405-nm wavelength.

Note: Set up parallel control experiments that do not contain conjugated substrate or caspase. Although the preferred cleavage sequences for individual caspases are different, all caspases can cleave the different sequences with varying efficiency (Fig. 4). Most of the recombinant caspases produced in bacteria cleave their substrates and do not need any other cellular factors for their activity. However, caspase-9 is known to form an active holoenzyme complex with Apaf-1 and cytochrome *c*, and this complex formation is necessary for its activity. The assay of caspase-9 is discussed separately later on in this chapter.

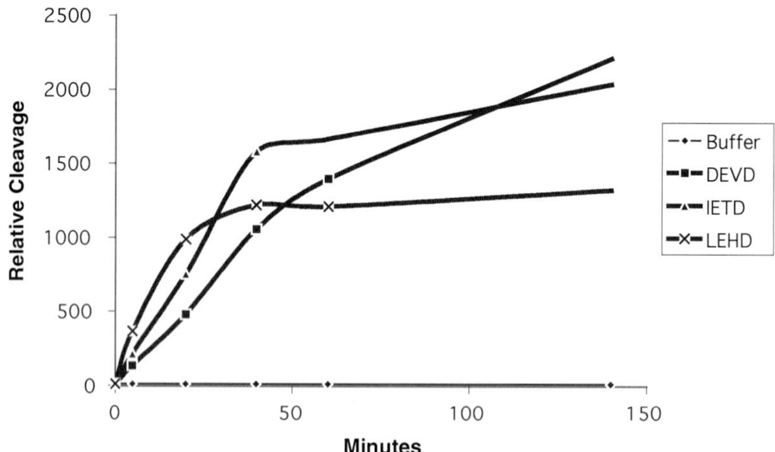

Fig. 4 Enzymatic activity of processed caspase-8 on various peptide substrates. Equal amounts of processed caspase-8 were incubated with various AMC-conjugated peptide substrates as described in the text. The relative cleavage of each substrate was plotted against time in minutes.

2. Protein Substrate Assay to Analyze Cleaved Products of Cellular Proteins of Interest *in Vivo* during Apoptosis

During apoptosis, caspases cleave a number of cellular polypeptides. To determine the effect of each caspase on various substrates, the following methods can be employed. Cleavage of polypeptides by caspases can be analyzed by polyacrylamide gel electrophoresis followed by immunoblotting or autoradiography.

a. Collect 1×10^6 normal or apoptosis-stimulated cells.

b. Centrifuge at 500*g* for 5 min at 4°C. Wash cell pellet once with 1 ml of cold PBS.

c. Resuspend the cell pellet in 50 μl of cold cell extraction buffer, pH 8 (see Section II). Vortex briefly and incubate on ice for 15 min.

d. Centrifuge at 20,000*g* for 15 min at 4°C and save the supernatant extract.

e. Use 30 μl of clear supernatant with sample buffer for a standard 10% polyacrylamide-glycine gel electrophoresis.

f. Transfer the fractionated proteins onto a nitrocellulose or PVDF membrane using routine Western blotting methods and immunostain the proteins using a specific antibody against the protein of interest.

Note: Detection of cleaved polypeptide bands only in apoptotic cell samples indicates the potential cleavage of the protein during apoptosis. However, it is necessary to confirm that the observed proteolytic cleavage is caspase specific. The specificity of cleavage can be checked using the pancaspase inhibitor ZVAD-*fmk*. Treat 1×10^6 cells with apoptotic stimuli in the presence or absence of 20 μM ZVAD-*fmk* in the cell media. Collect the cells and follow the procedure as described in Section III,F,2,b to f to detect the cleaved products. Disappearance of the cleaved polypeptide bands in ZVAD-*fmk*-treated samples confirms the requirement of active caspases for the observed cleavage.

3. Protein Substrate Assay to Analyze the Direct Cleavage of a Protein by Caspases under Cell-Free, *in Vitro* Conditions

The effect of a particular caspase on a specific polypeptide under cell-free conditions can be analyzed using either pure protein or radiolabeled protein. If a specific antibody against the protein is not available, T7 or any other epitope-tagged protein expressed in bacteria or mammalian cells can be used for this assay.

a. Set up the following reaction:

Pure/tagged recombinant protein	2 μg (or sufficient to detect on a blot)
Purified active caspase	10–100 n*M*
2x buffer X	10 μl
Water	to 20 μl

b. Incubate at 37°C for 60 min.

c. Add 7 μl of 3x sample buffer and run the samples on a standard 10% polyacrylamide–glycine gel. Include a control reaction minus the active caspase.

d. Transfer the proteins onto a nitrocellulose or PVDF membrane and immunostain using the specific antibody against the protein or the tag.

4. Cleavage of Radiolabeled Substrates

This method can be used if the cDNA of a potential substrate is available and is much simpler to use than the previous method, which requires prior purification of the protein substrate. For *in vitro* labeling of proteins with [^{35}S] methionine, we routinely use the TnT-coupled reticulocyte lysate systems from Promega corporation. cDNAs under a T7 or T3 or SP6 promoter can be transcribed and translated *in vitro*. We find that cDNAs cloned under the T7 promoter in pET vectors (Novagen) or pcDNA3 vectors (Invitrogen) give optimal translational products compared to other commercially available expression vectors. To purify plasmid DNA for use in coupled transcription and translation reactions, we recommend using a Qiaprep Spin Miniprep kit (Qiagen) or Wizard Plus Minipreps DNA purification system (Promega). We follow the standard reaction conditions recommended by the manufacturer, as described:

TnT rabbit reticulocyte lysate	25 μl
RNasin (RNase inhibitor) (40 U/μl)	2 μl
Amino acid mixture minus methionine	1 μl
TnT reaction buffer	2 μl
[^{35}S]Methionine (>1000 Ci/mmol at 10 μCi/μl)	3 μl
cDNA template (0.5 μg/μl)	2 μl
RNA polymerase (T3, T7, or Sp6)	1 μl
Nuclease-free water	to 50 μl

Incubate the reactions at 30°C for 60–90 min.

For preliminary experiments, this TnT reaction can be used directly for caspase cleavage assay as follows:

TnT reaction mix	2 μl
Pure caspase	10–100 n*M*
2x buffer X	10 μl
Water	to 20 μl

Incubate at 37°C for 60 min.

Add 7 μl of 3x sample buffer and fractionate the samples on a standard 12.5% polyacrylamide–glycine gel. Fix the gel in fixing solution for 1 h on a platform shaker. Rinse the gel in water for 15 min. Dry the gel onto a 3M Whatman paper using a gel drier followed by autoradiography overnight at room temperature.

Note: In this method, the signal strength depends on the number of methionine residues present in the polypeptide. Cleaved products that do not contain methionine residues will not be detected. A control lane of radiolabeled protein that is not incubated with caspases should always be included.

5. Caspase-9 and Apaf-1 Apoptosome Assay

Caspase-9 and Apaf-1 play central roles (Earnshaw *et al.,* 1999) in apoptosis. In apoptotic cells, procaspase-9 forms an oligomeric complex with Apaf-1 and cytochrome *c* released from mitochondria in the presense of dATP/ATP. This "apoptosome" activates downstream procaspases such as procaspase-3 and procaspase-7. Formation of the apoptosome is a very critical process and can be assayed as follows.

a. Apoptosome assay in normal cell extracts using fluorogenic substrate:

S100 extract (from normal cells) in buffer A	50 μg
Reduced bovine cytochrome *c* (0.5 μg/μl)	1 μl
100 m*M* dATP	1 μl
1 m*M* DEVD-AMC/AFC substrate in buffer A	5 μl
Buffer A	to 100 μl

Incubate the reaction for 60 min at 30°C. Read the value using a fluorometer as described in Section III,F,1, steps c and d.

b. Apoptosome assay in normal cell extracts using labeled procaspase-9:

S-100 extract (from normal cells) in buffer A	50 μg
Reduced bovine cytochrome *c* (0.5 μg/μl)	1 μl
100 m*M* dATP	1 μl
^{35}S-labeled procaspase-9 (TnT reaction)	10 μl
Buffer A	to 100 μl

Incubate the reaction for 60 min at 30°C. Take a 20-μl reaction mixture at different time points, stop the reaction, and analyze samples as described in Section III,F, 4.

Figure 5 shows the cleavage of labeled procaspase-9 in an apoptosome assay for 60 min using S-100 extracts from 293 cells.

dATP/Cyt c	−	−	+
293 S-100	−	+	+

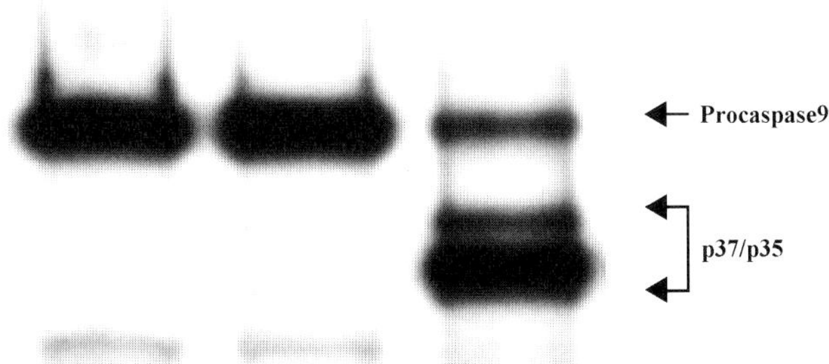

Fig. 5 Processing of labeled procaspase-9 in an apoptosome assay. ^{35}S-labeled procaspase-9 was used for the apoptosome assay as described in the text. The processing of procaspase-9 was analyzed by 10% SDS–PAGE and autoradiography.

c. Apoptosome assay using pure recombinant caspase-9 and Apaf-1: The processed caspase-9 purified from bacteria exhibits its full activity only in the presence of Apaf-1, cytochrome c, and dATP (Rodriguez and Lazebnik, 1999; Stennicke *et al.,* 1999). To assay for caspase-9 activity, the following procedures can be followed.

Assay using S-100 extracts depleted of procaspase-9:

S-100 extract depleted of procaspase-9 in buffer A	50 μg
Processed caspase-9	100 n*M*
Reduced bovine cytochrome *c* (0.5 μg/μl)	1 μl
100 m*M* dATP	1 μl
1 m*M* DEVD-AMC/AFC substrate in buffer A	5 μl
Buffer A	to 100 μl

Incubate the reaction at 30°C for 60 min. Read the value in a fluorometer as described in Section III,F,1, steps c and d.

Assay using S-100 extracts depleted of procaspase-9 and Apaf-1:

S-100 extract depleted of procaspase-9 and Apaf-1, in buffer A	50 μg
Processed caspase-9	100 n*M*
Pure Apaf-1 (see later)	100 n*M*
Reduced bovine cytochrome *c* (0.5 μg/μl)	1 μl
100 m*M* dATP	1 μl
1 m*M* DEVD-AMC/AFC substrate in buffer A	5 μl
Buffer A	to 100 μl

Incubate the reaction at 30°C for 60 min and read the value in a fluorometer as described in Section III,F,1, steps c and d.

Assay using LEHD substrates:

Processed caspase-9	100 n*M*
Pure Apaf-1	100 n*M*
Reduced bovine cytochrome *c* (0.5 μg/μl)	1 μl
100 m*M* dATP	1 μl
1 m*M* LEHD-AMC/AFC substrate in buffer A	5 μl
Buffer A	to 100 μl

Incubate the reaction for 1 h at 30°C and read the value in a fluorometer as described in Section III,F,1, steps c and d.

Assay of Apaf-1 and caspase-9 holoenzyme complex on substrates: Apaf-1 and caspase-9 holoenzyme can be used to assay caspase-9 activity or to identify its substrates.

Processed caspase-9	100 n*M*
Pure Apaf-1	100 n*M*
Reduced bovine cytochrome *c* (50 ng/μl)	2 μl
10 m*M* dATP	2 μl
Labeled protein	2 μl
(or) pure unlabeled protein	2 μg
Buffer A	to 20 μl

Incubate the reaction for 60 min at 30°C.

Run the reaction samples on a 12.5% polyacrylamide–glycine gel and process the gels either for autoradiography of the labeled proteins or for immunoblotting of pure unlabeled protein as described in Section III,F,4. Figure 6 shows the cleavage of T7-tagged procaspase-3 by the Apaf-1/caspase-9 complex in the presence of cytochrome *c* and dATP. Here we use dominant-negative procaspase-3 (active site C to A mutant) to avoid the possible autocatalytic cleavage of the protein.

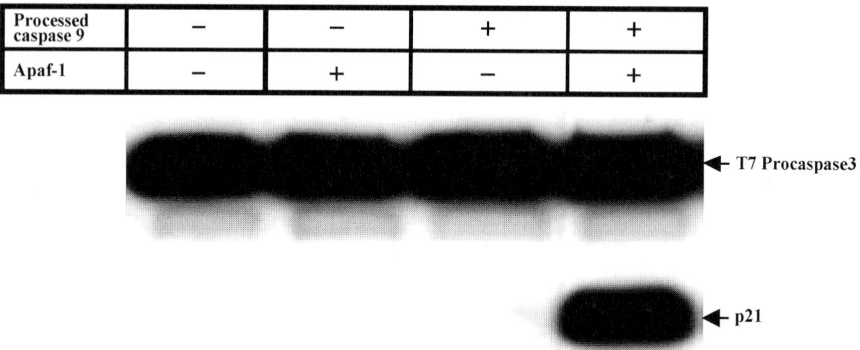

Fig. 6 Cleavage of T7-tagged procaspase-3 by Apaf-1/caspase-9 complex in the presence of cytochrome *c* and dATP. The T7-tagged purified procaspase-3 (C to A) mutant protein was incubated in the presence of cytochrome *c* and dATP with pure Apaf-1 and/or processed caspase-9. The processing of procaspase-3 was analyzed by 10% SDS–PAGE and immunoblotting using the T7-HRP antibody (Novagen).

G. Purification of Recombinant Human Apaf-1 Using Baculovirus Expression System

Full-length Apaf-1 can be expressed easily in Sf9 cells using the baculovirus expression system. The baculovirus transfer vector pVL1393 can be used to subclone the full-length Apaf-1 cDNA with a C-terminal His(6) tag and used for the generation of recombinant baculovirus.

For details on the generation of baculovirus transfer vectors and obtaining recombinant baculovirus, refer to Alnemri and Litwack (1994) and Piwnica-Worms (1989).

1. Expressing Recombinant Baculovirus in Insect Sf9 Cells

a. Infect 100×10^6 Sf9 cells with recombinant Apaf-1 virus (10 pfu/cell) using standard procedures.

b. Collect the cells by centrifugation at 500*g* for 10 min at 4°C. Wash the cell pellet with 10 ml of ice-cold (PBS).

c. Centrifuge at 500*g* for 10 min at 4°C. Remove the supernatant and freeze the cell pellet in liquid nitrogen (cell pellet can be stored at −80°C at this point).

d. Prepare S-100 cell extracts as described in Section III,E.

e. Incubate the S-100 extract with 250 U of Benzoaze (Novagen) for 15 min at 30°C. (This step removes any existing DNA in the S-100 extract.) Perform one more freeze and thaw as described earlier. (You may see some precipitate at this stage. However, we do not detect any Apaf-1 in this precipitate.)

f. Centrifuge the supernatant again at 100,000*g* for 30 min at 4°C to clear the S-100 extract. Measure the protein concentration. At this stage, the S-100 extract can be stored at −80°C after quick freezing in liquid nitrogen.

To purify Apaf-1 to homogeneity, we run the S-100 extract through successive columns of a Ni^+–NTA affinity column (Novagen), Mono Q ion-exchange column (Amerham-Pharmacia), and Superose 12 gel-filtration column (Amerham-Pharmacia) using protein liquid chromatography system FPLC (Amersham-Pharmacia).

The elution of Apaf-1 has to be monitored by immunoblotting of all fractions after each run using a specific Apaf-1 antibody (Pharmingen). All the purification steps have to be carried out at 4°C as quickly as possible to get the functionally active Apaf-1 protein.

2. Affinity Chromatography Using Ni^+–NTA Affinity Column

a. Carefully pack 2 ml of Talon metal affinity (Clontech) or Ni^+–NTA affinity resin (Novagen), without any air bubbles, onto an XQ FPLC column by gravity using standard procedures.

b. Wash the column with 10 ml of water to remove all traces of ethanol from the resin.

Note: At this stage the column can be recharged with nickel sulfate after stripping the resin of Ni ions, as described later. We found that this fresh recharging step enhances the column-binding capacity.

c. Wash the column with 10 ml of 6.0 *M* guanidine–HCl in 0.2% acetic acid to strip the resin. Rinse the column with 20 ml of water to remove strip solution.

d. Recharge the resin by passing 10 ml of freshly prepared 0.2% nickel sulfate solution through the column.

e. Hook the column to the FPLC machine and equilibrate with 20 ml of buffer B at a flow rate of 0.1 ml/min. This step also removes any excess, unbound nickel from the column.

f. Load the S-100 extract onto the recharged column at a flow rate of 0.5 ml/min.

g. Wash the column once with 10 ml of buffer B at a flow rate of 0.5 ml/min and twice with 20 ml of buffer B containing 20 m*M* imidazole at a flow rate of 0.15 ml/min.

h. Elute the bound protein from the column using a 30-ml gradient of 50–350 m*M* imidazole in buffer B at a flow rate of 0.15 ml/min.

i. Collect 0.5-ml fractions and check 25 μl of each fraction by immunoblotting with the Apaf-1 antibody to determine the elution profile of Apaf-1. Usually the protein elutes at a range of 90–150 m*M* imidazole.

j. Pool the Apaf-1-containing fractions and concentrate the pool using a Centricon-30 (Amicon) filter.

k. Adjust the final concentration of NaCl to 20 m*M* in a final volume of 2–4 ml.

3. Ion-Exchange Chromatography Using Mono Q Column

a. Equilibrate an FPLC Mono Q column (volume 1.0 ml) with buffer C.

b. Load the concentrated sample of Apaf-1 from step 3*k* onto the Mono Q column at a flow rate of 0.1 ml/min. Wash the column with 10 ml of buffer C.

c. Elute the bound protein from the column with a 20-ml gradient of buffer C containing 20–300 m*M* NaCl, 200 μg/ml bovine serum albumin (BSA, Sigma), and 0.1 m*M* DTT at a flow rate of 0.25 ml/min.

d. Collect 0.5-ml fractions, and blot 20 μl of each fraction with the Apaf-1 antibody to determine the elution profile of Apaf-1 protein. Usually the protein elutes at a range of 120–200 m*M* NaCl.

e. Pool the Apaf-1-containing fractions and concentrate the pool using a Centricon-30 (Amicon) filter. Adjust the final concentration of NaCl to 50 m*M* in a final volume of 0.5 ml.

4. Gel-Filtration Chromatography on Superose 12 Column

a. Preequilibrate the Superose 12 gel filtration column with buffer D at a flow rate of 0.2 ml/min.

b. Load 250 μl of the concentrated sample of Apaf-1 from step 3*e* onto the equilibrated gel filtration column and collect 0.25-ml fractions. Based on Bio-Rad gel filtration standards, pure Apaf-1 protein elutes at a size of approximately 125 kDa.

c. Repeat the gel filtration run using the rest of the 250-μl sample as described earlier.

d. Blot 20 μl of each fraction with Apaf-1 antibody to determine the elution profile of Apaf-1 protein. Pool the Apaf-1-containing fractions and concentrate using a Centricon-30 (Amicon) filter.

e. Store the pure protein as 50-μl aliquots at −80°C.

5. Assaying Apaf-1 Function by Activation of Procaspase-9

The activity of recombinant Apaf-1 can be assayed by the method described in Section III,F,5,b for its ability to form an active enzyme complex with caspase-9 in the presence of cytochrome *c* and dATP. Alternatively, the following method can also be used to measure its ability to autoactivate procaspase-9.

Pure Apaf-1	20 n*M*
^{35}S-labeled procaspase-9	2 μl
Reduced bovine cytochrome *c* (50 ng/μl)	2 μl
10 m*M* dATP	2 μl
Buffer A	to 20 μl

a. Incubate the *above* reaction for 60 min at 30°C.

b. Fractionate the samples on a 12.5% polyacrylamide–glycine gel, followed by autoradiography as described in Section III,F,d. Cleavage of the 46-kDa procaspase-9 to the processed caspase-9 subunits of 35 and 11 kDa is a result of autoactivation of procaspase-9 by active Apaf-1.

6. Assaying Apaf-1 Function by Oligomerization

The function of pure Apaf-1, or Apaf-1 in cytosolic extracts, can also be assayed for its ability to form a multimer in the presence of cytochrome *c* and dATP. These oligomers can be analyzed by gel-filtration chromatography.

a. Formation of Apaf-1 Apoptosome Using pure Apaf 1

Pure Apaf-1	4.0 μg
10x buffer D	10 μl
Reduced bovine cytochrome *c* (500 ng/μl)	2 μl
10 m*M* dATP	5 μl
Water	to 100 μl

Incubate the reaction at 4°C for 60 min.

b. Formation of Apaf-1 Apoptosome in Cytosolic S-100 extracts

Apaf-1 in cytosolic extracts can be activated by the addition of cytochrome *c* and dATP.

Cytosolic S-100 extract in buffer D	30 mg
10x buffer D	10 μl
Reduced bovine cytochrome *c* (500 ng/μl)	2 μl
100 m*M* dATP	5 μl
Water	to 100 μl

Incubate at 25°C for 30 min.

c. Fractionation of Apoptosome by Superose 6 Column

a. Equilibrate a Superose 6 HR 10/30 column with 3 column volumes of buffer D.

b. Dilute the reaction mix to 250 μl with buffer D.

c. Load the clear reaction sample to an equilibrated Superose 6 column at a flow rate of 0.2 ml/min and collect 500-μl fractions.

d. Analyze the Apaf-1 elution profile by running 50 μl of each fraction on a 8% acrylamide gel and immunoblotting using the anti-Apaf-1 antibody.

e. Estimate the molecular mass of the Apaf-1 eluted in different fractions by running the calibration protein standards through the column.

f. The size of the Apaf-1 oligomer should be approximately 1.4 MDa (Saleh *et al.*, 1999; Zou *et al.*, 1999).

d. Coupling of Apaf-1 CARD to Affi-Gel 10

The Affi-Gel 10 (Bio-Rad) provides an efficient and rapid method for the covalent attachment of the primary amino group of polypeptides to the reactive *N*-hydroxysuccinimide groups of the matrix.

a. Prepare 200 μl of reactive Affi-Gel 10 according to the manufacturer's instructions.

b. Add 2–3 mg of pure Apaf-1 CARD in a final volume of 1 ml under continuous and gentle agitation on a rocker.

c. Incubate for 1 h at 4°C.

d. Spin the mixture down at 500 rpm for 2–3 min at 4°C.

e. Carefully aspirate the supernatant and put it aside for subsequent measurement to determine the percentage of coupling (see later).

f. Wash the resin twice with 1 ml of buffer A. In each wash step, resuspend the resin gently in the buffer, centrifuge at 500 rpm, and remove the supernatant carefully.

g. Block excess reactive *N*-hydroxysuccinimide groups of the Affi-Gel matrix that did not form an amide bond with the primary amino group of the CARD polypeptide by the addition of 0.2 ml of 0.1 *M* ethanolamine HCl (pH 8).

h. Wash the resin twice with 1 ml of buffer A containing 100 m*M* NaCl.

i. Resuspend the resin in 0.2 ml of buffer A + 100 m*M* NaCl. The CARD-Affi-Gel can be stored at 4°C in this buffer up to 1 month after the addition of sodium azide to a final concentration of 0.1%.

To determine the coupling efficiency, measure the total protein before and after incubation with the Affi-Gel material. The coupling efficiency can also be determined by comparing the intensity of the Coomassie-stained band of the Apaf-1 CARD in a 15% SDS gel with 25 μl of the supernatants before and after the coupling reaction. Ideally, one observes 90% efficiency of binding of CARD to the Affi-Gel matrix. As a control, we couple the same amount of BSA protein to Affi-Gel following similar conditions to those described earlier for the Apaf-1 CARD.

e. Depletion of Caspase-9 from S-100 Extracts

Prepare S-100 extracts from the appropriate cells in buffer D as described in Section III,E.

a. Incubate 0.5 mg of the S-100 extract with 20 μl bed volume of the Affi-Gel-bound Apaf-1 CARD or -bound BSA (control) under continuous gentle agitation at 4°C for 1 h.

b. Spin down the reaction mixture tube at 1000 rpm for 2 min and recover the supernatant.

c. Repeat the just-described step by incubating the recovered supernatants with another 20 μl of Affi-Gel-bound CARD and BSA control for 40 min.

d. The depletion efficiency of caspase-9 from the S-100 extract incubated with CARD resin and from the control sample incubated with the BSA resin can be determined by Western blotting of 100 μg of these samples with a caspase-9-specific antibody (Pharmingen). Alternatively, 50 μg of the depleted sample can be assayed for the caspase-9-dependent activation of downstream caspases by measuring the caspase activity using a peptide substrate after stimulation of the extract by cytochrome *c* and dATP as described in Section III,F,5,a.

References

Ahmad, M., Srinivasula, S. M., Wang, L., Talanian, R. V., Litwack, G., Fernandes-Alnemri, T., and Alnemri, E. S. (1997). CRADD, a novel human apoptotic adaptor molecule for caspase-2 and Fas/TNF receptor interacting protein RIP. *Cancer Res.* **57,** 615–619.

Alnemri, E. S. (1997). Mammalian cell death proteases, a family of highly conserved aspartate-specific cysteine proteases. *J. Cell. Biochem.* **64,** 33–42.

Alnemri, E. S., Fernandes, T. F., Haldar, S., Croce, C. M., and Litwack, G. (1992). Involvement of BCL-2 in glucocorticoid-induced apoptosis of human pre-B leukemias. *Cancer Res.* **52,** 491–495.

Alnemri, E. S., and Litwack, G. (1994). Baculovirus-Mediated Overexpression of Glucocorticoid and Mineralocorticoid Receptors and Related Proteins. (K. W. Adolph, ed.), Vol. 5, Academic Press, San Diego.

Alnemri, E. S., Livingston, D. J., Nicholson, D. W., Salvesen, G., Thornberry, N. A., Wong, W. W., and Yuan, J. (1996). Human ICE/CED-3 protease nomenclature. *Cell* **87,** 171.

Ashkenazi, A., and Dixit, V. M. (1998). Death receptors: Signaling and modulation. *Science* **281,** 1305–1308.

Boldin, M. P., Goncharov, T. M., Goltsev, Y. V., and Wallach, D. (1996). Involvement of MACH, a novel MORT1/FADD-interacting protease, in Fas/Apo-1- and TNF Receptor-induced cell death. *Cell* **85,** 803–815.

Cohen, G. M. (1997). Caspases: The executioners of apoptosis. *Biochem. J.* **326,** 1–16.

Darmon, A. J., Nicholson, D. W., and Bleackley, R. C. (1995). Activation of the apoptotic protease CPP32 by cytotoxic T-cell-derived granzyme B. *Nature* **377,** 446–448.

Deiss, L. P., Galinka, H., Berissi, H., Cohen, O., and Kimchi, A. (1996). Cathepsin D protease mediates programmed cell death induced by interferon-gamma, Fas/APO-1 and TNF-alpha. *EMBO J.* **15,** 3861–3870.

Duan, H., and Dixit, V. M. (1997). RAIDD is a new "death" adaptor molecule. *Nature* **385,** 86–89.

Earnshaw, W. C., Martins, L. M., and Kaufmann, S. H. (1999). Mammalian caspases: Structure, activation, substrates, and functions during apoptosis. *Annu. Rev. Biochem.* **68,** 383–424.

Fernandes-Alnemri, T., Takahashi, A., Armstrong, R., Krebs, J., Fritz, L., Tomaselli, K. J., Wang, L., Yu, Z., Croce, C. M., Salvesen, G., Earnshaw, W. C., Litwack, G., and Alnemri, E. S. (1995). Mch3, a novel human apoptotic cysteine protease highly related to CPP32. *Cancer Res.* **55,** 6045–6052.

Garcia-Calvo, M., Peterson, E. P., Rasper, D. M., Vaillancourt, J. P., Zamboni, R., Nicholson, D. W., and Thornberry, N. A. (1999). Purification and catalytic properties of human caspase family members. *Cell Death Differ.* **6,** 362–369.

Hirsch, T., Dallaporta, B., Zamzami, N., Susin, S. A., Ravagnan, L., Marzo, I., Brenner, C., and Kroemer, G. (1998). Proteasome activation occurs at an early, premitochondrial step of thymocyte apoptosis. *J. Immunol.* **161,** 35–40.

Howard, A. D., Kostura, M. J., Thornberry, N., Ding, G. J., Limjuco, G., Weidner, J., Salley, J. P., Hogquist, K. A., Chaplin, D. D., Mumford, R. A., Schmidt, J. A., and Tocci, M. J. (1991). IL-1-converting enzyme requires aspartic acid residues for processing of the IL-1B precursor at two distinct sites and does not cleave 31-kDa IL-1a. *J. Immunol.* **147,** 2964–2969.

Kurucz, I., Titus, J. A., Jost, C. R., and Segal, D. M. (1995). Correct disulfide pairing and efficient refolding of detergent- solubilized single-chain Fv proteins from bacterial inclusion bodies. *Mol. Immunol.* **32,** 1443–1452.

Los, M., Wesselborg, S., and Schulze-Osthoff, K. (1999). The role of caspases in development, immunity, and apoptotic signal transduction: Lessons from knockout mice. *Immunity* **10,** 629–639.

Martin, D. A., Siegel, R. M., Zheng, L., and Lenardo, M. J. (1998). Membrane oligomerization and cleavage activates the caspase-8 (FLICE/MACHalpha1) death signal. *J. Biol. Chem.* **273,** 4345–4349.

Martins, L. M., Kottke, T. J., Kaufmann, S. H., and Earnshaw, W. C. (1998). Phosphorylated forms of activated caspases are present in cytosol from HL-60 cells during etoposide-induced apoptosis. *Blood* **92,** 3042–3049.

Mukhopadhyay, A. (1997). Inclusion bodies and purification of proteins in biologically active forms. *Adv. Biochem. Eng. Biotechnol.* **56,** 61–109.

Muzio, M., Chinnaiyan, A. M., Kischkel, F. C., O'Rourke, K., Shevchenko, A., Ni, J., Scaffidi, C., Bretz, J. D., Zhang, M., Gentz, R., Mann, M., Krammer, P. H., Peter, M. E., and Dixit, V. M. (1996). FLICE, a novel FADD-homologous ICE/CED-3-like protease, is recruited to CD95 (Fas/Apo-1) death-inducing signaling complex. *Cell* **85,** 817–827.

Muzio, M., Stockwell, B. R., Stennicke, H. R., Salvesen, G. S., and Dixit, V. M. (1998). An induced proximity model for caspase-8 activation. *J. Biol. Chem.* **273,** 2926–2930.

Piwnica-Worms, H. (1989). "Expression of Proteins in Insect Cells Using Baculoviral Vectors" (F. M. Ausubel, R. Brent, R. E. Knigston, D. D. Moore, J. G. Seidman, J. A. Smith, and K. Struhl, eds.), Vol. 2, Wiley, New York.

Rodriguez, J., and Lazebnik, Y. (1999). Caspase-9 and APAF-1 form an active holoenzyme. *Genes Dev.* **13,** 3179–3184.

Rudolph, R., and Lilie, H. (1996). In vitro folding of inclusion body proteins. *FASEB J.* **10,** 49–56.

Saleh, A., Srinivasula, S. M., Acharya, S., Fishel, R., and Alnemri, E. S. (1999). Cytochrome *c* and dATP-mediated oligomerization of Apaf-1 is a prerequisite for procaspase-9 activation. *J. Biol. Chem.* **274,** 17941–17945.

Salvesen, G. S., and Dixit, V. M. (1997). Caspases: Intracellular signaling by proteolysis. *Cell* **91,** 443–446.

Shresta, S., MacIvor, D. M., Heusel, J. W., Russell, J. H., and Ley, T. J. (1995). Natural killer and lymphokine-activated killer cells require granzyme B for the rapid induction of apoptosis in susceptible target cells. *Proc. Natl. Acad. Sci. USA* **92,** 5679–5683.

Simon, M. M., Hausmann, M., Tran, T., Ebnet, K., Tschopp, J., ThaHla, R., and Mullbacher, A. (1997). In vitro- and ex vivo-derived cytolytic leukocytes from granzyme A x B double knockout mice are defective in granule-mediated apoptosis but not lysis of target cells. *J. Exp. Med.* **186,** 1781–1786.

Sleath, P. R., Hendrickson, R. C., Kronheim, S. R., March, C. J., and Black, R. A. (1990). Substrate specificity of the protease that processes human interleukin-1B. *J. Biol. Chem.* **265,** 14526–14528.

Squier, M. K., Miller, A. C., Malkinson, A. M., and Cohen, J. J. (1994). Calpain activation in apoptosis. *J. Cell. Physiol.* **159,** 229–237.

Srinivasula, S. M., Ahmad, M., Fernandes-Alnemri, T., and Alnemri, E. S. (1998). Autoactivation of procaspase-9 by Apaf-1-mediated oligomerization. *Mol. Cell.* **1,** 949–957.

Srinivasula, S. M., Ahmad, M., Fernandes-Alnemri, T., Litwack, G., and Alnemri, E. S. (1996). Molecular ordering of the Fas-apoptotic pathway: The Fas/APO-I protease Mch5 is a CrmA-inhibitable protease that activates multiple Ced-3/ICE-like cysteine proteases. *Proc. Natl. Acad. Sci. USA* **93,** 13706–13711.

Stennicke, H. R., Deveraux, Q. L., Humke, E. W., Reed, J. C., Dixit, V. M., and Salvesen, G. S. (1999). Caspase-9 can be activated without proteolytic processing. *J. Biol. Chem.* **274,** 8359–8362.

Stennicke, H. R., Jurgensmeier, J. M., Shin, H., Deveraux, Q., Wolf, B. B., Yang, X., Zhou, Q., Ellerby, H. M., Ellerby, L. M., Bredesen, D., Green, D. R., Reed, J. C., Froelich, C. J., and Salvesen, G. S. (1998). Pro-caspase-3 is a major physiologic target of caspase-8. *J. Biol. Chem.* **273,** 27084–27090.

Stennicke, H. R., and Salvesen, G. S. (1997). Biochemical characteristics of caspases-3, -6, -7, and -8. *J. Biol. Chem.* **272,** 25719–25723.

Thornberry, N. A., Bull, H. G., Calaycay, J. R., Chapman, K. T., Howard, A. D., Kostura, M. J., Miller, D. K., Molineaux, S. M., Weidner, J. R., Aunins, J., Ellison, K. O., Ayala, J. M., Casano, F. J., Chin, J., Ding, G. J.-F., Egger, L. A., Gaffney, E. P., Limjuco, G., Palyha, O. C., Raju, S. M, Rolando, A. M., Salley, J. P., Yamin, T.-T., Lee, T. D., Shively, J. E., MacCross, M., Mumford, R. A., Schmidt, J. A., and Tocci, M. J. (1992). A novel heterodimeric cysteine protease is required for interleukin-1 beta processing in monocytes. *Nature* **356,** 768–774.

Thornberry, N. A., and Lazebnik, Y. (1998). Caspases: Enemies within. *Science* **281,** 1312–1316.

Thornberry, N. A., Rano, T. A., Peterson, E. P., Rasper, D. M., Timkey, T., Garcia-Calvo, M., Houtzager, V. M., Nordstrom, P. A., Roy, S., Vaillancourt, J. P., Chapman, K. T., and Nicholson, D. W. (1997). A combinatorial approach defines specificities of members of the caspase family and granzyme B: Functional relationships established for key mediators of apoptosis. *J. Biol. Chem.* **272,** 17907–17911.

Wilson, K. P., Black, J. F., Thompson, J. A., Kim, E. E., Griffith, J. P., Navia, M. A., Murcko, M. A., Chambers, S. P., Aldape, R. A., Raybuck, S. A., and Livingston, D. J. (1994). Structure and mechanism of interleukin-1beta converting enzyme. *Nature* **370,** 270–275.

Wolf, B. B., and Green, D. R. (1999). Suicidal tendencies: Apoptotic cell death by caspase family proteinases. *J. Biol. Chem.* **274,** 20049–20052.

Wright, S. C., Wei, Q. S., Zhong, J., Zheng, H., Kinder, D. H., and Larrick, J. W. (1994). Purification of a 24-kD protease from apoptotic tumor cells that activates DNA fragmentation. *J. Exp. Med.* **180,** 2113–2123.

Yamin, T. T., Ayala, J. M., and Miller, D. K. (1996). Activation of the native 45-kDa precursor form of interleukin-1-converting enzyme. *J. Biol. Chem.* **271,** 13273–13282.

Yang, X., Chang, H. Y., and Baltimore, D. (1998). Autoproteolytic activation of pro-caspases by oligomerization. *Mol. Cell.* **1,** 319–325.

Zhou, Q., and Salvesen, G. S. (1997). Activation of pro-caspase-7 by serine proteases includes a noncanonical specificity. *Biochem. J.* **324,** 361–364.

Zou, H., Li, Y., Liu, X., and Wang, X. (1999). An APAF-1.cytochrome *c* multimeric complex is a functional apoptosome that activates procaspase-9. *J. Biol. Chem.* **274,** 11549–11556.

CHAPTER 2

Cloning and Analysis of Bcl-2 Family Genes

Enrique Cepero, Bryan W. Johnson, and Lawrence H. Boise

Department of Microbiology and Immunology
University of Miami School of Medicine
Miami, Florida 33101

I. Introduction

This chapter has been designed to aid those interested in finding new Bcl-2 homologs or Bcl-2 interacting proteins. We have approached this by highlighting the different strategies employed for isolating Bcl-2 homologs and describing selected experimental protocols used to assay for Bcl-2 family member function. However, many of these same principles, techniques, and protocols can be applied to Bcl-2 interacting proteins that are not related to Bcl-2. The following chapter is divided into three sections. The first section is a brief review of the Bcl-2 family describing characteristics that have been

METHODS IN CELL BIOLOGY, VOL. 66

0091-679X/01 $35.00

used for the isolation of new members. The second section is an overview of how several Bcl-2 homologs were isolated, cloned, and characterized. A detailed description of the experimental protocols used to isolate and clone these Bcl-2 family members will not be given as most members of the Bcl-2 family have been cloned by common molecular techniques such as the polymerase chain reaction (PCR), immunoprecipitation, and the yeast two-hybrid system. These techniques are all described in detail in standard laboratory manuals. Therefore, we have highlighted the unique aspects of how each family member was cloned to demonstrate how these cloning techniques can be applied to the Bcl-2 family. The third section describes functional assays used to test Bcl-2 function and regulation. This experimental protocol section contains several common protocols, from our laboratory and others, that have been used to assay the function and regulation of the Bcl-2 family. The experimental section is designed to allow the reader to functionally characterize novel members of the Bcl-2 family.

A. The Bcl-2 Family

The protooncogene Bcl-2 was found at the t(14 : 18) chromosomal translocation in follicular lymphomas (reviewed by Boise *et al.,* 1995; Adams and Cory, 1998; Gross *et al.,* 1999a). It was determined subsequently that Bcl-2 functions primarily by inhibiting apoptosis. In a separate series of experiments, genetic studies in the nematode *Caenorhabditis elegans* revealed the presence of a death-inhibiting gene, CED-9, which appeared to function like Bcl-2. Taken together, these findings suggested that Bcl-2 may be part of a complex genetic pathway. This initiated a search for other genes involved in cell death and resulted in the cloning and characterization of several Bcl-2-like genes. Using a variety of different methods of identification, the Bcl-2 family has grown to at least 15 members. The Bcl-2 family of proteins regulates cell death by either inducing (proapoptotic) or inhibiting (antiapoptotic) apoptosis.

The Bcl-2 family is divided into two subfamilies. The antiapoptotic subfamily includes Bcl-2, Bcl-x_L, Bcl-w, Mcl-1, A1, NR-13, and CED-9 (Reed *et al.,* 1996; Adams and Cory, 1998). When overexpressed, these proteins have been shown to protect cells from apoptotic stimuli such as growth factor withdrawal, irradiation, DNA-damaging agents, and some death receptor signals. The proapoptotic subfamily can be further divided into two groups. The Bax subfamily includes Bax, Bak, and Bok. The BH3-only subfamily includes Bik, Blk, Hrk, Bim_L, Bad, Bid, and EGL-1. Both the Bax and BH3-only subfamilies have been shown to induce apoptosis when overexpressed (Reed *et al.,* 1996; Adams and Cory, 1998).

The molecular mechanism by which Bcl-2 family members regulate cell death is still unknown and is an area of intense research and much controversy. The following sections describe how several Bcl-2 family members were identified and cloned, and some of the experiments used to determine their function. This is by no means an exhaustive list, and we have tried to highlight the different techniques used to characterize Bcl-2 family members. However, before such a list can be made, several characteristic features of the Bcl-2 family used to identify and clone them must be described.

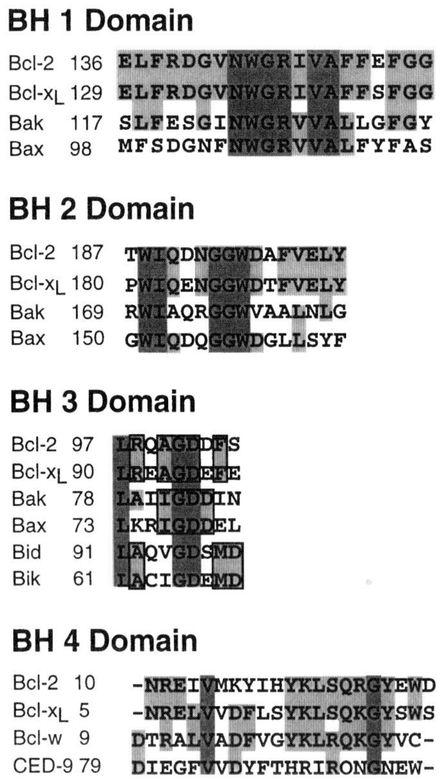

BH 1 Domain

```
Bcl-2  136 ELFRDGVNWGRIVAFFEFGG
Bcl-xL 129 ELFRDGVNWGRIVAFFSFGG
Bak    117 SLFESGINWGRVVALLGFGY
Bax     98 MFSDGNFNWGRVVALFYFAS
```

BH 2 Domain

```
Bcl-2  187 TWIQDNGGWDAFVELY
Bcl-xL 180 PWIQENGGWDTFVELY
Bak    169 RWIAQRGGWVAALNLG
Bax    150 GWIQDQGGWDGLLSYF
```

BH 3 Domain

```
Bcl-2  97 LRQAGDDFS
Bcl-xL 90 LREAGDEFE
Bak    78 LAIIGDDIN
Bax    73 LKRIGDDEL
Bid    91 LAQVGDSMD
Bik    61 LACIGDEMD
```

BH 4 Domain

```
Bcl-2  10 -NREIVMKYIHYKLSQRGYEWD
Bcl-xL  5 -NRELVVDFLSYKLSQKGYSWS
Bcl-w   9 DTRALVADFVGYKLRQKGYVC-
CED-9  79 DIEGFVVDYFTHRIRONGNEW-
```

Fig. 1 Amino acid sequences of the BH1, BH2, BH3, and BH4 domains of several Bcl-2 family members. Dark shades indicate residues conserved by all indicated members and light shades indicate conserved residues. Boxed residues in the BH3 domain represent regions of identity found within Bcl-2 subfamilies.

B. Bcl-2 Homology (BH) Domains

Several conserved domains have been identified at the amino acid level among the various Bcl-2 family members. A carboxy-terminal transmembrane (TM) domain is present in several but not all Bcl-2 family members and determines the intracellular localization of the proteins by targeting them to intracellular membranes, i.e., mitochondria, endoplasmic reticulum, and nuclear envelope (Reed *et al.,* 1996; Adams and Cory, 1998). Comparison at the amino acid level reveals four Bcl-2 homology (BH) domains shared by the Bcl-2 family (Fig. 1) (Reed *et al.,* 1996; Adams and Cory, 1998). Although not all Bcl-2 family members contain all four domains (see later), all Bcl-2 homologs contain at least one BH domain.

Among the Bcl-2 family members identified to date, only the antiapoptotic members Bcl-2, Bcl-x_L, and Bcl-w contain all four BH domains. Aside from these three members, there is variation in the number of BH domains that pro- and antiapoptotic Bcl-2 family

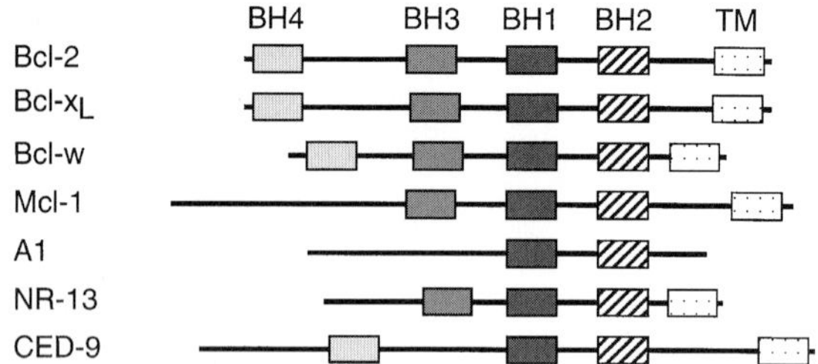

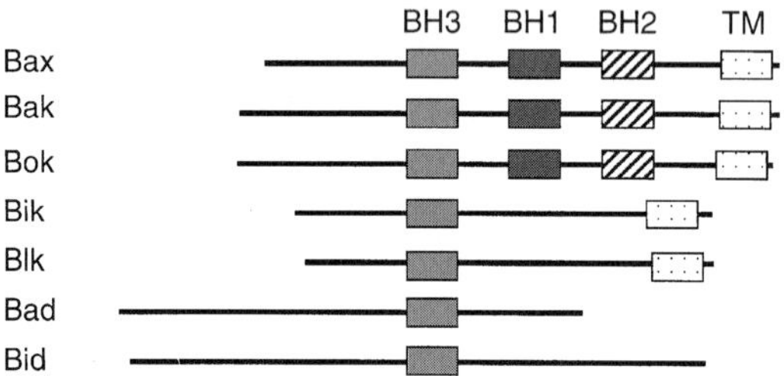

Fig. 2 Representative schematic diagrams of selected Bcl-2 family members are shown. Shaded boxes indicate the BH domains and the transmembrane domain (TM).

members contain. Figure 1 shows the amino acid sequence of the BH domains among several Bcl-2 family members (Muchmore *et al.,* 1996; Reed *et al.,* 1996). Antiapoptotic members contain at least a BH1 and a BH2 domain. CED-9 contains BH1, BH2, and BH4 domains but no BH3 domain. Because CED-9 is the *C. elegans* Bcl-2 homolog, and both Bcl-2 and Bcl-x_L contain a BH4 domain, this led many to believe that the presence of a BH4 domain was required for antiapoptotic function. However, there are antiapoptotic members that do not contain a BH4 domain, such as Mcl-1, A1, and NR-13. Among non-BH4 members, Mcl-1 and NR-13 contain BH1, BH2, and BH3 domains and A1 has only BH1 and BH2 domains (Fig. 2). All of the antiapoptotic family members mentioned previously except A1 contain a transmembrane domain (Reed *et al.,* 1996; Adams and Cory, 1998).

The proapoptotic Bcl-2 family can be subdivided into two separate groups based on their BH domains. Bax, Bok, and Bak contain BH1, BH2, and BH3 domains in addition to a transmembrane domain and are grouped into the Bax family. In contrast, Bik, Bid, Bad, Bim_L, EGL-1, Blk, and Hrk contain a BH3 domain only, and no other region in their peptide sequence is homologous to Bcl-2 (Reed *et al.,* 1996; Adams and Cory, 1998). Interestingly, despite these differences in BH domains, the crystal structure of Bid has been solved and shows a high degree of similarity with Bcl-x_L (Chou *et al.,* 1999; McDonnell *et al.,* 1999). As with the antiapoptotic Bcl-2 subfamily, the transmembrane domain is not found in all of the BH3-only subfamily. Bik, Blk, Hrk, and Bim_L contain a transmembrane domain, whereas Bid, Bad, and EGL-1 do not. Figure 2 is a schematic representation of the Bcl-2 family with the BH domains and the transmembrane domain indicated.

The presence of conserved BH domains among Bcl-2 family members has allowed identification and cloning of several Bcl-2 family members at the DNA level by low stringency DNA hybridization, PCR, and database searches. By understanding the differences in BH domains between Bcl-2 subfamilies, future searches for novel Bcl-2 family members can be directed toward a specific activity (pro- vs antiapoptotic). While all family members contain BH3 domains, the degree of identity at the amino acid level within the BH3 domains is greatest among subfamily members, as shown in Fig. 1.

C. Bcl-x_L Structure

Due to the conserved BH domains, the Bcl-2 family of proteins can homo- and heterodimerize with both pro- and antiapoptotic family members. The molecular mechanism for such binding was revealed by the crystal structure of Bcl-x_L (Muchmore *et al.,* 1996). Folding of the BH1, BH2, and BH3 domains forms an elongated hydrophobic cleft that can serve as a binding site for the BH3 domain of other Bcl-2 family members. As a result, site-directed mutagenesis of critical amino acids in BH1, BH2, or BH3 domains can abrogate dimerization (Muchmore *et al.,* 1996; Adams and Cory, 1998). However, it has been shown that dimerization-deficient mutants of Bcl-x_L are still able to prevent cell death (Cheng *et al.,* 1996; Minn *et al.,* 1999).

Because of the ability of Bcl-2 family members to homo- and heterodimerize, many laboratories have employed a protein–protein interaction strategy to isolate new Bcl-2 family members. Coimmunoprecipitation, the yeast two-hybrid system, and protein interactive cloning have been the most popular techniques to date. One advantage of this strategy is the potential for isolating Bcl-2 interacting proteins that do not have a high degree of similarity at the DNA level, yet interact at the protein level, such as BH3-only family members.

Most Bcl-2 family members have been cloned by either comparison at the DNA level or by protein–protein interactions. The following is a list of several Bcl-2 family members and how they were identified and cloned. The intent of this section is not to provide detailed protocols, but rather to demonstrate that many techniques can be employed when searching for Bcl-2 homologs. If a detailed protocol of the techniques described is required, the reader is referred to the original references.

II. Cloning of Bcl-2 Family Genes

A. DNA Analysis

1. Hybridization

a. Bcl-x_L/Bcl-x_S (Boise et al., 1993)

Bcl-x_L was first identified in chickens by low stringency hybridization of a mouse Bcl-2 cDNA to chicken genomic DNA and cDNA libraries. Subsequent screening of human libraries identified human Bcl-x_L. Overexpression of Bcl-x_L in an IL-3-dependent cell line provided protection against cytokine withdrawal-induced apoptosis. Bcl-x_L contains all four BH domains and can be alternatively spliced into a proapoptotic protein, Bcl-x_S.

2. EST Screen

a. Blk (Hedge et al., 1998)

Blk (Bik-like-killer) is a BH3-only proapoptotic Bcl-2 family member. Blk was identified by screening a GenBank-expressed sequence tag database for sequence similarity to the BH3 domain. After identification of a mouse partial cDNA, primers were generated based on the 3′ end and the full-length gene was cloned by PCR. Overexpression of Blk in MCF-7 cells established this protein as a death agonist.

3. PCR

a. Bak (Chittenden et al., 1995; Farrow et al., 1995; Kiefer et al., 1995)

Bak (Bcl-2 homologous antagonist killer) is a proapoptotic member of the Bax group and contains BH1, BH2, and BH3 domains. Bak was identified and cloned independently by three different groups. Chittenden *et al.* (1995) and Kiefer *et al.* (1995) cloned Bak by PCR using degenerate primers to the BH1 and BH2 domains of Bcl-2, whereas Farrow *et al.* (1995) used a yeast two-hybrid screen using the adenovirus Bcl-2 homolog E1B 19K as bait. Overexpression of Bak accelerated death induced by cytokine withdrawal, placing it in the proapoptotic Bcl-2 subfamily.

b. Bcl-w (Gibson et al., 1996)

Bcl-w is an antiapoptotic family member that, like Bcl-x_L and Bcl-2, contains all four BH domains. Bcl-w was cloned by using low stringency PCR with degenerate primers encoding parts of the BH1 and BH3 domains. These primers were then used to amplify sequences from mRNA of a mouse macrophage and a mouse brain cell line. The amplified sequence was used to probe a cDNA library and the full-length gene was cloned. Upon overexpression, Bcl-w protected cells from cytokine withdrawal-induced apoptosis.

B. Protein–Protein Interactions

1. Coimmunoprecipitation

a. Bax (Oltvai et al., 1993)

Bax (Bcl-2 associated X protein) is the prototypic member of the proapoptotic Bax subfamily. It contains BH1, BH2, and BH3 domains and was identified in a search for Bcl-2 interacting proteins. Bax was isolated by coimmunoprecipitation using a monoclonal antibody against Bcl-2. Bax was then amino acid sequenced and the full-length gene cloned. When overexpressed, Bax accelerates cell death induced by cytokine withdrawal, making it the first proapoptotic Bcl-2 family member described.

2. Protein Interactive Cloning

a. Bid (Wang et al., 1996)

Bid is a BH3-only proapoptotic family member that was identified by protein interactive cloning. Fusion proteins of Bcl-2 and Bax without their carboxy termini were made with GST and HMK tags. The HMK domain allowed phosphorylation with ^{32}P of the fusion protein. The radiolabeled fusion proteins were then used to screen an expression library made from the 2B4 cell line. Bid was pulled out as both Bcl-2- and Bax-interacting protein. Both the mouse and the human full-length genes were subsequently cloned and, when overexpressed, were shown to enhance cell death induced by cytokine withdrawal.

b. Bad (Yang et al., 1995)

Bad is a proapoptotic BH3-only Bcl-2 family member that was identified by both a yeast two-hybrid screen and a Bcl-2 fusion protein screen. In the yeast two-hybrid screen, a Bcl-2 fusion protein missing the carboxy terminus was used as bait to screen an oligo(dT)-primed mouse embryonic day 14.5 cDNA fusion library. Bad was also identified while screening a bacteriophage λ expression library using a GST-HMK-Bcl-2 fusion protein. The full-length Bad cDNA was then isolated from mouse libraries and shown to accelerate cell death induced by cytokine withdrawal.

c. Bim_L (O'Connor et al., 1998)

Bim_L (Bcl-2 interacting mediator of cell death) is a BH3-only proapoptotic Bcl-2 family member. A ^{32}P-labeled recombinant human Bcl-2 protein was used to screen a bacteriophage λ cDNA expression library from a $p53^{-/-}$ T cell line, KO52DA20. Bim_L was identified as a Bcl-2-interacting protein and shown to induce apoptosis when overexpressed.

3. Yeast Two-Hybrid

a. Bik (Boyd et al., 1995)

Bik (Bcl-2 interacting killer) is a BH3-only proapoptotic Bcl-2 family member that was identified and cloned using the yeast two-hybrid system. A Gal4-Bcl-2 fusion protein was used as bait to screen a human B-cell cDNA library. Upon overexpression, Bik has been shown to induce apoptosis.

b. Bok (Hsu et al., 1997)

Bok (Bcl-2 related ovarian killer) is a proapoptotic Bcl-2 family member in the Bax group. Bok contains BH1, BH2, and BH3 domains and was isolated in a yeast two-hybrid screen. The yeast two-hybrid screen was performed using Mcl-1 as bait to screen an ovarian cDNA library. Upon isolation of positive clones, the full-length gene was generated by PCR from an ovarian and brain library. Bok was shown to induce cell death when overexpressed.

c. Hrk (Inohara et al., 1997)

Hrk, which stands for harakiri (the Japanese suicide ritual), is a BH3-only proapoptotic Bcl-2 family member. Hrk was identified in a yeast two-hybrid screen using the Gal4-Bcl-2 fusion protein as bait to screen a HeLa cDNA library. Overexpression of Hrk induces apoptosis.

C. Miscellaneous Methods

1. Differentiation-Induced Genes

a. NR-13 (Gillet et al., 1995)

NR-13 is an antiapoptotic Bcl-2 family member containing BH1, BH2, and BH3 domains. NR-13 was identified as a gene whose expression was upregulated on Rous sarcoma virus infection. NR-13 was shown to have homology to Bcl-2 at the amino acid level.

b. Mcl-1 (Kozopas et al., 1993)

Mcl-1 is an antiapoptotic Bcl-2 family member containing BH1, BH2, and BH3 domains. Mcl-1 was identified and cloned by differential screening and hybridization when ML-1 cells were induced with TPA and then compared to uninduced cells.

c. A1 (Lin et al., 1993)

A1 is an antiapoptotic Bcl-2 family member containing BH1 and BH2 domains. A1 was identified and cloned as a gene induced in macrophages following stimulation with granulocyte macrophage-colony stimulating factor (GM-CSF).

III. Analysis of Bcl-2 Family Protein Function and Regulation

A. Transfections

A common mechanism for testing the function of Bcl-2 family members is to over-express these genes, either transiently or stably, in cell lines. The following transfection protocol is currently used in our laboratory for stable transfection of several hematopoietic cell lines (including FL5.12, Baf3, Jurkat, H9, WEHI-231, and SKW6.4) (Johnson and Boise, 1999). Several Bcl-2 family members were initially characterized using a version of this protocol (Oltvai *et al.,* 1993; Boise *et al.,* 1993; Yang *et al.,* 1995).

1. Pellet 10^7 cells by centrifugation at 250*g* for 5 min. Best results are obtained using cells that are in log phase growth.
2. Aspirate the medium and resuspend the cells in 0.4 ml of fresh medium.
3. Place the cells in a 0.4-cm electroporation cuvette (Bio-Rad).
4. Add 15 μg DNA containing the gene of interest and mix well. (Always use an empty vector as a control as this will alleviate some troubleshooting if stable transfectants do not grow out or if the protein of interest is not expressed in the transfectants; see later.)

Table I
Antibodies Currently Used in Our Laboratory to Detect Bcl-2 Family Member Proteins

Epitope	Host species[a]	Antibody name	Source	Applications[b]	Species cross-reactivity[c]
Bad	Goat pAb	C-20	Santa Cruz	WB, IP, IHC	M, R
Bax	Mouse mAb	B-9	Santa Cruz	WB, IHC	M, R, H
Bax	Rabbit pAb	N-20	Santa Cruz	WB, IP, IHC	H, M, R
Bcl-2	Hamster mAb	6C8	Pharmingen	WB, IP, IHC, FC	H
Bcl-2	Hamster mAb	3F11	Pharmingen	WB, IP, IHC, FC	M
Bcl-2	Goat pAb	N-19	Santa Cruz	WB, IP, IHC	M, R, H
Bcl-x	Mouse mAb	7B2	Southern Biotech.	IP, FC	M, H
Bcl-x	Rabbit pAb	S-18	Santa Cruz	WB, IP, IHC	M, R, H

[a]mAb, monoclonal antibody; pAb, polyclonal antibody.
[b]WB, Western blot; IP, immunoprecipitations; IHC, immunohistochemistry; FC, flow cytometry analysis.
[c]M, mouse; R, rat; H, human.

5. Electroporate the cells by pulsing on a Bio-Rad Gene Pulser at 250 V and 960 μF. Electroporation settings may vary between the cell line and the instrument used.

6. Incubate the cuvette on ice for 10 min. Some cells (such as FL5.12) can be incubated at room temperature without affecting transfection efficiency.

7. Add the electroporated cells to 10 ml fresh medium and incubate at 37°C for 48 h.

8. Select stable transfectants by adding 50 ml of medium containing the appropriate concentration of the selection drug (G418, L-histidinol, Puromycin, etc.) and incubating at 37°C until transfected cells grow out.

9. Determination of protein expression can be achieved by Western (immunoblot) analysis (see Table I for a list of selected commercially available antibodies useful for detecting Bcl-2 family member proteins).

Notes: Other transfection procedures can also be utilized for expression of Bcl-2 family members, including Lipofectamine or Lipofectin, calcium phosphate, or DEAE. When selecting for stable cells, it is essential to know beforehand the appropriate concentration of the selecting drug that is required to kill nontransfected cells. Based on our experience with G418, this can range from 0.4 to 2.0 mg/ml. This can be determined by incubating untransfected cells in varying concentrations of the selection drug to see what concentration will kill the parental cells. This protocol can also be used for transient assays if proapoptotic Bcl-2 family members do not allow for the generation of stable transfectants. We have seen transfection efficiencies up to 60% with this protocol. However, for subsequent death assays, a reporter construct should be utilized to assure that analyses are only performed on transfected cells. We have used a CD20 construct for transient assays in FL5.12 cells. β-Galactosidase and GFP are common reporter molecules used for transfection of adherent cells such as COS and 293.

Troubleshooting: If no stable transfectants grow out (including vector control cells):

1. Plasmid DNA is contaminated with either RNA or genomic DNA. Try repurifying DNA if this is the case.
2. Concentration of selection drug is too high. Try lower concentrations of selection drug as described earlier.
3. Some vectors display differing expression levels based on the cell line utilized. For instance, we have found that CMV promoter-based constructs do not express very well in FL5.12 cells and we thus use the vector pSFFV-Neo (Fuhlbrigge *et al.*, 1988).

If stable transfectants grow out, but do not express the protein of interest:

1. The vector may have been disrupted on integration. If using a circular vector, linearize it prior to transfection.
2. The gene transfected was toxic to the cells and only those cells that either fail to express the protein product or express it at very low levels grow out. In this case, transient transfections must be employed.

B. Cell Death Assays

Cell lines can be induced to undergo apoptosis by numerous stimuli, including growth factor withdrawal, chemotherapeutic drugs, irradiation, activation-induced cell death (AICD), and members of the tumor necrosis factor receptor (TNFR) family of death receptors. Three protocols utilized by our laboratory to induce apoptosis are (1) signaling with recombinant human TNFα (Johnson and Boise *et al.*, 1999; Gross *et al.*, 1999b), (2) use of anti-Fas antibodies (Boise and Thompson, 1997), and (3) cytokine withdrawal (Boise *et al.*, 1993; Oltvai *et al.*, 1993).

1. Interleukin-3 (IL-3) Withdrawal Assay

1. Harvest cells by spinning at 250*g* for 5 min.
2. Wash cells three times with unsupplemented RPMI 1640 medium.
3. Resuspend cells (5×10^5 cells/ml) in medium lacking IL-3 but containing all the other supplements (FBS, L-glutamine, pen/strep, etc.).
4. Determine viability of cells at various time points (common time points for IL-3 withdrawal assays are 0, 24, 48, 96 h). Two common techniques are used to determine cell viability after IL-3 withdrawal:

a. Trypan blue staining. Multiple fields are analyzed to determine standard deviations.
b. Propidium iodide (PI) staining. Triplicate analysis of each test group is achieved by adding 3×10^5 cells per tube in three separate FACS tubes (Falcon 2052 or 2054). Cells are harvested by centrifugation at 250*g* for 5 min. Medium is aspirated and cells are resuspended in 500 μl of filter-sterilized FACS buffer

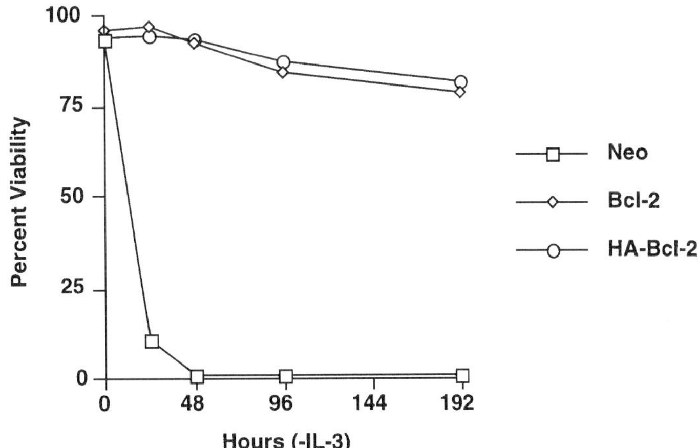

Fig. 3 Bcl-2 expression inhibits IL-3 withdrawal-induced cell death in the IL-3-dependent cell line FL5.12. FL5.12 cells were transfected with either the control vector pSFFV-Neo (Neo) or constructs encoding either wild-type Bcl-2 (Bcl-2) or Bcl-2 containing an influenza hemagglutinin tag on its N terminus (HA-Bcl-2). Cells were withdrawn from IL-3 as described, and viability was determined by analysis of propidium iodide staining on a FACScan flow cytometer.

[1X phosphate-buffered saline (PBS), pH 7.4, 1% bovine serum albumin (BSA), 0.01% sodium azide, stored at 4°C] containing 2 ng/ml propidium iodide. Viable (PI-negative) and nonviable (PI-positive) cells are analyzed on a flow cytometer and mean and standard deviations are derived from triplicate determinations. Figure 3 is a representative graph showing the protective ability of Bcl-2 in response to IL-3 withdrawal in FL5.12 cells as determined by PI staining.

2. Tumor Necrosis Factor-α Cytotoxicity Assay

1. Harvest cells by spinning at 250*g* for 5 min.

2. Resuspend cells at a concentration of 5×10^5 cells/ml.

3. Add the appropriate drugs to the cells. Many cell lines are not sensitive to TNFα alone and must also be treated with transcription or translation inhibitors such as actinomycin D (Act D) or cycloheximide (CHX), respectively. We generally treat FL5.12 cells with 2 ng/ml recombinant human TNFα (Calbiochem) and 10 μg/ml CHX (Sigma) for 6 h. This concentration of CHX is relatively nontoxic in these cells for up to 24 h. However, CHX toxicity can vary from cell line to cell line and must be determined prior to TNFα/CHX testing.

4. At the indicated time points, harvest 1 ml (because CHX or Act D inhibits cell division, the cell concentration will remain at 5×10^5 cells/ml throughout the time course of the experiment) by spinning at 250*g* for 5 min.

5. Wash cells twice with 1X PBS, pH 7.4.

6. Stain cells with Annexin V–FITC and PI according to the manufacturer's protocol (we use Annexin V–FITC from either Pharmingen or Biovision).

7. Analyze cells by flow cytometry. Live cells are negative for both stains, and apoptotic cells are positive for either Annexin V–FITC alone or will undergo secondary necrosis and be positive for both Annexin V–FITC and PI.

3. Anti-Fas (CD95/Apo-1)-Induced Cytotoxicity Assay

1. Harvest cells by spinning at 250*g* for 5 min.
2. Resuspend cells at a concentration of 5×10^5 cells/ml.
3. Add anti-Fas antibodies to induce apoptosis. In some cell lines, no Act D or CHX is required to kill cells by anti-Fas, although this must be determined before initiating anti-Fas experiments. We utilize the anti-Fas antibody Jo2 (Pharmingen) to kill mouse cells and anti-Fas antibody CH-11 (MBL) to kill human cells.
4. If this is the first time using a given cell line, perform a dose–response analysis with varying concentrations of anti-Fas antibody at different time points to optimize killing.
5. At the indicated time points, harvest 5×10^5 cells.
6. Stain cells with Annexin V–FITC and PI as described earlier.

Note: Jo2 does not kill all mouse cell lines. In contrast, it has been reported previously to inhibit activation-induced cell death in the 2B4 hybridoma (Yang *et al.,* 1995; Memon *et al.,* 1995). Therefore, Jo2 killing must be determined prior to testing for Bcl-2 function.

C. Detection of Bcl-2 Interactions

Antiapoptotic Bcl-2 family members have been reported to have multiple modes of action and can be regulated by many mechanisms. These regulatory mechanisms include, but are not restricted to, dimerization with other Bcl-2 family members, phosphorylation, and caspase-mediated cleavage (Gross *et al.,* 1999a). Also, as mentioned in the previous section, some Bcl-2 family member genes were discovered by screening for Bcl-2 interacting proteins. This can be achieved through the yeast two-hybrid assay or by coimmunoprecipitation assays. While the yeast two-hybrid system has been used by many laboratories in this respect, it does have drawbacks, including the frequent occurrence of false-positive interactions. For this reason, interactions found in the yeast two-hybrid system should always be confirmed by other biochemical analyses. We study interactions of Bcl-2 family members by coimmunoprecipitation assays, either by metabolically labeling cells and analyzing samples by autoradiography or by coimmunoprecipitating proteins and analyzing samples by Western blot analysis. Two protocols, both of which are adapted from Gottschalk *et al.* (1996), are provided that outline these procedures. Because Bcl-2 family members appear to function, at least in part, by regulating mitochondrial events during apoptosis, assays that measure mitochondrial function and cytochrome *c* distribution are an important aspect of studying Bcl-2 activity. Assays to study mitochondrial function during apoptosis are described elsewhere.

1. Coimmunoprecipitation of Metabolically Labeled Proteins

1. Resuspend cells at a concentration of 5×10^5 cells/ml.
2. Label cells by adding 0.1 mCi/ml of Trans-label [^{35}S]methionine to the wells overnight.
3. Harvest cells by centrifuging at 250*g* for 5 min and aspirate the media.
4. Lyse cells in 500 μl NET-N buffer (100 m*M* NaCl, 1 m*M* EDTA, 20 m*M* Tris, 0.2% NP-40, pH 8.0) supplemented with 8 μg/ml aprotinin, 2 μg/ml leupeptin, and 170 μg/ml phenylmethylsulfonyl fluoride (PMSF).
5. Centrifuge lysates at 14,000*g* for 10 min at 4°C and transfer supernatant to a fresh tube.
6. Preclear lysates by adding 50 μl of a protein G–agarose slurry to the lysates and rocking for 1 h at 4°C.
7. Centrifuge lysates at 14,000*g* for 10 min at 4°C and transfer supernatant to a fresh tube.
8. Add 1–4 μg of antibody (see Table I for a list of acceptable antibodies for IP) to the lysates and rock for at least 1–2 h at 4°C.
9. Add 50 μl protein G–agarose to the tubes to bind the protein/antibody complexes. Mix well and rock samples at 4°C for 1 h.
10. Centrifuge at 14,000*g* for 10 min at 4°C, aspirate, and remove supernatant.
11. Wash the protein G–agarose beads three times with NET-N buffer supplemented with protease inhibitors.
12. Add 60 μl of 1X Laemmli SDS–PAGE sample buffer to beads and mix well.
13. Boil samples for 5 min, pulse spin, and run on a 12 or 15% SDS–PAGE gel.
14. Incubate gel in EN^{3}HANCE (Dupont/NEN) or Amplify (Amersham) to enhance the signal.
15. Dry gel and expose to film at −80°C to detect interacting proteins.

2. Coimmunoprecipitation Assay by Western Blot Analysis

1. Harvest cells by spinning at 250*g* for 5 min, then aspirate, and remove the medium.
2. Wash cells with 1X PBS, pH 7.4, by resuspending cells and spinning again at 250*g*.
3. Aspirate PBS and lyse cells in NET-N buffer supplemented with protease inhibitors (as described earlier).
4. Transfer lysate to a 1.5-ml centrifuge tube.
5. Spin at 14,000*g* for 10 min at 4°C.
6. Transfer supernatant to a fresh 1.5-ml centrifuge tube.
7. Preclear lysate by incubating on a rocker for 60 min at 4°C with 50 μl protein A–agarose or protein G–agarose [we use GIBCO; for mouse IgG, protein G binds with greater affinity (Harlow and Lane, 1988)].
8. Spin down the slurry at 14,000*g* and transfer supernatant to a fresh 1.5-ml centrifuge tube.

9. Add 1–4 μg of antibody directed against one of the two proteins of interest and incubate 4 h to overnight at 4°C with shaking.
10. Add 50 μl of protein A/G–agarose to the lysate/antibody mix and incubate for 1 h at 4°C with shaking.
11. Spin down at 14,000*g* for 10 min at 4°C.
12. Remove the supernatant and wash beads three times with NET-N buffer supplemented with protease inhibitors.
13. Add 60 μl 1X Laemmli SDS–PAGE sample buffer.
14. Gently mix samples and boil for 5 min.
15. Load samples onto an SDS–PAGE gel and transfer to nitrocellulose by standard methods (Harlow and Lane, 1988).
16. Perform a Western (immunoblot) analysis using an antibody directed against the other protein involved in the hypothesized protein–protein interaction.

D. Analysis of Bcl-2 Posttranslational Modifications

Bcl-2 function is regulated by posttranslational modifications, such as phosphorylation and cleavage (Gross *et al.,* 1999a). Indeed, the proapoptotic Bcl-2 family member Bad becomes sequestered in the cytosol by 14-3-3 proteins when phosphorylated on serine residues in a PI-3K/Akt- and protein kinase A (PKA)-dependent fashion (Zha *et al.,* 1996; Harada *et al.,* 1999). Phosphorylation of Bcl-2 can also modulate its function (May *et al.,* 1994; Cheng *et al.,* 1997). We have tested the ability of Bcl-2 to become phosphorylated in response to TNFα-induced cell death (Fig. 4), and the following protocol is currently used in our laboratory.

1. Bcl-2 Phosphorylation Assay

1. Count cells and resuspend at a concentration of 1×10^6 cells/ml in phosphate-free medium (Life Technologies). We normally add 5×10^6 cells in 5 ml medium for each treatment condition. (We normally resuspend at 5×10^5 cells/ml for TNFα experiments, but the increased cell concentration has no effect on cell death and will give more lysate for the amount of ^{32}P used for subsequent steps in the protocol.)

2. Preincubate cells for 4 h at 37°C in [^{32}P]orthophosphate (0.1 mCi/ml).

3. Add the appropriate stimulus to flasks and incubate an additional 3 h at 37°C. We additionally add the serine/threonine phosphatase inhibitor okadaic acid (OA) at a concentration of 1 μ*M* to our flasks, as Bcl-2 phosphorylation is difficult to detect in its absence (Chang *et al.,* 1997). OA has no effect on cell viability over short time courses.

4. Harvest cells by spinning at 250*g* for 5 min.

5. Remove the supernatant and wash cells once with cold 1X PBS, pH 7.4.

6. Remove the supernatant and lyse cells in RIPA buffer (150 m*M* NaCl, 1.0% NP-40, 0.5% DOC, 0.1% SDS, and 50 m*M* Tris, pH 8.0) supplemented with protease inhibitors (2 μg/ml aprotinin, 1 μg/ml leupeptin, and 170 μg/ml PMSF) and phosphatase

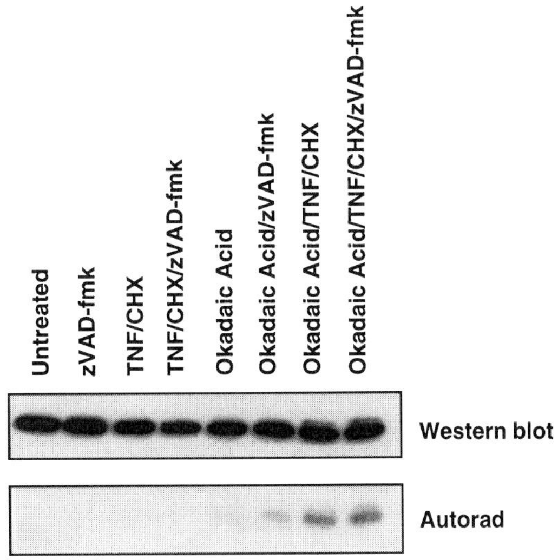

Fig. 4 Bcl-2 is phosphorylated in response to TNFα treatment in FL5.12 cells. FL5.12 cells transfected stably with wild-type Bcl-2 were subjected to the Bcl-2 phosphorylation assay as described in either the absence (first four lanes) or the presence (last four lanes) of the serine/threonine phosphatase inhibitor okadaic acid. (Top) After SDS–PAGE and transfer to nitrocellulose, the blot was subjected to Western analysis using the anti-Bcl-2 antibody 6C8 and protein was detected by enhanced chemiluminescence (ECL). The Bcl-2 band is shown and indicates equal protein loading in each lane. (Bottom) After allowing the ECL signal to decay, the blot was exposed to film at $-80°C$ for 7 h to test for ^{32}P incorporation by autoradiography. Upon longer exposure, ^{32}P incorporation is seen in the absence of okadaic acid, albeit with a much weaker signal.

inhibitors [10 m*M* sodium pyrophosphate, 1 m*M* sodium orthovanadate (Na_3VO_4), and 10 m*M* sodium fluoride (NaF)]. Lysates can be frozen at this step overnight at $-80°C$.

7. Preclear lysates by adding 50 μl of a protein A– or protein G–agarose slurry to the lysates and incubating for 1 h at 4°C rocking.
8. Spin tubes at 14,000*g* for 10 min at 4°C.
9. Transfer supernatant to a fresh 1.5-ml centrifuge tube.
10. Add 2 μg of an anti-Bcl-2 antibody (we use clone 6C8 from Pharmingen) to the lysates and incubate for 1–2 h at 4°C rocking.
11. Add 50 μl of a protein A– or protein G–agarose slurry to the lysate/antibody solution and incubate for 1 h at 4°C rocking.
12. Spin down slurry at 14,000*g* for 10 min at 4°C.
13. Aspirate supernatant and wash protein A– or protein G–agarose bead conjugates with RIPA buffer containing protease and phosphatase inhibitors.
14. Aspirate supernatant and add 60 μl 1X Laemmli SDS–PAGE sample buffer.
15. Boil samples for 5 min and load onto a 15% SDS–PAGE gel.
16. Analyze by Western blot analysis to ensure equal protein loading [see Fig. 4 (top) for a representative Western blot].

17. After the ECL signal has decayed, detect ^{32}P incorporation by autoradiography [see Fig. 4 (bottom) for a representative autorad].

2. Bcl-2 Cleavage Assay

It has been shown previously by several laboratories that death receptor signaling can induce cleavage of antiapoptotic Bcl-2 family members such as Bcl-2 (Cheng *et al.*, 1997; Johnson and Boise, 1999) and Bcl-x_L (Clem *et al.*, 1998), as well as proapoptotic members such as Bid (Li *et al.*, 1998; Luo *et al.*, 1998; Gross *et al.*, 1999b). The importance of this event in the regulation of Bcl-2 function appears to be both cell and stimulus specific. The following protocol is used in our laboratory to test Bcl-2 cleavage.

1. Harvest cells by spinning at 250*g* for 5 min.
2. Resuspend cells at a concentration of 5×10^5 cells/ml. To have enough cells for the preparation of lysates, we use at least 5×10^6 cells per treatment for each time point.
3. Add the appropriate drugs to the cells as described earlier.
4. At the appropriate time points, harvest 5×10^5 cells for Annexin V–FITC and PI staining and harvest another 4.5×10^6 cells in another tube for the preparation of whole cell lysates.
5. Stain cells with Annexin V–FITC and PI as described previously.
6. For whole cell lysates, wash cells once with ice-cold 1X PBS, pH 7.4.

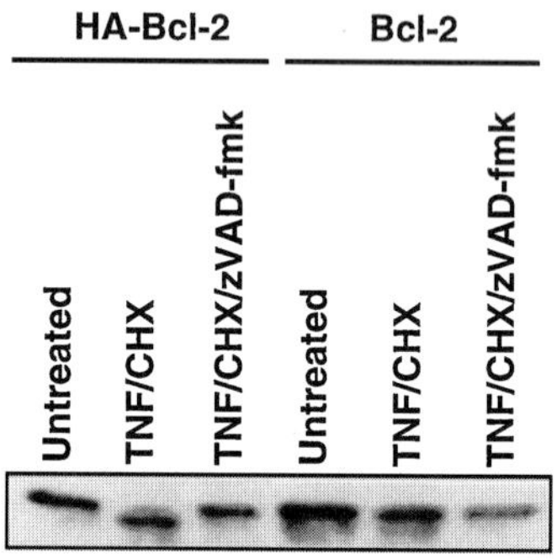

Fig. 5 The HA epitope tag is cleaved by caspases activated by TNFα. FL5.12 cells stably expressing an HA-tagged version of Bcl-2 (HA-Bcl-2) or wild-type Bcl-2 (Bcl-2) were subjected to a TNFα-induced cytotoxicity assay as described. Whole cell lysates were generated, and 50 μg total protein was run per lane on a 15% SDS–PAGE gel. The gel was transferred to nitrocellulose and probed with the anti-Bcl-2 antibody 6C8. The presence of a faster migrating band in the TNF/CHX lane of HA-Bcl-2 cells might be mistaken for cleavage of Bcl-2; however, this cleavage event is in fact the result of cleavage of the HA epitope tag, and the size of the band does not correlate with the predicted size of the caspase-cleaved fragment recognized by 6C8. Wild-type Bcl-2 cells also undergo TNFα-induced cleavage, but no faster migrating band is present due to the lack of the epitope tag.

7. Lyse cells in 1X RIPA buffer supplemented with protease inhibitors (for 4.5×10^6 cells, we normally lyse cells in approximately 50 μl to concentrate the protein samples for Western blot analysis) and spin at 14,000*g* for 5 min at 4°C.
8. Transfer supernatant to a fresh 1.5-ml tube.
9. Determine protein concentration and analyze samples by Western blot analysis.

Notes: It has become a common practice to transfect cells with Bcl-2 family members tagged on their N- or C-terminal regions to facilitate detection and manipulation of these proteins. While this "epitope tagging" usually has no effect on the ability of these proteins to function (see Fig. 3), it can be a problem when examining the caspase-mediated cleavage of pro- or antiapoptotic Bcl-2 family members. An influenza hemagglutinin (HA) tag, for example, has the amino acid sequence YPYDVPDYA. The sequences YPYD and DVPD are putative caspase cleavage sites. As shown in Fig. 5, the addition of the HA tag to the N-terminus of Bcl-2 leads to a faster migrating band in the presence of TNFα/CHX. While this band might be mistaken for a cleavage fragment of Bcl-2, this band is actually generated by cleavage of the HA tag. More importantly, this could result in loss of detection if the epitope tag is being utilized for IP or Western blotting purposes. Similarly, the commonly used FLAG epitope (amino acid sequence DYKDDDDK) has the potential caspase cleavage sites DYKD and DDDD.

References

Adams, J. M., and Cory, S. (1998). The Bcl-2 protein family: Arbiters of cell survival. *Science* **281,** 1322–1326.

Boise, L. H., Gonzalez-Garcia, M., Postema, C. E., Ding, L., Lindsten, T., Turka, L. A., Mao, X., Nunez, G., and Thompson, C. B. (1993). Bcl-x, a Bcl-2-related Gene that functions as a dominant regulator of apoptotic cell death. *Cell* **74,** 597–608.

Boise, L. H., Gottschalk, A. R., Quintans, J., and Thompson, C. B. (1995). Bcl-2 and Bcl-2-related proteins in apoptosis regulation. *In* "Apoptosis in Immunology" (G. Kroemer and C. Martinez-A., eds.), pp. 107–123. Springer-Verlag.

Boise, L. H., and Thompson, C. B. (1997). Bcl-x_L can inhibit apoptosis in cells that have undergone Fas-induced protease activation. *Proc. Natl. Acad. Sci. USA* **94,** 3759–3764.

Boyd, J. M., Gallo, G. J., Elangovan, B., Houghton, A. B., Malstrom, S., Avery, B. J., Ebb, R. G., Subramanian, T., Chittenden, T., Lutz, R. J., and Chinnadurai, G. (1995). Bik, a novel death-inducing protein shares a distinct sequence motif with Bcl-2 family proteins and interacts with viral and cellular survival-promoting proteins. *Oncogene* **11,** 1921–1928.

Chang, B. S., Minn, A. J., Muchmore, S. W., Fesik, S. W., and Thompson, C. B. (1997). Identification of a novel regulatory domain in Bcl-x_L and Bcl-2. *EMBO J.* **16,** 968–977.

Cheng, E. H.-Y., Levine, B., Boise, L. H., Thompson, C. B., and Hardwick, J. M. (1996). Bax-independent inhibition of apoptosis by Bcl-x_L. *Nature* **379,** 554–556.

Cheng, E. H.-Y., Kirsch, D. G., Clem, R. J., Ravi, R., Kastan, M. B., Bedi, A., Ueno, K., and Hardwick, J. M. (1997). Conversion of Bcl-2 to a Bax-like death effector by caspases. *Science* **278,** 1966–1968.

Chittenden, T., Harrington, E. A., O'Conner, R., Flemington, C., Lutz, R. J., Evan, G. I., and Guild, B. C. (1995). Induction of apoptosis by the Bcl-2 homologue Bak. *Nature* **374,** 733–736.

Chou, J. J., Li, H., Salvesen, G. S., Yuan, J., and Wagner, G. (1999). Solution structure of Bid, an intracellular amplifier of apoptotic signaling. *Cell* **96,** 615–624.

Clem, R. J., Cheng, E. H.-Y., Karp, C. L., Kirsch, D. G., Ueno, K., Takahashi, A., Kastan, M. B., Griffin, D. E., Earnshaw, W. C., Veliuona, M. A., and Hardwick, J. M. (1998). Modulation of cell death by Bcl-x_L through caspase interaction. *Proc. Natl. Acad. Sci. USA* **95,** 554–559.

Conradt, B., and Horvitz, H. R. (1998). The *C. elegans* protein EGL-1 is required for programmed cell death and interacts with the Bcl-2-like protein CED-9. *Cell* **93,** 519–529.

Farrow, S. N., Whaite, J. H. M., Martinou, I., Raven, T., Pun, K., Grinham, C. J., Martinou, J., and Brown, R. (1995). Cloning of a bcl-2 homologue by interaction with adenovirus ElB 19K. *Nature* **374,** 731–733.

Fuhlbrigge, R. C., Fine, S. M., Unanue, E. R., and Chaplin, D. D. (1988). Expression of membrane interleukin 1 by fibroblast transfected with pro-interleukin 1α. *Proc. Natl. Acad. Sci. USA* **85,** 5649–5653.

Gibson, L., Holmgreen, S. P., Huang, D. C., Bernard, O., Copeland, N. G., Jenkins, N. A., Sutherland, G. R., Baker, E., Adams, J. M., and Cory, S. (1996). Bcl-w, a novel member of the bcl-2 family, promotes cell survival. *Oncogene* **13,** 665–675.

Gillet, G., Guerin, M., Trembleau, A., and Brun, G. (1995). A Bcl-2-related gene is activated in avian cells transformed by the Rous sarcoma virus. *EMBO J.* **14,** 1372–1381.

Gottschalk, A. R., Boise, L. H., Oltvai, Z. N., Accavitti, M. A., Korsmeyer, S. J., Quintans, J., and Thompson, C. B. (1996). The ability of Bcl-x_L and Bcl-2 to prevent apoptosis can be differentially regulated. *Cell Death Differ.* **3,** 113–118.

Gross, A., McDonnell, J. M., and Korsmeyer, S. J. (1999a). BCL-2 family members and the mitochondria in apoptosis. *Genes Dev.* **13,** 1899–1911.

Gross, A., Yin, X. M., Wang, K., Wei, M. C., Jockel, J., Milliman, C., Erdjument-Bromage, H., Tempst, P., and Korsmeyer, S. J. (1999b). Caspase cleaved BID targets mitochondria and is required for cytochrome *c* release, while BCL-X_L prevents this release but not tumor necrosis factor-R1/Fas death. *J. Biol. Chem.* **274,** 1156–1163.

Harada, H., Becknell, B., Wilm, M., Mann, M., Huang, L. J., Taylor, S. S., Scott, J. D., and Korsmeyer, S. J. (1999). Phosphorylation and inactivation of BAD by mitochondria-anchored protein kinase A. *Mol. Cell* **3,** 413–422.

Harlow, E., and Lane, D. (1988). "Antibodies; A Laboratory Manual." Cold Spring Harbor Laboratory, Cold Spring Harbor, NY.

Hegde, R., Srinivasula, S. M., Ahmad, M., Fernandes-Alnemri, T., and Alnemri, E. S. (1998). Blk, a BH3-containing mouse protein that interacts with Bcl-2 and Bcl-x_L, is a potent death agonist. *J. Biol. Chem.* **273,** 7783–7786.

Hsu, S. Y., Kaipia, A., McGee, E., Lomeli, M., and Hsueh, J. W. (1997). Bok is a pro-apoptotic Bcl-2 protein with restricted expression in reproductive tissues and heterodimerizes with selective anti-apoptotic Bcl-2 family members. *Proc. Natl. Acad. Sci. USA* **94,** 12401–12406.

Inohara, N., Ding, L., Chen, S., and Núñez, G. (1997). Harakiri, a novel regulator of cell death, encodes a protein that activates apoptosis and interacts selectively with survival-promoting proteins Bcl-2 and Bcl-X_L. *EMBO J.* **16,** 1686–1694.

Johnson, B. W., and Boise, L. H. (1999). Bcl-2 and caspase inhibition cooperate to inhibit tumor necrosis factor-α-induced cell death in a Bcl-2 cleavage-independent fashion. *J. Biol. Chem.* **274,** 18552–18558.

Kiefer, M. C., Brauer, M. J., Powers, V. C., Wu, J. J., Umansky, S. R., Tomei, L. D., and Barr, P. J. (1995). Modulation of apoptosis by the widely distributed Bcl-2 homologue Bak. *Nature* **374,** 736–739.

Kozopas, K. M., Yang, T., Buchan, H. L., Zhou, P., and Craig, R. W. (1993). MCL1, a gene expressed in programmed myeloid cell differentiation, has sequence similarity to BCL2. *Proc. Natl. Acad. Sci. USA* **90,** 3516–3520.

Li, H., Zhu, H., Xu, C., and Yuan, J. (1998). Cleavage of Bid by caspase 8 mediates the mitochondrial damage in the Fas pathway of apoptosis. *Cell* **94,** 491–501.

Lin, E. Y., Orlofsky, A., Berger, M. S., and Prystowsky, M. B. (1993). Characterization of A1, a novel hemopoietic-specific early-response gene with sequence similarity to bcl-2. *J. Immunol.* **151,** 1979–1988.

Luo, X., Budihardjo, I., Zou, H., Slaughter, C., and Wang, X. (1998). Bid, a Bcl-2 interacting protein, mediates cytochrome *c* release from mitochondria in response to activation of cell surface death receptors. *Cell* **94,** 481–490.

May, W. S., Tyler, P. G., Ito, T., Armstrong, D. K., Qatsha, K. A., and Davidson, N. E. (1994). Interleukin-3 and bryostatin 1 mediate hyperphosphorylation of Bcl 2 alpha in association with suppression of apoptosis. *J. Biol. Chem.* **269,** 26865–26870.

McDonnell, J. M., Fushman, D., Milliman, C. L., Korsmeyer, S. J., and Cowburn, D. (1999). Solution structure of the proapoptotic molecule Bid: A structural basis for apoptotic agonist and antagonist. *Cell* **96,** 625–634.

Memon, S. A., Moreno, M. B., Petrak, D., and Zacharchuk, C. M. (1995). Bcl-2 blocks glucocorticoid-but not Fas- or activation-induced apoptosis in a T cell hybridoma. *J. Immunol.* **155,** 4644–4652.

Minn, A. J., Kettlun, C. S., Liang, H., Kelekar, A., Vander Heiden, M. G., Chang, B. S., Fesik, S. W., Fill, M., and Thompson, C. B. (1999). Bcl-x_L regulates apoptosis by heterodimerization-dependent and -independent mechanisms. *EMBO J.* **18,** 632–643.

Muchmore, S. W., Sattler, M., Liang, H., Meadows, R. P., Harlan, J. E., Yoon, H. S., Nettesheim, D., Chang, B. S., Thompson, C. B., Wong, S., Ng, S., and Fesik, S. W. (1996). X-ray and NMR structure of human Bcl-x_L, an inhibitor of programmed cell death. *Nature* **381,** 335–341.

O'Connor, L., Strasser, A., O'Reilly, L. A., Hausmann, G., Adams, J. M., Cory, S., and Huang, D. C. S. (1998). Bim: A novel member of the Bcl-2 family that promotes apoptosis. *EMBO J.* **17,** 384–395.

Oltvai, Z. N., Milliman, C. L., and Korsmeyer, S. J. (1993). Bcl-2 heterodimerizes in vivo with a conserved homolog, Bax, that accelerates programmed cell death. *Cell* **74,** 609–619.

Reed, J. C., Zha, H., Aime-Sempe, C., Takayama, S., and Wang, H.-G. (1996). Structure-function analysis of Bcl-2 family proteins. In "*Mechanisms of Lymphocyte Activation and Immune Regulation,*" Vol. I. New York, Plenum Press.

Wang, K., Yin, X. M., Chao, D. T., Milliman, C. L., and Korsmeyer, S. J. (1996). BID: A novel BH3 domain-only death agonist. *Genes Dev.* **10,** 2859–2869.

Yang, E., Zha, J., Jockel, J., Boise, L. H., Thompson, C. B., and Korsmeyer, S. J. (1995). Bad, a heterodimeric partner for Bcl-X_L and Bcl-2, displaces Bax and promotes cell death. *Cell* **80,** 285–291.

Yang, Y., Marcep, M., Ware, C. F., and Ashwell, J. D. (1995). Fas and activation-induced Fas ligand mediate apoptosis of T cell hybridomas: Inhibition of Fas ligand expression by retinoic acid and glucocorticoids. *J. Exp. Med.* **181,** 1673–1682.

Zha, J., Harada, H., Yang, E., Jockel, J., and Korsmeyer, S. J. (1996). Serine phosphorylation of death agonist Bad in response to survival factor results in binding to 14-3-3 not Bcl-x_L. *Cell* **87,** 619–628.

CHAPTER 3

Flow Cytometric Analysis of Cell Shrinkage and Monovalent Ions during Apoptosis

Carl D. Bortner and John A. Cidlowski

The Laboratory of Signal Transduction
National Institute of Environmental Health Sciences
National Institutes of Health
Research Triangle Park, North Carolina 27709

I. Introduction

Apoptosis is a physiological mode of cell death that affects many aspects of natural life from embryonic development through cellular homeostasis and disease. An explicit set of morphological and biochemical characteristics, which include the loss of cell volume, internucleosomal DNA degradation, and the formation of apoptotic bodies, sets apoptosis apart from the accidental cell death process known as necrosis. The loss of

0091-679X/01 $35.00

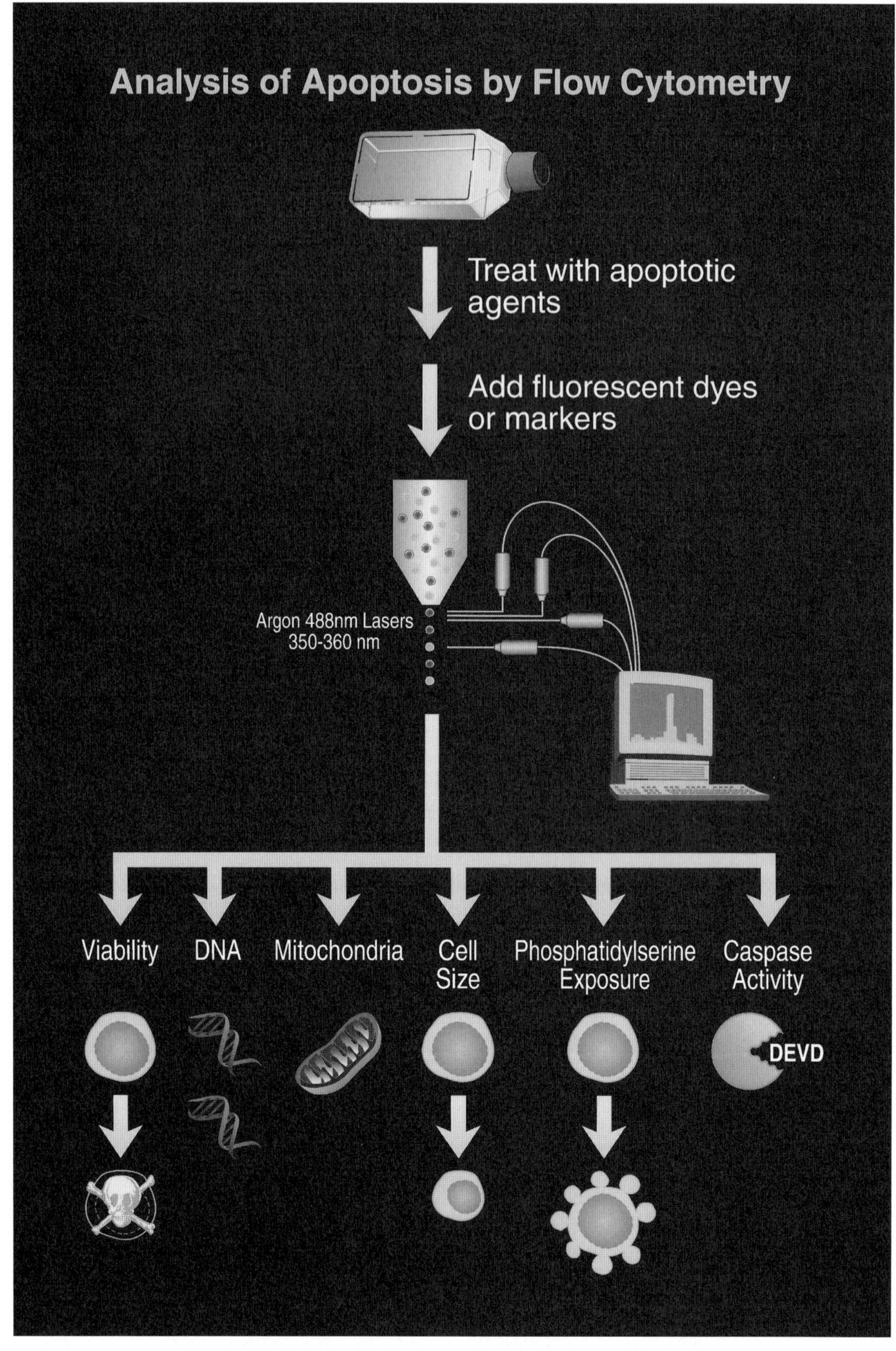
Analysis of Apoptosis by Flow Cytometry
Treat with apoptotic agents
Add fluorescent dyes or markers
Argon 488nm Lasers
350-360 nm
Viability
DNA
Mitochondria
Cell Size
Phosphatidylserine Exposure
Caspase Activity
DEVD

cell volume, or cell shrinkage, has been a defining feature in all well-characterized cases of apoptosis and is in stark contrast to the cellular swelling that occurs during necrosis (Kerr *et al.*, 1972; Wyllie, 1980; Thomas and Bell, 1981; Wyllie and Morris, 1982). The regulation of cell shrinkage during apoptosis has become an area of intense research focus. A loss of intracellular ions, particularly sodium and potassium, has been shown to be associated with the shrunken population of apoptotic cells (Klassen *et al.*, 1993; Jonas *et al.*, 1994; Barbiero *et al.*, 1995; Beauvais *et al.*, 1995; Benson *et al.*, 1996; Bortner *et al.*, 1997; Hughes *et al.*, 1997; McCarthy and Cotter, 1997; Yu *et al.*, 1997). Additionally, this decrease in intracellular ionic strength has been shown to be required for apoptotic nuclease activity and effector caspase activation (Hughes *et al.*, 1997), suggesting that the normal concentration of intracellular ions can inhibit the apoptotic machinery, and thus plays a more extensive role than simply facilitating the loss of cell volume during cell death.

Normal cellular volume is tightly controlled within very narrow limits in relation to changes in the extracellular environment (for review, see Hoffmann, 1987; Al-Habori, 1994). Upon exposure to a hypertonic environment, cells initially shrink through an obligatory loss of intracellular water. However, most cells compensate for this decrease in cell volume by the activation of specific ion transport pathways known collectively as a regulatory volume increase (RVI) response. During the RVI response, ions are transported into cells, resulting in an increase in intracellular water and in cell size. In contrast, exposure of cells to a hypotonic environment results in initial cellular swelling. Again, most cells compensate for this increase in cell volume by the activation of a regulatory volume decrease (RVD) response. The RVD response transports ions out of the cells, resulting in an obligatory efflux of intracellular water and a decrease in cell size. While these volume regulatory mechanisms protect cells from anisotonic changes, they appear to be either inactivated or overridden during cell death, as cells that possess an RVI response can still undergo apoptosis.

When a cell receives an apoptotic signal, a series of morphological and biochemical events initiate the cell death program. These early events lead to the activation of effector processes that carry out the apoptotic program and ensure the death of the cell. Currently, the interrelationship among these various apoptotic characteristics and events is not completely understood. Because numerous apoptotic characteristics can be examined by flow cytometry, it has been employed as a valuable tool in the study of apoptosis. Flow cytometry has many advantages in the examination of such a dynamic physiological

Fig. 1 Analysis of apoptosis by flow cytometry. Primary or cultured cell lines are treated with various apoptotic agents. After a period of time, numerous fluorescent dyes or markers can be added to the cells and the cells immediately or subsequently examined. During flow cytometry, the cells individually pass by a laser that excites the dye or marker. The fluorescent emission then passes through various filters and is detected by photomultiplier tubes. Several different cellular characteristics can be determined for a single cell and analyzed by various computer programs. In the study of apoptosis, characteristics such as cellular viability, cell size, DNA content, changes in the mitochondrial membrane potential, phosphatidylserine exposure, and caspase activity can be determined using flow cytometry. (See Color Plate.)

process as programmed cell death, including the ability to examine thousands of cells in a relatively short period of time at the single cell level. In addition, at any point during the cell death process, subpopulations of apoptotic cells can be isolated and examined, either by electronic or by physical sorting, to isolate apoptotic cells with characteristics distinct from those of the normal cell population.

The use of various fluorescent dyes and markers in cells treated with an apoptotic agent allows for the detection of specific programmed cell death characteristics by flow cytometry (Fig. 1, see Color Plate). For example, changes in cell size, DNA content, mitochondrial membrane potential, plasma membrane asymmetry [such as translocation of phosphatidylserine (PS) from the inner to the outer leaflet of the plasma membrane], and caspase activity can be examined individually by flow cytometry with different fluorescent indicators. Additionally, flow cytometry can be used to assist in defining the relationship between these complex series of events by comparing one cellular characteristic to another throughout the apoptotic process at the single cell level. This type of analysis permits the ordering of cell death characteristics into either early or late apoptotic events. This chapter focuses on the use of flow cytometry to examine cell shrinkage, a fundamental characteristic of apoptosis, and how in combination with other fluorescent markers, cell size can be related to other features of the cell death process.

II. Analysis of Cell Size by Flow Cytometry

Flow cytometry is a relatively fast and easy way to determine changes in cell size during apoptosis if one has access to a suitable instrument. Variations in the physical characteristics of a cell are assessed with flow cytometry by examining changes in its light-scattering properties. A change in the light-scattering property of a cell in the forward direction (forward scatter) is a direct measure of cell size; a decrease or increase in forward scatter indicates a decrease or increase in cell size, respectively. Additionally, a change in light scatter at a 90° angle (side scatter) is a measure of cell density or granularity. When a cell undergoes apoptosis, the cell shrinks, with an observed decrease in forward-scattered light. In contrast, when a cell swells, as occurs during an accidental cell death process known as necrosis, an increase in forward-scattered light can be observed.

A. Data Acquisition for Changes in Cell Size during Apoptosis

1. Place approximately 5×10^5 to 1×10^6 control cells in normal media (for example, we use RPMI 1640 for lymphoid cells) into a 6-ml round-bottom polystyrene tube (Falcon 2058).

2. Examine the cells by flow cytometry with excitation at 488 nm on a forward-scatter (FSC) versus side-scatter (SSC) dot plot and individual FSC and SSC histograms.

3. With the FSC instrument setting in the linear mode, adjust the FSC voltage and amp gain to initially position the main population of cells in the center of the FSC histogram.

4. With the SSC instrument setting in either linear or log mode, adjust the SSC voltage and amp gain to initially position the main population of cells in the center of the SSC histogram.

Note: Depending on the changes in cell size and cell density that may occur during apoptosis in a given model system, adjustment of these initial settings to keep the population of cells within the dot plot and histogram boundaries may be necessary. For example, because apoptotic cells shrink and some cell types display an increase in cell density, the initial cell population may be positioned to the right of center on the FSC histogram and to the left of center on the SSC histogram. The optimal position of the cells on the flow cytometric dot plots and histograms can only be determined empirically for each cell type being investigated. However, once instrument settings are determined for a particular study, these settings should not be changed during the course of the experiment.

5. Examine each experimental cell sample for changes in cell size by flow cytometry by placing the cells into individual 6-ml round-bottom, polystyrene tubes (Falcon 2058) and determining their position on a FSC versus SSC dot plot or a FSC histogram.

B. Data Analysis

There are several ways to examine data obtained from the light-scattering properties to assess changes in cell size. An increase in forward-scattered light indicates that cells have increased in size, whereas a decrease in forward-scattered light indicates cells have decreased in size. Additional information about the cells can be obtained by examining changes in their ability to scatter light at a 90° angle or side scatter. An increase in side-scattered light indicates that cells have increased cellular density or granularity, whereas a decrease in side-scattered light indicates that cells have decreased cellular density. As shown in Fig. 2A (see Color Plate), these light-scattering properties of the cells can be observed on a FSC versus SSC dot plot. S49 Neo cells, a murine immature T-cell line treated with the synthetic glucocorticoid dexamethasone (Dex), have been shown to undergo apoptosis in both a time- and a concentration-dependent manner. After 24 h of Dex treatment, the entire population of cells had a reduction in forward-scattered light, indicating a loss in cell size. Additionally, a separate or distinct population of cells that have an increase in cell density (as indicated by an increase in side-scattered light) along with a further decrease in cell size is observed. This latter population becomes more evident after 48 h of Dex treatment, when a significant number of cells have a decrease in cell size and an increase in cell density. In contrast, treatment of S49 Neo cells with 250 m*M* mannitol for 1 min results in an immediate and complete decrease in cell size without the observed increase in cell density as observed during apoptosis (Fig. 2A), demonstrating the ability of flow cytometry to detect rapid changes in cell size.

One can also examine the cells on individual histograms to compare control cells to cells treated with various apoptotic agents (Fig. 2A). A margin can be set on these histograms to give a clear indication of subtle changes in cell size, which may occur under various apoptotic conditions. Additionally, these histograms can be overlaid onto a single plot to directly examine differences in the forward-scattered light property of the cells with various apoptotic stimuli. Other ways to assess changes in the light-scattering properties of a cell include contour plots, which have the advantage of giving a three-dimensional view of the cells on a two-dimensional medium (Fig. 2B, see Color Plate). Additionally, the cells can be analyzed directly on a three-dimensional plot to simultaneously examine changes in the number of cells when comparing two different cellular characteristics (Fig. 2B).

If only the viable population of cells is of interest for a particular study, one can also incorporate propidium iodide (PI) into the sample. PI functions as a vital dye and will only enter cells and stain DNA + RNA if a loss of plasma membrane integrity has occurred. Thus, one can gate out cells that have lost their membrane integrity. PI can be excited at 488 nm and determined using either a 585-nm filter (FL-2) or a 650-nm long-pass filter (FL-3), which are standard on most flow cytometers. Because apoptosis in a population of cells is a very dynamic process, individual cells in a population can be found at different stages of cell death at any given time. Therefore, the appearance of cells that have lost their membrane integrity can occur at any point during the analysis. Depending on the percentage of cells that comprises this nonviable population, subtle changes in cell size may not be evident when examining the entire population of cells. Thus, the inclusion of PI, prior to flow cytometric examination, may be beneficial or even essential. The procedure for using PI in the study of cell size by flow cytometry is described.

C. Data Acquisition for Changes in Cell Size during Apoptosis with Propidium Iodide

1. Place approximately 5×10^5 to 1×10^6 control cells into a 6-ml round-bottom polystyrene tube (Falcon 2058).

2. Add 1 μl of a 10-mg/ml stock solution of PI (in 1X PBS; Sigma) for each milliliter of cells in the flow tubes. *Note:* PI will immediately enter cells that have compromised membrane integrity and will be excluded from cells with an intact plasma membrane. No specific incubation time or washes are required.

3. Examine the cells by flow cytometry with excitation at 488 nm on a FSC versus SSC dot plot and individual FSC and SSC histograms.

Fig. 2 Changes in cell size can be determined by flow cytometry. (A) S49 Neo cells treated with dexamethasone or mannitol were initially examined on a forward-scatter versus side-scatter dot plot. In this example, 10,000 cells were examined on a FACSort flow cytometer (Becton-Dickinson, San Jose, CA). Following 48 h of Dex treatment, we observed an increase in a population of cells that had a decrease in forward-scattered light, indicating a loss of cell size. Additionally, an increase in side scatter, indicating an increase in cellular granularity, was also observed during the course of apoptosis. In contrast, cells placed in a hypertonic environment using mannitol displayed an immediate loss of cell volume, without the increase in cell granularity. Forward-scatter histograms can be used to determine the position of the cells at any time during the cell death process. (B) Numerous flow cytometric plots can be employed to examine changes in cell size. Cells can be examined on a forward-scatter versus side-scatter contour plot to give a three-dimensional view of the cell population on a two-dimensional medium. Additionally, the cells can be examined on a three-dimensional plot to observed specific changes in cell number at various times during the apoptotic process. (See Color Plate.)

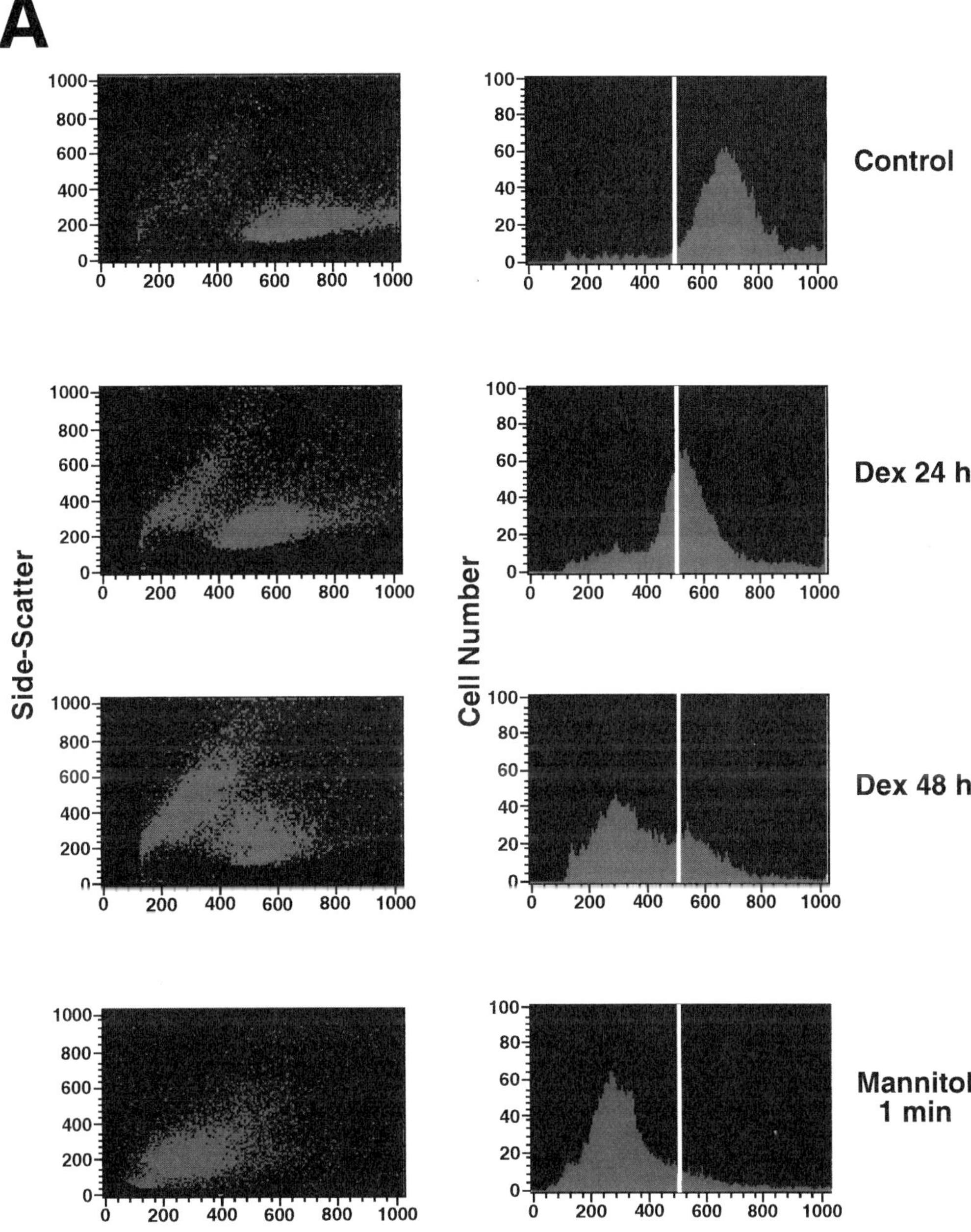
A
Control
Dex 24 h
Dex 48 h
Mannitol 1 min
Side-Scatter
Cell Number
Forward-Scatter
0
200
400
600
800
1000
0
20
40
60
80
100

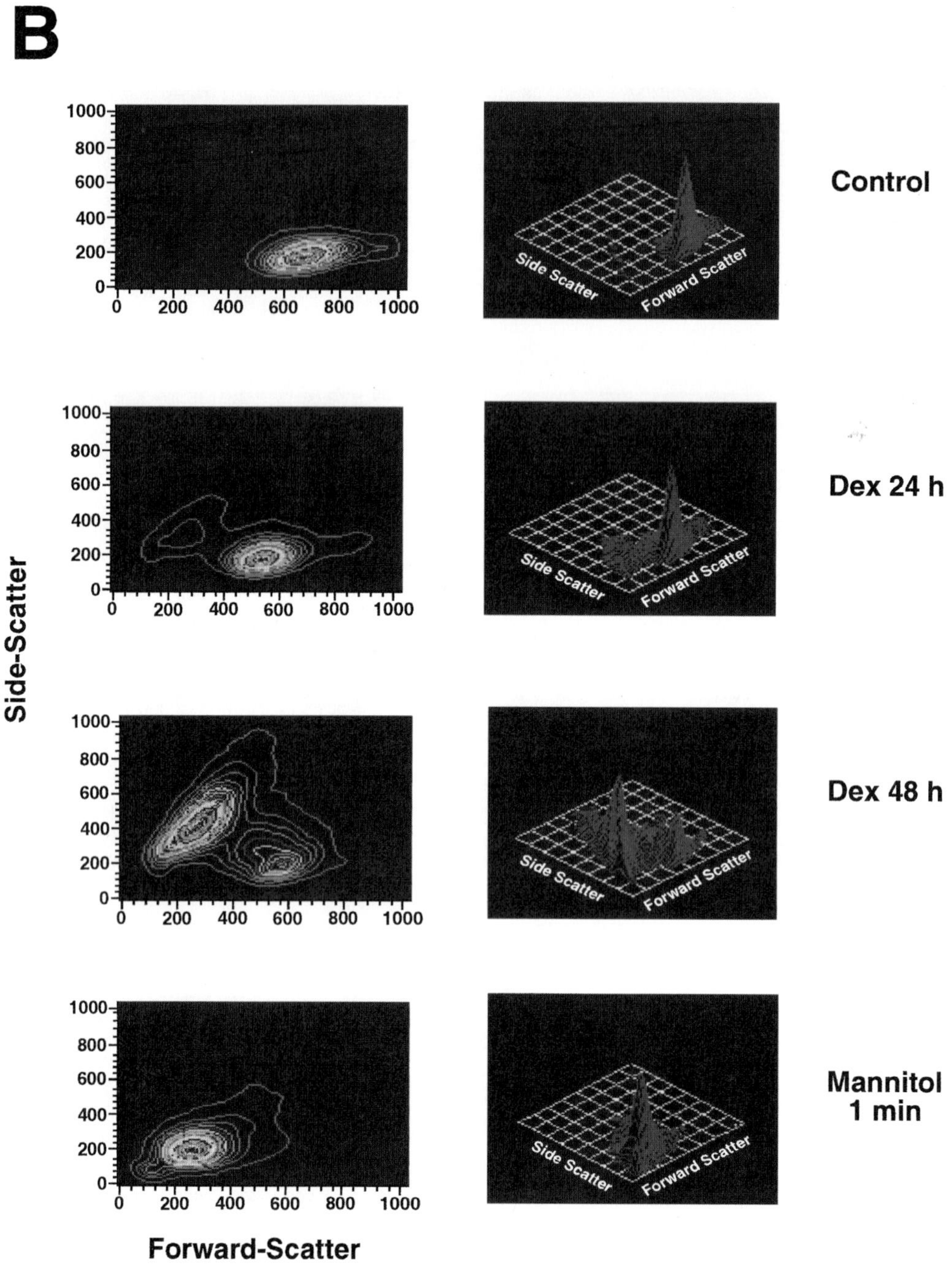

Fig. 2 *(continued).*

4. With the FSC instrument setting in the linear mode, adjust the FSC voltage and amp gain to initially position the main population of cells in the center of the FSC histogram.

5. With the SSC instrument setting in either linear or log mode, adjust the SSC voltage and amp gain to initially position the main population of cells in the center of the SSC histogram.

Note: Once again, as with the examination of cell size in the absence of PI, adjustments to the FSC and SSC settings of the cells may be necessary depending on the particular model system used in order to keep the population of cells within the dot plot and histogram boundaries. However, the instrument settings that are chosen for a particular study should not be changed during the course of the experiment.

6. Examine the cells on a FSC versus PI dot plot and a PI histogram. With the PI instrument setting in log mode, adjust the PI voltage and amp gain to position the main population of cells in the first decade of the PI scale. An acquisition gate can be set on the FSC versus PI dot plot to examine only cells that are PI negative and have therefore retained their membrane integrity. Set all other dot plots and histograms to examine only the PI-negative (viable) population of cells.

7. Examine each experimental cell sample for changes in cell size and PI fluorescence by flow cytometry by placing the cells into individual 6-ml round-bottom, polystyrene tubes (Falcon 2058), adding PI to a final concentration of 10 μg/ml, and observing their position on a FSC versus SSC dot plot, a FSC versus PI dot plot, or a FSC histogram.

There are many advantages to using flow cytometry to determine changes in cell size during apoptosis. First, flow cytometry allows for a very rapid and simple analysis of cell size, based on the light-scattering properties of the cell. Thousands of cells can be analyzed in a relatively short period of time with very little sample preparation. Cells can be examined in the media in which they are cultured, eliminating any laborious manipulation of the sample and permitting a large number of samples to be examined at any given time. Second, changes in cell size can be examined in relation to other cellular characteristics, such as cell density and membrane integrity. Finally, these distinct populations of apoptotic cells can be isolated and examined at various times during the cell death process.

III. Analysis of Monovalent Intracellular Ions by Flow Cytometry

Flow cytometry can be used to examine cellular characteristics other than changes in cell size during apoptosis, depending on the specific fluorescent dye or indicator that is incorporated into the experiment. As mentioned earlier, the advantage of using flow cytometry to study a dynamic process such as apoptosis is the ability to simultaneously examine multiple cellular characteristics in thousands of cells at the single cell level in a short period of time. Therefore, we have utilized flow cytometry to examine changes in intracellular monovalent ions and cell size during apoptosis. It has been determined previously that changes in cell size, specifically the loss of cell volume during apoptosis, occur through the loss of both intracellular sodium and potassium. In our flow cytometric studies, we have used two commercially available ionic dyes: a potassium indicator designated PBFI-AM and a sodium indicator designated SBFI-AM (Molecular Probes, Eugene, OR). Both of these dyes are acetoxymethyl (AM) ester derivatives that can be loaded into live cells. The dye is retained in the cell by nonspecific cellular esterases that cleave the lipophilic-blocking AM ester and restore the charged free acid form of the dye. Both PBFI-AM and SBFI-AM are UV (340–360 nm) excitable dyes and can be detected at 425 nm. The following section describes how these fluorescent ionic indicators can be used in the examination of apoptotic cells by flow cytometry.

A. Instrument Setup

Many fluorescent ionic dyes require the use of both UV excitation (laser line of 340–360 nm) and specific filters in front of the photomultiplier tubes (PMT). Therefore, a flow cytometer with multiple laser lines and exchangeable filters in front of the PMTs is the best choice for this type of analysis. We use a FACSVantage SE flow cytometer (Becton-Dickinson), which has both a 488-nm laser line and UV excitation. Both PBFI-AM and SBFI-AM are UV excitable and can be examined using a 425-nm filter in front of the PMT for the fluorescent channel connected to the UV line. Standard filters can be used in the fluorescence channels associated with the 488-nm laser.

B. Cell Preparation and Loading of Fluorescent Ionic Dyes

1. Prepare a 2.5 m*M* stock of PBFI-AM and/or SBFI-AM by adding equal volumes of dimethyl sulfoxide (DMSO) and 20% (w/v) Pluronic F-127 (Molecular Probes) to 50 μg of prepackaged dye. The addition of Pluronic F-127 is used to facilitate cellular loading and to limit dye compartmentalization within cells.

2. One hour prior to examination, place 1 ml of cells (approximately 5×10^5 cells/ml) into a 6-ml round-bottom polystyrene tube (Falcon 2058).

3. Add 2 μl of 2.5 m*M* stock of PBFI-AM or SBFI-AM (final dye concentration of 5 μM) to each individual tube of cells. Keep the tubes out of direct light as much as possible.

4. Incubate the cells for 1 h at 37°C under the appropriate CO_2 condition. *Note:* Different cell types may require different concentrations of the dyes and loading conditions. The concentrations and times mentioned earlier for loading the cells with either PBFI-AM or SBFI-AM have worked well in experiments involving lymphoid cells in media, even in the presence of 10% FCS. Each individual cell type should be examined for optimal loading by varying the initial dye concentration incubation period and temperature in a control set of cells prior to further experimentation.

5. Add 1 μl of a 10-mg/ml stock solution of PI (Sigma) to each tube of cells immediately prior to flow cytometric examination.

C. Data Acquisition for Changes in Intracellular Ions during Apoptosis

1. After the appropriate incubation period for loading the cells with the fluorescent ionic indicators, examine the control cells initially on a FSC versus SSC dot plot to determine the light-scattering properties of the cells. Make the appropriate adjustments to the voltage and amp gain for these individual parameters.

2. Examine these cells for their ionic concentration by UV excitation (laser line 340–360 nm). Changes for both PBFI-AM and SBFI-AM fluorescence can be detected using a 425-nm filter in front of the detector positioned for the UV laser. On our instrument we use FL-4 for UV excitation. Examine the cells on a forward-scatter versus FL-4 (425 nm) dot plot and an FL-4 histogram. Set the FL-4 signal amplification mode to log and adjust the FL-4 detector voltage and fine gain to position the control population of cells on the various plots. We usually position the population of control cells to the right of center for the PBFI-AM indicator (as apoptotic cells lose intracellular potassium during the cell death process) and to the left of center for the SBFI-AM indicator on an FL-4 histogram (as the normal concentration of intracellular sodium is not very high compared to that of potassium).

3. Examine these cells for a loss of membrane integrity with 488 nm excitation, detecting the PI fluorescence with a 575-nm filter in front of the FL-3 detector on a forward-scatter versus FL-3 (PI) dot plot and an FL-3 histogram. Set the FL-3 signal amplification mode to log and adjust the FL-3 detector voltage and fine gain to position the control population of cells in the first decade on an FL-3 (PI) histogram. Cells that have lost their membrane integrity will dislay an increase in PI fluorescence and can be gated out on the forward-scatter versus FL-3 (PI) dot plot.

4. Examine the cells on an FL-4 (425 nm) versus FL-3 (PI) dot plot. If a gate was set on the forward-scatter versus FL-3 (PI) dot plot to gate out the cells that have lost their membrane integrity (cells that are PI positive), the FL-4 (425 nm) versus FL-3 (PI) dot plot can be set to show only viable (PI negative) cells. Once all instrument settings have been adjusted, they can be saved and used for all subsequent experimental samples.

5. Examine each experimental sample by flow cytometry using the previously established instrument settings and observe the change for the various fluorescent indicators on either dot plots or histograms.

D. Data Analysis

There are a variety of ways to analyze changes in the fluorescent properties of the cells. Similar to analyzing changes in light scattering of a cell, various dot plots and histograms can be employed to determine the relationship between the characteristics of interest. For example, the viable population of cells, as determined on the forward-scatter versus FL-3 (PI) dot plot, can be examined on a forward-scatter versus side-scatter dot plot. Experimentally treated cells that have a change in intracellular ion concentration can be compared to control cells on a forward-scatter versus FL-4 (425 nm) dot plot or a FL-4 (425 nm) histogram. Additionally, various types of plots such as contour or three-dimensional plots can be used to compare changes in the fluorescent and light-scattering properties of the cells.

Figure 3 (see Color Plate) shows an example of the use of flow cytometry in determining the relationship between cell size and intracellular potassium during apoptosis. Jurkat cells induced to undergo apoptosis with an antibody to Fas (CD95) were initially examined on a forward-scatter versus PI fluorescence dot plot to isolate the viable population (data not shown). Examination of the viable population of cells on a forward-scatter versus side-scatter three-dimensional plot revealed a time-dependent increase in the number of cells that have decreased ability to scatter light in the forward direction, indicative of cell shrinkage (Fig. 3A, see Color Plate). Interestingly, this same cell population has a slight increase in cell granularity, as indicated by an increase in side-scattered light. When these cells are examined for a loss of intracellular potassium by flow cytometry using the fluorescent potassium indicator dye PBFI-AM on a forward-scatter versus PBFI-AM fluorescence three-dimensional plot, a decrease in PBFI-AM fluorescence is observed only in cells that have a decrease in forward-scattered light, suggesting that only the shrunken population of cells have lost intracellular potassium. This relationship between cell size and intracellular potassium is observed with a variety of apoptotic stimuli. For example, when Jurkat cells were treated with either the calcium ionophore A23187 or the calcium ATPase inhibitor thapsigargin, again only the shrunken population of cells have a decrease in intracellular potassium (Fig. 3B, see Color Plate), similar to the observation with anti-Fas-treated cells.

Caspase activity is an additional apoptotic characteristic that can be examined in relation to other features of programmed cell death using flow cytometry. A commercially available fluorescent caspase substrate known as PhiPhiLux (OncoImmunin, Inc., College Park, MD) can be used to examine caspase-3-like enzyme activity in live cells. PhiPhiLux is a 33 amino acid peptide that has a consensus DEVD sequence in the center and rhodamine conjugated to each end. The peptide is constructed in such a way that the two rhodamines are in close proximity to each other, thus quenching their fluorescence. However, when the peptide is cleaved by a caspase-3-like enzyme the rhodamines are separated and the increase in fluorescence can be detected at 585 nm (FL-2). Although the standard 488-nm excitation is not optimal for rhodamine dyes, PhiPhiLux can be reasonably excited at this wavelength. A 514-nm excitation line is superior for this substrate, as is a 560 nm +/−10 filter for detecting rhodamine fluorescence.

When S49 Neo cells are induced to undergo apoptosis in response to A23187 or thapsigargin and subsequently examined on a forward-scatter versus PhiPhiLux contour plot, only the shrunken cells, or cells that have a decrease in their ability to scatter light in the forward direction, have an increase in PhiPhiLux fluorescence (Fig. 4, see Color Plate). Interestingly, when Jurkat cells are induced to undergo apoptosis in response to either anti-Fas or A23187, the cells shrink prior to the detection of caspase-3-like activity (Fig. 4, see Color Plate). Therefore, similar to the observations with S49 Neo cells, caspase-3-like activity is only observed in the shrunken population of apoptotic Jurkat cells, suggesting that the loss of cell volume occurs prior the activation of caspase-3-like enzymes during programmed cell death.

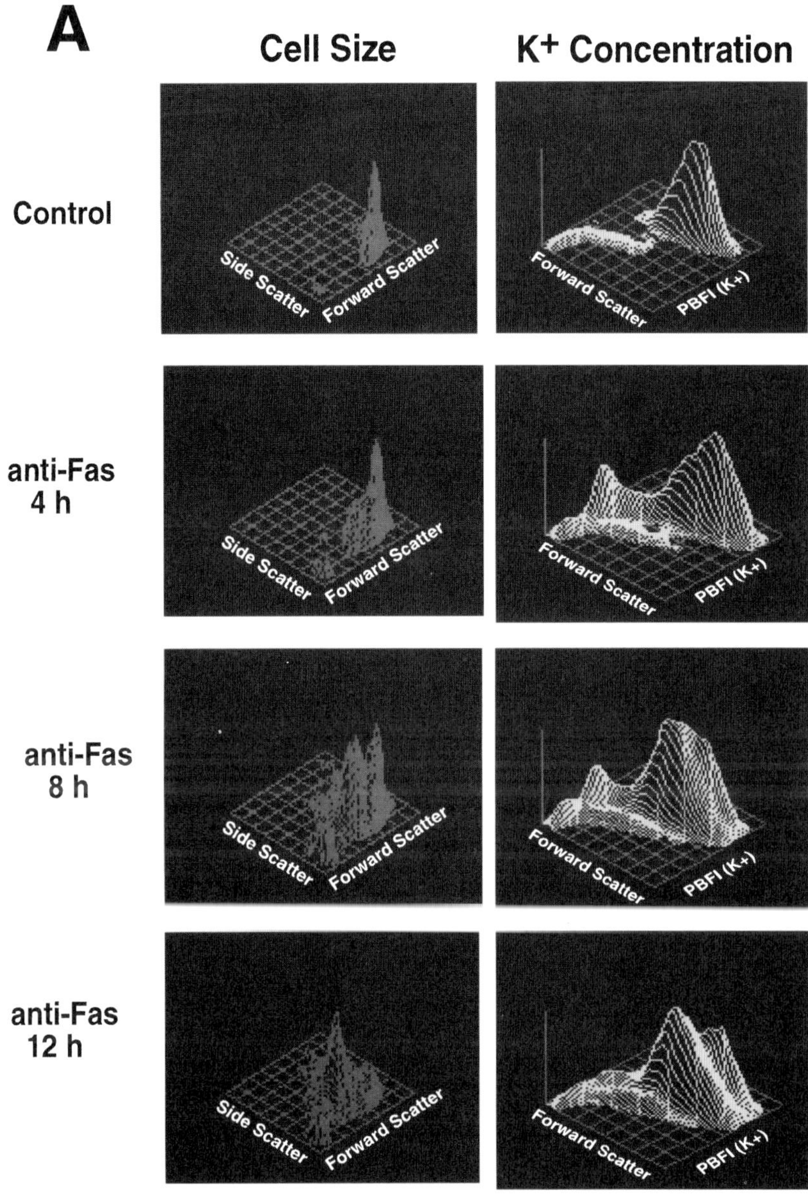

Fig. 3 Changes in cell size can be compared to those in intracellular potassium during apoptosis by flow cytometry. (A) Jurkat cells were treated with an anti-Fas antibody for 4, 8, and 12 h. Flow cytometry was used to assess the light-scattering properties of the cell and the loss of intracellular potassium. Ten thousand cells were examined on a FACSVantage flow cytometer (Becton-Dickinson, San Jose, CA). A loss in cell size, as indicated by a decrease in forward-scattered light, was observed in a time-dependent manner after the addition of anti-Fas.

(*continued*)

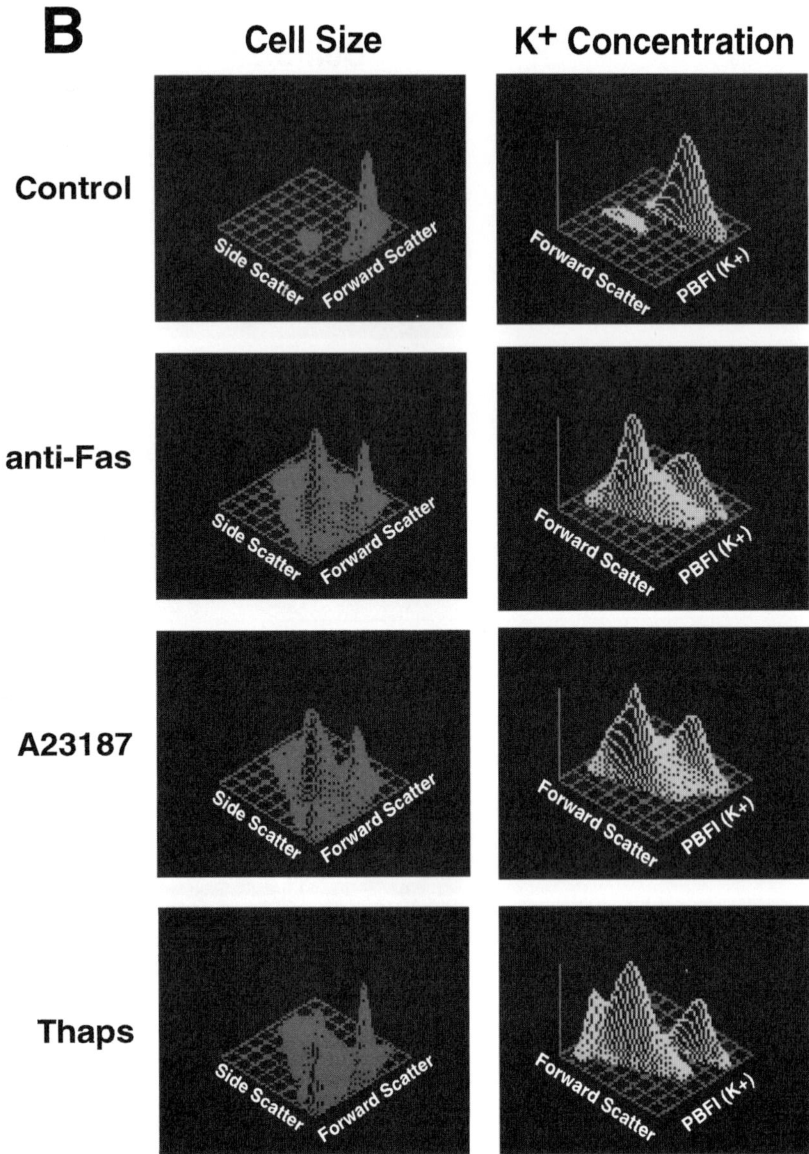

Fig. 3 *(continued)*. Additionally, a slight increase in side-scattered light, indicating an increase in cellular granularity, was also observed in a time-dependent manner. Gating on only viable or PI-negative anti-Fas-treated cells in the presence of the potassium indicator PBFI-AM showed that only the shrunken population of apoptotic cells had a decrease in intracellular potassium. (B) Jurkat cells treated with anti-Fas, A23187, or thapsigargin were examined for changes in cell size and intracellular potassium. A loss in cell size occurred with each apoptotic stimulus as a decrease in the ability of a population of cells to scatter light in the forward direction was observed. Comparing the loss of cell size to changes in intracellular potassium showed that only the shrunken population of cells had a decrease in intracellular potassium. Reprinted with permission from *J. Biol. Chem.* 274, 21953–21962 (1999). Copyright (1999) The American Society for Biochemistry and Molecular Biology. (See Color Plate.)

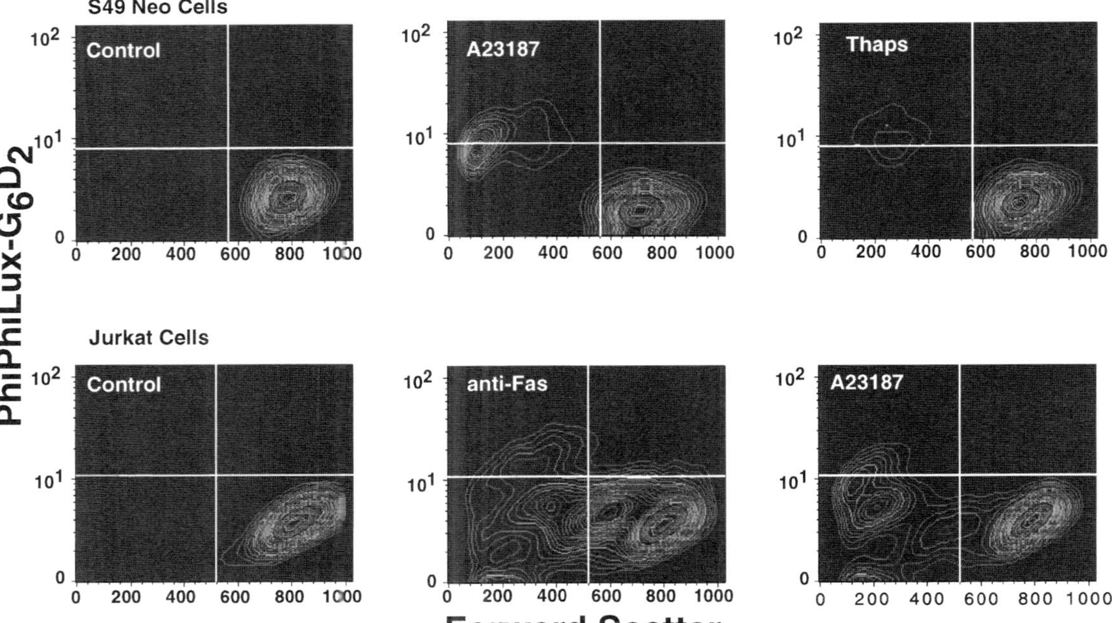

Fig. 4 Changes in cell size can be compared to changes in caspase-3-like activity during apoptosis by flow cytometry. S49 Neo cells were treated with either A23187 or thapsigargin (top) or Jurkat cells were treated with either anti-Fas or A23187 (bottom). Cells were exposed to the fluorescent caspase-3-like substrate (PhiPhiLux) for 1 h prior to flow cytometric examination. Ten thousand cells were examined on a FACSort flow cytometer (Becton-Dickinson, San Jose, CA). When these cells were examined on a FSC vs PhiPhiLux fluorescence contour plot, only the shrunken population of cells had an increase in caspase-3-like activity. Interestingly, a loss of cell volume was clearly observed prior to detection of caspase-3-like activity in the Jurkat cell model system, supporting the conclusion that cell shrinkage must occur prior to effector caspase activation. (See Color Plate.)

IV. Sorting Apoptotic Cells by Flow Cytometry

One of the most unique advantages of flow cytometry in the study of cell death is the ability to physically separate distinct populations of cells. Isolation of distinct populations of apoptotic cells by cell sorting permits further biochemical and morphological analysis on a homogeneous cell population. This sorting can be accomplished using one or more of the various characteristics used to detect cell death. For example, one can sort shrunken cells from normal ones during apoptosis. Additionally, one can sort cells that display a change in their mitochondrial membrane potential, caspase activity, or external phosphatidylserine exposure from those cells that remain normal for these parameters. The following procedure describes how apoptotic cells can be sorted in order to achieve the highest yield and best purity of the population of interest.

A. Instrument Setup

Many different flow cytometers are capable of physically sorting distinct populations of cells. We use a FACSVantage SE flow cytometer (Becton-Dickinson) equipped with Turbo Sort for sorting apoptotic cells. The Turbo Sort option permits a high sorting speed, which we have found to be advantageous in isolating apoptotic cells. Because our interest is in a specific population of apoptotic cells at a given point during the cell death process, the ability to recover the highest number of cells in the least amount of time is of great importance. Turbo Sort allows us to examine up to 15,000 cells per second, thus recovering a large number of cells in a relatively short period of time.

In using Turbo Sort, certain precautions need to be addressed in order to obtain the best sort possible.

After the standard optimization of the flow cytometer using either beads or chicken red blood cells, it is essential that a profile for the drop delay be obtained if one is performing less than a 3-drop sort envelop. The drop delay profile ensures that the flow cytometer is set at the optimal settings to achieve the greatest cell yield at the highest purity. We have had great success using Turbo Sort at a 1.5-drop sort envelop with a drive level of 72 in the Normal-R sort mode. The appropriate excitation (laser) and filters can then be set up for the particular apoptotic characteristic of interest.

B. Cell Preparation and Sorting

The best control of the flow cytometer during a sort is achieved by starting with a high concentration of cells in the sample to be sorted (approximately 5×10^6 cells/ml). This allows for the core stream to remain as tight as possible, which decreases the chances of two cells passing by the laser side by side, which would in turn abort these cells. Again, keeping the abort rate as low as possible permits the greatest number of cells to be recovered at the highest purity in the shortest amount of time. After a given period of time following the apoptotic stimulus, cells will be harvested and resuspended in either growth medium or phosphate-buffered saline to a final concentration of approximately 5×10^6 cells/ml. The following simple protocol is used for sorting apoptotic cells.

1. Examine a control sample of cells on a FSC vs SSC dot plot to determine the light-scattering properties of the cells. Make the appropriate adjustments to the voltage and amp gain for these individual parameters.
2. Examine the control cells on the individual plots or histograms of interest for the specific apoptotic characteristic. For example, if the shrunken but viable population of cells is the population to be sorted, examine the cells on a FSC vs PI dot plot. Examination of the control cells is required to ensure that the appropriate instrument settings are employed, i.e., that this sample does not contain any cells that have an increase in PI fluorescence.
3. Once the experimental cells have been resuspended to a concentration of 5×10^6 cells/ml, examine the cells on a FSC vs SSC dot plot and on the other plots of interest.
4. Draw a sort gate around the cells to be collected.
5. Set the sort gate selection to the region defining the population of cells to be collected.
6. Once the sort gate selection has been loaded, the cells of interest will accumulate in the appropriate collection device.

Because programmed cell death is a very dynamic process, several important considerations should be addressed prior to sorting apoptotic cells. For the best recovery of apoptotic cells, we have found it advantageous to chill the collection device during the sort. Depending on the number of cells required for further biochemical or morphological analysis, the sorted cells may be on the instrument for an extended period of time. Collection speed is enhanced by concentrating the initial cell sample, which in turn allows the desired number of cells to be collected in the shortest period of time. Keeping the collection device chilled to approximately 4°C also helps maintain cell viability, especially during long sorts. Additionally, because it is important to collect only viable cells, incorporation of PI in the sample and gating on only the viable or PI-negative cell population will also permit the best recovery, independent of the apoptotic characteristic of interest. In our experience, the isolation of DNA or proteins from cells that have lost their membrane integrity is virtually impossible. Finally, processing the cells as soon as possible after the sort increases the ability to recover viable cells for further biochemical or morphological analysis.

C. Data Analysis

Depending on the type of flow cytometer used for the sort, data can be collected and saved to the computer at various times to monitor the progression of apoptosis in the cell population. This is similar to collecting flow cytometry data in the absence of sorting. For example, primary rat thymocytes treated with Dex for 2 h were examined on a FSC vs SSC dot plot (Fig. 5A, see Color Plate). Sort gates were drawn around the shrunken and normal population of cells, which were simultaneously sorted into collection tubes,

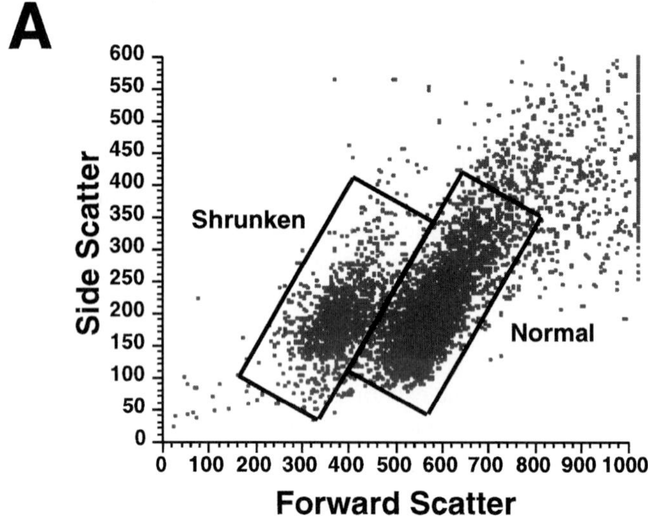

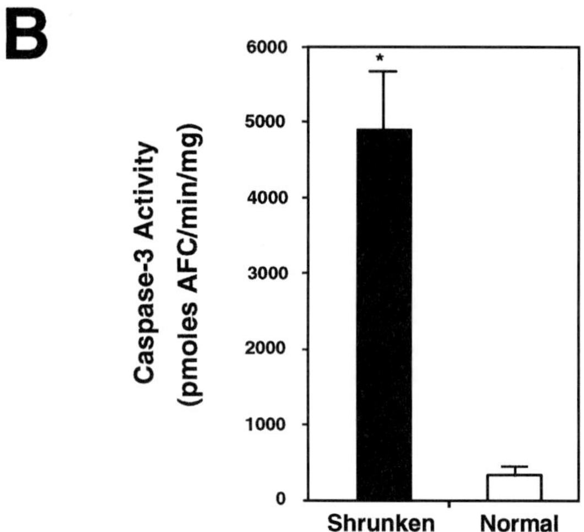

Fig. 5 Simultaneous sorting of normal and shrunken Dex-treated primary rat thymocytes for additional biochemical analysis. (A) Primary rat thymocytes were isolated and placed in culture in the presence of dexamethasone to induce apoptosis. These cells were initially examined by flow cytometry on a FSC vs SSC dot plot where two distinct populations of cells were observed. Sort gates were drawn around the normal and shrunken population of cells, which were sorted simultaneously into collection tubes. (B) Each individual sorted population of cells was examined for caspase-3-like activity using a fluorometric caspase assay. Only the shrunken population of apoptotic cells had an increase in caspase-3-like activity. Reprinted with permission from *J. Biol. Chem.* 272, 30567–30576 (1997). Copyright (1997) The American Society for Biochemistry and Molecular Biology. (See Color Plate.)

harvested, and examined individually for caspase-3-like activity using a fluorometric assay (Hughes *et al.,* 1997). Similar to the results observed in S49 Neo and Jurkat cells using the fluorescent caspase-3-like substrate (Fig. 4), only the shrunken population of Dex-treated primary rat thymocytes had an increase in caspase-3-like activity (Fig. 5B). These results suggest that similar conclusions can be reached using multiple methods for examining cells by flow cytometry.

In summary, flow cytometry is a very rapid method for examining numerous apoptotic characteristics at any given time during the programmed cell death process. With the incorporation of various fluorescent dyes and markers in the sample, the relationship between one apoptotic characteristic and another can be assessed to determine if these events occur in the same or different cells during the cell death process. The advantage of physically sorting cells by flow cytometry permits further biochemical or morphological analysis of a discrete population of apoptotic cells. With the continual advancement of specific substrates and reagents to be used in fluorescent experiments, flow cytometry will continue to remain a valuable tool to study apoptosis.

References

Al-Habori, M. (1994). Cell volume and ion transport regulation. *Int. J. Biochem.* **26,** 319–334.

Barbiero, G., Duranti, F., Bonelli, G., Amenta, J. S., and Baccino, F. M. (1995). Intracellular ionic variations in the apoptotic death of L cells by inhibitors of cell cycle progression. *Exp. Cell Res.* **217,** 410–418.

Beauvais, F., Michel, L., and Dubertret, L. (1995). Human eosinophils in culture undergo a striking and rapid shrinkage during apoptosis. Role of K^+ channels. *J. Leukoeyte Biol.* **57,** 851–855.

Benson, R. S. P., Heer, S., Dive, C., and Watson, A. J. M. (1996). Characteristics of cell volume loss in CEM-C7A cells during dexamethasone-induced apoptosis. *Am. J. Physiol.* **270,** C1190–C1203.

Bortner, C. D., Hughes, F. M., Jr., and Cidlowski, J. A. (1997). A primary role for K^+ and Na^+ efflux in the activation of apoptosis. *J. Biol. Chem.* **272,** 32436–32442.

Hoffmann, E. K. (1987). Volume regulation in cultured cells. *Curr. Top. Membr. Transp.* **30,** 125–180.

Hughes, F. M., Jr., Bortmer, C. D., Purdy, G. D., and Cidlowski, J. A. (1997). Intracellular K^+ suppresses the activation of apoptosis in lymphocytes. *J. Biol. Chem.* **272,** 30567–30576.

Jonas, D., Walev, I., Berger, T., Liebetrau, M., Palmer, M., and Bhakdi, S. (1994). Novel path to apoptosis: Small transmembrane pores created by Staphylococcal alpha-toxin in T lymphocytes evokes internucleosomal DNA degradation. *Infect. Immun.* **62,** 1304–1312.

Kerr, J. F. R., Wyllie, A. H., and Currie, A. R. (1972). Apoptosis: A basic biological phenomenon with wide-ranging implications in tissue kinetics. *Br. J. Cancer* **26,** 239–257.

Klassen, N. V., Walker, P. R., Ross, C. K., Cygler, J., and Lach, B. (1993). Two-stage cell shrinkage and the OER for radiation-induced apoptosis of rat thymocytes. *Int. J. Radiat. Biol.* **64,** 571–581.

McCarthy, J. V., and Cotter, T. G. (1997). Cell shrinkage and apoptosis: A role for potassium and sodium ion efflux. *Cell Death Differ.* **4,** 756–770.

Thomas, N., and Bell, P. A. (1981). Glucocorticoid-induced cell-size changes and nuclear fragility in rat thymocytes. *Mol. Cell. Endocrinol.* **22,** 71–84.

Wyllie, A. H. (1980). Glucocorticoid-induced thymocyte apoptosis is associated with endogenous endonuclease activation. *Nature (Lond.)* **284,** 555–556.

Wyllie, A. H., and Morris, R. G. (1982). Hormone-induced cell death Purification and properties of thymocytes undergoing apoptosis after glucocorticoid treatment. *Am. J. Pathol.* **109,** 78–87.

Yu, S. P., Yeh, C.-H., Sensi, S. L., Gwag, B. J., Canzoniero, L. M. T., Farhangrazi, Z. S., Ying, H. S., Tian, M., Dugan, L. L., and Choi, D. W. (1997). Mediation of neuronal apoptosis by enhancement of outward potassium current. *Science* **278,** 114–117.

CHAPTER 4

Use of Flow and Laser-Scanning Cytometry in Analysis of Cell Death

Zbigniew Darzynkiewicz,* Xun Li,* and Elzbieta Bedner*,†

*Brander Cancer Research Institute
New York Medical College
Hawthorne, New York 10532

† Pomeranian School of Medicine
Szczecin, Poland

METHODS IN CELL BIOLOGY, VOL. 66

0091-679X/01 $35.00

I. Introduction

Applications of flow cytometry in cell necrobiology (reviews: Darzynkiewicz *et al.*, 1992, 1997a; Ormerod, 1998; van Engeland *et al.*, 1998) can be subdivided into two groups. In one group are applications aimed at revealing molecular and functional mechanisms associated with cell death, primarily by apoptosis. In such studies, flow cytometry is often used to measure cellular levels of the immunocytochemically detected components that are involved directly or indirectly in the regulation and/or execution of apoptosis. The most prominent among them are members of the Bcl-2 protein family, caspases, the protooncogenes (i.e., c-*myc* or *ras*), or tumor suppressor genes (i.e., p53 or pRB). Flow cytometry is also used widely to study functional attributes of the cell such as mitochondrial metabolism, oxidative stress, intracellular pH, or ionized calcium, all closely associated with mechanisms regulating cell sensitivity to apoptosis. The major advantage of flow cytometry in these applications is that it offers the possibility of multiparametric measurements of a multitude of cell attributes. Multivariate analysis of such data allows one to study quantitative correlations between the measured cell constituents. For example, when one of the measured attributes is cellular DNA content, the parameter that reports the cell cycle position or DNA ploidy, an expression of the other measured attribute can be then directly correlated with the cell cycle position (or cell ploidy) without a need for cell synchronization or separation by elutriation. Furthermore, because individual cells are measured, intercellular variability can be assessed, cell subpopulations identified, and rare cells easily detected.

The second group of cytometry applications presented here comprises the methods to identify and quantify dead cells and discriminate between apoptotic vs necrotic modes of death. Dead cell recognition is generally based on the presence of a particular biochemical or molecular marker that is characteristic for apoptosis, necrosis, or both. A plethora of methods have been developed, especially for the identification of apoptotic cells. Apoptosis-associated changes in the gross physical attributes of cells, such as cell size and granularity, can be detected by analysis of laser light scattered by the cell in forward and side directions (Ormerod *et al.,* 1995; Swat *et al.,* 1981). Some of the methods rely on apoptosis-associated changes in the distribution of plasma membrane phospholipids (Fadok *et al.,* 1992; Koopman *et al.,* 1994). Others detect the loss of active transport function of the plasma membrane. Still other methods probe the mitochondrial transmembrane potential, which dissipates early during apoptosis (Cossarizza *et al.,* 1995; Kroemer, 1998; Zamzani *et al.,* 1998). The detection of DNA fragmentation provides another convenient marker of apoptosis: apoptotic cells are then recognized either by their fractional ("subdiploid," "sub-G_1") DNA content due to extraction of low molecular weight (MW) DNA from the cell (Nicoletti *et al.,* 1991; Umansky *et al.,* 1981) or by the presence of DNA strand breaks, which can be detected by labeling their 3′OH termini with fluorochrome-conjugated nucleotides in a reaction utilizing exogenous terminal deoxynucleotidyl transferase (TdT) (Gorczyca *et al.,* 1992, 1993; Li and Darzynkiewicz, 1995; Li *et al.,* 1996).

The drawback of all flow cytometric methods stems from the fact that the identification of apoptotic or necrotic cells relies on a single attribute of the cell, the attribute that is assumed to represent a characteristic feature of apoptosis or necrosis. However, this attribute may be absent when apoptosis is atypical, as was shown in many cases of death of epithelial and fibroblast lineage cells (e.g., Catchpoole and Stewart, 1993; Collins *et al.,* 1992; Ormerod *et al.,* 1996; Zamai *et al.,* 1996). Likewise, apoptosis caused by agents that inhibit specific apoptotic effectors may also lack certain characteristic attributes. As examples, induction of apoptosis by an endonuclease inhibitor results in death without DNA fragmentation, whereas inhibitors of proteases suppress the degradation of particular proteins ("death substrates") such as nuclear lamin, thereby preventing nuclear breakdown (e.g., Hara *et al.,* 1996). The characteristic changes in cell morphology, therefore, still remain the gold standard for the recognition of apoptotic cell death, particularly in such cases (Kerr *et al.,* 1972; Majno and Joris, 1995).

A laser-scanning cytometer (LSC) is a microscope-based cytofluorometer manufactured in the United States (CompuCyte Corp., Cambridge, MA) and Japan (Olympus Co., Tokyo), which offers the combined advantages of flow cytometry and image analysis (reviews: Darzynkiewicz *et al.,* 1999; Kamentsky *et al.,* 1991, 1997). This newly developed instrument is finding wide applicability in many disciplines of biology and medicine. LSC measures cell fluorescence rapidly and with accuracy comparable to that obtained by flow cytometry. In addition, because the cell position on the slide is recorded in a list-mode fashion, together with other measured cell parameters, cells can be relocated after the measurement. The relocated cells can be then examined visually or subjected to image analysis to correlate the observed change in the measured parameter with the change in their morphology (e.g., Bedner *et al.,* 1998, 1999; Deptala *et al.,* 1998). This is of particular value in the case of atypical apoptosis. Furthermore, morphometric analysis allows one to study translocations of cell constituents, e.g., from cytoplasm to nucleus, such as seen with the activation of NF-κB (Deptala *et al.,* 1998). Because analysis on slides eliminates cell loss, which otherwise occurs during repeated centrifugations in sample preparation for flow cytometry, LSC is suitable for hypocellular specimens.

Perhaps the most attractive feature of LSC for studies of apoptosis is the possibility of sequential measurements of the same set of cells and recording the sequential data as a list-mode single file ("file merge"). This feature allows one to directly correlate, on the same cells, the results of functional cell assays such as mitochondrial transmembrane potential ($\Delta\Psi_m$), change in pH, or generation of reactive oxygen intermediates (ROIs) (Hedley and McCullogh, 1996) with cell attributes that can be measured only after cell fixation and permeabilization (e.g., DNA strand breaks, cell cycle position) (Li and Darzynkiewicz, 1999). Thus, the sequence of both functional and structural changes occurring during apoptosis can be mapped and it is possible to determine if a particular death-associated event is a prerequisite for the latter steps.

Several flow cytometric methods developed for the identification of apoptotic and necrotic cells have been modified to adapt them to LSC (described later) (Darzynkiewicz and Bedner, 2000; Darzynkiewicz *et al.,* 1998). The unique features of LSC that are of special value in the analysis of apoptosis are emphasized. Methods for the use of flow cytometry are also described, although we have focused primarily on novel approaches as general methods have been well documented in the literature (Darzynkiewicz and Li, 1996; Darzynkiewicz *et al.,* 1994, 1997b). Finally, some problems and analysis strategies are also provided.

A variety of kits are available commercially to identify apoptotic cells using the methods presented in this chapter. Because the reagents are already prepackaged in these kits and the procedures are described in a cook-book format, the kits offer the advantage of simplicity. The price of these kits, however, is manyfold higher than the cost of the individual reagents. Furthermore, the kits do not allow one the flexibility that is often required to optimalize procedures for a particular cell system. In many situations, therefore, the preparation of samples for analysis by flow cytometry or LSC, as described in this chapter, may be preferred.

II. Principles of Cell Measurement by Laser-Scanning Cytometry (LSC)

A diagrammatic scheme of the LSC is presented in Fig. 1 (see Color Plate). The microscope (Olympus Optical Co.) is the central part of the instrument. The

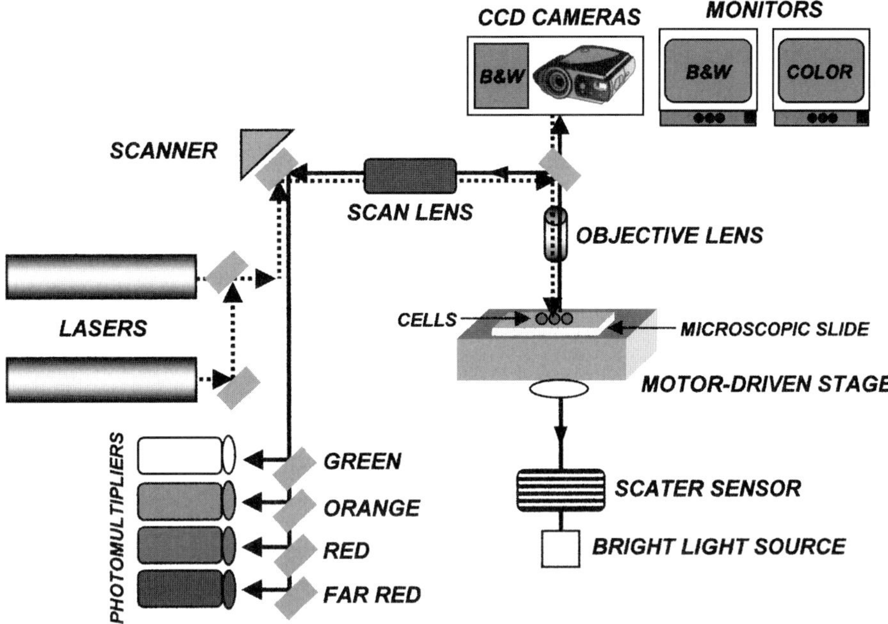

Fig. 1 Scheme representing major components of the LSC. See text for explanation. (See Color Plate.)

fluorochrome-stained specimen on the microscope slide is placed on the stage of the microscope and its fluorescence is excited by a laser beam that scans the microscope slide rapidly. Actually, in the current instruments, beams from two lasers (Ar ion and He–Ne) spatially merged by a set of dichroic mirrors are directed onto the computer-controlled oscillating (350 Hz) mirror, which in turn directs the beams through the epi-illumination port of the microscope and images them through the objective lens onto the slide. The laser beams, therefore, sweep the area of microscope slide under the lens rapidly. Depending on the lens magnification, the beam spot size varies from 2.5 (at 40×) to 10.0 μm (at 10× magnification). The position of the slide on *xy* coordinates is monitored by sensors located on the computer-controlled motorized microscope stage and the slide is moved, with the stage, at 0.5-μm steps per each laser scan, perpendicularly to the scan. Light scattered by the cells is imaged by the condenser lens and recorded by scatter sensors. A portion of the fluorescence emitted by the specimen is collected by the objective lens and is directed to a charge-coupled device (CCD) camera for imaging. Another portion of emitted fluorescence is directed through the scan lens to the scanning mirror. Upon reflection, it passes through a series of dichroic mirrors and optical interference filters to reach one of the four photomultipliers. Each photomultipler records fluorescence at a specific wavelength range, defined by the combination of filters and dichroic mirrors. In addition to the lasers, a light source provides transmitted illumination that is used to visualize the objects through an eyepiece or the CCD camera. The measurement of cell fluorescence (or light scatter) is computer controlled and triggered

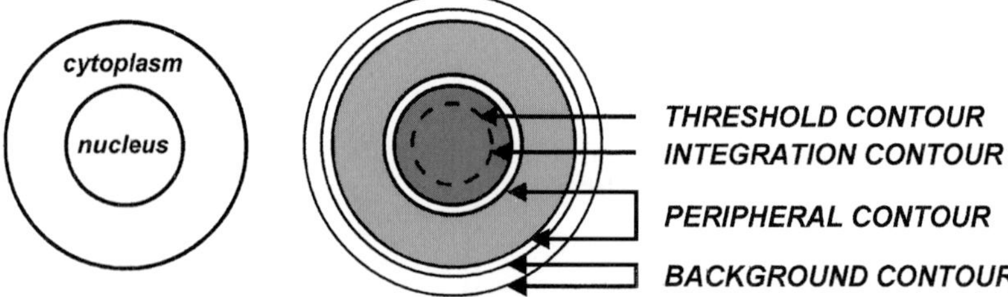

Fig. 2 The principle of analysis of nuclear and/or cytoplasmic fluorescence by LSC. When nuclear DNA is stained with the red fluorescing dye (e.g., 7-AAD) the threshold contour is set on red signal to detect the nucleus. To measure nuclear fluorescence the integration contour is set at a desired number of pixels outside of the threshold to ensure that all fluorescence emitted from the nucleus is measured and integrated. When fluorescence from the whole cell is measured the integration contour is set far from the threshold to ensure that fluorescence emitted from the cytoplasm is integrated as well. It is also possible to measure nuclear and cytoplasmic fluorescence separately. The peripheral contours are then set at the desired number of pixels outside of the nuclear integration contour, and fluorescence intensities emitted from the integration boundary (nuclear) and from the peripheral torus of the desired width (cytoplasmic) are measured and integrated separately. The background contour is automatically set outside the cell and background fluorescence is subtracted from nuclear, cytoplasmic, or total cell fluorescence. (See Color Plate.)

by setting a threshold contour for the cell above the background level of emission. For each measured object the following parameters are recorded by LSC.

a. Integrated fluorescence intensity over the integration contour, which can be adjusted to a desired width with respect to the threshold contour, represents the sum of intensities of all pixels within the area (Fig. 2, see Color Plate).

b. The value of maximal pixel intensity within this area, the so-called "peak" or "max pixel value."

c. The area within the integration contour that represents the number of pixels within the contour area.

d. The perimeter of the contour (in micrometers).

e. Fluorescence intensity integrated over the area of a torus of desired width defined by the peripheral contour located around (outside) the primary integration contour. Thus, if the integration contour is set for the nucleus based on red fluorescence (DNA stained by propidium), then the integrated (or maximal pixel) green fluorescence of FITC-stained cytoplasm can be measured separately, within the integration contour (i.e., over the nucleus) and within the peripheral contour, i.e., over the rim of cytoplasm of desired width outside the nucleus. It should be noted that all of these values of integrated fluorescence are automatically corrected for background, measured locally outside the cell, within the background contour (Fig. 2).

f. The slide position on *X* and *Y* coordinates of the maximal pixel.

g. The computer clock time at the moment of measurement.

Ratios of the respective parameters are preset easily as new parameters, and ratiometric data are then collected or calculated during data analysis. The spectrum overlap measured by individual photodetectors can be compensated electronically during data analysis.

There are many similarities between LSC and FC. The measurements by LSC are rapid and, with optimal cell density, up to 100 cells can be measured per second. The accuracy and sensitivity of cell fluorescence measurements by LSC are comparable to the most advanced flow cytometers (Kamentsky and Kamentsky, 1991; Kamentsky *et al.,* 1997). Other features that can be measured, such as integrated fluorescence intensity of the cells, time of measurement, and forward light scatter, are also identical for both instruments. However, right angle (side) light scatter, common to FC, cannot be measured by LSC. LSC, however, measures individual pixel values, which cannot be measured by FC. This parameter reflects inhomogeneity of the fluorochrome distribution within the analyzed object, and the peak pixel value represents the maximal concentration per area imaged on a single pixel. In contrast, peak fluorescence measured by FC represents the peak value of the analog electronic signal from fluorescence integration of the cellular fluorescence. The possibility of differential analysis of fluorescence emitted from nucleus vs cytoplasm is another feature provided by LSC that is not available with FC.

The most characteristic feature of LSC that distinguishes it from FC is that cell analysis is performed on a slide. This allows visual cell examination to assess morphology and to correlate it with the measured parameters. It also allows cell image capture, analysis, and/or display. Furthermore, additional cytofluorometric analysis of the same cell is possible using new sets of markers or other contouring thresholds. Results of the sequential measurements can be then integrated in list-mode fashion using the "merge" capability of the instrument. Applications for studies of apoptosis that descend from these unique features of LSC are presented further in this chapter.

III. Specifics of Cell Preparation for Analysis by LSC

Many assays of apoptosis by LSC are performed on fixed cells. In these assays, the cells are attached to microscope slides by standard methods that include smear films, tissue sections, "touch" preparations from the tissues, or cytocentrifuging cell suspensions. Cytocentrifugation is often preferred over "touch" or smear preparations because it flattens cells on the slides so that their geometry is favorable and more morphological details can be revealed. The method of attaching cells (to be subsequently fixed) by cytocentrifugation is described later.

However, several cytometric methods designed to identify apoptotic cells or to study molecular or metabolic events associated with apoptosis require the cells to be alive with preservation of vital functions. Among them are analyses of plasma membrane transport function (e.g., Darzynkiewicz *et al.,* 1994), detection of phosphatidylserine on the cell surface (Fadok *et al.,* 1992; Koopman *et al.,* 1994), probing $\Delta\Psi_m$ (Cossarizza *et al.,* 1994; Zamzani *et al.,* 1996), intracellular pH, ROIs (Hedley and McCullogh, 1996), and level of calcium ions. Suspensions of live cells in appropriately prepared reaction media are generally used when such analyses are performed by flow cytometry. In the case of LSC, however, the measured cells often have to be attached to a microscope slide. The attachment is required if one intends to relocate the measured cells for their subsequent morphologic examination or to additionally probe by another fluorochrome(s). The relocation then allows one to correlate the initial measurement with cell morphology or with the secondary analysis involving another fluorochrome. The following methods are used to attach live cells to be studied by LSC.

A. Attachment of Cells to Slides by Cytocentrifugation

1. Prepare cell suspension in tissue culture medium (with serum) at a density of 5–10 × 10^3 cells per 1 ml.

2. Transfer 300 μl of this suspension into a cytospin chamber (e.g., Shandon Scientific, Pittsburgh, PA).

3. Cytocentrifuge at 1000 rpm for 6 min.

4. Without allowing the cytospun cells to dry completely in air, fix them by immersing the slides in a Coplin jar containing fixative [e.g., 1% formaldehyde in phosphate-buffered saline (PBS) or 70% ethanol]. For most applications (e.g., immunocytochemistry, see later in the chapter), cells may be fixed in 1% formaldehyde at 0–4°C for 15–30 min, rinsed in PBS, and postfixed and/or stored for up to several days in 70% ethanol at −20°C.

B. Attachment of Live Cells to Microscope Slides

A variety of cells, including cells of epithelial or fibroblast lineage, nerve cells, and macrophages, adhere readily to flasks in culture. Such cells can be attached easily to microscope slides by culturing them on slides or coverslips. Culture vessels that have a microscope slide at the bottom of the chamber are available commercially (e.g., "Chamberslide," Nunc, Inc., Naperville, IL). Cells growing in these chambers spread and attach to the slide surface in a few hours after suspending them in full culture medium (with serum) and incubation at 37°C. Chambers with glass rather than plastic slides are preferred as the latter have high autofluorescence that interferes with measurements by LSC. Alternatively, cells can be grown on coverslips, e.g., placed on the bottom of petri dishes. The coverslips are then inverted over shallow (<1 mm) wells on microscope slides. The wells can be prepared by constructing the well walls (~2 × 1 cm^2) with a pen that deposits a hydrophobic barrier ("Isolator," Shandon Scientific), nail polish, or melted paraffin. Alternatively, slides with wells can also be made by preparing a strip of Parafilm "M" (American National Can, Greenwich, CT) of the size of the slide, cutting a hole ~2 × 1 cm in the middle of this strip, placing the strip on the microscope slide, and heating the slide on a warm plate until the Parafilm starts to melt. It should be stressed, however, that because the cells detach during the late stages of apoptosis, these cells may be selectively lost if the analysis is limited to attached cells.

Cells that grow in suspension can be attached to glass slides by electrostatic forces. This is due to the fact that sialic acid on the cell surface has a net negative charge. Glass surface, however, is positively charged. Incubation of cells on microscope slides in the absence of any serum or serum proteins (which otherwise neutralize the charge) thus leads to their attachment. Cells taken from culture should be rinsed in PBS in order to remove serum proteins and then resuspended in PBS at a concentration of $2 \times 10^5 - 10^6$ cells/ml. An aliquot (50–100 μl) of this suspension should be deposited within a shallow well (prepared as described earlier) on the horizontally placed microscope slide. To prevent drying, a small piece (~2 × 2 cm) of thin polyethylene foil or Parafilm may be placed on top of the cell suspension drop. A short (15–20 min) incubation of such a cell suspension at room temperature in a closed box providing 100% humidity is adequate to ensure that most cells will become firmly attached to the slide surface. Cells attached in this manner remain viable for several hours and can be subjected to surface immmunophenotyping (Clatch *et al.,* 1998), viability tests, or intracellular enzyme kinetic assays (Bedner *et al.,* 1998). Such preparations can be fixed (e.g., in formaldehyde) without a significant loss of cells from the slide. Thus, when such preparations are restained, the large majority of the cells (>95%) are still attached and can be relocated by LSC (Clatch *et al.,* 1998). However, as in the case of cell growth on glass, late apoptotic cells have a tendency not to attach or may even detach after the initial attachment.

IV. Chromatin Condensation as a Marker of Apoptotic Cells Detected by LSC

Measurement of total nuclear or cellular fluorescence is performed by LSC via the integration of light intensity of individual pixels over the area of nucleus and/or cytoplasm (Kamentsky *et al.,* 1997). In addition, intensity of the maximal pixel within the measured area is also recorded. Because of the high degree of chromatin condensation in apoptotic cells, DNA stains with greater intensity per unit of the projected nuclear area in these cells (hyperchromasia). The maximal pixel value of the DNA-associated fluorescence measured in the chromatin of apoptotic cells, therefore, is greater than in nonapoptotic nuclei (Bedner *et al.,* 1999; Furuya *et al.,* 1997). This situation is similar to chromatin staining in mitotic cells, which is also strongly condensed. Apoptotic cells, therefore, in analogy to mitotic cells (Luther and Kamentsky, 1996; Kawasaki *et al.,* 1997) or other cells with condensed chromatin such as lymphocytes (Bedner *et al.,* 1997), can be identified by high values of the maximal pixel of DNA-associated fluorescence (Fig. 3, see Color Plate). Propidium iodide (PI) is used as the DNA fluorochrome in the following method.

A. Materials

1. Prepare 1% formaldehyde solution in PBS. This solution should be made fresh.
2. A stock solution of PI (Molecular Probes, Eugene, OR) is prepared in distilled water at a concentration of 1 mg/ml and stored at 4°C in the dark.
3. A stock solution of DNase-free RNase A (Sigma Chemical Co., St Louis, MO) is prepared by dissolving RNase A in distilled water (5 mg/ml). If DNase-free RNase is

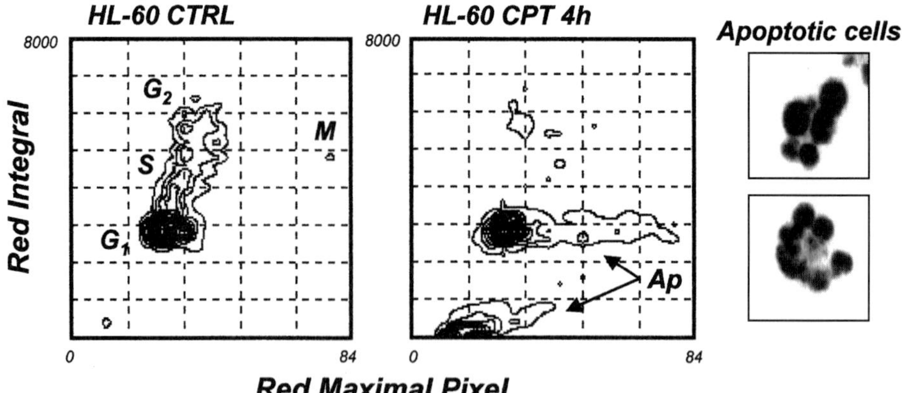

Fig. 3 Identification of apoptotic cells by LSC based on high values of maximal pixel detecting red fluorescence or fractional DNA content of propidium iodide (PI)-stained cells. Exponentially growing HL-60 cells, untreated (CTRL) or induced to undergo apoptosis by treatment with 0.15 μM camptothecin (CPT) for 4 h, were stained with PI in the presence of RNase A as described in the text. Contour maps represent bivariate distributions of cells with respect to their integrated red fluorescence (proportional to DNA content) vs maximal red fluorescence pixel value. Only mitotic cells (M) have a high maximal pixel value in the untreated culture. Apoptotic cells (Ap) that are present in CPT-treated cultures are characterized either by the increased intensity of maximal pixel of red fluorescence or by a low ("sub-G_1") DNA content. The relocation feature of LSC allows one to observe morphology of the cells selected from particular regions of the bivariate distributions. Upon relocation, cells with a high maximal pixel value or with a fractional DNA content show chromatin condensation and nuclear fragmentation typical of apoptosis (two panels on right). (See Color Plate.)

unavailable, DNase activity is destroyed (but RNase is preserved) by boiling the stock solution for 3–5 min. Aliquots can be stored at −20°C.

4. Staining solution of PI/RNase: Add the proper volumes of stock solutions of PI and RNase A to PBS to obtain final concentrations of 10 μg/ml PI and 0.1 mg/ml RNase A. This solution should be prepared fresh.

B. Cell Staining and Measurement

1. Deposit cells on the microscope slide by cytocentrifugation, electrostatically, or by growing them on the slide, as described earlier.

2. Without allowing the attached cells to dry completely, fix them by immersing the slides in a Coplin jar containing 1% formaldehyde in PBS on ice for 15 min.

3. Wash the slides briefly by immersing in PBS and transfer them into Coplin jars containing 70% ethanol. The cells may be stored in ethanol for several days at 4°C.

4. After fixation, rinse slides in PBS for 5 min. Immerse slides in a Coplin jar containing the PI-staining solution. Keep slides immersed in PI solution for 30 min at room temperature in the dark. (*Note:* For best results, whole slides should be immersed in the staining solution. However, to save the reagents, a small volume (∼0.5 ml) of PI-staining

solution may be deposited onto the horizontally placed slide at the area containing the cells and the slide maintained within in the box at 100% humidity for 30 min at room temperature in the dark.)

5. Mount the cells under a coverslip by adding a drop of a solution containing 90% glycerol in PBS in which PI was dissolved at the same concentration as in the staining solution and seal the preparation with melted paraffin or nail polish. Alternatively, a commercial antifade mounting medium (e.g., Vectashield, Vector Laboratories Inc., Burlingame, CA) can be used. Preventing specimen drying during measurement is essential for high-quality DNA content analysis. The slides should be kept in the dark until measurement on LSC.

6. Measure cellular red fluorescence (>600 nm; integrated fluorescence and maximal pixel) by LSC, illuminating the cells at 488 nm.

Apoptotic cells are identified by LSC as cells with high maximal pixel values of the PI fluorescence as shown in Fig. 3 (see Color Plate). Because cellular DNA content is also measured, cell ploidy and/or cell cycle position of nonapoptotic cells can be determined at the same time. This staining procedure is simple and can be combined with analysis of other constituents of the cell when they are probed with fluorochromes of another color. It should be stressed, however, that because DNA undergoes fragmentation during apoptosis and because fragmented DNA of low molecular weight may be lost from the cells (e.g., extracted during the staining procedure or lost as a result of shedding of apoptotic bodies that may contain fragments of nuclear chromatin), the DNA content of apoptotic cells may not always be a reliable marker of their cell cycle position or ploidy.

The drawback of this approach is that it cannot discriminate between mitotic and apoptotic cells. In addition, early G_1 (postmitotic) cells may have high fluorescence intensity of the maximal pixel (Kawasaki *et al.,* 1997). The distinction between apoptotic and mitotic cells is critical after treatment with agents such as taxol or other mitotic blockers, i.e., when mitotic cells undergo apoptosis. Visual examination of the cells or analysis of other morphometric features such as nucleus-to-cytoplasm ratio, nuclear or cellular area or circumference, and forward light scatter, as offered by LSC, however, can be helpful in these instances. It should also be noted that LSC may not discriminate between individual nuclear fragments of the late apoptotic cells and single nuclei, of apoptotic cells with fractional ("subdiploid") DNA content. Because individual fragments of broken up nuclei, as well as whole nuclei of apoptotic cells, have subdiploid DNA content, they are represented by the "sub-G_1" peak on DNA content frequency histograms.

V. Gross Changes in Cell Structure during Apoptosis Measured by Laser Light Scattering

The intersection of cells with the laser light beam in a flow cytometer results in light scattering. Analysis of the light scattered in different directions reveals information about cell size and structure. Forward light scatter correlates with cell size, whereas side scatter yields information on cell light refractive and reflective properties and reveals optical inhomogeneity of the cell structure, such as that resulting from the condensation of cytoplasm or nucleus and granularity. Sensors measuring light scatter in the forward and right angle ("side scatter"; 90° angle) direction are a built-in feature of every commercially available flow cytometer utilizing laser illumination. It should be noted, however, that side scatter cannot be measured by LSC.

As a consequence of cell shrinkage, a decrease in forward light scatter is observed at a relatively early stage of apoptosis (Ormerod *et al.,* 1995; Swat *et al.,* 1981). Initially, there is little change in side scatter during apoptosis, although in some cell systems an increase in intensity of the side scatter signal is seen, reflecting perhaps chromatin and cytoplasm condensation and nuclear fragmentation. When apoptosis is more advanced and the cells become small, the intensity of side scatter decreases, similar to forward scatter. Late apoptotic cells, therefore, are characterized by a markedly diminished intensity of both forward and side scatter signals. In contrast to apoptosis, cell swelling, which occurs early during cell necrosis, is detected by a transient increase in forward light scatter. Rupture of the plasma membrane and leakage of the cytosol during subsequent steps of necrosis correlate with a marked decrease in intensity of both forward and side scatter signals.

Because there is a great variability in light scatter properties of individual cells, an exponential scale (logarithmic amplifiers) should be used during scatter measurements. Analysis of light scatter is often combined with other assays, most frequently surface immunofluorescence (e.g., to identify the phenotype of the dying cell), or another marker of apoptosis. It should be mentioned, however, that the change in light scatter alone is not a specific marker of apoptosis or necrosis. Mechanically broken cells, isolated nuclei, cell debris, and individual apoptotic bodies may also display diminished light scatter properties. Therefore, the analysis of light scatter should be combined with measurements that can provide a more definite identification of apoptotic or necrotic cells.

VI. Dissipation of Mitochondrial Transmembrane Potential ($\Delta\Psi_m$) during Apoptosis

The critical role of mitochondria during apoptosis is associated with the release of two intermembrane proteins, cytochrome *c* and apoptosis-inducing factor (AIF), which are essential for the sequential activation of procaspase-9 and procaspase-3 (Liu *et al.,* 1996; Yang *et al.,* 1997). AIF is also involved in the protelytic activation of apoptosis-associated endonuclease (Susin *et al.,* 1997). Dissipation (collapse) of mitochondrial transmembrane potential ($\Delta\Psi_m$), also called the "permeability transition" (PT), also occurs early during apoptosis (Cossarizza *et al.,* 1995; Kroemer, 1998; Zamzani *et al.,* 1998). However, a growing body of evidence suggests that this event may not always be correlated with the release of cytochrome *c* or AIF or activation of caspases (e.g., Scorrano *et al.,* 1999; Finucane *et al.,* 1999; Li *et al.,* 2000).

Membrane-permeant lipophilic cationic fluorochromes such as rhodamine 123 (Rh123) or 3,3′-dihexiloxa-dicarbocyanine [$DiOC_6(3)$] can serve as probes of $\Delta\Psi_m$

(Darzynkiewicz *et al.*, 1981, 1982; Johnson *et al.*, 1982). When live cells are incubated in their presence, the probes accumulate in mitochondria and the extent of their uptake, as measured by the intensity of cellular fluorescence, reflects $\Delta\Psi_m$. A combination of Rh123 and PI was introduced as a viability assay that discriminates among live cells that only stain with Rh123 (green fluorescence), dead or dying cells whose plasma membrane integrity is compromised (cells with damaged plasma membrane, late apoptotic, and necrotic cells) that stain only with PI (red fluorescence), and early apoptotic cells that show somewhat increased staining with PI but still take up Rh123 (Darzynkiewicz *et al.*, 1994). The specificity of Rh123 and $DiOC_6(3)$ as $\Delta\Psi_m$ probes is increased when they are used at low concentrations (<0.5 μg/ml). Still another probe of $\Delta\Psi_m$ is the J-aggregate forming lipophilic cationic fluorochrome 5,5′,6,6′-tetrachloro-1,1′,3,3′-tetraethylbenzimidazolcarbocyanine iodide (JC-1). Its uptake by charged mitochondria driven by the transmembrane potential is detected by the shift in color of fluorescence from green, which is characteristic of its monomeric form, to orange, which reflects its aggregation in mitochondria (Cossarizza *et al.*, 1995, 1997).

A. Materials

1. Stock solution of PI: Dissolve 1 mg of PI (Molecular Probes) in 1 ml of distilled water.
2. Stock solution of Rh123: Prepare 0.1 m*M* solution of Rh123 (Molecular Probes) by dissolving 3.8 mg of the dye in 10 ml of methanol. Store in small aliquots at −20°C in the dark. Prior to use dilute 10-fold with PBS to obtain 10 μ*M* concentration.
3. Stock solution of $DiOC_6(3)$: Prepare 0.1 m*M* solution of $DiOC_6(3)$ (Molecular Probes) by dissolving 5.7 mg of the dye in 10 ml of dimethyl sulfoxide (DMSO). Store in small aliquots at −20°C in the dark. Prior to use dilute 10-fold with PBS to obtain 10 μ*M* concentration.
4. Stock solution of JC-1: Prepare 0.2 m*M* solution of JC-1 (Molecular Probes) by dissolving 13 mg of the dye in 10 *ml* *N*,*N*-dimethylformamide (Sigma). Store in small aliquots at −20°C in the dark.

Each of the just-described stock solutions is stable and can be stored at 0–4°C in the dark for weeks.

B. Staining with Rh123 or $DiO_6(3)$ and PI and Analysis by Flow Cytometry

1. Add either 20 μl of a 10 μ*M* solution of Rh123 or 5 μl of $DiO_6C(3)$ to approximately 10^6 cells suspended in 1 ml of full tissue culture medium (with serum) to obtain the final concentration (200 or 50 n*M*, respectively) and incubate for 20 min at 37°C in the dark.
2. Add 10 μl of the PI stock solution and incubate for 5 min at room temperature in the dark.
3. Analyze cell fluorescence by flow cytometry: excite fluorescence with blue light (e.g., 488-nm line of the argon ion laser), trigger cell measurement on light scatter signal,

measure green [Rh123 or $DiO_6C(3)$] fluorescence at 530 ± 20 nm, and measure red (PI) fluorescence at >600 nm.

C. Staining with Rh123 or $DiO_6C(3)$ and PI and Analysis by LSC

The principle of cell labeling with Rh123 or $DiOC_6(3)$ and PI for analysis by LSC is essentially the same as for flow cytometry except that the cells are attached to microscope slides or coverslips. The methods of attachment of live cells and of their incubation were described earlier in this chapter. The cells are incubated with the appropriately diluted stock solution of either Rh123 or $DiO_6C(3)$ at 200 or 50 nM final concentration of the respective dye. Cells grown on slides may be incubated with Rh123 or $DiOC_6(3)$ in the same culture medium in which they are normally maintained, with full serum content. Electrostatically attached cells, however, have a tendency to detach in the presence of serum. Therefore, they have to be incubated with Rh123 or $DiO_6C(3)$ in a serum-free medium or in Hanks' buffered salt solution (HBSS). Following a 15-min incubation with Rh123 or $DiO_6C(3)$, an aliquot of PI stock solution is added to obtain a final concentration of 10 μg/ml. The cells are then incubated for an additional 5 min, mounted under a coverslip in the same medium in which they were incubated with Rh123 or $DiO_6C(3)$ and PI, and subjected to fluorescence measurement by LSC. Cell fluorescence is excited at 488 nm and is measured at green and red wavelengths as described earlier for flow cytometry.

D. Staining with JC-1 and Analysis by Flow Cytometry

1. Suspend cell pellet (~10^6 cells) in 1 ml of tissue culture medium with 10% serum.
2. Add 10 μl of the stock solution of JC-1. Vortex cells intensely during the addition of JC-1 and for the next 20 s. Wash twice with PBS, centrifuging each time at 300g.
3. Incubate cells for 15 min at room temperature in the dark.
4. Measure cell fluorescence by flow cytometry: excite fluorescence with blue light (488 nm line of argon laser), measure green fluorescence at 530 ± 20 nm, and measure orange fluorescence at 570 ± 20 or above 570 nm (long-pass filter).

E. Staining with JC-1 and Analysis by LSC

JC-1 fluorochrome is poorly soluble in aqueous media, particularly in the absence of serum. It is difficult therefore to apply it to electrostatically attached cells as they detach in the presence of serum. One can use LSC as a cytometer, however, to measure cell fluorescence of the nonattached cells, as follows.

1. Prepare cells and incubate them with JC-1 in suspension as described earlier.
2. Place 50 μl of this suspension into a well on the microscope slide. Cover with a coverslip.
3. Analyze cells by LSC: use excitation in blue light (488-nm), use forward scatter as a contouring (triggering) parameter, measure green fluorescence at 530 ± 20 nm, and measure orange fluorescence at 570 ± 20 or above 570 nm (long-pass filter).

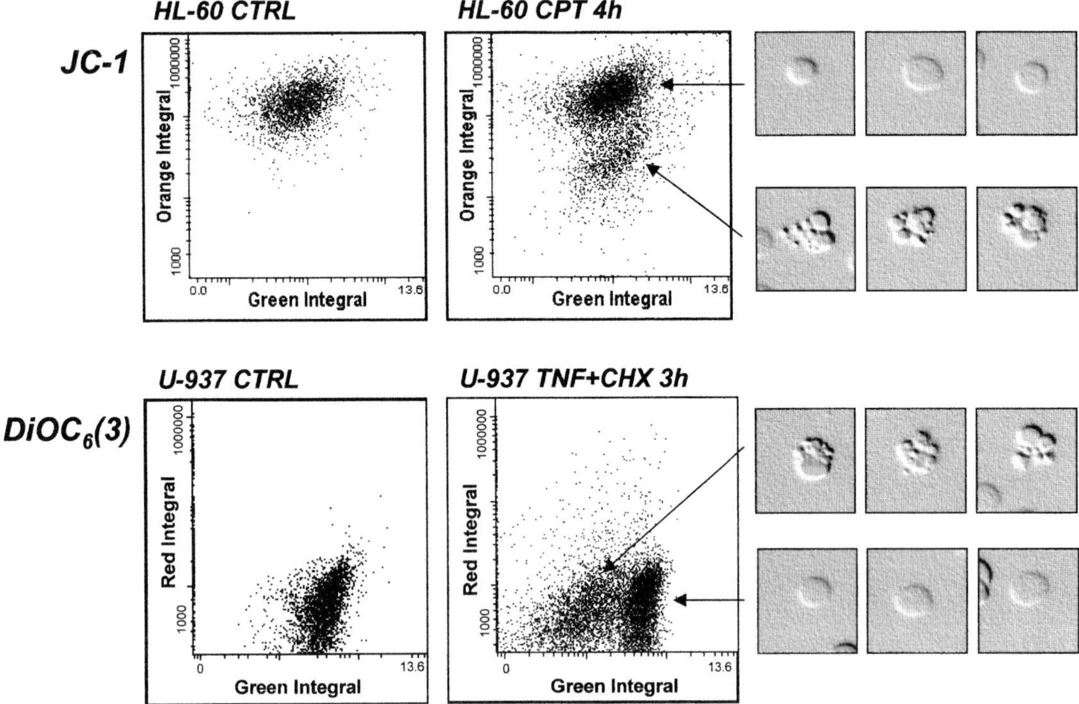

Fig. 4 Detection of the collapse of mitochondrial electrochemical potential ($\Delta\Psi_m$) by LSC after cell staining with JC-1 (top) or $DiOC_6(3)$ (bottom). Exponentially growing HL-60 cells, untreated (CTRL) or induced to undergo apoptosis by treatment with 0.15 μM camptothecin (CPT) for 4 h or with TNFα + cycloheximide (CHX) for 3 h, were stained with JC-1 or with a combination of $DiOC_6(3)$ and PI as described in the respective protocols. A decrease in the orange fluorescence of JC-1 stained cells reflecting dissociation of the JC-1 dye aggregates characterizes apoptotic cells stained with this dye. Apoptotic cells stained with $DiOC_6(3)$ show decreased green fluorescence. A few late apoptotic cells are stainable with PI. Cell galleries show representative cells relocated and based on their $\Delta\Psi_m$.

F. Data Analysis and Interpretation

As shown in Fig. 4, the decrease in $\Delta\Psi_m$ that occurs during apoptosis can be measured by different markers. A combination of PI and $DiOC_6(3)$ identifies nonapoptotic cells that stain only green, early apoptotic cells whose green fluorescence is diminished markedly and late apoptotic or necrotic cells that stain with PI and have red fluorescence only. Likewise, a combination of Rh123 and PI labels live nonapoptotic cells green, early apoptotic cells dim green, and late apoptotic and necrotic cells red. The change in binding of JC-1 is manifested by a loss of orange fluorescence, which represents the aggregate binding of this dye and characterizes only charged mitochondria. Green fluorescence of JC-1 also decreases during apoptosis, although to a lesser degree than orange.

It has been reported that other mitochondrial probes, 10-nonyl acridine orange, MitoFluor Green, and MitoTracker Green, are markers of mitochondrial mass and are not sensitive to $\Delta\Psi_m$ (Ratinaud *et al.,* 1988; Poot *et al.,* 1997). It was proposed, therefore, to measure both $\Delta\Psi_m$ and mitochondrial mass by using a combination of $\Delta\Psi_m$-sensitive and $\Delta\Psi_m$-nonsensitive mitochondrial probes (e.g., Petit *et al.,* 1995). Observations, however, indicate that with 10-nonyl acridine orange, MitoFluor Green and MitoTracker Green are quite sensitive to changes in $\Delta\Psi_m$ and therefore, either alone or in combination with $\Delta\Psi_m$-sensitive probes, cannot be used as markers of mitochondrial mass (Keiji *et al.,* 2000).

It should be stressed that $\Delta\Psi_m$, as is observed with other functional markers, is sensitive to any change in cell environment. The samples to be compared, therefore, should be incubated and measured under identical conditions, taking into an account temperature, pH, time elapsed between the onset of incubation and actual fluorescence measurement, and other potential variables. If possible (e.g., the microscope stage of LSC is controlled thermostatically), the measurements should be performed at 37°C. Otherwise, the samples should be equilibrated to ambient temperature.

Another point to be considered in measuring $\Delta\Psi_m$ is that most mitochondrial potential probes lack absolute specificity and also accumulate in cytosol. Their specificity toward mitochondria is increased when used at low concentration. It is advisable, therefore, to use these probes at a minimal concentration, if possible even below that given in the protocols described earlier. The limit for the minimal dye concentration that still provides an adequate signal-to-noise ratio during the measurement is dictated by sensitivity of the instrument (laser power, optics, photomultiplier sensitivity) and by the mitochondrial mass per cell; the latter varies depending on the cell type or on mitogenic stimulation (Darzynkiewicz *et al.,* 1981).

A series of MitoTracker dyes (chloromethyltetramethylrosamine analogs) of different color was introduced by Molecular Probes Inc. as new mitochondrial $\Delta\Psi_m$-sensitive probes and some of these dyes remain attached to mitochondria following cell fixation using a cross-linking agent (Haugland, 1998; Poot *et al.,* 1997). It should be noted, however, that their retention after fixation may not be correlated with the transmembrane potential because of the binding to thiol moieties within mitochondria (Ferlini *et al.,* 1998; Gilmore and Wilson, 1999). Furthermore, they are potent inhibitors of respiratory chain I and may themselves induce dissipation of $\Delta\Psi_m$ (Scorrano *et al.,* 1999). Because it is likely that other $\Delta\Psi_m$ probes may also either induce or predispose the cells to the permeability transition, one has to be cautious in interpreting data on their use in the analysis of apoptosis.

VII. Annexin V as a Marker of Apoptotic Cells

Phospholipids are distributed asymmetrically between inner and outer leaflets of the plasma membrane of live cells: phosphatidylcholine and sphingomyelin are exposed on the external leaflet of the lipid bilayer whereas phosphatidylserine is located on the inner surface. Early during apoptosis this asymmetry is disrupted and phosphatidylserine becomes exposed on the outside surface of the plasma membrane (Fadok *et al.,* 1992; Koopman *et al.,* 1994; van Engeland *et al.,* 1998). Because the anticoagulant protein Annexin V binds with high affinity to phosphatidylserine, fluorochrome-conjugated Annexin V has found an application as a marker of apoptotic cells, particularly for their detection by flow cytometry (van Engeland *et al.,* 1998). The cells become reactive with Annexin V prior to the loss of the ability of the plasma membrane to exclude cationic dyes such as PI. Therefore, by staining cells with a combination of Annexin V–FITC and PI, it is possible to detect unaffected, nonapoptotic cells (Annexin V negative/PI negative), early apoptotic cells (Annexin V positive/PI negative), and late apoptotic cells ("necrotic stage" of apoptosis), as well as necrotic cells (PI positive).

A. Materials

1. Dissolve fluorescein-conjugated Annexin V (1 : 1 stoichiometric complex, available from BRAND Applications, AW Maastricht, The Netherlands, in binding buffer [10 m*M* Hepes (*N*-2-hydroxyethylpiperazine-*N*-2-ethanesulfonic acid) -NaOH, pH. 7.4, 140 m*M* NaCl, 2.5 m*M* $CaCl_2$] at a concentration of 1.0 μg/ml. This solution has to be prepared fresh prior to use.

2. PI stock solution: Dissolve 1 mg of PI in 1 ml of distilled water. The solution is stable for months when stored in the dark at 0–4°C.

B. Cell Staining and Analysis by Flow Cytometry

1. Suspend 10^5–10^6 cells in 1 ml of fluorescein-conjugated Annexin V in binding buffer for 5 min at room temperature in the dark.

2. Add appropriate volume of stock solution of PI to the cell suspension prior to analysis to have a final PI concentration of 1.0 μg/ml. Incubate for 5 min at room temperature in the dark.

3. Analyze cells by flow cytometry: use excitation in blue light (e.g., 488-nm line of the argon ion laser), use light scatter (forward vs side) to trigger cell measurements, measure green fluorescein–Annexin V fluorescence at 530 ± 20 nm, and measure red (PI) fluorescence at >600 nm.

C. Cell Staining and Analysis by LSC

The methods of attachment of live cells and of their incubation were described earlier in this chapter. Cells are incubated for 5 min in the same solution of fluorescein-conjugated Annexin V in binding buffer as for analysis by flow cytometry, to which the PI stock solution is added to obtain its final concentration of 1 μg/ml. The cells are mounted under a coverslip and their fluorescence excited at 488 nm is measured in the green and red wavelengths, as described earlier for flow cytometry. Because live nonapoptotic cells stain with neither Annexin V–fluorescein nor PI and thus have no fluorescence at all, triggering (contouring) should be set on the light scatter signal. It is critical that cells not dry or be damaged mechanically during the attachment and staining procedure.

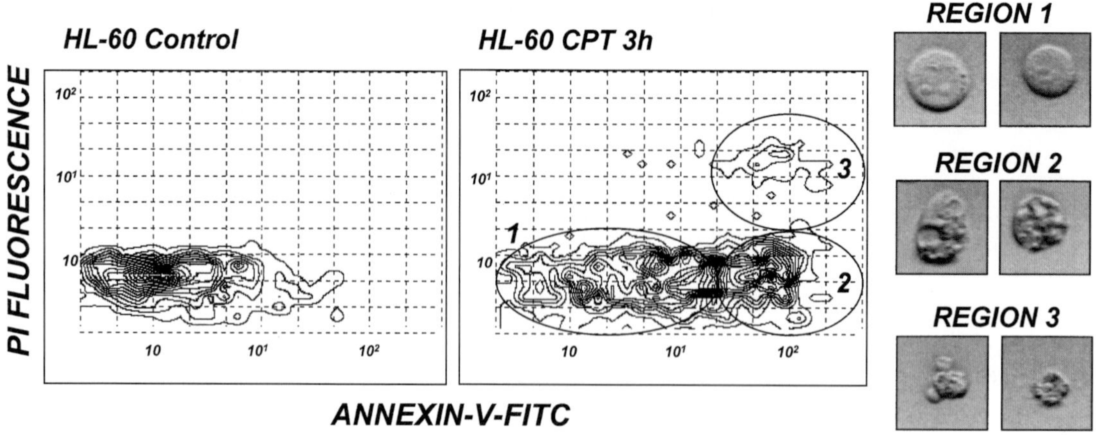

Fig. 5 Detection of early and late apoptotic cells by LSC after staining with Annexin V–FITC conjugate and PI. To induce apoptosis, HL-60 cells were treated with CPT (see legend to Fig. 3), attached to a microscope slide, and subjected to the procedure of labeling with Annexin V–FITC and PI. Their fluorescence was measured by LSC as described in the text. Following the relocation for visual inspection, cells that bind neither Annexin V nor PI (region 1) appear to have normal morphology, cells that bind Annexin V but exclude PI (region 2, early apoptosis) display undulations of the plasma membrane and granularity in chromatin and cytoplasm, and cells that bind Annexin V and stain with PI (late apoptosis, "necrotic" stage of apoptosis) are shrunk and some show the formation and detachment of apoptotic bodies ("budding"). (See Color Plate.)

D. Data Analysis

Live nonapoptotic cells, when stained according to the procedures described previously, have minimal green (fluorescein–Annexin V) fluorescence and also minimal or undetectable red (PI) fluorescence (Fig. 5, see Color Plate). At early stages of apoptosis, cells stain green but still exclude PI and therefore continue to have no significant red fluorescence; analysis by flow cytometry reveals that these cells often display decreased forward and either unchanged or increased side light scatter. At late stages of apoptosis, the cells show intense green and red fluorescence. It should be noted that isolated nuclei, cells with severely damaged membranes, and very late apoptotic cells stain rapidly and intensely with PI and may not bind Annexin V. Stainability of DNA with PI in isolated nuclei is stoichiometric and therefore their frequency histograms of DNA content may have patterns characteristic of cell cycle distribution.

Interpretation of the results may be complicated by the presence of nonapoptotic cells with damaged membranes. Such cells may have phosphatidylserine exposed on the plasma membrane and, therefore, similar to apoptotic cells, bind Annexin V. Mechanical disaggregation of tissues to isolate individual cells; extensive use of proteolytic enzymes to disrupt cell aggregates, remove adherent cells from cultures, or to isolate cells from tissue; mechanical removal of the cells from tissue culture flasks (e.g., by "rubber policeman"); or cell electroporation all affect the binding of Annexin V. Such treatments, therefore, may introduce experimental bias in the subsequent analysis of apoptosis by this method.

Even intact and live cells take up PI on prolonged incubation. Therefore, fluorescence measurement, whether by flow cytometry or LSC, should be performed rather shortly after addition of the dye.

VIII. DNA Fragmentation Assay: Detection of Cells with Fractional ("Subdiploid") DNA Content

Endonuclease(s) activated during apoptosis target(s) internucleosomal DNA sections and cause(s) extensive DNA fragmentation (Arends *et al.,* 1990; Halenbeck *et al.,* 1998; Kerr *et al.,* 1972; Mukae *et al.,* 1998). The fragmented, low molecular weight DNA can be extracted from the cells following their fixation in precipitating fixatives such as ethanol. Conversely, fixation with cross-linking fixatives such as formaldehyde results in the retention of low MW DNA in the cell and therefore should be avoided. Generally, the extraction occurs during the process of cell staining in aqueous solutions after transfer from the fixative. Apoptotic cells, thus, often end up with a deficit in DNA content and when stained with a DNA-specific fluorochrome can be recognized by cytometry as cells having less DNA than G_1 cells. On DNA content frequency histograms, they form a characteristic "sub-G_1" peak (Gong *et al.,* 1994; Nicoletti *et al.,* 1991; Umansky *et al.,* 1981). It should be noted that loss of DNA may also occur as a result of the shedding of apoptotic bodies containing fragments of nuclear chromatin.

The degree of DNA degradation varies depending on the stage of apoptosis, cell type, and often the nature of the apoptosis-inducing agent. Hence, the extractability of DNA during the staining procedure also varies. It has been noted that a high molarity phosphate–citrate buffer enhances extraction of the fragmented DNA (Gong *et al.,* 1994). With some limitations (see later), this approach can be used to extract DNA from apoptotic cells to the desired level in order to achieve their optimal separation by flow cytometry.

A. Materials

1. DNA extraction buffer: Mix 192 ml of 0.2 *M* Na_2HPO_4 with 8 ml of 0.1 *M* citric acid; pH 7.8.
2. DNA staining solution: Dissolve 200 μg of PI and 2 mg of DNase-free RNase A in 10 ml of PBS. Prepare fresh staining solution before each use.

B. Cell Staining with PI and Analysis by Flow Cytometry

1. Fix cells in suspension in 70% ethanol by adding 1 ml of cells suspended in PBS ($1–5 \times 10^6$ cells) into 9 ml of 70% ethanol in a tube on ice. Cells may be stored in fixative at −20°C for several weeks.

2. Centrifuge cells (200*g*, 3 min), decant ethanol, suspend the cell pellet in 10 ml of PBS, and centrifuge (300*g*, 5 min).

3. Suspend cells in 0.5 ml of PBS. To facilitate extraction of low MW DNA, add 0.2–1.0 ml of the DNA extraction buffer.

4. Incubate at room temperature for 5 min and centrifuge.

5. Suspend cell pellet in 1 ml of DNA staining solution.

6. Incubate cells for 30 min at room temperature in the dark.

7. Analyze cells by flow cytometry: use 488-nm laser line (or a mercury arc lamp with a BG12 filter) for excitation and measure red fluorescence (>600 nm) and forward light scatter.

C. Cell Staining with DAPI and Analysis by Flow Cytometry

Cellular DNA may be stained with other fluorochromes instead of PI, and other cell constituents may be counterstained in addition to DNA. The following procedure is used to stain DNA with DAPI.

1. After step 4 in Section VIII,B, suspend the cell pellet in 1 ml of a staining solution containing DAPI (Molecular Probes) at a final concentration of 1 μg/ml in PBS. Keep on ice for 20 min.

2. Analyze cells by flow cytometry: use excitation with UV light (e.g., 351-nm line from an argon ion laser, or mercury lamp with a UG1 filter) and measure the blue fluorescence of DAPI in a band from 460 to 500 nm.

D. Cell Analysis by LSC

To be analyzed by LSC, cells should be attached to the microscope slide as described earlier in this chapter. The slides should then be fixed in 70% ethanol rinsed with phosphate–citrate buffer and stained with PI step by step, as described earlier for cells in suspension. Fixation, subsequent rinses in PBS and in extraction buffer, and staining with PI should be performed in Coplin jars. Following staining with PI, the cells should be mounted under a coverslip in a drop of a solution containing 90% glycerol in PBS into which PI has been dissolved at the same concentration as in the staining solution and sealed with melted paraffin or nail polish.

E. Data Analysis

Apoptotic cells have decreased PI (or DAPI) fluorescence and diminished forward light scatter relative to cells in the main peak (G_1) [Fig. 6, see also Fig. 3]. Optimally, the "sub-G_1 peak" representing apoptotic cells should be separated from the G_1 peak of the nonapoptotic cell population with little or no overlap between the two. It should be stressed, however, that the degree of extraction of low MW DNA, and consequently the content of DNA remaining in apoptotic cells for flow cytometric analysis, varies markedly depending on the extent of DNA degradation (duration of apoptosis), the number of cell washings, and the pH and molarity of the washing and staining buffers. Often DNA fragmentation is so extensive that most DNA is removed during the postfixation rinse

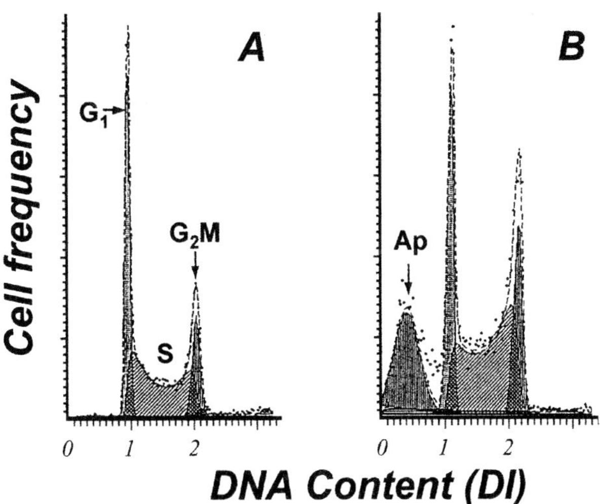

Fig. 6 Detection of apoptotic cells by flow cytometry based on cellular DNA content analysis. To induce apoptosis, HL-60 cells were treated with DNA topoisomerase II inhibitor fostriecin (Hotz *et al.*, 1994). The cells were then fixed in 70% ethanol and stained with PI according to the procedure described in the text. Cell fluorescence was measured with FACScan (Becton-Dickinson, San Jose, CA). The subpopulation of apoptotic cells (Ap) with fractional ("subdiploid") DNA content, i.e., less than DNA index (DI) 1.0 ("sub-G_1" cells), is quite distinct. Also noted is an increase in the proportion of S-phase cells in the nonapoptotic cell population.

with PBS and in the staining solution and therefore the extraction step is unnecessary. Conversely, when DNA degradation does not proceed to internucleosomal regions but stops after generating 50- to 300-kb fragments (Oberhammer *et al.*, 1993), little DNA can be extracted and this method may fail to detect such apoptotic cells. It also should be noted that if G_2, M, or even late S phase cells undergo apoptosis, the loss of DNA from these cells may not be adequate to place them at the "sub G_1, peak" as they may end up with a DNA content equivalent of that of G_1 or early S phase cells and therefore be indistinguishable from the latter.

It is a common practice to use detergents or hypotonic solutions instead of fixation in the process of DNA staining for flow cytometry (Nicoletti *et al.*, 1991). Such treatments cause lysis of plasma membrane and release of the nucleus. Although this approach is simple and yields excellent resolution for DNA content analysis, it introduces bias when used to quantify apoptotic cells. This bias is due to the fact that nuclei of apoptotic cells are often fragmented. Lysis of cells with fragmented nuclei releases nuclear fragments rather than individual nuclei. Thus, several fragments can be released from a single cell. Likewise, lysis of mitotic cells that happen to be in the specimen releases individual chromosomes or chromosome aggregates. In the case of micronucleation (e.g., after cell irradiation), micronuclei are released on cell lysis. Each nuclear fragment, chromosome, or micronucleus is then recorded by flow cytometer as an individual object with a sub-G_1 DNA content. Such objects are then classified erronously as individual apoptoptic cells. This bias is increased if the logarithmic scale is used to display DNA content. Such a scale allows one to record the objects with as little DNA content as 1 or even 0.1% of that of G_1 cells, which certainly cannot be individual apoptotic cells.

IX. DNA Fragmentation: Detection of DNA Strand Breaks ("TUNEL" Assay)

DNA fragmentation during apoptosis, particularly when it progresses to internucleosomal regions (Arends *et al.,* 1990; Oberhammer *et al.,* 1993), generates a multitude of DNA strand breaks in the nucleus. The 3′OH ends of the breaks can be detected by attaching a fluorochrome to them. This is generally done directly or indirectly (e.g., via biotin or digoxygenin) using fluorochrome-labeled deoxynucleotides in a reaction catalyzed preferably by exogenous terminal deoxynucleotidyltransferase (TdT; Gorczyca *et al.,* 1992, 1993; Li and Darzynkiewicz, 1995). This reaction is commonly known as TUNEL from "TDT-mediated dUTP-biotin nick-end labeling" (Gavrieli *et al.,* 1992). This acronym is a misnomer, however, because DNA double-strand breaks and not single-strand nicks are labeled. Of all the markers used to label DNA breaks, BrdUTP appears to be the most advantageous with respect to sensitivity, low cost, and simplicity of the reaction (Li and Darzynkiewicz, 1995). When attached to DNA strand breaks in the form of poly-BrdU, this deoxynucleotide can be detected with an FITC-conjugated, anti-BrdU Ab; the same Ab is commonly used to detect BrdU incorporated during DNA replication. Poly-BrdU attached to DNA strand breaks by TdT, however, is accessible to the Ab without a need for DNA denaturation, which otherwise is required to detect the precursor incorporated during DNA replication.

It should be stressed that the detection of DNA strand breaks by this method requires prefixation of cells with a cross-linking agent such as formaldehyde. Unlike ethanol, formaldehyde prevents the extraction of small pieces of fragmented DNA. Thus, despite cell permeabilization and the subsequent cell washings required by the procedure, the DNA content of early apoptotic cells (and the number of DNA strand breaks) is not diminished markedly due to extraction. Labeling DNA strand breaks in this procedure, which utilizes fluorescein-conjugated, anti-BrdU Ab, can be combined with staining of DNA with the fluorochrome of another color (PI, red fluorescence). Cytometry of cells that are stained differentially for DNA strand breaks and for DNA allows one to distinguish apoptotic from nonapoptotic cell subpopulations and reveals the cell cycle distribution in these subpopulations (Gorczyca *et al.,* 1992, 1993).

A. Materials

1. Prepare fixatives:

Primary fixative: 1% methanol-free formaldehyde (available from Polysciences Inc., Warrington, PA) in PBS, pH 7.4

Secondary fixative: 70% ethanol

2. The TdT reaction buffer (5× concentrated) contains 1 *M* potassium (or sodium) cacodylate, 125 m*M*, Tris–HCl, pH 6.6, and 1.25 mg/ml bovine serum albumin (BSA).

3. Cobalt chloride ($CoCl_2$), 10 mM.

4. TdT in storage buffer, 25 units in 1 μl. The buffer, TdT, and $CoCl_2$ are available from Boehringer Mannheim (Indianapolis, IN).

5. BrdUTP stock solution: BrdUTP (Sigma) 2 mM (100 nmol in 50 μl) in 50 m*M* Tris–HCl, pH 7.5.

6. FITC-conjugated, anti-BrdU mAb solution (per 100 μl of PBS): 0.3 μg of anti-BrdU FITC-conjugated mAb (available from Becton-Dickinson, San Jose, CA), 0.3% Triton X-100, and 1% BSA.

7. For rinsing buffer, dissolve 0.1% (v/v) Triton X-100, and 5 mg/ml BSA in PBS.

8. For PI staining buffer, dissolve 5 μg/ml PI, and 200 μg/ml DNase-free RNase A in PBS.

B. Cell Fixation, Staining, and Analysis by Flow Cytometry

1. Fix cells in suspension in 1% formaldehyde for 15 min on ice.

2. Centrifuge (300*g*, 5 min), resuspend cell pellet ($\sim 2 \times 10^6$ cells) in 5 ml of PBS, centrifuge (300*g*, 5 min), and resuspend cells in 0.5 ml of PBS.

3. Add the 0.5-ml aliquot of cell suspension into 5 ml of ice-cold 70% ethanol. The cells can be stored in ethanol at −20°C for several weeks.

4. Centrifuge (300*g*, 3 min), remove ethanol, resuspend cells in 5 ml of PBS, and centrifuge (300*g*, 5 min).

5. Resuspend the pellet (not more than 10^6 cells) in 50 μl of a solution containing 10 μl of the reaction buffer, 2.0 μl of BrdUTP stock solution, 0.5 μl (12.5 units) of TdT in storage buffer, 5 μl of $CoCl_2$ solution, and 33.5 μl of distilled H_2O.

6. Incubate cells in this solution for 40 min at 37°C (alternatively, incubation can be carried out overnight at 22–24°C).

7. Add 1.5 ml of the rinsing buffer and centrifuge (300*g*, 5 min).

8. Resuspend cells in 100 μl of FITC-conjugated, anti-BrdU mAb solution.

9. Incubate at room temperature for 1 h or at 4°C overnight. Add 2 ml of rinsing buffer and centrifuge (300*g*, 5 min).

10. Resuspend the cell pellet in 1 ml of PI staining solution containing RNase.

11. Incubate for 30 min at room temperature in the dark.

12. Analyze cells by flow cytometry: excite cell fluorescence with blue light (488-nm laser line or BG12 excitation filter), measure green fluorescence of FITC–anti BrdU mAb at 530 ± 20 nm, and measure red fluorescence of PI at >600 nm.

C. Cell Analysis by LSC

1. Attach the cells to the microscope slide preferably either by cytocentrigation or by growth on the slide, as described earlier in this chapter.

2. Without allowing the cytospins to dry completely, prefix cells by immersing the slide in 1% methanol-free formaldehyde (Polysciences) in PBS in a Coplin jar for 15 min on ice.

3. Transfer the slides to 70% ethanol and fix for at least 1 h; the cells can be stored in Coplin jars in ethanol for several days.

4.–9. Remove the slide from ethanol, rinse in PBS, and place it horizontally. Follow steps 4 to 9 as described for flow cytometry, but layer small volumes (50–100 μl) of the respective buffers, rinses, or staining solutions carefully on the area of the slide where the cells are deposited. At appropriate times these solutions are removed with a Pasteur pipette or vacuum suction pipette. Place small pieces (2.5 × 2.5 cm) of thin polyethylene foil on slides atop the drops to prevent drying. Carry out all incubations at 100% humidity (e.g., in a closed box with wet paper towels) to prevent drying at any step of the reaction.

10. Rinse the slide in PBS and apply a drop or two of the PI-staining solution containing RNase A over the area with cells. Cover with a strip of polyethylene foil and incubate for 20 min at 100% humidity in the dark at room temperature. Replace the PI staining solution with a drop of a mixture of glycerol and PI-staining solution (9:1) and mount under the coverslips. To preserve the specimen for a longer period of time or transport, seal the coverslip with nail polish or melted paraffin.

11. Measure cell fluorescence by LSC: excite fluorescence with 488-nm laser line, measure green fluorescence of FITC–anti BrdU mAb at 530 ± 20 nm, and measure red fluorescence of PI at >600 nm.

D. Commercial Kits

A plethora of kits designed to label DNA strand breaks applicable to flow cytometry are available from different vendors. For example, Phoenix Flow Systems, PharMingen Inc., and ALEXIS (all from San Diego, CA) all provide kits to identify apoptotic cells based on a single-step procedure utilizing TdT and FITC-conjugated dUTP (APO-DIRECT) or TdT and BrdUTP, as described earlier (APO-BRDU). A description of the method, which is nearly identical to the one presented in this chapter, is included with the kit. Another kit (ApopTag), based on two-step DNA strand break labeling with digoxygenin-16-dUTP by TdT, is provided by ONCOR Inc., (Gaithersburg, MD; now owned by Intergen, Purchase, NY).

E. Data Analysis

Apoptotic cells are strongly labeled with fluoresceinated anti-BrdU Ab, which distinguishes them from nonapoptotic cells (Fig. 7). Because of the high intensity of their green fluorescence, an exponential scale (logarithmic photomultipliers) often must be used for data acquisition and display. Simultaneous measurement of DNA content makes it possible to identify the cell cycle position of cells in apoptotic and nonapoptotic populations. It should be noted, however, that late apoptotic cells may have diminished DNA content because of prior shedding of apoptotic bodies (which may contain nuclear

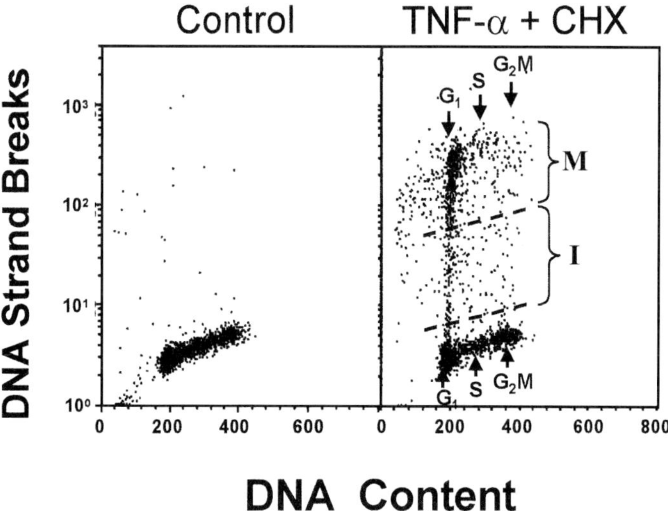

Fig. 7 Detection of apoptotic cells by flow cytometry based on the presence of DNA strand breaks. To induce apoptosis, HL-60 cells were treated with TNF-α in the presence of cycloheximide (CHX) for 60 min as described (Li and Darzynkiewicz, 2000). DNA strand breaks were labeled with BrdUTP as described in the text. The cells are very heterogeneous in terms of the frequency of DNA strand breaks: the cells most advanced in the process of apoptosis display maximal labeling (M; note that the DNA strand break coordinate is exponential) and the cells that are less advanced have intermediate labeling (I). The cell cycle distribution of both the apoptotic and the nonapoptotic cell subpopulations can be estimated based on the DNA content of individual cells. Cell fluorescence was measured in a FACScan cytometer (Becton-Dickinson).

fragments) or due to such massive DNA fragmentation that small DNA fragments cannot be retained in the cell even after fixation with formaldehyde. Such late apoptotic cells thus may have sub-G_1 DNA content (Fig. 7).

In some instances of apoptosis, DNA fragmentation stops after the initial DNA cleavage to 50- to 300-kb fragments (Collins *et al.*, 1992; Oberhammer *et al.*, 1993). The frequency of DNA strand breaks in nuclei of these cells is low and therefore they may not be detected easily by the TUNEL method.

X. Analysis of Caspase Activation by the Labeled Inhibitor Method

Activation of cysteine aspartic acid-specific proteases (caspases) is the critical event of apoptosis, initiating the irreversible ("execution") steps of the cell demise (Alnemri *et al.*,1996; Kaufmann *et al.*, 1992; Lazebnik *et al.*, 1994). Several approaches are used to detect and study the process of activation of these enzymes. Immunoprecipitation and immunoblotting techniques using caspase-specific antibodies are the most common but cannot be applied to cytometry. One approach that is potentially useful for cytometry utilizes fluorogenic (or chromogenic) substrates of caspases. The peptide substrates are colorless or not fluorescent but upon caspase-induced cleavage they generate colored or fluorescing products (Gorman *et al.,* 1999; Liu *et al.,* 1999). Thus far the utility of this approach was tested primarily on tissue extracts whose color or fluorescence changes were measured in 96-well format microtiter plates. Evidence shows, however, that the substrates penetrate through the plasma membrane into apoptotic cells and therefore can be used in studies of intact cells by flow cytometry (Hug *et al.,* 1999). Many kits designed to measure the activity of caspases using fluorometric or colorimetric assays are available commercially (e.g., from BIOMOL Research Laboratories, Inc., Plymouth Meeting, PA, or Calbiochem, La Jolla, CA). Some of these kits can be used to detect the activation of multiple caspases, whereas others are based on substrates that are specific for caspase-1, caspase-3, or caspase-8.

The second approach in studies of activation of caspases, applicable to cytometry, is based on the immunocytochemical detection of the epitope of these enzymes that is characteristic of their active form. This epitope appears as a result of conformational changes that occur during the activation of caspases, e.g., such as associated with the transcatalytic cleavage of the zymogen procaspases (reviews: Budihardijo *et al.,* 1999; Earnshow *et al.,* 1999). Antibodies developed to react only with the activated caspases have become available from commercial sources (e.g., from Promega, Madison, WI). These antibodies can be used in standard immunocytochemical assays. In principle, the protocol that is provided in this chapter for the immunocytochemical detection of the cleaved poly(ADP-ribose) polymerase (PARP, PARP p89) (see further) can be used for their detection. The cells may be counterstained with another color fluorochrome (e.g., for analysis of DNA content), and their fluorescence can be measured by multiparameter flow cytometry or LSC. However, no published data are yet available about the cytometric analysis of active caspases by this approach. Furthermore, one has to remember that often antibodies that are applicable to immunoblotting or immunoprecipitation may not be useful in the immunocytochemical assays and vice versa.

The third approach used to discern the activation of caspases relies on the use of labeled inhibitors of these enzymes. This methodology has a long history and was applied initially in studies of proteases in mast cells utilizing the radioisotope-labeled inhibitors that were detected by autoradiography (Darzynkiewicz and Barnard, 1966). The essence of the methodology is covalent binding of the specific inhibitor to the active center of the enzyme. In the case of caspases, one of several ligands that may bind specifically to their active center is a complex conjugate consisting of a fluorochrome, recognition peptide, and ketone moiety. One such ligand [carboxyfluorescein derivative of benzyloxycarbonyl-valylalanylaspartic acid fluoromethyl ketone (FAM-VAD-FMK)] has become available commercially (Intergen Co., Purchase, NY). Another (FITC-VAD-FMK) is provided by Promega (Madison, WI). These ketone reagents penetrate through the plasma membrane of live cells and are relatively nontoxic to the cell. Their irreversible binding to active centers of caspases ensures that only cells with activated enzymes become labeled. Because these inhibitors consist of three amino acid recognition sequence instead of four (which is characteristic for the individual caspases; Thornberry *et al.,* 1997), they are rather generic for most caspases than specific for the particular ones. The protocol given in this section is a modification of the protocol provided with the kit by the vendor (Intergen Co.).

A. Materials

1. Dissolve lyophilized FAM-VAD-FMK (available as a component of the CaspaTag fluorescein caspase activity kit from Intergen) in dimethyl sulfoxide (DMSO) as specified in the kit to obtain a 150× concentrated (stock) solution of this inhibitor. Aliquots of this solution may be stored at −20°C in the dark for several months.

2. Just prior to use, prepare a 30× working solution of FAM-VAD-FMK by diluting the stock solution 1:5 in PBS. Mix the vial until the solution becomes transparent and homogeneous. This solution should not be stored. Protect all FAM-VAD-FMK solutions from light.

3. Stock solution of PI: Dissolve 1 mg of PI (Molecular Probes) in 1 ml of distilled water.

B. Cell Staining and Analysis by Flow Cytometry

1. Suspend approximately 10^6 cells in 0.3 ml of medium [with 10 or 1% (w/v) serum albumin].

2. Add 10 μl of the working solution of FAM-VAD-FMK to this cell suspension. Mix gently and incubate for 1 h at 37°C.

3. Add 2 ml of PBS, mix gently, and centrifuge (300*g*, 5 min, room temperature).

4. Resuspend the cell pellet in 2 ml of PBS and centrifuge as in step 3.

5. Resuspend the cell pellet in 1 ml of PBS. Add 10 μl of stock solution of PI. *Note:* Protect samples from light at all times.

6. Measure cell fluorescence by flow cytometry: excite cell fluorescence with blue light (488-nm laser line or BG12 excitation filter), measure green fluorescence of FAM-VAD-FMK at 530 ± 20 nm, and measure red fluorescence of PI at >600 nm.
Note: Apoptotic cells are fragile and tend to disintegrate during centrifugations. Addition of bovine serum albumine (2%; w/v) to PBS used for rinsing provides some protection.

C. Cell Staining and Analysis by LSC

1. Attach the cells to the microscope slide electrostatically (within the shallow wells) or by growing them on the slide or coverslip, as described earlier in this chapter. Keep the cells immersed in the culture medium by adding 100 μl of the medium (with 10% serum) into the well to cover the area with the cells.

2. Prepare the staining solution by adding 3 μl of the FAM-VAD-FMK working solution into 100 μl of culture medium. Replace the medium from over the cells with the staining solution of FAM-VAD-FMK.

3. Place a thin polyethylene foil or parafilm strip (2.5 × 2.5 cm) atop the staining solution to prevent drying. Incubate the slides horizontally for 1 h at 37°C in a closed box with wet cotton balls or tissue paper to prevent drying.

4. Remove the incubation medium by vacuum suction or by a Pasteur pipette. Rinse three times with PBS each time by adding new PBS solution, mixing gently, and, after 2 min, replacing with the next rinse.

5. Prepare the PI-staining solution by adding 20 μl of PI stock solution to 1 ml of PBS. Apply one or two drops of this solution on top of the cells deposited on the slide. Cover with a coverslip and seal the edges to prevent drying. *Note:* Protect cells from light throughout the procedure.

6. Measure cell fluorescence on LSC: excite fluorescence with 488-nm laser line, measure green fluorescence of FAM-VAD-FMK at 530 ± 20 nm, measure red fluorescence of PI at >600 nm, and use light scatter signal for contouring.

Note: Staining with PI is optional. It allows to distinguish cells that have the integrity of plasma membrane compromised to the extent that they cannot exclude PI (necrotic and late apoptotic cells, cells with mechanically damaged membranes, isolated cell nuclei).

D. Data Analysis

Figure 8 illustrates the increasing propensity of HL-60 cells treated with CPT to induce apoptosis, to react with FAM-VAD-FMK, as measured by LSC according to the protocol presented earlier. A distinct subpopulation of cells with increased green fluorescence (integrated values) is apparent in the CPT-treated cultures. The intensity of their maximal pixel of green fluorescence is even more increased. The frequency of these cells was increased with time of treatment with CPT. Upon relocation and visual

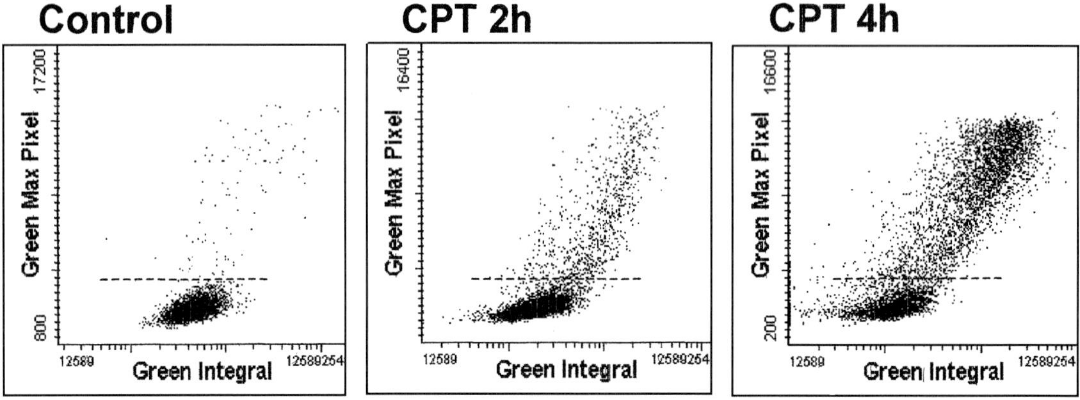

Fig. 8 Activation of caspases detected by the fluorochrome-labeled caspase inhibitor method measured by LSC. Exponentially growing HL-60 cells were untreated (control) or, to induce apoptosis, were treated in culture with 0.15 μ*M* camptothecin (CPT) for 2 or 4 h as described (Li and Darzynkiewicz, 2000). The cells were then attached to microscope slides electrostatically, incubated with a staining solution of FAM-VAD-FMK as described in the text (steps 1 to 4), and their green fluorescence (integrated value and pixel of maximal intensity) measured by LSC. Note the appearance of the apoptotic cell subpopulation characterized by the increased green fluorescence (above the threshold level marked by the dashed line), reflecting activation of caspases that bind FAM-VAD-FMK, in cultures treated with CPT.

examination, these cells showed changes in morphology typical of apoptosis, such as extensive "blebbing," shrinkage, and nuclear fragmentation (Bedner *et al.*, 2000b).

XI. Detection of Apoptotic Cells Based on Cleavage of Poly(ADP-ribose) Polymerase

Poly(ADP-ribose) polymerase is a nuclear enzyme that is involved in DNA repair and is activated in response to DNA damage (de Murcia, 1994). Early in apoptosis, PARP is cleaved by caspases, primarily by caspase-3 (Alnemri *et al.*, 1996; Kaufmann *et al.*, 1992; Lazebnik *et al.*, 1994). The specific cleavage of this protein resulting in distinct 89- and 24-kDa fragments, usually detected electrophoretically, is considered to be a hallmark of the apoptotic mode of cell death.

The development of antibodies that recognize the cleaved PARP products prompted their use as immunocytochemical markers of apoptotic cells. The antibody that recognizes the 89-kDa fragment (PARP p89) was initially used to score the frequency of apoptosis in tissue sections (Kockx *et al.*, 1998; Sallman *et al.*, 1997). This antibody has been adapted to label apoptotic cells for detection by flow cytometry and LSC (Li and Darzynkiewicz, 2000). A good correlation was observed between the frequency of apoptosis detected cytometrically with PARP p89 Ab and that detected by the DNA strand break (TUNEL) assay. However, at least in some cell systems, the cleavage of PARP occurs prior to the onset of DNA fragmentation (Li and Darzynkiewicz, 2000). In these instances the correlation may not be apparent at early stages of apoptosis because the apoptotic index estimated based on PARP cleavage may exceed the estimate based on the TUNEL reaction. Cytometric analysis of cells stained differentially for PARP p89 and DNA, similar as the TUNEL assay, makes it possible not only to idenitify and score apoptotic cell populations, but also to correlate apoptosis with the cell cycle position or DNA ploidy.

A. Materials

1. Cell fixatives:
   1% methanol-free formaldehyde (Polysciences, Inc., Warrington, PA) in PBS 70% ethanol
2. Anti-PARP p89 antibody (Promega Corp., MI; reported by the vendor as "anti-PARP-85 fragment," rabbit polyclonal).
3. Fluorescein-conjugated antirabbit immunoglobulin Ab (DAKO Corporation, Carpintiera, CA).
4. Prepare 0.25% solution of Triton X-100 (Sigma) in PBS.
5. Prepare 1% (w/v) solution of bovine serum albumin (Sigma) in PBS (PBS/BSA solution).
6. Stock solution of PI: Dissolve 1 mg of PI (Molecular Probes) in 1 ml of distilled water.

7. Stock solution of RNase: Dissolve 2 mg of DNase-free RNase A (Sigma) in 1 ml of distilled water. If RNase is not DNase free, boil this solution for 3 min.

Solutions 3–6 may be stored at 4°C for several weeks.

B. Cell Staining and Analysis by Flow Cytometry

1. Suspend approximately 10^6 cells in 1 ml of PBS. Fix cells by transferring this suspension to a centrifuge tube containing 10 ml of 1% formaldehyde on ice. After 15 min, centrifuge the cells (300*g*), rinse them once with PBS (300*g*), and resuspend the cell pellet in 1 ml of PBS. With a Pasteur pipette, transfer this cell suspension into a centrifuge tube containing 10 ml of 70% ethanol. The cells may be stored in ethanol at −20°C for several days.

2. Centrifuge cells (300*g*, 5 min) and resuspend the cell pellet in 2 ml of PBS; repeat this step again.

3. Resuspend the cells in 0.25% Triton X-100/PBS solution for 10 min.

4. Centrifuge cells (300*g*, 5 min) and resuspend the cells in 2 ml of BSA/PBS solution for 10 min.

5. Centrifuge cells (300*g*, 5 min) and resuspend the cells in 100 μl of BSA/PBS containing anti-PARP p89 pAb diluted 1:200. Incubate for 2 h at room temperature or at 4°C overnight.

6. Add 5 ml of BSA/PBS solution for 5 min and centrifuge (300*g*, 5 min).

7. Resuspend cell pellet in 100 μl of PBS/BSA containing fluorescein-conjugated secondary Ab $F(ab')_2$ fragment (swine antirabbit immunoglobulin) diluted 1:30. Incubate for 1 h in the dark at room temperature.

8. Add 5 ml of BSA/PBS and centrifuge (300*g*, 5 min). Resuspend cell pellet in 1 ml of PBS. Add 20 μl of PI and 50 μl of RNase stock solutions. Incubate for 20 min in the dark at room temperature.

9. Measure cell fluorescence by flow cytometry: excite cell fluorescence with blue light (488-nm laser line or BG12 excitation filter), measure green fluorescence of FITC-anti PARP p89 at 530 ± 20 nm, and measure red fluorescence of PI at >600 nm.

C. Cell Staining and Analysis by LSC

1. Attach the cells to the microscope slide either by cytocentrigation or by growth on the slide, as described earlier in this chapter.

2. Without allowing the cytospins to dry completely, prefix the cells by immersing the slide in 1% formaldehyde in Coplin jar, cooled to ice temperature. After 15 min transfer the slide into 70% ethanol, also on ice, for at least 2 h. The slides can be stored at −20°C for several days.

3. Remove the slide from ethanol, rinse in PBS, and place horizontally. Follow steps 2 to 6 as described for flow cytometry but by layering small volumes (50–100 μl) of the respective buffers, rinses, or staining solutions carefully on the area of the slide where the cells are deposited. At appropriate times, remove the solutions with a Pasteur pipette or

vacuum suction pipette. Place thin polyethylene foil (2.5 × 2.5 cm) on top of the drops of incubation solutions to prevent drying during incubations. Carry out all incubations at 100% humidity (e.g., in a closed box with wet cotton balls or tissue paper) to prevent drying.

4. Prepare the PI-staining solution by adding 20 μl of PI and 50 μl of RNase stock solutions to 1 ml of PBS. Apply one or two drops of this solution on top of the cells deposited on the slide. Cover with polyethylene foil and incubate for 30 min at 100% humidity in the dark at room temperature. Mount under the coverslips in a drop of a mixture of glycerol and PI-staining solution (9:1). To preserve the specimen for longer periods of time or to transport, seal the coverslip with nail polish or melted paraffin and store at 4°C in the dark.

5. Measure cell fluorescence on LSC: excite fluorescence with 488-nm laser line, measure green fluorescence of FITC-anti PARP p89 Ab at 530 ± 20 nm, and measure red fluorescence of PI at >600 nm.

D. Data Analysis

The bivariate distributions of HL-60 cells treated with TNFα or CPT representing immunofluorescence of PARP p89 vs DNA content (Fig. 9) are strikingly similar to these representing DNA strand breaks vs DNA content (Fig. 7). The distinction between

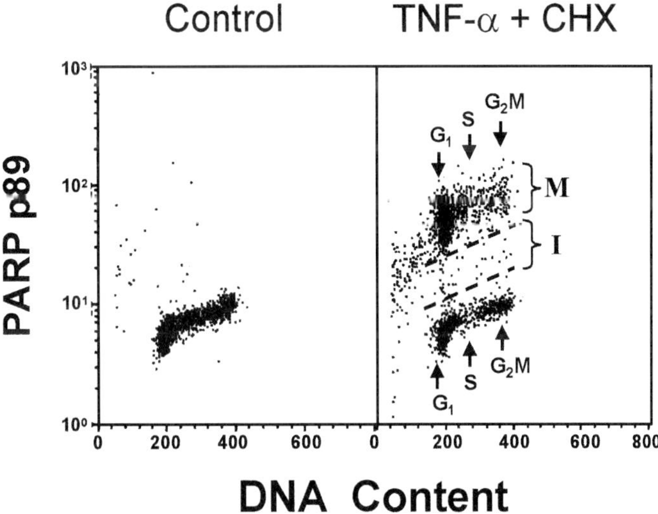

Fig. 9 Identification of apoptotic cells by flow cytometry based on the immunocytochemical detection of the 89-kDa PARP cleavage fragment. To induce apoptosis, HL-60 cells were treated with TNF-α in the presence of cycloheximide (CHX) for 60 min (Li and Darzynkiewicz, 2000) identically as for the assay based on DNA fragmentation (see Fig. 7). In contrast to the DNA strand break assay, apoptotic cells detected by PARP p89 immunofluorescence are more uniform: most cells are labeled maximally (M) and only a few have an intermediate (I) level of PARP p89 fluorescence. Cell fluorescence was measured by a FACScan flow cytometer (Becton Dickinson).

apoptotic and nonapoptotic cells is equally good in both assays. Both assays also offer the opportunity to assess the cell cycle distribution in apoptotic and nonapoptotic cell subpopulations. Late apototic cells that have lost DNA via shedding of apoptotic bodies are characterized by a sub-G_1 DNA content. Compared with early apoptotic cells, they also have diminished PARP p89 immunofluorescence.

Comparison of the DNA fragmentation pattern (Fig. 7) with that of PARP cleavage (Fig. 9) of identically treated cells reveals the differences in the kinetics of these processes. As is evident there are more cells with an intermediate level of DNA strand breaks than with intermediate PARP p89 immunofluorescence. With a "snapshot" measurement, the probability of finding intermediate-step cells of any kinetic event is proportional to the duration of this step. It is apparent that PARP cleavage, from its onset to maximal level, is more rapid than DNA fragmentation.

Although the TUNEL assay offers similar resolution as well as an opportunity to investigate the cell cycle phase specificity of apotosis, the procedure of identifying apoptotic cells based on PARP p89 immunofluorescence is simpler. It may be preferred, therefore, particularly in studies trying to correlate apoptosis with cell cycle progression. It may also be of special value in instances when DNA cleavage during apoptosis does not progress to internucleosomal regions and thus is undetectable by the TUNEL assay.

XII. Unique Possibilities Offered by LSC in Studies of Apoptosis

Three attributes of LSC make it particularly useful in studies of apoptosis. The first is the possibility of examining the cells of interest morphologically. Thus, several thousand cells can be measured per sample, with rates approaching 100 cells per second to quantify the frequency of apoptotic cells based on a particular marker. Morphology of the presumed apoptotic cells can then be discerned following their relocation and microscopic examination, as shown in Fig. 3. This attribute of LSC made it possible, for example, to identify "false-positive" apoptotic cells in the bone marrow of leukemic patients undergoing chemotherapy (Bedner *et al.*, 1999). These cells, showing presence of DNA strand breaks ("TUNEL positive"), were actually nonapoptotic monocytes and macrophages that engulfed apoptotic bodies containing fragmented nuclei, the products of disintegration of apoptotic cells. Such cells are classified erroneously by flow cytometry as genuine apoptotic cells. Thus LSC, allows one to quantify the frequency of apoptotic cells at rates approaching that of flow cytometry and to confirm the accuracy of classification based on the gold standard of apoptosis, morphology.

The second attribute of LSC that distinguishes it from flow cytometry and is of great utility is the possibility LSC offers in studying the spatial location of a fluorochrome vis-a-vis a particular component of the cell structure. The instrument allows, for example, measurement of nuclear and cytoplasmic fluorescence separately (Fig. 2). Activation of many cell constituents (e.g., transcription factors and signal transduction molecules) correlates with their movement between the cytoplasm and the nucleus. Analysis of their translocation, therefore, reveals activation. Translocation from the cytoplasm to the nucleus, or vice versa, of any component that can be detected immunocytochemically is measured conveniently by LSC as a change in the ratio of cytoplasmic to nuclear immunofluorescence. For example, LSC was used to measure the nuclear translocation of the ubiquitous transcription factor NF-κB induced by TNF-α (Deptala *et al.*, 1998). Translocation of Bax from cytosol to mitochondria can also be measured by LSC (Darzynkiewicz *et al.*, 1999). In cytosol, Bax is distributed uniformly and, when detected immunocytochemically, its immunofluorescence is diffuse. Its local concentration is increased when translocated into mitochondria, which is manifested by strong, localized foci of immunofluorescence. This change in local inhomogeneity of fluorescence distribution within the cell is detected as an increase in maximal pixel fluorescence by LSC (Bedner *et al.*, 2000a). It is likely that the translocation of cytochrome *c* from mitochondria to cytoplasm, the event presumed to initiate the "execution" phase of apoptosis (Reed, 1997), may also be detected in a similar fashion.

The third unique attribute of LSC that has found an application in studies of apoptosis relates to the possibility of repeated measurements of the same set of cells and the subsequent integration ("merging") of the results from all sequential measurements into a single file (Kamentsky *et al.*, 1997). This feature allows one to study enzyme kinetics and other time-resolved events in individual cells (Bedner *et al.*, 1998). It can thus be applied to study the activity of caspases, cathepsin, and other enzymes involved in apoptosis. Furthermore, multivariate analysis of integrated data reveals correlations between cell features that are measured sequentially. This capability of LSC can be employed to combine analysis of functional features of live cells with cell attributes that can be probed only after fixation (Li and Darzynkiewicz, 1999). Specifically, the functional changes that occur during apoptosis, dissipation of the mitochondrial transmembrane potential ($\Delta\Psi_m$; Petit *et al.*, 1995; Zamzani *et al.*, 1998), and oxidative stress (increase in reactive oxygen intermediates, ROIs; Hedley and McCulloch, 1996; Sheng-Tanner *et al.*, 1998)

Table I
Apoptosis Associated Changes in Cell Attributes Measured Supravitally or after Cell Fixation[a]

Cell attributes studied supravitally	Cell attributes measured after fixation
Plasma membrane integrity	DNA content/cell cycle position
Asymmetry of plasma membrane phospholipids	DNA fragmentation/strand breaks
Intracellular pH	Chromatin condensation
Intracellular Ca^{2+} and other ions	Activation of NF-κB
Mitochondrial transmembrane potential ($\Delta\Psi_m$)	Bax translocation (cytosol–mitochondria)
Reactive oxygen intermediates	Cytochrome *c* translocation (mitochondria–cytosol)
Intracellular level of glutathione	Translocation of other molecules (AIF, APAF-1, etc.)
Activation of caspases (detected by cleavage of the substrate)	Activation of caspases (detected immunocytochemically, e.g., PARP cleavage)
Activity of many intracellular enzymes	Changes in cell morphology

[a]The "merge" feature of LSC as described allows one to correlate, within the same cell, the attributes from both groups. It is possible, therefore, to disclose whether a dissipation of $\Delta\Psi_m$ or oxidative stress precedes and/or is a prerequisite for DNA fragmentation or activation of caspases.

can be correlated with the attributes measured in fixed cells, the cell cycle position, and the presence of DNA strand breaks. The cells are first measured when alive to assess their $\Delta\Psi_m$ or ROIs are then fixed and subjected to analysis of DNA content and/or the presence of DNA strand breaks. The results of both measurements can then be integrated into a single file for multivariate analysis. Using this procedure it is possible to determine the status of DNA fragmentation or cell cycle position of the same cell whose $\Delta\Psi_m$ or ROIs has been measured. This approach appears to be of particular utility in mapping the sequence of intracellular events that occur during apoptosis, as well as in assessing the cause–effect correlation between these events (Table I).

XIII. Strategies in Analysis of Apoptosis by Flow Cytometry or LSC

A. Which Method to Choose?

The choice of a particular method depends on the cell system, the nature of the inducer of apoptosis, the desired information (e.g., specificity of apoptosis with respect to the cell cycle phase or DNA ploidy), and technical restrictions (e.g., the need for sample transportation, type of flow cytometer available). Positive identification of apoptotic cells is not always easy. One of the most specific markers appears to be the presence of DNA strand breaks. The number of DNA strand breaks in apoptotic cells is so large that the intensity of their labeling in the TUNEL reaction ensures their positive identification (Gorczyca *et al.,* 1992). As mentioned, however, the situation is complicated in those instances of apoptosis where internucleosomal DNA degradation does not occur (Catchpoole and Stewart, 1993; Collins *et al.,* 1992; Ormerod *et al.,* 1994; Knapp *et al.,* 1999). The number of DNA strand breaks in such atypical apoptotic cells may be inadequate to distinguish them by the TUNEL method. Although necrotic cells have many fewer DNA strand breaks relative to apoptotic cells (Gorczyca *et al.,* 1992), in some instances of necrosis DNA fragmentation may be extensive and such necrotic cells may be undistinguishable from apoptotic cells. It is unknown, however, how frequent such cases are.

The activation of caspases is also considered to be a specific marker of apoptosis. It can be detected by variety of methods, some presented in this chapter. It should be stressed, however, that caspases may be activated, and several "death substrates" such as PARP cleaved, in nonapoptotic cells. During the mitogenic stimulation of lymphocytes, for example, the extensive activation of caspases appears to be a physiological step associated with the mitogenic stimulation process and is unrelated to apoptosis (Miossec *et al.,* 1997; Kennedy *et al.,* 1999; Zapata *et al.,* 1998). One has to be careful, therefore, in identifying apoptotic cells based on the activation of caspases alone, particularly in studies of lymphocytes.

The ability of cells to bind Annexin V is still another marker considered to be specific to apoptosis. It should be kept in mind, however, that the use of the Annexin V binding assay is hindered in some situations, as discussed earlier in this chapter, when the plasma membrane may be damaged during cell preparation or storage, leading to the loss of asymmetry in the distribution of phosphatidylserine across the membrane. Furthermore, macrophages and other cells engulfing apoptotic bodies may also be positive in the Annexin V assay (Marguet *et al.,* 1999).

Apoptosis can be recognized with greater certainty when the cells are subjected to several assays probing different viability features. For example, the assay of plasma membrane integrity (exclusion of PI) and Annexin V binding, combined with the analysis of PARP cleavage or DNA fragmentation, may provide a more definitive assessment of the mode of cell death than can be determined when each of these methods is used alone. It should be stressed that in light of the evidence that the collapse of $\Delta\Psi_m$ may not be a prerequisite for the release of cytochrome *c*, AIF, and other apoptotic events (Li *et al.,* 2000; Finucano *et al.,* 1999; Scorrano *et al.,* 1999), one should be cautious in interpreting the collapse as a marker of apoptosis.

B. Controls

The *lack of evidence* of apoptosis detected by a particular assay is not necessarily *evidence of the lack of apoptosis*. As already mentioned in discussing particular methods, there are instances when cells die by a process resembling apoptosis that lacks one or more typical apoptotic features. It is also possible, however, that for technical reasons the assay used to identify apoptotic cells malfunctions. For example, the enzyme TdT used in the TUNEL assay may be inactive due to its improper storage or an error can occur during the staining procedure. It is therefore essential to distinguish between the genuine lack of apoptosis and the inability to detect it due to technical causes. A positive control consisting of cells known to be apoptotic (confirmed by a standard method and inspection of cell morphology) is therefore necessary. Such control cells have to be processed in parallel with the investigated sample through all the steps prescribed in the protocol. Some vendors provide positive and negative control cells with their kits (e.g., Phoenix Flow Systems, Inc., APO-BRDU).

It is convenient to have positive control cells prepared in large quantity that can be stored in aliquots to be used during each experiment. Exponentially growing HL-60 or U937 cells treated in cultures for 3–6 h with 0.2 μM CPT to induce apoptosis can serve as a useful control. Cells so treated consist of a subpopulation of apoptotic (S phase) and nonapoptotic (G_1 phase) cells present in the same sample. To induce the apoptosis of S-phase cells it is critical that the cells are in an exponential phase of their growth, at a relatively low cell density (<800,000 cells per 1 ml) during the treatment with CPT; subconfluent cultures are quite resistant to this drug. A large batch of cells treated in such manner can be fixed appropriately and then stored in aliquots in 70% ethanol at $-20°C$ to be used as a positive and negative control for each assay that utilizes fixed cells. For assays that require live cells, the controls should be made fresh and cannot be fixed.

C. Live Cells Engulfing Apoptotic Bodies Can be Mistakenly Identified as Genuine Apoptotic Cells

The exposure of phosphatidylserine on cell surfaces that occur during apoptosis (Fadok *et al.,* 1992; Koopman *et al.,*1994) makes apoptotic cells and apoptotic bodies "tasty" to neighboring cells, which phagocytize them. The ability to engulf apoptotic bodies is not solely the property of professional phagocytes, but also of cells from fibroblast or epithelial lineages. It is frequently observed, especially in solid tumors, that the cytoplasm of both nontumor and tumor cells located in the neighborhood of apoptotic cells contains inclusions typical of apoptotic bodies. The remains of apoptotic cells engulfed by neighboring cells contain an altered plasma membrane, fragmented DNA, and other constituents with attributes characteristic of apoptosis. Thus, if the distinction is based on any of these attributes, the live, nonapoptotic cells that phagocytized apoptotic bodies cannot be distinguished from genuine apoptotic cells by flow cytometry. Nonapoptotic cells that engulfed apoptotic bodies, for example, become reactive with Annexin V (Marguet *et al.,* 1999). Most likely this is due to the fact that during engulfment the plasma membrane of apoptotic bodies fuses with the plasma membrane of the phagocytizing cell. It has also been shown by LSC that nonapoptotic cells in the bone marrow of patients undergoing chemotherapy (primarily monocytes and macrophages) have large quantities of apoptotic bodies in their cytoplasm and are strongly positive in the TUNEL reaction (Bedner *et al.,* 1999).

Regardless of the cytometric assay used to identify apoptosis, the mode of cell death should be confirmed by the inspection of cells by light or electron microscopy. Morphological changes during apoptosis have a very characteristic pattern (Kerr *et al.,* 1972; Majno and Joris, 1995) and should be the deciding factor in situations where any ambiguity arises regarding the mechanism of cell death. The ability of LSC to relocate the cells for inspection of their morphology makes this instrument of particular value in studies of apoptosis.

D. Apoptotic Index and Incidence of Cell Death

An assumption is often made that the frequency of apoptotic cells (apoptotic index, AI), *in vivo* or in cultures, is a reflection of the incidence of cell death. This may not always be the case because apoptosis is a kinetic event. The entire apoptotic process, from the initiation to the total disintegration of the cell, is of short and variable duration. The time window during which individual apoptotic cells demonstrate the characteristic features (markers) that allow them to be recognizable varies depending on the method of detection used, the cell type, and the nature of the inducer of apoptosis. Different methods differ in the width of the time window through which they recognize apoptosis. For example, PARP cleavage precedes DNA fragmentation; as a result, early in apoptosis the AI estimated based on PARP cleavage is higher relative to that estimated by the TUNEL assay (Li *et al.,* 2000). Furthermore, some inducers may slow down or accelerate the apoptotic process. This may occur if the rate of either formation and/or shedding of apoptotic bodies, endonucleolysis, or proteolysis is affected by the inducer or by the growth conditions (e.g. temperature, pH). Thus, when the duration of apoptosis is shortened, the frequency of apoptotic cells (AI) may be diminished even if the *incidence of cell death* remains the same. Conversely, the prolongation of apoptosis in the absence of any change in the incidence of cell death is manifested by an increase in AI. Protease inhibitors, for example, including inhibitors of serine proteases, delay nuclear fragmentation and prolong the process of apoptosis (Hara *et al.*, 1996). Apoptosis induced by these agents, or in their presence, is therefore reflected by an increased AI relative to that measured in parallel cultures with the same incidence of cell death but where apoptosis is of shorter duration. To estimate the *incidence* of cell death, the *absolute number* (not the percentage) of live cells should be measured in the treated culture and compared with the appropriate untreated control. A correction should also be made to account for the rate of cell proliferation. These issues are discussed in more detail elsewhere (Darzynkiewicz *et al.,* 2000).

Acknowledgments

Supported by NCI Grant CA RO1 28704, the Chemotherapy Foundation, and Robert A. Welke Cancer Research Foundation. EB was on leave from the Department of Pathology, Pomeranian School of Medicine, Szczecin, Poland.

References

Alnemri, E. S., Livingston, D. I., Nicholson, D. W., Salvesen, G., Thornberry, N. A., Wong, W. W., and Yuan, J. (1996). Human ICE/CED-4 protease nomenclature. *Cell* **87,** 171–173.

Arends, M. J., Morris, R. G., and Wyllie, A. H. (1990). Apoptosis: The role of endonuclease. *Am. J. Pathol.* **136,** 593–608.

Bedner, E., Burfeind, P., Gorczyca, W., Melamed, M. R., and Darzynkiewicz, Z. (1997). Laser scanning cytometry distinguishes lymphocytes, monocytes and granulocytes by differences in their chromatin structure. *Cytometry* **29,** 191–196.

Bedner, E., Li, X., Gorczyca, W., Melamed, M. R., and Darzynkiewicz, Z. (1999). Analysis of apoptosis by laser scanning cytometry. *Cytometry* **35,** 181–195.

Bedner, E., Li, X., Kunicki, J., and Darzynkiewicz, Z. (2000a). Translocation of Bax to mitochondria during apoptosis measured by laser scanning cytometry. *Cytometry,* **41,** 83–88.

Bedner, E., Melamed, M. R., and Darzynkiewicz, Z. (1998). Enzyme kinetic reactions and fluorochrome uptake rates measured in individual cells by laser scanning cytometry (LSC). *Cytometry* **33,** 1–9.

Bedner, E., Smolewski, P., Amstad, P., and Darzynkiewicz, Z. (2000b). Activation of caspases measured in situ by binding of fluorochrome-labeled inhibitors of caspases (FLICA): Correlation with DNA fragmentation. *Exp. Cell Res.* **259,** 308–313.

Budihardjo, I., Oliver, H., Lutter, M., and Luo, X. (1999). Biochemical pathways of caspase activation during apoptosis. *Annu. Rev. Cell Dev. Biol.* **15,** 269–290.

Catchpoole, D. R., and Stewart, B. W. (1993). Etoposide-induced cytotoxicity in two human T-cell leukemic lines: Delayed loss of membrane permeability rather than DNA fragmentation as an indicator of programmed cell death. *Cancer Res.* **53,** 4287–4296.

Clatch, R. J., Foreman, J. R., and Walloch, J. L. (1998). Simplified immunophenotypic analysis by laser scanning cytometry. *Cytometry* **34,** 3–16.

Collins, R. J., Harmon, B. V., Gobe, G. C., and Kerr, J. F. R. (1992). Internucleosomal DNA cleavage should not be the sole criterion for identifying apoptosis. *Int. J. Radiat. Biol.* **61,** 451–453.

Cossarizza, A., Kalashnikova, G., Grassilli, E., Chiappelli, F., Salvioli, S., Capri, M., Barbieri, D., Troiano, L., Monti, D., and Franceschi, C. (1994). Mitochondrial modifications during rat thymocyte apoptosis: A study at a single cell level. *Exp. Cell Res.* **214,** 323–330.

Darzynkiewicz, Z., and Barnard, E. A. (1966). Specific proteases of mast cells. *Nature* **212,** 1198–1203.

Darzynkiewicz, Z., and Bedner, E. (2000). Analysis of apoptotic cells by flow- and laser scanning-cytometry. *Methods Enzymol.* **322,** 18–39.

Darzynkiewicz, Z., Bedner, E., Burfeind, P., and Traganos, F. (1998). Analysis of apoptosis by flow and laser scanning cytometry. *In* "Apoptosis Detection and Assay Methods" (L. Zhu and J. Chun, eds.), pp. 75–91. BioTechniques Books, Natick, MA.

Darzynkiewicz, Z., Bedner, E., Li, X., Gorczyca, W., and Melamed, M. R. (1999). Laser scanning cytometry: A new instrumentation with many applications. *Exp. Cell Res.* **249,** 1–12.

Darzynkiewicz, Z., Bruno, S., Del Bino, G., Gorczyca, W., Hotz, M. A., Lassota, P., and Traganos, F. (1992). Features of apoptotic cells measured by flow cytometry. *Cytometry* **13,** 795–808.

Darzynkiewicz, Z., Juan, G., Li, X., Murakami, T., and Traganos, F. (1997a). Cytometry in cell necrobiology: Analysis of apoptosis and accidental cell death (necrosis). *Cytometry* **27,** 1–20.

Darzynkiewicz, Z., and Li, X. (1996). Measurements of cell death by flow cytometry. *In* "Techniques in Apoptosis: A User's Guide" (T. G. Cotter and S. J. Martin, eds.), pp. 71–106, Portland Press Ltd, London.

Darzynkiewicz, Z., Li, X., and Gong, J. (1994). Assays of cell viability: Discrimination of cells dying by apoptosis. *Methods Cell Biol.* **41,** 16–39.

Darzynkiewicz, Z., Li, X., Gong, J., and Traganos, F. (1997b). Methods for analysis of apoptosis by flow cytometry. *In* "Manual of Clinical Laboratory Immunology" (N. R. Rose, E. C. de Macario, J. D. Folds, H. C. Lane, and R. Nakamura, eds.), 5th Ed., pp. 334–344, ASM Press, Washington, DC.

Darzynkiewicz, Z., Staiano-Coico, L., and Melamed, M. R. (1981). Increased mitochondrial uptake of rhodamine 123 during lymphocyte stimulation. *Proc. Natl. Acad. Sci. USA* **78,** 2383–2387.

Darzynkiewicz, Z., Traganos, F., Staiano-Coico, L., Kapuscinski, J., and Melamed, M. R. (1982). Interactions of rhodamine 123 with living cells studied by flow cytometry. *Cancer Res.* **42,** 799–806.

de Murcia, G., and Menissier-de Murcia, J. M. (1994). Poly(ADP-ribose) polymerase: A molecular nick sensor. *Trends Biochem. Sci.* **19,** 172–176.

Deptala, A., Bedner, E., Gorczyca, W., and Darzynkiewicz, Z. (1998). Activation of nuclear factor kappa B (NF-κB) assayed by laser scanning cytometry. *Cytometry* **33,** 376–382.

Earnshaw, W. C., Martins, L. M., and Kaufmann, S. H. (1999). Mammalian caspases: Structure, activation, substrates, and functions during apoptosis. *Annu. Rev. Biochem.* **68,** 383–424.

Fadok, V. A., Voelker, D. R., Campbell, P. A., Cohen, J. J., Bratton, D. L., and Henson, P. M. (1992). Exposure of phosphatidylserine on the surface of apoptotic lymphocytes triggers specific recognition and removal by macrophages. *J. Immunol.* **148,** 22.

Ferlini, C., Scambia, G., and Fattorossi, A. (1998). Is chlorometyl-X-rosamine useful in measuring mitochondrial transmembrane potential. *Cytometry* **31,** 74.

Finucane, D. M., Waterhouse, N. J., Amaranto-Mendes, G. P., Cotter, T. G., and Green, D. R. (1999). Collapse of the inner mitochondrial transmembrane potential is not required for apoptosis of HL-60 cells. *Exp. Cell Res.* **251,** 166–174.

Furuya, T., Kamada, T., Murakami, T., Kurose, A., and Sasaki, K. (1997). Laser scanning cytometry allows detection of cell death with morphological features of apoptosis in cells stained with PI. *Cytometry* **29,** 173–177.

Gavrieli, Y., Sherman, Y., and Ben-Sasson, A. (1992). Identification of programmed cell death in situ via specific labeling of nuclear DNA fragmentation. *J. Cell Biol.* **119,** 493–501.

Gilmore, K., and Wilson, M. (1999). The use of chloromethyl-X-rosamine (Mitotracker Red) to measure loss of mitochondrial membrane potential in apoptotic cella is incompatible with cell fixation. *Cytometry* **36,** 355–358.

Gong, J., Traganos, F., and Darzynkiewicz, Z. (1994). A selective procedure for DNA extraction from apoptotic cells applicable for gel electrophoresis and flow cytometry. *Anal. Biochem.* **218,** 314–319.

Gorczyca, W., Bigman, K., Mittelman, A., Ahmed, T., Gong, J., Melamed, M. R., and Darzynkiewicz, Z. (1993). Induction of DNA strand breaks associated with apoptosis during treatment of leukemias. *Leukemia* **7,** 659–670.

Gorczyca, W., Bruno, S., Darzynkiewicz, R., Gong, J., and Darzynkiewicz, Z. (1992). DNA strand breaks occuring during apoptosis: Their early in situ detection by the terminal deoxynucleotydil transferase and nick translation assays and prevention by serine protease inhibitors. *Int. J. Oncol.* **1,** 639–648.

Gorman, A. M., Hirt, U. A., Zhivotovsky, B., Orrenius, S., and Ceccatelli, S. (1999). Application of a fluorometric assay to detect caspase activity in thymus tissue undergoing apoptosis in vivo. *J. Immunol. Methods* **226,** 43–48.

Halenbeck, R., MacDonald, H., Roulston, A., Chen, T. T., Conroy, L., and Williams, L. T. (1998). CPAN, a human nuclease regulated by the caspase-sensitive inhibitor DFF45. *Curr. Biol.* **8,** 537–540.

Hara, S., Halicka, H. D., Bruno, S., Gong, J., Traganos, F., and Darzynkiewicz, Z. (1996). Effect of protease inhibitors on early events of apoptosis. *Exp. Cell Res.* **232,** 372–384.

Haugland, R. P. (1996). "Handbook of Fluorescent Probes and Research Chemicals," 6th Ed. Molecular Probes, Eugene, OR.

Hedley, D. W., and McCulloch, E. A. (1996). Generation of oxygen intermediates after treatment of blasts of acute myeloblastic leukemia with cytosine arabinoside: Role of bcl-2. *Leukemia* **10,** 1143–1149.

Hotz, M. A., Gong, J., Traganos, F., and Darzynkiewicz, Z. (1994). Flow cytometric detection of apoptosis: Comparison of the assays of in situ DNA degradation and chromatin changes. *Cytometry* **15,** 237–244.

Hug, H., Los, M., Hirt, W., and Debatin, K. M. (1999). Rhodamine 110–linked amino acids and peptides as substrates to measure caspase activity upon apoptosis induction in intact cells. *Biochemistry* **38,** 13906–13911.

Johnson, L. V., Walsh, M. L., and Chen, L. B. (1980). Localization of mitochondria in living cells with rhodamine 123. *Proc. Natl. Acad. Sci. USA* **77,** 990–994.

Kamentsky, L. A., Burger, D. E., Gershman, R. J., Kamentsky, L. D., and Luther, E. (1997). Slide-based laser scanning cytometry. *Acta Cytol.* **41,** 123–143.

Kamentsky, L. A., and Kamentsky, L. D. (1991). Microscope-based multiparameter laser scanning cytometer yielding data comparable to flow cytometry. *Cytometry* **12,** 381–387.

Kaufmann, S. H., Desnoyers, S., Ottaviano, Y., Davidson, N. E., and Poirier, G. G. (1993). Specific proteolytic cleavage of poly(ADP-ribose) polymerase: An early marker of chemotherapy-induced apoptosis. *Cancer Res.* **53,** 3976–3985.

Kawasaki, M., Sasaki, K., Satoh, T., Kurose, A., Kamada, T., Furuya, T., Murakami, T., and Todoroki, T. (1997). Laser scanning cytometry (LSC) allows detailed analysis of the cell cycle in PI stained human fibroblasts (TIG-7). *Cell Prolif.* **30,** 139–147.

Keij, J. F., Bell-Prince, C., and Steinkamp, J. A. (2000). Staining of mitochondrial membranes with 10–nonyl acridine orange MitFluor Green, and MitoTracker Green is affected by mitochondrial membrane potential altering drugs. *Cytometry* **39,** 203–210.

Kennedy, N. J., Kataoka, T., Tschopp, J., and Budd, R. C. (1999). Caspase activation is required for T-cell proliferation. *J. Exp. Med.* **190,** 1891–1896.

Kerr, J. F. R., Wyllie, A. H., and Curie, A. R. (1972). Apoptosis: A basic biological phenomenon with wide-ranging implications in tissue kinetics. *Br. J. Cancer* **26,** 239–257.

Knapp, P. E., Bartlett, W. P., Williams, L. A., Yamada, M., Ikenaka, K., and Skoff, R. P. (1999). Programmed cell death without DNA fragmentation in the jimpy mouse: Secreted factors can enhance survival. *Cell Death Differ.* **6,** 136–145.

Kockx, M. M., De Meyer, G. R., Muhring, J., Jacob, W., Bult, H., and Herman, A. G. (1998). Apoptosis and related proteins in different stages of human atherosclerotic plaques. *Circulation* **97,** 2307–2315.

Koopman, G., Reutelingsperger, C. P. M., Kuijten, G. A. M., Keehnen, R. M. J., Pals, S. T., and van Oers, M. H. J. (1994). Annexin V for flow cytometric detection of phosphatidylserine expression of B cells undergoing apoptosis. *Blood* **84,** 1415–1420.

Kroemer, G. (1998). The mitochondrion as an integrator/coordinator of cell death pathways. *Cell Death Diff.* **5,** 547–548.

Lazebnik, Y. A., Kaufmann, S. H., Desnoyers, S., Poirier, G. G., and Earnshaw, W. C. (1994). Cleavage of poly(ADP-ribose) polymerase by proteinase with properties like ICE. *Nature* **371,** 346–347.

Li, X., and Darzynkiewicz, Z. (1995). Labelling DNA strand breaks with BdrUTP: Detection of apoptosis and cell proliferation. *Cell Prolif.* **28,** 571–579.

Li, X., and Darzynkiewicz, Z. (1999). The Schrödinger's cat quandary in cell biology: Integration of live cell funtional assays with measurements of fixed cells in analysis of apoptosis. *Exp. Cell Res.* **249,** 404–412.

Li, X., and Darzynkiewicz, Z. (2000). Cleavage of poly(ADP-ribose) polymerase measured *in situ* in individual cells: Relationship to DNA fragmentation and cell cycle position during apoptosis. *Exp. Cell Res.* **255,** 125–132.

Li, X., Du, L., and Darzynkiewicz, Z. (2000). Caspases are activated during apoptosis independent on dissipation of mitochondrial electrochemical potential. *Exp. Cell Res.* **257,** 290–297.

Li, X., Melamed, M. R., and Darzynkiewicz, Z. (1996). Detection of apoptosis and DNA replication by differential labeling of DNA strand breaks with fluorochromes of different color. *Exp. Cell Res.* **222,** 28–37.

Liu, J., Bhalgat, M., Zhang, C., Diwu, Z., Hoyland, B., and Klaubert, D. H. (1999). Fluorescent molecular probes V: A sensitive caspase-3 substrate for fluorometric assays. *Bioorg. Med. Chem. Lett.* **9,** 3231–3236.

Liu, X., Kim, C. N., Yang, J., Jemmerson, R., and Wang, X. (1996). Induction of apoptotic program in cell-free extracts: Requirements for dATP and cytochrome *c*. *Cell* **86,** 147–157.

Luther, E., and Kamentsky, L. A. (1996). Resolution of mitotic cells using laser scanning cytometry. *Cytometry* **23,** 272–278.

Majno, G., and Joris, I. (1995). Apoptosis, oncosis, and necrosis: An overview of cell death. *Am. J. Pathol.* **146,** 3–16.

Marguet, D., Luciani, M.-F., Moynault, A., Williamson, P., and Chimini, G. (1999). Engulfment of apoptotic cells involves the redistribution of membrane phosphatidylserine on phagocyte and pray. *Nature Cell Biol.* **1,** 454–456.

Miossec, C., Dutilleul, V., Fassy, F., and Diu-Hercend, A. (1997). Evidence for CPP32 activation in the absence of apoptosis during T lymphocyte stimulation. *J. Biol. Chem.* **272,** 13459–13462.

Mukae, N., Enari, M., Sakahira, H., Fukuda, Y., Inazawa, J., Toh, H., and Nagata, S. (1998). Molecular cloning and characterization of human caspase-activated DNase. *Proc. Natl. Acad. Sci. USA* **95,** 9123–9128.

Nicoletti, I., Migliorati, G., Pagliacci, M. C., Grignani, F., and Riccardi, C. (1991). A rapid and simple method for measuring thymocyte apoptosis by propidium iodide staining and flow cytometry. *J. Immunol. Method* **139,** 271–280.

Oberhammer, F., Wilson, J. M., Dive, C., Morris, I. D., Hickman, J. A., Wakeling, A. E., Walker, P. R., and Sikorska, M. (1993). Apoptotic death in epithelial cells: Cleavage of DNA to 300 and/or 50 kb fragments prior to or in the absence of internucleosomal fragmentation. *EMBO J.* **12,** 3679–3684.

Ormerod, M. G. (1998). The study of apoptotic cells by flow cytometry. *Leukemia* **12,** 1013–1025.

Ormerod, M. G., Cheetham, F. P. M., and Sun, X.-M. (1995). Discrimination of apoptotic thymocytes by forward light scatter. *Cytometry* **21,** 300–304.

Ormerod, M. G., O'Neill, C. F., Robertson, D., and Harrap, K. R. (1994). Cisplatin induced apoptosis in a human ovarian carcinoma cell line without a concomitant internucleosomal degradation of DNA. *Exp. Cell Res.* **211,** 231–237.

Petit, J. M., Ratinaud, M. H., Cordelli, E., Spano, M., and Julien, R. (1995). Mouse testis cell sorting according to DNA and mitochondrial changes during spermatogenesis. *Cytometry* **19,** 304–312.

Poot, M., Gibson, L. L., and Singer, V. L. (1997). Detection of apoptosis in live cells by MitoTracker^{TM} Red CMXRos and SYTO dyes flow cytometry. *Cytometry* **27,** 358–364.

Ratinaud, M. H., Leprat, P., and Julien, R. (1988). In situ cytometric analysis of nonyl acridine orange-stained mitochondria from splenocytes. *Cytometry* **9,** 206–212.

Reed, J. C. (1997). Cytochrome *c*: Can't live with it—can't live without it. *Cell* **91,** 559–562.

Sallman, F. R., Bourassa, S., Saint-Cyr, J., and Poirier, G. G. (1997). Characterization of antibodies specific for the caspase cleavage site on poly(ADP-ribose) polymerase: Specific detection of apoptotic fragments and mapping of the necrotic fragments of poly(ADP-ribose) polymerase. *Biochem. Cell Biol.* **75,** 451–458.

Scorrano, L., Petronilli, V., Colonna, R., Di Lisa, F., and Bernard, P. (1999). Cloromethyltetramethylrosamine (Mitotracker Orange^{TM}) induces the mitochondrial permeability transition and inhibits respiratory complex I: Implications for the mechanism of cytochrome *c* release. *J. Biol. Chem.* **274,** 24657–24663.

Susin, S. A., Zamzani, N., Larochette, N., Dallaporta, B., Marzo, I, Brenner, C., Hirsch, T., Petit, P. X., Geuskens, M., and Kroemer, G. (1997). A cytofluorometric assay of nuclear apoptosis induced in a cell-free system: Application to ceramide-induced apoptosis. *Exp. Cell Res.* **236,** 397–403.

Swat, W., Ignatowicz, L., and Kisielow, P. (1981). Detection of apoptosis of immature $CD4^{+}8^{+}$ thymocytes by flow cytometry. *J. Immunol. Methods* **137,** 79–87.

Thornberry, N. A., Rano, T. A., Peterson, E. P., Rasper, D. M., Timkey, T., Garcia-Calvo, M., Houtzager, V. M., Nordstrom, P. A., Roy, S., Vaillancourt, J. P., Chapman, K. T., and Nicholson, D. W. (1997). A combinatorial approach defines specificities of members of the caspase family and granzyme B: Functional relationships established for key mediators of apoptosis. *J. Biol. Chem.* **272,** 17907–17911.

Umansky, S. R., Korol, B. R., and Nelipovich, P. A. (1981). In vivo DNA degradation in the thymocytes of gamma-irradiated or hydrocortisone-treated rats. *Biochim. Biophys. Acta* **655,** 281–290.

van Engeland, M., Nieland, L. J. W., Ramaekers, F. C. S., Schutte, B., and Reutelingsperger, P. M. (1998). Annexin V-affinity assay: A review on an apoptosis detection system based on phosphatidylserine exposure. *Cytometry* **31,** 1–9.

Yang, J., Liu, X., Bhalla, K., Ibrado, A. M., Cai, J., Peng, T. I., Jones, D. P., and Wang, X. (1997). Prevention of apoptosis by Bcl-2: Release of cytochrome *c* from mitochondria blocked. *Science* **275,** 1129–1132.

Zamai, L., Falcieri, E., Marhefka, G., and Vitale, M. (1996). Supravital exposure to propidium iodide identifies apoptotic cells in the absence of nucleosomal DNA fragmentation. *Cytometry* **23,** 303–311.

Zamzani, N., Brenner, C., Marzo. I., Susin, S. A., and Kroemer, G. (1998). Subcellular and submitochondrial mode of action of Bcl-2–like oncoproteins. *Oncogene* **16,** 2265–2282.

Zapata, J. M., Takahashi, R., Salvesen, G. S., and Reed, J. C. (1998). Granzyme release and caspase activation in activated human T-lymphocytes. *J. Biol. Chem.* **273,** 6916–6920.

CHAPTER 5

Analysis of Protein Transglutamylation in Apoptosis

Zoltán Nemes,* András Mádi,† Lyuben N. Marekov,§ Mauro Piacentini,‡ Peter M. Steinert,§ and László Fésüs†

*Departments of Psychiatry and
†Biochemistry and Molecular Biology
University of Debrecen Medical and Health Sciences Center
Debrecen, Hungary H-4012

‡Department of Biology
University of Rome "Tor Vergata" and
Laboratory of Cell Biology and E. M., IRCCS "L. Spallanzani"
Rome, Italy I-00133

§Laboratory of Skin Biology
NIAMS, National Institutes of Health
Bethesda, Maryland 20892

METHODS IN CELL BIOLOGY, VOL. 66

0091-679X/01 $35.00

I. Introduction

There has been a good deal of recent innovation in the field of transglutaminase (TGase) research. Although remarkable progress has been achieved in the characterization of TGase promoter sequences (Nagy *et al.,* 1996), the bulk of apoptosis-related TGase research is still accomplished by biochemical methods aimed at identifying the targets of TGase activity and characterizing structures created by TGase-mediated cross-linking. Molecular biological methods applicable for the characterization of TGase expression are routine. This chapter focuses on protein isolation and characterization techniques developed specially for TGase research. Several of these techniques were invented and introduced for the analysis of keratinocyte-cornified envelope (CE), a specialized 15- to 20-nm-thick structure that is tightly cross-linked by three different TGases directly below the cell membrane of cornifying keratinocytes. As the terminal differentiation of keratinocytes can be regarded as a specialized form of apoptosis (Melino *et al.,* 1998), and such highly cross-linked, insoluble, and (supra)colloidal size protein meshworks ("apoptotic bodies") are also formed frequently in apoptotic cells (Fesus *et al.,* 1989; Knight *et al.,* 1993), the CE research techniques need little if any modification to be used in apoptosis model systems.

TGases are Ca^{2+}-dependent enzymes that catalyze an acyl transfer reaction between the γ-carboxamido group of protein-bound glutamine to the ε amino group of protein bound lysyl residues or other primary amines. Whereas the glutamyl substrate specificity of TGases has been studied intensively (Kahlem *et al.,* 1996), little is known about the amine substrate specificity as simple diamines are used as *in vitro* amine substrates (Grootjans *et al.,* 1995). Because of their common reaction mechanism, data have revealed that, TGases are not only capable of forming ε-(γ-glutamyl)-lysine isodipepitide or N^1,N^8-bis(γ-glutamyl) spermidine bonds between peptide chains, but can also create ester bonds (Nemes *et al.,* 1999) or deamidate substrate glutamine residues to glutamate under appropriate conditions (Nemes *et al.,* 2000).

Seven phylogenetically related human TGase enzymes (TGase1, 2, 3, 4, X, Band 4.2, and Factor XIIIa) have been identified in humans thus far, and several others have been isolated lately from lower vertebrates, invertebrates, plants, fungi, and bacteria (Makarova *et al.,* 1999; Madi *et al.,* 1998).

Several lines of evidence, including the isolation of highly polymerized protein products from apoptotic cells and immunohistochemical studies, have established the ubiquitous TGase2 (or "tissue" transglutaminase) as a useful biochemical marker of programmed cell death without defining its precise role in apoptosis (Thomazy and Fesus, 1989; Nemes *et al.,* 1996). It was hypothesized that the TGase-mediated covalent cross-linking of cellular proteins could be useful for hindering the release of intracellular antigens and DNA from dying cells and thus prevent the immunological and mutagenic sequelae that are often seen in necrotic tissue (Fesus *et al.,* 1989, 1991). However, it is still disputed whether the enzyme is involved in the execution of the death program as part of the killing mechanism or, alternatively, is involved in modulating the course of apoptosis by temporarily stabilizing dying cells (Melino and Piacentini, 1998).

Emerging data suggest that intracellular TGases other than TGase2 (e.g., TGase1, TGase3, TGase X) may be expressed in unanticipated cell types and may exert both overlapping functions within the cells and substitute for one another in apoptosis (Kim *et al.,* 1999; V. DeLaurenzi, personal communication). Therefore, several apoptotic phenomena previously attributed solely to TGase2 may actually result from the action of other intracellular TGases. Elucidation of the role of protein transglutamylation in apoptosis, and characterization of the thusly assembled protein structures, is going to remain a promising field of cell biology and protein biochemistry.

Some crucial issues need to be considered before attempting to identify and characterize proteins that become cross-linked in specific apoptotic model systems. First, is transglutaminase activity induced during cell death? If there is no detectable TGase induction in response to the specific apoptotic stimulus, then the execution of cell death is incomplete, which is often the case with malignant cell lines and nonphysiological death stimuli where living cells do not benefit from the removal of apoptotic ones. Second, if there is apoptosis-associated TGase expression or activity in the experimental system of interest, does it really participate in cross-link formation?

In addition to its enzymatic activity, TGase2 also functions as a regulatory G protein and its appearance does not unequivocally mean *in situ* activation (Im *et al.,* 1997). Therefore it is necessary to determine the amount of cross-linking in cells that occurs after the initiation of apoptosis. Third, if some proteins are cross-linked during the cell death process, they must be identified by *in situ* labeling techniques or sequence identification as components of larger cross-linked complexes. Once a protein is identified as a real and physiologically relevant *in vivo* substrate, its cross-linking properties may be identified. This is accomplished by locating TGase-reactive glutamine and lysine residues via cross-linking to small traceable substrates or, alternatively, by locating isodipeptide cross-links via sequence analysis.

The methods described in this chapter provide tools to perform a systematic analysis of protein transglutamylation and cross-linking in the CE as well as appropriately chosen apoptotic cell culture systems. It is hoped that these methods will make it easier for other researchers to harvest valuable data in this very promising and unexploited aspect of cell suicide.

II. Transglutaminase (TGase) Activity Assays

Methods for measuring specific TGase activity in tissue or cell homogenates have been available since the 1960s. Initial macro methods from the late 1960s utilized the esterase activity of TGases in liberating *P*-nitrophenol from its esters or colorimetrically measured the decrease of free amine substrates added far below saturating concentrations (Folk and Finlayson, 1977). The 1980s brought the introduction of isotopically labeled diamines and the widespread use of the glutamine substrate dimethylated casein as standard assay components (Folk and Chung, 1985). This so-called "filter paper" or "putrescine isotope" method is still the "gold standard" of kinetic measurements in TGase biochemistry. Lately, a plated ELISA method [first described by Slaughter *et al.* (1992) and developed further by Lilley *et al.* (1997)] has also gained popularity as a simple, productive method that does not produce large amounts of radioactive waste. Nevertheless, the ELISA method does not directly yield absolute turnover rates and has to be calibrated, possibly by means of the [^{14}C]putrescine method.

A. [^{14}C]Putrescine Method

1. Combine 10 μl putrescine substrate buffer, 10× (see Section X), 10 μl [^{14}C]putrescine (Amersham Pharmacia, 1 mCi/ml), 20 μl 2% dimethylated casein solution (Sigma), 1–60 μl biological sample of known protein content (20–200 μg cell protein), and water to 100 μl.
2. Incubate at 37°C.
3. Take 20 μl aliquots at 0, 10, 20, and 30 min.
4. Pipette aliquots onto 25-mm Whatmann C glass fiber filter circles wetted with 20% trichloroacetic acid (TCA).
5. Place filter circles on a filtration mainfold (e.g., Hoefer FH225, Amersham Pharmacia) and apply vacuum.
6. Wash with 5 ml 5% TCA containing 0.2 *M* KCl.
7. Dehydrate with 3 × 5 ml acetone.
8. Air dry filter circles.
9. Place circles into scintillation vials and overlay with scintillation fluid.
10. Measure disintegrations per minute (DPM).
11. Calculate specific activity from DPM increase (per minute), sample protein content and concentration, and the specific activity of the given isotope batch. Multiply by the dilution of cold putrescine

Note: To obtain reliable activities, DPM increase/min should be in the 10–200 DPM/min range. Incomplete drying of filter circles can make the scintillation fluid turbid, resulting in quenching and artificially low DPM values.

B. Plated ELISA Method

1. Pipette 200 μl 2% dimethylated casein solution into each well of a flat-bottom 96-well ELISA plate (Pierce).
2. Incubate the plate for 1 h at 37°C.
3. Discard solution from wells.
4. Add 200 μl 5% nonfat dry milk solution to block free protein-binding sites. Incubate at room temperature for 30 min.
5. Wash wells twice with 350 μl TBS (20 m*M* Tris–Cl, 150 m*M* NaCl, pH 8.0, at 25°C) buffer. Coated plates can be stored for several moths at −20°C.
6. Add known amounts of sample protein (5–100 μg) in 50 μl TBS (pH 8.0) containing 1 m*M* dithiothreitol (DTT) and 0.5 m*M* EDTA, 175 μl TBS, and 25 μl X-biotin reaction buffer (see Section X) into the wells.
7. Incubate the plate at 25°C for 30 min.
8. Discard the reaction mixture. Wash twice with 350 μl TBS-EDTA (TBS + 1 m*M* EDTA-Na, pH 8.0) solution.
9. Wash twice with 350 μl TBS.
10. Dilute streptavidin–alkaline phosphatase solution (Jackson Immunoresearch, 1.5 mg/ml) with 1% bovine serum albumin (BSA), 15 m*M* NaN_3 in TBS to 0.1 μg/ml. Pipette 200 μl into each well and incubate the plate for an hour at room temperature.
11. Wash wells twice with 350 μl TBS + 0.1% Triton X-100.
12. Wash four times with 350 μl TBS.
13. Add 250 μl phosphatase substrate solution (see Section X) into each well.
14. Immediately start the OD reading at 405 nm.
15. Read OD every 5 min for 30 min.

TGase activity can be assessed by calculating Δ OD/min. Measure activity in triplicates or more. Subtract sample-free blank from every measurement if the reader does not do so automatically.

III. Determination of ε-(γ-Glutamyl)-lysine Cross-Link Density in Apoptotic Tissue

Cross-linked tissue proteins harbor inter- or intrachain ε-(γ-glutamyl)-lysine linkages between α-aminoacyl peptide chains. The determination of protein cross-link density within the tissue relies on the liberation of free ε-(γ-glutamyl)-lysine isodipeptide from larger protein structures by fragmentation into smaller (soluble) peptides and exhaustive digestion by ectopeptidases (aminopeptidases and carboxypeptidases) that do not split the isopeptide bond. Because TGase cross-linking characteristically results in supracolloidal aggregates of proteins, as exemplified in the case of apoptotic bodies or keratinocyte-cornified envelopes, these tightly cross-linked macropolymeric structures often have to be freed from non-cross-linked proteins. Removal of non-cross-linked proteins is achieved by boiling in chaotropic solutions to solubilize protein structures held together by ionic or hydrophobic interactions or disulfide bonds. Nonspecific proteases such as proteinase K represent a major problem for the ε-(γ-glutamyl)-lysine isopeptide bond, whereas pronase is thought to be much less aggressive and still efficient enough to fragment and solubilize large cross-linked protein meshes. Large complex protein structures are first solubilized and broken into smaller oligopeptides by pronase or Proteinase K and then digested by leucine aminopeptidase and carboxypeptidase A and Y to free amino acids and isodipeptides.

Several methods are applicable for -chromatographic separation of the isodipeptides from free amino acids. In order to detect and quantitate the eluted isopeptide, amino acids are converted into light-absorbing derivatives. This derivatization step can be performed either before ("precolumn derivatization") or after ("post-column derivatization") separation. Precolumn derivatization is followed by reversed-phase HPLC separation; post-column methods separate by cation-exchange chromatography. While automated amino acid analyzers are widely available and are the tools of choice for measuring isopeptide concentrations, precise results can also be achieved by manual derivatization and simple HPLC chromatography as well.

While first-step phenythiourea (PTU) derivatization followed by C18 reversed-phase HPLC chromatography (Tarcsa and Fésüs) is regarded as the "gold standard" for determining the concentration of ε-(γ-glutamyl)-lysine, automated techniques are less tedious. Two protocols are given: one for manual sample processing using PITC derivatization and another for automated analysis using a Beckman 6300 amino acid analyzer employing postcolumn ninhydrine conversion.

A. Isolation of Keratinocyte-Cornified Envelopes from Human Foreskin

1. Wash flattened foreskins thoroughly with cold phosphate-buffered saline (PBS) containing 0.1% Triton X-100.
2. Overlay foreskins with 10–20 volumes cold PBS containing 4 *M* urea.
3. Incubate on ice for 1 h.
4. Gently peel off stratum corneum sheets from the skin surface with forceps.
5. Place the cornified layers onto a dense mesh wire filter.
6. Wash thoroughly with water.
7. Boil cornified layers for 15 min in 50–100 volumes keratin lysis buffer A (see Section X) in a 35-ml Oak Ridge (polyethylene) centrifuge tube.
8. Collect sedimentable material by centrifuging 60 min at 40,000*g* at room temperature.

9. Resuspend and boil the pellet in keratin lysis buffer B (see Section X) as described earlier.
10. While hot, sonicate sample for 4 × 30 s with a micro tip, using maximal intensity.
11. Centrifuge as before.
12. Repeat steps 7–11 two more times.
13. Extract SDS by resuspending the pellet and centrifuging twice in 90% acetone, 5% acetic acid, and 5% triethylamine.
14. Evaporate the solvent in vacuum.

Note: This method (steps 7–14) is also applicable for the preparation of apoptotic bodies from most cultured cells. A typical yield is 50–80 mg CE mass per 25 newborn foreskins or 5–25 mg apoptotic bodies per 10^{10} DMSO and retinate (all-*trans* retinoic acid)-treated apoptotic HL-60 cells (Nemes *et al.,* 1997). However, determinating cross-link density from cultured apoptotic cells usually does not require isolation of the cross-linked (supra)colloidal size protein structures, but instead can be performed by direct protease digestion of total cell protein after a mild detergent lysis.

B. Breakdown of Cornified Envelope Proteins into Amino Acids and ε-(γ-Glutamyl)-lysine Isodipeptides

1. Suspend 1–5 mg of dried CE in 0.5 ml 0.1 *M* NH_4HCO_3 (pH 8.3).
2. Add 10 μg pronase mg CE mass (from 20 mg/ml aliquoted frozen stock solution in TBS).
3. Digest overnight at 56°C.
4. Determine if digestion is complete by centrifuging for 15 min at 15,000*g*. If there is a visible pellet, repeat steps 2–4.
5. Autoclave samples at 121°C for 20 min.
6. Lyophilize samples and dissolve in 0.5 ml 0.1 *M* NH_4HCO_3.
7. Add 1.5 μg leucine-aminopeptidase, 0.4 μg carboxypeptidase A, and 0.3 μg carboxypeptidase Y (from 10 mg/ml aliquoted frozen stocks) mg CE and incubate at 37°C for 36 h.
8. Lyophilize samples.

Note: For exhaustive enzymatic proteolysis of total cell homogenates, cells should be washed by centrifugation and resuspended with buffer, lysed by sonication, detergent lysis or French press, and defatted with phenol-chloroform or ether extraction. Chromosomal DNA should be subsequently digested with DNase I (10–20 Kunitz units/10^6 cells) to decrease viscosity.

C. Determination of ε-(γ-Glutamyl)-lysine by PTU Derivatization

1. Dissolve lyophilized protein digest (corresponding to approximately 0.1–10 nmol total amino acid) in 0.5 ml 100 m*M* *N*-ethylmorpholine-acetate, pH 7.5 (NEM-Ac), buffer.

2. Pass through a 0.5 × 2-cm DOWEX 50X4 (200–400 mesh, equilibrated with the specified NEM-Ac buffer) column (suitably assembled by pouring the resin into Pasteur pipettes stoppered with glass wool) to free samples of arginine and lysine.

3. Lyophilize and dissolve in 100 μl NEM-Ac buffer.

4. Inject samples onto a Waters Nova-Pak silica 3.9 × 150-mm HPLC column and desalt sample by elution with distilled water. Monitor A_{254} and collect the largest peak harboring the amino acids. This step frees the sample from buffer, salts, proteolytic enzymes, and undigested higher molecular weight peptides.

5. After eluting the peak of interest, wash column with a gradient to 60% methanol.

6. Lyophilize and dissolve in 80 μl ethanol:water:triethylamine (2 : 2 : 1).

7. Evaporate solvents under vacuum.

8. Flush sample tubes with N_2 and then dissolve in 100 μl ethanol : water : triethylamine : phenylisothiocyanate (7 : 1 : 1 : 1) (work under N_2 stream) to convert amino acids into their PTU derivatives.

9. Close lid and incubate for 1 h at room temperature.

10. Lyophilize and dissolve in 100 μl 100 m*M* acetate-Na buffer, pH 6.35.

11. Inject sample onto a Waters Nova-Pak C18 3.9 × 150-mm HPLC column equilibrated with 86% 100 m*M* acetate-Na buffer, pH 6.35 (solvent A), and 14% acetonitrile (solvent B). Elute at 1 ml/min with 14% solvent B and monitor A_{268}. The di-PTU derivative of ε-(γ-glutamyl)-lysine is eluted as a distinct peak between alanine and tyrosine. The exact elution time can be established by running the PITC-modified ε-(γ-glutamyl)-lysine standard [prepared from ε-(γ-glutamyl)-lysine (Sigma) via steps 7–9].

12. Calculate the amount of ε-(γ-glutamyl)-lysine by integrating peak areas and comparing to standards.

Note: Automated amino acid analyzer systems using precolumn derivatization and reversed-phase separation will also detect the ε-(γ-glutamyl)-lysine elution peak between alanine and tyrosine, regardless of whether PTU, FMOC, OPA-$SHCH_2CH_2SH$, or other derivatization is used. Poor separation from neighboring peaks may necessitate changes in chromatography parameters. (Usually a slight change of column thermostation temperature is enough to separate the isopeptide peak.)

D. Determination of ε-(γ-Glutamyl)-lysine by Automated Amino Acid Analysis

1. Dissolve proteolytically digested (see Section III,B) samples in a 3 : 1 mixture of formic acid (88%) : H_2O_2 (30%).
2. Cap tightly and incubate at room temperature for 3 h.
3. Evaporate solvent in vacuum.
4. Dissolve in 100 μl 0.1 *M* citrate-Na buffer, pH 2.2.
5. Process samples in a Beckman 6300 amino acid analyzer.

6. Read data for methionine content and factor by the values established by measuring known amounts of an ε-(γ-glutamyl)-lysine standard.

Note: Amino acid analyzers using post-column derivatization will detect ε-(γ-glutamyl)-lysine isodipeptide between valine and isoleucine, about where methionine is expected to elute.

IV. Immunoprecipitation of TGase-Interacting Proteins from Apoptotic Cells

In order to identify and characterize cellular proteins that specifically interact with and eventually act as regulators or substrates of cellular TGases during the commitment of the cell to apoptosis, a screening of such proteins is accomplished by isolation and analysis of proteins that bind to the TGase of interest (Piredda *et al.,* 1999). Proteins isolated on the basis of their *in vivo* coimmunoprecipitation with TGase can be consecutively subjected to *in vivo* biochemical analysis in order to determine if they are real substrates of the TGase of interest.

1. Wash 1.5×10^7 cells by centrifuging at 200*g* for 10 min and resuspending in cold PBS plus 1 m*M* EDTA, pH 7.6, three times.
2. Lyse cell pellet in 4 ml RIPA buffer with freshly added protease inhibitors (see Section X) by vortexing at 4°C.
3. Centrifuge for 30 min at 15,000*g*.
4. Add 5–10 μg anti-TGase antibody (mouse monoclonal AB clone TG100 from Lab Vision Corp., Fremont, CA) to the supernatant and agitate overnight in the cold.
5. Add 250 μl protein A-Sepharose (Amersham Pharmacia) to the sample and shake for 2 h at 4°C.
6. Collect protein A-Sepharose beads by centrifugation and wash several times with RIPA buffer (see Section X).
7. Boil beads in Laemmli buffer (see Section X).
8. Separate (co)immunoprecipitated proteins by SDS–PAGE and blot onto PVDF membrane.
9. Stain membrane with Coomassie dye and identify novel bands that are distinct from TGase and antibody heavy and light chains as detailed in Section V,E.

V. *In Situ* Identification of TGase Glutamyl Donor Substrate Proteins in Apoptotic Cells

Glutamyl donor substrate proteins can be identified if reacted with a suitable primary amine harboring an immunologically ("hapten") or photometrically detectable moiety. Pioneering innovation in TGase glutamine substrate identification was accomplished by

Daniel Aeschlimann, who introduced *in vivo* dansylcadaverine labeling to the TGase research community in the early nineties. Using his own antidansyl antibody, these elegant experiments revealed amine incorporation into osteonectin and osteopontin *in vivo* as well as in *ex vivo* cartilage ECM (Aeschlimann and Paulsson, 1991; Aeschlimann *et al.*, 1992). The use of dansylcadaverine has been subsequently replaced by 5-biotinamidopentylamine ("X- biotin") due to the versatility of readily available biotin/(strept)avidin labeling kits and affinity columns. The major drawbacks of X-biotin labeling are due to the endogenous background of intracellular biotinylated proteins and the low affinity of simple diamines. An advance in TGase amine substrate development was the introduction of DALP (3-[N^{α}[N^{ε}-[2′,4′-dinitrophenyl]-amino-*n*-hexanoyl-L-lysylamido]-propane-1-ol), an αN, αC peptide-bounded lysine attached to a dinitrophenyl haptenic group, to which very specific and high-affinity IgE class commercial antibodies are available (e.g., from Sigma). K_{cat}/K_m values for DALP are a magnitude of order higher than those of simple diamine derivatives (Folk and Finlayson, 1977; Slaughter *et al.*, 1992; Nemes *et al.*, 1997). Nevertheless, haptenized primary amines are too big and—at physiological pH—too charged to penetrate cellular membranes. Cells need to be permeabilized by methods that often result in cell death (e.g., by electroporation, Smethurst and Griffin, 1996), to take up labeled amine substrates, or the amines themselves need to be modified to transiently lose their charge in order to enter the cells and be converted to the amine substrate. A splendid way for such derivatization is to convert the primary amines into alkyloxy carbonyl products (R-NH-CO-O-R2), which then undergo hydrolysis by cellular esterase activity and release carbon dioxide, liberating the amine substrate compound (Nemes *et al.*, 1997). The free amine substrate is thus trapped within the cells. Although any amine substrate (including dansylcadaverine or X-biotin) can be carboxyalkylated for intracellular labeling, DALP-methylcarbamate (DALP-MC) is the first choice due to its better substrate properties. X-biotin is also convenient if the substrate concentration is not limiting, as in the case of *in vivo*-labeling experiments using broken cells or tissue. This section describe methods for *in situ* TGase glutamyl substrate identification in apoptotic cell cultures by DALP (Nemes *et al.*, 1997) and for labeling putative substrates of *Caenorhabditis elegans* TGase (Madi *et al.*, 1998).

A. Preparation of DALP-MC

1. Dissolve 5 mg DALP (trifluoroacetate salt, from L. Fésüs, Debrecen, Hungary) (Nemes *et al.*, 1997) in 250 μl freshly distilled dry triethylamine in an Eppendorff tube. Flush with a weak stream of argon.
2. Add 10 μl methyl chloroformate while under argon.
3. Cap tightly and let stand overnight at room temperature.
4. Evaporate solvents in vacuum.
5. Redissolve pellet in 1 ml 70% ethanol.
6. Equilibrate a 5 × 20-mm column of TMD-8 mixed bed resin (Sigma) column (suitably filled into a glass wool-stoppered Pasteur pipette) with 70% ethanol.

7. Apply the sample on top of the column and elute with 70% ethanol. Collect the yellow flow through.
8. Aliquot samples and lyophilize.

Note: Aliquots of DALP-MC can be stored at −70°C for several weeks, however, it is prudent to prepare a fresh batch before every experiment in order to prevent spontaneous isomerization.

B. Loading and *in Situ* Labeling of TGase Glutamyl Substrates in Apoptotic Cells by DALP-MC

1. Place thoroughly defatted and acid-washed glass coverslips into plastic cell culture dishes.
2. Culture cells in these vessels to facilitate cell attachment to the glass.
3. Add 25 μ*M* freshly prepared DALP-MC (13 μg/liter culture fluid, diluted from a 1-mg/ml ethanol stock solution) to cultures 16–24 h before apoptosis induction.
4. Proceed with apoptosis induction.
5. Remove coverslips and wash with a gentle stream of PBS.
6. Place coverslips into cold (−20°C) St. Marie's fixative (2% acetic acid in 96% ethanol) for 5 min.
7. Air dry.
8. Wash coverslips with TBS containing 0.1% Triton X-100 (TBS-TX, pH 8.0).
9. Block nonspecific binding with a solution of 5% nonfat milk powder, 2% BSA, and 5% nonimmune donkey serum in TBS-TX for 30 min at room temperature.
10. Apply monoclonal antidinitrophenyl mouse antibody (Sigma) at a 1 : 250 dilution in TBS-TX + 1% BSA for 4 h at room temperature.
11. Wash with TBS-TX.
12. Apply biotinylated donkey anti-mouse immunglobulins (Amersham Pharmacia) at a 1 : 200 dilution in TBS-TX-BSA overnight at 4°C.
13. Wash with TBS-TX.
14. Apply streptavidin-fluorescein isothiocyanate (Roche), 10 μg/ml in TBS-TX-BSA, for 1 h at room temperature.
15. Wash with TBS-TX.
16. Mount with 2% Mowiol (Calbiochem) containing 5% diazabicyclooctane (DABCO) onto microscopic slides.
17. Observe FITC fluorescence by confocal image microscopy.

Note: Not all cell types nor all apoptotic stimuli produce detectable levels of TGase induction and activation. Prior to planning substrate labeling the experimenter should check the apoptotic model system for detectable TGase2 induction in order to avoid daunting results. In our experience, DALP labeling by TGase2 targets the cytoplasmic

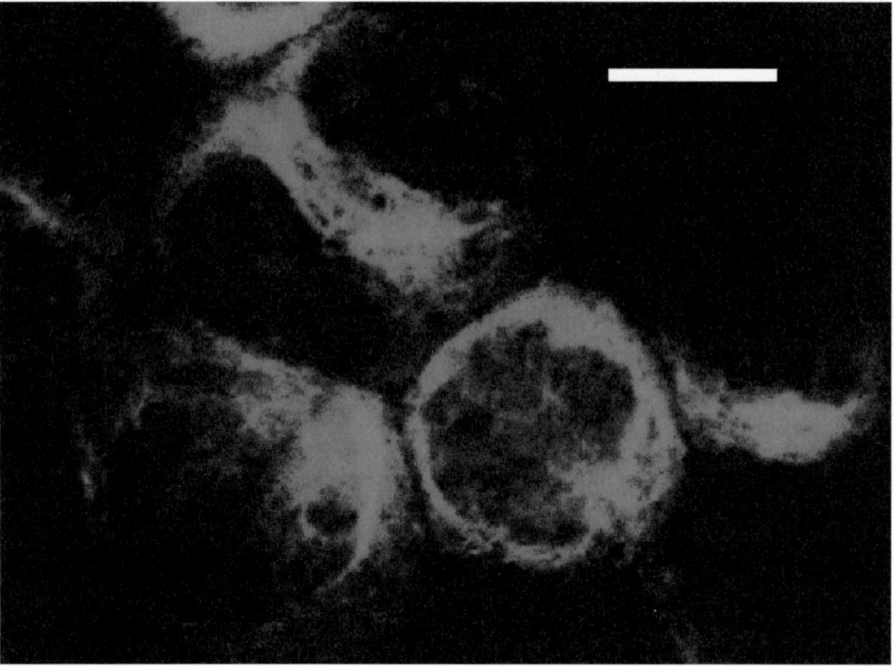

Fig. 1 Confocal image of apoptotic HL-60 cells. Intracellular glutamyl substrates were labeled by TGgase by the haptenized amine substrate DALP. The dinitrophenyl group of DALP was detected by immunofluorescence. Bar : 10 nm. (See Color Plate).

microfilamental cytoskeleton (Fig. 1, see Color Plate), which can only be observed by confocal imaging (Nemes *et al.*, 1997). If the subcellular localization of DALP incorporation is not investigated, standard fluorescent microscopy can also be used to visualize DALP-incorporating cells.

C. Affinity Isolation of TGase Glutamyl Substrate Proteins from Cultured Apoptotic Cells after Cross-Linking to DALP

1. Grow as many cells as possible (5–10 or more liters of floating cell cultures). *Note:* Isolation and biochemical identification of TGase glutamine substrate proteins may need up to several grams of crude cell protein.
2. Load cells with DALP-MC and induce apoptosis as in Section V,B steps 3 and 4.
3. Harvest cells and wash them with cold PBS.
4. Pellet cells by centrifugation and lyse with a French press at 4°C.
5. Add 1 m*M* aminoethybenzenesulfonyl fluoride (AEBSF), 1 mg/liter pepstatin A, 0.5 mg/liter leupeptin, and 1 m*M* $MgCl_2$ and add 10 Kunitz units of DNase I/10^6 cells. Stir until viscous at room temperature.
6. Add 2 m*M* EDTA-Na, pH 8.0.

7. Boil sample in 2x PL buffer (see Section X) for 5 min.

8. Add iodoacetamide to 0.1 *M* after cooling to room temperature and then incubate for 30 min.

9. Precipitate samples with 4 volumes of cold acetone.

10. Remove supernatant after centrifugation at 20,000*g* for 30 min.

11. Redissolve pellet by boiling in Laemmli buffer (see Section X).

12. Fill a continuous elution preparative SDS–PAGE electrophoresis column (e.g., Prep-Cell 491 Bio-Rad) with a 4% pH 8.4 gel.

13. Run sample through the electrophoresis column, discard blue SDS-rich front fractions, and collect column eluate in the column elution buffer (the same as the gel running buffer, see Section X) for a period at least 10-fold longer than the passage time of the blue dye.

14. Cool column eluate to 4°C.

15. Suspend 1–2 ml antidinitrophenyl-linked beads (see Section X) in the cold column eluate and stir overnight in the cold.

16. Recover affinity beads by centrifugation and wash several times with cold column elution buffer.

17. Release antibody-captured proteins by stirring 1 ml of bead volume in 10 ml DNP affinity elution buffer (see Section X) overnight at 37°C.

18. Reduce sample volume to 0.5–1 ml by centrifugal concentration or dialysis against carboxymethylcellulose (Aquacide III, Calbiochem).

D. Labeling of Putative TGase Glutamyl Substrate Proteins in *Caenorhabditis elegans* by X-Biotin

1. Purify *C. elegans* from *E. coli*-rich broth by flotation on 1 *M* sucrose at 1500*g* for 30 min at 4°C.

2. Lyse worms in 0.5 ml worm lysis buffer (see Section X) per 100 mg worm mass by sonication (at maximum output) on ice for 1 min.

3. Remove insoluble worm chitin and other debris by centrifugation at 1500*g* for 15 min at 4°C.

4. Measure sample protein content by any detergent-compatible protein essay (e.g., Bio-Rad DC protein assay kit).

5. Dilute samples to 200 μg protein/ml with TBS.

6. Add 0.1 volume X-biotin substrate buffer (10X, see Section X) and incubate for 30 min at 25°C.

7. Add an equal volume of 2X Laemmli buffer (see Section X) and boil samples for 5 min.

8. Perform SDS–PAGE and blot proteins onto PVDF membrane.

9. Block the membrane with 5% nonfat milk powder in TBS plus 0.1% NaN_3 for 60 min at room temperature.

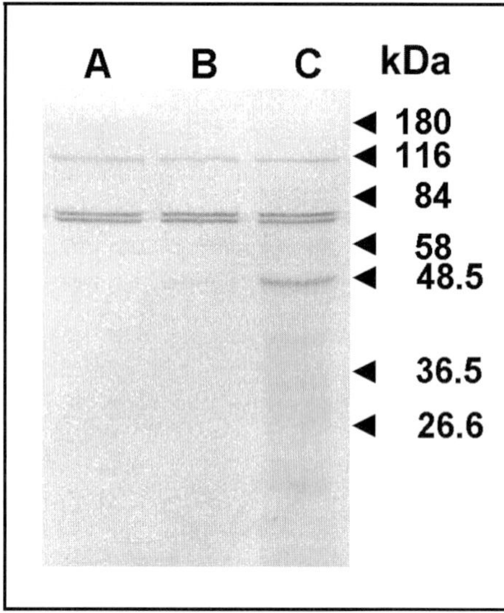

Fig. 2 Affinity-isolated TGase glutamyl substrate proteins from *Caenorhabditis elegans*. Nematode proteins were labeled with TGase amine substrate X-biotin and detected with streptavidin-coupled alkaline phosphatase. Lane A is an unlabeled blank. Lane B is a control containing amine substrate and TGase active site inhibitor iodoacetamide. Lane C reveals a band at approximately 48 kDa, the major substrate protein for *C. elegans* TGase.

10. Add streptevidin-conjugated alkaline phosphatase (Jackson Immunoresearch) to 0.1 μg/ml and incubate with shaking for 2 h at room temperature.

11. Wash membrane five times with TBS plus 0.1% Triton X-100.

12. Develop color reaction with 100 m*M* triethanolamine–Cl, pH 9.5, buffer containing 0.33 mg/ml nitroblue tetrazolium, 0.165 mg/ml X-phosphate (BCIP, 5-bromo-4-chloro-3-indolyl phosphate) and 5 m*M* $MgCl_2$. Biotinylated proteins appear as bluish-brown bands on a yellow background (Fig. 2).

Note: X-biotin-labeled TGase substrate proteins can be isolated by preparative SDS–PAGE followed by avidin affinity chromatography (similar to the method described in Section, IV and V,C) and subsequently by two-dimensional electrophoresis.

VI. Identification of Labeled Substrate Proteins

Identification of affinity-labeled TGase glutamine substrate proteins is performed by trial-and-error experimentation. The first approach is to separate affinity-labeled substrate proteins by one- or two-dimensional electrophoresis, transfer them onto a PVDF membrane, visualize them by protein staining, and attempt direct N-terminal microsequencing. Most intracellular proteins, however, are blocked at their N termini. Therefore, purified (denatured) proteins may be fragmented either by chemical methods (e.g., CNBr fragmentation) or by protease (e.g., tryptic) digestion. Peptide fragments of 7–30 amino acid, can be separated by reversed-phase HPLC and subjected to sequencing. As a rule of thumb, two to three accurately identified sequences of five to eight consecutive amino acids suffice to identify the glutamyl substrate protein of interest from sequence databases. Alternatively, proteins blotted onto PVDF are suitable for MALDI-TOF analysis after protease digestion, assuming that they have been identified previously.

VII. Identification of TGase Reactive Glutamine Residues by Protein Sequencing

Glutamine is a common amino acid; nearly all natural proteins contain several glutamines within their primary sequence. Nevertheless, not all of these residues are suitably aligned to serve as substrates for TGases, the substrate specificities of which are remarkably different from one another. Whether or not a glutamine residue is utilized efficiently by a specific TGase depends on its neighboring residues within the primary amino acid sequence as well as its sterical position within the secondary or perhaps tertiary structure. To date, no reliable modeling method is available to predict whether a specific glutamine residue is utilized for transglutamylation. The exact location of Tgase reactive glutamine residues can only be determined empirically by protein sequencing and identifying TGase-modified glutamine residues (Nemes *et al.,* 1999). The following method involves labeling of purified protein by [^{14}C] putrescine and subsequent fragmentation into peptides that are then separated by HPLC. Peptides harboring incorporated isotope are subsequently sequenced in a gas-phase microsequencer, which identifies γ-glutamylputerescine as a distinct peak. The relative peak intensity, as compared to the unmodified glutamine peak, serves as the indicator of the individual glutamine's substrate quality.

1. Mix 10–20 nmol from the protein of interest, 200–500 atamol/s (10^{-15}mol/s) TGase activity (as measured by the [^{14}C]putrescine method described in Section II,A), 100 μl putrescine substrate buffer 10X (Section X), 200 μl [^{14}C]putrescine di-HCl (Amersham Pharmacia, 1 mCi/ml), and water to 1 ml.
2. Incubate reaction at 37°C.
3. Take 250 μl aliquots at 0, 5, 60, and 240 min.
4. Terminate TGase reaction by heating samples to 100°C for 5 min.
5. Precipitate protein by 1 ml of cold acetone. Recover precipitate and resuspend in 0.1 *M* NH_4HCO_3 (pH 8.3) and add 20 μg of modified trypsin (Roche) into each tube.
6. Incubate overnight at 37°C.

Or

a. 70% formic acid and add 100 μmol cyanogen bromide. Shake overnight at room temperature.

Or

b. Other enzymes of choice may be used on the basis of primary protein structure.

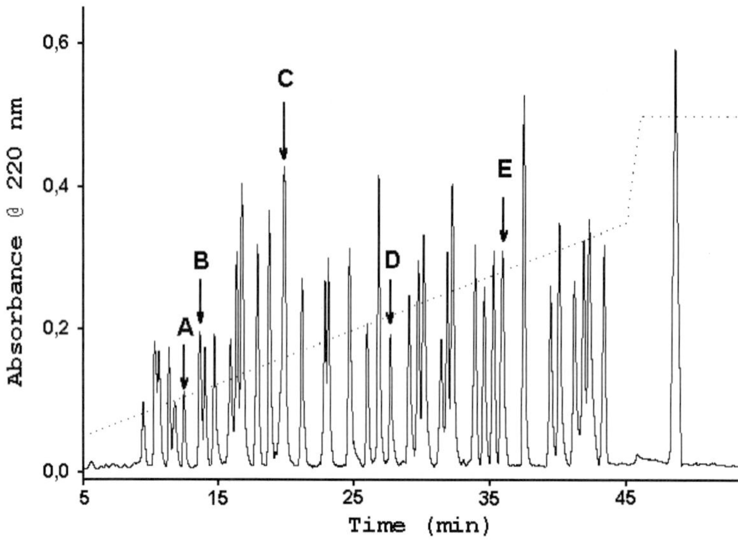

Fig. 3 HPLC separation profile of involucrin peaks reacted with membrane-bound TGase1 enzyme after [^{14}C]putrescine labeling. Arrows denote the peptide peaks showing incorporation of radioactivity.

7. Dry in vacuum. Dissolve in 200 μl 0.08 % TFA in water.

8. Separate peptides by C18 HPLC on a 4.6 × 250-mm Beckman ultrasphere C18 HPLC column using a gradient from 95% 0.08% TFA in water (solvent A) and 5% 0.08% TFA in acetonitrile (solvent B) to 35–50% buffer B for 50–60 min at a 1-ml/min flow rate. Collect peptide peaks by monitoring $A_{208-220}$.

9. Identify peaks harboring radioactivity (Fig. 3).

Note: γ-glutamyl-putrescine modification has little effect on the retention of longer peptides; usually, putrescine-labeled peptides are eluted together or overlap with unmodified ones.

10. Sequence peptides.

11. Examine the HPLC elution profile from sequencing cycles where a glutamine residue is expected. The PITC (PTH + PTU) derivative of γ-glutamyl-putrescine is eluted as a novel peak between valine and DPTU (Fig. 4). Comparing the γ-glutamyl-putrescine peak area to that of glutamine (apply the same absorption coefficient as for lysine) gives the yield of utilization for a given glutamine residue.

VIII. Location of Cross-Links in Protein Complexes

Although cross-links are formed easily by any transglutaminase when added to an arbitrary mixture of proteins harboring substrate glutamine and lysine residues, evidence suggests that cross-linking *in vivo* is more specific and accomplishes organized structures by sequentially layering distinct protein components (Steinert, 1995). A well-characterized experimental model for cross-link identification is the keratinocyte-cornified envelope. The cross-linking patterns of CE structural proteins have been studied extensively by sequencing of cross-linked peptide fragments obtained after sequential limited proteolysis by trypsin and proteinase K (Steinert and Marekov, 1997). In contrast, the organization of cross-linked protein structures formed within the apoptotic cells has not yet been characterized systematically. This section describes a method that has been used successfully to identify cross-linking sites in recombinant involucrin protein by the membrane-bound TGase 1 enzyme (Nemes *et al.,* 2000). Whereas locating a few cross-links between one or two known proteins is an uncomplicated exercise, the fragmentation of larger cross-linked protein masses (cornified envelopes, apoptotic bodies, Japanese fish paste, "surimi," etc.) requires aggressive proteolysis, results in poorly separated peaks, and yields tremendous sequence ambiguity.

1. React TGase substrate proteins under similar conditions as in Section V (steps 1–5) with the omission of (cold or ^{14}C) putrescine from the reaction mixture.
2. Check the cross-linking reaction by SDS–PAGE and Coomassie blue staining. Both native and cross-linked products should be present in similar amounts (Fig. 5).
3. Perform peptide fragmentation and HLPC separation as described in Section V (steps 5–8). Compare C18 HPLC profile of cross-linked peptides to that of uncross-linked ingredients (compare Figs. 3 and 6).

Note: If there is no difference between the two elution patterns, alternative fragmentation procedures must be found, as the peptides harboring the cross-link are either too big or too small to be resolved by HPLC.

4. Collect peaks (or peak shoulders) that appear in the course of TGase reaction (such as P1 and P2 in Fig. 6).

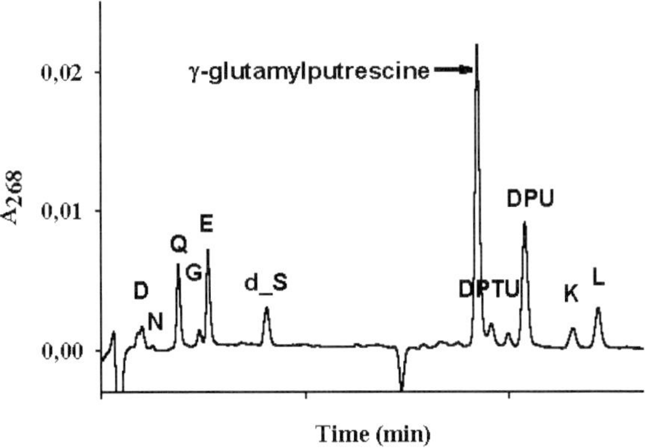

Fig. 4 Sequencing chromatogram of residue Gln^{133} from peak D of Fig. 3. The chromatogram was obtained by a Beckman Porton 3000 microsequencer.

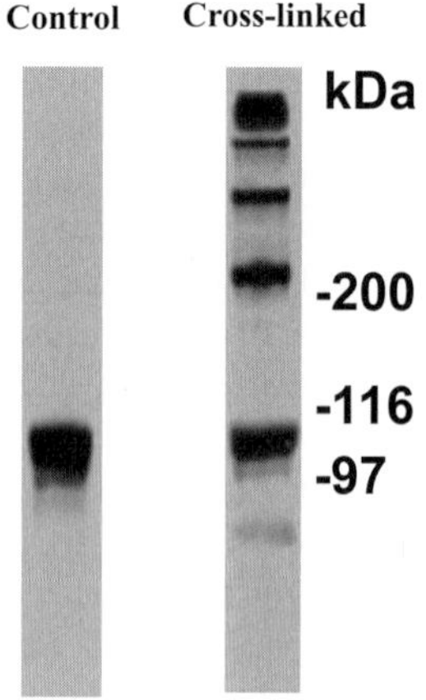

Fig. 5 *In vitro* cross-linking of recombinant involucrin protein by membrane-bound TGase I. Oligomerization of the protein is indicated by the appearance of multiple bands on a Coomassie-stained SDS–PAGE gradient gel.

5. Subject newly emerged peaks to protein sequencing.
6. Identify cross-linking sites by

a. The sequence overlap of the identified peptide sequences.

b. The appearance of the PTH-derivatized γ-glutamyl-lysine isodipeptide in the appropriate cycle.

c. The sequencing gap: amino acid residues of two (or more) cross-linked peptides are sequenced in the same cycle until sequencing gets to the cross-link of the peptide

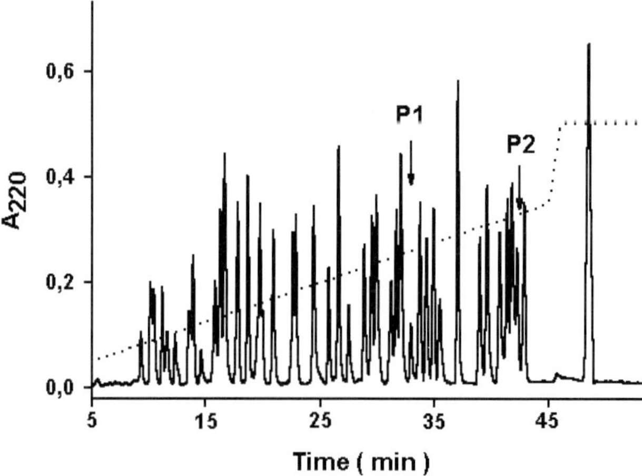

Fig. 6 HPLC elution profile of tryptic involucrin peptides cross-linked by TGase 1. Comparing the peak profile to that on Fig. 3, newly emerged cross-linked peaks can be identified.

Table I
Amino Acids Sequenced from Cross-Linked Peptide P2 of Fig. 6[a]

Cycle	1	2	3	4	5	6	7	8	9	10	11	12	13	14	15
Amino acids	Q	Q	E	Q	Q	E	G	Q	E	Q	H	H	G	T	Q
	G	E	A	E	L	P	V	P	A	E	A	P	E	H	A
	H	L	L				L	F	L	P	**CL**	V	M	Q	P
												L			V

[a]CL denotes an ε(γ-glutamyl)lysine cross-link. A single letter amino acid code is used. Involucrin tryptic peptides are deconvoluted from the sequencer output (cross-linked residues are boxed).

Cross-linked peptides:

Q E E [K] H M T A (V K)

Q E A Q L E L P E Q [Q] V G Q P (K)

Contaminating not cross-linked peptides:

G Q L E Q P V F A P A P G Q V (Q D I Q P A L P T K)

H L E Q Q E G Q L E H L E H L (E H Q E G Q L G L P E Q Q V L Q L K)

that harbors the cross-link closest to its amino terminus. There the peptide fails to release amino acids as long as the amino terminus of the cross-linked counterpart is degraded and the cross-link is released. Then the residues from the two sequences appear together again. (Table I)

Note: The peaks of cross-linked peptides may overlap with background peptides (Table I). This necessitates sequencing of the corresponding non-cross-linked peptide sequences (Fig. 7). This rarely causes sequence evaluation ambiguity in case of a parsimonious choice of protein substrate with a known amino acid sequence.

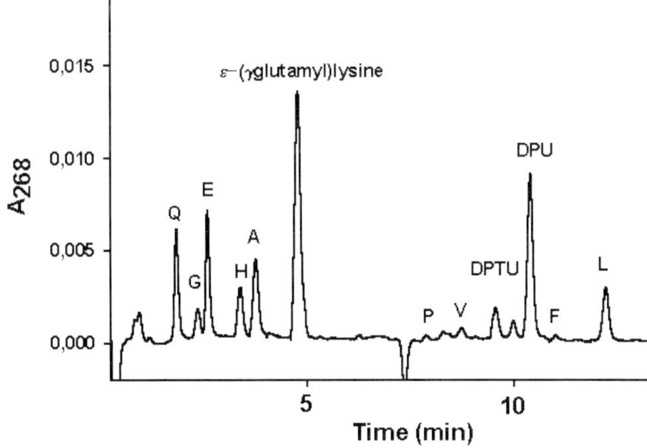

Fig. 7 Sequencing profile of cycle 11 of Table I. The C18 HPLC chromatogram of PTH-converted amino acids was obtained by a Beckman Porton 3000 automated gas-phase microsequencer.

IX. Analysis of TGase-Mediated Glutamine Deamidation by Peptide Sequencing

In addition to amines and alcohols, water can also serve as a glutamine acceptor for TGases. Although a weak nucleophyl and thus a poor substrate, water is present in vast molar concentration and has significant propensity to occur in the otherwise hydrophobic acceptor–substrate binding pocket of TGases. Utilization of water as substrate results in a net deamidation of reactive glutamine residues, which prevents TGase from being trapped as a glutamyl-enzyme intermediate in the absence of amine acceptors and thus contributes to higher turnover rates (Nemes *et al.*, 2000). Deamidation by TGase is involved in bacterial toxin pathogenesis (Schmidt *et al.*, 1998) and can drastically alter the outcome of TGase action on the same proteins if the conformation of the enzyme is distorted by environmental interactions (Tarcsa *et al.*, 1997). This section describes a method for assessing TGase-mediated deamidation. The rates of deamidation are assessed by protein sequencing using the equations described later.

1. Determine the exact molar amount of used protein substrate.

Note: Protein assays are unreliable. Automated amino acid analysis after acid hydrolysis is the most precise method available.

2. Identify glutamine moieties serving as substrates for TGase in the experimental system of interest as described in Section VI.

3. Perform TGase reaction, limited proteolysis, and HPLC separation of the proteolytic fragments as described in Section VI.

4. Collect the peaks that correspond to those showing [^{14}C]putrescine incorporation in step 1 as well as cross-linked peptides (if there are any).

5. Quantify the amount of these peaks by amino acid analysis.

Note: Overlapping elution of background peptides requires sequence determination of the contaminating peptides and correction of molar peptide amounts by subtraction of background. Background peptide amounts can be derived from residues not present in the peptides of interest or by suitable mathematical formulae.

6. Perform peptide sequencing of the peaks that harbor TGase reactive glutamine residues.

7. Evaluate peak areas for glutamine and glutamate in the sequencing cycles corresponding to, preceding, and following TGase substrate glutamine moieties.

8. The ratio of TGase-deamidated glutamines can be assessed by correcting the amounts for carryover from preceding cycles, spontaneous deamidation, and incomplete Edman's degradation by the following equation:

$$\mathrm{Q/E} = \frac{[Q_n - Q_{n-1} \cdot (1 - Q_{n+1}/Q_n)][1 + E_m/(Q_m + E_m)]}{[E_n - E_{n-1} \cdot (1 - E_{n+1}/E_n)][1 - E_m/(Q_m + E_m)]},$$

where Q_n and E_n denote the amount of glutamine and glutamate released from the sequencing cycle corresponding to the reactive glutamine residue position, Q_{n-1}, Q_{n+1} and E_{n-1}, E_{n+1} denote the yield of the same amino acid in the previous or consecutive

sequencing cycle. Q_m and E_m denote the yield from any glutamine cycle where this Q has not been preceded by E within the previous four cycles.
Note: This formula is not applicable if there is a Q or E in $n + 1$ cycle. Here calculate carryover from the last Q or E of the sequence.

9. Calculate residue deamidation by the molar amount of substrate protein and cross-linked residues and the *Q/E* ratio determined by sequencing. Scrupulous work and reliable technique can yield ±1–2% reproducibility of deamidation results. Nevertheless, biologically relevant glutamine deamidation should convert >50% of the given reactive glutamine moiety to glutamate.

X. Reagents

A. Buffer

Putrescine substrate buffer (10×): 500 m*M* Tris–Cl, pH 8.0, 20 m*M* DTT, 20 m*M* $CaCl_2$, and 200 m*M* putrescine–diHCl
Note: Store in frozen aliquots up to 3 months.

X-biotin substrate buffer (10×): 50 m*M* Tris–Cl, pH 8.0, 10 m*M* DTT, 20 m*M* $CaCl_2$, 10 m*M* 5-biotinamido-pentylamine ("X-biotin," Pierce), and 1% NaN_3
Note: Store in frozen aliquots up to 3 months.

Keratinocyte lysis buffer A: 100 m*M* Tris–Cl, pH 8.0, 1 m*M* EDTA–Na, pH 8.0, 2% β-mercaptoethanol, and 2% SDS

Keratinocyte lysis buffer B: 1 *M* NaOH, 6 *M* guanidine–HCl, and 2% β-mercaptoethanol

RIPA buffer with protease inhibitors: 150 m*M* NaCl, 1% Nonidet P-40, 0.5% sodium deoxycholate, 0.1% SDS, 1 m*M* EDTA, and 50 m*M* Tris–Cl, pH 7.5, at 25°C, Add the following prior to use: 1 m*M* AEBSF 10 m*M* benzamidine–HCl, 10 m*M* chymostatin, 0.7 μg/ml pepstatin A, and 1 μg/ml leupeptin

Laemmli buffer (1×): 1% SDS, 62.5 m*M* Tris–Cl, pH 6.8, 100 m*M* DTT, and 0.005% bromphenol blue

PL buffer 2×: 2% SDS, 100 m*M* DTT, and 100 m*M* Tris–Cl, pH 8.0

Column elution buffer: 192 m*M* glycine, 25 m*M* Tris, and 0.05% SDS (pH approximately 8.4)

Affinity elution buffer: 50 m*M* Tris–Cl, pH 7.5, 120 m*M* NaCl, 15 m*M* NaN_3, 100 μ*M* DALP, and 0.05% SDS

Worm lysis buffer: 100 m*M* Tris–Cl, pH 8.0, 1% CHAPS, 1 m*M* PMSF, 2 m*M* EDTA, and 10 m*M* DTT

B. Preparation of Antidinitrophenyl Antibodies Immobilized on Hydrazine-Linked Beads

1. Precipitate globulins from 2 ml rabbit antidinitrophenyl serum (Sigma) by adding 4 ml two-thirds saturated ammonium sulfate.

2. Pellet precipitate and redissolve in 2 ml PBS.
3. Dialyze globulins against PBS overnight at 4°C.
4. Add 10 m*M* $NaIO_4$ at room temperature and wait 30 min.
5. Apply globuline solution to hydrazine-linked beads (Affi-Gel Hz Immunoaffinity Kit, Bio-Rad) according to the kit instructions.
6. Stabilize Schiff's base bonds between immunoglobulin sugar moieties and hydrazine linkages of the gel matrix by reduction with 25 m*M* $NaBH_3$ in PBS according to the kit protocol.
7. Wash column with TBS. Beads may be removed from the column for batch techniques.

Acknowledgments

This work was partially supported by grants from Hungarian Science foundation OTKA F0032545, ETT07-454 EU "Copernicus" program, "AIDS" Ricerca Corrente from Ministerio Sanitá AIRC and MURST Cofin 97.

References

Aeschlimann, D., and Paulsson, M. (1991). Cross-linking of laminin-nidogen complexes by tissue transglutaminase: A novel mechanism for basement membrane stabilization. *J. Biol. Chem.* **266,** 15308–15317.

Aeschlimann, D., Paulsson, M., and Mann, K. (1992). Identification of Gln726 in nidogen as the amine acceptor in transglutaminase-catalyzed cross-linking of laminin-nidogen complexes. *J. Biol. Chem.* **267,** 11316–11321.

Fesus, L., Davies, P. J., and Piacentini, M. (1991). Apoptosis: Molecular mechanisms in programmed cell death. *Eur. J. Cell Biol.* **56,** 170–177.

Fesus, L., Thomazy, V., Autuori, F., Ceru, M. P., Tarcsa, E., and Piacentini, M. (1989). Apoptotic hepatocytes become insoluble in detergents and chaotropic agents as a result of transglutaminase action. *FEBS Lett.* **245,** 150–154.

Folk, J. E., and Chung, S. I. (1985). Transglutaminases. *Methods Enzymol.* **113,** 358–375.

Folk, J. E., and Finlayson, J. S. (1977). The epsilon-(gamma-glutamyl)lysine crosslink and the catalytic role of transglutaminases. *Adv. Protein Chem.* **31,** 1–133.

Grootjans, J. J., Groenen, P. J., and de Jong, W. W. (1995). Substrate requirements for transglutaminases: Influence of the amino acid residue preceding the amine donor lysine in a native protein. *J. Biol. Chem.* **270,** 22855–22858.

Im, M. J., Russell, M. A., and Feng, J. F. (1997). Transglutaminase II: A new class of GTP-binding protein with new biological functions. *Cell Signal* **9,** 477–482.

Kahlem, P., Terre, C., Green, H., and Djian, P. (1996). Peptides containing glutamine repeats as substrates for transglutaminase-catalyzed cross-linking: Relevance to diseases of the nervous system. *Proc. Natl. Acad. Sci. USA* **93,** 14580–14585.

Kim, S. Y., Grant, P., Lee, J. H., Pant, H. C., and Steinert, P. M. (1999). Differential expression of multiple transglutaminases in human brain: Increased expression and cross-linking by transglutaminases 1 and 2 in Alzheimer's disease. *J. Biol. Chem.* **274,** 30715–30721.

Knight, C. R., Rees, R. C., Platts, A., Johnson, T., and Griffin, M. (1993). Interleukin-2-activated human effector lymphocytes mediate cytotoxicity by inducing apoptosis in human leukaemia and solid tumour target cells. *Immunology* **79,** 535–541.

Lilley, G. R., Griffin, M., and Bonner, P. L. (1997). Assays for the measurement of tissue transglutaminase (type II) mediated protein crosslinking via epsilon-(gamma-glutamyl) lysine and N',N'-bis (gamma-glutamyl) polyamine linkages using biotin labelled casein. *J. Biochem. Biophys. Methods* **34,** 31–43.

Madi, A., Punyiczki, M., di Rao, M., Piacentini, M., and Fesus, L. (1998). Biochemical characterization and localization of transglutaminase in wild-type and cell-death mutants of the nematode Caenorhabditis elegans. *Eur. J. Biochem.* **253,** 583–590.

Makarova, K. S., Aravind, L., and Koonin, E. V. (1999). A superfamily of archaeal, bacterial, and eukaryotic proteins homologous to animal transglutaminases. *Protein Sci.* **8,** 1714–1719.

Melino, G., De Laurenzi, V., Catani, M. V., Terrinoni, A., Ciani, B., Candi, E., Marekov, L., and Steinert, P. M. (1998). The cornified envelope: A model of cell death in the skin. *Results Probl. Cell Differ.* **24,** 175–212.

Melino, G., and Piacentini, M. (1998). "Tissue" transglutaminase in cell death: A downstream or a multifunctional upstream effector? *FEBS Lett.* **430,** 59–63.

Nagy, L., Saydak, M., Shipley, N., Lu, S., Basilion, J. P., Yan, Z. H., Syka, P., Chandraratna, R. A., Stein, J. P., Heyman, R. A., and Davies, P. J. (1996). Identification and characterization of a versatile retinoid response element (retinoic acid receptor response element-retinoid X receptor response element) in the mouse tissue transglutaminase gene promoter. *J. Biol. Chem.* **271,** 4355–4365.

Nemes, Z., Marekov, L. N., Fesus, L., and Steinert, P. M. (1999). A novel function for transglutaminase 1: Attachment of long-chain omega-hydroxyceramides to involucrin by ester bond formation. *Proc. Natl. Acad. Sci. USA* **96,** 8402–8407.

Nemes, Z., Marekov, L. N., Fésüs, L., and Steinert, P. M. (2000). Cholesterol-3-sulfate interferes with cornified envelope assembly by diversion of transglutaminase 1 activity from crosslink and ester formation to glutamine hydrolysis. *J. Biol. Chem.* **275,** 2636–2646.

Nemes, Z., Marekov, L. N., and Steinert, P. M. (1999). Involucrin cross-linking by transglutaminase 1: Binding to membranes directs residue specificity. *J. Biol. Chem.* **274,** 11013–11021.

Nemes, Z., Jr., Adany, R., Balazs, M., Boross, P., and Fesus, L. (1997). Identification of cytoplasmic actin as an abundant glutaminyl substrate for tissue transglutaminase in HL-60 and U937 cells undergoing apoptosis. *J. Biol. Chem.* **272,** 20577–20583.

Nemes, Z., Jr., Friis, R. R., Aeschlimann, D., Saurer, S., Paulsson, M., and Fesus, L. (1996). Expression and activation of tissue transglutaminase in apoptotic cells of involuting rodent mammary tissue. *Eur. J. Cell Biol.* **70,** 125–133.

Piredda, L., Farrace, M. G., Lo Bello, M., Malorni, W., Melino, G., Petruzzelli, R., and Piacentini, M. (1999). Identification of "tissue" transglutaminase binding proteins in neural cells committed to apoptosis. *FASEB J.* **13,** 355–364.

Schmidt, G., Selzer, J., Lerm, M., and Aktories, K. (1998). The Rho-deamidating cytotoxic necrotizing factor 1 from *Escherichia coli* possesses transglutaminase activity: Cysteine 866 and histidine 881 are essential for enzyme activity. *J. Biol. Chem.* **273,** 13669–13674.

Smethurst, P. A., and Griffin, M. (1996). Measurement of tissue transglutaminase activity in a permeabilized cell system: Its regulation by Ca^{2+} and nucleotides. *Biochem. J.* **313,** 803–808.

Slaughter, T. F., Achyuthan, K. E., Lai, T. S., and Greenberg, C. S. (1992). A microtiter plate transglutaminase assay utilizing 5-(biotinamido)pentylamine as substrate. *Anal. Biochem.* **205,** 166–171.

Steinert, P. M. (1995). A model for the hierarchical structure of the cornified cell envelope. *Cell Death Differ.* **2,** 33–40.

Steinert, P. M., and Marekov, L. N. (1997). Direct evidence that involucrin is a major early isopeptide cross-linked component of the keratinocyte cornified cell envelope. *J. Biol. Chem.* **272,** 2021–2030.

Tarcsa, E., and Fésüs, L. (1990). Determination of epsilon (gamma-glutamyl)lysine crosslink in proteins using phenylisothiocyanate derivatization and high-pressure liquid chromatographic separation. *Anal. Biochem.* **186,** 135–140.

Tarcsa, E., Marekov, L. N., Andreoli, J., Idler, W. W., Candi, E., Chung, S. I., and Steinert, P. M. (1997). The fate of trichohyalin: Sequential post-translational modifications by peptidyl-arginine deiminase and transglutaminases. *J. Biol. Chem.* **272,** 27893–27901.

Thomazy, V., and Fesus, L. (1989). Differential expression of tissue transglutaminase in human cells: An immunohistochemical study. *Cell Tissue Res.* **255,** 215–224.

CHAPTER 6

Analysis of Sphingomyelin and Ceramide Levels and the Enzymes Regulating their Metabolism in Response to Cell Stress

Rick T. Dobrowsky* **and Richard N. Kolesnick**†

* Department of Pharmacology and Toxicology
University of Kansas
Lawrence, Kansas 66045

† Laboratory of Signal Transduction
Memorial Sloan-Kettering Cancer Center
New York, New York 10021

METHODS IN CELL BIOLOGY, VOL. 66

0091-679X/01 $35.00

I. Introduction

Ceramide is the common moiety forming the hydrophobic backbone of glycosphingolipids and sphingomyelin (SM). Over the last decade ceramide has been implicated as a mediator of diverse biologic responses such as cell cycle arrest, terminal cell differentiation, neurotransmitter and neurotrophin secretion, and insulin resistance, to list a few (Hannun, 1996; Mathias and Kolesnick, 1998). These observations have led to the proposal that ceramide production serves as a general signal of cell stress in organisms from yeast to humans (Hannun, 1996; Mathias and Kolesnick, 1998). However, ceramide has gained the greatest recognition due to extensive data supporting its critical signaling role in apoptosis (Ariga *et al.,* 1998; Kolesnick and Kronke, 1998).

Cellular ceramide levels may be regulated by the concerted action of many different enzymes involved in its formation and metabolism. The goal of this chapter is to provide clear and well-documented protocols for investigators who are interested in assessing the role of ceramide and ceramide-metabolizing enzymes in their particular cell model. A prime emphasis of this chapter is on basic methods of lipidology with which many cell biologists may be unfamiliar and clarifying several critical misconceptions present in the literature regarding methods for the measurement of this bioactive lipid metabolite. Additionally, we provide a brief discussion of our current understanding on the regulation of SM and ceramide-metabolizing enzymes and the availability of molecular probes, antibodies, and inhibitors where applicable.

II. Metabolic Labeling

Metabolic radiolabeling of cellular lipids is one of the easiest and most manageable methods for measuring SM and ceramide levels. Depending on the particular goal of the experiment, either [9,10-^{3}H]palmitate (60 Ci/mmol, American Radiolabeled Chemicals) or L-[3-^{14}C]serine (58 mCi/mmol, American Radiolabeled Chemicals) may be used to label the fatty acyl chain or sphingolipid backbone of ceramide, respectively (Fig. 1).

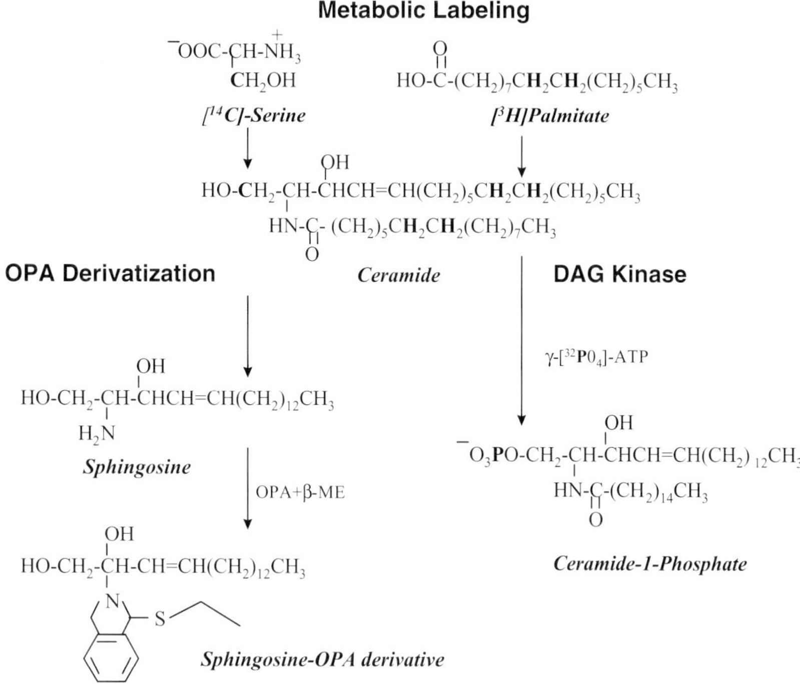

Fig. 1 Schematic overview of methods for qualitative and quantitative analysis of ceramide. Bold lettering indicates the position of the radioisotope. OPA, *o*-phthaldehyde; β-ME, β-mercaptoethanol.

[*methyl*-^{3}H]Choline chloride (80 Ci/mmol, American Radiolabeled Chemicals) may be used to radiolabel cellular SM. Importantly, metabolic labeling does not provide an accurate measure of mass and therefore can only give qualitative results. Nonetheless, this approach is very useful in determining if particular experimental treatments may have an impact on the SM pathway.

Prior to a description of the protocols, a few considerations about this approach are warranted. Although SM is concentrated primarily in the outer leaflet of the plasma membrane, the location of the signaling-sensitive pool in response to given agonists or cellular stresses may vary. For example, SM is hydrolyzed primarily from caveolar membranes in response to neurotrophins (Bilderback *et al.,* 1997), whereas tumor necrosis factor (TNF) induces hydrolysis from pools of SM localized to the inner leaflet of the plasma membrane (Linardic and Hannun, 1994) and hepatectomy may induce the turnover of SM present in nuclear membranes (Albi *et al.,* 1997). Because the different cellular pools of SM label at different rates (Linardic and Hannun, 1994), cells should be labeled with the radioactive precursor for sufficient duration to ensure that all the cellular pools of SM are labeled to metabolic equilibrium. Because the incorporation of the radiolabeled precursor into all SM pools is related to the doubling time of the cells, cellular pools of SM should be at isotopic steady state after incubating the cells with radiolabel for two to three population doublings. Failing to label all cellular pools of SM may diminish the detection of SM hydrolysis, and therefore apparent ceramide production, in response to treatment.

A second consideration when labeling ceramide metabolically is whether the radiolabeled precursor is utilized effectively by the different metabolic pathways that produce ceramide. For example, some *in vivo* studies have erringly used a short 2-h incubation of cells with lysophosphatidylcholine containing [^{14}C]palmitate in the *sn*-1 position of glycerol to label ceramide (Schutze *et al.,* 1992). Because the acylation of sphingomyelin and ceramide does not directly involve donation of a fatty acyl group from lysophosphatidylcholine (Merrill and Jones, 1990), the amount of label incorporated into these molecules would be expected to be rather low and would undoubtedly require a very long incubation to reach equilibrium. In the absence of more convincing evidence on the biochemical mechanisms justifying this strategy, any radiolabeled species of lysophosphatidylcholine is an extremely poor choice as a precursor. Finally, when labeling ceramide metabolically it is important to consider the effect of availability of the precursor in the medium. Thus, when using radiolabeled serine as a precursor, incorporation will be greatest in serine-deficient media.

A. Cell Culture

A critical consideration for effective SM labeling is the initial cell density. Cells should be seeded so that at the required density for experimental treatment, the cells will have incubated for 48–72 h or three doubling times in radiolabeled medium. This time may need to be increased if the doubling time for the cells is particularly long. We generally incubate fibroblast and PC12 cells lines for 48–72 h in the presence of the radiolabel. Freshly split cells are labeled with [^{3}H]palmitate or [^{14}C]serine in complete or serine-deficient medium that contains 0.5 to 1 μCi/ml of the radioisotope. Following incubation, the medium should be removed, the cells washed with fresh serum-free medium or phosphate-buffered saline, and then placed in fresh serum-free medium lacking the radiolabel for a period of 2–6 h. The 2- to 6-h rest period is to allow the cells to recover from the wash step, as previous studies have demonstrated that simple medium changes increase the level of sphingolipid metabolites dramatically (Smith *et al.,* 1997). The levels of these metabolites typically return to baseline levels after 2–6 h although this may need to be determined empirically for any given cell type.

B. Lipid Isolation

Following experimental treatment, the cells may be harvested by adding 1 ml of any standard lysis buffer (detergent is not absolutely necessary) and scraping the cells from the plate; nonadherent cells should be pelleted by centrifugation. An aliquot of this lysate should be reserved for analysis of total protein if desired. Alternatively, if a protein measurement is not required, the cells may be scraped directly into 2 ml of ice-cold methanol. Transfer the solution to a 13 $\times$ 100-mm screw cap test tube and add one 1 ml of $CHCl_3$. Secure the tube with a Teflon-lined cap and shake or vortex the solution vigorously. Lipids are extracted using a modification of the method of Bligh and Dyer (1959).

1. Add 0.8 ml of the cell lysate to a 13 × 100-mm screw cap test tube containing 3 ml of $CHCl_3 : CH_3OH$ (1 : 2, v/v), secure the tube with a Teflon-lined cap, and shake well. Alternatively, if the cells were scraped directly into methanol, add 1 ml of $CHCl_3$ and about 0.6–0.7 ml of water as some liquid is always recovered with the cells. At this point the mixture should be a monophase.

2. After 5 min, phase separation is achieved by the addition of 1 ml of $CHCl_3$ and 1 ml of 1% perchloric acid or water. Acidification primarily increases the extraction of more polar phospholipids, but is not absolutely necessary to efficiently extract SM or ceramide. Mix the samples well by vigorous shaking and centrifuge for 5–10 min at room temperature. Centrifugation in a table-top clinical centrifuge at 2000 to 3000*g* is sufficient.

3. Aspirate the upper aqueous layer. Transfer 1.5 ml of the 2 ml (theoretical) lower organic phase to a fresh tube. Evaporate the solvent under a stream of nitrogen gas in either a Speed Vac or an oven at 45°C.

Note: Prior to phase separation, the solution should be a clear monophase. If the solution is not forming a clear monophase or has prematurely partitioned, carefully add a few drops of methanol and shake to achieve a clear monophase before proceeding with step 2.

C. Alkaline Methanolysis

It is advisable to base hydrolyze the samples prior to chromatography as glycerophospholipids will also incorporate the various radiolabeled precursors to differing extents and make the separation and identification of ceramide or SM more difficult. To accomplish this:

1. Resuspend the dried lipid samples in 1 ml of $CHCl_3$ and add 0.1 ml of 1 or 2 *N* KOH dissolved in CH_3OH. Incubate the tubes for 1 h at 37°C. During this incubation, a white precipitate of glycerophosphate will form due to the hydrolysis of the acyl groups.

2. Neutralize the samples with 0.1 ml of 2 *N* HCl in CH_3OH and add 0.3 ml of CH_3OH. Add 0.2 ml of water, vortex, and centrifuge the sample. Carefully aspirate the upper aqueous layer and quantitatively transfer an aliquot of the 1-ml organic phase to a fresh tube.

3. Add 5 μg of cold ceramide or SM standard and evaporate the solvent. The lipid residue is dissolved in 30 μl of cold $CHCl_3$, and a 20-μl aliquot is spotted on a TLC plate about 2 cm from the bottom of the plate. The plate may be developed in a solvent system consisting of $CHCl_3 : CH_3OH : NH_4OH$ [200 : 25 : 2.5 (v/v)] for separating ceramide or $CHCl_3 : CH_3OH$: glacial acetic acid : H_2O [50 : 30 : 8 : 5, (v/v)] for separating SM (Liu and Anderson, 1995).

D. Thin-Layer Chromatography

A few cautionary notes on aspects of thin-layer chromatography (TLC) that are often overlooked by many newcomers to lipidology.

1. To optimize chromatography, be sure that the TLC chamber is fully equilibrated. To equilibrate the TLC chamber, place a sheet of Whatman No. 1 paper inside the tank to serve as a wick. Pour the solvent in the tank and allow the solvent to fully saturate the paper. This usually takes several hours and should be prepared before the samples are spotted on the TLC plate.

2. When preparing solvent, do not individually add the solvent components directly to the TLC tank, especially in the presence of the wick. Add the solvent components to a flask, mix them well, and then pour the mixed solvent solution into the TLC chamber.

3. Do not overfill the TLC chamber. Use only enough solvent so that when the TLC plate is placed in the chamber, the solvent covers about the bottom 0.5 to 1 cm of the plate. It is important to be sure that the solvent will not cover the area of the plate spotted with the sample when the plate is placed in the chamber.

4. Once the plate has been inserted into the chamber, avoid opening the chamber until the solvent has migrated about 2–4 cm below the top of the plate. Two TLC plates may be developed at the same time in one chamber, but it is advisable to place both plates in the chamber together in such a way as to ensure that they do not touch and that they are both sufficiently immersed in the solvent.

5. Once the solvent front has reached the desired position, remove the plate from the chamber, rapidly mark the position of the solvent front with a pencil, and place the plate in a fume hood to dry.

E. Visualization of Lipids

1. The isolated lipids may be visualized using iodine vapors. In a fume hood, place some crystalline iodine in a clean TLC chamber and allow the vapors to accumulate for several hours. Do not open the tank with your face near the opening, even with the fume hood operating, because iodine vapors are very toxic. Place the plate in the tank and after 2–5 min, depending on the amount of lipid, iodine-positive spots can be visualized as yellow bands. Remove the plate from the chamber, mark the bands with a pencil, and allow the iodine to dissipate from the plate in the fume hood.

2. Radioactive bands may be visualized by fluorography. To help orient the film after development, mark the plate with three drops of radioactive dye (add 1 μl of L-[3-^{14}C] serine stock solution (58 mCi/mmol) to 200 μl of bromphenol blue or India ink). Spray the plate liberally with En^{3}Hance (New England Nuclear) without oversaturating the silica gel. Wrap the plate in a single layer of plastic wrap and expose it to film in an autoradiography cassette with intensifying screens for 2–3 days at −80°C.

3. After developing the film, place it on top of the TLC plate and line up the radioactive dye spots on the film with the blue dye spots on the plate. While firmly holding the film in place, mark the radioactive bands with a pencil.

4. Spray the silica gel with a water mister and lightly saturate the gel with water without causing an excess of water to run off. Wetting the silica minimizes the generation of radioactive silica dust when scraping the silica from the plate. Using a sharp razor, scrape each band outlined by pencil onto a piece of weigh paper and transfer the moist

silica to a scintillation vial. If concerns exist about water acting as a quenching agent, which is extremely minimal in our experience, dry the silica in an oven prior to adding 8 ml of scintillation fluid.

5. Alternatively, a densitometer or phosphoimager may be used to quantitate the radioactive bands on the film. For accurate quantitation, be sure that the intensity of the bands is in the linear range of the densitometer.

F. Protein and Phospholipid Measurement

To account for variations in the number of cells extracted, data from scintillation counting or densitometry may be normalized to protein, cell number, or phospholipid content. Protein levels may be measured using any standard protein assay. Using the specific activity of the starting isotope, the moles (this value will typically be in the picomole to nanomole range) of labeled lipid can be calculated and normalized to the milligram of protein. Alternatively, an aliquot of the organic extract may be saved *prior* to base hydrolysis and used to determine the total phospholipid content; do not use aliquots of the organic extract following base hydrolysis as most of the glycerophospholipids have been lost. Each mole of phospholipid contains 1 mol of phosphate, which can be measured directly after liberation from the lipid backbone. Although several procedures are available for phosphate analysis, we prefer the method of Ames and Dubin (1960).

1. Duplicate aliquots of sample from the organic layer of the initial lipid extraction, as well as standards of NaH_2PO_4 (0–80 nmol), are aliquoted into inexpensive 13 × 100-mm test tubes. The tubes do not need to be capped at any point so plain borosilicate test tubes are sufficient. (*Note:* a 1 m*M* phosphate solution may be prepared easily by making a 1 : 50 dilution of 50 m*M* NaH_2PO_4, pH 7.0, buffer used to calibrate pH meters.)

2. Add 100 μl of ashing buffer [10% $Mg(NO_3)_2$ in ethanol (w/v)] and evaporate the solvents at 80°C in an oven.

3. Ash the samples in a strong flame just until the generation of the brown gas is complete. Ashing too long will char the sample, producing a black residue that will interfere with the spectrophotometry. Transfer the tubes to a metal test tube rack to avoid melting of plastic racks.

4. After the tubes have cooled, add 0.3 ml of 0.5 *N* HCl and boil for 15 min to hydrolyze pyrophosphates. Cover the tops of individual tubes with marbles or fit the entire rack of tubes with a sheet of aluminum foil to avoid evaporation.

5. Add 0.7 ml of a solution containing 6 parts of 0.42% acid ammonium molybdate (4.2 g ammonium molybdate in 1 liter of 1 *N* H_2SO_4) and 1 part 10% ascorbic acid [10% (w/v), made fresh in water], i.e., 30 ml 0.42% acid ammonium molybdate plus 5 ml of 10% ascorbic acid.

6. Mix the tubes thoroughly and incubate for either 30 min at 45°C or 60 min at 37°C. The absorbance at 820 nm is read to determine the nanomoles of phosphate. A standard curve is constructed and the nanomoles of phosphate in the samples are extrapolated from the curve.

G. Bacterial Sphingomyelinase Assay

An alternative approach to quickly determine the extent of SM hydrolysis from cells labeled metabolically with [^{3}H]choline without using chromatography is to use bacterial SMase. The principle of the assay relies on the ability of bacterial SMase from *Streptomyces* sp. to quantitatively hydrolyze SM present in a lipid extract. The enzyme cleaves [^{3}H]phosphocholine from SM releasing the radiolabeled head group, which can then be recovered in the aqueous phase following lipid extraction (Jayadev *et al.*, 1995). Hydrolysis of labeled cellular SM by *in vivo* experimental treatment leads to less SM available for hydrolysis *in vitro* by the bacterial SMase. Conversely, an *in vivo* increase in SM synthesis would provide more substrate for *in vitro* hydrolysis by the bacterial enzyme. Thus, in comparison to a control, the relative effect of a given treatment on SM levels may be determined.

1. Evaporate the lipid sample in an inexpensive 13 × 100-mm test tube and resuspend the dried lipid residue in 50 μl of 200 m*M* Tris–HCl, pH 7.5, 20 m*M* $MgCl_2$ containing 1% Triton X-100 with vigorous vortexing.
2. Place the tubes in a 37°C water bath and initiate the reaction by the addition of 100 mU *Streptomyces* sp. SMase in 50 μl 100 m*M* Tris–HCl, pH 7.5.
3. After 2 h, quench the reaction by the addition of 1.5 ml of $CHCL_3 : CH_3OH$ (2 : 1). The released [^{3}H]choline phosphate is recovered in the upper aqueous phase after the addition of 0.2 ml H_2O. Following centrifugation, transfer 0.4 ml of the 0.8-ml upper aqueous phase (theoretical) to a scintillation vial for quantitation.

Note: To ensure quantitative hydrolysis of the SM in the sample by the bacterial SMase, samples should contain no more than 10–15 nmol total phospholipid phosphate. Under these reaction conditions, the bacterial SMase does not hydrolyze phosphatidylcholine (Jayadev *et al.*, 1995). The amounts of SM hydrolyzed may be normalized to protein or phospholipid phosphate content as described earlier. The extent of SM hydrolysis determined by this method is identical to that determined by chromatographic separation (Jayadev *et al.*, 1995).

III. Determination of Sphingomyelin Mass Levels

A. Phosphate Analysis

The mass of SM in a lipid extract may be determined following TLC by extracting the lipid from the silica matrix and determining the phosphate content. Because SM typically accounts for about 5% of the total phospholipid phosphate in most cell types, it is best to start with as much cellular material as possible, i.e., at least 1×10^7 cells.

1. Extract the lipids as described in Section II,B and remove duplicate aliquots of the recovered organic phase for the determination of total phospholipid phosphate prior to performing the next base hydrolysis step.

2. Base hydrolyze the glycerophospholipids as described in Section II,C, quantitatively transfer 0.9 ml of the organic phase to a fresh tube, and evaporate the solvent. Chill the tube on ice briefly and resuspend the lipid residue in 40 μl of cold $CHCl_3$ to minimize evaporation of the solvent. Apply 35 μl of the sample to a TLC plate and separate the SM from the remaining lipids by developing the plate in one of the aforementioned solvent systems.

3. Visualize the lipid using the I_2 vapors as described in Section II,E and carefully mark the area. Mist the plate with water, scrape the silica onto a piece of weighing paper, and transfer the silica to a test tube.

4. Elute the SM from the silica by sequential extraction with 2 × 1-ml aliquots of $CHCl_3 : CH_3OH$ (2 : 1) and 1 ml of $CHCl_3 : CH_3OH$ (1 : 2). Vortex the solution vigorously and sediment the silica gel by centrifugation. Pool the eluates in a fresh tube.

5. Evaporate the eluate and proceed with the phosphate assay described in Section II,F.

Two drawbacks with determining the phosphate content of SM after TLC are that (1) the recovery of the lipid from the silica gel can vary between samples, especially at low lipid concentrations; this may be accounted for by adding a known trace amount of radioactive SM directly to the SM spot on the TLC plate prior to extraction, and (2) the phosphate assay described earlier does not reliably determine less than 1 nmol of inorganic phosphate [other phosphate assays with picomole sensitivity have been developed using Malachite green (Petitou *et al.*, 1978)].

B. Cupric Sulfate Charring

SM may be quantitated directly from a TLC plate by charring the lipid samples along with known amounts of SM standards. Using a densitometer, a standard curve can be constructed and the amount of SM in the samples extrapolated. We have found that the technique is sensitive to about 0.5 nmol of SM, but the intensity of charring shows the best linearity between 2 and 10 nmol of SM. The use of copper sulfate is based on the procedure of Rustenbeck and Lenzen (1990).

1. Prepare 500 ml of a solution of 10% copper sulfate in 8% aqueous phosphoric acid and prepare the samples as described in steps 1 and 2 in the procedure described previously.

2. Prior to chromatography, prepare a 1 m*M* SM standard (MW ~ 731.09, Avanti Polar Lipids) in $CHCl_3$ and make a serial dilution to prepare a 0.1 m*M* SM standard. Apply standard amounts of SM in triplicate ranging from 0.25 to 10 nmol along with the samples. Be sure to include a blank lane for background. After the chromatography step is completed, dry the plate in a 180°C oven for 10 min, remove, and then allow the plate to cool.

3. Pour the copper sulfate solution into a shallow dish that is large enough to accommodate the size of the TLC plate. To dip the plate into the solution efficiently, angle the bottom of the plate into the dish and slide the plate along the bottom quickly while immersing the remaining area of the plate. Allow the plate to sit in the solution for 15 s.

Remove the plate from the solution and allow any excess solution to drip off the plate; avoid contact of the solution with your skin and clothes.

4. Char the lipids by heating the plate initially for 2 min at 110°C and then for 10 min at 175°C. The lipids appear as brownish-black spots and the SM is identified by comigration with the SM standards.

5. Wrap the plate tightly with one layer of plastic wrap and use a laser densitometer at 530 nm to scan the plate. Construct a standard curve of relative absorption units versus nanomoles of SM to determine the amount of SM in samples. This technique is also useful for analyzing ceramide if desired. We have found more intense charring with ceramide relative to SM with the limit of sensitivity being about 0.1 nmol.

IV. Enzymatic Method for the Quantitation of Ceramide-Diacylglycerol (DAG) Kinase Assay

The *Escherichia coli* enzyme diacylglycerol kinase phosphorylates ceramide to generate ceramide-1-phosphate (Fig. 1) (Priess *et al.,* 1987). This reaction occurs in a mixed micelle and is performed in the presence of excess enzyme, allowing the reaction to go to completion. The kinetic conditions for this assay have been well documented; when performed properly, the assay can yield reliable quantitative data on ceramide levels in biological samples. Radiolabeled [^{32}P]ceramide-1-phosphate is produced in the presence of DAG kinase and [γ-^{32}P]ATP as the phosphate donor. The reaction products are extracted, separated by TLC, and quantitated by liquid scintillation counting in comparison with a standard curve.

A. Materials

Type III ceramide standard is purchased from Sigma. [γ-^{32}P]ATP (3000 Ci/mmol) is purchased from New England Nuclear. The Easytide solution of [γ-^{32}P]ATP from this supplier works fine in the DAG kinase assay and is more convenient than preparations requiring freeze–thaw cycles. *n*-Octyl-β-D-glucopyranoside (β-octylglucoside) is purchased from Calbiochem. L-α-Dioleoylphosphatidylglycerol is obtained from Avanti Polar Lipids. *E. coli* DAG kinase may be purchased from Biomol as a turbid membrane suspension of 1 mg/ml protein. Lyophilized enzyme preparations are not advised (see later).

B. Lipid Extraction

Lipids are extracted from cell samples by the modified method of Bligh and Dyer as described in Section II,B. Importantly, because serum contains significant amounts of ceramide, the cells should be washed prior to extraction if necessary. Because the DAG kinase assay is very sensitive (Priess *et al.,* 1987), we typically use 0.5 ml of the 2-ml (theoretical) organic layer for the determination of ceramide from lipid extracts prepared from whole cell lysates (~1–2 × 10^6 cells). Aliquots of the remaining organic phase can be used for the determination of total phospholipid phosphate (Section II,F).

C. DAG Kinase Assay Reagents

2× buffer: 100 m*M* imidazole (pH 6.6), 100 m*M* LiCl, 25 m*M* $MgCl_2$, 2 m*M* EGTA (pH 6.6)

Dilution buffer: 10 m*M* imidazole (pH 6.6), 1 m*M* diethylenetriaminepentaacetic acid (DTPA)

Mixed micelles are prepared by drying 0.97 ml of 20 mg/ml L-α-dioleoylphosphatidylglycerol (DOPG) under nitrogen. To the dried DOPG, add 1 ml of 7.5% β-octylglucoside. Vortex and sonicate the mixture until the DOPG is completely dissolved. The solution should be only slightly cloudy. Alternatively, store the samples at 4°C overnight to hydrate the lipid and vortex vigorously the next day. The lipid should solubilize easily with minimal or no sonication. A 10-ml batch of mixed micelles may be conveniently prepared and stored in 1-ml aliquots at −20°C for up to a year.

Note: The indicated grade of β-octylglucoside can be used without further purification. Less pure grades of detergent may give spurious results and may require recrystallization from acetone at −20°C. Recover the crystals by filtration through a chilled fine sintered glass funnel, wash the crystals with 200–500 ml of ice cold ethyl ether, and dry in a vacuum dessicator.

D. DAG Kinase Assay Method

After the lipid extraction, transfer an aliquot of the sample to a fresh 13 × 100-mm screw cap test tube. Similarly, aliquot ceramide type III standards (0–1000 pmol) in duplicate to prepare the external standard curve. The assay is typically linear up to at least 1.2 nmol of ceramide.

1. Evaporate the solvent and resuspend the lipid residue in 20 μl of the mixed micelles by vortexing. Add 70 μl of the reaction mixture to each sample. The needed amount of reaction mixture may be prepared by following Table I. Always prepare enough for a few extra assays to account for pipetting errors.

2. Each DAG kinase assay should contain a final [γ-^{32}P]ATP concentration of 1 m*M* at a specific activity of 40 μCi/μmol. We typically prepare a stock solution of 40 m*M* ATP in 10 m*M* Tris buffer, pH 7.4. To prepare sufficient amounts of 10 m*M* [γ-^{32}P]ATP for 20 assays, combine 50 μl of the 40 m*M* ATP stock solution, 80 μCi [γ-^{32}P]ATP, and bring the total volume to 0.2 ml. The assay is initiated by adding 10 μl of the 10 m*M* [γ-^{32}P]ATP to the reaction mix. After 30 min at room temperature, stop the reaction by adding 3 ml of $CHCl_3:CH_3OH$ (1 : 2). Next, add 0.7 ml of H_2O and mix the solution. After at least 5 min, phase partitioning is achieved by the addition of 1 ml of $CHCl_3$ and 1 ml of 1% perchloric acid or water. Centrifuge the samples for 10 min at 2–4000*g* and aspirate the upper aqueous phase into a radioactive waste container. The lower phase

Table I
Preparation of DAG Kinase Reaction Mixture[a]

Solution	Volume/assay	Number of assays	Total volume
2× buffer	50 μl		
1*M* DTT	0.2 μl		
5.0 mg/ml DGK	1.0 μl		
Dilution buffer	18.8 μl		

[a]To prepare the complete reaction buffer, enter the number of assays to be performed in column 3 and multiply column 2 by column 3 to get the total volume of reagent needed. Mix all the reagents thoroughly and add 70 μl per assay. The DAG kinase enzyme concentration should be 3.5–5 μg/assay.

may be washed again if desired, although this is not necessary. If an emulsion forms on washing, add a few drops of methanol.

3. Transfer 1.5 ml of the lower phase to a fresh tube and evaporate the solvent. When dry, resuspend the residue in 30 μl of cold $CHCl_3$ and immediately apply 20 μl to a TLC plate. The lipids are resolved by development in chloroform : acetone : methanol : acetic : acid : water (10 : 4 : 3 : 2 : 1) or chloroform : methanol : glacial acetic acid (65 : 15 : 5). Mark the plate with radioactive ink and expose the plate to film overnight at −80°C. No En^3Hance spraying is necessary. Develop the film and locate the ceramide phosphate spots by comparison to the standards.

4. The radioactive products are scraped from the plate and the radioactivity quantitated as detailed in Section II,E. Alternatively, the ceramide may be quantitated using a phosphoimager. The amount of ceramide per sample is extrapolated from the standard curve and normalized to total phospholipid phosphate, protein, or cell number. The efficiency of the mass conversion of ceramide to its phosphorylated derivative should be determined using the specific activity of the [γ-^{32}P]ATP. In most instances this conversion should exceed 90% (Priess *et al.,* 1987).

E. Kinetic and Enzymatic Considerations in Performing the DAG Kinase Assay

Typical Michaelis–Menten conditions dictate that a limiting amount of enzyme should be used to study kinetic mechanisms or to explore the effects of activators and inhibitors on the enzyme. However, the DAG kinase assay may be viewed essentially as a derivatization reaction; the reaction quantitatively converts the ceramide substrate to the phosphorylated derivative. Therefore, to allow an accurate quantitation of substrate and to avoid any kinetic effects of endogenous activators (Watts *et al.,* 1997), it is necessary that an excess DAG kinase enzyme be used and the reaction allowed to go to completion (Van Veldhoven *et al.,* 1989; Perry and Hannun, 1999). Moreover, attention should also be given to the nature of the DAG kinase enzyme preparation. Lyophilized DAG kinase enzyme preparations may be more subject to lipid activators relative to enzyme obtained from a glycerol solution of a bacterial membrane preparation (Perry and Hannun, 1999). When performed appropriately, the DAG kinase assay gives results similar to both metabolic labeling and mass measurements obtained by other physical derivatization techniques (Tepper *et al.,* 1997; Bose *et al.,* 1998; Garzotto *et al.,* 1998).

V. Chromatographic Quantitation of Ceramide Using HPLC and Fluorescence Spectroscopy

A nonenzymatic alternative to quantitating ceramide employs the strategy of derivatizing the sphingoid base of ceramide. This approach necessitates the degradation of ceramide to sphingosine and requires the addition of an internal standard to the lipid extract. The internal standard is chemically similar to natural ceramide and is expected to undergo the same amount of loss during the analytic workup but has a distinct chromatographic migration enabling separation from endogenous ceramide. By calculating the ratio of the analyte signal (ceramide) to that of the internal standard, the amount of ceramide in the sample can be determined by extrapolation from a standard curve constructed using the internal standard and an authentic ceramide standard.

Derivatization of the sphingoid base of ceramide is accomplished with *o*-phthalaldehyde (OPA) (Merrill *et al.,* 1988). OPA derivatizes the primary amine of the sphingoid base forming a fluorescent derivative, which can be separated by high-performance liquid chromatography (HPLC) and detected fluorometrically. This procedure has low picomole sensitivity and permits a true mass measurement. The major drawback to OPA derivatization is that it is necessary to purify ceramide and perform acid or base hydrolysis prior to derivatization. This increases the difficulty of handling a large number of samples. However, we also describe an alternative procedure that does not require prior chromatographic isolation of ceramide.

A. Chromatographic Isolation of Ceramide and Acid Hydrolysis

The cells are recovered as described earlier, aliquots are saved for protein determination, and the lipids are extracted using the method of Bligh and Dyer (1959) (Section II,B).

1. Transfer a 1.5-ml aliquot of the lipid extract to a fresh tube and evaporate the solvent. Resuspend the residue in 40 μl of cold $CHCl_3$ and apply 30 μl of each sample to a TLC plate. A series of type III ceramide standards (0–2 nmol) are spiked into a similar matrix that been extracted from control cells. Ceramide is resolved from other lipids by developing the plate in $CHCl_3$: CH_3OH : triethylamine : 2-propyl alcohol : 0.25% potassium chloride (30 : 9 : 25 : 18 : 6) (Nikolova-Karakashian *et al.,* 1997). The location of ceramide is determined by comparison to the ceramide standards using iodine vapors.
2. Prepare a 50 μ*M* solution of *N*-acetyl-C20-sphinganine (Matreya Biochemicals) in $CHCl_3$ for use as an internal standard. Do not use SM as the internal standard. Carefully add 10 μl (0.5 nmol) of the internal standard directly to the ceramide spot of each sample

visualized on the TLC plate. Mist the TLC plate with water and scrape the ceramide spots into a fresh 13×100-mm screw cap tube. The lipid is eluted from the silica gel with 2×1 ml of $CHCl_3:CH_3OH$ (2 : 1) followed by 1 ml of methanol. Add each aliquot of solvent to the silica gel, vortex the samples vigorously, and sediment the silica by brief centrifugation. Combine the eluates in a fresh tube and evaporate the solvent.

3. To deacylate, add 1 ml of 0.5 *M* HCl in methanol and incubate the samples at 65°C for 15 h. Be sure to cap the tube tightly with a Teflon-lined screw cap. Do not use a rubber-lined screw cap. After cooling, neutralize the sample with 1 ml of 1 *M* KOH in methanol and add 1 ml of $CHCl_3$.

4. Induce phase separation with the addition of 1 ml of 1 *M* NaCl and 1 ml of $CHCl_3$. Vortex the solution, centrifuge, and aspirate the upper aqueous phase. Wash the organic phase several times with water containing 50 μl of 1 *N* NH_4OH per 15 ml of water. Evaporate the lower phase organic solvent containing the free long chain bases. Resuspend the lipid residue in 0.1 ml of CH_3OH.

B. Direct Deacylation of Ceramide from Lipid Extracts by Base Hydrolysis

This procedure is a quicker alternative to that described earlier and does not require prior chromatographic isolation of ceramide (Bose *et al.,* 1998).

1. Standards of 0–2 nmol of ceramide type III are prepared fresh as described previously. An internal standard of 0.5 nmol of C20-sphinganine is added to both samples and standards to estimate recovery. Following the lipid extraction, evaporate the lower phase organic solvent and resuspend the lipid residue in 0.5 ml of 1 *M* KOH in 90% CH_3OH. Seal the tubes and heat at 90°C for 1 h to quantitatively convert ceramide into sphingosine. This digestion procedure does not convert complex sphingolipids, such as sphingomyelin, galactosylceramide, or glucosylceramide, into sphingosine (Bose *et al.,* 1998).

2. Extract the samples and standards with 0.5 ml of 1 *M* HCl in CH_3OH, 1.0 ml of $CHCl_3$, and 0.75 ml of 1 *M* aqueous NaCl. Recover the lower organic phase, evaporate the solvent, and dissolve the residue in 0.1 ml of CH_3OH.

C. Formation of the OPA Derivative of Long Chain Bases and HPLC

1. The OPA solution must be prepared fresh daily. Dissolve 10 mg OPA in 0.2 ml ethanol and add 10 μl of 2-mercaptoethanol followed by 19.8 ml of 3% aqueous boric acid (pH adjusted to 10.5 with KOH). Some researchers allow the solution to sit in the dark for 10 min before use.

2. Add 0.1 ml of the OPA solution to the methanol solution of long chain bases from step B,2 and incubate at room temperature for 5 min. Next, add 0.5 ml of CH_3OH : 5 m*M* aqueous potassium phosphate (pH 7.0) (90 : 10, v/v) (solvent A). The samples may be cleared prior to chromatography by centrifugation if desired.

3. The derivatized long chain sphingoid bases may be resolved by reversed-phase HPLC using a Beckman octadecylsilyl (C18) column with isocratic elution using solvent A. The following settings are employed: injection volume, 20 μl; solvent flow rate, 0.6 ml/min; cycle time, 30 min; fluorescence excitation wavelength, 340 nm; and emission wavelength, 455 nm. Retention times of various sphingosines are directly related to alkyl chain length. Although specific elution times may vary somewhat, the elution times should follow the order of C14 < C16 < C18 < C20 and must be determined empirically for each system. Typically, the 18 carbon sphingoid base is the predominant molecular species forming the backbone of mammalian ceramides.

4. To quantitate the ceramide in the original sample, first calculate the peak area ratio of the ceramide standards to the internal standard and construct a standard curve by plotting the peak area ratio versus the amount of ceramide. For each sample, perform a similar calculation and extrapolate the amount of ceramide in the sample from the standard curve. Correct for any dilutions and volume transfers to quantitate the amount of ceramide in the original extract. The amount of ceramide per sample should be normalized to the milligram of protein extracted; samples may be normalized to total phospholipid phosphate, but be sure to use aliquots of the original lipid extract prior to any chromatographic or deacylation steps. If the direct deacylation procedure is employed, the fluorescence signal represents the sum of ceramide and free sphingosine in the sample as the ceramide was not first isolated chromatographically. However, the contribution of free sphingosine is minor as ceramide levels are 10- to 20-fold greater than free sphingosine in mammalian cells.

VI. Assays for Enzymes Regulating Cellular Levels of Sphingomyelin and Ceramide

A. Acid Sphingomyelinase

1. Background

Acid SMase is a ubiquitous type C phosphodiesterase that catalyzes the hydrolysis of phosphocholine from SM (Fig. 2). Both lysosomal and secretory forms of the enzyme have been described and represent posttranslationally processed products of the same gene (Schissel *et al.*, 1996; Lin *et al.*, 1998). The two forms of the enzymes may be distinguished based on the Zn^{2+} dependence of the secretory form for *in vitro* activity (Schissel *et al.*, 1996).

2. Regulation of Signaling

Although long considered a lysosomal housekeeping enzyme, interest in acid SMase has undergone a renaissance due to the observations that this enzyme can undergo regulated activation in response to cytokines and cell stress (Mathias and Kolesnick, 1998). Indeed, genetic evidence strongly supports that acid SMase plays a central role

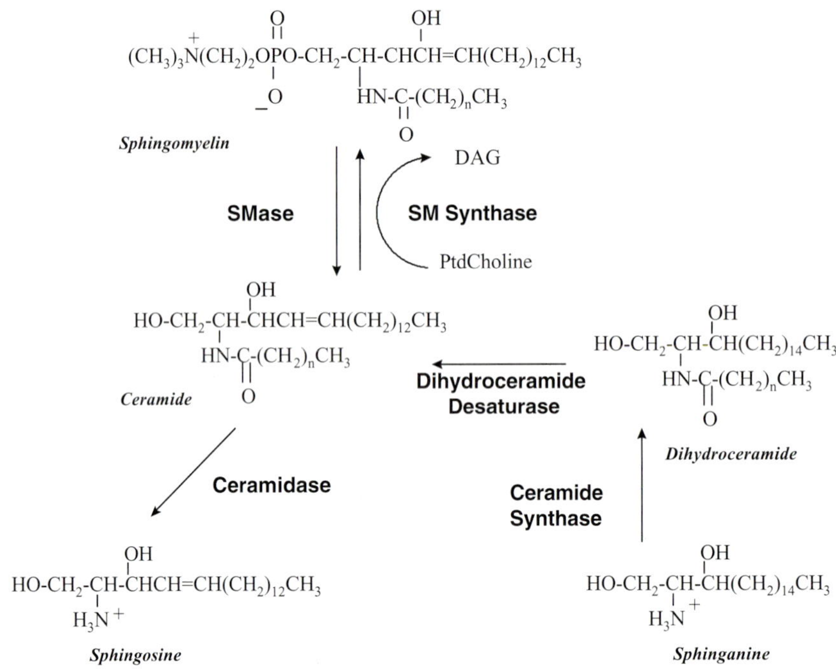

Fig. 2 Schematic overview of enzymatic interconversion of ceramide and sphingomyelin. SMase, sphingomyelinase; SM synthase, sphingomyelin synthase.

in radiation-induced apoptosis in specific cell types (Santana *et al.,* 1996; Haimovitz-Friedman *et al.,* 1997). Significantly, a novel pool of acid SMase that may be involved in signal transduction has been localized to caveolae and caveolae-related domains (Liu and Anderson, 1995; Dobrowsky and Gazula, 2000; Zundel *et al.,* 2000).

To date, the mechanisms that control ligand or stress-induced activation of acid SMase are not well defined. Several reports suggest that accessory/adapter proteins may be critical in mediating receptor coupling to acid SMase. For example, the interleukin-1 receptor accessory protein is required for the activation of acid SMase by interleukin-1α (Hofmeister *et al.,* 1997). However, the role of adapter proteins in acid SMase activation have been best characterized using the TNF ligand-receptor pair as the model system. Kronke and colleagues have demonstrated that the TNF-mediated activation of acid SMase utilizes the adapter protein TRADD but not other TNF receptor-associated proteins such as TRAF2 and RIP (Schwandner *et al.,* 1998). Moreover, coexpression of FADD, which interacts with TRADD, further increased TNF-induced acid SMase activity (Wiegmann *et al.,* 1999). Interestingly, expression of caspase-8, whose activation is mediated by binding to FADD, had no effect on TNF-induced acid SMase activation (Wiegmann *et al.,* 1999). These results suggest that FADD recruitment likely leads to acid SMase activation via a noncaspase 8-dependent mechanism.

3. Materials

[*choline-N-methyl*-^{14}C]Sphingomyelin (54.5 mCi/mmol) is purchased from American Radiolabeled Chemicals. Cold SM standard should be purchased from Avanti Polar Lipids. SM micelles are prepared by evaporating a total of 400 nmol of cold and radiolabeled SM in a screw cap glass tube. Add 1 ml of 0.2% Triton X-100 in 100 m*M* sodium acetate, pH 5.0, 2 m*M* EDTA, vortex, and sonicate in a bath sonicator. The specific activity of the [*choline-N-methyl*-^{14}C]SM should be adjusted to at least 2–2.5 $\times$ 10^3 dpm/nmol.

4. Principle of the Assay

The acid SMase assay measures the hydrolysis of [*choline-N-methyl*-^{14}C]phosphocholine from [*choline-N-methyl*-^{14}C]SM at pH 5.0 in the absence of added cations. Nonhydrolyzed [*choline-N-methyl*-^{14}C]SM is separated from the water-soluble [^{14}C] phosphocholine by simple lipid extraction. Enzymatic activity is assessed by quantitating the amount of [^{14}C]phosphocholine recovered in the upper aqueous phase.

5. Assay Procedure

Because immunoprecipitating antibodies to acid SMase are not readily available, most investigations have assessed this enzyme activity in whole cell lysates following ligand stimulation.

1. The cells are collected into a lysis buffer consisting of 0.2% Triton X-100, 100 m*M* sodium acetate, pH 5.0, 2 m*M* EDTA, 0.1 m*M* Na_3VO_4, 1 m*M* phenylmethylsulfonylfluoride (PMSF), and 10 μg/ml each of aprotinin and leupeptin. Adherent cells may be scraped directly into lysis buffer, whereas nonadherent cells should be centrifuged at 500*g* for 3 min and the cell pellet resuspended in buffer. After 15 min on ice, pulse sonicate the cells with a fine tip probe and centrifuge at 10,000*g* for 10 min. The resulting supernatant is used as a source for acid SMase activity (see later for a caveat to this procedure).
2. In a total volume of 0.1 ml the assay contains 20 nmol of [*choline-N-methyl*-^{14}C]SM and is initiated by the addition of up to 0.05 ml of the cell lysate. The assay is performed at 37°C and is terminated by the addition of 1.5 ml of $CHCl_3 : CH_3OH$ (2 : 1).
3. Phase partitioning is induced by adding 0.2 ml of water. Vortex and centrifuge the tube and transfer 0.4 ml of the upper aqueous layer to a scintillation vial for the quantitation of hydrolyzed [^{14}C]choline phosphate.

Because acid SMase activity varies dramatically between cell types and subcellular fractions, the duration of the assay is variable and should be adjusted such that no more than 10 to 15% of the substrate is hydrolyzed. During the collection of the aqueous phase, care should be taken to avoid contamination with the lower phase. To measure secreted forms of acid SMase, add 0.1 m*M* Zn^{2+} to the assay and delete EDTA from the buffers, assay conditions are similar to those described earlier. It should also be noted that a signal-activated pool of acid SMase may also localize to detergent-resistant and/or caveolar domains of the plasma membrane, which in turn would be sedimented following high-speed centrifugation of the whole cell lysate. To avoid this, either centrifuge the lysate at 800*g* or prepare detergent-resistant domains. Procedures for the analysis of SM hydrolysis and SMase activation in these domains have been described in detail elsewhere (Bilderback *et al.,* 1997; Dobrowsky and Gazula, 2000).

6. Molecular Probes, Antibodies, and Inhibitors

The gene for acid SMase has been cloned (Quintern *et al.,* 1989; Schuchman *et al.,* 1991; Lin *et al.,* 1998), and transgenic mice lacking the acid SMase have been generated (Otterbach and Stoffel, 1995; Santana *et al.,* 1996). Additionally, Niemann–Pick fibroblasts, which lack acid SMase, represent a potentially powerful cell model for studying the role of acid SMase (Andrieu *et al.,* 1994; Santana *et al.,* 1996).

Antibodies against lysosomal and secretory acid SMase have been generated but are not available commercially (Hurwitz *et al.,* 1994b). Moreover, no highly specific inhibitors of acid SMase are readily available. Although originally designed as a L-type Ca^{2+} channel blocker, SR33557,[(2-isopropyl-1-(4-[3-*N*-methyl-(3,4-dimethoxy-phenethyl) amino]propyloxy) benezenesulphonyl)]indolizine, is reportedly a specific inhibitor of acid versus neutral SMase and blocks TNF-induced apoptosis and NF-κB activation (Higuchi *et al.,* 1996). However, this molecule has not seen widespread use as a specific acid SMase inhibitor (Lee *et al.,* 1998). Desipramine may be considered an *in vivo* inhibitor of acid SMase as it induces the proteolytic degradation of the enzyme (Lee *et al.,* 1998; Hurwitz *et al.,* 1994a). However, use of this compound raises obvious concerns regarding specificity and is not appropriate for *in vitro* enzyme assays.

Recombinant human acid SMase has been shown to be expressed and secreted effectively from either *Sf*21 insect cells (Bartelsen *et al.,* 1998) or Chinese hamster ovary cells (He *et al.,* 1999). In both cases, the recombinant enzymes had physical and kinetic properties similar to those observed for the native mammalian enzyme. Thus, these expression systems may serve as excellent sources for large amounts of acid SMase for *in vitro* biochemical studies.

B. Neutral Sphingomyelinase

1. Background

Neutral SMase catalyzes the identical reaction as described for acid SMase. Because neutral SMase is inactive at low pH, this activity can be discriminated from acid SMase by performing the reaction at pH 7.4. Neutral SMase has been classically described as a membrane-associated, Mg^{2+}-dependent activity (Spence *et al.,* 1982), although a Mg^{2+}-independent cytosolic neutral SMase has also been described (Okazaki *et al.,* 1994). In most cells and tissues, the specific activity of neutral SMase is generally about 10-fold less than acid SMase. In contrast, neutral SMase activity is about 2- to 2.5-fold greater than acid SMase activity in gray and white matter of the cerebral cortex (Spence and Burgess, 1978).

2. Regulation of Signaling

Numerous cytokines and growth factors have been shown to activate the membrane-associated, Mg^{2+}-dependent neutral SMase. The localization of Mg^{2+}-dependent neutral SMase to the plasma membrane suggests a pivotal role in the SM pathway and has led to the intense study of potential mechanisms for its regulation. In general, at least three mechanisms may be involved in regulating the Mg^{2+}-dependent neutral SMase involving (1) a novel adapter protein, (2) manipulation of the cellular redox levels, or (3) protease activation. It should also be noted that the tyrosine phosphorylation of neutral SMase reportedly suppresses its activity (Nikolova-Karakashian *et al.,* 1997).

The WD repeat protein FAN has been implicated in coupling neutral SMase to the p55 TNF receptor (Adam-Klages *et al.,* 1996). Indeed, genetic knockout of FAN totally uncoupled TNF from the activation of neutral SMase and led to increased trans-epidermal water loss and a delay in cutaneous barrier repair (Kreder *et al.,* 1999).

Unfortunately, neutral SMase has been rather intractable to purification to homogeneity, rendering biochemical analysis problematic. A highly enriched enzyme preparation has been purified from rat brain that is dependent on phosphatidylserine for activity and is stimulated by dithiothreitol (Liu *et al.,* 1998b). Importantly, in agreement with *in vitro* results (Liu *et al.,* 1998a), the partially purified enzyme was also inhibited by glutathione. These observations have led to the hypothesis that the regulation of cellular glutathione levels may be a key control mechanism of neutral SMase (Liu *et al.,* 1998a,b; Levade and Jaffrezou, 1999).

Neutral SMase may also be subject to regulation by caspases and/or serine proteases [for a more comprehensive review, see Levade and Jafferzou (1999)]. For example, nitric oxide-induced activation of caspase-3 was required to stimulate neutral SMase in HL-60 cells (Takeda *et al.,* 1999). Interestingly, the addition of caspase-3, but not caspase-6, to cell extracts stimulated neutral SMase activity (Takeda *et al.,* 1999). Alternatively, environmental stimuli appear to increase neutral SMase and ceramide in a caspase-independent manner (Mathias *et al.,* 1998; Yoshimura *et al.,* 1998).

3. Materials

Radiolabeled SM micelles are prepared as described earlier except that the lipid residue is dissolved in 1 ml of 0.2% Triton X-100 in 100 m*M* Tris–HCl, pH 7.4, and 10 m*M* $MgCl_2$.

4. Principle of the Assay

This principle is identical to that described for acid SMase except that the reaction is performed at pH 7.4 in the presence of Mg^{2+} ions.

5. Assay Procedure

Similar to acid SMase, the enzymatic activity is typically assessed in whole cell lysates following ligand stimulation.

1. Because neutral SMase is sensitive to proteolysis and potentially to the phosphorylation state (Schutze *et al.*, 1992), the cells are resuspended in up to 0.5 ml of an ice-cold solution of 0.2% Triton X-100 in 20 m*M* Hepes (pH 7.4), 2 m*M* EDTA, 10 m*M* $MgCl_2$, 5 m*M* dithiothreitol, 0.1 m*M* Na_3VO_4, 10 m*M* β-glycerophosphate (prepare fresh), 0.75 m*M* ATP, 1 m*M* PMSF, and 10 μg/ml each of leupeptin and aprotinin. Prepare the lysate as described for acid SMase, keeping in mind that since the Mg^{2+}-dependent neutral SMase is a membrane-associated activity, it is advisable to perform only a low-speed centrifugation to remove nuclei and unbroken cells. Membrane fractions may then be prepared if desired by centrifugation at 100,000*g* for 30–60 min at 4°C.
2. In a total volume of 0.1 ml, the assay contains 20 nmol of [^{14}C]SM and is initiated by the addition of up to 0.05 ml of the cell lysate. The assay is performed at 37°C and is terminated by the addition of 1.5 ml of $CHCl_3 : CH_3OH$ (2 : 1).
3. Phase partitioning is induced by adding 0.2 ml of water. Vortex and centrifuge the tube and transfer 0.4 ml of the upper aqueous layer to a scintillation vial for the quantitation of hydrolyzed [^{14}C]choline phosphate.

Because neutral SMase activity is usually low in most cell extracts, assay times are between 1 and 3 h at 37°C. Attention should be given to assure the linearity of the reaction with respect to time and protein concentration such that no more than 10 to 15% of the substrate is hydrolyzed.

6. Molecular Probes, Antibodies, and Inhibitors

Putative murine and human homologs of Mg^{2+}-dependent neutral SMase have been cloned (Tomiuk *et al.*, 1998). However, the cloned enzyme was not activated by TNF in transfected HEK 293 cells, similar to other negative findings from this group with acid SMase (Zumbansen and Stoffel, 1997). Additionally, the cloned enzyme exhibited several characteristics not associated with the Mg^{2+} dependent neutral SMase purified from rat brain, e.g., lack of stimulation by dithiothreitol and an ability to utilize phosphatidylcholine as substrate (Liu *et al.*, 1998b; Tomiuk *et al.*, 1998). Moreover, Sawai *et al.* (1999) reported that the cloned enzyme does not act as a neutral SMase *in vivo* but that the protein functions as a lyso-platelet-activating factor-phospholipase C. Thus, it is unlikely that this enzyme is the ligand-activated neutral SMase.

Although antibodies have been generated against a neutral SMase isolated from human urine, the relatedness of this preparation to the signal-activated enzyme is also unclear. Furthermore, these antibodies are not generally available (Chatterjee and Ghosh, 1989).

Specific inhibitors of neutral SMase have not been well characterized. Gentamicin is known to inhibit this enzyme (Chatterjee, 1993), and 3-*O*-methyl-SM (Biomol) reportedly inhibits Mg^{2+}-dependent neutral SMase with an IC_{50} of about 50 μ*M* (Lister *et al.*, 1995). Scyphostatin (Sankyo Inc, Tokyo, Japan) is a specific inhibitor of Mg^{2+}-dependent neutral SMase with an IC_{50} of about 10 μ*M*; this drug has no effect on acid

SMase activity (Tanaka *et al.*, 1997). Interestingly, 1 μM sycphostatin blocked NGF-induced axonal outgrowth in developing hippocampal neurons (Brann *et al.*, 1999).

C. Ceramide Synthase (Sphinganine-*N*-acyltransferase)

1. Background

Although the apoptotic effect of ceramide produced by SM catabolism is well appreciated, apoptotic pools of ceramide may also be produced through *de novo* synthesis (Fig. 2). Ceramide synthase is required for the *de novo* synthesis of ceramide and catalyzes the acylation of sphinganine (dihydrosphingosine) to produce dihydroceramide (Merrill and Jones, 1990). Because dihydroceramide is not apoptogenic, it must be oxidized to ceramide by the introduction of a *trans*-4,5 double bond by the enzyme dihydroceramide desaturase (Michel *et al.*, 1997).

Enzymatically, ceramide synthase activity is typically determined using microsomal membrane preparations coincubated with sphinganine and radiolabeled palmitoyl-coenzyme A. The resulting radiolabeled dihydroceramide is isolated by TLC and quantified.

2. Regulation of Signaling

The ceramide synthase pathway can be stimulated by drugs and ionizing radiation, typically over a period of hours (Mathias and Kolesnick, 1998). However, mechanisms controlling the activation of the enzyme by any stimuli have not yet been determined, although a signal from damaged DNA may be involved. In this regard, the metabolic incorporation of ^{125}I-labeled 5-iodo-2′deoxyuridine ([^{125}I]dURd), which produces DNA double strand breaks, signaled *de novo* ceramide synthesis by the posttranslational activation of ceramide synthase (Liao *et al.*, 1999).

Ceramide synthase activity has been detected in the endoplasmic reticulum (Mandon *et al.*, 1992) and mitochondria (Shimeno *et al.*, 1995). Although the endoplasmic reticulum localization likely relates to ceramide synthesis for the generation of higher sphingolipids, the role of ceramide synthase in mitochondrial function is unknown. Whether ceramide synthase may provide a localized stress signal to mitochondria, which in turn play a central role during the commitment phase of the apoptotic process (Zamzami *et al.*, 1998), is presently under investigation. It will be important to determine if the mitochondrial pool of enzyme is regulated by either oxidative modifications and/or interactions with Bcl-2 family members. A better understanding of the mechanisms regulating ceramide synthase action should help elucidate the role of ceramide in some forms of stress-induced apoptosis.

3. Materials

Ceramide type III, palmitoyl-coenzyme A, and fatty acid free bovine serum albumin are obtained from Sigma Chemical Co. Sphinganine is from Matreya Biochemicals. [1-^{14}C]palmitoyl-coenzyme A (55 mCi/mmol) is from American Radiolabeled

Chemicals. The chemical nomenclature notes that the radiolabel is in the C-1 carboxyl group of palmitate. Ensure that the label is located in the fatty acyl moiety and not the coenzyme A, as this portion of the molecule is lost during the transferase reaction. Prepare a solution of 1 m*M* palmitoyl CoA containing 2 μCi/ml of [1-^{14}C]palmitoyl CoA. A stock solution of 2 m*M* sphinganine in 100% ethanol needs to be prepared fresh every 3–7 days.

4. Principle of the Assay

The ceramide synthase assay quantitates the fatty acyl CoA-dependent formation of [^{14}C]dihydroceramide from sphinganine and [1-^{14}C]palmitoyl CoA. Following lipid extraction, dihydroceramide is separated from other reaction components by TLC. Based on comigration with authentic standard, the dihydroceramide is scraped from the TLC plate and quantitated by scintillation counting.

5. Assay Procedure

The following assay of ceramide synthase activity is adapted from the procedures of Harel and Futerman (1993) and Merrill and Wang (1992).

1. Cells are homogenized in a buffer of 25 m*M* Hepes, pH 7.4, 5 m*M* EGTA, 50 m*M* NaF, 10 μg/ml leupeptin, and 10 μg/ml soybean trypsin inhibitor. Remove unbroken cells and nuclei by centrifugation at 800*g* for 5 min at 4°C.

2. Microsomal membranes are prepared by centrifuging the postnuclear supernatant at 100,000*g* for 30 min. Resuspend the microsomal membrane pellet in up to 1 ml of cold homogenization buffer and determine the protein concentration. Membranes should be prepared fresh on the day of the experiments.

3. In a final volume of 1 ml, the reaction mixture contains 0.2–20 μ*M* of sphinganine, obtained by evaporating aliquots of the 2 m*M* sphinganine stock solution under N_2 gas. As quickly as possible after evaporation of the solvent, resuspend the lipid residue in 2 m*M* $MgCl_2$, 20 m*M* Hepes, pH 7.4, 0.5 m*M* dithiothreitol, 20 μ*M* defatted (fatty acid free) bovine serum albumin by mixing and/or sonication. Add 75–200 μg of microsomal membrane protein to the reaction mixture and initiate the reaction by the addition of 0.1 ml of 1 m*M* [1-^{14}C]palmitoyl CoA.

4. After 1 h at 37°C, the reaction is terminated by the addition of 2 ml of chloroform : methanol (1 : 2). The upper phase is aspirated carefully and 500 μl of the lower phase is removed and the solvent evaporated. The resulting lipid film is resuspended in 50 μl of chloroform : methanol (1 : 1) containing 1 mg/ml of ceramide (type III from bovine brain) and 1 mg/ml diacylglycerol. Forty microliters of this solution is loaded onto a silica gel thin-layer chromatography plate.

5. Dihydroceramide is resolved from free radiolabeled fatty acid using a solvent system of chloroform : methanol : 3.5 *N* aqueous ammonium hydroxide (85 : 15 : 1), identified by iodine vapor staining based on comigration with ceramide type III standards, and quantified by liquid scintillation counting. Relevant R_f values in this system include

palmitoyl-CoA, 0.0; sphingosine, 0.17; palmitic acid, 0.1–0.2; ceramide, 0.66; and diacylglycerol, 0.78.

It should be noted that important controls include microsomes that have been boiled for 20 min and samples containing enzyme but lacking sphinganine. The velocity of the reaction is linear for at least 2 h and the amount of palmitoyl-coenzyme A consumed should not exceed 5–10% of the total.

6. Molecular Probes, Antibodies, and Inhibitors

Ceramide synthase is a membrane-bound enzyme that has not been purified to homogeneity and has not yet been identified molecularly. Thus, antibodies are unavailable. The most powerful tool for identifying the biologic role of ceramide generated by ceramide synthase is to inhibit the enzyme specifically with the fungal toxin fumonisin B1 (Merrill *et al.,* 1993). Fumonisin B1 is a sphingoid base analog that acts as a potent competitive inhibitor of ceramide synthase with respect to both sphinganine and fatty acyl CoA; the IC_{50} is 100-fold less than the K_m for sphingosine (Merrill *et al.,* 1993). Additionally, it has also been determined that an acylated analog of the fumonisin B1 backbone, *N*-palmitoyl-aminopentol$_1$ (PAP_1), is a potent inhibitor of ceramide synthase with an IC_{50} of about 10 μM (Humpf *et al.,* 1998).

D. Sphingomyelin Synthase

1. Background

One of the most enigmatic enzymes involved in regulating intracellular ceramide levels is phosphatidylcholine : ceramide phosphocholine transferase or SM synthase. This enzyme catalyzes the transfer of the phosphocholine head group from phosphatidylcholine (PtdCholine) to ceramide generating SM and diacylglycerol (Fig. 2). The enzyme is localized primarily in the cis/medial Golgi and the plasma membrane (Miro *et al.,* 1997; van Helvoort *et al.,* 1997).

2. Regulation of Signaling

Overall, little is known regarding the specific regulation of SM synthase, although it may be responsible for resynthesizing SM following TNF treatment of human fibroblasts (Andrieu-Abadie *et al.,* 1998).

3. Materials

[1-^{14}C]*N*-Hexanoyl-D-*erythro*-sphingosine is purchased from American Radiolabeled Chemicals. The chemical nomenclature indicates that the radiolabel is in the C-1 carboxyl group of the hexanoic acid. Prepare a 50 m*M* stock solution in ethanol to give a specific activity of about 1×10^4 cpm/nmol. Prepare a ceramide : BSA complex in a 1 : 1 molar ratio by injecting 10 μl of the ethanol solution into 1 ml of 0.5 m*M* fatty acid-free BSA, vortex vigorously, and incubate the solution at 37°C for 30 min (Nikolova-Karakashian *et al.*, 1997). Alternatively, L-α-dipalmitoyl [*choline-N–methyl-*^{3}H]phosphatidylcholine may be used as the phosphocholine donor (Luberto and Hannun, 1998).

4. Principle of the Assay

The assay quantitates the formation of radiolabeled SM following the transfer of phosphocholine from PtdCholine. The reaction mix is subjected to a lipid extraction and the radiolabeled SM is separated from the [^{14}C]ceramide and other components by TLC. The radiolabeled SM is recovered from the TLC plate and quantitated by scintillation counting.

5. Assay

The following assay is based on the procedure of Luberto and Hannun (1998).

1. Prepare a cell lysate by homogenization in 25 m*M* Tris–HCl, pH 7.4, 5 m*M* EDTA, 1 m*M* PMSF, and 10 μg/ml each of leupeptin and aprotinin. Spin the lysate at 800*g* for 10 min at 4°C and subject the postnuclear supernatant to centrifugation at 100,000*g* for 1 h at 4°C. Resuspend the membrane pellet in lysis buffer and determine the protein concentration.
2. In a total volume of 0.5 ml, preincubate 50–200 μg of protein in 50 m*M* Tris–HCl, pH 7.4, 25 m*M* KCl, and 0.5 m*M* EDTA at 37°C. Some authors have found that the addition of up to 5 m*M* $MnCl_2$ enhances activity (Marggraf *et al.*, 1981). Initiate the reaction by the addition of 40 μl of the radiolabeled substrate (20 nmol) and continue the incubation for 1 h.
3. Terminate the reaction by adding 3 ml of $CHCl_3 : CH_3OH$ (1 : 2) and 0.3 ml of water. Complete the extraction by the addition of 1 ml of $CHCl_3$ and 1 ml of water. Vortex, centrifuge, and recover the lower organic phase. Quantitatively transfer an aliquot to a fresh tube and evaporate the solvent.
4. Resuspend the lipid residue in 40 μl of cold $CHCl_3$ and apply 30 μl to a TLC plate. Resolve the radioactive lipid products by developing the plate in $CHCl_3 : CH_3OH$: 15 m*M* anhydrous $CaCl_2$ (60 : 35 : 8). Visualize the lipids by autoradiography after spraying the plate with En^{3}Hance. If using [^{3}H]*N*-acetylsphingosine as the substrate the radiolabeled short chain SM can be identified by comparison to the migration of a C_2-SM standard (Matreya).

Similar to the ceramide synthase assay, a boiled protein control is used for the determination of background radioactivity.

6. Molecular Probes, Antibodies, and Inhibitors

SM synthase has not been purified to homogeneity and specific antiserum is not available. The putative phosphatidylcholine phospholipase C inhibitor, D609, also inhibits

SM synthase with an IC_{50} of about 100 μg/ml (~0.4 mM). Thus, data generated with the use of this inhibitor must be interpreted cautiously, as the biochemical and molecular identities of both of these enzymes are poorly characterized.

E. Acid and Alkaline Ceramidase

1. Background

Ceramidases catalyze the deacylation of ceramide to free fatty acid and sphingosine (Merrill and Jones, 1990). Several forms of ceramidase have been biochemically identified based on their pH optima for hydrolysis, acidic, pH 4.5; neutral, pH 7.4; and alkaline, pH 9.0 (Coroneos *et al.,* 1995; Nikolova-Karakashian *et al.,* 1997). Interestingly, the expression of alkaline ceramidase may be tissue specific as this activity has been detected in cerebellum but not kidney extracts (Sugita *et al.,* 1975b). Additionally, a novel membrane-bound nonlysosomal ceramidase with a broad neutral to alkaline pH optimum has been purified from rat brain (Bawab *et al.,* 1999).

2. Regulation of Signaling

Acid ceramidase is known to be a lysosomal enzyme that functions in the biodegradation of sphingolipids. However, unpublished information indicates that acid ceramidase, like acid SMase, may exist in a secreted form. Deficiency in the activity of this enzyme, which results from a single amino acid substitution, leads to the lysosomal storage disorder known as Farber's disease (Koch *et al.,* 1996). Emerging evidence suggests that acid ceramidase may be specifically activated or inhibited by cytokines and growth factors and, in turn, regulate cellular ceramide levels. For example, interleukin-1β-induced ceramide elevation is attenuated by increased acid ceramidase activity. The products of ceramidase action, sphingosine and its phosphorylated form sphingosine-1-phosphate, are also sphingolipid second messengers (Nikolova-Karakashian *et al.,* 1997). The activation of alkaline ceramidase by growth factors, but not cytokines, has similarly been implicated in promoting the mitogenic and pro-survival potential of these proteins by keeping ceramide levels low (Coroneos *et al.,* 1995). Indeed, activation of acid ceramidase by epidermal growth factor inhibited increases in cellular ceramide and blocked apoptosis in primary placental trophoblasts cotreated with TNF and interferon-γ (Payne *et al.,* 1999). However, inhibition of ceramidase activity by cell stress signals may also reduce survival. Nitric oxide-induced apoptosis in mesangial cells was associated with the activation of SMase activity (increasing ceramide synthesis) and an inhibition of both acid and neutral ceramidase activity (decreasing ceramide degradation), leading to prolonged increases in steady-state ceramide levels (Huwiler *et al.,* 1999). Alternately, TNF elevated ceramide synthesis rapidly by stimulating SMase activity but also stimulated its degradation by increasing ceramidase activity. The concurrent increases in SMase and ceramidase activities lead to essentially a steady state in ceramide levels; this correlated with the inability of TNF to induce apoptosis in mesangial cells (Huwiler *et al.,* 1999). The importance of acid ceramidase activation in protecting cells from ceramide-induced apoptosis is underscored by the observation that an inhibitor of acid ceramidase, *N*-oleoylethanolamine, blocked the TNF-induced increase in acid ceramidase activity and enabled TNF to induce apoptosis in mesangial cells. Collectively, these data suggest that ceramidases are likely to be under tight regulation and that these enzymes are critical to affecting the cellular responses to ceramide production.

3. Materials

[1-^{14}C]Palmitoyl sphingosine may be purchased from American Radiolabeled Chemicals. The chemical nomenclature indicates that the radiolabel is in the C-1 carboxyl group of the palmitic acid. Prepare a 1 m*M* stock solution of the substrate in $CHCl_3$ and adjust the specific activity to about 2–3 × 10^3 cpm/nmol. Prepare separate stock solutions of 0.1% Triton X-100 and 0.2% sodium cholate both in $CHCl_3 : CH_3OH$ (2 : 1).

4. Principle of the Assay

The assay measures the release of radiolabeled fatty acid from ceramide at acidic, neutral, or alkaline pH in the presence of Mg^{2+} cations. Using a modified Dole extraction, the released fatty acid is recovered by partitioning into heptane following protonation. Therefore, it is critical to use a ceramide substrate labeled in the fatty acyl moiety.

5. Assay

The assay is based on the procedure of Yavin and Gatt (1969) as modified by Bielawska *et al.* (1996).

1. Prepare a cell lysate by homogenization in 50 m*M* Hepes, pH 9.5 (alkaline), 50 m*M* Tris–HCl, pH 7.4 (neutral), or 50 m*M* sodium acetate, pH 4.5 (acid), containing 1 m*M* EDTA, 1 m*M* PMSF, and 10 μg/ml each of leupeptin and aprotinin. Spin the lysate at 800*g* for 10 min at 4°C and subject the postnuclear supernatant to centrifugation at 100,000*g* for 1 h at 4°C. Resuspend the membrane pellet in the same buffer and determine the protein concentration.
2. To each assay tube, add 0.1 ml each of 0.1% Triton X-100 and 0.2% sodium cholate detergent stock solutions. Thereafter, add 10 nmol of radiolabeled substrate, mix the solution well, and evaporate the solvent.
3. Resuspend the residue in 0.1 ml of 50 m*M* Hepes, pH 9.5 (alkaline), 50 m*M* Tris–HCl, pH 7.4 (neutral), or 50 m*M* sodium acetate, pH 4.5 (acid), containing 10 m*M* $MgCl_2$ and vortex the tubes vigorously.
4. Initiate the reaction by the addition of up to 0.1 mg of protein and incubate for 1 h at 37°C. Terminate the reaction by the addition of 2 ml of isopropyl alcohol : heptane : 2 *N* NaOH (78 : 20 : 2) and mix well.
5. Add 50 μg of cold palmitic acid as carrier lipid and induce phase partitioning by the addition of 1.2 ml of heptane and 1 ml of water. After centrifugation, discard the upper heptane phase. Wash the lower phase with 2 × 1-ml aliquots of heptane and discard the

upper heptane phase after each centrifugation. Acidify the lower phase with 1 ml of 1 *N* H_2SO_4 to protonate the released fatty acid. Add 2 ml of heptane, vortex, and centrifuge. Transfer 1.5 ml of the upper phase to a scintillation vial and count. Heptane is not a quenching agent and scintillation cocktail may be added directly without evaporating the solvent.

6. Molecular Probes, Antibodies, and Inhibitors

The gene encoding acid ceramidase has been cloned (Koch *et al.,* 1996). The protein is composed of α and β subunits and contains a potential tyrosine phosphorylation site at Tyr 307 (Koch *et al.,* 1996). Sandhoff and colleagues have generated antibodies against acid ceramidase, although they are not available commercially (Bernardo *et al.,* 1995).

N-Oleoylethanolamine is an inhibitor of acid ceramidase and has been used in both *in vivo* and *in vitro* assays despite the fact that the IC_{50} is about 0.5 m*M* (Sugita *et al.,* 1975a; Bielawska *et al.,* 1996; Wiesner and Dawson, 1996).

A new and potent inhibitor of alkaline ceramidase has been characterized: (1*S*,2*R*-D-erythro-2-(*N*-myristoylamino)-1-phenyl-1-propanol (D-*e*-MAPP) (Bielawska *et al.,* 1996). The IC_{50} for inhibition of alkaline ceramidase *in vitro* is 1–5 μM with little effect on acid ceramidase; the effect of D-MAPP on neutral ceramidase was not reported. Treatment of cells with D-*e*-MAPP increased intracellular ceramide levels up to threefold, providing a powerful tool for examining the role of endogenous ceramide in signal transduction events. Importantly, the stereoisomer, L-*e*-MAPP, is inactive in the inhibition of ceramidase and can serve as an excellent negative control when using D-*e*-MAPP to investigate the role of endogenous ceramide in signaling. All the just-mentioned inhibitors are available commercially (Matreya Biochemicals).

References

Adam-Klages, A., Adam, D., Weigmann, K., Struve, S., Kolanus, W., Schneider-Mergener, J., and Kronke, M. (1996). FAN, a novel WD-repeat protein, couples the p55 TNF-receptor to neutral sphingomyelinase. *Cell* **86,** 937–947.

Albi, E., Peloso, I., and Magni, M. P. (1997). Nuclear membrane sphingomyelin-cholesterol changes in rat liver after hepatectomy. *Biochem. Biophys. Res. Commun.* **236,** 29–33.

Ames, B. N., and Dubin, D. T. (1960). The role of polyamines in the neutralization of bacteriophage deoxyribonucleic acid. *J. Biol. Chem.* **235,** 769–775.

Andrieu, N., Salvayre, R., and Levade, T. (1994). Evidence against involvement of the acid lysosomal sphingomyelinase in the tumor-necrosis-factor- and interleukin-1-induced sphingomyelin cycle and cell proliferation in human fibroblasts. *Biochem. J.* **303,** 341–345.

Andrieu-Abadie, N., Carpentier, S., Salvayre, R., and Levade, T. (1998). The tumor necrosis factor-sensitive pool of sphingomyelin is resynthesized in a distinct compartment of the plasma membrane. *Biochem. J.* **333,** 91–97.

Ariga, T., Jarvis, W. D., and Yu, R. K. (1998). Role of sphingolipid-mediated cell death in neurodegenerative diseases. *J. Lipid. Res.* **39,** 1–16.

Bartelsen, O., Lansmann, S., Nettersheim, M., Lemm, T., Ferlinz, K., and Sandhoff, K. (1998). Expression of recombinant human acid sphingomyelinase in insect *Sf*21 cells: Purification, processing and enzymatic characterization. *J. Biotechnol.* **63,** 29–40.

Bernardo, K., Hurwitz, R., Zenk, T., Desnick, R. J., Ferlinz, K., Schuchman, E. H., and Sandhoff, K. (1995). Purification, characterization, and biosynthesis of human acid ceramidase. *J. Biol. Chem.* **270,** 11098–11102.

Bielawska, A., Greenberg, M. S., Perry, D., Jayadev, S., Shayman, J. A., McKay, C., and Hannun, Y. A. (1996). (1S,2R)-D-erythro-2-(N-myristoylamino)-1-phenyl-1-propanol as an inhibitor of ceramidase. *J. Biol. Chem.* **271,** 12646–12654.

Bilderback, T. R., Grigsby, R. J., and Dobrowsky, R. T. (1997). Association of p75^{NTR} with caveolin and localization of neurotrophin-induced sphingomyelin hydrolysis to caveolae. *J. Biol. Chem.* **272,** 10922–10927.

Bligh, E. G., and Dyer, W. J. (1959). A rapid method of total lipid extraction and purification. *Can. J. Biochem. Physiol.* **37,** 911–917.

Bose, R., Chen, P., Loconti, A., Grullich, C., Abrams, J. M., and Kolesnick, R. N. (1998). Ceramide generation by the Reaper protein is not blocked by the caspase inhibitor, p35. *J. Biol. Chem.* **273,** 28852–28859.

Bose, R., Verheji, M., Haimovitz-Friedman, A., Scotto, K., Fuks, Z., and Kolesnick, R. N. (1995). Ceramide synthase mediates danorubicin-induced apoptosis: An alternative mechanism for generating death signals. *Cell* **82,** 405–414.

Brann, A. B., Scott, R., Neuberger, Y., Abulafia, D., Boldin, S., Fainzilber, M., and Futerman, A. H. (1999). Ceramide signaling downstream of the p75 neurotrophin receptor mediates the effects of nerve growth factor on outgrowth of cultured hippocampal neurons. *J. Neurosci.* **19,** 8199–8206.

Chatterjee, S. (1993). Neutral sphingomyelinase. *Adv. Lipid Res.* **26,** 25–48.

Chatterjee, S., and Ghosh, N. (1989). Neutral sphingomyelinase from human urine: Purification and preparation of monospecific antibodies [published erratum appears in *J. Biol. Chem.* 1990 Jan 15;**265**(2):1231]. *J. Biol. Chem.* **264,** 12554–12561.

Coroneos, E., Martinez, M., McKenna, S., and Kester, M. (1995). Differential regulation of sphingomyelinase and ceramidase activities by growth factors and cytokines. *J. Biol. Chem.* **270,** 23305–23309.

Dobrowsky, R. T., and Gazula, V.-R. (2000). Analysis of sphingomyelin hydrolysis in caveolar membranes. *Methods Enzymol.* **311,** 184–193.

Garzotto, M., White-Jones, M., Jiang, Y., Ehleiter, D., Liao, W. C., Haimovitz-Friedman, A., Fuks, Z., and Kolesnick, R. (1998). 12-O-tetradecanoylphorbol-13-acetate-induced apoptosis in LNCaP cells is mediated through ceramide synthase. *Cancer Res.* **58,** 2260–2264.

Haimovitz-Friedman, A., Cordon-Carlo, C., Bayoumy, S., Garzotto, M., McLoughlin, M., Gallily, R., Edwards, C. K., III, Schuchman, E. H., Fuks, Z., and Kolesnick, R. N. (1997). Lipopolysaccharide induces disseminated endothelial apoptosis requiring ceramide formation. *J. Exp. Med.* **186,** 1831–1834.

Hannun, Y. A. (1996). Functions of ceramide in coordinating cellular response to stress. *Science* **274,** 1855–1859.

Harel, R., and Futerman, A. H. (1993). Inhibition of sphingolipid synthesis affects axonal outgrowth in cultured hippocampal neurons. *J. Biol. Chem.* **268,** 14476–14481.

He, X., Miranda, S. R. P., Xiong, X., Dagan, A., Gatt, S., and Schuchman, E. H. (1999). Characterization of human acid sphingomyelinase purified from the media of overexpressing Chinese hamster ovary cells. *Biochim. Biophys. Acta* **1432,** 251–264.

Higuchi, M., Singh, S., Jaffrezou, J. P., and Aggarwal, B. B. (1996). Acidic sphingomyelinase-generated ceramide is needed but not sufficient for TNF-induced apoptosis and nuclear factor-kappa B activation. *J. Immunol.* **157,** 297–304.

Hofmeister, R., Wiegmann, K., Korherr, C., Bernado, K., Kronke, M., and Falk, W. (1997). Activation of acid sphingomyelinase by interleukin-1 (IL-1) requires the IL-1 receptor accessory protein. *J. Biol. Chem.* **272,** 27730–27736.

Humpf, H. U., Schmelz, E. M., Meredith, F. I., Vesper, H., Vales, T. R., Wang, E., Menaldino, D. S., Liotta, D. C., and Merrill, A. H. J. (1998). Acylation of naturally occurring and synthetic 1-deoxysphinganines by ceramide synthase: Formation of N-palmitoyl-aminopentol produces a toxic metabolite of hydrolyzed fumonisin, AP1, and a new category of ceramide synthase inhibitor. *J. Biol. Chem.* **273,** 19060–19064.

Hurwitz, R., Ferlinz, K., and Sandhoff, K. (1994a). The tricyclic antidepressant desipramine causes proteolytic degradation of the lysosomal sphingomyelinase in human fibroblasts. *Biol. Chem. Hoppe-Seyler* **375,** 447–450.

Hurwitz, R., Ferlinz, K., Vielhaber, G., Moczall, H., and Sandhoff, K. (1994b). Processing of human acid sphingomyelinase in normal and I-cell fibroblasts. *J. Biol. Chem.* **269,** 5440–5445.

Huwiler, A., Pfeilschifter, J., and van den Bosch, H. (1999). Nitric oxide donors induce stress signaling via ceramide formation in rat mesangial cells. *J. Biol. Chem.* **274,** 7190–7195.

Jayadev, S., Liu, B., Bielawska, A. E., Lee, J. Y., Nazaire, F., Pushkareva, M. Yu., Obeid, L. M., and Hannun, Y. A. (1995). Role for ceramide in cell cycle arrest. *J. Biol. Chem.* **270,** 2047–2052.

Koch, J., Gartner, S., Li, C. M., Quintern, L. E., Bernardo, K., Levran, O., Schnabel, D., Desnick, R. J., Schuchman, E. H., and Sandhoff, K. (1996). Molecular cloning and characterization of a full-length complementary DNA encoding human acid ceramidase: Identification of the first molecular lesion causing Farber disease. *J. Biol. Chem.* **271,** 33110–33115.

Kolesnick, R. N., and Kronke, M. (1998). Regulation of ceramide production and apoptosis. *Annu. Rev. Physiol.* **60,** 643–665.

Lee, J. Y., Leonhart, L. G., and Obeid, L. M. (1998). Cell-cycle-dependent changes in ceramide levels preceding retinoblastoma protein dephosphorylation in G_2/M. *Biochem. J.* **334,** 457–461.

Levade, T., and Jaffrezou, J.-P. (1999). Signalling sphingomyelinases: Which, where, how and why? *Biochim. Biophys. Acta* **1438,** 1–17.

Lin, X., Hengartner, M. O., and Kolesnick, R. (1998). *Caenorhabditis elegans* contains two distinct acid sphingomyelinases. *J. Biol. Chem.* **273,** 14374–14379.

Linardic, C. M., and Hannun, Y. A. (1994). Identification of a distinct pool of sphingomyelin involved in the sphingomyelin cycle. *J. Biol. Chem.* **269,** 23530–23537.

Liao, W. C., Haimovitz-Friedman, A., Persaud, R. S., McLoughlin, M., Ehleiter, D., Zhang, N., Gatei, M., Lavin, M., Kolesnick, R., and Fuks, Z. (1999). Ataxia teleangiectasia-mutated gene product inhibits DNA damage-induced apoptosis via ceramide synthase. *J. Biol. Chem.* **274,** 17908–17917.

Lister, M. D., Ruan, Z. S., and Bittman, R. (1995). Interaction of sphingomyelinase with sphingomyelin analogs modified at the C-1 and C-3 positions of the sphingosine backbone. *Biochim. Biophys. Acta* **1256,** 25–30.

Liu, B., Andrieu-Abadie, N., Levade, T., Zhang, P., Obeid, L. M., and Hannun, Y. A. (1998a). Glutathione regulation of neutral sphingomyelinase in tumor necrosis factor-alpha-induced cell death. *J. Biol. Chem.* **273,** 11313–11320.

Liu, B., Hassler, D. F., Smith, G. K., Weaver, K., and Hannun, Y. A. (1998b). Purification and characterization of a membrane bound neutral pH optimum magnesium-dependent and phosphatidylserine-stimulated sphingomyelinase from rat brain. *J. Biol. Chem.* **273,** 34472–34479.

Liu, P., and Anderson, R. G. W. (1995). Compartmentalized production of ceramide at the cell surface. *J. Biol. Chem.* **270,** 27179–27185.

Luberto, C., and Hannun, Y. A. (1998). Sphingomyelin synthase, a potential regulator of intracellular levels of ceramide and diacylglycerol during SV40 transformation. *J. Biol. Chem.* **273,** 14550–14559.

Mandon, E. C., Ehses, I., Rother, J., van Echten, G., and Sandhoff, K. (1992). Subcellular localization and membrane topology of serine palmitoyltransferase, 3-dehydrosphinganine reductase, and sphinganine N-acyltransferase in mouse liver. *J. Biol. Chem.* **267,** 11144–11148.

Marathe, S., Schissel, S. L., Yellin, M. J., Beatini, N., Mintzer, R., Williams, K. J., and Tabas, I. (1998). Human vascular endothelial cells are a rich and regulatable source of secretory sphingomyelinase: Implications for early atherogenesis and ceramide-mediated cell signaling. *J. Biol. Chem.* **273,** 4081–4088.

Marggraf, W.-D., Anderer, F. A., and Kanfer, J. N. (1981). The formation of sphingomyelin from phosphatidylcholine in plasma membrane preparations from mouse fibroblasts. *Biochim. Biophys. Acta* **664,** 61–73.

Mathias, S., and Kolesnick, R. N. (1998). Signal transduction of stress via ceramide. *Biochem. J.* **335,** 465–480.

Merrill, A. H. J., and Wang, E. (1992). Enzymes of ceramide biosynthesis. *Methods Enzymol.* **51,** 427–437.

Merrill, A. H. J., Wang, E., Mullins, R. E., Jamison, W. C., Nimkar, S., and Liotta, D. C. (1988). Quantitation of free sphingosine in liver by high-performance liquid chromatography. *Anal. Biochem.* **171,** 373–381.

Merrill, A. H., Jr., and Jones, D. D. (1990). An update of the enzymology and regulation of sphingomyelin metabolism. *Biochim. Biophys. Acta* **1044,** 1–12.

Merrill, A. H., Jr., van Echten, G., Wang, E., and Sandhoff, K. (1993). Fumonisin B_1 inhibits sphingosine (sphinganine) *N*-acyltransferase and *de novo* sphingolipid biosynthesis in cultured neurons *in situ*. *J. Biol. Chem.* **268,** 27299–27306.

Michel, C., van Echten-Deckert, G., Rother, J., Sandhoff, K., Wang, E., and Merrill, A. H., Jr. (1997). Characterization of ceramide synthesis. *J. Biol. Chem.* **272,** 22432–22437.

Miro, O. M., Sillence, D., Howitt, S., and Allan, D. (1997). The subcellular sites of sphingomyelin synthesis in BHK cells. *Biochim. Biophys. Acta* **1359,** 1–12.

Nikolova-Karakashian, M., Morgan, E. T., Alexander, C., Liotta, D. C., and Merrill, A. H., Jr. (1997). Bimodal regulation of ceramidase by interleukin-1β: Implications for the regulation of cytochrome p450 2C11. *J. Biol. Chem.* **272,** 18718–18724.

Okazaki, T. O., Bielawska, A., Domae, N., Bell, R. M., and Hannun, Y. A. (1994). Characteristics and partial purification of a novel cytosolic, magnesium-independent, neutral sphingomyelinase activated in the early signal transduction of 1α,25-dihydroxyvitamin D_3-induced HL-60 cell differentiation. *J. Biol. Chem.* **269,** 4070–4077.

Otterbach, B., and Stoffel, W. (1995). Acid sphingomyelinase-deficient mice mimic the neurovisceral form of human lysosomal storage disease (Niemann-Pick disease). *Cell* **81,** 1053–1061.

Payne, S. G., Brindley, D. N., and Guilbert, L. J. (1999). Epidermal growth factor inhibits ceramide-induced apoptosis and lowers ceramide levels in primary placental trophoblasts. *J. Cell Physiol.* **180,** 263–270.

Perry, D. K., and Hannun, Y. A. (1999). The use of diglyceride kinase in quantitating ceramide: Is it really valid? *TIBS* **24,** 226–227.

Petitou, M., Tuy, F., and Rosenfeld, C. (1978). A simplified procedure for organic phosphorus determination from phospholipids. *Anal. Biochem.* **91,** 350–353.

Priess, J., Loomis, C. R., Bell, R. M., and Neidel, J. E. (1987). Quantitative measurement of sn-12, diacylglycerols. *Methods Enzymol.* **141,** 294–299.

Quintern, L. E., Schuchman, E. H., Levran, O., Suchi, M., Ferlinz, K., Reinke, H., Sandhoff, K., and Desnick, R. J. (1989). Isolation of cDNA clones encoding human acid sphingomyelinase: Occurrence of alternatively processed transcripts. *EMBO J.* **8,** 2469–2473.

Rustenbeck, I., and Lenzen, S. (1990). Quantitation of hexadecylphosphocholine by high performance thin layer chromatography with densitometry. *J. Chromatogr.* **525,** 85–91.

Santana, P., Pena, L. A., Haimovitz-Friedman, A., Martin, S., Green, D., McLoughlin, E. H., Cordon-Cardo, C., Schuchman, E. H., Fuks, Z., and Kolesnick, R. (1996). Acid sphingomyelinase-deficient human lymphoblasts and mice are defective in radiation-induced apoptosis. *Cell* **86,** 189–199.

Sawai, H., Domae, N., Nagan, N., and Hannun, Y. A. (1999). Function of the cloned putative neutral sphingomyelinase as lyso-platelet activating factor-phospholipase C. *J. Biol. Chem.* **274,** 38131–38139.

Schissel, S. L., Schuchman, E. H., Williams, K. J., and Tabas, I. (1996). Zn^{2+}-stimulated sphingomyelinase is secreted by many cell types and is a product of the acid sphingomyelinase gene. *J. Biol. Chem.* **271,** 18431–18436.

Schuchman, E. H., Suchi, M., Takahashi, T., Sandhoff, H., and Desnick, R. J. (1991). Human acid sphingomyelinase. *J. Biol. Chem.* **266,** 8531–8539.

Schwandner, R., Wiegmann, K., Bernardo, K., Kreder, D., and Kronke, M. (1998). TNF receptor death domain-associated proteins TRADD and FADD signal activation of acid sphingomyelinase. *J. Biol. Chem.* **273,** 5916–5922.

Shimeno, H., Soeda, S., Yasukouchi, M., Okamura, N., and Nagamatsu, A. (1995). Fatty acyl-Co A: sphingosine acyltransferase in bovine brain mitochondria: Its solubilization and reconstitution onto the membrane lipid liposomes. *Biol. Pharm. Bull.* **18,** 1335–1339.

Smith, E. R., Jones, P. L., Boss, J. M., and Merrill, A. H. J. (1997). Changing J774A.1 cells to new medium perturbs multiple signaling pathways, including the modulation of protein kinase C by endogenous sphingoid bases. *J. Biol. Chem.* **272,** 5640–5646.

Spence, M. W., and Burgess, J. K. (1978). Acid and neutral sphingomyelinases of rat brain: Activity in developing brain and regional distribution in adult brain. *J. Neurochem.* **30,** 917–919.

Spence, M. W., Wakkary, J., and Cook, H. W. (1982). Localization of neutral, magnesium-stimulate sphingomyelinase in plasma membrane of cultured neuroblastoma cells. *Biochim. Biophys. Acta* **719,** 162–164.

Sugita, M., Williams, M., Dulaney, J., and Moser, H. (1975a). Ceramidase deficiency in Farber's disease. *Biochim. Biophys. Acta* **398,** 125–133.

Sugita, M., Williams, M., Dulaney, J. T., and Moser, H. W. (1975b). Ceramidase and ceramide synthesis in human kidney and cerebellum: Description of a new alkaline ceramidase. *Biochim. Biophys. Acta* **398,** 125–131.

Takeda, Y., Tashima, M., Takahashi, A., Uchiyama, T., and Okazaki, T. (1999). Ceramide generation in nitric oxide-induced apoptosis. *J. Biol. Chem.* **274,** 10654–10660.

Tanaka, M., Nara, F., Suzuki, K., Hosoya, T., and Ogita, T. (1997). Structural elucidation of scyphostatin, an inhibitor of membrane-bound neutral sphingomyelinase. *J. Am. Chem. Soc.* **119,** 7871–7872.

Tepper, A. D., Boesen-de Cock, J. G. R., deVries, E., Borst, J., and van Blitterswijk, W. J. (1997). CD95/FAS-induced ceramide formation proceeds with slow kinetics and is not blocked by caspase-3/CPP32 inhibition. *J. Biol. Chem.* **272,** 24308–24312.

Tomiuk, S., Hofmann, K., Nix, M., Zumbansen, M., and Stoffel, W. (1998). Cloned mammalian neutral sphingomyelinase: Functions in sphingolipid signaling? *Proc. Natl. Acad. Sci. USA* **95,** 3638–3643.

van Helvoort, A., Stoorvogel, W., van Meer, G., and Burger, N. J. (1997). Sphingomyelin synthase is absent from endosomes. *J. Cell Sci.* **110,** 781–788.

Van Veldhoven, P. P., Bishop, W. R., and Bell, R. M. (1989). Enzymatic quantification of sphingosine in the picomole range in cultured cells. *Anal. Biochem.* **183,** 177–189.

Watts, J. D., Gu, M., Polverino, A. J., Patterson, S. D., and Aebersold, R. (1997). Fas-induced apoptosis of T cells occurs independently of ceramide generation. *Proc. Natl. Acad. Sci. USA* **94,** 7292–7296.

Weigmann, K., Schwander, R., Krut, O., Yeh, W.-C., Mak, T. W., and Kronke, M. (1999). Requirement of FADD for tumor necrosis factor-induced activation of acid sphingomyelinase. *J. Biol. Chem.* **274,** 5267–5270.

Wiesner, D. A., and Dawson, G. (1996). Staurosporine induces programmed cell death in embryonic neurons and activation of the ceramide pathway. *J. Neurochem.* **66,** 1418–1425.

Yavin, E., and Gatt, S. (1969). Enzymatic hydrolysis of sphingolipids. 8. Further purification and properties of rat brain ceramidase. *Biochemistry* **8,** 1692–1698.

Yoshimura, S.-I., Banno, Y., Nakashima, S., Takenaka, K., Sakai, H., Nishimura, Y., Sakai, N., Shimizu, S., Eguchi, Y., Tsujimoto, Y., and Nozawa, Y. (1998). Ceramide formation leads to caspase-3 activation during hypoxic PC12 cell death. *J. Biol. Chem.* **273,** 6921–6927.

Zamzami, N., Brenner, C., Marzo, I., Susin, S. A., and Kroemer, G. (1998). Subcellular and submitochondrial mode of action of Bcl-2-like oncoproteins. *Oncogene* **16,** 2265–2282.

Zundel, W., Swiersz, I. M., and Giaccia, A. (2000). Caveolin 1-mediated regulation of receptor tyrosine kinase-associated phosphatidylinositol 3-kinase activity by ceramide. *Mol. Cell. Biol.* **20,** 1507–1514.

CHAPTER 7

Cell-Free Systems to Study Apoptosis

Howard O. Fearnhead

Apoptosis Section
Regulation of Cell Growth Laboratory
NCI-FCRDC Frederick, Maryland 21702

I. Introduction

Use of cell-free systems to investigate apoptosis is now widespread, and different systems have been developed to study different aspects of the apoptotic process. Cell-free systems have been used to study how nuclear and mitochondrial changes of apoptosis are brought about, to identify caspases that are activated during apoptosis, to investigate mechanisms of caspase activation, and to identify caspase substrates.

In the first cell-free systems used to study apoptosis, extracts were prepared from either overtly apoptotic cells or cells committed to die by exposure to an apoptotic stimulus (Lazebnik *et al.*, 1993, 1994, 1995). These extracts were used to study nuclear changes of apoptosis. As a result the proteases responsible for poly-(ADP-ribose) polymerase (PARP) and lamin cleavage (caspase-3 and caspase-6, respectively) were identified (Lazebnik *et al.*, 1994; Nicholson *et al.*, 1995; Takahashi *et al.*, 1996). Similar

METHODS IN CELL BIOLOGY, VOL. 66

0091-679X/01 $35.00

systems were later used to purify and identify caspase substrates whose cleavage causes DNA fragmentation (Enari *et al.,* 1998; Liu *et al.,* 1997).

The obvious limitation of systems derived from apoptotic cells and containing active caspases is that the caspase activation step itself cannot be studied. This limitation was overcome by the development of *Xenopus laevis* (Newmeyer *et al.,* 1994) and mammalian (Liu *et al.,* 1996; Fearnhead *et al.,* 1997) cell-free systems in which caspases were inactive in the extract, but became active on incubation. As a result, it became clear that cell-free systems could be used to investigate how bcl-2 blocked apoptosis (Newmeyer *et al.,* 1994), and cytochrome *c* and Apaf-1 were identified as cofactors for the autoactivation of caspase-9 (Liu *et al.,* 1996, 1997; Zou *et al.,* 1997).

The identification of cytochrome *c* as a cofactor focused attention on its release from mitochondria (Green and Reed, 1998). Mammalian (Luo *et al.,* 1998) and *Xenopus* (Kluck *et al.,* 1997a,b cell-free systems have been used to identify proteins (Luo *et al.,* 1998; Thress *et al.,* 1998, 1999) involved in regulating this step of the apoptotic process. A cell-free system was also used to identify another pro-apoptotic mitochondrial protein, apoptosis inducing factor (AIF) (Susin *et al.,* 1999).

Besides being used to identify proteins necessary for apoptosis, cell-free systems have also been used to investigate how inhibitors of apoptosis function. As examples, the role of bcl-2 in controlling cytochrome *c* release (Kluck *et al.,* 1997a) and the interaction of IAPs with caspases (Deveraux *et al.,* 1997) were demonstrated.

Even from this brief and incomplete list it is clear that cell-free systems have been instrumental in identifying new components of the apoptotic machinery and ordering the sequence of events that constitute apoptosis. This chapter describes protocols for extract preparation and methods for assessing nuclear changes of apoptosis as well as more direct assays for detecting active caspases.

II. Preparation of Extracts

The following detailed methods are used for making extracts from both suspension and adherent cell lines.

A. Requirements for Extract Preparation

1. Buffers

Extract buffer: 50 m*M* PIPES, pH 7.0, 50 m*M* KCl, 5 m*M* EGTA, 2 m*M* $MgCl_2$, 1 m*M* dithiothireitol (DTT) supplemented immediately before use with 10 μg/ml of cytochalasin B and protease inhibitor cocktail [0.1 m*M* phenylmethylsulfonyl fluoride (PMSF) and 2 μg/ml each of chymostatin, pepstatin, leupeptin, and antipain]

DTT: 1 *M* in water, stored at −20°C.

PMSF: 0.1 *M* in ethanol, stored at −20°C.

Protease inhibitor cocktail: Stocks of chymostatin, pepstatin, leupeptin, and antipain at 10 mg/ml in dimethyl sulfoxide (DMSO) mixed to give 2 mg/ml of each in DMSO, stored at −20°C.

Cytochalasin B: 10 mg/ml in DMSO and stored at −20°C.

Mineral oil

Reagents for Bradford assay (Bio-Rad)

2. Equipment

Sorvall RC-3B centrifuge

1-liter centrifuge bottles

15-ml Falcon tubes

250-ml conical bottom Falcon tubes

Low-speed centrifuge (Beckman Accuspin)

Ultracentrifuge (Beckman Optima TLX or a floor standing model)

Appropriate rotor for the ultracentrifuge (TLA 110 rotor for the Optima, SW 50.1 or SW 55 Ti for floor models)

Polycarbonate centrifuge tubes (Beckman, number 362305 for TLA 110 or number 349622 for SW 50.1 or SW 55 Ti)

Dewar or ice bucket for holding liquid nitrogen; container should be capable of holding 5–10 15-ml Falcon tubes

Spectrophotometer and disposable cuvettes

B. Suspension Cells

1. Grow 293 (human embryonic kidney) or HeLa (human cervical carcinoma cells) adapted to suspension culture in Jokliks medium (JRH Biosciences) supplemented with 5% calf serum. Cells are grown in flasks with a magnetic stirrer placed inside a 37°C incubator.

2. Transfer one or more liters of 293 or HeLa cells (5–7 × 10^5 cells/ml) to 1-liter bottles and centrifuge (Sorvall RC-3B centrifuge, 1400 rpm for 15 min at 4°C).

3. Carefully pour off three-quarters of the medium and resuspend the cells in the residual medium by swirling the bottle gently.

4. Transfer the resuspended cells to 250-ml conical bottomed bottles and centrifuge again (Beckman Accuspin, 300*g*, for 10 min at 4°C). Remove the medium.

All subsequent steps are performed on ice using prechilled buffers.

5. Gently but quickly resuspend the pellet in 1 ml of extract buffer using a 1-ml pipette tip with the end cut off.

6. Add extract buffer to 10 pellet volumes. The author typically uses 100 ml of extract buffer for the pellet from 8 liters of 293 cells. Speed is important at this stage as

the buffer is hypotonic and the cells become progressively more fragile. If the cells are allowed to sit, they lyse and the protein concentration of the final extract will be reduced drastically.

7. Transfer the cell suspension to 15-ml Falcon tubes and centrifuge immediately (300*g*, 5 min in Beckman Accuspin at 4°C). The pellet will have a loose consistency, so carefully remove as much extract buffer as possible.

8. Lyse the cells by three cycles of freezing in liquid nitrogen and thawing in cold running water with a brief vortex mix prior to each freezing. Once thawed, keep pellets on ice.

Extracts can also be prepared using a Dounce homogenizer instead of freeze thawing. By adding sucrose (250 m*M*) to the extract buffer and using a Dounce homogenizer, extracts can be prepared without breaking mitochondria, preventing the release of cytochrome *c* into the extract (Liu *et al.*, 1996). This experiment was originally performed to demonstrate the cytochrome *c* dependence of caspase-9 activation.

9. Transfer the lysate to polycarbonate centrifuge tubes and overlay this with 0.3 ml of mineral oil. Take care not to overfill the tubes. Centrifuge the lysate in an ultracentrifuge for 1 h at 100,000*g* and 4°C.

10. Carefully remove and discard the mineral oil. The yellowish material at the interface of oil and supernatant is also discarded. Carefully collect the supernatant (S-100 extract) and store it on ice until snap freezing in liquid nitrogen. Approximately 1.5 ml of supernatant is recovered per tube.

11. Take a 1- to 10-μl aliquot of extract and dilute 1 : 20, 1 : 10, and 1 : 5 with water. Use 1–2 μl of these dilutions plus 1 ml of Bradford reagent to determine the protein concentration.

12. Extracts can be stored at −70°C in aliquots of 10–100 μl for several months with no loss in activity.

C. Adherent Cells

There are two approaches for preparing extracts from adherent cells, for convenience called A and B. Method A is more reliable but requires many more cells.

1. Method A

1. Grow adherent cells on 15-cm dishes. Expand until 20–25 confluent plates are available.
2. Detach cells by trypsinization. Process plates in batches of 5, pooling the cells in a 50-ml tube containing 10 ml of chilled culture medium (50% fetal bovine serum to inactivate trypsin).
3. Wash the cells once in phosphate-buffered saline (PBS) and then resuspend in extract buffer.
4. Prepare extracts as described for suspension cells.

The major drawback of this technique is the number of plates required. Some success was achieved, particularly with 293 cells, preparing extracts from a smaller number of plates using Method B (described next).

2. Method B

1. Grow cells in a 10-cm dish until confluent.
2. Wash plate with 10 ml of ice-cold extract buffer.
3. Add 10 ml of ice-cold extract buffer and place plate on ice for 20 min to allow the cells to swell.
4. Remove the buffer and tip the plate 90° to allow the residual buffer to drain. Leave the plate in this position for 15–20 min. The plate must be kept cold, either by packing ice around it or by draining the plate in a fridge. Check the plate every 5 min. Care must be taken that the plate does not dry completely, but it is important to remove as much buffer as possible.
5. Aspirate residual buffer.
6. Scrape the cells from the plate using a cell scraper (or a razor blade). Approximately 30–50 μl of cells can be collected per plate.
7. Prepare extract from the lysate by freeze thawing as described earlier.

D. Expected Protein Concentrations of Extracts

Typically a protein extract from 1 liter of 293 cells contains 30–35 mg/ml protein. This varies slightly from cell type to cell type, with Jurkat and HeLa giving less (20–30 mg/ml) concentrated extracts. Adherent cells also yield protein extracts of 20–30 mg/ml. The effect of protein concentration on caspase activation is discussed briefly in Section IV.

III. *In Vitro* Apoptosis Assays

All of the following assays exploit apoptotic changes that depend on caspase activity. Some of these assays use isolated nuclei as a readout for apoptotic changes whereas others measure caspase activity more directly. Direct assays of caspase activity can be further divided into those using caspase substrates and those using caspase processing as an indicator of activation. Assays using isolated nuclei will be described first, beginning with the preparation of nuclei.

A. Nuclei Assays

1. Nuclei Preparation

Nuclei for the following assays can be prepared from suspension cultures of 293, Jurkat, or HeLa cells.

a. Buffers

Nuclei buffer: 10 m*M* PIPES, pH 7.0, 10 m*M* KCl, 1.5 m*M* $MgCl_2$, 1 m*M* DTT, 10 μg/ml cytochalasin B, and the protease inhibitor cocktail (see extract preparation described earlier)

Nuclei storage buffer: 10 m*M* PIPES, pH 7.0, 80 m*M* KCl, 20 m*M* NaCl, 250 m*M* sucrose, 5 m*M* EGTA, 1 m*M* DTT, 0.5 m*M* spermidine, 0.2 m*M* spermine, the protease inhibitor cocktail, and 50% glycerol

Sucrose cushion: 10 m*M* PIPES, pH 7.0, 10 m*M* KCl, 1.5 m*M* $MgCl_2$, 30% sucrose, 1 m*M* DTT, 10 μg/ml cytochalasin B, and the protease inhibitor cocktail (see extract preparation)

Extract dilution buffer: 10 m*M* Hepes, pH 7.0, 50 m*M* NaCl, 2 m*M* $MgCl_2$, 5 m*M* EGTA, and 1 m*M* DTT

b. Equipment

15-ml Dounce homogenizer

Low-speed centrifuge (Beckman Accuspin)

Compound microscope

Hemocytometer

Slides and coverslips

1. Grow 1 liter of HeLa Jurkat, or 293 cells in suspension.
2. Pellet the cells (300*g* for 10 min in a Beckman Accuspin at 4°C) and resuspend in 15 ml of nuclei buffer.
3. Allow the cells to swell on ice for 20 min in nuclei buffer and then gently lyse with a Dounce homogenizer. Monitor lysis by microscopy. Twenty passes with a tight pestle is usually sufficient.
4. Layer the lysate over a 10-ml sucrose cushion and centrifuge (800*g* for 10 min in a Beckman Accuspin at 4°C).
5. Discard the supernatant and resuspend the nuclei in nuclei buffer.
6. Wash the nuclei in nuclei buffer (300–400*g* in a bench-top microcentrifuge for 4 min at 4°C).
7. Resuspend the nuclei in nuclei buffer and use a hemocytometer to assess their density.
8. Either use the nuclei immediately or store them at −20°C in nuclei storage buffer at 1×10^8 nuclei/ml. If nuclei have been stored, wash twice in extract dilution buffer before use.

2. Chromatin Condensation Assay

Chromatin condensation is one of the archetypal apoptotic changes and provides an easily detectable end point for a cell-free assay (Lazebnik *et al.,* 1993).

a. Buffers

ATP regenerating system: ATP (40 m*M*) and phosphocreatine (200 m*M*) in Hepes (10 m*M*, pH 7.0), stored at −70°C

Creatine kinase (1 mg/ml) in Hepes (10 m*M*, pH 7.0), NaCl (50 m*M*), and DTT (1 m*M*), stored at −70°C

Paraformaldehyde: 4% paraformaldehyde. To prepare this, heat 90 ml of PBS to 80°C and them add 4*g* of paraformaldehyde while stirring. After cooling adjust the pH to 7.4, filter the solution, and store at −20°C

4,6-Diamidino-2-phenylindole (DAPI) (Sigma): 10 mg/ml in water. Store at −20°C and protect from light

b. Equipment

Microcentrifuge

Fluroescence microscope

Slides and coverslips

1. Add isolated nuclei (5×10^5) to an extract (10 μl) supplemented with an ATP regeneration system.
2. Incubate the reaction for 60 min at 37°C.
3. Stop the reaction by adding an equal volume of 4% paraformaldehyde.
4. Incubate on ice for 20 min.
5. Stain nuclei with DAPI (final concentration 1 μg/ml) for 5 min and examine by fluorescence microscopy. Determine the percentage of nuclei with an apoptotic morphology.

After an hour incubation with a 293 cell extract, approximately 80% of isolated nuclei will show obvious apoptotic signs; typically the chromatin will have condensed into brightly staining "pebbles" (see Fig. 1). The remaining 20% of nuclei may appear altered.

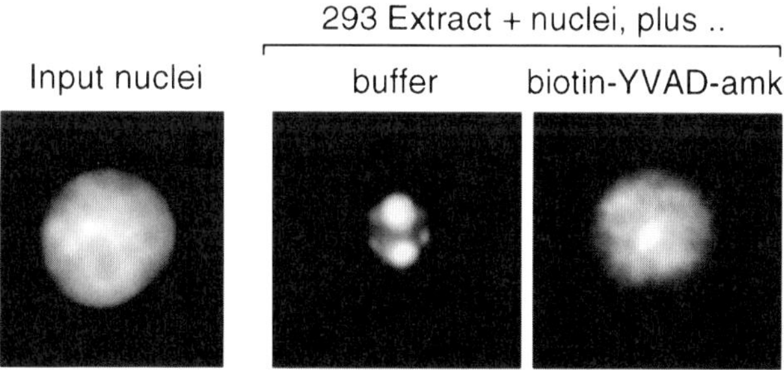

Fig. 1 Nuclear condensation in isolated nuclei. Nuclei (1×10^5) were incubated with 293 extract for 60 min at 37°C as described in the text. After this time, nuclei were fixed and stained with DAPI. To test caspase dependence of the chromatin condensation, 10 μ*M* biotin-YVAD-cmk was added to the reaction before incubation. Reproduced with permission from Fearnhead *et al.* (1997).

For example, they may increase in size and decrease in staining intensity or appear "moth eaten." The rate at which nuclei undergo chromatin condensation will vary from extract to extract, but in the fastest reactions, 30 min is required for 60–80% of nuclei to have undergone chromatin condensation.

In an extract without active caspases, no chromatin condensation is observed even after 60 min. However, some of these nuclei will increase in size and decrease in DNA staining as mentioned earlier. With longer incubations more and more nuclei will undergo these "nonspecific" changes.

3. DNA Fragmentation Assay

The cleavage of DNA into a characteristic ladder pattern is an easily detectable apoptotic event. The following method is based on that described by Liu *et al.* (1997) and was used to purify a caspase-activated nuclease, DFF. The same protein was also purified using plasmid rather than isolated nuclei DNA as an endonuclease substrate.

a. Buffers

Proteinase buffer: 0.1 *M* Tris–HCl, pH 8.5, 5 m*M* EDTA, 0.2 *M* NaCl, 0.2% (w/v) sodium dodecyl sulfate (SDS), and 0.2 mg/ml proteinase K

5 *M* NaCl

Isopropanol

70% ethanol

RNase buffer: 10 m*M* Tris–HCl, pH 7.5, 1 m*M* EDTA, and 0.2 mg/ml RNase

2× DNA sample buffer: 10 m*M* Tris–HCl, pH 5.5, 1 m*M* EDTA, 40% (w/v) Ficoll, and 0.005% (w/v) bromphenol blue

Ethidium bromide: 50 mg/ml in water. Store protected from light

TBE: 90 m*M* Tris–borate and 2 m*M* EDTA

b. Equipment

Agarose gel electrophoresis system

Microcentrifuge

1. Incubate nuclei (5×10^5) for 60 min with an extract as described for the chromatin condensation assay.
2. Add 240 μl proteinase buffer to each reaction and incubate the samples overnight at 37°C.
3. Add 84 μl of 5 *M* NaCl and spin out the debris (14,000 rpm for 5 min in a microcentrifuge at room temperature).
4. Precipitate the DNA in the supernatant by adding an equal volume of ice-cold isopropanol (333 μl) and centrifuge immediately at 14,000 rpm for 10 minutes. Mark the outside of the tubes before spinning so the pellet can be found more easily after centrifugation.

5. Wash the DNA once with ice-cold 70% ethanol.

6. Resuspend the precipitated DNA in 20 μl of RNase buffer and incubate for 1 h at 37°C.

7. Add an equal volume of 2× DNA sample buffer and load on an 1.8% agarose gel (0.5× TBE) containing 2.5 μg/ml ethidium bromide. Carry out electrophoresis at 100 V.

8. Assess DNA laddering under UV light.

The apoptotic changes induced by 293 cell extracts described previously are dependent on active caspases. However, this may not be true for all extracts, and some apoptotic changes may be caspase independent [e.g., the chromatin condensation induced by AIF (Susin *et al.,* 1999)]. Testing whether peptide-based inhibitors of caspases (e.g., the aldehyde or chloromethyl ketone derivatives of YVAD or DEVD) prevent nuclear changes in isolated nuclei is straightforward as there is no plasma membrane to limit the efficacy of these compounds. Typically, incubation of an extract with 10 μ*M* of inhibitor (available from a number of suppliers including Biomol and Calbiochem) will prevent all chromatin condensation in isolated nuclei (see Fig. 1). Details about the specificity of peptide based-caspase inhibitors can be found elsewhere (Garcia-Calvo *et al.,* 1998; Margolin *et al.,* 1997).

The dependence of chromatin condensation and DNA fragmentation on caspase activity brings us to assays that directly measure caspase activity in the extract.

B. Caspase Assays

1. PARP Cleavage Assay

Poly-(ADP-ribose)polymersase is a 116-kDa nuclear protein and one of the first caspase substrates to be described (Kaufmann *et al.,* 1993). Caspase cleavage generates an 85-kDa fragment that is recognized by the anti-PARP antibody C2-10 (Pharmingen). Immunoblotting for PARP cleavage is indicative of caspase activity in the extract.

a. Equipment

SDS–PAGE system

Western blotting apparatus

1. Nuclei are incubated as for the chromatin condensation assay. Alternatively, recombinant PARP can be incubated with an extract.
2. After incubation, SDS–PAGE sample buffer containing 4 *M* urea is added and the samples are heated at 37°C for 10 min.
3. Standard SDS–PAGE and immunoblotting protocols are then used to detect PARP using the C2-10 antibody (1/1000).

A similar strategy can be used to detect cleavage of other nuclear substrates for caspases, e.g., lamins (Oberhammer *et al.,* 1994; Takahashi *et al.,* 1996). However, while proteins such as lamins and PARP are bona fide caspase substrates, using them to assess caspase activity has several drawbacks. First, detecting their cleavage by immunoblotting is time-consuming. Second, this approach is unsuitable for quantifying caspase activity. There are a number of techniques using synthetic substrates that are faster, allow the identification of the active caspases, and are better suited to quantifying caspase activity.

2. Affinity Labeling Assay

Affinity labeling of caspases depends on tetrapeptides that mimic caspase substrates (Thornberry *et al.,* 1994). For example, YVAD- mimics a caspase-1 cleavage site in interleukin-1β, whereas DEVD mimics a caspase-3 cleavage site in PARP. More details about caspase substrate specificity may be found elsewhere (Garcia-Calvo *et al.,* 1998; Margolin *et al.,* 1997; Talanian *et al.,* 1997; Thornberry *et al.,* 1997). In addition to the tetrapeptide, the affinity label has a reactive alkylating group [acyloxyfluoromethyl ketone (amk), chloromethyl ketone (cmk), or fluoromethyl ketone (fmk)] and is biotinylated. Caspase binding of the tetrapeptide results in alkylation of the caspase's active site cysteine, inhibiting the enzyme irreversibly. The biotin group allows inhibited caspases to be detected. In this way caspases that are active in an extract can be labeled and subsequently identified (Faleiro *et al.,* 1997).

Two protocols are presented. The first is derived from Falerio *et al.* (1997) and allows labeling of active caspases present in a cell lysate. The second is a modification used to label caspases in cell extracts. Either Biotin-YVAD-amk or Biotin-DEVD-fmk was used as affinity labels. At the concentrations used, both compounds label caspase-3 and -6 (Faleiro *et al.,* 1997). The lower concentration of Biotin-DEVD-fmk is necessary due to higher background labeling.

3. Affinity Labeling Caspases in Lysates from Apoptotic Cells

a. Buffers

Cell lysate labeling buffer: 20 μ*M* biotin-YVAD-amk or 2 μ*M* DEVD-fmk in 10 m*M* Hepes, pH 7.0, 50 m*M* NaCl, 2 m*M* $MgCl_2$, and 5 m*M* EGTA

Etoposide: 50 m*M* in DMSO

Biotin-YVAD-acyl-oxyfluoromethyl ketone (YVAD-amk) or biotin-DEVD-fluoromethyl ketone

(DEVD-fmk) (Biosyn): 10 m*M* stock solution in DMSO, which can be aliquoted and stored at −70°C. Immediately before use, aliquots are diluted to 20 μM with labeling buffer

b. Equipment

Hemocytometer

Fluorescence microscope

SDS–PAGE system

Blotting apparatus

1. Treat 5–10 ml of Jurkat cells (5×10^5 cells/ml) with 50 μM etoposide or DMSO (0.1% final concentration for 6 h.

2. Take an aliquot of cells and stain with Hoechst 33342 (1-μg/ml final concentration) for 5 min. Assess cell density using a hemocytometer and determine the percentage of cells that have apoptotic nuclear morphology by fluorescence microscopy. By 6 h, 70–80% of Jurkat cells will be apoptotic.

3. Pellet cells at 300*g* for 10 min in a Beckman Accuspin centrifuge.

4. Resuspend cells at 2×10^6 cells/10 μl in 10 μM biotin-YVAD-amk in labeling buffer supplemented with the protease inhibitor cocktail and 10 mg/ml cytochalasin B.

5. Lyse the cells by three cycles of freezing and thawing.

6. Centrifuge the lysate in a microcentrifuge (14,000 rpm, 10 min, 4°C).

7. Collect the supernatant and take an aliquot to determine the protein concentration using the by Bradford assay.

8. Incubate 10 ml of the supernatant at 37°C for 4 min to label active caspases. This incubation is usually sufficient to achieve maximal caspase labeling but some optimization may be necessary.

9. Stop labeling by adding 2X SDS sample buffer (10 μl) and boiling for 4 min. At this stage samples can be stored at −20°C.

10. Subject the samples to SDS–PAGE (15% gel) with 10 μg of total protein loaded per lane.

11. While the gel is running, a piece of PVDF membrane (Millipore) is prepared by soaking in methanol for 30 s, followed by 1 min in water and then 30 min in transfer buffer.

12. Transfer proteins onto the membrane using the mini-Protean system at 100 V constant voltage in a cold room for 1 h.

13. Soak the membrane in methanol for 1 min and dry; first air for 30 min and then at 80°C for 15 min.

14. Incubate the membrane for 1 h with 1 μg/ml of avidin in Tris-buffered saline (TBS) (3% milk powder) and then wash in TBS (3×10 min).

15. Incubate membrane with 50 ng/ml of biotin-peroxidase in TBS (3% milk powder) and then wash in TBS (3×10 min).

16. Visualize labeled caspases by ECL (Amersham Inc).

4. Labeling Caspases in Cell Extracts

a. Buffers

ATP or dATP: 10 mM in 10 mM Hepes, pH 7.0, stored at −20°C

Extract labeling buffer: 10 μM biotin-YVAD-amk or 1 μM DEVD-fmk in 10 mM Hepes, pH 7.0, 50 mM NaCl, 2 mM $MgCl_2$, and 5 mM EGTA

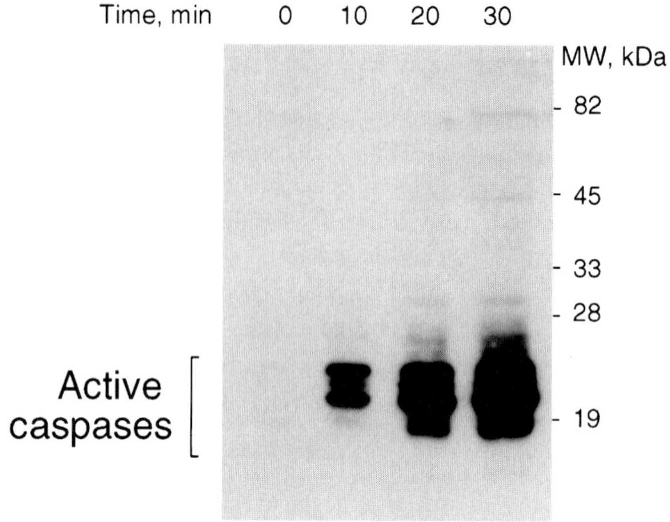

Fig. 2 Affinity-labeled caspases. 293 cell extract (300 μg) was incubated with 1 m*M* ATP for up to 60 min. at 37°C. At various times, an aliquot was taken and active caspases affinity labeled as described. Reproduced with permission from Fearnhead *et al.* (1997).

1. Add ATP or dATP to an extract (final concentration 1m*M*) and incubate one aliquot (10 μl) at 37°C and another aliquot (10 μl) on ice for up to 60 min. This incubation allows caspase activation.
2. Add an equal volume of extract labeling buffer to each aliquot and incubate at 37°C for 4 min.
3. Stop labeling by adding 2× SDS sample buffer (10 μl) and boiling the samples for 4 min.
4. Labeled caspases are visualized as described earlier for caspases labeled in cells.

A typical labeling pattern of a 293 cell extract is shown in Fig. 2. Initially, no caspase activity is detectable. Over a period of 30 min, a number of bands appear representing the large subunit of an active caspase. The number of bands reflects the activation in the extract of different caspases and also that a caspase may be processed to different sized active forms (Faleiro *et al.,* 1997).

5. Protease Assay

While affinity labeling may allow the active caspase to be identified, it is unsuited for quantitating caspase activity. Accurate quantitation can be achieved using colorimetric or fluorimetric caspase substrates. There are now a range of different fluorimetric and colorimetric caspase substrates available from a number of suppliers (e.g., Calbiochem, Biomol, Enzyme Systems). Typically these are described as substrates for a particular caspase. However, care should be taken when ascribing a particular activity in the extract to a particular caspase based only on the use of tetrapeptide substrates. For more detailed information on the specifities of caspases, a number of papers should be consulted (Garcia-Calvo *et al.*, 1998; Margolin *et al.*, 1997; Stennicke and Salvesen, 1997; Talanian *et al.*, 1997; Thornberry *et al.*, 1997).

The following fluorimetric assay uses acetyl-DEVD-7-amido-4-(trifluoromethyl) coumarin (DEVD-afc), which is based on the caspase-3 cleavage site in PARP. In 293 cell extracts, immunodepletion of caspase-3 removes almost all DEVDase activity and so cleavage of DEVD-afc in this case reflects caspase-3 activity. This is the most rapid and least labor-intensive assay described in this chapter, allowing many samples to be assayed simultaneously, and is essentially as described by Thornberry (1994).

a. Buffers

7-amido-4-methylcoumarin (amc) or 7-amido-4-(trifluoromethyl)coumarin (afc) (Sigma): 1 m*M* solution in DMSO stored at −20°C

Assay buffer: 50 m*M* PIPES/KOH, pH 7.0, 0.1 m*M* EDTA, 1 m*M* DTT, and 10% glycerol

DEVD-afc or amc (Biomol): DEVD-afc, 10 m*M* stock in DMSO, stored at −20°C

b. Equipment

Fluorimeter capable of reading 96-well plates (Cytofluor 4000, PE Biosystems)

Excitation and emission filters appropriate for detecting amc (excitation, 380 nm; emission, 460 nm) or afc (excitation, 400 nm; emission, 505 nm)

Black 96-well plates with clear bottoms (Costar)

1. Use a series of amc or afc solutions (1 pmol–1 nmol/reaction volume of 200 μl) to contruct a standard curve.
2. Incubate extract with ATP or dATP (1 m*M* final concentrations) as described earlier to activate caspases.
3. Pipet 1–2 μl (25–50 μg) of extract into a well of a 96-well plate and add 200 μl of assay buffer containing 20 μ*M* of the fluorescent substrate DEVD-afc or -amc.
4. Place the plate in the Cytofluor preequilibrated at 30°C. Measure the amount of free every 30 s over a period of 15 min and determine the rate of substrate cleavage (express as picomoles of afc generated per minute per milligram of the total extract protein at 30°C).

A typical 293 cell extract will have a maximal caspase activity of 300–400 pmol/min/mg, although the rate at which extracts activate will vary from extract to extract. The time course of caspase activation in a typical 293 cell extract is shown in Fig. 3.

6. Detection of Caspase Processing

Caspases can be activated by proteolytic processing of the pro-enzyme (Thornberry and Lazebnik, 1998). Thus, an additional indicator of caspase activation is

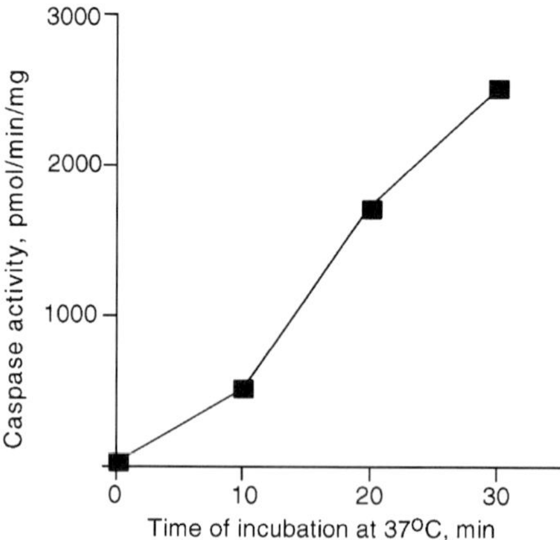

Fig. 3 Caspase activity using a fluorimetric assay. 293 cell extract (300 μg) was incubated with 1 m*M* ATP for up to 60 min at 37°C. At various times, 60 μg of extract was assayed for caspase activity as described. Reproduced with permission from Fearnhead *et al.* (1997).

immunoblotting to detect the disappearance of the pro-enzyme and the appearance of the polypeptides that make up the active caspase. The drawbacks of this approach are that (1) it is limited by the availability of suitable antibodies and (2) it has been reported that caspase-9 activity is not necessarily dependent on proteolytic cleavage (Stennicke *et al.*, 1999). The following protocol is used for the detection of caspase-3 processing.

a. Buffers

Buffers necessary for SDS–PAGE and immunoblotting

b. Equipment

SDS–PAGE system

Blotting apparatus

1. Incubate extract to activate caspases as described earlier.
2. Stop activation by adding an equal volume of 2× SDS sample buffer.
3. Load 10 μg of extract per lane on a 15% SDS–polyacrylamide gel.
4. Resolve proteins by electrophoresis.
5. While the gel is running, prepare a piece of PVDF membrane (Millipore) by soaking in methanol for 30 s, followed by 1 min in water and then 30 min in transfer buffer.
6. Transfer proteins onto the membrane using the mini-Protean system at 100 V constant voltage in a cold room for 1 h.
7. Stain the transferred protein with Ponceau-S; confirm even transfer. After washing away the Ponceau-S in running distilled water, the membrane can be blocked and probed in the conventional manner. However, it can also be probed using a fast protocol, described next.

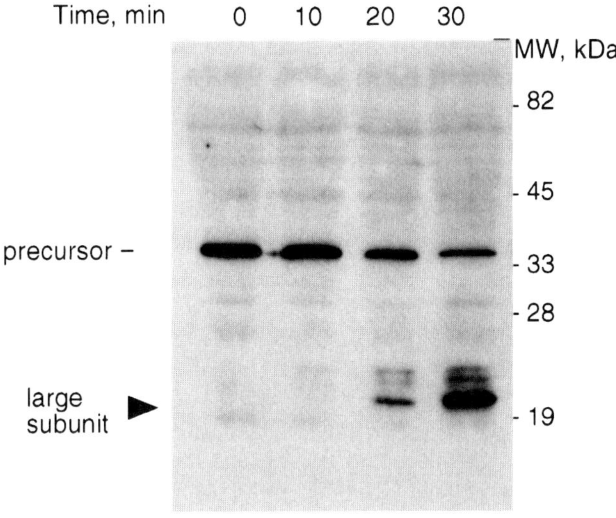

Fig. 4 Caspase-3 processing. 293 cell extract (300 μg) was incubated with 1 m*M* ATP for up to 60 min at 37°C. At various times, an aliquot was taken and caspase-3 processing was assessed by immunoblotting affinity as described. Reproduced with permission from Fearnhead *et al.* (1997).

8. Soak the membrane in methanol for 1 min and dry; first air dry for 30 min and then at 80°C for 15 min.

9. Incubate the membrane for 1 h with a monoclonal antibody to caspase-3 (Clone 19, from Transduction Laboratories) (1/1000 in Tris-buffered saline, 1% BSA) for 30 min followed by three 10-min washes in Tris-buffered saline.

10. Incubate the membrane for 30 min with a secondary antibody conjugated to horseradish peroxidase (1/10,000 in Tris-buffered saline, 1% BSA) followed by another three 10-min washes.

11. Caspase-3 on the blot is visualized by ECL (Amersham Inc). An example is shown in Fig. 4.

IV. Factors That Affect Caspase Activation in Extracts

This section considers some factors that affect caspase activation *in vitro* and describes some basic manipulations to separate the components involved in caspase activation.

The methods described in this chapter were developed for investigating oncogene-dependent caspase activation in extracts from mammalian cells. In brief, extracts prepared from mouse embryo fibroblasts (MEFs) expressing the E1A oncogene do not contain active caspases. However, upon incubation, caspases in the extract activate and induce apoptotic changes in isolated nuclei. In contrast, caspases in extracts from MEFs lacking E1A do not activate (Fearnhead *et al.*, 1997). These data suggest that extracts from E1A-expressing cells contain an oncogene-generated activity capable of activating caspases. This is particularly interesting as E1A is a pro-apoptotic oncogene, capable of inducing apoptosis and sensitizing cells to other apoptotic stimuli (Rao *et al.,* 1992; Lowe and Ruley, 1993).

Apaf-1 was subsequently purified from E1A expressing cells (293 cells) as a protein that triggered caspase activation when added to MEF or human primary lung fibroblasts (IMR90 cells) (Fearnhead *et al.,* 1998). Despite associating Apaf-1 with the oncogene-dependent activity that activates caspases, how E1A triggers death is still unclear. Apaf-1, cytochrome *c*, caspase-9, and caspase-3 are present in extracts from cells lacking E1A, although caspase activation does not occur. At present it is unclear whether E1A expression alters Apaf-1 activity or an inhibitor of Apaf-1-dependent caspase activation.

The activation of caspase-3 is a multistep process. Apaf-1 and cytochrome *c* bind, which allows Apaf-1 to interact with caspase-9. Once bound to Apaf-1, the activity of the caspase-9 zymogen increases, triggering autoactivation (Li *et al.,* 1997; Saleh *et al.,* 1999; Zou *et al.,* 1999) and the formation of an Apaf-1 : caspase-9 holoenzyme (Rodriguez and Lazebnik, 1999). Removal of cytochrome *c*, Apaf-1, or caspase-9 from the extract prevents caspase-3 activation. Cytochrome *c* can be removed by either immunodepleting using a commercially available antibody (clone 6H2.B4, available from Pharmingen) or incubating an extract with 1/10 the extract volume of S-Sepharose. Cytochrome *c*, but not caspases-3 or 9 nor Apaf-1, binds to S-Sepharose. Adding back equine cytochrome *c*, to an extract treated in this way is sufficient to reestablish caspase activation.

Apaf-1 contains Walker A and B motifs (Zou *et al.,* 1997) and hydrolyses ATP (Saleh *et al.,* 1999). Consistent with this, ATP or dATP and magnesium are required for caspase activation in the extracts described earlier (Fearnhead *et al.,* 1997, 1998). In a number of cell types, substituting dATP for ATP will produce a more rapid caspase activation (Table I). A more detailed investigation of the nucleotide dependence of Apaf-1-mediated

Table I
Caspase Activation[a]

Cell type	Activation with ATP	Activation with dATP	Notes
293	++	+++	Prepared from adherent and suspension cells
HeLa	+	++	Prepared from adherent and suspension cells
Jurkat	−	++	Prepared from suspension cells
K562	+	++	Prepared from suspension cells
U2-OS	−	+	Prepared from adherent cells
MEF	−	+	Prepared from adherent cells
MEF+E1A/E1B	+	++	Prepared from adherent cells

[a]Extracts were prepared from a number of different cell types and were then incubated with ATP (1 m*M*) or dATP (1 m*M*) for 15 min at 37°C. After this time caspase activity was measured using DEVD-afc (H. Fearnhead, unpublished results).

caspase activation, including chemotherapeutic deoxyadenosine nucleotides, was carried out by Genini *et al.* (2000). Care should also be taken with magnesium concentrations in the extract as maximal activation is seen in the range of 0.25–2.0 m*M* magnesium, but 4 m*M* inhibited activation in 293 cell extracts approximately 90% (H. Fearnhead, unpublished results).

The salt concentration of the extract is also important: 100 m*M* KCl blocks activation completely, whereas reducing the KCl concentration accelerates the rate of caspase activation (Genini *et al.,* 2000). KCl concentration is also important when considering the relationship between the rate of extract activation and the protein concentration of the extract. At 50 m*M* KCl, a 50% dilution of the extract inhibits activation completely. However, at 10 m*M* KCl, activation is proportional to protein concentration in the range of 3–12 mg/ml (H. Fearnhead, unpublished results).

V. Concluding Remarks

Work in a large number of laboratories has demonstrated the power of cell-free systems in investigating many aspects of the apoptotic process. Their power comes from the precise way the biochemical processes of apoptosis can be dissected. However, the price of this precision may be the relevance of the cell-free system to the workings of a cell, tissue, or organism. Thus, verification that cell-free data reflect *in vivo* biology is essential.

References

Deveraux, Q. L., Takahashi, R., Salvesen, G. S., and Reed, J. C. (1997). X-linked IAP is a direct inhibitor of cell-death proteases. *Nature* **388,** 300–304.

Enari, M., Sakahira, H., Yokoyama, H., Okawa, K., Iwamatsu, A., and Nagata, S. (1998). A caspase-activated DNase that degrades DNA during apoptosis, and its inhibitor ICAD. *Nature* **391,** 43–50.

Faleiro, L., Kobayashi, R., Fearnhead, H. O., and Lazebnik, Y. A. (1997). Multiple species of CPP32 and Mch2 are the major active caspases present in apoptotic cells. *EMBO J.* **16,** 2271–2281.

Fearnhead, H. O., McCurrach, M. E., O'Neill, J., Zhang, K., Lowe, S. W., and Lazebnik, Y. A. (1997). Oncogene-dependent apoptosis in extracts from drug-resistant cells. *Genes Dev.* **11,** 1266–1276.

Fearnhead, H. O., Rodriguez, J., Govek, E. E., Guo, W., Kobayashi, R., Hannon, G., and Lazebnik, Y. A. (1998). Oncogene-dependent apoptosis is mediated by caspase-9. *Proc. Natl. Acad. Sci. USA* **95,** 13664–13669.

Garcia-Calvo, M., Peterson, E. P., Leiting, B., Ruel, R., Nicholson, D. W., and Thornberry, N. A. (1998). Inhibition of human caspases by peptide-based and macromolecular inhibitors. *J. Biol. Chem.* **273,** 32608–32613.

Genini, D., Budihardjo, I., Plunkett, W., Wang, X., Carrera, C. J., Cottam, H. B., Carson, D. A., and Leoni, L. M. (2000). Nucleotide requirements for the in vitro activation of the apoptosis protein-activating factor-1-mediated caspase pathway. *J. Biol. Chem.* **275,** 29–34.

Green, D. R., and Reed, J. C. (1998). Mitochondria and apoptosis. *Science* **281,** 1309–1312.

Kaufmann, S. H., Desnoyers, S., Ottaviano, Y., Davidson, N. E., and Poirier, G. G. (1993). Specific proteolytic cleavage of poly(ADP-ribose) polymerase: An early marker of chemotherapy-induced apoptosis. *Cancer Res.* **53,** 3976–3985.

Kluck, R. M., Bossy-Wetzel, E., Green, D. R., and Newmeyer, D. D. (1997a). The release of cytochrome *c* from mitochondria: A primary site for Bcl-2 regulation of Apoptosis. *Science* **275,** 1132–1136.

Kluck, R. M., Martin, S. J., Hoffman, B. M., Zhou, J. S., Green, D. R., and Newmeyer, D. D. (1997b). Cytochrome *c* activation of CPP32-like proteolysis plays a critical role in a Xenopus cell-free apoptosis system. *EMBO J.* **16,** 4639–4649.

Lazebnik, Y. A., Cole, S., Cooke, C. A., Nelson, W. G., and Earnshaw, W. C. (1993). Nuclear events of apoptosis in vitro in cell-free mitotic extracts: A model system for analysis of the active phase of apoptosis. *J. Cell Biol.* **123,** 7–22.

Lazebnik, Y. A., Kaufmann, S. H., Desnoyers, S., Poirier, G. G., and Earnshaw, W. C. (1994). Cleavage of poly(ADP-ribose) polymerase by a proteinase with properties like ICE. *Nature* **371,** 346–347.

Lazebnik, Y. A., Takahashi, A., Poirier, G. G., Kaufmann, S. H., and Earnshaw, W. E. (1995). Characterization of the execution phase of apoptosis in vitro using extracts from condemned-phase cells. *J. Cell Sci.* (Suppl. 19), 41–49.

Li, P., Nijhawan, D., Budihardjo, I., Srinivasula, S. M., Ahmad, M., Alnemri, E. S., and Wang, X. (1997). Cytochrome *c* and dATP-dependent formation of Apaf-1/caspase-9 complex initiates an apoptotic protease cascade. *Cell* **91,** 479–489.

Liu, X., Kim, C. N., Yang, J., Jemmerson, R., and Wang, X. (1996). Induction of apoptotic program in cell-free extracts: Requirement for dATP and cytochrome *c*. *Cell* **86,** 147–157.

Liu, X., Zou, H., Slaughter, C., and Wang, X. (1997). DFF, a heterodimeric protein that functions downstream of caspase-3 to trigger DNA fragmentation during apoptosis. *Cell* **89,** 175–184.

Lowe, S. W., and Ruley, H. E. (1993). Stabilization of the p53 tumor suppressor is induced by adenovirus 5 E1A and accompanies apoptosis. *Genes Dev.* **7,** 535–545.

Luo, X., Budihardjo, I., Zou, H., Slaughter, C., and Wang, X. (1998). Bid, a Bc12 interacting protein, mediates cytochrome *c* release from mitochondria in response to activation of cell surface death receptors. *Cell* **94,** 481–490.

Margolin, N., Raybuck, S. A., Wilson, K. P., Chen, W., Fox, T., Gu, Y., and Livingston, D. J. (1997). Substrate and inhibitor specificity of interleukin-1 beta-converting enzyme and related caspases. *J. Biol. Chem.* **272,** 7223–7228.

Newmeyer, D. D., Farschon, D. M., and Reed, J. C. (1994). Cell-free apoptosis in Xenopus egg extracts: Inhibition by Bcl-2 and requirement for an organelle fraction enriched in mitochondria. *Cell* **79,** 353–364.

Nicholson, D. W., Ali, A., Thornberry, N. A., Vaillancourt, J. P., Ding, C. K., Gallant, M., Gareau, Y., Griffin, P. R., Labelle, M., Lazebnik, Y. A., Munday, N., Raju, S., Smulson, M., Yamin, T., Yu, V., and Miller, D. (1995). Identification and inhibition of the ICE/CED-3 protease necessary for mammalian apoptosis. *Nature* **376,** 37–43.

Oberhammer, F. A., Hochegger, K., Froschl, K., Tiefenbacher, R., and Pavelka, M. (1994). Chromatin condensation during apoptosis is accompanied by degradation of lamin A+B without enhanced activation of cdc2 kinase. *J. Cell Biol.* **126,** 827–837.

Rao, L., Debbas, M., Sabbatini, P., Hockenbery, D., Korsmeyer, S., and White, E. (1992). The adenovirus E1A proteins induce apoptosis, which is inhibited by the E1B 19-kDa and bcl-2 proteins. *Proc. Natl. Acad. Sci. USA* **89,** 7742–7746.

Rodriguez, J., and Lazebnik, Y. (1999). Caspase-9 and APAF-1 form an active holoenzyme. *Genes Dev.* **13,** 3179–3184.

Saleh, A., Srinivasula, S. M., Acharya, S., Fishel, R., and Alnemri, E. S. (1999). Cytochrome *c* and dATP-mediated oligomerization of Apaf-1 is a prerequisite for procaspase-9 activation. *J. Biol. Chem.* **274,** 17941–17945.

Stennicke, H. R., Deveraux, Q. L., Humke, E. W., Reed, J. C., Dixit, V. M., and Salvesen, G. S. (1999). Caspase-9 can be activated without proteolytic processing. *J. Biol. Chem.* **274,** 8359–8362.

Stennicke, H. R., and Salvesen, G. S. (1997). Biochemical characteristics of caspases-3, -6, -7, and -8. *J. Biol. Chem.* **272,** 25719–25723.

Susin, S. A., Lorenzo, H. K., Zamzami, N., Marzo, I., Snow, B. E., Brothers, G. M., Mangion, J., Jacotot, E., Costantini, P., Loeffler, M., Larochette, N., Goodlett, D. R., Aebersold, R., Siderovski, D. P., Penninger, J. M., and Kroemer, G. (1999). Molecular characterization of mitochondrial apoptosis-inducing factor. *Nature* **397,** 441–446.

Takahashi, A., Alnemri, E. S., Lazebnik, Y. A., Fernandesalnemri, T., Litwack, G., Moir, R. D., Goldman, R. D.,

Poirier, G. G., Kaufmann, S. H., and Earnshaw, W. C. (1996). Cleavage of lamin A by Mch2 alpha but not CPP32: Multiple interleukin 1 beta-converting enzyme-related proteases with distinct substrate recognition properties are active in apoptosis. *Proc. Natl. Acad. Sci. USA* **93,** 8395–8400.

Talanian, R. V., Quinlan, C., Trautz, S., Hackett, M. C., Mankovich, J. A., Banach, D., Ghayur, T., Brady, K. D., and Wong, W. W. (1997). Substrate specificities of caspase family proteases. *J. Biol. Chem.* **272,** 9677–9682.

Thornberry, N. A. (1994). Interleukin-1 beta converting enzyme. *Methods Enzymol.* **244,** 615–631.

Thornberry, N. A., and Lazebnik, Y. (1998). Caspases: Enemies within. *Science* **281,** 1312–1316.

Thornberry, N. A., Peterson, E. P., Zhao, J. J., Howard, A. D., Griffin, P. R., and Chapman, K. T. (1994). Inactivation of interleukin-1 beta converting enzyme by peptide (acyloxy)methyl ketones. *Biochemistry* **33,** 3934–3940.

Thornberry, N. A., Rano, T. A., Peterson, E. P., Rasper, D. M., Timkey, T., Garcia-Calvo, M., Houtzager, V. M., Nordstrom, P. A., Roy, S., Vaillancourt, J. P., Chapman, K. T., and Nicholson, D. W. (1997). A combinatorial approach defines specificities of members of the caspase family and granzyme B: Functional relationships established for key mediators of apoptosis. *J. Biol. Chem.* **272,** 17907–17911.

Thress, K., Evans, E. K., and Kornbluth, S. (1999). Reaper-induced dissociation of a Scythe-sequestered cytochrome *c*-releasing activity. *EMBO J.* **18,** 5486–5493.

Thress, K., Henzel, W., Shillinglaw, W., and Kornbluth, S. (1998). Scythe: A novel reaper-binding apoptotic regulator. *EMBO J.* **17,** 6135–6143.

Zou, H., Henzel, W. J., Liu, X., Lutschg, A., and Wang, X. (1997). Apaf-1, a human protein homologous to *C. elegans* CED-4, participates in cytochrome *c*-dependent activation of caspase-3. *Cell* **90,** 405–413.

Zou, H., Li, Y., Liu, X., and Wang, X. (1999). An APAF-1.cytochrome *c* multimeric complex is a functional apoptosome that activates procaspase-9. *J. Biol. Chem.* **274,** 11549–11556.

CHAPTER 8

Role of c-Jun N-terminal Kinase in Apoptosis

Zheng-gang Liu,[*] Joseph Lewis,[*] Tzu-Hao Wang,[†] and Amy Cook[*]

[*] Department of Cell and Cancer Biology
Medicine Branch, Division of Clinical Sciences
National Cancer Institute, National Institutes of Health
Bethesda, Maryland 20892

[†] Department of Obstetrics and Gynecology
Chang-Gung Memorial Hospital, Lin-Kou Medical Center
Taoyuan, Taiwan

I. Introduction

Signal transduction plays a crucial role in the regulation of cell growth, differentiation, and cell death. Protein kinases are important mediators in transducing signals from the cell surface to the nucleus (Hill and Treisman, 1995; Marshall, 1995). Mitogen-activated protein kinases (MAPKs) are one group of signal-transducing enzymes that play important roles in mediating responses to a variety of extracellular stimuli from yeast to mammals (Herskowitz, 1995; Marshall, 1995). Three subgroups of MAPKs have been identified: ERKs (extracellular signal-regulated kinases), JNKs (c-Jun N-terminal kinases), and p38 MAPKs (Karin, 1998; Davis, 1999). JNKs are also known as

METHODS IN CELL BIOLOGY, VOL. 66

0091-679X/01 $35.00

SAPKs (stress-activated kinases) (Kyriakis *et al.,* 1994). Among these three subgroups of MAPKs, ERKs were identified first and their functions have been studied extensively (Marshall, 1995). This subgroup of MAPKs is thought to mediate many of the biological effects of growth factors that bind to tyrosine kinase receptors (Marshall, 1995). The antiapoptotic effect of ERKs has been discovered (Xia *et al.,* 1996). In contrast, JNK and p38 subgroups of MAPKS are thought to be important apoptotic mediators because they are activated by many apoptosis-inducing stimuli (Verheij *et al.,* 1996; Huang *et al.,* 1997). Studies with generating different JNK null mice have demonstrated that the physiological functions of JNK are much more complicated and diverse (Yang *et al.,* 1997, 1998; Dong *et al.,* 1998; Kuan *et al.,* 1999; Sabapathy *et al.,* 1999). In addition to apoptosis, it is now believed that JNK activation is involved in many other biological processes, such as cell proliferation, embryogenesis, and immunological responses (Davis, 1999). These biological functions of JNK have been confirmed by the molecular genetic studies of JNK in *Drosophila* (Sluss *et al.,* 1996).

The JNKs are distantly related to the ERKs, to which they exhibit about 40% identity. Three genes that encode JNKs have been identified as *jnk1, jnk2,* and *jnk3* by molecular cloning (Derijard *et al.,* 1994; Kallunki *et al.,* 1994; Sluss *et al.,* 1994; Mohit *et al.,* 1995). While jnk1 and jnk2 are widely expressed in most types of tissues, jnk3 expression is detected mainly in the brain and heart (Mohut *et al.,* 1995; Yang *et al.,* 1997). As shown in Table I, the alternative splicing of the transcripts of these three genes generates at least 10 JNK isoforms with molecular masses of 46 and 55 kDa (Gupta *et al.*, 1996). No functional differences between the 46- and 55-kDa isoforms from each *jnk* gene have been found (Gupta *et al.,* 1996). In contrast, the α and β isoforms of *jnk1* and *jnk2* have distinct substrate specificities (Gupta *et al.,* 1996). All of these different JNK isoforms can be potently activated by ultraviolet (UV) radiation, proinflammatory cytokines, and environmental stress. More modest activation is found in response to growth factors (Kullunki *et al.,* 1994; Karin, 1998). As shown in Fig. 1, JNKs are activated through

Table I
JNK Isoforms

Protein	Isoforms	Molecular mass (kDa)	Homology[a]
JNK1	JNK1-α1	46	—
	JNK1-α2	55	—
	JNK1-β1	46	98%
	JNK1-β2	55	97%
JNK2	JNK2-α1	46	83%
	JNK2-α2	55	80%
	JNK2-β1	46	87%
	JNK2-β2	55	81%
JNK3	JNK3-α1	51	84%
	JNK3-α2	60	83%

[a]All of the α1 and β1 isoforms are compared to JNK1-α1 and all of the α2 and β2 isoforms are compared to JNK1-α2.

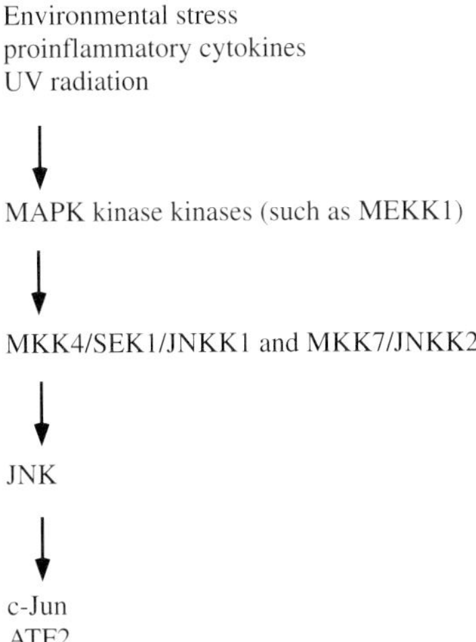

Fig. 1 The signaling pathway of JNK activation in response to different stimuli.

a signaling kinase cascade. In response to different stimuli, a MAPK kinase kinase (MAP3K), such as MEKK1, is activated and then activates the immediate upstream kinases of JNK (MAPK kinases) known as MKK4/SEK1/JNKK1 (Sanchex *et al.,* 1994; Derijard *et al.,* 1995; Lin *et al.,* 1995) and MKK7/JNKK2 (Lu *et al.,* 1997; Tournier; *et al.,* 1997; Wu *et al.,* 1997). In turn, MKK4/SEK1/JNKK1 and MKK7/JNKK2 elevate the activity of JNKs by phosphorylating them on both Thr and Tyr in the Thr-Pro-Tyr (T-P-Y) motif (Karin, 1998; Davis, 1999). In turn, the activated JNKs phosphorylate their specific substrates such as transcription factors c-Jun and ATF2 (Karin *et al.,* 1997; Davis 1999). Both c-Jun and ATF2 are important components of the transcription factor AP-1, whose activation is essential for many biological events (Karin *et al.*, 1997). Phosphorylation of either c-Jun or ATF2 by JNKs will increase AP-1 activity. It is believed that AP-1 is the major target of the JNK pathway and that most biological functions of JNKs are achieved by regulating AP-1 activity (Karin *et al.,* 1997; Davis, 1999).

Apoptosis removes harmful or unwanted cells from organisms during development (Ellis *et al.,* 1991). Apoptosis is also involved in eliminating transformed cells, and inhibition of apoptosis may indirectly promote oncogenic transformation by allowing neoplasms to develop (Korsmeyer, 1995). Because JNK is activated by many stimuli that induce apoptosis, it has been thought to play an important role in many types of apoptosis (Verheij *et al.,* 1996; Xia *et al.,* 1996). However, accumulating evidence indicates that the activation of JNK is not linked to cell death for certain types of apoptosis and, in some circumstances, the activation of JNK actually protects cells against apoptosis (Liu *et al.,* 1996; Lenczowski *et al.*, 1997; Roulston *et al.,* 1998). Because the major effect of JNK activation is to increase AP-1 activity, it seems that JNK activation is only required for those forms of apoptosis in which *de novo* protein synthesis is pivotal (Xia *et al.,* 1996; Kasibhatla *et al.,* 1998). For these types of apoptosis, JNK activation may be needed to induce the expression of some critical factors that are essential for initiating cell death. For instance, JNK activation is crucial for NGF withdrawal-induced apoptosis in PC12 cells, in which it mediates the expression of the death factor FasL (Le-Niculescu *et al.*, 1999). However, JNK activation is not required in other forms of apoptosis in which *de novo* protein synthesis is not essential, such as death factors Fas- and TNF-induced apoptosis (Liu *et al.*, 1996; Lenczowski *et al.*, 1997). Moreover, it was also reported that JNK activation had a protective effect against TNF-induced apoptosis (Roulston *et al.*, 1998). Therefore, the role of JNK activation in different types of apoptosis can be completely different. Instead of always inducing or accelerating apoptosis, JNK activation may play the opposite role in certain cases.

II. Methods for the Detection of c-Jun N-terminal Kinase (JNK) Activation during Apoptosis

A. Measurement of JNK Activation by *in Vitro* Kinase Assay

1. Using Endogenous JNK

1. Cells are induced to undergo apoptosis and then are harvested at different time points, such as 0, 5, 15, 30, 60, 90, 120, and 240 min, after the death stimulus. For each time point, 5×10^5 cells are lysed in 300 μl cold lysis buffer [20 m*M* Tris, pH 7.6, 250 m*M* NaCl, 3 m*M* EDTA, 3 m*M* EGTA, 0.5% Nonidet P-40 (NP-40), 2 m*M* dithiothreitol (DTT), 0.5 m*M* phenylmethylsulfonyl fluoride (PMSF), 10 m*M* PNPP, 20 m*M* β-glycerol phosphate, 1 m*M* sodium vanadate, and 1 μg/ml leupeptin] (the last six components need to be added fresh). Samples are rotated gently at 4°C for 30 min and then spun at 15,000*g* for 15 min. Supernatants are collected as cell extracts.

2. One hundred micrograms of cell extract from each sample is mixed with 1 μg of the mouse anti-JNK1 antibody (Pharmingen) and 20 μl 50% protein A-Sepharose. Adjust the total volume to 500 μl with lysis buffer and incubate by rotation for 2 h at 4°C.

3. Spin down the Sepharose beads at low speed (<2000*g*) and wash the beads twice with 500 μl lysis buffer and twice with 500 μl of kinase buffer (20 m*M* Hepes, pH 7.5, 20 m*M* β-glycerol phosphate, 10 m*M* PNPP, 10 m*M* $MgCl_2$, 1 m*M* DTT, and 50 μ*M* sodium vanadate). Incubate the beads in 30 μl of kinase buffer containing 20 μ*M* ATP, 5 μCi [^{32}P]-γ-ATP (Amersham), and 2 μg of GST-c-Jun(1-79) for 30 min at 30°C (Derijard *et al.*, 1994).

4. The reaction is stopped by the addition of 8 μl 5 × sodium dodecyl sulfate (SDS) loading buffer (62.5 m*M* Tris, pH 6.8, 2% SDS, 0.02% bromphenol blue, and 10% glycerol) and boiled for 5 min. The supernatant of each kinase assay is applied to 10% SDS–PAGE. When the front dye (bromphenol blue) reaches the bottom of the gel, stop electrophoresis. After the gel is dried, the JNK activity is displayed by autoradiography.

As shown in Fig. 2, JNK activity is barely detected in nontreated cells as represented by the low level of GST-c-Jun phosphorylation. When cells are treated with UV, IL-1, or TNF, all of which are potent inducers of JNK activation, JNK activity is elevated greatly

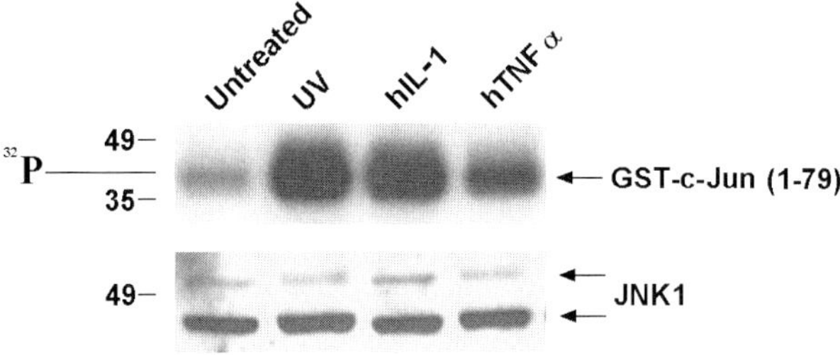

Fig. 2 Detection of JNK activation by *in vitro* kinase assay. Cells are collected in lysis buffer at 15 min after UV (20 J/m^2), hIL-1 (4 μg/ml), or hTNFα (20 μg/ml) treatment. Nontreated cells are used as a control. One hundred micrograms of cell extract from each treatment is used for immune-precipitating JNK1 with anti-JNK1 antibody, and an *in vitro* kinase assay is performed with GST-c-Jun(1-79) as the substrate as described in the text. JNK activity is represented by the level of GST-c-Jun phosphorylation (top). JNK protein content in each sample is measured by Western blotting with an anti-JNK1 antibody (bottom).

as represented by intensive GST-c-Jun phosphorylation (Fig. 2). Therefore, the untreated cells can be used as a negative control whereas TNF-, IL-1-, or UV-treated cells can be used as a positive control. To be sure that the same amount of JNK from each sample is used for the kinase assay, a Western blotting with the same cell extracts as used in the kinase assay is performed to measure the JNK protein content. Accordingly, JNK activity displayed by an *in vitro* kinase assay should be normalized to the JNK protein level (Fig. 2). For certain types of cells, such as NIH3T^3 cells, JNK may have high basal activity. In order to measure JNK activation in those cells accurately, serum starvation may be necessary (Lin *et al.*, 1995).

2. Using Transfected JNK

1. Cells (3 × 10^5 cells per 35-mm dish) are transfected with 0.5 μg of epitope-tagged JNK1 such as HA-JNK1 or Flag-JNK1 (Derijard *et al.*, 1994). Twenty-four hours after transfection, cells are induced to undergo apoptosis with the desired treatment and are collected in lysis buffer at a variety of time points following treatment. Cell extracts are generated as described in Section II,A,1, and typically 3 × 10^5 cells are lysed in 250 ml lysis buffer.

2. To detect the expression of transfected HA-JNK1 or Flag-JNK1, a 50-μg cell extract from each sample is used to perform a Western blotting with an anti-HA or anti-Flag antibody. Then, according to the result of the Western blotting, the same amount of HA-JNK1 or Flag-JNK1 from each sample will be used for a JNK kinase assay as described earlier.

3. The *in vitro* JNK kinase assay is performed as described in Section II,A,1 except that 1 μg of anti-HA or anti-Flag antibody will be used instead of anti-JNK antibody.

Note: In these *in vitro* kinase assays, Western blotting for JNK protein content in each sample is always necessary to normalize JNK activity.

It should be pointed out that measuring the activity of either endogenous or transfected JNK is sufficient to determine whether JNK is activated during apoptosis when all of the cells are treated in the same manner. However, in some circumstances when cells need to be manipulated with the transfection of certain genes prior to be induced to undergo apoptosis, the method with transfected JNK should be used. In those cases, HA- or Flag-JNK1 is normally cotransfected with those genes (Liu *et al.,* 1996). One example for such cases is described in Section III.

B. Determination of JNK Activation by Western Blotting with Antiphospho-JNK Antibody

Because JNK activation requires the phosphorylation of both Thr and Tyr in the Thr-Pro-Tyr (T-P-Y) motif, this unique feature of JNK activation has been used to develop antiphospho-JNK antibodies for detecting JNK activation (Karin, 1998; Davis, 1999).

Cell extracts are generated as described in Section II,A,1. Use a 50-μg cell extract from each sample for Western blotting. Probe the blot with antiphospho-JNK1 polyclonal antibody (New England Biolabs) overnight at 4°C. The dilution of the antiphospho-JNK1 antibody is 1 : 1000 in 5% nonfat dry milk, 1 × phosphate-buffered saline (PBS), 0.05% Tween 20. After the secondary antibody incubation, the phospho-JNK1 is visualized by enhanced chemiluminescence (ECL), according to the manufacturer's (Amersham) instructions. As shown in Fig. 3, no phospho-JNK1 is observed in nontreated cells, yet both of its isoforms (46 and 55 kDa) are detected after UV, IL-1, and TNF treatments. To be sure that the same amount of JNK1 is present in each sample, probe the same blot with anti-JNK1 antibody (Pharmingen) (Fig. 3, bottom).

Note: When using antiphospho-JNK antibody to detect JNK activation, it is critical to collect cell extracts in the lysis buffer containing phosphatase inhibitors as described in Section II,A,1. Otherwise, the signal of phospho-JNK will be weak and the measurement will not be accurate. In addition, although detection of JNK activation with the antiphospho-JNK antibody has been a commonly used method for determining JNK activation, one has to keep in mind that the phosphorylation of JNK is just an indication of JNK activation. Measurement of the phosphorylation of JNK is not equivalent to JNK activity.

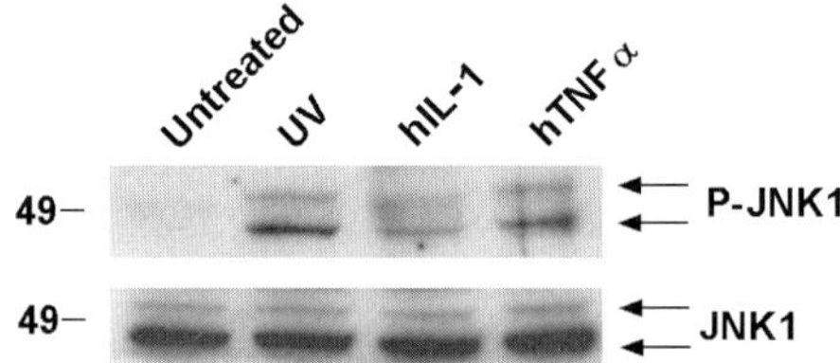

Fig. 3 Detection of JNK activation by Western blotting with antiphospho-JNK1 antibody. Cells are collected in lysis buffer at 15 min after UV (20 J/m^2), hIL-1 (4 μg/ml), or hTNFα (20 μg/ml) treatment. Nontreated cells are used as a control. Fifty micrograms of cell extract from each sample is used for Western blotting with either antiphospho-JNK1 (top) or anti-JNK1 antibody (bottom).

III. Evaluation of the Role of JNK Activation in Apoptosis

After determination of JNK activation during apoptosis, the obvious question is what kind of role the JNK activation plays in apoptosis in the system of interest. In other words, what happens to the process of apoptosis in that system when JNK activity is abolished needs to be investigated. Inhibition of JNK activation can be achieved either by overexpression of the dominant-negative mutants of JNK upstream kinases, such as MKK4/JNKK1 and MKK7/JNKK2 (Lenczowski *et al.,* 1997; Chen *et al.,* 1998), or by using JNK inhibitors, such as SB202190 (Le-Niculescu *et al.,* 1999). The advantage of using dominant-negative mutant MKK7/JNKK2, JNKK2(AA), is that the effect of JNKK2(AA) is very specific and overexpression of it only blocks JNK activation. However, because of the presence of the endogenous wt JNKK2 in cells, the disadvantage of using the JNKK2(AA) mutant is that JNKK2(AA) usually cannot completely block JNK activation. Therefore, a high expression of JNKK2(AA) is essential for blocking JNK activation efficiently. JNKK2(AA) can be introduced into cells either transiently or stably. Transient transfection of JNKK2(AA) is relatively easy and fast, but can only be used with *in vitro* systems. Studies with stable expression of JNKK2(AA) are much cleaner compared to those with transient expression but take much longer. Typically, when JNKK2(AA) is used, 1 μg of Xpress-tagged JNKK2(AA) (Wu *et al.,* 1997; Chen *et al.,* 1998) and 0.5 μg of HA- or Flag-JNK1 are cotransfected with 3×10^5 cells. Twenty-four hours after transfection, cells are treated and collected for *in vitro* kinase as described in Section II,B. However, the usage of JNK inhibitors such as SB202190 (Calbiochem) to block JNK activity is simple and efficient, but less specific. In the case of SB202190, the activity of p38 MAP kinase is also blocked. Normally, 50 μM of SB202190 is added into the culture medium 30 min prior to the apoptotic treatment (Le-Niculescu *et al.,* 1999). Moreover, because those inhibitors are often toxic to cells (Nemoto *et al.,* 1998; Le-Niculescu *et al.,* 1999), one has to be cautious when working with them. Because efficient and specific inhibition of JNK activation is critical for concluding the role of JNK activation in a certain type of apoptosis, both approaches for inhibiting JNK activity should be applied. As reported in many previous studies, the role of JNK activation during apoptosis in a given system could be promotive, protective, or irrelevant.

References

Chen, C. Y., Del Gatto-Konczak, F., Wu, Z., and Karin, M. (1998). Stabilization of interleukin-2 mRNA by the c-Jun NH2-terminal kinase pathway. *Science* **280,** 1945–1949.

Davis, R. J. (1999). Signal transduction by the c-Jun N-terminal kinase. *Biochem. Soc. Symp.* **64,** 1–12.

Derijard, B., Hibi, M., Wu, I. H.,, Barrett, T., Su, B., Deng, T., Karin, M., and Davis, R. J. (1994). JNK1: A protein kinase stimulated by UV light and Ha-Ras that binds and phosphorylates the c-Jun activation domain. *Cell* **76,** 1025–1037.

Derijard, B., Raingeaud, J., Barrett, T., Wu, I. H., Han, J., Ulevitch, R. J., and Davis, R. J. (1995). Independent human MAP-kinase signal transduction pathways defined by MEK and MKK isoforms. *Science* **267,** 682–685.

Dong, C., Yang, D. D., Wysk, M., Whitmarsh, A. J., Davis, R. J., and Flavell, R. A. (1998). Defective T cell differentiation in the absence of Jnk1. *Science* **282,** 2092–2095.

Ellis, R. E., Yuan, J. Y., and Horvitz, H. R. (1991). Mechanisms and functions of cell death. *Annu. Rev. Cell Biol.* **7,** 663–698.

Gupta, S., Barrett, T., Whitmarsh, A. J., Cavanagh, J., Sluss, H. K., Derijard, B., and Davis, R. J. (1996). Selective interaction of JNK protein kinase isoforms with transcription factors. *EMBO J.* **15,** 2760–2770.

Herskowitz, I. (1995). MAP kinase pathways in yeast: For mating and more. *Cell* **80,** 187–197.

Hill, C. S., and Treisman, R. (1995). Transcriptional regulation by extracellular signals: Mechanisms and specificity. *Cell* **80,** 199–211.

Huang, S., Jiang, Y., Li, Z., Nishida, E., Mathias, P., Lin, S., Ulevitch, R. J., Nemerow, G. R., and Han, J. (1997). Apoptosis signaling pathway in T cells is composed of ICE/Ced-3 family proteases and MAP kinase kinase 6b. *Immunity* **6,** 739–749.

Kallunki, T., Su, B., Tsigelny, I., Sluss, H. K., Derijard, B., Moore, G., Davis, R., and Karin, M. (1994). JNK2 contains a specificity-determining region responsible for efficient c-Jun binding and phosphorylation. *Genes Dev.* **8,** 2996–3007.

Karin, M. (1998). Mitogen-activated protein kinase cascades as regulators of stress responses. *Ann. N. Y. Acad. Sci.* **851,** 139–146.

Karin, M., Liu, Z-G., and Zandi, E. (1997). AP-1 function and regulation. *Curr. Opin. Cell Biol.* **9,** 240–246.

Kasibhatla, S., Brunner, T., Genestier, L., Echeverri, F., Mahboubi, A., and Green, D. R. (1998). DNA damaging agents induce expression of Fas ligand and subsequent apoptosis in T lymphocytes via the activation of NF-kappa B and AP-1. *Mol. Cell* **1,** 543–551.

Korsmeyer, S. J. (1995). Regulators of cell death. *Trends Genet.* **11,** 101–105.

Kuan, C. Y., Yang, D. D., Samanta Roy, D. R., Davis, R. J., Rakic, P., and Flavell, R. A. (1999). The Jnk1 and Jnk2 protein kinases are required for regional specific apoptosis during early brain development. *Neuron* **22,** 667–676.

Kyriakis, J. M., Banerjee, P., Nikolakaki, E., Dai, T., Rubie, E. A., Ahmad, M. F., Avruch, J., and Woodgett, J. R. (1994). The stress-activated protein kinase subfamily of c-Jun kinases. *Nature* **369,** 156–160.

Lenczowski, J. M., Dominguez, L., Eder, A. M., King, L. B., Zacharchuk, C. M., and Ashwell, J. D. (1997). Lack of a role for Jun kinase and AP-1 in Fas-induced apoptosis. *Mol. Cell Biol.* **17,** 170–181.

Le-Niculescu, H., Bonfoco, E., Kasuya, Y., Claret, F. X., Green, D. R., and Karin, M. (1999). Withdrawal of survival factors results in activation of the JNK pathway in neuronal cells leading to Fas ligand induction and cell death. *Mol. Cell Biol.* **19,** 751–763.

Lin, A., Minden, A., Martinetto, H., Claret, F. X., Lange-Carter, C., Mercurio, F., Johnson, G. L., and Karin, M. (1995). Identification of a dual specificity kinase that activates the Jun kinases and p38-Mpk2. *Science* **268,** 286–290.

Liu, Z.-G., Hsu, H., Goeddel, D. V., and Karin, M. (1996). Dissection of TNF receptor 1 effector functions: JNK activation is not linked to apoptosis while NF-kappaB activation prevents cell death. *Cell* **87,** 565–576.

Lu, X., Nemoto, S., and Lin, A. (1997). Identification of c-Jun NH2-terminal protein kinase (JNK)-activating kinase 2 as an activator of JNK but not p38. *J. Biol. Chem.* **272,** 24751–24754.

Marshall, C. J. (1995). Specificity of receptor tyrosine kinase signaling: Transient versus sustained extracellular signal-regulated kinase activation. *Cell* **80,** 179–185.

Mohit, A. A., Martin, J. H., and Miller, C. A. (1995). p493F12 kinase: A novel MAP kinase expressed in a subset of neurons in the human nervous system. *Neuron*, **14,** 67–78.

Nemoto, S., Xiang, J., Huang, S., and Lin, A. (1998). Induction of apoptosis by SB202190 through inhibition of p38beta mitogen-activated protein kinase. *J. Biol. Chem.* **273,** 16415–16420.

Roulston, A., Reinhard, C., Amiri, P., and Williams, L. T. (1998). Early activation of c-Jun N-terminal kinase and p38 kinase regulate cell survival in response to tumor necrosis factor alpha. *J. Biol. Chem.* **273,** 10232–10239.

Sabapathy, K., Hu, Y., Kallunki, T., Schreiber, M., David, J. P., Jochum, W., Wagner, E. F., and Karin, M. (1999). JNK2 is required for efficient T-cell activation and apoptosis but not for normal lymphocyte development. *Curr. Biol.* **9,** 116–125.

Sanchez, I., Hughes, R. T., Mayer, B. J., Yee, K., Woodgett, J. R., Avruch, J., Kyriakis, J. M., and Zon, L. I. (1994). Role of SAPK/ERK kinase-1 in the stress-activated pathway regulating transcription factor c-Jun. *Nature* **372,** 794–798.

Sluss, H. K., Barrett, T., Derijard, B., and Davis, R. J. (1994). Signal transduction by tumor necrosis factor mediated by JNK protein kinases. *Mol. Cell Biol.* **14,** 8376–8384.

Sluss, H. K., Han, Z., Barrett, T., Davis, R. J., and Ip, Y. T. (1996). A JNK signal transduction pathway that mediates morphogenesis and an immune response in Drosophila. *Genes Dev.* **10,** 2745–2758.

Tournier, C., Whitmarsh, A. J., Cavanagh, J., Barrett, T., and Davis, R. J. (1997). Mitogen-activated protein kinase kinase 7 is an activator of the c-Jun NH2-terminal kinase. *Proc. Natl. Acad. Sci. USA* **94,** 7337–7342.

Verheij, M., Bose, R., Lin, X. H., Yao, B., Jarvis, W. D., Grant, S., Birrer, M. J., Szabo, E., Zon, L. I., Kyriakis, J. M., Haimovitz-Friedman, A., Fuks, Z., and Kolesnick, R. N. (1996). Requirement for ceramide-initiated SAPK/JNK signalling in stress-induced apoptosis. *Nature* **380,** 75–79.

Wu, Z., Wu, J., Jacinto, E., and Karin, M. (1997). Molecular cloning and characterization of human JNKK2, a novel Jun NH2-terminal kinase-specific kinase. *Mol. Cell Biol.* **17,** 7407–7416.

Xia, Z., Dickens, M., Raingeaud, J., Davis, R. J., and Greenberg, M. E. (1996). Opposing effects of ERK and JNK-p38 MAP kinases on apoptosis. *Science* **270,** 1326–1331.

Yang, D. D., Conze, D., Whitmarsh, A. J., Barrett, T., Davis, R. J., Rincon, M., and Flavell, R. A. (1998). Differentiation of $CD4^+$ T cells to Th1 cells requires MAP kinase JNK2. *Immunity* **9,** 575–585.

Yang, D. D., Kuan, C. Y., Whitmarsh, A. J., Rincon, M., Zheng, T. S., Davis, R. J., Rakic, P., and Flavell, R. A. (1997). Absence of excitotoxicity-induced apoptosis in the hippocampus of mice lacking the Jnk3 gene. *Nature* **389,** 865–870.

CHAPTER 9

Methods for Studying Pro- and Antiapoptotic Genes in Nonimmortal Cells

Mila E. McCurrach and Scott W. Lowe

Cold Spring Harbor Laboratory
Cold Spring Harbor, New York 11724

METHODS IN CELL BIOLOGY, VOL. 66

0091-679X/01 $35.00

I. Introduction

During the last decade, basic cancer research has produced remarkable advances in our understanding of cancer biology and cancer genetics. Among the most important of these advances is the realization that apoptosis, and the genes that control it, has a profound effect on the malignant phenotype (for review, see Lowe and Lin (2000). It is now clear that some oncogenic mutations (e.g., Bcl-2 overexpression or p53 mutations) disrupt apoptosis, leading to tumor initiation, progression, or metastasis. Conversely, other oncogenic changes can promote apoptosis (e.g., c-Myc activation), thereby producing selective pressure to override apoptosis during multistage carcinogenesis. Finally, it is now well documented that most cytotoxic anticancer agents induce apoptosis, raising the intriguing possibility that defects in apoptotic programs contribute to treatment failure. Because the same mutations that suppress apoptosis during tumor development also reduce treatment sensitivity, apoptosis provides a conceptual framework to link cancer genetics with cancer therapy.

As a result of the profound relationship between the control of apoptosis and malignancy, an intense research effort has been directed toward understanding the genes that modulate apoptosis during tumorigenesis and therapy, as well as toward exploiting apoptotic programs for therapeutic benefit. Many studies that characterize the apoptotic activities of known or novel genes involve overexpression of the pro- or antiapoptotic activity in established cell lines (usually human tumor-derived cell lines or immortalized rodent fibroblasts). Although such studies have produced important insights into apoptotic gene function, they suffer from inherent limitations. First, tumors acquire genetic alterations that affect apoptosis, and establishment of permanent cell lines in culture provides a strong antiapoptotic selection. These changes are generally not defined and can confound data interpretation. Second, studies using permanent cell lines often rely on either transient or stable overexpression following DNA transfection. Transient transfections can produce protein levels that are extremely unphysiological, and stable cell lines suffer from clonal variation and acquire mutations during clonal expansion. Both methods are particularly problematic when assessing the activity of putative proapoptotic genes, which may simply induce apoptosis by toxic overexpression or may be masked by secondary mutations that arise during the generation of stable cell lines. As a consequence, these methods can produce misleading information concerning gene function.

Nonimmortal cells provide an excellent alternative for studying genes that control apoptosis and oncogenic transformation, as experimental conclusions are not limited by the unknown alterations prevalent in immortal cell lines. Moreover, the availability of null mutant cells derived from "knockout" mice has increased the power and versatility of nonimmortal cells; hence, it is possible to analyze loss-of-function mutations in a manner not previously possible in mammalian systems. However, nonimmortal cells have been an impractical system for most gene function studies because nonimmortal cells invariably display extremely poor transfection efficiencies and a relatively short life span in culture. Thus, stable cell lines cannot be generated in the absence of immortalizing oncogenes, which in turn can disrupt normal apoptotic programs.

Improvements in retroviral vectors and packaging systems have made recombinant retroviruses the method of choice in many gene transfer studies and it is particularly suitable for genetic manipulation in nonimmortal cells. Among the salient features of retroviral-mediated gene transfer are that (i) genes can be introduced synchronously and stably into whole cell populations; (ii) genes are expressed shortly after gene transfer; and (iii) the retrovirus itself has no cytopathic effects (Pear *et al.,* 1996). Many cell types, including nonimmortal cells, can be infected to near 100% efficiency and, in principle, the consequences of gene expression can be examined within hours of gene transfer and subsequently for many weeks. Because genes are introduced into whole cell populations, gene function can be analyzed rapidly without substantial expansion of the cell population in culture. This virtually eliminates problems associated with clonal variation and minimizes selection for variants that might otherwise occur upon clonal expansion. This is particularly important when studying proapoptotic genes. High-efficiency retroviral gene transfer acts as an "inducible" system, with retroviral infection itself acting as the "on" switch. Finally, although retroviral vectors are overexpressed, generally only one or at most several proviruses integrate into the target cells, resulting in more physiological expression levels than are observed with standard transfection techniques.

This chapter describes procedures used in our laboratory for studying gene function in early passage mouse embryo fibroblasts or other nonimmortalized cell types (see, e.g., McCurrach *et al.,* 1997; Samuelson and Lowe, 1997; Serrano *et al.,* 1997; Schmitt *et al.,* 1999; Soengas *et al.,* 1999). Our approach is to combine high-titer retroviral vectors with various null mutant cells to assess combinations of genes in a defined genetic background. A typical experiment involves four stages: (i) isolation of fibroblasts from normal and/or knockout mice; (ii) generation of a retroviral vector; (iii) infection of cells and selection for infected populations; (iv) analysis of cell death, transformation, etc. This approach allows one to mix and match dominant and recessive activities to build model "tumor" cells from normal counterparts without clonal expansion in culture. As a result, one can have strict control over genetic background, thereby avoiding complications arising from unknown mutations that accumulate in immortal or tumor-derived lines. Although the power of this approach is in its well-defined genetic nature, it is important to recognize that these systems are necessarily artificial. Therefore, it is recommended that, having understood pro- or antiapoptotic gene function in normal cells, one tests predictions of these systems in animal models or by using patient material.

II. Mouse Embryo Fibroblasts (MEFs)

Mouse embryonic fibroblasts are an excellent system for studying genes in nonimmortal cells. Importantly, MEFs are isolated easily from most "knockout" mice, allowing analysis of loss of gene function in cultured cells. In addition, MEFs are relatively hardy cells, adapt well to culture conditions, and can be passaged 7–10 times without difficulty. Eventually, MEFs will undergo crisis (although some cells will undergo genetic mutations and immortalize) (Tadaro and Green, 1963). Immortalized MEFs often acquire mutations in p53 or ARF (Harvey and Levine, 1991; Kamijo *et al.,* 1997), which can disrupt some apoptotic programs. The longer MEFs are cultured, the more likely they will acquire genetic changes. Although the expression of certain oncogenes can facilitate MEF immortalization, these changes may also affect the pathway being analyzed. Hence, it is advisable to use cells that have been passaged minimally (less than five times). Some MEF types (e.g., p53 null) are genetically unstable (Livingstone *et al.,* 1992), again making it crucial to conduct experiments as rapidly as possible.

A. Isolation of Cells

i. Sacrifice the pregnant mouse at day 13.5 of gestation (or 13 days after observing a mucus plug). Refer to your animal facility for the proper humane procedure.

ii. Wash the mouse torso with 70% EtOH to disinfect.

iii. Using sterile technique, extract the two uterine horns containing the embryos and place them in a 10-cm dish containing phosphate-buffered saline (PBS).

iv. Using a scalpel and tweezers, cut the uterus between each embryo, dividing them into individual segments. Place each embryo into a separate 6-cm dish containing PBS.

v. Dissect away the decidua (maternally derived). Save the yolk sac (embryonic) for genotyping.

vi. Examine the embryo. Note whether large, small, live or dead.

vii. Clamp down on the neck with tweezers and cut off the head with scalpel or sharp scissors. Freeze it for genotyping (the head provides a secondary and more reliable source of embryonic DNA).

viii. Dissect out red organs (e.g., heart, liver, kidney) with hooked watchmaker forceps.

ix. Transfer the torso to a 6-cm plate with 1.0 ml trypsin.

x. Sterilize all instruments between the dissection of each embryo to avoid cross-contamination (wash in PBS, dip in EtOH, and flame).

xi. Mince the embryos with fine scissors or scalpel to approximately 1-mm^3 pieces (this should take only 1–2 min, no need to overdo). Again, flame-sterilize the instruments between embryos.

xii. Place each dish in a 37°C incubator on an angle (e.g., on a 5-ml glass pipette) such that the trypsin and the embryos collect at one end of the plate.

xiii. Incubate the plates for 45 min total. Halfway through the incubation, gently pipette each cell suspension up and down through a Pasteur pipette 10–15 times to homogenize.

xiv. Prepare 75-cm^2 flasks with 15 ml growth medium.

xv. After the incubation, add 5 ml growth medium to each plate, pipette up and down 15 times, and transfer each embryo to a 75-cm^2 flask. Record these flasks as passage 1.

xvi. Allow the cells to grow to confluence. This typically takes 48 h.

xvii. Split the cells from the 75-cm^2 flask to 2 × 150-cm^2 flasks.

xviii. Grow the cells until very confluent (e.g., 1 day after they are all touching).

xix. Freeze down the cells from both flasks into 15–18 vials. See Section II,B,2.

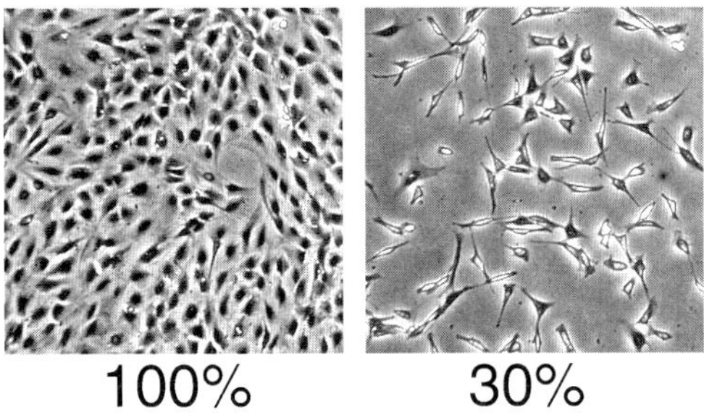

Fig. 1 Mouse embryo fibroblasts plated at either 100 or 30% confluence.

Notes:

a. When comparing the impact of gene deletion of the cellular phenotype, genetic background differences are minimized by comparing wild-type (+/+) and null (−/−) littermate embryos. This is done by using intercrosses between mice heterozygous (+/−) for a gene of interest. In this situation, rapid genotyping can reduce the total number of MEFs to freeze by identifying the +/−, which can be discarded.

b. On average, MEFs double every 18 h. Figure 1 shows MEFs at 30 and 100% confluence. (Approximately 5×10^6 MEFs per 10-cm dish equals 100% confluence.)

c. MEFs that have been isolated well should appear to be a homogeneous cell population by passage 2 with few, if any, floating cells.

d. Using this protocol it is rare to have contamination of cell cultures. Should contamination occur, it is best to dispose of the contaminated plates and start over. If reagents are limiting, MEFs can be grown in the presence of fungizone (GIBCO-BRL) or 100 μg/ml gentamycin (GIBCO-BRL).

e. It is possible to isolate MEFs as early as day 9.5 (Yeh *et al.,* 1998), but this is difficult due to the size of the embryo and cell yields are reduced dramatically. These mutant MEF populations must be compared to wild-type MEFs isolated at the same gestational age.

B. Maintenance of MEFs

1. MEF Culture

MEFs are best maintained between 30 and 80% confluent at 37°C in 7% CO^2. They should be passaged every 3 to 4 days and split between 1 : 3 and 1 : 6 depending on growth rate. To split a 10-cm dish, wash the cells twice with PBS and then incubate 2–5 minutes in 1 ml of trypsin at 37°C. The passage number should always be recorded.

Solutions

MEF growth media: Dulbecco's modified Eagle medium (DMEM) (GIBCO-BRL) supplemented with 10% fetal bovine serum (FBS) (v/v), 50 U/ml penicillin G, and 100 μg/ml streptomycin sulfate

Trypsin: 0.125% trypsin (GIBCO) and 1 m*M* EDTA in Hepes-buffered saline. Sterile filter and store at −20°C

Hepes-buffered saline (per liter): 7.07 g NaCl, 0.4 g KCl, 0.043 g Na_2HPO_4, 1.0 g D-glucose, and 4.77 g Hepes. Combine ingredients and bring volume up to 1 liter with dH_2O. pH to 7.3 with NaOH. Sterile filter and store at 4°C.

Phosphate-buffered saline (per liter): 7.07 g NaCl, 0.4 g KCl, and 0.06 g KH2PO4. Combine ingredients and bring volume up to 1 liter with dH_2O.

2. Freezing MEFs

Trypsinize the cells and collect them in a 50-ml Falcon tube. Pellet cells by centrifugation (1500 rpm, 5 min) and discard supernatant. Resuspend cell pellet in ice cold freezing media (1 ml per vial to be frozen) and aliquot into cold freezing vials. **Once cells are in freezing media they must be kept cold**. Cells should be frozen gradually by placing them in a styrofoam rack at −70°C over night. Cells should be transferred to −180°C within 2 weeks.

Solutions

Freezing media: DMEM supplemented with 10% dimethyl sulfoxide (DMSO)(v/v), 30% FBS (v/v), 50 U/ml penicillin G, and 100 μg/ml streptomycin sulfate

III. Retroviral-Mediated Gene Transfer

A recombinant retrovirus transduces a nonviral gene into cells (for review, see Cepko, 1997). These vectors utilize the same machinery of naturally occurring retroviruses to replicate and integrate into a host genome. Naturally occurring retroviruses infect their target cells via interaction of a viral envelope glycoprotein with a host cell receptor. The viral RNA is injected into the host cell and a DNA copy is generated by reverse transcriptase. The DNA copy (provirus) integrates into the host cell genome, usually requiring M phase (Roe *et al.*, 1993). Once integrated, the host cell transcribes many RNA copies of the viral genome. Envelope and capsid proteins are translated and assembled into new virus particles, each containing reverse transcriptase and the viral genomic RNA.

Recombinant retrovirus particles are essentially identical to those occurring naturally. They are generated by introducing a retroviral vector into a retrovirus packaging cell line. Retroviral vectors contain the DNA sequences intended for gene transduction flanked by the signals present at the ends of the retroviral genome [long terminal repeats (LTR)] and a retroviral packaging site (Ψ), which harbor all the integration and expression sequences. The packaging cells contain the retroviral structural gene products [the retroviral core (*gag*), polymerase (*pol*) and envelope (*env*)] but lack a retroviral packaging site and LTR. These proteins function in *trans* with the elements supplied in the retroviral vector to package retroviral RNA into virions. Thus, retroviral vector particles produced using packaging cells contain all the components necessary to infect cells and integrate into their genome, but contain none of the structural genes that are required to package new virus; hence, they are often referred to as replication defective.

The range of cell types from different species that can be infected by a particular retrovirus is called the "host range" of the virus and is mainly determined by the envelope protein and the proliferation state of the target cell (Weiss *et al.,* 1985). Packaging cell lines have been developed from various retroviruses with different host ranges, including avian leukosis virus, lentivirus and ecotropic (mouse and rat infecting), amphotropic (human, feline, canine, guinea pig, rabbit, rat, and mouse infecting), and pleiotropic (all species infecting) Moloney murine leukemia viruses (Miller, 1990; Burns *et al.,* 1993; Williams, 1995; Mulligan, 1993; Verma, 1994; Leiden, 1995; Naldini *et al.,* 1996).

Each packaging system has distinct advantages. The ecotropic packaging lines allow for the relatively safe generation of oncogene expressing viruses, as they will not infect human cells. Pleiotropic murine systems have been generated by replacing the retroviral envelope proteins with the G glycoprotein of vesicular stomatitis virus (VSV-G) (Burns *et al.,* 1993). Because VSV-G interacts with a phospholipid component of the cell membrane to mediate viral entry rather than a specific protein receptor, it has an extremely broad host range. In addition, VSV-G retroviral vectors are highly stable and can thus be concentrated by ultracentrifugation without a loss of infectivity. Finally, lentivirus-based vectors can integrate into the host chromosome without the need for M phase and can thus be used to infect senescent, arrested, or terminally differentiated cells (Naldini *et al.,* 1996). Lentivirus-based systems with a VSV-G envelope have also been developed (Kafri *et al.,* 1999).

Two types of packaging strategies have been used: stable and transient. Stable systems require the generation of a packaging line (murine or avian cells), which contains both the packaging signals and the retroviral vector, a procedure that takes 1–2 months under optimal circumstances. However, once the line is generated, it will always produce virus. Transient systems were a step forward in retrovirus technology and exploit highly transfectable packaging cell lines in which the retroviral plasmid is introduced by transient transfection. The virus is produced by the cells within 48 h and can be collected and used for infections immediately. The one drawback is that viral production decreases rapidly over time after the transfection. Hence, one must continue to repeat the packaging line transfection to continuously produce the virus.

Our laboratory has found that we can achieve >80% infection efficiency using the Phoenix packaging cells generated by Gary Nolan (http://www.stanford.edu/group/nolan/NL-Homepage.html). Phoenix cells were generated from 293T cells (an adenoviral-transformed human embryonic kidney cell line that expresses a temperature-sensitive SV40 large T antigen)(DuBridge *et al.,* 1987). The Phoenix system is highly transfectable and, depending on the cell line, can generate either ecotropic or amphotropic retrovirus. The retroviral structural genes are expressed from two different stably integrated plasmids that contain multiple mutations (Danos and Mulligan, 1988); hence, at least three recombination events are necessary for the generation of replication-competent virus. Its major advantage for studying pro- and antiapoptotic genes in nonimmortal cells is the rapidity with which high-titer virus can be produced. In addition, because >80% of target cells can be infected, it is possible to analyze growth inhibitory genes shortly after

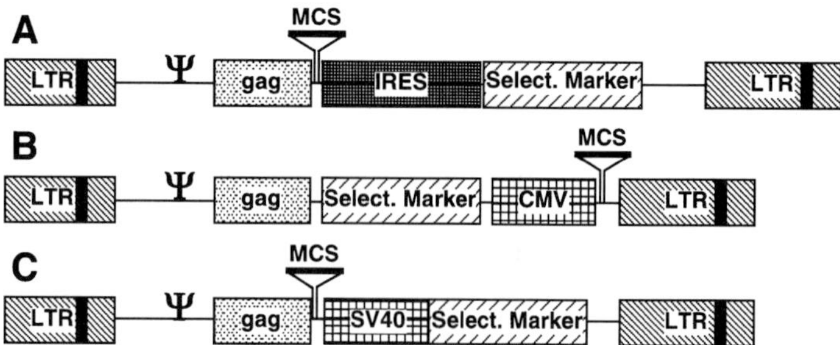

Fig. 2 Diagram of three retroviral vectors: (A) pWZL, (B) LPC (McCurrach *et al.,* 1997), and (C) pBABE (Morgenstern and Land, 1990). All contain the viral long-terminal repeat sequences (LTR), RNA packaging signal (Ψ), multicloning site (MCS), and mutagenized *gag* gene, which cannot generate a functional protein but whose sequence location is necessary for a high viral titer.

gene transfer. Due to the fact that 293 cells constitutively express the adenovirus E1B oncoproteins, which block apoptosis, they are protected from the effects of many of the proapoptotic genes we study. The vectors used in retroviral gene transfer are relatively similar (Fig. 2). The "gene of interest" is inserted in a multicloning site and can be driven from the LTR or an internal promoter such as the cytomegalovirus immediate early promoter (CMV). Many vectors contain an antibiotic resistance gene that is either driven by a separate promoter than the "gene of interest" or, in some cases, by the same promoter but separated by an "internal ribosome entry site" (IRES), allowing coexpression via a bicistronic message. Another type of selectable marker is fluorescence. These are either green fluorescence protein (GFP) based or cell surface markers, such as CD4 or CD8, which can be recognized by fluorescently conjugated antibodies or attached to antibody-coated magnetic beads (Gaines and Wojchowski, 1999). Finally, some vectors have introduced lox P sites in the LTR that allow for the deletion of the provirus once it has integrated (Bergemann *et al.,* 1995).

The choice of vector depends on the individual experiment. Some considerations are the selection options available for the target cells, the desired level of expression, the response of cells to the gene of interest, and the rapidity with which the cells must be analyzed postinfection. Cells that express a selectable marker, either because they were infected previously or are derived from a "knockout" mouse (which usually express the neomycin resistance gene), must use a different selectable marker.

Different promoters result in differing levels of protein. Typically, genes driven from CMV promoters result in higher protein levels than LTR-driven genes. These differences are not greatly significant and can vary with the "gene of interest." Nonetheless, it is advisable to use the same vector when comparing different genes of interest.

Proapoptotic genes can be introduced into nonimmortalized cells but may require immediate analysis after gene delivery. Puromycin is the most rapid drug selectable marker but to date has not been usable in IRES-type vectors. Fluorescence type markers can be detected visually for nonselected infected populations or FACS sorted within 24–48 h after infection and have the advantage of not requiring a drug to select infected cells. However, some oncogene-expressing populations or cells expressing proapoptotic genes may not survive FACS sorting due to the fact that the suspension culture can trigger apoptosis. Also keep in mind that cells that express proapoptotic genes can be selected against even during a short-term culture. When working with these types of genes, it may be advisable to use a vector with an IRES construction to ensure the "gene of interest" expression along with the selectible marker or an inducible system [e.g., RetroOn (Clonetech) or estrogen receptor fusions (Evan *et al.*, 1992)].

IV. Basic Infection Protocol for MEFs (Using the Phoenix System)

A. Culture of Packaging Cells (Phoenix)

Phoenix cells should be maintained in MEF growth media at 37°C in 7% CO_2 and passaged 1 : 5 every 3 days. A 10-cm dish is subcultured by washing the cells twice with PBS and then incubating 2–5 min in 1 ml of trypsin at 37°C. These cells readily detach from the plate so care must be taken not to wash them off in the PBS. Over time, Phoenix cells will discard one of their viral components and begin to generate diminished viral titers. It is therefore recommended to discard any cells higher than passage 20 and replace them with an earlier passage frozen stock. It is possible to reselect the packaging cells for the viral plasmids. However, this has not proven to be very successful in our hands.

B. Transfection of Packaging Cells

A variety of methods can be used to introduce viral vectors into the packaging cell line. The standard calcium phosphate coprecipitation method works extremely well for Phoenix cells (adapted from Sambrook *et al.*, 1989).

i. Plate 5×10^6 packaging cells (Phoenix) per 10-cm plate per transfection in MEF growth media (cells should be fresh and actively growing before splitting). Incubate for 12–24 h.

ii. Immediately prior to transfection, change the media on the packaging cells.

iii. DNA–calcium phosphate coprecipitates are formed by the mixing of two solutions. One contains 20 μg retroviral vector DNA in 0.25X TE, 62.5 μl 2 M $CaCl_2$, and H_2O to 500 μl and the other is 500 μL 2X BBS. Mix the DNA solution to the BBS dropwise. While the DNA solution is being added, constantly mix the BBS solution by bubbling air into it through a sterile Pasteur pipette attached to an automated pipette aid. The precipitate forms in 10–15 min when the solution turns slightly milky. This can be confirmed by looking at a drop on a hemocytometer. Add 1 ml to each 10-cm plate and mix thoroughly.

iv. Incubate the plates overnight at 37°C (no longer than 18 h).

v. Stop transfection by changing the media.

Notes:

a. Bubbling the BBS solution while mixing is important for a uniform precipitate.

b. The precipitate should be barely visible when first added to cells and should be uniform in size. Large precipitate can reduce transfection efficiency.

c. Some proapoptotic genes can cause apoptosis even in the packaging lines (e.g., bax). In these cases it is possible to reduce cell death and increase viral titers by cotransfecting a nonretroviral plasmid that expresses Bcl-2 along with the proapoptotic retroviral vector (15 μg of each vector) or by culturing the packaging lines with zVAD-fmk (Zhu *et al.*, 1995).

d. We have had success increasing the viral titer of later passage packaging cells (>15) by cotransfecting a virus helper plasmid along with the retroviral vector (15 μg of each vector) (Finer *et al.*, 1994; Soneoka *et al.*, 1995). The helper plasmid encodes all of the viral structural components contained in the Phoenix packaging cells. Hence, any of the packaging components that may have been lost due to selection in longer term culture are replaced. It is possible to generate a high-titer virus using the helper virus/retroviral vector cotransfection in 293T cells alone (Finer *et al.*, 1994; Soneoka *et al.*, 1995).

C. Collection of Viral Supernatant

Following DNA transfection of viral constructs, the virus is produced for several days. Strategies for collecting the virus supernatant are variable, but most work equally well. The main consideration is virus stability. Viruses have a half-life of 3–6 h at 37°C (Sanes *et al.*, 1986). Viral breakdown can be slowed significantly by maintaining the supernatant at 4°C (Le Doux *et al.*, 1999) or stopped by freezing at −80°C (the act of freezing and thawing reduces viral titers twofold) (Pear *et al.*, 1996). The highest viral titers are obtained by using a supernatant directly from the packaging lines to infect the target cells. In general, this is not necessary as modern packaging systems can produce manyfold more virus than is needed for a standard infection and much of virus infection is limited not by particle number but by the target cell (e.g., growth rate). Our laboratory uses a procedure adapted from Owen Witte and Charles Sawyers (University of California, Los Angeles, CA) and M. Roussel (St. Jude Children's Research Hospital, Memphis, TN).

The viral supernatant is collected every 5–6 h in 3–4 ml of media during the 24–72 h after transfection. The collected supernatant is pooled and kept on ice for the duration of the collection. The viral supernatant must be filtered through a 0.45-μm syringe filter or centrifuged twice at 1000 rpm (Sorvall tabletop or equivalent) for 5 min to eliminate any packaging cells that might contaminate the target cell populations. It can then either be used immediately or frozen at −70°C until needed.

A sample collection scheme for Phoenix cells per 10-cm plate is

9 AM—stop transfection

8 PM—replace media with 6 ml (more media is added for overnight collection)

9 AM—collect and replace with fresh 4 ml
2 PM—collect and replace with fresh 4 ml
7 PM—collect and replace with fresh 6 ml
9 AM—collect and replace with fresh 4 ml
2 PM—collect and replace with fresh 4 ml
7 PM—collect and replace with fresh 6 ml
9 AM—collect, filter pooled supernatant, and freeze

D. Infection

i. Twelve to 24 hours prior to infection, the target cells must be plated at 30% confluence (for MEFs 7.5×10^5 in 10-cm plates). The cells should be fresh and actively growing (remember that when using ecotropic or amphotropic viruses, retroviral integration requires M phase. This is not true for lentivirus-based systems.)

ii. Thaw 3–4 ml of frozen supernatant (per 10-cm plate) on ice and add polybrene to a final concentration of 10 μg/ml. Replace the media on the target cells with the viral supernatant and return to incubator for 3–4 h.

iii. Repeat step ii twice more, for a total of three infections.

iv. One hour after the last infection, add 10 ml media to the plate (no polybrene) and incubate overnight.

v. If the vector contains a drug-selectable marker, the cells are either split and selected or simply selected 32 h after the last infection, depending on growth. Table I shows target concentrations of several selection drugs. It is important, however, to always perform a drug titration on uninfected cells to establish the optimum concentration. Keep in mind that temperature and media composition can affect the efficiency of selection drugs.

vi. If the vector contains a sortable marker, the cells can be trypsinized and sorted by either flow cytometry or magnetic beads 48 h after infection (to ensure expression of the marker). For cell surface markers, trypsinization must be kept to a minimum.

Table I
Target Concentrations of Selection Drugs

Drug	Concentration	Days on selection	Advantages	Drawbacks
Puromycin	2–2.5 μg/ml	2–3	Inexpensive, rapid selection, consistent optimal dose range	Cannot be used with IRES vectors
Hygromycin	75–100 μg/ml	4–5	Inexpensive, can be used with IRES vectors	Widely varying optimal dose range
Neomycin (G418)	200 μg/ml	6–7	Can be used with IRES vectors	Widely varying optimal dose range, expensive, long selection time

Notes:

a. Polybrene is a polycation that increases infection efficiency over 100-fold (for review see, Burns *et al.,* 1993).

b. The amount of virus supernatant used and the number of infections can alter the percentage of transduced cells or the average virus copy number (multiplicity of infection). This can have an impact on expression levels, so the infection procedure should be optimized for each application and then kept constant.

Solutions

2X BBS: 50 m*M* *N,N*-bis(2-hydroxyethyl)-2 aminoethanesulfonic acid (BES), 280 m*M* NaCl, 1.5 m*M* Na_2HPO_4, pH 6.95, 800 ml H_2O, and pH to 6.95 with NaOH (pH is extremely important for good precipitate formation and high efficiency gene transfer. Use a BBS that is known to work as your pH standard or generate several batches at varying pHs and test transfection efficiency using a control plasmid). Filter sterilize solution through a 0.45-μm filter and freeze in aliquots at −20°C. BBS can be stored frozen for three months, and frozen and thawed once.

2M $CaCl_2$: 2 *M* $CaCl_2$ and 0.01 *M* Hepes, pH 5.5 Filter sterilize and store at −20°C.

E. Suggestions and Tips

1. Controls

All infections should be performed in parallel. This is particularly important when studying apoptotic genes for which there may be selection for or against high expressing members of the population. In addition, all experiments should include a control vector expressing selectable marker alone to infect parallel cultures.

A parallel transfection and infection using a control lacZ or GFP should be included with each experiment. By looking for β-Gal staining or green fluorescence in both the packaging and the infected cells one can readily evaluate both transfection and infection efficiency. Although this cannot directly monitor experimental viruses, low transduction efficiencies with these viruses will warn of general problems or inconsistencies.

Another useful control is a noninfected target plate that is put on selection. Waiting for all the cells to die on the control plate ensures a complete selection. The concentrations of selection drugs given in Table I are general guidelines for normal primary MEFs. For other cell types, the optimal drug concentration must be determined empirically.

2. Variables

The protocol described here should generate a virus that is capable of infecting exponentially growing MEFs at an efficiency of 75–100%. However, situations may arise where efficiencies are lower. If this is the case, several modifications to the protocol may improve transduction efficiency. In retroviral infections there are two variables that can be limiting: concentration of the viral particles and susceptibility of the target cells to infection.

The concentration of viral particles can be bolstered by either increasing the number of virus particles per volume of supernatant or slowing the rate of viral breakdown. Freezing viral supernatant decreases the titer by about twofold (Pear *et al.,* 1996). Hence, using a fresh supernatant should bolster infectivity. It is also possible to concentrate the virus by centrifugation. Centriprep-30 filters (Amicon) are recommended because they reduce virus loss during centrifugation. Keep in mind that the virus should be maintained at 4°C to slow breakdown.

Another technique that can be tried is centrifugation of the virus onto the target cells, called "spin infection" (Kotani *et al.,* 1994). After the viral supernatant is placed on the target cells, the plates are sealed with Parafilm and placed in a Sorvall tabletop centrifuge (or its equivalent) and spun at 1500 rpm for 45 min at room temperature. Round plates can be taped onto a 96-well swinging plate holder up to three plates high without breakage.

Supplementing the transfection medium with several factors may increase viral titers. For example, the addition of 25 μ*M* chloroquine during transfection may increase transfection efficiency and thus virus titers. (Note that chloroquine is toxic and should not be left on the cells for longer than 12 h.) The addition of 10 m*M* sodium butyrate (pH 7) 12–24 h after transfection has been reported to increase retroviral titers from severalfold to 10-fold. This treatment is thought to increase transcription from the Moloney LTR, as well as other promoters, including CMV (Soneoka *et al.,* 1995). Treatment should last for 12–14 h and cells should be washed with PBS prior to harvesting supernatant.

The final virus titer is determined by two opposing forces: production and breakdown. By increasing the stability of the virus, the effective concentration increases. Viruses have a relatively short half-life at 37°C (3–6 h) (Sanes *et al.,* 1986). Collection of the virus at 32–34°C has been reported to increase viral titers by 5- to 15-fold (Kotani *et al.,* 1994). When used with 293T-derived packaging lines, in conjunction with constructs that contain the SV40 origin of replication, this technique may further increase viral titers due to the fact that 293T cells contain a temperature-sensitive large T antigen whose permissive temperature is 32°C (DuBridge *et al.,* 1987). Thus, incubation at this temperature may result in plasmid replication and higher retroviral titers.

The second limiting factor in retroviral transduction is the susceptibility of the target cells to infection. For most viruses to infect and integrate into a host genome, they must bind the host cell receptor, be internalized, and remain intact long enough for the target cell to pass through M phase (5.5–7.5 h) (Andreadis *et al.,* 1997). One way to increase susceptibility of the target cell is to enhance the virus–receptor interaction. This can be achieved by overexpressing the virus receptor in the target cells or by increasing the polybrene concentration. Polybrene gives the best results at concentrations between 2 and 10 μg/ml but can be toxic in some cells types. A second strategy is to increase the probability that the infected cells will enter M phase prior to virus breakdown. This can be accomplished by increasing the proliferation rate of the target cells or by multiple infections. In a population with a doubling time of 18 h, only a fraction of the cells will be going through M phase during the window of time that the virus is active. By infecting three times at 3- to 4-h intervals, this fraction is increased. However, a slower or faster growing cell population may require more or differently spaced infection strategies.

3. Multiple Gene Infection

It is possible to infect one cell population with multiple genes. This can be accomplished by generating vectors that coexpress two genes (plus marker) or by either sequential or coinfection with two viruses. Sequential infections are more time-consuming but have the advantage of ensuring that a large enough population of cells receives both genes while avoiding the need to use two different selection agents simultaneously (the optimal concentration of selection drugs can vary in the presence of other drugs). Coinfections can be performed with equal success by either transfecting two plasmids onto the same packaging cells or mixing two viral supernatants. Drug interaction issues can be resolved by performing a sequential selections or by combining a drug-selectable vector and a fluorescent one.

4. Cautions

The generation of viral supernatants may, depending on the retroviral insert, contain potentially hazardous recombinant virus. Use caution and the appropriate protection when producing, using, and storing recombinant retroviruses, especially those with amphotropic and polytropic host ranges. Appropriate NIH and other institutional guidelines should be followed when using recombinant retrovirus production systems.

V. Analysis of Cells

With the exception of biochemical purifications, sufficient material can be obtained from infected cell populations for most applications. In general, the length of time that a given culture is maintained should be kept to a minimum to avoid selection for variants.

A. Confirmation of Transduced Levels of Gene

Transgene expression should be monitored in all cell populations. This can be accomplished by a variety of methods, including Northern and Western blotting. This is particularly important when analyzing several mutants of a gene of interest or when comparing the impact of transgene expression in cells with different genetic backgrounds—interpretation errors can arise if mutants have different stabilities or if the genetic background affects transgene expression directly or indirectly. Our laboratory has found that constructs made in different vectors give rise to markedly different levels of gene expression. In addition, we have found enormous differences in gene expression of some transduced genes after several passages in culture, probably due to selective pressure for or against the transduced gene.

Table II
Apoptosis Inducing Agents

Agent	Dose range for E1A/ras expressing MEFs	Time to collection (h)
Adriamycin	0.1–0.5 μg/ml	24
Etoposide	1–3 μ*M*	24
Tumor necrosis factor	2–100 ng/ml	24
Paclitaxel	50–1000 n*M*	48
Hypoxia	2–0.02% oxygen	24
Low serum	0.2–2%	48
γ radiation	10–20 Gy	24–48

B. Apoptosis Assays

A large number of qualitative and quantitative assays exist for measuring apoptosis (e.g., TUNEL, DNA ladders, electron microscopy, caspase assays), and virtually any can be applied to MEF populations. In general, cells are treated with an apoptotic stimulus (chemotherapeutic agents, serum deprivation, irradiation, low oxygen, or proapoptotic genes) (see Table II). We strongly recommend dose–response experiments, as studies using only single doses can often miss important aspects of apoptotic regulation. At various time intervals or at a fixed time point (24 h is convenient), the number of apoptotic bodies can be calculated. Note that for many stimuli, the appropriate time posttreatment is dependent on the dose of the stimulus—for new systems, this should be worked out empirically. Several assays are described.

1. Annexin V and Propidium Iodide Staining

Apoptotic cells lose membrane phospholipid asymmetry, causing phosphatidylserine to appear on the cell surface. Annexin V binds phosphatidylserine and, when conjugated fluorescently, can act as a marker of early apoptosis (Martin *et al.,* 1995). Several kits are available commercially on the market that use FITC-conjugated Annexin V along with propidium iodide (PI) staining to distinguish live, early apoptotic, late apoptotic, and necrotic cells (see Fig. 3A). The following protocol uses the system of Annexin V/PI from R&D Systems, which has worked well in our hands.

i. In a six-well dish, plate between 3×10^4 and 1×10^5 cells/well [depending on cell type (30% confluent)] approximately 24 h prior to adding stimulus.

ii. Prior to apoptotic treatment, check cells to ensure that most are alive.

iii. Wash each well with normal medium. Aspirate off as much medium as possible without drying out the cells.

iv. Treat cells with an apoptotic stimulus in 1 ml medium/well.

v. Incubate the cells for the appropriate period.

vi. Transfer medium and floating cells to a labeled 15-ml Falcon tube.

vii. Wash each well with 1 ml PBS. Pool the PBS wash, which contains more floating cells, with medium from the same well.

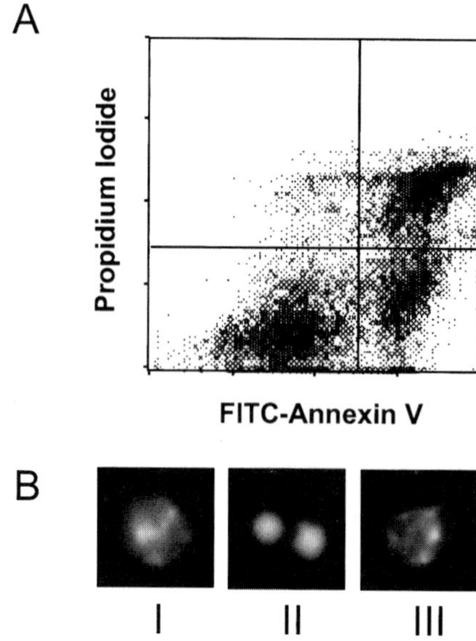

Fig. 3 (A) Dot plots from two-color flow cytometry on apoptotic E1A expressing MEFs stained with FITC-conjugated Annexin V and PI. Cells were treated for 24 h with 0.2 μg/ml doxorubicin. (Lower left) Viable cells, (lower right) early apoptotic cells, and (upper right) late apoptotic cells. (B) DAPI staining of (i) nonapoptotic, (ii) classically apoptotic, and (iii) dead but nonapoptotic cells viewed under a fluorescent microscope (100× objective).

viii. Add 1–2 drops fresh trypsin. Incubate briefly at 37°C. Use pooled media + wash from the appropriate well to detach and resuspend adherent cells. Transfer the entire suspension back to the original tube. After this point the cells must be kept at 4°C.

ix. Pellet cells at 1200 rpm for 3 min.

x. Resuspend cells in 100 μl Annexin binding buffer in a 5-ml culture tube.

xi. Add 10 μl of FITC-conjugated Annexin V (10 μg/ml stock) and 10 μl of PI (50 μg/ml stock).

xii. Incubate for 15 min at room temperature protected from light.

xiii. Add 400 μl Annexin binding buffer to each tube.

xiv. Perform flow cytometry with a laser, emitting excitation at 488 nm. FITC can be detected at 530 ± 20 nm and PI at >600 nm.

a. This technique can be problematic when used with adherent cells due to the fact that trypsinization alone can bring phosphatidylserine to the cell surface and produce false positives. For this reason, the time in trypsin should be minimized and, moreover, kept constant for all samples in the experiment.

b. Control cell populations of "all live" and "all dead," as well as cells stained only with Annexin V or PI, greatly facilitate identifying the appropriate flow cytometry parameters.

c. Apoptotic cells shrink substantially during the cell death process; therefore, it is important not to "gate out" the small cells that are often considered debris.

2. DAPI Staining

Apoptotic cells undergo a series of well-characterized cellular alterations, including cell shrinkage, membrane blebbing, and chromatin condensation. The DAPI method identifies apoptotic cells by fluorescently staining chromatin. Apoptotic cells display marked condensed chromatin (often with a "grape-like" appearance), and many of the nuclei are fragmented (see Fig. 3B).

i. Prepare cells as described in steps i–x of the Annexin V and propidium iodide-staining protocol (Section V,B,1).

ii. Resuspend the pellet in 100 μl 70% EtOH or 4% formaldehyde.

iii. Add 500 ng/ml DAPI (stock solution 1 mg/ml in water).

iv. Analyze at least 200 cells for chromatin condensation using a fluorescent microscope (use a UV filter emitting excitation at 360 nm and emitting at >400 with a UV-transparent objective).

v. Fixed cells can be maintained at 4°C in the dark for several weeks prior to microscopic analysis.

Notes:

a. This technique is a very easy way to obtain a ratio of apoptotic versus live cells in a population. The preparation of the cells is simple and rapid. Because the cells are fixed, they can be analyzed weeks after the experiment has been completed. However, counting statistically significant numbers of cells is very time-consuming. Our laboratory usually uses this technique as a confirmation of apoptotic cell death once viability assays have demonstrated differences in death rates.

b. Dead cells that do not fully condense their chromatin (Fig. 3B, III) may be indicative of necrosis or caspase-independent apoptosis (e.g., Woo *et al.,* 1998).

3. FACS Analysis of DNA Content

DNA content analysis is used extensively in cell cycle research because cells in different phases of the cell cycle have different amounts of DNA (G_0/G_1, G_2/M, and S phases have a $2N$, $4N$, and intermediate DNA content, respectively). It can also be extremely useful in studying apoptosis as apoptotic cells have a reduced DNA content due to degradation. DNA intercalating dyes, such as propidium iodide, fluoresce with an intensity proportional to bound DNA, making it relatively easy to estimate the DNA content/cell cycle distribution by using flow cytometry. Similarly, the percentage of apoptotic cells can be estimated by calculating the "sub-G_1" fraction of cells, i.e., cells with less then a $2N$ DNA content. The following method was adapted from Dlugosz *et al.* (1995) and Gorczyca *et al.* (1998).

i. Collect cells. When studying adherent cell types, detach the cells with trypsin and pool with the floating cells.

ii. Adjust the cell concentration to (10^6–5×10^6/ml) in PBS.

iii. Fix the cells by adding 1 ml of cells to 9 ml ice-cold 70% EtOH. (Cells can be stored at this point for several weeks at −20°C.)

iv. Pellet the cells by centrifugation (1500 rpm in a tabletop, centrifuge 5 min) and discard the ethanol.

v. Wash the cell pellet in 10 ml PBS and repeat the centrifugation step. Discard the supernatant.

vi. Resuspend the cell pellet in 0.5 ml of PBS.

vii. Incubate cells at room temperature for 5 min and centrifuge as in step iv.

viii. Resuspend the cell pellet in 1 ml of PBS containing 20 μg PI.

ix. Incubate for 30 min at room temperature.

x. Analyze cells by flow cytometry. Use a 488-nm laser line or BG12 filter for excitation and long-pass (>600 nm) fluorescence emission filter. Measure red fluorescence and forward light scatter.

Notes:

a. This assay can give an overview of the proliferation and cell death features of the same population.
b. Data are generally plotted as histograms of relative cell number (y axis) in each red channel (x axis, DNA content).
c. The sub-G_1 DNA content can be measured directly from the machine using the appropriate gates or by using curve-fitting programs such as MODFIT LT (Verity Software House Inc.)

C. Viability Assays

Viability assays determine the relative susceptibility of a cell population to death, but usually cannot distinguish between apoptosis and necrosis.

1. Trypan Blue

Trypan blue is a dye that crosses the plasma membrane but is actively pumped out of live cells. Dead cells cannot pump out the dye and stain blue.

i. In a 12-well dish, plate between 1×10^4 and 5×10^4 cells/well approximately 24 h prior to adding stimulus.

ii. Prior to apoptotic treatment, using a microscope, check cells to make sure that most are alive. This ensures that any loss of viability is due to the agent tested.

iii. Wash each well with normal media. Aspirate off as much media as possible without drying the cells.

iv. Treat the cells maintaining 0.75 ml media/well.

v. Incubate for the appropriate period.

vi. Transfer the medium and floating cells to a labeled Eppendorf tube.

vii. Wash each well with 0.5 ml PBS. Pool the PBS wash with medium from the same well. This collects all the floating cells.

viii. Add 1–2 drops of fresh trypsin. Incubate briefly. Use the pooled media + wash from the appropriate well to detach and resuspend adherent cells. Transfer the entire suspension back to the appropriate tube. After this point the cells must be kept at 4°C.

ix. Pellet the cells by centrifugation at 1200 rpm for 3 min.

x. Pour off the supernatant and shake each tube gently (this should leave the cell pellet and approximately 100 μl of media at the bottom of the tube).

xi. Add 100 μl trypan blue (0.4% stock) (GIBCO-BRL) to each tube.

xii. Using a Pipetman with a 200-μl tip, gently pipette the cells up and down to mix. Place a drop onto a hemocytometer.

xiii. Count 200 cells using a light microscope with a 20 × objective, scoring for white (live) and blue (dead) cells. Apoptotic cells often shrink, so blue cells are often smaller than white ones.

Notes:

a. The trypan blue method is a rapid, inexpensive assay that is reproducible. Because all cells are scored by eye, larger experiments are more tedious.
b. Cells scored for viability by Trypan blue can be fixed in formaldehyde (1 drop of formalin per tube) and stained with DAPI to confirm apoptosis (see Section V,B,2).

2. Crystal Violet

This assay measures the quantity of cells on a plate following an apoptotic stimulus. Floating cells (dead) are washed away and adherent cells (live) are quantified by crystal violet staining. This assay can only be used for adherent cells.

i. For MEFs (or other normal fibroblasts) in a 12-well plate, add 4000 cells/well in 150 μl of medium. It is advisable to make sixtuplicates for each point.

ii. After an overnight incubation, add the proapoptotic stimulus to each well.

iii. At the appropriate time, wash wells twice with 200 μl PBS (when washing with a tip attached to a vacuum, do not touch the bottom of the dish to avoid removing adherent cells).

iv. Fix the cells with 100 μl of 1% glutaraldehyde for 10 min on ice.

v. Wash the wells twice with 200 μl PBS.

vi. Stain the fixed cells with 100 μl of 0.1% crystal violet (in PBS or H_2O) for 30 min at room temperature.

vii. Wash once with 200 μl PBS.

viii. Wash twice with distilled water by submerging the plate in a tray full of water.

ix. Dry the plates at room temperature or in a 37°C incubator.

x. Solubilize the attached crystal violet by adding 200 μl of 10% acetic acid to each well and pipetting several times.

xi. Transfer the acetic acid to a 96-well plate and measure the OD in a plate reader at 590 nm.

xii. Readings should be normalized to a well that contained media alone and was processed along with the other wells.

Notes:

a. This method is very fast and requires relatively few cells.
b. Because this method does not give a ratio of live versus dead cells, it cannot distinguish growth inhibition from cell death nor can it demonstrate that cell death is apoptotic.
c. Another colorimetric assay that is widely used is the MTT assay (Mosmann, 1983), which can be set up in a similar way as just described.

D. Transformation Assays

Disruption of apoptosis can promote oncogenic transformation. Usually, nonimmortal cells require two or more mutations to become transformed. However, nonimmortal cells from certain "knockout" mice can be transformed by single oncogenes (Lowe *et al.,* 1994; Tanaka *et al.,* 1994; Kamijo *et al.,* 1997). It has become possible to transform nonimmortal human cells with defined genetic alterations (Hahn *et al.,* 1999). Thus, the methods described here should also be applicable to human cells.

A number of oncogene combinations are known to oncogenically transform normal MEFs. These include ras + E1A, ras + myc, and ras + mutant p53. Expression of some genes can either enhance or diminish transformation induced by these or other combinations. To examine the effect of loss of gene function on transformation, one can analyze cells from "knockout" mice (e.g., McCurrach *et al.,* 1997).

The transformation assays outlined here measure three traits associated with malignant cells: the ability to overcome contact inhibition (focus assay), anchorage independence (soft agar assay), and the ability to form tumors *in vivo* (transplantation to nude mice). Clearly, the ability of cells to form tumors in mice is the most solid evidence of transformation of a cell; however, this experiment is the most costly and, for laboratories not actively doing mouse work, the most time-consuming. In addition, all animal experiments require the approval of the institutional authorities.

1. Focus Assay

i. Prior to transfection, maintain the cell cultures at a subconfluent density.

ii. Plate 1×10^6 cells per 10-cm plate and incubate for 24 h.

iii. Change the medium 4 h before transfection.

iv. Introduce the genes of interest by $CaPO_4$ transfection (see Section III,B,1 for the protocol). Incubate for 16–18 h at 37°C. For E1A and ras transfection, use 5 μg experimental plasmid and 15 μg carrier plasmid (e.g., pBluescript).

v. Change medium and incubate for 24 h.

vi. Split each plate 1 : 4 (two plates for focus formation and two plates for selection).

vii. Change media on all plates every third day.

viii. Colonies on selected plates should start to appear by the third media change. Foci usually appear around the fourth media change.

ix. At day 15, add 1 ml formalin (37% formaldehyde solution) per 10 ml medium to the plates (do not remove medium).

x. Swirl to mix.

xi. Incubate the plates at room temperature for a minimum of 30 min (up to a couple of days).

xii. Pour off the medium and fixative and add 5 ml (per 10-cm plate) of a 0.02% Giemsa solution in PBS.

xiii. Incubate the plates for 30 min at room temperature. Remove the dye and wash each plate under running distilled water (taking care that the water falls smoothly onto the cells). Air dry the plates.

xiv. Foci will appear as dark spots on the plate (see Fig. 4A). Count the number of foci per plate, scoring only those that are 5 mm in diameter or larger.

xv. Plates can be photographed easily for records when placed on a light box.

Notes:

a. Formaldehyde is toxic—use in a fume hood. Giemsa is carcinogenic—wear gloves.

b. In order to assess transfection efficiency, it is advisable to place one series of the tranfected plates on selection. The number of cells remaining after selection is used to

Fig. 4 Representative plates from a focus assay and soft agar assay carried out using p53 −/− MEFs expressing both adenovirus-5 E1A and an activated *ras* oncogene. Plates were fixed and photographed 2 weeks after plating.

calculate transfection efficiency. This control rules out the possibility that the differences in focus formation are due to different transfection efficiencies.

c. Another useful control is carrier alone, as a negative control, and a transforming plasmid (e.g., ras plus T antigen or mutant p53) as a positive control.

d. A variation of this procedure is to mix infected cell populations with normal MEFs (e.g., 1 : 10,000, 1 : 1000, 1 : 100) and incubate them for 2 weeks. The advantage of this procedure is that the initial number of transduced cells is precise.

2. Soft Agar

i. The day before initiating the soft agar assay, plate the cells at 60–70% confluence, trying to prevent the formation of cell clusters.

ii. Prepare the plates by layering 3 ml of bottom agar (see later) per 6-cm plate.

iii. Let the agar solidify at room temperature for 10–15 min.

iv. Trypsinize and count the cells, making sure that they are not aggregated.

v. Prepare the top agar solution. When the temperature has reached 42°C, add 10^3–10^4 cells to 4 ml of top agar (see later). Swirl and add dropwise on top of the solidified bottom agar (work fast to prevent solidification of the agar).

vi. Feed the plates once a week by adding 1 ml of top agar solution.

vii. Colonies form usually within 2 weeks. To stain the colonies, add 1–2 ml of a 0.02% Giemsa solution (in PBS) for 10 min at room temperature (longer times will increase background staining).

viii. Remove staining solution with a pipette. The colonies are usually not visible unless they are >100 cells.

ix. Let the plates sit at 4°C for several hours (overnight). Colonies should stain dark blue. The agarose may turn slightly blue.

x. Photographs of a typical experiment are shown in Figs. 4B and 4C. Data can be quantified in two ways. Stained or unstained colonies can be counted under a light microscope and scored for size (<50 and >50 cells per colony). The percentage of cells forming colonies can be listed as % = [(colony no. >50/total) × 100%]. The second method is to count the total number of large (>50 cells) colonies on each plate. This is accomplished most easily with stained plates.

Note: The plates can be photographed using a light box. Counting colonies from photographs is often easier than from the plates themselves. Each colony can be marked to improve counting accuracy.

Solutions

Bottom agar: Melt 0.5% Nobel agar in growth medium (without FCS) by autoclaving. Cool to 65°C and add FCS to 10% (v/v).

Top agar: Melt 0.3% Nobel agar in growth medium (without FCS) by autoclaving. Cool to 65°C and add FCS to 10% (v/v). Cool to 42°C in a water bath before adding cells.

3. Transplantation in Nude Mice

i. The day before injection, plate cells at 60–70% confluence.

ii. Trypsinize cells and concentrate to 10^6 cells/0.25 ml cold PBS. Keep on ice.

iii. Inject 0.25 ml of the cell suspension subcutaneously into each rear flank of NSW athymic nude mouse using a 23-gauge needle. This is accomplished most easily by loosely lifting the skin over the target injection site with thumb, forefinger, and middle finger. This should form a "tent" of skin into which the cells are injected. The needle should be pushed completely through the skin. There should be virtually no resistance when injecting the cells. If there is, the needle is probably subdermal and not subcutaneous. Be careful not to push the needle out the back side of the "tent." Inject cells slowly and pull the needle out carefully to reduce the amount of leakage.

iv. Injection sites should be monitored for tumor formation three times per week by palpation. The length (L) and width (W) of tumors can be measured using a caliper and the volume of tumor is determined using the formula: $V = (L \times W^2)/2$ where the length equals the widest part of the tumor. The width is equal to the widest part of the tumor perpendicular to the length. Tumor latency (time to palpable tumor) and size at successive times are recorded.

v. Mice should be sacrificed before tumors become too large (generally ≤ 1 cm^3). Refer to your institutional guidelines.

Notes:

a. Begin the transplantation experiment within two passages of retroviral gene transfer to avoid selection for mutants *in vitro*.
b. One mouse can be injected at two sites.
c. Each cell line should be injected at least three times to ensure reproducibility.
d. Tumors can be excised and tissue studied *ex vivo* or recultured following trypsinizaton (see MEF protocol).
e. Material can be fixed in formalin for histological staining or immunohistochemistry.

E. Growth Assays

A large body of evidence indicates that the excessive proliferation of normal cells is coupled to cell death (Evan and Littlewood, 1998). A large number of assays can be used to assess cell proliferation. Many simply measure the rate of cell accumulation. Hence, it is important to perform cell death assays in conjunction with growth assays.

1. Growth Curves

This assay quantifies the amount of cell material on a plate after a given growth period. Cells are plated and allowed to grow for various intervals (usually between 12 and 72 h). They are then quantified by crystal violet staining, which is proportional to cell number.

This is a relatively simple technique, but it cannot distinguish slow growth from enhanced cell death.

i. Plate cells at a density of 2.5×10^4 per well in 12-well plates following completion of drug selection for infected cells. It is best to arrange the cells so that an entire plate is fixed per time point, as gluteraldehyde vapors in an incubator will kill the remaining cells in an experiment.

ii. Twelve hours later, fix one plate (time zero) using 500 μl of 1% glutaraldehyde. Incubate for 10 min on ice and wash the wells twice with 1 ml PBS. The plates can then be left in PBS at 4°C until the experiment is complete.

iii. Change medium on remaining plates every 3 days until the experiment is completed. Fix the other cells at appropriate intervals.

iv. Stain all the wells with 500 μl of 0.1% crystal violet (in PBS or H_2O) for 30 min at room temperature.

v. Wash with 1 ml PBS.

vi. Wash the plates twice with distilled water by submerging the plate in a tray full of water.

vii. Dry the plates (at room temperature or in a 37°C incubator).

viii. Solublize the attached crystal violet by adding 500 μl of 10% acetic acid to each well. Pipett the solution up and down several times.

ix. Transfer the solublized dye to a 24-well plate and measure optical density at 590 nm in a microplate reader.

x. Normalize the values to those of the time zero plate. Staining is proportional to the relative cell number.

Notes:

a. Each cell type and time point should be assessed in triplicate to correct for plating differences.
b. All samples from an experiment should be stained at the same time.

2. [^{3}H]Thymidine Incorporation

This assay measures the relative proliferation rates using the incorporation of [^{3}H]thymidine into DNA as a measure of the DNA synthesis rate of a cell population. This is usually proportional to the proliferation rate of the cell population. This assay is fast and easy, but provides less information than other assays and requires radioactive materials.

i. Plate 2×10^4 cells per well in 12-well plates. Each sample should be measured in triplicate.

ii. Incubate cells overnight and add 5 μCi/ml [*methyl*-^{3}H]thymidine. Incubate for between 3 and 24 h (the length of the [^{3}H]thymidine pulse will be dependent on the baseline growth rate of the cells and must be determined empirically).

iii. After the designated incubation time, add 5 μCi/ml [^{3}H]thymidine to three control wells containing cells that were not pulsed with [^{3}H]thymidine. Proceed immediately to the next step.

iv. Transfer the media and floating cells to a labeled 15-ml Falcon tube.

v. Wash each well with 1 ml PBS. Pool PBS wash with media from the same well.

vi. Add 1–2 drops of fresh trypsin. Incubate briefly to detach the cells. Use pooled media + wash from the appropriate well to detach and resuspend adherent cells. Transfer the entire suspension back to the appropriate tube. Pass the pooled cells through a glass fiber filter (Filtermat, Wallac).

vii. Wash each filter twice with 10% ice-cold trichloroacetic acid followed by one wash in 70% EtOH and one in 100% EtOH.

viii. Place the dried filters in marked scintillation vials. Add scintillation fluid and count radioactivity using the ^{3}H channel of a scintillation counter.

ix. Normalize counts to the control wells that serve as background.

Note: Once [^{3}H]thymidine is added to the cells, all media and plasticware must be disposed of in approved radioactive waste containers.

3. Fluorescent Bromodeoxyuridine (BrdU) Protocol

This assay is a modification of the DNA content protocol (Section V,B,3). The incorporated BrdU can subsequently be detected using a fluorescently conjugated antibody and measured using flow cytometry. Analysis of the cells requires a FACS scanner. This is a very powerful assay because it analyzes individual cells. Analysis can be carried out with the addition of propidium iodide (see Section V,B,1) to distinguish dividing and dead cells.

i. In a six-well dish, plate between 5×10^4 and 1×10^5 cells/well depending on cell type. Cells should be ~30% confluent.

ii. Allow cells to adhere overnight.

iii. Add BrdU (1 : 1000 dilution in media of a 3-mg/ml stock) in normal growth medium.

iv. Incubate for 4 h at 37°C. (This can be varied depending on the experiment.)

v. Transfer the medium and floating cells to a labeled 15-ml Falcon tube.

vi. Wash each well with 1 ml PBS and pool with media from the same well.

vii. Add 1–2 drops of fresh trypsin. Incubate briefly at 37°C. Use pooled media + wash from appropriate well, to detach and resuspend the adherent cells. Transfer the entire suspension back to the original tube.

viii. Pellet the cells by centrifuging at 1200 rpm for 3 min.

ix. Wash the cells by resuspending them in 1 ml PBS. Centrifuge the cells as described earlier.

x. Resuspend the cell pellet in 1 ml fixative (95% EtOH; 5% glacial acetic acid; precooled to −20°C). Incubate at −20°C for 15 min.

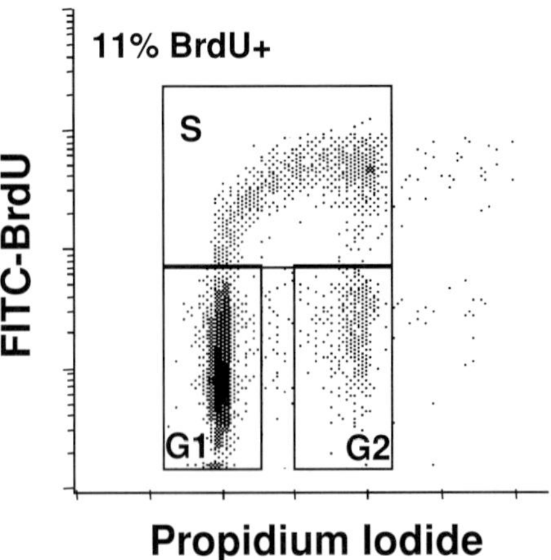

Fig. 5 Dot plot representation of FITC-conjugated BrdU incorporation and PI staining of growing MEFs expressing E1A. The three boxes represent G_1-arrested cells (G1), S-phase cells (S) that incorporate BrdU, and G_2-arrested/M-phase cells.

xi. Wash out fixative by centrifuging cells at 1200 rpm for 3 min. Resuspend the cell pellet in 1 ml PBS and centrifuge. Repeat twice more.

xii. After the final wash, denature the DNA by resuspending the cell pellet in 2 *N* HCl for 20 min at room temperature.

xiii. Wash out the denaturing solution as in step xi.

xiv. Neutralize the denaturing solution by resuspending the cells in 0.1 *M* sodium borate (pH 7.5) for 2 min at room temperature.

xv. Wash out neutralizing solution as in step xi.

xvi. Add 15 μl FITC-conjugated anti-BrdU antibody (Pharmingen) to the cells. Incubate for 1 h at room temperature.

xvii. Wash out nonbound antibody as in step xi.

xviii. After the last wash, resuspend the cell pellet in 1 ml PBS containing 10 μg/ml RNase A and 10 μg/ml PI. Incubate for 30 min at 37°C. (PI binds DNA, allowing it to be quantified by a flow cytometer. RNase A removes RNA, which causes background staining.)

xix. Perform flow cytometry with a laser emitting at 488 nm. FITC can be detected at 530 ± 20 nm and PI at >600 nm. Data are typically represented as dot plots showing the green channel (BrdU incorporation) versus the red chanel (PI, DNA content) (Fig. 5).

VI. Other Systems

The methods just described work well with early passage MEFs. With the right retroviral vectors and conditions, almost any cell type can be infected and analyzed as described.

When beginning any new system it is advisable to optimize infection conditions using control viruses. This section describes several other systems.

A. Mouse Skin Fibroblasts

The primary reason for choosing MEFs is to study recessive mutations *in vitro*. This is because MEFs are readily isolated from many strains of "knockout" mice. However, MEFs are heterogeneous (being derived from a whole embryo) and are short-lived in culture. In contrast, primary and early passage dermal fibroblasts are relatively homogeneous and, in our hands, can be passaged much longer than MEFs without showing overt signs of crises. Of course, dermal fibroblasts cannot be isolated from mice that die during embryogenesis; hence, it may not be possible to obtain dermal fibroblasts from some knockout strains.

Dermal fibroblasts can be isolated from the skin of newborn embryos and from adults (skin keratinocytes and melanocytes do not grown in MEF media). Our laboratory produces dermal fibroblasts from newborn mice as a biproduct of our keratinocyte isolation procedure (Dlugosz *et al.,* 1995). The dermis is simply minced and trypsinzied as described for MEFs (see Section II,A). Once cultures have been established, dermal fibroblasts are passaged and infected as described for MEFs (Sections II,B,1 and IV, respectively).

B. Thymocytes

Although detailed protocols are beyond the scope of this chapter, primary mouse thymocytes are used extensively in apoptosis research. Thymocytes are extremely short-lived and cannot be expanded in culture, but their strength as a system lies in their readiness to undergo apoptosis in response to a variety of highly diverse stimuli and in the ease of isolating relatively large numbers from mice. Because of their short life span and the fact they are not cycling, it is not possible to introduce genes into thymocytes using standard retroviral vectors. The adenoviral receptor has been directed to hematologic cells using transgenic technology, allowing T cells to be infected successfully with adenoviral vectors (Leon *et al.,* 1998).

C. Primary Epithelial Cells

Fibroblasts provide a powerful system for studying the relationship among apoptosis, proliferation, and transformation, largely because they are one of the few nonimmortal cell types that are relatively easy to culture. However, tumors arising from fibroblasts and other connective tissues (sarcomas) are uncommon in adults. In fact, most adult cancers arise from various epithelia (carcinomas). While the extent to which principles uncovered in fibroblasts can be extrapolated to epithelial cells is often unknown, there is considerable interest in studying cancer-related processes in epithelial cells. In contrast to fibroblasts, however, primary epithelial cells are notoriously difficult to isolate, culture, and maintain.

In mice, the most practical epithelial cell type to study is dermal keratinocytes, which can be obtained from normal and "knockout" newborn embryos (for protocol, see Dlugosz *et al.,* 1995). Keratinocytes derived from wild-type mice can be passaged only a few times in culture, but can be infected with reasonable efficiency using the protocols described for MEFs. Importantly, because keratinocytes differentiate in calcium-containing medium (e.g., DMEM), it is important that retrovirus collections are performed in the actual keratinocyte medium, which contains low calcium and dialyzed serum. Once infected, keratinocytes can be analyzed using the same death, proliferation, and transformation assays described for MEFs. We have found that it is difficult to get substantial numbers of normal keratinocytes for routine assays such as Western blots. This problem can be overcome using a variety of single cell assays to study proliferation (e.g., FITC-BrdU incorporation), gene expression (e.g., immunofluorescence), and apoptosis (e.g., TUNEL or DAPI). Human epithelial cells are somewhat easier to culture and many can be obtained commercially (e.g., Clonetics, http://www.clonetics.com/). Keep in mind that the infection of human cells requires amphotropic or VSV-G-based systems, and thus involves additional biosafety concerns.

D. Bone Marrow Stem Cells

A variety of hematologic cell types can be infected with retroviral vectors, and these cell types present powerful systems to study gene function *in vitro* and *in vivo*. Our laboratory has begun to study pro- and antiapoptotic genes by infecting murine bone marrow cells and using these cells to reconstitute the immune systems of lethally irradiated mice. Procedures to isolate and infect bone marrow stem cells are beyond the scope of this chapter, but have been optimized by a number of groups (Luskey *et al.,* 1992; Barker *et al.,* 1993; Moritz *et al.,* 1994; Kuefer *et al.,* 1997; Pear *et al.,* 1998). Most rely on the use of MSCV-derived viruses, which are a mutant form of the Moloney murine leukemia virus that has been optimized for long-term gene expression in stem cells (standard Moloney-based vectors are often silenced). By isolating bone marrow from genetically engineered mice, it is possible to combine retrovirus and knockout technology to study apoptosis and tumor development *in vitro* or *in vivo*, without the need for generating additional germline transgenic mice.

E. Introduction of Murine Ecotropic Retroviral Receptor into Human Cells

The methods described for MEFs can be applied to nonimmortal fibroblasts, epithelial cells, and immortalized human tumor-derived lines. The main advantage of using human cells is their relevance to human cancer; the disadvantage is the general inability to study loss of gene function using "knockouts." However, homologous recombination has been applied to nonimmortal human cells (e.g., Brown *et al.,* 1997). It is also possible to approximate the impact of gene inactivation using viruses expressing "dominant-negative" proteins.

Human cells can be infected to varying degrees with amphotropic retroviruses as described in Section III. However, we have used an alternative approach that makes the target cell competent for infection with murine ecotropic viruses (Serrano *et al.*, 1997). In this method, a retroviral vector expressing the murine ecotropic retroviral receptor is produced in an amphotropic packaging system and is then transduced into human cells. Subsequent genes can be introduced using viruses generated in ecotropic virus, packaging systems. This approach can be highly effective. Because the ecotropic retrovirus receptor is overexpressed, the transduction efficiency for subsequent genes increases. Also, the procedure reduces the biosafety concerns associated with viruses with a human host range. For this reason, this procedure is recommended for studies using potentially oncogenic genes.

Acknowledgments

The authors thank the members of the Lowe Laboratory, who developed and/or optimized many of the protocols described here. We also thank G. Hannon, M. Roussel, G. Nolan, W. Pear, and J. Morgenstern for helpful discussions and M. Soengas, E. de Stanchina, A. Lin, and C. Schmitt for editorial advice. S.W.L. is a Rita Allen Scholar. Many of the protocols described in this article were optimized with the support of Grants CA13106 and AG16379 from the N.I.H. and by Grant RPG-99-200-01-LBC from the A.C.S.

References

Andreadis, S. T., Brott, D., Fuller, A. O., and Palsson, B. O. (1997). Moloney murine leukemia virus-derived retroviral vectors decay intracellularly with a half-life in the range of 5.5 to 7.5 hours. *J. Virol.* **71,** 7541–7548.

Barker, J. E., Wolfe, J. H., Rowe, L. B., and Birkenmeier, E. H. (1993). Advantages of gradient vs. 5-fluorouracil enrichment of stem cells for retroviral-mediated gene transfer. *Exp. Hematol.* **21,** 47–54.

Bergemann, J., Kuhlcke, K., Fehse, B., Ratz, I., Ostertag, W., and Lother, H. (1995). Excision of specific DNA-sequences from integrated retroviral vectors via site-specific recombination. *Nucleic Acids Res.* **23,** 4451–4456.

Brown, J. P., Wei, W., and Sedivy, J. M. (1997). Bypass of senescence after disruption of p21CIP1/WAF1 gene in normal diploid human fibroblasts. *Science* **277,** 831–834.

Burns, J. C., Friedmann, T., Driever, W., Burrascano, M., and Yee, J. K. (1993). Vesicular stomatitis virus glycoprotein pseudotyped retroviral vectors: Concentration to very high titer and efficient gene transfer into mammalian and nonmammalian cells. *Proc. Natl. Acad. Sci. USA* **90,** 8033–8037.

Cepko, C. C., and Pears, W. (1997). Overview of the retrovirus transduction system. *In* "Current Protocols in Molecular Biology" (F. A. Ausubel, *et al.*, ed.), pp. 9.9.1–9.9.16. John Wiley & Sons, New York.

Danos, O., and Mulligan, R. C. (1988). Safe and efficient generation of recombinant retroviruses with amphotropic and ecotropic host ranges. *Proc. Natl. Acad. Sci. USA* **85,** 6460–6464.

Dlugosz, A. A., Glick, A. B., Tennenbaum, T., Weinberg, W. C., and Yuspa, S. H. (1995). Isolation and utilization of epidermal keratinocytes for oncogene research. *Methods Enzymol.* **254,** 3–20.

DuBridge, R. B., Tang, P., Hsia, H. C., Leong, P. M., Miller, J. H., and Calos, M. P. (1987). Analysis of mutation in human cells by using an Epstein-Barr virus shuttle system. *Mol. Cell. Biol.* **7,** 379–387.

Evan, G., and Littlewood, T. (1998). A matter of life and cell death. *Science* **281,** 1317–1322.

Evan, G. I., Wyllie, A. H., Gilbert, C. S., Littlewood, T. D., Land, H., Brooks, M., Waters, C., Penn, L. Z., and Hancock, D. C. (1992). Induction of apoptosis in fibroblasts by c-myc protein. *Cell* **69,** 119–128.

Finer, M. H., Dull, T. J., Qin, L., Farson, D., and Roberts, M. R. (1994). A high-efficiency retroviral transduction system for primary human T lymphocytes. *Blood* **83,** 43–50.

Gaines, P., and Wojchowski, D. M. (1999). pIRES-CD4t, a dicistronic expression vector for MACS- or FACS-based selection of transfected cells. *Biotechniques* **26,** 683–688.

Gorczyca, W., Melamed, M. R., and Darzynkiewicz, Z. (1998). "Analysis of Apoptosis by Flow Cytometry" (M. J. Jaroszeski and R. Heller, eds.), Vol. 91, Humana Press, Totowa, NJ.

Hahn, W. C., Counter, C. M., Lundberg, A. S., Beijersbergen, R. L., Brooks, M. W., and Weinberg, R. A. (1999). Creation of human tumour cells with defined genetic elements. *Nature* **400,** 464–468.

Harvey, D. M., and Levine, A. J. (1991). p53 alteration is a common event in the spontaneous immortalization of primary BALB/c murine embryo fibroblasts. *Genes Dev.* **5,** 2375–2385.

Kafri, T., van Praag, H., Ouyang, L., Gage, F. H., and Verma, I. M. (1999). A packaging cell line for lentivirus vectors. *J. Virol.* **73,** 576–584.

Kamijo, T., Zindy, F., Roussel, M. F., Quelle, D. E., Downing, J. R., Ashmun, R. A., Grosveld, G., and Sherr, C. J. (1997). Tumor suppression at the mouse INK4a locus mediated by the alternative reading frame product p19ARF. *Cell* **91,** 649–659.

Kotani, H., Newton, P. B., Zhang, S., Chiang, Y. L., Otto, E., Weaver, L., Blaese, R. M., Anderson, W. F., and McGarrity, G. J. (1994). Improved methods of retroviral vector transduction and production for gene therapy. *Hum. Gene Ther.* **5,** 19–28.

Kuefer, M. U., Look, A. T., Pulford, K., Behm, F. G., Pattengale, P. K., Mason, D. Y., and Morris, S. W. (1997). Retrovirus-mediated gene transfer of NPM-ALK causes lymphoid malignancy in mice. *Blood* **90,** 2901–2910.

Le Doux, J. M., Davis, H. E., Morgan, J. R., and Yarmush, M. L. (1999). Kinetics of retrovirus production and decay. *Biotechnol. Bioeng.* **63,** 654–662.

Leiden, J. M. (1995). Gene therapy: Promise, pitfalls, and prognosis. *N. Engl. J. Med.* **333,** 871–873.

Leon, R. P., Hedlund, T., Meech, S. J., Li, S., Schaack, J., Hunger, S. P., Duke, R. C., and DeGregori, J. (1998). Adenoviral-mediated gene transfer in lymphocytes. *Proc. Natl. Acad. Sci. USA* **95,** 13159–13164.

Livingstone, L. R., White, A., Sprouse, J., Livanos, E., Jacks, T., and Tlsty, T. D. (1992). Altered cell cycle arrest and gene amplification potential accompany loss of wild-type p53. *Cell* **70,** 923–935.

Lowe, S. W., Jacks, T., Housman, D. E., and Ruley, H. E. (1994). Abrogation of oncogene-associated apoptosis allows transformation of p53-deficient cells. *Proc. Natl. Acad. Sci. USA* **91,** 2026–2030.

Lowe, S. W., and Lin, A. W. (2000). Apoptosis in cancer. *Carcinogenesis* **21,** 485–495.

Luskey, B. D., Rosenblatt, M., Zsebo, K., and Williams, D. A. (1992). Stem cell factor, interleukin-3, and interleukin-6 promote retroviral-mediated gene transfer into murine hematopoietic stem cells. *Blood* **80,** 396–402.

Martin, S. J., Reutelingsperger, C. P., McGahon, A. J., Rader, J. A., van Schie, R. C., LaFace, D. M., and Green, D. R. (1995). Early redistribution of plasma membrane phosphatidylserine is a general feature of apoptosis regardless of the initiating stimulus: inhibition by overexpression of Bcl-2 and Abl. *J. Exp. Med.* **182,** 1545–1556.

McCurrach, M. E., Connor, T. M., Knudson, C. M., Korsmeyer, S. J., and Lowe, S. W. (1997). Bax-deficiency promotes drug resistance and oncogenic transformation by attenuating p53-dependent apoptosis. *Proc. Natl. Acad. Sci. USA* **94,** 2345–2349.

Miller, A. D. (1990). Retrovirus packaging cells. *Hum. Gene Ther.* **1,** 5–14.

Morgenstern, J. P., and Land, H. (1990). Advanced mammalian gene transfer: High titre retroviral vectors with multiple drug selection markers and a complementary helper-free packaging cell line. *Nucleic Acids Res.* **18,** 3587–3596.

Moritz, T., Patel, V. P., and Williams, D. A. (1994). Bone marrow extracellular matrix molecules improve gene transfer into human hematopoietic cells via retroviral vectors. *J. Clin. Invest.* **93,** 1451–1457.

Mosmann, T. (1983). Rapid colorimetric assay for cellular growth and survival: Application to proliferation and cytotoxicity assays. *J. Immunol. Methods* **65,** 55–63.

Mulligan, R. C. (1993). The basic science of gene therapy. *Science* **260,** 926–932.

Naldini, L., Blomer, U., Gallay, P., Ory, D., Mulligan, R., Gage, F. H., Verma, I. M., and Trono, D. (1996). In vivo gene delivery and stable transduction of nondividing cells by a lentiviral vector. *Science* **272,** 263–267.

Pear, W. S., Miller, J. P., Xu, L., Pui, J. C., Soffer, B., Quackenbush, R. C., Pendergast, A. M., Bronson, R., Aster, J. C., Scott, M. L., and Baltimore, D. (1998). Efficient and rapid induction of a chronic myelogenous leukemia-like myeloproliferative disease in mice receiving P210 bcr/abl-transduced bone marrow. *Blood* **92,** 3780–3792.

Pear, W. S., Scott, M. L., and Nolan, G. P. (1996). "Generation of High-Titer, Helper-Free Retroviruses by Transient Transfection" (P. Robbins, ed.). Humana Press, Totowa, NY.

Roe, T., Reynolds, T. C., Yu, G., and Brown, P. O. (1993). Integration of murine leukemia virus DNA depends on mitosis. *EMBO J.* **12,** 2099–2108.

Sambrook, J., Fritsch, E. F., and Maniatis, T. (1989). "Molecular Cloning: A Laboratory Manual," 2nd Ed. Cold Spring Harbor Laboratory Press, Cold Spring Harbor, NY.

Samuelson, A. V., and Lowe, S. W. (1997). Selective induction of p53 and chemosensitivity in RB-deficient cells by E1A mutants unable to bind the RB-related proteins. *Proc. Natl. Acad. Sci. USA* **94,** 12094–12099.

Sanes, J. R., Rubenstein, J. L., and Nicolas, J. F. (1986). Use of a recombinant retrovirus to study post-implantation cell lineage in mouse embryos. *EMBO J.* **5,** 3133–3142.

Schmitt, C. A., McCurrach, M. E., de Stanchina, E., Wallace-Brodeur, R. R., and Lowe, S. W. (1999). INK4a/ARF mutations promote lymphomagenesis and chemoresistance by disabling p53. *Genes Dev.* **13,** 2670–2677.

Serrano, M., Lin, A. W., McCurrach, M. E., Beach, D., and Lowe, S. W. (1997). Oncogenic ras provokes premature cell senescence associated with accumulation of p53 and p16INK4a. *Cell* **88,** 593–602.

Soengas, M. S., Alarcon, R. M., Yoshida, H., Giaccia, A. J., Hakem, R., Mak, T. W., and Lowe, S. W. (1999). Apaf-1 and caspase-9 in p53-dependent apoptosis and tumor inhibition. *Science* **284,** 156–159.

Soneoka, Y., Cannon, P. M., Ramsdale, E. E., Griffiths, J. C., Romano, G., Kingsman, S. M., and Kingsman, A. J. (1995). A transient three plasmid expression system for the production of high titer retroviral vectors. *Nucleic Acids Res.* **23,** 628–633.

Tadaro, G. J., and Green, H. (1963). Quantitative studies of the growth of mouse embryo cells in culture and their development into established cell lines. *J. Cell Biol.* **17,** 299–313.

Tanaka, N., Ishihara, M., Kitagawa, M., Hisashi, H., Kimura, T., Matsuyama, T., Lamphier, M. S., Alzawa, S., Mak, T. W., and Taniguchi, T. (1994). Cellular commitment to oncogene-induced transformation or apoptosis is dependent on the transcription factor IRF-1. *Cell* **77,** 829–839.

Verma, I. M. (1994). Gene therapy: Hopes, hypes, and hurdles. *Mol. Med.* **1,** 2–3.

Weiss, R., Teich, N., Varmus, H., and Coffin, J. (1985). "RNA Tumor Viruses." Cold Spring Harbor Laboratory, Cold Spring Harbor, NY.

Williams, R. S. (1995). Human gene therapy: Of tortoises and hares. *Nat. Med.* **1,** 1137–1138.

Woo, M., Hakem, R., Soengas, M. S., Duncan, G. S., Shahinian, A., Kagi, D., Hakem, A., McCurrach, M., Khoo, W., Kaufman, S. A., Senaldi, G., Howard, T., Lowe, S. W., and Mak, T. W. (1998). Essential contribution of caspase 3/CPP32 to apoptosis and its associated nuclear changes. *Genes Dev.* **12,** 806–819.

Yeh, W. C., Pompa, J. L., McCurrach, M. E., Shu, H. B., Elia, A. J., Shahinian, A., Ng, M., Wakeham, A., Khoo, W., Mitchell, K., El-Deiry, W. S., Lowe, S. W., Goeddel, D. V., and Mak, T. W. (1998). FADD: Essential for embryo development and signaling from some, but not all, inducers of apoptosis. *Science* **279,** 1954–1958.

Zhu, H., Fearnhead, H. O., and Cohen, G. M. (1995). An ICE-like protease is a common mediator of apoptosis induced by diverse stimuli in human monocytic THP. 1 cells. *FEBS Lett.* **374,** 303–308.

CHAPTER 10

Calcium Flux Measurements in Apoptosis

David J. McConkey and Leta K. Nutt

Department of Cancer Biology
U.T.M.D. Anderson Cancer Center
Houston, Texas 77030

I. Background

A. Early Studies Implicating Ca^{2+} in Apoptosis

Parallel efforts in whole cells and isolated nuclei provided some of the first evidence for the involvement of Ca^{2+} in apoptosis (reviewed in McConkey and Orrenius, 1997). Hewish and Burgoyne (1973) and other investigators (Vanderbilt *et al.*, 1982; Cohen and Duke, 1984; Wyllie *et al.*, 1984) demonstrated that oligonucleosomal DNA fragmentation ("DNA ladders") could be induced in isolated nuclei from a variety of tissues by incubating them in the presence of millimolar concentrations of Ca^{2+} and Mg^{2+}. With

METHODS IN CELL BIOLOGY, VOL. 66

0091-679X/01 $35.00

the demonstration of similar DNA fragmentation patterns in apoptosis (Wyllie, 1980), it was suspected that a Ca^{2+}/Mg^{2+}-dependent endonuclease could be involved (Cohen and Duke, 1984). This notion was further supported by independent work by Kaiser and Edelman (1977), who demonstrated that glucocorticoid-induced apoptosis in thymocytes was associated with enhanced Ca^{2+} influx. Furthermore, they showed that the effects of glucocorticoids could be mimicked by specific Ca^{2+} ionophores (Kaiser and Edelman, 1978). We subsequently used Ca^{2+}-sensitive fluorescent dyes to confirm that cytosolic Ca^{2+} elevations are involved in glucocorticoid-induced thymocyte apoptosis and that extracellular or intracellular Ca^{2+} chelators block the DNA fragmentation associated with cell death (McConkey *et al.,* 1989).

Studies in rat prostate provided further evidence for the involvement of Ca^{2+} fluxes in apoptosis. Early work by Kerr and Searle (1973) demonstrated that castration-induced prostatic involution was associated with a wave of programmed cell death in the gland. It was shown that this programmed cell death is associated with enhanced nuclear Ca^{2+}-dependent endonuclease activity (Kyprianou *et al.,* 1988) and can be inhibited by Ca^{2+} channel blockers (Martikainen and Isaacs, 1990). Subsequent studies also showed that Ca^{2+} ionophores (Martikainen and Isaacs, 1990) and the endoplasmic reticular Ca^{2+} ATPase inhibitor, thapsigargin, (Furuya *et al.,* 1994), can also induce apoptosis in normal prostatic epithelial cells and prostate cancer cell lines, respectively. Other studies have shown that Ca^{2+} ionophores and thapsigargin induce apoptosis in a variety of other cell types, suggesting that this pathway of apoptosis is well conserved. Finally, more recent work on excitatory toxicity in the central nervous system revealed a central role for Ca^{2+}-mediated apoptosis in the response. Sustained exposure to glutamate can overstimulate NMDA receptors on neurons and promotes a form of Ca^{2+}-mediated programmed cell death (Ankarcrona *et al.,* 1995; Choi, 1995). More recently, other evidence has been advanced implicating Ca^{2+}-dependent processes in neurodegeneration due to Alzheimer's disease (Vito *et al.,* 1996; Nakagawa *et al.,* 2000).

B. Alterations in Subcellular Ca^{2+} Compartmentalization

Initially, most investigators assumed that increases in cytosolic Ca^{2+} concentration would be the most significant consequences of the Ca^{2+}-associated death stimuli introduced earlier. However, scattered reports of instances where increases in cytosolic Ca^{2+} actually protected cells from death began to emerge, as in the cases of hematopoietic cells Rodriguez-Tarduchy *et al.,* 1990; Whyte *et al.,* 1993 or in neurons (Koike *et al.,* 1989) deprived of trophic factor support. Parallel, independent work (Baffy *et al.,* 1993; Lam *et al.,* 1994) suggested a possible explanation for these apparently disparate results. It was shown that apoptosis induced by growth factor withdrawal in hematopoietic BaF3 progenitors (Baffy *et al.,* 1993) or by glucocorticoid treatment in WEHI lymphocytes (Lam *et al.,* 1994) involves early depletion of the endoplasmic reticular (ER) Ca^{2+} store without an increase in cytosolic Ca^{2+}. Subsequent work showed that these changes are attenuated in cells overexpressing the antiapoptotic BCL-2 protein (Lam *et al.,* 1994) and that BCL-2 directly regulates Ca^{2+} uptake and release in the ER (He *et al.,* 1997). We

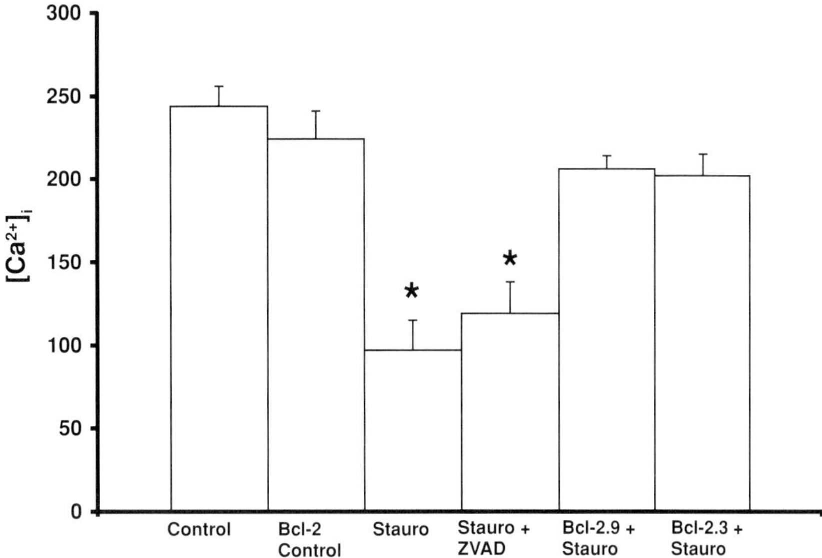

Fig. 1 Staurosporine-induced depletion of ER Ca^{2+} store PC-3 parental cells or BCL-2 transfectants (BCL2.3, BCL2.9) were plated on coverslips and allowed to adhere overnight. Cells were then treated with staurosporine (stauro, 1 μM), the pan-caspase inhibitor zVADfmk (zVAD, 20 μM), or both agents. Cells were loaded with 12.5 μM fura-2 AM in the presence of 1 mM probenicid for 1 h at 37°C. Endoplasmic reticular Ca^{2+} ($[Ca^{2+}]_i$) was quantified by measuring the difference between peak and basal Ca^{2+} levels after thapsigargin (5 μM) treatment. $^*p < 0.001$ versus control, BCL-2 control, BCL2.9 plus stauro, BCL-2.3 plus stauro as measured by ANOVA ordinary (unpaired) test and Student–Newmann–Keuls posthoc analysis.

have confirmed that most (but not all) proapoptotic agents stimulate ER Ca^{2+} depletion via a caspase-independent, BCL-2-sensitive mechanism; representative results with a human prostate cancer cell line are shown in Fig. 1.

How might release of ER Ca^{2+} precipitate the downstream events in apoptosis? The answer to this question is still not available, although several interesting possibilities have been advanced. Early work suggested that the Ca^{2+}-sensitive endonuclease DNase I might be responsible for chromatin cleavage in apoptotic cells (Peitsch *et al.,* 1993), and it was speculated that the Ca^{2+} increase occurring after ER pool depletion could promote activation of the enzyme. In addition, a pool of DNase I exists in the ER, and it is possible that this pool is released into the nucleus after pool depletion (Peitsch *et al.,* 1993). These explanations have fallen out of favor since the subsequent implication of caspase-activated endonucleases (DFF/CAD) in apoptotic DNA fragmentation (Liu *et al.,* 1997; Enari *et al.,* 1998; Sakahira *et al.,* 1998). However, an exciting alternative hypothesis has been put forward (Nakagawa *et al.,* 2000). It was shown that caspase-12 is localized to the ER in resting cells and that agents such as thapsigargin or ionophores that deplete ER Ca^{2+} stores initiate a caspase-12-dependent pathway of cell death (Nakagawa *et al.,* 2000). Thus, caspase-12 activation may represent a specific means of coupling ER Ca^{2+} pool depletion to the downstream events associated with apoptosis. Endoplasmic reticular Ca^{2+} pool depletion also serves as a specific signal for the TRAF-dependent activation of Jun kinase (JNK) via the transmembrane ER-resident protein serine/threonine kinase, IRE1 (Urano *et al.,* 2000). A causative role for JNK activation in apoptosis is well documented (Verheij *et al.,* 1996; Cuvillier *et al.,* 1996; Yang *et al.,* 1997; Adachi-Yamada *et al.,* 1999).

Another possibility is that ER pool depletion results in subsequent Ca^{2+} loading in mitochondria and that this mitochondrial Ca^{2+} loading results in the specific activation of apoptosis. Mitochondria are currently considered the point of convergence for many different upstream triggers of apoptosis (Kroemer *et al.,* 1997). The suggestion that this is true came from the observation that BCL-2 accumulates in the organelle (Hockenberry *et al.,* 1990) and subsequent work by Newmeyer *et al.* (1994) showing that the heavy mitochondrial fraction of cells contains a factor(s) required for efficient caspase activation. In other studies, it was demonstrated that mitochondrial membrane depolarization (a drop in $\Delta\Psi_{mito}$) occurs at an early stage during glucocorticoid-induced apoptosis in thymocytes (Zamzami *et al.,* 1995a,b), providing the first evidence for the disruption of mitochondrial function as a trigger for cell death. Finally, identification of the mitochondrial electron transport chain intermediate, cytochrome *c*, as an apoptogenic caspase activating factor (Liu *et al.,* 1996) suggested that membrane depolarization may promote the release of cytochrome *c* from mitochondria. These results have prompted an intensive search for factors that can regulate mitochondrial cytochrome *c* release, a drop in $\Delta\Psi$, or both. Proapoptotic members of the BCL-2 family can directly trigger both events via a Ca^{2+}-sensitive mechanism (Narita *et al.,* 1998; Antonsson *et al.,* 1997), but how they do so remains unclear.

Exogenous Ca^{2+} is one of the best characterized stimuli for opening of the mitochondrial permeability transition pore (PT pore), an event that has been implicated in cellular injury induced by oxidants. This led Richter (1993) to suggest that mitochondrial Ca^{2+} loading might precipitate MPT pore opening and cytochrome *c* release in apoptotic cells. Indeed, exogenous Ca^{2+} can promote the release of cytochrome *c* from isolated mitochondria via a BCL-2-inhibitable mechanism (Narita *et al.,* 1998; Shimizu *et al.,* 1999). Exciting work indicates that the proapoptotic agents, ceramide and staurosporine, sensitize mitochondria to Ca^{2+}-induced PT pore opening (Szalai *et al.,* 1999). Where studied, BCL-2 appears to enhance mitochondrial Ca^{2+} uptake capacity (Murphy *et al.,* 1996), and it is possible that this Ca^{2+} tolerance underlies the inhibition of BCL-2 of cytochrome *c* release. Indeed, cyclosporin A and bongkreic acid, which inhibit the MPT pore, can block cytochrome *c* release and the drop in $\Delta\Psi$ in some systems (Kroemer *et al.,* 1997; Narita *et al.,* 1998). However, other studies suggest that cytochrome *c* release occurs prior to detectable changes in $\Delta\Psi$ (Green and Reed, 1998; Kluck *et al.,* 1997). Whether or not Ca^{2+} (or other factors) can promote cytochrome *c* release without triggering MPT pore opening is the subject of current active investigation. Preliminary results confirm that ER Ca^{2+} pool emptying occurs prior to cytochrome *c* release in prostate cancer cells (Fig. 2).

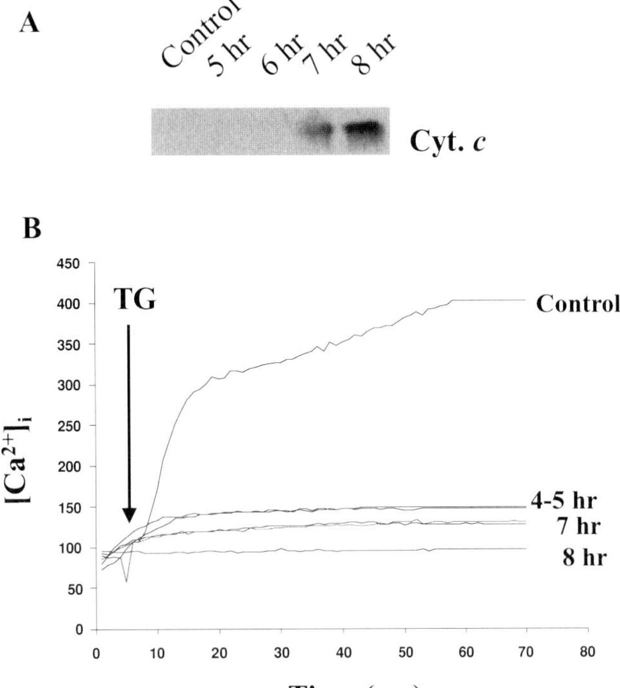

Fig. 2 Relative kinetics of staurosporine-induced ER Ca^{2+} pool depletion and cytochrome *c* release. (A) Kinetics of cytochrome *c* release. PC-3 cells were treated with staurosporine (1 μM) for the times indicated. Cytosols were prepared, and cytochrome *c* content was measured by immunoblotting. (B) Kinetics of ER Ca^{2+} pool depletion. Examples of several typical Ca^{2+} traces measured at different times after staurosporine treatment. Endoplasmic reticular Ca^{2+} content was released by thapsigargin (T6) stimulation (5 μM).

C. Regulation of Nuclear Ca^{2+} Fluxes

The observation that Ca^{2+} promoted endogenous endonuclease activation in isolated nuclei suggested that nuclear Ca^{2+} accumulation might play an important role in regulating apoptosis. Early work had established that nuclei maintain a Ca^{2+} concentration gradient across the nuclear envelope, in that IP_3-mediated Ca^{2+} spikes are associated with the dramatic reorganization of the nuclear matrix that occurs in cells during anaphase (Poenie *et al.*, 1985, 1986; Ciapa *et al.*, 1994). Furthermore, Ca^{2+} gradients form between the nucleus and the cytosol in contracting smooth muscle cells (Williams *et al.*, 1985, 1987). We found that exogenous ATP promoted endogenous endonuclease activation at submicromolar Ca^{2+} concentrations in isolated rat liver nuclei (Jones *et al.*, 1989), and subsequent analysis of the mechanisms involved revealed that ATP promotes nuclear Ca^{2+} uptake (Nicotera *et al.*, 1989). More recently, we have shown that nuclear Ca^{2+}

uptake precedes overt signs of apoptosis in whole cells exposed to proapoptotic stimuli *in vitro* (Bellomo *et al.*, 1992; Marin *et al.*, 1996). Importantly, like the ER, the nuclear envelope store contains receptors for IP_3 and is regulated by the second messenger (Nicotera *et al.*, 1990; Gerasimenko *et al.*, 1995). However, unlike the ER, the nuclear store is sensitive to calmodulin antagonists (Nicotera *et al.*, 1989), suggesting that it represents a distinct subcellular Ca^{2+} pool.

D. Communication among Intracellular Ca^{2+} Pools

For years the existence of a mitochondrial "Ca^{2+} pool" has been the subject of intense controversy. On the one hand, some investigators presented evidence that mitochondria can act as a Ca^{2+} "sink" under conditions of Ca^{2+} overload (Richter, 1993) and that levels of intramitochondrial Ca^{2+} could approach the micromolar levels thought to be present within the lumen of the ER. However, other studies reported that the Ca^{2+} pool released by mitochondrial uncoupling agents (cyanide, azide, CCCP) was insignificant under either control conditions or in cells exposed to oxidative stress (Lemasters *et al.*, 1990). This controversy has been dissipated somewhat by studies of subcellular Ca^{2+} mobilization following hormonal stimulation, which have demonstrated extensive communication between the ER and mitochondrial Ca^{2+} pools (Rizzuto *et al.*, 1998; Hajnoczky *et al.*, 1995). The ER and mitochondrial membranes appear to be juxtaposed, and IP_3-mediated release of ER Ca^{2+} leads to a nearly immediate and somewhat prolonged increase in the concentration of Ca^{2+} present in the mitochondrial matrix (Rizzuto *et al.*, 1998; Hajnoczky *et al.*, 1995; McCormack *et al.*, 1990). Under normal circumstances these Ca^{2+} signals may promote the activation of Ca^{2+}-dependent metabolic enzymes located in the mitochondrial matrix, leading to increased cellular activation and respiration (McCormack *et al.*, 1990). However, it is also possible that this same communication may promote transient opening of a mitochondrial cytochrome *c* release channel(s) during apoptosis (Szalai *et al.*, 1999; Xia *et al.*, 1995; Sadoul *et al.*, 1996), without global depolarization of the mitochondrial membrane (Szalai *et al.*, 1999). As discussed earlier, BCL-2 may act at both the ER and mitochondrion to prevent the effects of Ca^{2+} on cytochrome *c* release.

Together, the work introduced earlier strongly suggests that alterations in subcellular Ca^{2+} localization, and not global increases in intracellular Ca^{2+}, are more generally involved in the regulation of apoptosis. Therefore, techniques capable of visualizing particular subcellular Ca^{2+} pools appear to be yielding the most important information in this area of investigation.

II. Ca^{2+}-Specific Probes

A. Synthetic Fluorescent Dyes

The first Ca^{2+}-selective fluorescent dyes were synthesized by Tsien and colleagues (1980, 1982) and Grynkiewicz *et al.* (1985), and these agents remain the most common Ca^{2+} indicators in use today. All of the dyes are based on the structure of the Ca^{2+}-selective chelator EGTA. Modifications include decreased affinity for Ca^{2+} and improved fluorescent properties that exhibit dramatic changes upon interaction with Ca^{2+} (Grynkiewicz *et al.,* 1985). In their active forms, all of the dyes possess a high negative charge, and they must therefore be either microinjected into single cells or coupled with a hydrophobic moiety to allow for transplasma membrane passage. Most investigators load their cells with the acetoxymethyl ester (AM) forms of the dyes. These forms pass freely into cells where they are cleaved by cytosolic esterase(s), yielding highly charged species that are trapped primarily in the cytosol. Using this method, low micromolar extracellular dye concentrations can be used to accumulate near-millimolar intracellular concentrations.

Each synthetic Ca^{2+} probe has specific properties that can be exploited in specific applications. The first EGTA analog synthesized (BAPTA) exhibits fluorescence properties (very high-energy excitation) that make it unsuitable for use as a Ca^{2+} indicator in viable cells, but it is commonly used as an intracellular Ca^{2+} buffering agent due to its relatively high affinity (50 n*M*) for Ca^{2+}. The second-generation dye, quin-2, has much more preferable, single-wavelength fluorescence properties (excitation = 325 nm, emission = 500 nm), but it is not particularly bright and it bleaches fairly rapidly. The third dye synthesized (fura-2) has the advantage that it exhibits opposite fluorescence properties when it is excited at either of two wavelengths: at 340 nm, fura-2 fluorescence increases upon binding Ca^{2+}, whereas at 380 its fluorescence decreases (emission = 510 nm). Thus, in fluorimeters capable of measuring both wavelengths simultaneously, the fluorescence signals can be quantified in "ratio" (340/380), resulting in much greater sensitivity. In addition, results obtained in ratio mode are independent of the specific intracellular concentration of dye and are therefore less subject to artifacts generated by differences in dye loading. Finally, a fourth dye (indo-1) has properties (single-wavelength excitation, dual–wavelength emission) that allow it to be used to measure Ca^{2+} in "ratio" mode, and its optical characteristics make it ideal for measuring Ca^{2+} in mixed populations of cells by flow cytometry (FACS). A comparison of the chemical properties of several commonly used dyes can be found in Table I.

More recent generations of dyes are superior for use in the sophisticated imaging applications required for the visualization of subcellular Ca^{2+} Compartmentalization. The relatively short excitation wavelengths required to excite quin-2, fura-2, and indo-1 make them difficult to use to image subcellular Ca^{2+} changes by fluorescence or confocal microscopy, as they tend to bleach very rapidly. This prompted the development of dyes that are excited at longer wavelengths (400 nm and greater) such as fluo-3 and calcium green that do not bleach as readily. Parallel staining with fluo-3 and a mitochondrial marker (such as MitoTracker, TMRE, or rhodamine B) allows for direct measurement of global and intramitochondrial Ca^{2+} changes by confocal microscopy. A more detailed discussion of this approach has been provided elsewhere (Lemasters *et al.,* 1995).

Subcellular dye localization is a major factor that must be controlled for in experiments that employ the fluorescent dyes outlined earlier. Early work demonstrated that dye accumulation within organelles was highly sensitive to temperature. In one example, human skin fibroblasts loaded with fura-2 AM at 37°C displayed heterogeneous dye uptake concentrated in acidic organelles, whereas loading at 15°C produced homogeneous distribution throughout the cytosol (Malgaroli *et al.,* 1987). The increased interest

Table I
Calcium-Binding and Fluorescence Properties of Common Ca^{2+} Dyes[a]

		Absorbance max (nm)		Emission max (nm)	
Type of indicator	K_d (nM)	Ca^{2+} free	Bound	Ca^{2+} free	Bound
Single wavelength, intensity—modulating					
Quin-2	115	352	332	492	498
Fluo-3	400	503	506	526	526
Rhod-2	1000	556	553	576	576
Dual wavelength, ratiometric					
Fura-2	224	362	335	512	505
Indo-1	250	349	331	485	410

[a] K_d, dissociation constants for Ca^{2+} binding. Excitation and emission wavelength maxima are presented in nanometers. For use of fura-2 in ratiometric measurements, the dye is excited at 340 and 380 nm, and fluorescence is measured at 495–510 nm emission. The signal obtained at 340 increases and the signal at 380 decreases on Ca^{2+} binding. For indo-1, the dye is generally excited at 360 nm and fluorescence is monitored at 410 and 485 nm emission. (Calcium binding results in an increase in the former and a decrease in the latter.)

in imaging Ca^{2+} fluxes within organelles requires that a significant fraction of the acetoxymethyl ester form of the dye pass through the cytoplasm without first being cleaved (and trapped) by cytosolic esterases. Studies suggest that loading cells at 4°C can help accomplish this goal (Lemasters *et al.,* 1995). Nonetheless, it is extremely important to confirm homogeneous dye distribution, especially when semiquantitative measurements are made. Dye flux from one cellular compartment to another can produce artifacts that over- or underestimate actual Ca^{2+} levels.

B. Aequorin

Aequorin is a 22-kDa photoprotein product of jellyfish and certain other marine organisms. The aequorin complex contains the luminophore coelenterazine, which emits blue light (466 nm) on Ca^{2+}-dependent oxidation. Although its K_d for Ca^{2+} is fairly high (15 μM), aequorin has a broad detection range (from 100 nM to over 100 μM), allowing for accurate measurements to be made across all physiological Ca^{2+} concentrations.

Aequorin is not normally compartmentalized within organelles, which is a strength of using the microinjected compound for cytosolic Ca^{2+} measurements. However, Rizzuto and colleagues (1995) have generated recombinant aequorin fusion proteins that contain organelle-targeting motifs directing localization to the nucleus, mitochondrion, and endoplasmic reticulum. The proteins are also tagged with an epitope marker (HA) that allows for detection of the transfected aequorin with a monoclonal antibody. Cells are transfected with the construct of choice (now available commercially through Molecular

Probes). Once transfectants have been isolated, the cells are incubated for 1–2 h in coelenterazine (or one of the improved analogs that are now available from Molecular Probes), thereby reconstituting the aequorin complex. This strategy has been used successfully to measure intranuclear (Brini *et al.*, 1993), intramitochondrial (Rizzuto *et al.*, 1992), intra-Golgi (Pinton *et al.*, 1998), or intraendoplasmic reticular (Montero *et al.*, 1995) Ca^{2+} concentrations in intact cells.

Although measuring organelle Ca^{2+} levels with targeted aequorin is extremely appealing, there are serious pitfalls associated with using this approach in apoptotic cells. Aequorin is very good for measuring acute Ca^{2+} changes, but it is destroyed irreversibly by Ca^{2+} binding, which means that its light output is dependent on its entire past exposure to Ca^{2+}. Therefore, the intracellular half-life of aequorin is especially short in organelles that contain high resting Ca^{2+} concentrations, such as the ER. To circumvent this problem, the targeted aequorin molecules are reconstituted with coelenterazine after organelles are depleted of Ca^{2+} using ionomycin and EGTA (Pinton *et al.*, 1998; Montero *et al.*, 1995). Given that ER pool depletion is currently thought to be a trigger for apoptosis, the reconstitution procedure could have serious adverse effects on downstream components of the response. The procedure is also not amenable to long-term kinetic analyses.

C. Cameleons

Cameleons were developed by Tsien and colleagues to address the problems associated with the imaging of organelle Ca^{2+} fluxes with synthetic dyes or aequorin. Cameleons are chimeric proteins consisting of a blue or cyan mutant form of the green fluorescent protein (GFP), calmodulin, the calmodulin-binding domain of myosin light chain kinase (MLCK), and a green or yellow version of GFP (Miyawaki *et al.*, 1997). Calcium binding to calmodulin causes an intramolecular binding of calmodulin to MLCK, which increases the efficiency of fluorescence resonance energy transfer between the two GFP subunits. Cameleons have been targeted to specific organelles by strategies analogous to those used for the aequorin derivatives discussed earlier. Initial problems associated with pH-dependent effects on cameleon fluorescence have been rectified by the development of new structural derivatives (Miyawaki *et al.*, 1997).

III. Calcium Imaging

A. Spectrofluorimetric Measurements in Bulk Populations

Cells are typically loaded with low micromolar (1–5 μM) concentrations of the acetoxymethyl ester forms of quin-2 or fura-2 (the dyes of choice for this application) for 15–60 min, cells are washed, and fluorescence is measured at appropriate excitation and emission wavelengths. The advantage of quin-2 in these assays is that it can be used very effectively in the single-wavelength mode (excitation = 325 nm, emission = 500 nm), and it tends to leak out of cells less rapidly than fura-2. However, its relatively low fluorescence requires that higher intracellular concentrations of dye be used, which can lead to significant intracellular Ca^{2+} buffering. (The K_d value for quin-2 for Ca^{2+} is about 100 n*M*.) For these reasons, fura-2 has largely replaced quin-2 for most applications today. The improved fluorescence properties of fura-2 allow for greater sensitivity, especially when used in the ratio mode. In addition, its lower K_d for Ca^{2+} (225 n*M*) results in less interference with intracellular Ca^{2+} fluxes. However, it is especially important to control for dye leakage in experiments with fura-2.

Dye leakage can generate serious artifacts when cells are analyzed by spectrofluorimetry. Extracellular dyes are exposed to the millimolar concentrations of Ca^{2+} present in medium, and the high background fluorescence that results can dramatically interfere with Ca^{2+} measurements. Cells are therefore typically washed just prior to analysis to minimize extracellular dye. Another control for dye leakage is the application of extracellular Zn^{2+} or Mn^{2+}, which quench dye fluorescence. Alternatively, the addition of extracellular EGTA will generate a decrease in fluorescence if extracellular dye is present.

The availability of compounds capable of selectively emptying the endoplasmic reticular Ca^{2+} store has allowed investigators to obtain indirect measurements of ER pool depletion in apoptotic cells (Baffy *et al.*, 1993; Lam *et al.*, 1993; He *et al.*, 1997). Cells are loaded with fura-2 AM under conditions that promote cytoplasmic retention of the dye. Cells are then placed in Ca^{2+}-free medium and are exposed to either thapsigargin or 2,5-ditertbutyl-1,4-benzohydroquinone (DBHQ) (Moore *et al.*, 1987; Kass *et al.*, 1989). (Both compounds are available from Calbiochem, La Jolla, CA.) The magnitude of the increase in cytosolic Ca^{2+} that follows is directly proportionate to ER Ca^{2+} content, as capacitative Ca^{2+} entry is prevented by the absence of extracellular Ca^{2+}. Once the ER pool has been emptied, other intracellular pools can be evaluated by incubating the cells with ionomycin. Alternatively, some investigators have used mitochondrial uncouplers to obtain indirect measurements of mitochondrial Ca^{2+} pools (Baffy *et al.*, 1993). However, it is important to ensure that proapoptotic triggers do not alter dye localization in the cells, which can produce artifactual results (Tombal *et al.*, 1999).

1. Protocol: Spectrofluorimetric Measurement of Cytosolic Ca^{2+} in Human PC-3 Prostate Adenocarcinoma Cells [Adapted from Nutt and O'Neil (2000)]

Confluent primary cultures of cells are rinsed twice with Dulbecco's phosphate-buffered saline (D-PBS) and detached by mild trypsinization (0.05% trypsin plus 0.53 m*M* EDTA) for 2–3 min. Digestion is stopped by the addition of RPMI 1640 medium containing 10% fetal calf serum (FCS). Cells are washed once with isotonic bathing solution containing 140 m*M* NaCl, 4.2 m*M* KCl, 0.4 m*M* $NaxHPO_4$, 0.5 m*M* HaH_2PO_4, 0.3 m*M* $MgCl_2$, 0.4 m*M* $MgSO_4$, 1 m*M* $CaCl_2$, and 20 m*M* *N*-2-hydroxymethylpiperazine-*N*′-2-ethanesulfonic acid (HEPES), pH 7.4, at 37°C. Just before use, 0.2% bovine serum albumin (BSA) and 5 m*M* glucose are added to the solution. Cells are resuspended in bathing solution containing 10 μ*M* fura-2 acetoxymethyl ester (fura-2 AM) and are incubated for 30 min at 37°C, with periodic gentle mixing. At the end of this incubation, the mixture of cells and fura-2 is diluted 1 : 4 with isotonic bathing solution and is incubated for an additional 60 min at 37°C. The cells are then washed in bathing medium without fura-2, resuspended in the medium, and maintained at room temperature until use. Just before analysis, cells are washed twice in bathing medium by centrifugation for 30 s in a microcentrifuge (1200*g*). An aliquot of cells (usually about 1×10^5 to 1×10^6) is added to a quartz cuvette containing 2 ml of isotonic bathing medium with or without 5 m*M* EGTA.

Intracellular Ca^{2+} concentrations are estimated from fura-2 fluorescence by monitoring emission at 511 nm following dye excitation at 340 and 380 nm (Delta Scan dual-excitation fluorometer, Photon Technology International, South Brunswick, NJ). All fluorescence measurements are initiated when cells are added to the cuvette. Fura-2 fluorescence ratios are converted to intracellular Ca^{2+} concentrations according to the formula described by Grynkiewicz *et al.* (1985) as follows

$$[Ca^{2+}]_i = \beta K_d[R - R_{min}/R_{max} - R]$$

where R is the ratio at any time and β is the ratio of the fluorescence emission intensity at 380 nm excitation in Ca^{2+}-depleting and Ca^{2+}-saturating conditions, K_d is the Ca^{2+} dissociation constant for fura-2 (Table I) (225 n*M*), R_{min} is the minimum ratio in Ca^{2+}-depleting conditions (i.e., upon addition of 5 m*M* EGTA to lysed cells), and R_{max} is the maximum ratio in Ca^{2+}-saturating conditions (i.e., lysed cells in the presence of the 1 m*M* $CaCl_2$ contained in the bath medium).

B. Flow Cytometry (FACS)

Major advantages of flow cytometry for Ca^{2+} analysis are the ability to detect population heterogeneity in response and to measure multiple biochemical end points simultaneously. The newer FACS machines are generally capable of identifying up to seven fluorescent signals simultaneously. A multicolor approach has been used successfully to demonstrate that cytosolic Ca^{2+} increases occur after decreases in intracellular reduced glutathione levels, decreases in $\Delta\Psi$mito, and increases in oxygen radical production in thymocytes (Macho *et al.*, 1997).

Generally, FACS machines are not equipped to excite fluorescent dyes at more than one wavelength. Therefore, the dye of choice for FACS-based Ca^{2+} analyses has been indo-1, which is used in the ratio mode by measuring fluorescence at 400 and 480 nm simultaneously. Extracellular dye does not interfere with these measurements, as the machine is typically gated on forward-and side-scatter properties typical to whole cells. Likewise, Ca^{2+} alterations that occur during secondary necrosis ("postmortem") can be gated out by staining cells with propidium iodide or another vital dye that does not interfere with indo-1 fluorescence.

C. Luminescence Measurements (Aequorin)

Another minor disadvantage associated with using aequorin for Ca^{2+} imaging is the need to measure responses in a fairly unique apparatus. The investigators who developed the targeted aequorin constructs have devised a special unit consisting of a low-noise photomultiplier and an amplifier/discriminator connected to a power supply and a computer-controlled photon-counting board. For maximum efficiency the photomultiplier is maintained at 4°C in a dedicated, light-protected refrigerator. Generally speaking, such units are dedicated to Ca^{2+} measurement and are therefore not usually shared departmental instrumentation.

D. Fluorescence Microscopy

Quantitation of Ca^{2+} at the single-cell level is now a routine procedure with the development of the epifluorescence/phase-contrast microscope. A wide range in hardware is available from the top-end Photon Technology Inc., delta scan apparatus to the moderately priced but very efficient Intracellular Imager. Fura-2 AM is the preferred intracellular Ca^{2+} dye for this application. Cells are grown on coverslips, and fluorescence is monitored using a 20× fluorescence objective and equipped with a video camera. The calculated free Ca^{2+} is determined using a cell-free calibration curve. This technique affords the convenience of measuring Ca^{2+} without disrupting adhesion. Examples of Ca^{2+} measurements obtained by this method can be found in Fig. 2.

The development of fluorescent dyes that selectively distribute to particular organelles would represent a major technological advance. However, as discussed earlier, control of intracellular dye localization continues to pose a challenge to investigators in the field, and none of the conventional (chemical) dyes can be used to directly measure intraorganellar Ca^{2+} levels without first localizing the organelle of interest with another fluorescent dye. In our hands, standard fluorescence microscopy does not usually provide the resolution required for such studies, and a confocal microscope is required. However, ER and non-ER Ca^{2+} pools can be estimated indirectly using thapsigargin, DBHQ, mitochondrial uncouplers, and ionomycin as described previously. Alternatively, methods have been developed for measuring intracellular Ca^{2+} pools in digitonin-permeabilized cells (Szalai *et al.*, 1999; Csordas *et al.*, 1999).

Alternatively, confocal microscopy is a powerful way to measure qualitative changes in intracellular Ca^{2+} compartmentalization. Cells are loaded with a long-wavelength dye such as fluo-3 or calcium green under conditions that promote optimal distribution of the dye to the organelle of interest. Usually the particular conditions necessary have to be worked out empirically, but important variables include temperature, time of incubation, and buffer composition. Cells are then counterstained with a fluorescent dye specific for a particular organelle with an emission wavelength that does not overlap with the Ca^{2+} probe. Commonly used tracking dyes include rhodamine derivatives or Mitotracker (from Molecular Probes, and specific for mitochondria), DAPI (nuclear membranes), and Bodipy (Golgi apparatus). Images are obtained at each emission wavelength, and overlays are produced to measure Ca^{2+} indicator fluorescence intensity at organelle sites within the cell. Some of the newer microscopes are also equipped with hardware and software packages that allow for semiquantitative measurements to be made. Examples of confocal images obtained from human PC-3 prostate adenocarcinoma cells loaded with a Ca^{2+} probe (fluo-3 AM) and a

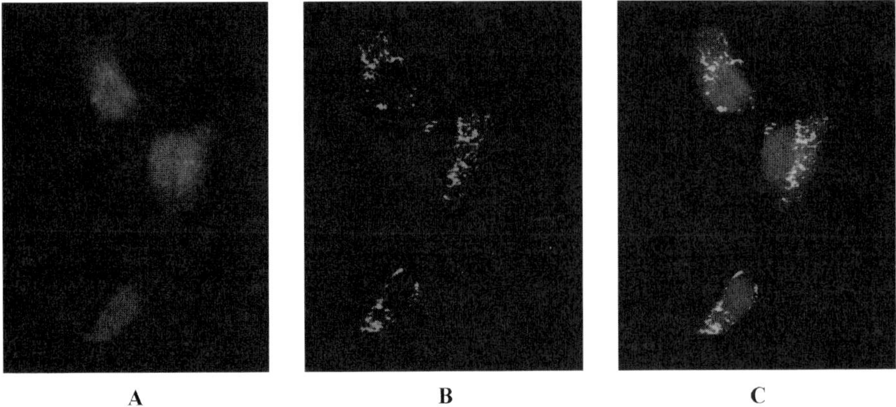

Fig. 3 Dual labeling of cells with a Ca^{2+} probe and a mitochondrial tracer. (A) Mitochondrial staining. PC-3 cells were labeled with the mitochondrial tracer rhodamine B (1 : 10^6 dilution) for 2 min. (B) Calcium dye fluorescence. The same cells presented in A were loaded with 5 μM fluo-3 AM for 30 min. (C) Images presented in A and B were overlaid. The intensity of orange/yellow fluorescence corresponds to levels of intramitochondrial Ca^{2+} concentrations. (See Color Plate.)

mitochondrial probe (rhodamine B)(both from Molecular Probes) are presented in Fig. 3 (see Color Plate).

1. Protocol: Single-Cell Measurements of Intracellular Ca^{2+} in Human PC-3 Prostate Adenocarcinoma Cells

For measurement of cytosolic Ca^{2+} concentrations, cells are plated on 22 × 30 glass coverslips. On culture day 2, cells are incubated for 1 h at 37°C with 10 μM fura-2 AM in RPMI 1640 medium containing 10% FCS and 1 mM probenecid (Molecular Probes, Eugene, OR). The coverslips are washed thoroughly with PBS and mounted onto a 1.5-ml chamber, with the cells facing upward. The chamber is placed on an epifluorescence/phase-contrast microscope (INCA workstation, Intracellular Imaging, Inc., Cincinnati, OH). Cells are illuminated alternatively at excitation wavelengths of 340 and 380 nm using a xenon arc lamp. The emitted fluorescence is monitored at 511 nm with a video camera, and the calculated free [Ca^{2+}] is determined using a cell-free calibration curve. Data are collected with INCA software (Win 3.1 version). See Fig. 3 for representative results.

For qualitative assessment of Ca^{2+} fluxes in mitochondria, cells plated on 22 × 30 glass coverslips are incubated for 1 h at 37°C with 10 μM fluo-3 AM in RPMI 1640 medium containing 10% FCS and 1 mM probenecid. Dye uptake into mitochondria can be enhanced in some cell types by lowering the loading temperature (to as low as 4°C), which apparently allows for lower hydrolysis (and cytosolic trapping) by cytosolic esterases. To visualize mitochondria, cells are washed with PBS and incubated for 5 min at 37°C with rhodamine B, Mitotracker Red, or another appropriate mitochondrial marker. Confocal analysis of red fluorescence should reveal punctate cytoplasmic fluorescence consistent with dye uptake by mitochondria. Colocalization of fluo-3 and rhod-B results in orange or yellow fluorescence, depending on the amount of mitochondrial fluo-3 loading and the intramitochondrial Ca^{2+} concentration. Control experiments should be conducted (with excess Ca^{2+} or EGTA) to confirm that alterations in fluo-3 fluorescence are not due to alterations in intracellular dye localization. Alternatively, we have found that rhod-2 AM preferentially accumulates in mitochondria in PC-3 cells. Cells may therefore be loaded with rhod-2 and a green mitochondrial marker ($DiOC_6$ or Mitotracker Green), and fluorescence overlays can be generated. These approaches must be considered qualitative at best, but semiquantitative results can be obtained if the confocal microscope is equipped with software that allows for pseudo-color image analysis. The results obtained are then presented as relative increases in fluorescence using a color scale as a legend.

IV. Summary and Conclusions

Early studies in apoptosis implicated an increase in cytosolic Ca^{2+} as a direct mediator of DNA fragmentation. However, efforts to delineate targets for this increase in Ca^{2+} have been slow in evolving. Several previous studies have implicated ER Ca^{2+} pool depletion in the initiation of apoptosis. Our own preliminary studies confirm that many (but not all) apoptotic stimuli empty the ER store via a mechanism that is blocked by BCL-2 expression. Furthermore, ER pool depletion is not affected by broad spectrum caspase inhibitors, indicating that it occurs via a caspase-independent mechanism. Finally, our data demonstrate that ER pool depletion occurs prior to release of cytochrome *c* from mitochondria. Given previous work demonstrating close coordination of ER and mitochondrial Ca^{2+} levels, we speculate that ER-dependent changes in mitochondrial Ca^{2+} serve as important signals for cytochrome *c* release. Alternative mechanisms include activation of caspase-12 and/or the JNK pathway, both of which can be directly stimulated by depletion of the ER Ca^{2+} pool.

Although substantial improvements in intracellular Ca^{2+} imaging have emerged, compelling answers to many of the present questions related to the role of Ca^{2+} in apoptosis await future technical improvements. The development of organelle-specific, recombinant Ca^{2+} probes (targeted aequorins and cameleons) certainly should facilitate some of this work, although the target cell of interest must be amenable to molecular manipulation (transfection), which precludes straightforward analysis of primary cells. Pharmacological tools (i.e., thapsigargin and DBHQ) can provide conclusive data on ER pool status without requiring an overly sophisticated image analysis system. However, confocal microscopy allows for the effective analysis of Ca^{2+} pools as long as dye localization is homogeneous and properly controlled. However, current techniques should be considered semiquantitative at best and will remain so until specific organelle-targeted fluoroescent dyes are developed and widely available.

References

Adachi-Yamada, T., Fujimura-Kamada, K., Nishida, Y., and Matsumoto, K. (1999). Distortion of proximodistal information causes JNK-dependent apoptosis in Drosophila wing. *Nature* **400,** 166–169.

Ankarcrona, M., Dypbukt, J. M., Bonfoco, E., Zhivotovsky, B., Orrenius, S., Lipton, S. A., and Nicotera, P. (1995). Glutamate-induced neuronal death: A succession of necrosis or apoptosis depending on mitochondrial function. *Neuron* **15,** 961–973.

Antonsson, B., Conti, F., Ciavatta, A., Montessuit, S., Lewis, S., Martinou, I., Bernasconi, L., Bernard, A., Mermond, J., Mazzei, G., Maundrell, K., Gambale, F., Sadoul, R., and Martinou, J. C. (1997). Inhibition of BAX channel-forming activity by BCL-2. *Science* **277,** 370–372.

Baffy, G., Miyashita, T., Williamson, J. R., and Reed, J. C. (1993). Apoptosis induced by withdrawal of interleukin-3 (IL-3) from an IL-3-dependent hematopoietic cell line is associated with repartitioning of intracellular calcium and is blocked by enforced BCL-2 oncoprotein production. *J. Biol. Chem.* **268,** 6511–6519.

Bellomo, G., Perotti, M., Taddei, F., Mirabelli, F., Finardi, G., Nicotera, P., and Orrenius, S. (1992). Tumor necrosis factor a induces apoptosis in mammary adenocarcinoma cells by an increase in intranuclear free Ca^{2+} concentration and DNA fragmentation. *Cancer Res.* **52,** 1342–1346.

Brini, M., Murgia, M., Pasti, L., Picard, D., Pozzan, T., and Rizzuto, R. (1993). Nuclear Ca^{2+} concentration measured with specifically targeted recombinant aequorin. *EMBO J.* **12,** 4813–4819.

Choi, D. W. (1995). Calcium: Still center-stage in hypoxic-ischemic neuronal cell death. *Trends Neurosci.* **18,** 58–60.

Ciapa, B., Pesando, D., Wilding, M., and Whitaker, M. (1994). Cell-cycle calcium transients driven by cyclic changes in inositol trisphosphate levels. *Nature* **368,** 875–878.

Cohen, J. J., and Duke, R. C. (1984). Glucocorticoid activation of a calcium-dependent endonuclease in thymocyte nuclei leads to cell death. *J. Immunol.* **132,** 38–42.

Csordas, G., Thomas, A. P., and Hajnoczky, G. (1999). Quasi-synaptic calcium signal transmission between endoplasmic reticulum and mitochondria. *EMBO J.* **18,** 96–108.

Cuvillier, O., Pirianov, G., Kleuser, B., Vanek, P. G., Coso, O. A., Gutkind, J. S., and Spiegel, S. (1996). Suppression of ceramide-mediated programmed cell death by sphingosine-1-phosphate. *Nature* **381,** 800–803.

Enari, M., Sakahira, H., Yokoyama, H., Okawa, K., Iwamatsu, A., and Nagata, S. A. (1998). Caspase-activated DNase that degrades DNA during apoptosis, and its inhibitor ICAD. *Nature* **391,** 43–50.

Furuya, Y., Lundmo, P., Short, A. D., Gill, D. L., and Isaacs, J. T. (1994). The role of calcium, pH, and cell proliferation in the programmed (apoptotic) death of androgen-independent prostatic cancer cells induced by thapsigargin. *Cancer Res.* **54,** 6167–6175.

Gerasimenko, O. V., Gerasimenko, J. V., Tepikin, A. V., and Petersen, O. H. (1995). ATP-dependent accumulation and inositol trisphosphate- or cyclic ADP-ribose-mediated release of Ca^{2+} from the nuclear envelope. *Cell* **80,** 439–444.

Green, D. R., and Reed, J. C. (1998). Mitochondria and apoptosis. *Science* **281,** 1309–1312.

Grynkiewicz, G., Poenie, M., and Tsien, R. Y. (1985). A new generation of Ca^{2+} indicators with greatly improved fluorescence properties. *J. Biol. Chem.* **260,** 3440–3450.

Hajnoczky, G., Robb-Gaspers, L. D., Seitz, M. B., and Thomas, A. P. (1995). Decoding of cytosolic calcium oscillations in the mitochondria. *Cell* **82,** 415–424.

He, H., Lam, M., McCormick, T. S., and Distelhorst, C. W. (1997). Maintenance of calcium homeostasis in the endoplasmic reticulum by Bcl-2. *J. Cell Biol.* **138,** 1219–1228.

Hewish, D. R., and Burgoyne, L. A. (1973). Chromatin sub-structure. The digestion of chromatin DNA at regularly spaced sites by a nuclear deoxyribonuclease. *Biochem. Biophys. Res. Commun.* **52,** 504–510.

Hockenbery, D. M., Nunez, G., Milliman, C., Schreiber, R. D., and Korsmeyer, S. J. (1990). BCL-2 is an inner mitochondrial membrane protein that blocks programmed cell death. *Nature* **348,** 334–336.

Jones, D. P., McConkey, D. J., Nicotera, P., and Orrenius, S. (1989). Calcium-activated DNA fragmentation in rat liver nuclei. *J. Biol. Chem.* **264,** 6398–6403.

Kaiser, N., and Edelman, I. S. (1977). Calcium dependence of glucocorticoid-induced lymphocytolysis. *Proc. Natl. Acad. Sci. USA* **74,** 638–642.

Kaiser, N., and Edelman, I. S. (1978). Further studies on the role of calcium in glucocorticoid-induced lymphocytolysis. *Endocrinology* **103,** 936–942.

Kass, G. E., Duddy, S. K., Moore, G. A., and Orrenius, S. (1989). 2,5-Di-(tert-butyl)-1,4-benzohydroquinone rapidly elevates cytosolic Ca^{2+} concentration by mobilizing the inositol 1,4,5-trisphosphate-sensitive Ca^{2+} *pool. J. Biol. Chem.* **264,** 15192–15198.

Kerr, J. F. R., and Searle, J. (1973). Deletion of cells by apoptosis during castration-induced involution of the rat prostate. *Virch. Arch. Abt. B Zellpath.* **13,** 87–102.

Kluck, R. M., Bossy-Wetzel, E., Green, D. R., and Newmeyer, D. D. (1997). The release of cytochrome *c* from mitochondria: A primary site for bcl-2 regulation of apoptosis. *Science* **275,** 1132–1136.

Koike, T., Martin, D. P., and Johnson, E. M., Jr. (1989). Role of Ca^{2+} channels in the ability of membrane depolarization to prevent neuronal death induced by trophic-factor deprivation: Evidence that levels of internal Ca^{2+} determine nerve growth factor dependence of sympathetic ganglion cells. *Proc. Natl. Acad. Sci. USA* **86,** 6421–6425.

Kroemer, G., Zamzami, N., and Susin, S. A. (1997). Mitochondrial control of apoptosis. *Immunol. Today* **18,** 44–52.

Kyprianou, N., English, H. F., and Isaacs, J. T. (1988). Activation of a Ca^{2+}-Mg^{2+}-dependent endonuclease as an early event in castration-induced prostatic cell death. *Prostate* **13,** 103–117.

Lam, M., Dubyak, G., Chen, L., Nunez, G., Miesfeld, R. L., and Distelhorst, C. W. (1994). Evidence that bcl-2 represses apoptosis by regulating endoplasmic reticulum-associated Ca^{2+} fluxes. *Proc. Natl. Acad. Sci. USA* **91,** 1994.

Lam, M., Dubyak, G., and Distelhorst, C. W. (1993). Effect of glucocorticoid treatment on intracellular calcium homeostasis in mouse lymphoma cells. *Mol. Endocrinol.* **7,** 686–693.

Lemasters, J. J., Chacon, E., Ohata, H., Harper, I. S., Nieminen, A. L., Tesfai, S. A., and Herman, B. (1995). Measurement of electrical potential, pH, and free calcium ion concentration in mitochondria of living cells by laser scanning confocal microscopy. *Methods Enzymol.* **260,** 428–444.

Lemasters, J. J., Nieminen, A. L., and Herman, B. (1990). Is there release of mitochondrial calcium in toxic injury? *Hepatology* **11,** 902–903.

Liu, X., Kim, C. N., Yang, J., Jemmerson, R., and Wang, X. (1996). Induction of the apoptotic program in cell-free extracts: Requirement for dATP and cytochrome c. *Cell* **86,** 147–157.

Liu, X., Zou, H., Slaughter, C., and Wang, X. (1997). DFF, a heterodimeric protein that functions downstream of caspase-3 to trigger DNA fragmentation during apoptosis. *Cell* **89,** 175–184.

Macho, A., Hirsch, T., Marzo, I., Marchetti, P., Dallaporta, B., Susin, S. A., Zamzami, N., and Kroemer, G. (1997). Glutathione depletion is an early and calcium elevation is a late event of thymocyte apoptosis. *J. Immunol.* **158,** 4612–4619.

Malgaroli, A., Milani, D., Meldolesi, J., and Pozzan, T. (1987). Fura-2 measurement of cytosolic free Ca^{2+} in monolayers and suspensions of various types of animal cells. *J. Cell Biol.* **105,** 2145–2155.

Marin, M. C., Fernandez, A., Bick, R. J., Brisbay, S., Buja, M., Snuggs, M., McConkey, D. J., Eschenbach, A. C. v., Keating, M. J., and McDonnell, T. J. (1996). Apoptosis suppression by Bcl-2 is correlated with the regulation of nuclear and cytosolic Ca^{2+}. *Oncogene* **12,** 2259–2266.

Martikainen, P., and Isaacs, J. (1990). Role of calcium in the programmed cell death of rat ventral prostatic glandular cells. *Prostate* **17,** 175–187.

McConkey, D. J., and Orrenius, S. (1997). The role of calcium in the regulation of apoptosis. *Biochem. Biophys. Res. Commun.* **239,** 357–366.

McConkey, D. J., Nicotera, P., Hartzell, P., Bellomo, G., Wyllie, A. H., and Orrenius, S. (1989). Glucocorticoids activate a suicide process in thymocytes through an elevation of cytosolic Ca^{2+} concentration. *Arch. Biochem. Biophys.* **269,** 365–370.

McCormack, J. G., Halestrap, A. P., and Denton, R. M. (1990). Role of calcium ions in regulation of mammalian intramitochondrial metabolism. *Physiol. Rev.* **70,** 391–425.

Miyawaki, A., Griesbeck, O., Heim, R., and Tsien, R. Y. (1999). Dynamic and quantitative Ca^{2+} measurements using improved cameleons. *Proc. Natl. Acad. Sci. USA* **96,** 2135–2140.

Miyawaki, A., Llopis, J., Heim, R., McCaffery, J. M., Adams, J. A., Ikura, M., and Tsien, R. Y. (1997). Fluorescent indicators for Ca^{2+} based on green fluorescent proteins and calmodulin. *Nature* **388,** 882–887.

Montero, M., Brini, M., Marsault, R., Alvarez, J., Sitia, R., Pozzan, T., and Rizzuto, R. (1995). Monitoring dynamic changes in free Ca^{2+} concentration in the endoplasmic reticulum of intact cells. *EMBO J.* **14,** 5467–5475.

Moore, G. A., McConkey, D. J., Kass, G. E., O'Brien, P. J., and Orrenius, S. (1987). 2,5-Di(tert-butyl)-1,4-benzohydroquinone: A novel inhibitor of liver microsomal Ca^{2+} sequestration. *FEBS Lett.* **224,** 331–336.

Murphy, A. N., Bredesen, D. E., Cortopassi, G., Wang, E., and Fiskum, G. (1996). BCL-2 potentiates the maximal calcium uptake capacity of neural cell mitochondria. *Proc. Natl. Acad. Sci. USA* **93,** 9893–9898.

Nakagawa, T., Zhu, H., Morishima, N., Li, E., Xu, J., Yankner, B. A., and Yuan, J. (2000). Caspase-12 mediates endoplasmic-reticulum- specific apoptosis and cytotoxicity by amyloid-beta. *Nature* **403,** 98–103.

Narita, M., Shimizu, S., Ito, T., Chittenden, T., Lutz, R. J., Matsuda, H., and Tsujimoto, Y. (1998). Bax interacts with the permeability transition pore to induce permeability transition and cytochrome *c* release in isolated mitochondria. *Proc. Natl. Acad. Sci. USA* **95,** 14681–14686.

Newmeyer, D. D., Farschon, D. M., and Reed, J. C. (1994). Cell-free apoptosis in *Xenopus* extracts: Inhibition by bcl-2 and requirement for an organelle fraction enriched in mitochondria. *Cell* **79,** 353–364.

Nicotera, P., McConkey, D. J., Jones, D. P., and Orrenius, S. (1989). ATP stimulates Ca^{2+} uptake and increases the free Ca^{2+} concentration in isolated rat liver nuclei. *Proc. Natl. Acad. Sci. USA* **86,** 453–457.

Nicotera, P., Orrenius, S., Nilsson, T., and Berggren, P. O. (1990). An inositol 1,4,5-trisphosphate-sensitive Ca^{2+} pool in liver nuclei. *Proc. Natl. Acad. Sci. USA* **87,** 6858–6862.

Nutt, L. K., and O'Neil, R. G. (2000). Effect of elevated glucose on endothelin-induced store-operated and non-store-operated Ca^{2+} influx in renal cells. *J. Am. Soc. Nephrol.* **11,** 1225–1235.

Peitsch, M. C., Polzar, B., Stephan, H., Crompton, T., MacDonald, H. R., Mannherz, H. G., and Tschopp, J. (1993). Characterization of the endogenous deoxyribonuclease involved in nuclear DNA degradation during apoptosis (programmed cell death). *EMBO J.* **12,** 371–377.

Pinton, P., Pozzan, T., and Rizzuto, R. (1998). The Golgi apparatus is an inositol 1,4,5-trisphosphate-sensitive Ca^{2+} store, with functional properties distinct from those of the endoplasmic reticulum. *EMBO J.* **17,** 5298–5308.

Poenie, M., Alderton, J., Tsien, R. Y., and Steinhardt, R. A. (1985). Changes of free calcium levels with stages of the cell division cycle. *Nature* **315,** 147–149.

Poenie, M., Alderton, J., Steinhardt, R., and Tsien, R. (1986). Calcium rises abruptly and briefly throughout the cell at the onset of anaphase. *Science* **233,** 886–889.

Richter, C. (1993). Pro-oxidants and mitochondrial Ca^{2+}: Their relationship to apoptosis and oncogenesis. *FEBS Lett.* **325,** 104–107.

Rizzuto, R., Brini, M., Bastianutto, C., Marsault, R., and Pozzan, T. (1995). Photoprotein-mediated measurement of calcium ion concentration in mitochondria of living cells. *Methods Enzymol.* **260,** 417–428.

Rizzuto, R., Pinton, P., Carrington, W., Fay, F. S., Fogarty, K. E., Lifshitz, L. M., Tuft, R. A., and Pozzan, T. (1998). Close contacts with the endoplasmic reticulum as determinants of mitochondrial Ca^{2+} responses. *Science* **280,** 1763–1766.

Rizzuto, R., Simpson, A. W., Brini, M., and Pozzan, T. (1992). Rapid changes of mitochondrial Ca^{2+} revealed by specifically targeted recombinant aequorin [published erratum appears in *Nature* 1992 Dec 24-31;360(6406):768]. *Nature* **358,** 325–327.

Rodriguez-Tarduchy, G., Collins, M., and Lopez-Rivas, A. (1990). Regulation of apoptosis in interleukin-3-dependent hemopoietic cells by interleukin-3 and calcium ionophores. *EMBO J.* **9.**

Sadoul, R., Fernandez, P. A., Quiquerez, A. L., Martinou, I., Maki, M., Schroter, M., Becherer, J. D., Irmler, M., Tschopp, J., and Martinou, J. C. (1996). Involvement of the proteasome in the programmed cell death of NGF-deprived sympathetic neurons. *EMBO J.* **15,** 3845–3852.

Sakahira, H., Enari, M., and Nagata, S. (1998). Cleavage of CAD inhibitor in CAD activation and DNA degradation during apoptosis. *Nature* **391,** 97–99.

Shimizu, S., Narita, M., and Tsujimoto, Y. (1999). Bcl-2 family proteins regulate the release of apoptogenic cytochrome *c* by the mitochondrial channel VDAC. *Nature* **399,** 483–487.

Szalai, G., Krishnamurthy, R., and Hajnoczky, G. (1999). Apoptosis driven by IP(3)-linked mitochondrial calcium signals. *EMBO J.* **18,** 6349–6361.

Tombal, B., Denmeade, S. R., and Isaacs, J. T. (1999). Assessment and validation of a microinjection method for kinetic analysis of $[Ca^{2+}]_i$ in individual cells undergoing apoptosis. *Cell Calcium* **25,** 19–28.

Tsien, R. Y. (1980). New calcium indicators and buffers with high selectivity against magnesium and protons: Design, synthesis, and properties of prototype structures. *Biochemistry* **19,** 2396–2404.

Tsien, R. Y., Pozzan, T., and Rink, T. J. (1982). T-cell mitogens cause early changes in cytoplasmic free Ca^{2+} and membrane potential in lymphocytes. *Nature* **295,** 68–71.

Urano, F., Wang, X., Bertolotti, A., Zhang, Y., Chung, P., Harding, H. P., and Ron, D. (2000). Coupling of stress in the ER to activation of JNK protein kinases by transmembrane protein kinase IRE1. *Science* **287,** 664–666.

Vanderbilt, J. N., Bloom, K. S., and Anderson, J. N. (1982). Endogenous nuclease: Properties and effects on transcribed genes in chromatin. *J. Biol. Chem.* **257,** 13009–13017.

Verheij, M., Bose, R., Lin, X. H., Yao, B., Jarvis, W. D., Grant, S., Birrer, M. J., Szabo, E., Zon, L. I., Kyriakis, J. M., Haimovitz-Friedman, A., Fuks, Z., and Kolesnick, R. N. (1996). Requirement for ceramide-initiated SAPK/JNK signalling in stress-induced apoptosis. *Nature* **380,** 75–79.

Vito, P., Lacana, E., and D'Adamio, L. (1996). Interfering with apoptosis: Ca(2+)-binding protein ALG-2 and Alzheimer's disease gene ALG-3. *Science.* **271,** 521–525.

Whyte, M. K. B., Hardwick, S. J., Meagher, L. C., Savill, J. S., and Haslett, C. (1993). Transient elevations of cytosolic free calcium retard subsequent apoptosis in neutrophils in vitro. *J. Clin. Invest.* **92,** 446–455.

Williams, D. A., Becker, P. L., and Fay, F. S. (1987). Regional changes in calcium underlying contraction of single smooth muscle cells. *Science* **235,** 1644–1648.

Williams, D. A., Fogarty, K. E., Tsien, R. Y., and Fay, F. S. (1985). Calcium gradients in single smooth muscle cells revealed by the digital imaging microscope using Fura-2. *Nature* **318,** 558–561.

Wyllie, A. H. (1980). Glucocorticoid-induced thymocyte apoptosis is associated with endogenous endonuclease activation. *Nature* **284,** 555–556.

Wyllie, A. H., Morris, R. G., Smith, A. L., and Dunlop, D. (1984). Chromatin cleavage in apoptosis: Association with condensed chromatin morphology and dependence on macromolecular synthesis. *J. Pathol.* **142,** 67–77.

Xia, Z., Dickens, M., Raingeaud, J., Davis, R. J., and Greenberg, M. E. (1995). Opposing effects of ERK and JNK-p38 MAP kinases on apoptosis. *Science* **270,** 1326–1331.

Yang, D. D., Kuan, C. Y., Whitmarsh, A. J., Rincon, M., Zheng, T. S., Davis, R. J., Rakic, P., and Flavell, R. A. (1997). Absence of excitotoxicity-induced apoptosis in the hippocampus of mice lacking the Jnk3 gene. *Nature.* **389,** 865–870.

Zamzami, N., Marchetti, P., Castedo, M., Decaudin, D., Macho, A., Hirsch, T., Susin, S. A., Petit, P. X., Mignotte, B., and Kroemer, G. (1995a). Sequential reduction of mitochondrial transmembrane potential and generation of reactive oxygen species in early programmed cell death. *J. Exp. Med.* **182,** 367–377.

Zamzami, N., Marchetti, P., Castedo, M., Zanin, C., Vayssiere, J. L., Petit, P. X., and Kroemer, G. (1995b). Reduction in mitochondrial potential constitutes an early irreversible step of programmed lymphocyte death in vivo. *J. Exp. Med.* **181,** 1661–1672.

CHAPTER 11

Proteinase Families and Their Inhibitors

Donald L. Mykles

Department of Biology
Cell and Molecular Biology Program
and Molecular, Cellular, and Integration
Neurosciences Program
Colorado State University
Fort Collins, Colorado 80523

I. Introduction

Proteolytic cascades play a central role in the initation and execution of the cell death program. Much of the work has focused on caspases, but it is now clear that other proteinases are involved, including calpains, proteasome, serine proteinases, and metalloproteinases (for reviews, see Patel *et al.,* 1996; Grimm and Osborne, 1998; Lee

0091-679X/01 $35.00

and Goldberg, 1998b; Solary *et al.*, 1998; Chan and Mattson, 1999; Earnshaw *et al.*, 1999; Kass and Orrenius, 1999; Orlowski, 1999). The availability of cell-permeable inhibitors has greatly facilitated our ability to elucidate the functions, as well as identify protein substrates, of proteinases in intact cells.

The purpose of this chapter is to review the classification and biochemical properties of proteinases and their inhibitors. Emphasis is placed on three proteolytic systems that mediate apoptosis in various cell types: caspases, calpains, and proteasome. It is intended to be a practical guide on the proper use of proteinase inhibitors. Most of the information is organized into tables. This enables the reader to compare and select inhibitors most appropriate for a particular application. Limited space does not permit a complete listing of experimental conditions (e.g., concentrations, incubation intervals) used for each compound. However, the reader is encouraged to consult the references cited in the tables for these details. Additional information is usually available from suppliers on their web sites (see Table VII).

II. Enzyme Classification

There are five general types of proteinases: serine, threonine, cysteine, aspartic, and metallo (Table I). All proteinases require a nucleophile for hydrolyzing the peptide bond. The catalytic mechanism, however, is fundamentally different between enzymes of the first three types and those of the last two types. For serine, threonine, and cysteine proteinases, the nucleophile is part of the side chain of serine, threonine, and cysteine residues, respectively. For aspartic and metallo proteinases, the nucleophile is an activated water molecule. In addition to a nucleophile, catalysis also requires a proton donor or general base in the active site. This is often a histidine residue in serine and cysteine proteinases. A lysine residue is essential for proteasome activity (Groll *et al.,* 1999). Information on specific proteinases is available in Barrett *et al.* (1998) and on the web sites listed in Table II.

The specificity of inhibitors is determined by the catalytic mechanism and the conformation in and around the active site. Compounds used commonly to characterize

Table I
Proteinase Classification Based on Catalytic Properties[a]

Protease class	Selected examples
Aspartic	Cathepsins D and E; pepsin; renin
Cysteine	Calpains; caspases; cathepsins B, H, K, L, and S; deubiquitinases; papain
Metallo	Astacin; carboxypeptidase; collagenase; gelatinase; hedgehog protein; meprin; stromelysin; thermolysin; tolloid protein
Serine	Cathepsin G; chymotrypsin; granzyme; elastase, kallikrein; subtilisin; proteinase K; thrombin; tricorn protease; tripeptidyl-peptidase II; trypsin
Threonine	Proteasome

[a]From Barrett *et al.* (1998).

Table II
Web Sites for Information on Proteinases and Proteinase Inhibitors

Name	Address	Comments
Merops	www.bi.bbsrc.ac.uk/Merops/Merops.htm	Lists over 900 peptidases on individual cards; entries linked to sequence, taxonomic, and structural databases
Peptidase Nomenclature	www.chem.qmw.ac.uk/iubmb/enzyme/EC34	Enzyme nomenclature recommended by the International Union of Biochemistry and Molecular Biology; contains links to related databases
PROLYSIS	delphi.phys.univ-tours.fr/Prolysis	Lists peptidases and peptidase inhibitors; includes assays and enzyme kinetics; contains links to protease sites; uses "CUTTER" program to generate and analyze proteolytic fragments
Ubiquitin	cbweb.med.harvard.edu/ubiquitin	Web site for "Ubiquitin and the Biology of the Cell" (Peters *et al.*, 1998); contains links to database, software, and protease sites
Bioinformatic Links	www.cbs.dtu.dk/janhan/protlink.html	Contains comprehensive listing of bioinformatic resources

purified proteinases *in vitro* are listed in Tables III–VI. Reagents that react with hydroxyl or thiol groups are often potent inhibitors of serine and threonine or cysteine proteinases, respectively. However, they react with any available group on the protein and thus may inhibit activity indirectly by altering higher order structure. Greater specificity is usually achieved with compounds that mimic a polypeptide substrate. The residue at the P1 position, which is located on the N-terminal side to the scissile bond, usually determines substrate binding and cleavage. All caspases, for example, cleave after an Asp in a polypeptide. Additional residues toward the N terminus (P2, P3, P4, etc.) and/or toward the C terminus (P1', P2', P3', etc.) from the scissile bond often influence substrate binding, therefore modifying proteinase specificity (Barrett *et al.*, 1998).

III. Proteinase Inhibitors

A. Caspase

All caspases prefer an Asp at P1 and a Glu at P3 positions in the cleavage site, which has facilitated the synthesis of relatively specific inhibitors. Further selectivity is obtained by exploiting differences between isozymes in binding residues at the P2 and P4 positions in protein and peptide substrates (Margolin *et al.*, 1997; Talanian *et al.*, 1997; Villa *et al.*, 1997; Garcia-Calvo *et al.*, 1998, 1999; Chan and Mattson, 1999). A variety of cell-permeable inhibitors are available from commercial suppliers (Tables VIII and IX). Most are N-blocked tetrapeptides with either an aldehyde or a ketone at the C terminus. Peptide ketones are more effective than peptide aldehydes at low concentrations, as

Table III
Inhibitors of Serine Proteinases[a]

Inhibitor[b]	Comments	Commercial sources[c]
AEBSF	Irreversible; stable in aqueous solutions	Cal, MP, Roc, Sig
Antipain	Reversible; also inhibits cysteine proteases	Bac, Cal, ICN, Roc, Sig
APMSF	Irreversible; unstable in aqueous solutions	Bac, Cal, Fl, Roc, Sig
Aprotinin	Reversible protein inhibitor (6.5 kDa); very stable	Cal, Fl, ICN, Roc, Sig
Benzamidine	Irreversible; inhibits trypsin-like serine proteases	Cal, Fl, ICN, MP, Sig
Chymostatin	Reversible; also inhibits chymotrypsin-like serine proteases, cysteine proteases and proteasome	Bac, Cal, Roc, Sig
DCI	Irreversible	Cal, Roc, Sig
DFP	Irreversible; neurotoxin; volatile; requires special handling	Ald, Cal, Fl, ICN, Sig
Ecotin	Reversible protein inhibitor (32 kDa)	
Elastinal	Reversible; inhibits elastase-like proteases	Cal, ICN, Sig
Leupeptin	Reversible; also inhibits cysteine proteases and proteasome	Bac, Cal, Fl, ICN, MP, Roc, Sig
PMSF	Irreversible; unstable in aqueous solutions	Ald, Bac, Cal, ICN, MP, Roc, Sig
Soybean trypsin inhibitor	Reversible protein inhibitor (20 kDa)	Cal, Fl, Roc, Sig
TLCK	Irreversible; inhibits trypsin-like serine proteases	Aff, Bac, Cal, Fl, ICN, MP, Roc, Sig
TPCK	Irreversible; inhibits chymotyrypsin-like serine proteases	Aff, Ald, Bac, Cal, Fl, ICN, MP, Roc, Sig

[a]From Barrett *et al.* (1998) and Beynon and Bond (1989).

[b]AEBSF, 4-(2-aminoethyl)benzenesulfonyl flouride; APMSF, (4-amidinophenyl)methanesulfonyl fluoride; Ata, N^{α}-[(acetylthio)acetyl]; DCI, 3,4-dichloroisocoumarin; DFP, diisopropylfluorophosphate; PMSF, phenylmethanesulfonyl fluoride; TLCK, Tosyl-Lys-CH_2Cl; TPCK, Tosyl-Phe-CH_2Cl.

[c]See Table VII for abbreviations used for commercial sources.

peptide ketones bind irreversibly and are more permeable than peptide aldehydes. For peptide chloromethyl ketones and fluoromethyl ketones, the acidic residues (Glu and/or Asp) are methylated to improve cell permeability; intracellular esterase activity removes the methyl groups to produce a biologically active compound. This also "traps" the inhibitor inside cells, enhancing the effectiveness of these compounds even after cells are returned to inhibitor-free media.

There are few reports of the effects of caspase inhibitors on other intracellular proteinases. Given the restricted cleavage site specificity of caspases, it is likely that inhibitors with these sequences would be fairly specific. However, granzyme B, a serine proteinase, is inhibited by Z-IETD-FMK and Ac-IETD-CHO (Thornberry *et al.*, 1997). The trypsin- and chymotrypsin-like activities of the 26S proteasome are partially inhibited by Z-VAD-FMK, whereas these activities in the 20S proteasome are stimulated by Z-VAD-FMK (Beyette *et al.*, 1998). Furthermore, proteasomes can cleave substrates after Asp residues (Rivett *et al.*, 1994), suggesting that they could bind caspase inhibitors at higher concentrations. Z-VAD-FMK also inhibits calpain (Waterhouse *et al.*, 1998; Wolf *et al.*, 1999).

Table IV
Inhibitors of Cysteine Proteinases[a]

Inhibitor[b]	Comments	Commerical sources[c]
Antipain	Reversible; also inhibits trypsin-like serine proteases	Bac, Cal, ICN, Roc, Sig
Chymostatin	Reversible; also inhibits chymotrypsin-like proteases	Bac, Cal, Roc, Sig
Cystatin	Reversible protein inhibitor (12 kDa)	Cal, Sig
E-64	Irreversible; highly specific	Cal, MP, Roc, Sig
Leupeptin	Reversible; also inhibits trypsin-like serine proteases and proteasome	Bac, Cal, Fl, ICN, MP, Roc, Sig
NCO-700	Irreversible; highly specific	Cal
Z-Leu-Leu-Tyr-CHN_2	Irreversible	MP
Z-Leu-Val-Gly-CHN_2	Irreversible	Sig
Z-Phe-Ala-CHN_2	Irreversible	Bac

[a]From Barrett *et al.* (1998) and Beynon and Bond (1989).
[b]E-64, L-*trans*-epoxysuccinyl-leucylamido(4-guanidino)butane; NCO-700, bis[ethyl(2*R*,3*R*)-3-[(*S*)-methyl-1-[4-(2,3,4-trimethoxyphenylmethyl) piperazin-1-ylcarbonyl]butyl-carbonyl] oxiran-2-carboxylate] sulfate; Z, carbobenzoxy.
[c]See Table VII for abbreviations used for commercial sources.

B. Calpain

Calpains consist of a diverse family of Ca^{2+}-dependent cysteine proteinases in animal cells (for reviews, see Croall and DeMartino, 1991; Mykles, 1998; Wang and Yuen, 1999). In mammals, two calpains (μ- and m-calpain) are ubiquitously expressed and have a heterodimeric subunit composition. The 80-kDa catalytic subunit (CAPN1 or CAPN2 for μ- or m-calpain, respectively) has a cysteine proteinase domain and a Ca^{2+}-binding domain; the 30-kDa regulatory subunit (CAPN4) also has a Ca^{2+}-binding domain. Each Ca^{2+}-binding domain has five EF hand motifs, which also occur in calmodulin and other Ca^{2+}-binding proteins. Other calpain genes often have a tissue-specific distribution. p94 calpain (CAPN3) is expressed primarily in skeletal muscle and is responsible for limb girdle muscular dystrophy type 2 (Spencer *et al.*, 1997; Baghdiguian *et al.*, 1999); Lp82, an alternatively spliced form of p94, is expressed in lens (Ma *et al.*, 1998); nCL-2 is

Table V
Inhibitors of Aspartic Proteinases[a]

Inhibitor[b]	Comments	Commercial sources[c]
EPNP	Highly specific	MP
Pepstatin A	Reversible; inhibits some aspartic proteases	Bac, Cal, Fl, MP, Roc, Sig

[a]From Barrett *et al.* (1998) and Beynon and Bond (1989).
[b]EPNP, 1, 2-epoxy-3-(4-nitrophenoxy)propane.
[c]See Table VII for abbreviations used for commercial sources.

Table VI
Inhibitors of Metallo-Proteinases[a]

Inhibitor	Comments	Commercial sources[b]
Bestatin	Reversible; inhibits aminopeptidases	Roc, Sig
EDTA	Reversible; chelator; also inhibits calpains	Cal, Roc, Sig
Epibestatin	Reversible; inhibits aminopeptidases	Sig
1,10-Phenanthroline	Reversible; chelator	MP, Sig
Phosphoramidon	Reversible; does not inhibit all metallo-proteases	Cal, Roc, Sig

[a]From Barrett *et al.* (1998) and Beynon and Bond (1989).
[b]See Table VII for abbreviations used for commercial sources.

expressed in stomach (Sorimachi *et al.,* 1993a); nCL-4 is expressed in digestive organs (Lee *et al.,* 1998); CAPN6 is expressed in placenta (Dear *et al.,* 1997); and CANP11 is expressed in testis (Dear *et al.,* 1999). Other calpain genes (CAPN5 and CANP8) have wider distributions (Dear *et al.,* 1997; Braun *et al.,* 1999). An endogenous inhibitor, calpastatin, is expressed in all tissues.

A large number of effective cell-permeable inhibitors of calpains have been developed (for reviews, see Crawford, 1990; Hayes *et al.,* 1998; Wells and Bihovsky, 1998; Wang, 1999). These are primarily directed toward the ubiquitous calpains (Table X). There is extensive literature on the biochemical properties of μ- and m-calpains, including the effects of inhibitors on purified enzymes *in vitro*. Most compounds inhibit calpains by binding reversibly or irreversibly to the active site. Calpeptin and epoxysuccinyl peptides

Table VII
Commercial Sources of Proteinase Inhibitors

Supplier	Abbreviation	Web site
Affiniti Research Products	Aff	www.proteasome.com
Alexis Biochemicals	AB	www.alexis-corp.com
AnaSpec	AS	www.anaspec.com
Aldrich	Ald	www.sigma-aldrich.com
Bachem	Bac	www.bachem.com
BIOMOL	BI	www.biomol.com
Calbiochem	Cal	www.calbiochem.com
CHEMICON International	CH	www.chemicon.com
CLONTECH	CL	www.clontech.com
Enzyme Systems Products	ESP	www.enzymesys.com
Fluka	Fl	www.sigma-aldrich.com
ICN Pharmaceuticals	ICN	www.icnpharm.com
Molecular Probes	MP	www.probes.com
Oncogene Science	OS	www.oncogene.com
Peptide Institute (Osaka, Japan)	PI	
Roche	Roc	biochem.roche.com
Sigma	Sig	www.sigma-aldrich.com

Table VIII
Cell-Permeable Irreversible Inhibitors of Caspases[a]

Inhibitor[b]	Sequences and modifications	Inhibitor specificity (caspase isozyme) 1	2	3	4	5	6	7	8	9	10	Comments and references	Commercial sources[c]
Z-D-DCB	Z-Asp-CH_2-DCB	All caspases										Wang *et al.* (1998a,b), Zhao *et al.* (1999)	Cal, BI
Boc-D-FMK	Boc-Asp(OMe)-CH_2F	All caspases											Cal, ESP
Ac-DEVD-CMK	Ac-Asp-Glu-Val-Asp-CH_2Cl			X			X	X	X		X	Talanian *et al.* (1997), Thornberry *et al.* (1997)	Cal, CH
Z-DEVD-FMK	Z-Asp-Glu-Val-Asp-CH_2F			X			X	X	X		X	Denning *et al.* (1998), Talanian *et al.* (1997), Thornberry *et al.* (1997)	Cal, CL, ESP
Z-DQMD-FMK	Ac-Asp-Gln-Met-Asp-CH_2F			X			X					Talanian *et al.* (1997)	ESP
Z-IETD-FMK	Z-Ile-Glu(OMe)-Thr-Asp(OMe)-CH_2F								X			Also inhibits granzyme B (Pörn-Ares *et al.*, 1998; Thornberry *et al.*, 1997)	AB, Cal, CH, CL, ESP
Z-LEED-FMK	Z-Leu-Glu(OMe)-Glu(OMe)-Asp(OMe)-CH_2F											Inhibits caspase-13	ESP
Z-LEHD-FMK	Z-Leu-Glu(OMe)-His-Asp(OMe)-CH_2F				X	X				X		Thornberry *et al.* (1997)	AB, Cal
Z-LETD-FMK	Z-Leu-Glu(OMe)-Thr-Asp(OMe)-CH_2F								X				ESP
Z-LEVD-FMK	Z-Leu-Glu(OMe)-Val-Asp(OMe)-CH_2F	X											ESP
Z-VAD-FMK	Z-Val-Ala-Asp(OMe)-CH_2F	X		X	X			X				Denning *et al.* (1998), Knepper-Nicolai *et al.* (1998), Wu *et al.* (1999), Sun *et al.* (1999). Also inhibits proteasome (Beyette *et al.*, 1998) and calpain (Waterhouse *et al.*, 1998; Wolf *et al.*, 1999)	BI, Cal, CH, CL, ESP
Z-VDVAD-FMK	Z-Val-Asp(OMe)-Val-Ala-Asp(OMe)-CH_2F		X									Talanian *et al.* (1997)	Cal, ESP
Z-VEID-FMK	Z-Val-Glu(OMe)-Ile-Asp(OMe)-CH_2F						X					Talanian *et al.* (1997)	AB, Cal, CH, ESP
Z-WEHD-FMK	Z-Trp-Glu(OMe)-His-Asp(OMe)-CH_2F	X			X	X						Thornberry *et al.* (1997)	Cal, ESP
Ac-YVAD-AOM	Ac-Tyr-Val-Ala-Asp-AOM	X			X								Cal
Ac-YVAD-CMK	Ac-Tyr-Val-Ala-Asp-CH_2Cl	X			X							Denning *et al.* (1998), Wu *et al.* (1999)	Cal, CL
Ac-YVAD-FAOM	Ac-Tyr-Val-Ala-Asp-FAOM	X			X								Cal
Z-YVAD-FMK	Z-Tyr-Val-Ala-Asp(OMe)-CH_2F	X			X							Talanian *et al.* (1997)	Cal, CH, ESP

[a]For the fluoromethyl ketones, the acidic groups in Glu and/or Asp are methylated to improve cell permeability; endogenous esterases remove the methyl group to produce the biologically active compound [see Cohen (1997) and Villa *et al.* (1997) for additional references].

[b]Ac, acetyl; AOM, 2,6-dimethylbenzoyloxymethyl ketone; CMK, chloromethyl ketone; DCB, 2,6-dichlorobenzoyloxy; FAOM, 2,6-bis(trifluoromethyl)benzoyloxymethyl ketone; FMK, fluoromethyl ketone; Z, carbobenzoxy.

[c]See Table VII for abbreviations used for commercial sources.

Table IX
Cell-Permeable Reversible Inhibitors of Caspases[a]

Inhibitor[b]	Sequences and modifications	Inhibitor specificity (caspase isozyme) 1	2	3	4	5	6	7	8	9	10	Comments and references	Commercial sources[c]
Ac-DEVD-CHO[d,e]	Ac-Asp-Glu-Val-Asp-CHO			X				X	X			Weaker inhibitor of caspases-1, -2, -4, -5, -6, -9, and -10 (Garcia-Calvo *et al.*, 1998; Margolin *et al.*, 1997; Han *et al.*, 1972; Talanian *et al.*, 1997; Thornberry *et al.*, 1997; Vanags *et al.*, 1996)	AS, BI, Cal, CL, ESP, PI
Ac-DQMD-CHO	Ac-Asp-Gln-Met-Asp-CHO			X			X						BI
Ac-DMQD-CHO	Ac-Asp-Met-Gln-Asp-CHO			X								Hirata *et al.* (1998)	AS, Bac
Ac-ESMD-CHO	Ac-Glu-Ser-Met-Asp-CHO											Han *et al.* (1972)	Sig
Ac-IETD-CHO[d]	Ac-Ile-Glu-Thr-Asp-CHO	X					X		X			Weaker inhibitor of capsases-3, -4, -5, -9, and -10; also inhibits granzyme B (Garcia-Calvo *et al.*, 1998; Thornberry *et al.*, 1997)	AB, AS, BI, Cal, ESP, Sig
Ac-LEHD-CHO[d]	Ac-Leu-Glu-His-Asp-CHO				X	X				X		Thornberry *et al.* (1997)	AB, Bac, BI, Cal
Ac-LEVD-CHO[d]	Ac-Leu-Glu-Val-Asp-CHO				X							Talanian *et al.* (1997)	AB, Bac, Cal
Ac-VAD-CHO	Ac-Val-Ala-Asp-CHO	X		X	X			X					Cal, ESP
Ac-VEID-CHO[d]	Ac-Val-Glu-Ile-Asp-CHO						X					Talanian *et al.* (1997), Hirata *et al.* (1998)	AB, AS, BI, Cal, ESP, Sig
Ac-WEHD-CHO	Ac-Trp-Glu-His-Asp-CHO	X										Weaker inhibitor of caspases-4, -5, -8, -9, and -10 (Garcia-Calvo *et al.*, 1998; Rano *et al.*, 1997; Thornberry *et al.*, 1997)	Sig
Ac-YVAD-CHO[d]	Ac-Tyr-Val-Ala-Asp-CHO	X										Weaker inhibitor of caspases-4, -5, -8, -9, and -10 (Garcia-Calvo *et al.*, 1998; Margolin *et al.*, 1997; Han *et al.*, 1972; Talanian *et al.*, 1997)	AS, BI, Cal, ESP, OS, PI

[a]Generally the aldehydes are not as permeable than the irreversible inhibitors in Table 8 [see Cohen (1997) and Villa *et al.* (1997) for additional references].
[b]Ac, acetyl.
[c]See Table VII for abbreviations used for commercial sources.
[d]Longer peptide versions with greater cell permeability are available from Calbiochem.
[e]Longer peptide versions with greater cell permeability are available from BIOMOL.

(E-64c and E-64d) are effective inhibitors of calpains, while having no effect on caspases and proteasome. A derivative of these compounds, NCO-700, has been introduced, but its effects on caspases and proteasome are unknown. Two α-mercaptoacrylate compounds, PD150606 and PD151746, are highly specific for calpains and do not inhibit other proteinases, as they bind to the Ca^{2+}-binding domain rather than to the active site (Wang *et al.,* 1996a). Another highly specific inhibitor is the calpain inhibitory peptide, which consists of a 27 residue inhibitory sequence from calpastatin. Despite its large size, it can penetrate cells and inhibit calpain activity (Maki *et al.,* 1989; Eto *et al.,* 1995). Several potent and selective calpain inhibitors have been developed, but are not yet available commercially; these include dipeptidyl α-ketoamides [AK275, AK295 (Li *et al.,* 1993, 1996)], a quinoline carboxamide [compound 8 (Graybill *et al.,* 1995)], and a peptide acycloxymethyl ketone [compound 69 (Harris *et al.,* 1995)].

The effects of these calpain inhibitors on tissue-specific calpains are largely unknown, as they were initially characterized from deduced sequences from cDNA and genomic DNA libraries. Few have been characterized biochemically. Calpain inhibitors may not suppress all calpains that are expressed in a tissue. For example, p94 undergoes rapid autolysis even in the absence of Ca^{2+}. Leupeptin, calpastatin, and E-64, which are effective inhibitors of μ- and m-calpains, do not block p94 autolysis (Sorimachi *et al.,* 1993b). A novel 74-kDa calpain-like proteinase from rat brain is insensitive to E-64 and leupeptin (Yoshihara *et al.,* 1990). Consequently, the reagents listed in Table X can be used with reasonable confidence to determine the role of ubiquitous calpains, but not tissue-specific calpains, in cellular processes.

C. Proteasome

The 26S proteasome is a multisubunit, multicatalytic proteinase complex in eukaryotic cells (for reviews, see Mykles, 1998; Peters *et al.,* 1998; Bochtler *et al.,* 1999; DeMartino and Slaughter, 1999; Voges *et al.,* 1999). It is composed of a 20S proteasome and two 19S regulatory complexes and mediates the ATP-dependent degradation of ubiquitinated proteins. The 20S proteasome is a threonine proteinase with three different active sites (two of each in each complex), corresponding to trypsin-like, chymotrypsin-like, and peptidylglutamyl peptide hydrolase activities. Other peptidase activities, such as BrAAP and SNAAP, are associated with one or more of these sites (Mykles, 1997; McCormack *et al.,* 1998; Cardozo *et al.,* 1999; Covi *et al.,* 1999).

An extensive array of proteasome inhibitors has been developed, and many are available commercially (for reviews, see Wilk and Figueiredo-Pereira, 1993; Goldberg *et al.,* 1997a; Lee and Goldberg, 1998b; Groettrup and Schmidtke, 1999). Lactacystin, a *Streptomyces* metabolite, is a selective and potent inhibitor of the proteasome (Table XI). It binds tightly to all active sites, although with different kinetics, and thus inhibits all catalytic activities (Fenteany *et al.,* 1995; Craiu *et al.,* 1997; Corey *et al.,* 1999). Peptidyl-vinyl sulfones (NLVS and ZL_3VS) are also strong inhibitors of all three activities (Bogyo *et al.,* 1997; Elofsson *et al.,* 1999). However, there are no published reports determining the sensitivities of calpains and caspases to lactacystin, NLVS, or ZL_3VS. Both cathepsin A and tripeptidyl peptidase II are inhibited by lactacystin (Ostrowska *et al.,* 1997; Geier

Table X
Cell-Permeable Inhibitors of Calpains

Inhibitor[a]	Other names	Comments/references	Commercial sources[b]
α-mercaptoacrylate derivatives		Reversible; highly specific for calpains at low concentrations; binds to Ca^{2+}-binding domain	
3-(4-iodophenyl)-2-mercapto-(Z)-2-PA	PD150606	Similar inhibition of μ- and m-calpain (Edelstein *et al.*, 1996; Nath *et al.*, 1996a; Wang *et al.*, 1996a,b,c; Lin *et al.*, 1997; Schnellmann and Williams, 1998; Squier and Cohen, 1997; Debiasi *et al.*, 1999; Squier *et al.*, 1999). Inhibits apoptosis (Nath *et al.*, 1996a,b; Squier and Cohen, 1997; Schnellmann and Williams, 1998; Debiasi *et al.*, 1999; Squier *et al.*, 1999)	AB, Cal
3-(5-fluoro-3-indolyl)-2-mercapto-(Z)-2-PA	PD151746	20-fold greater inhibition of μ-calpain over m-calpain (Nath *et al.*, 1996a; Wang *et al.*, 1996a,c; Hajimohammadreza *et al.*, 1997; Squier *et al.*, 1999). Inhibits apoptosis (Squier *et al.*, 1999)	
Calpain inhibitor peptide		Reversible; highly specific; derived from calpain inhibitory domain of calpastatin (Maki *et al.*, 1989; Eto *et al.*, 1995; Yamazaki *et al.*, 1997; Kilic and Trevithick, 1998)	Cal, Sig
Dipeptidyl α-ketoamides		Reversible; highly specific for calpains (Li *et al.*, 1993)	
Z-Leu-Abu-CONH-CH_2CH_3	AK275	Low solubility in aqueous media (Bartus *et al.*, 1994a, 1999; Harbeson *et al.*, 1994; Li *et al.*, 1996; James *et al.*, 1998)	
Z-Leu-Abu-CONH$(CH_2)_3$-morpholine	AK295	More soluble in aqueous media than AK275 (Bartus *et al.*, 1994b, 1999; Harbeson *et al.*, 1994; Saatman *et al.*, 1996; James *et al.*, 1998; Blomgren *et al.*, 1999)	
Dipeptidyl phosphorus derivatives			
Z-Leu-Leu-P(O)$(OCH_3)_2$		Reversible; effects on other proteases not known (Tao *et al.*, 1998)	
Epoxysuccinyl peptides		Irreversible; inhibit other cysteine proteases (Suzuki *et al.*, 1981; Barrett *et al.*, 1982; Suzuki, 1983; Parkes *et al.*, 1985; Huang *et al.*, 1992; Posner *et al.*, 1995; Okamura-Oho *et al.*, 1997; Kudo *et al.*, 1999; Matsumoto *et al.*, 1999)	
E-64c	Ep-475	Does not inhibit caspases (Juin *et al.*, 1998) or proteasome (Barrett *et al.*, 1982; Tanaka *et al.*, 1986; Mykles, 1989, 1990). Inhibits (Sarin *et al.*, 1994), does not inhibit (Sadoul *et al.*, 1996), or does not induce (Drexler, 1997) apoptosis	ICN, Sig
E-64d	EST, loxastatin	More permeable than E-64c; converted to E-64c intracellularly by endogenous esterase activity (McGowan *et al.*, 1989; Ishii *et al.*, 1990; Mykles, 1990; Azuma *et al.*, 1992; Wilcox and Mason, 1992; Song *et al.*, 1994; Squier and Cohen, 1997; Barnoy *et al.*, 1998; Meredith *et al.*, 1998). Inhibits (Sarin *et al.*, 1994; Squier and Cohen, 1997) or does not induce (Dou *et al.*, 1999) apoptosis	Cal, ICN, Sig
Epoxysuccinyl acid derivative			
NCO-700		Irreversible; inhibits other cysteine proteases; an antioxidant (Toyo-Oka *et al.*, 1986a,b; Carrasco-Marín *et al.*, 1998)	AB, Cal
Hydroxyoxazolidines			
MDL 104,903		Derived from MDL 28170; cell permeability not established (Peet *et al.*, 1999)	

N-Methyltyrosine-containing peptides		Reversible; cell-permeability not established (Alvarez *et al.*, 1994)	
NmTyr-diketopiperazine	WIN-68100		
NmTyr-NmTyr-Leu-Ala	WIN-69211		
Peptide aldehydes		Reversible	
Ac-Leu-Leu-Arg-CHO	Leupeptin	Inhibits serine and other cysteine proteases and proteasome (T-L activity); not as permeable as other peptide aldehydes (Azanza *et al.*, 1979; Sugita *et al.*, 1980; Croall *et al.*, 1986; Mykles and Skinner, 1986; Mehdi *et al.*, 1988; Ishii *et al.*, 1990; Mykles, 1990; Saitoh *et al.*, 1991; Pinter *et al.*, 1992; Rami and Krieglstein, 1993; Iqbal *et al.*, 1997; Kudo *et al.*, 1999). Inhibits (Sarin *et al.*, 1994; Belaud-Rotureau *et al.*, 1999), does not induce (Drexler, 1997) or inhibit (Tan *et al.*, 1998) apoptosis, or induces apoptosis by inhibiting cathepsins (Maeda *et al.*, 1996)	AB, BI, Cal, ICN, MP, Roc, Sig
Ac-Leu-Leu-nLeu-CHO	Calpain inhibitor I	Also inhibits cathepsins and proteasome (ChT-L activity) (Sasaki *et al.*, 1990; Vinitsky *et al.*, 1992; Rami and Krieglstein, 1993; Sherwood *et al.*, 1993; Figueiredo-Pereira *et al.*, 1994a; Squier *et al.*, 1994; Griscavage *et al.*, 1996; Milligan *et al.*, 1996; Wang *et al.*, 1996b; Hajimohammadreza *et al.*, 1997; Iqbal *et al.*, 1997; Rami *et al.*, 1997; Squier and Cohen, 1997; Mellgren, 1997; Xie and Johnson, 1998; Debiasi *et al.*, 1999). Does not inhibit caspases (Juin *et al.*, 1998). Induces (Zhu *et al.*, 1995; Lu and Mellgren, 1996; Vanags *et al.*, 1996; Drexler, 1997; Monti *et al.*, 1998; Herrmann *et al.*, 1998; Wu *et al.*, 1999; Dou *et al.*, 1999), inhibits (Zheng *et al.*, 1994; Gressner *et al.*, 1997; Knepper-Nicolai *et al.*, 1998; Schnellmann and Williams, 1998; Spinedi *et al.*, 1998; Villa *et al.*, 1998; Debiasi *et al.*, 1999; Iwamoto *et al.*, 1999; Squier *et al.*, 1999; Stempien-Otero *et al.*, 1999), or does not inhibit (Lu and Mellgren, 1996; Vanags *et al.*, 1996; Lynch *et al.*, 1997) apoptosis	Aff, AB, BI, Cal, ICN, Roc
Ac-Leu-nLeu-CHO	Calpeptin	Highly specific (Kwak *et al.*, 1993; Figueiredo-Pereira *et al.*, 1994a; Barnoy *et al.*, 1997, 1998; Iqbal *et al.*, 1997; Schoenwaelder *et al.*, 1997; Gitler and Spira, 1998; James *et al.*, 1998; Knepper-Nicolai *et al.*, 1998; Potter *et al.*, 1998; Wood *et al.*, 1998). Also inhibits protein-tyrosine phosphatase (Schoenwaelder and Burridge, 1999). Does not inhibit caspases (Vanags *et al.*, 1996). Induces (Vanags *et al.*, 1996; Monti *et al.*, 1998), inhibits (Behrens *et al.*, 1995; Vanags *et al.*, 1996; Chen *et al.*, 1998), does not inhibit (Sadoul *et al.*, 1996; Vanags *et al.*, 1996; Knepper-Nicolai *et al.*, 1998; Tan *et al.*, 1998; Wood *et al.*, 1998; Belaud-Rotureau *et al.*, 1999; Wood and Newcomb, 1999), or does not induce (Stoklosa *et al.*, 1999) apoptosis	Cal, ICN
Ac-Leu-Leu-Met-CHO	Calpain inhibitor II	Also inhibits cathepsins and proteasome (ChT-L activity) (Sasaki *et al.*, 1990; Vinitsky *et al.*, 1992; Figueiredo-Pereira *et al.*, 1994a; Iqbal *et al.*, 1997; Mellgren, 1997; Xie and Johnson, 1998). Does not inhibit caspases (Juin *et al.*, 1998). Induces (Herrmann *et al.*, 1998), inhibits (Sarin *et al.*, 1994; Nath *et al.*, 1996a,b; Gressner *et al.*, 1997; Pike *et al.*, 1998b; Schnellmann and Williams, 1998; Villa *et al.*, 1998), does not inhibit (McGinnis *et al.*, 1999b; Stempien-Otero *et al.*, 1999), or does not induce (Drexler, 1997; Dou *et al.*, 1999; Glockzin *et al.*, 1999; Wu *et al.*, 1999) apoptosis	Aff, AB, BI, Cal, ICN, Roc

continues

Table X
continued

Inhibitor[a]	Other names	Comments/references	Commercial sources[b]
Fps-Val-Leu-CHO	SJA6017	Effects on other proteases not known (Fukiage *et al.*, 1997, 1998)	
Z-Val-Phe-CHO	MDL 28170, calpain inhibitor III	Also inhibits cathepsin B (Mehdi *et al.*, 1988; Song *et al.*, 1994; Chard *et al.*, 1995; Harris *et al.*, 1995; Wang *et al.*, 1996b; Rami *et al.*, 1997; Squier and Cohen, 1997; Chatterjee *et al.*, 1998; Potter *et al.*, 1998; Steiner *et al.*, 1998). Inhibits (Squier *et al.*, 1994) apoptosis.	Cal
Peptide chloromethyl ketones		Irreversible	
Z-Leu-Leu-Phe-CH_2Cl	Z-LLF-CH_2Cl	Inhibits serine proteases and other cysteine proteases (Sasaki *et al.*, 1986)	
Peptide diazomethyl ketones		Irreversible	
Z-Leu-Leu-Tyr-CHN_2	Z-LLY-CHN_2	Also inhibits cathepsins B and L, proteasome (Crawford *et al.*, 1988; Angliker *et al.*, 1992; Rosser *et al.*, 1993; Savory *et al.*, 1993; Mellgren *et al.*, 1994, 1996; Mellgren, 1997; Xie and Johnson, 1997, 1998; Potter *et al.*, 1998; Kohli *et al.*, 1999); inhibits caspase-1 and -3 at higher concentrations (Sindram *et al.*, 1999). Induces (Zhu *et al.*, 1995; Lu and Mellgren, 1996), inhibits (Squier and Cohen, 1997; Kohli *et al.*, 1999), or does not inhibit (Lu and Mellgren, 1996) apoptosis	EPS, MP
Peptide disulfides		Irreversible	
Leu-Leu-Cys(Npys)-NH_2		(Matsueda *et al.*, 1990)	
Phe-Gln-Val-Val-Cys(Npys)-Gly-NH_2	P1, compound 3	(Matsueda *et al.*, 1990; Puri *et al.*, 1993a,b)	
Peptide fluoromethyl ketones		Irreversible	
Tq-Leu-Leu-D,L-Phe-CH_2F	Compound 4f	Also inhibits cathepsins B and L (Chatterjee *et al.*, 1996, 1997)	
Z-Leu-Leu-Tyr-CH_2F	Calpain inhibitor IV	Also inhibits cathepsin L (Angliker *et al.*, 1992; Anagli *et al.*, 1993)	Cal, ESP
Mu-Val-HPh-CH_2F	Calpain inhibitor V	Also inhibits cathepsins B and L (Esser *et al.*, 1994)	Cal, ESP
Peptide methyl esters			
Z-Val-Phe-methyl ester		Inhibits caspase-1 and -3 at higher concentrations; inhibits apoptosis (Sindram *et al.*, 1999).	
Peptide acycloxymethyl ketones			
Z-(D)Ala-Leu-Phe-(OCO-2,6-F_2)-Ph	Compound 69	Reversible; highly specific for calpains (Harris *et al.*, 1995)	
Quinoline carboxamides			
Compound 8		Reversible; highly specific for calpains (Graybill *et al.*, 1995)	

[a]Abu, aminobutyric acid; Ac, acetyl; Boc, *t*-butyloxycarbonyl; E-64c, L-*trans*-epoxysuccinyl-leucylamido(3-methyl)butane; E-64d, ethyl(+)-(2*S*,3*S*)-3-[(*S*)-methyl butylcarbamoyl)-butylcarbamoyl]-2-oxiranecarboxylate; Fps, 4-fluorophenylsulfonyl; HPh, homophenylalanyl; Mu, morpholinoureidyl; NCO-700, bis[ethyl(2*R*,3*R*)-3-[(*S*)-methyl-1-[4-(2,3,4-trimethoxyphenyl-methyl) piperazin-1-ylcarbonyl]butyl-carbonyl] oxiran-2-carboxylate] sulfate; Nm, *N*-methyl; Npys, 3-nitro-2-sulfenylpyridine; PA, propenoic acid; Tq, (1, 2, 3, 4-tetrahydroisoquinolin-2-yl)carbonyl.
[b]See Table VII for abbreviations used for commercial sources.

et al., 1999), raising some concerns that it is not as selective as commonly thought. Epoxomicin appears to be a specific inhibitor of the proteasome; it does not affect calpain, cathepsin B, and other proteinases (caspases were not examined) (Sin *et al.,* 1999; Meng *et al.,* 1999b; Kim *et al.,* 1999b). Other broad-spectrum proteasome inhibitors, such as calpain inhibitor I, MG115, and MG132, are frequently used, but they lack the specificity of lactacystin, epoxomicin, and perhaps the peptidyl-vinyl sulfones.

Some inhibitors preferentially suppress one catalytic site, allowing one to determine the importance of a particular activity of the proteasome in a specific cellular process. The trypsin-like activity is more sensitive to leupeptin (Savory and Rivett, 1993; Figueiredo-Pereira *et al.,* 1995; Mykles and Haire, 1995; Haire *et al.,* 1995). The chymotrypsin-like activity is inhibited preferentially by PSI, Z-LLY-CHN_2, Ac-hFLFL-epoxide, Z-LLY-COCHO, CVT-634, aclacinomycin A, cyclosporin A, gliotoxin, lovastatin, or Ritonavir, although many of these compounds inhibit other proteinases and/or enzymes (Table XI). Z-LLE-CHO and Z-GPFL-CHO selectively inhibit a single active site that apparently mediates the PGPH and BrAAP activities (Cardozo *et al.,* 1996, 1999; McCormack *et al.,* 1998; Covi *et al.,* 1999). A recommended strategy is to use low concentrations (1 μM or less) and short incubations (no more than 1 h) to exploit differences between active sites in binding an inhibitor. Peptidyl boronic acids (PS-262 and PS-341), epoxomicin, NLVS, and lactacystin, for example, are at least 10-fold more effective inhibitors of the chymotrypsin-like activity than the trypsin-like and PGPH activities (Elofsson *et al.,* 1999; Wang *et al.,* 1999).

IV. Considerations for Inhibitor Selection and Application

The proteolytic pathways within eukaryotic cells are complex and versatile. Analogous to signal transduction, proteolytic pathways converge on common protein substrates, allowing integration, or "cross-talk," between proteolytic activities that regulate the levels and/or activities of enzymes and regulatory proteins. The number of caspase substrates is growing rapidly (for reviews, see Cryns and Yuan, 1998; Chan and Mattson, 1999); many of these are also substrates of calpains and/or proteasome (Table XI). Spectrin/fodrin, a cytoskeletal protein that stabilizes the cell membrane, is hydrolyzed by both calpain and caspase during apoptosis (Wang *et al.,* 1998b). Spectrin is also ubiquitinated (Corsi *et al.,* 1997, 1999), suggesting that it is degraded by the proteasome under certain conditions. Furthermore, a component of one proteolytic system can be a substrate for another, thus modulating the activity and/or specificity of the other system. Both caspase-3 and -9 are cleaved by calpain, preventing caspase activation (Wolf *et al.,* 1999; McGinnis *et al.,* 1999a). The ubiquitin-protein ligase Nedd4 is cleaved by several caspases during the apoptosis of Jurkat cells (Harvey *et al.,* 1998). Calpastatin is hydrolyzed by caspases during apoptosis (Wang *et al.,* 1998a; Pörn-Ares *et al.,* 1998). BRUCE, a giant ubiquitin-conjugating enzyme, contains a baculovirus inhibitor of an apoptosis repeat (BIR) motif, which occurs in IAP apoptosis inhibitor proteins (Hauser *et al.,* 1998; Budihardjo *et al.,* 1999). These examples indicate potential interactions among the caspase, calpain, and proteasome systems, but the nature of these interactions is poorly understood.

Table XI
Cell-Permeable Inhibitors of Proteasomes

Inhibitor[a]	Other names	Comments/references	Commercial sources[b]
Aclacinomycin A		Inhibits ChT-L activity; also inhibits chymotrypsin and calpain; stimulates trypsin; does not inhibit cathepsin B (Isoe *et al.*, 1992; Figueiredo-Pereira *et al.*, 1996; Mimnaugh *et al.*, 1997)	Aff, Cal
Gliotoxin		Reversible; inhibits ChT-L activity; does not inhibit calpain (Elofsson *et al.*, 1999; Kroll *et al.*, 1999a). Induces apoptosis (Ward *et al.*, 1999)	ICN, Sig
Lactacystin and *clasto*-Lactacystin β-lactone		Irreversible; inhibits T-L, ChT-L, PGPH, and BrAAP activities (Fenteany *et al.*, 1995; Mori *et al.*, 1995; Bogyo *et al.*, 1997; Craiu *et al.*, 1997; Fenteany and Schreiber, 1998; McCormack *et al.*, 1998; Corey and Li, 1999; Corey *et al.*, 1999; Elofsson *et al.*, 1999); also inhibits cathepsin A (Ostrowska *et al.*, 1997) and tripeptidyl peptidase II (Geier *et al.*, 1999). A more potent derivative (PS-519) has been synthesized (Campbell *et al.*, 1999; Elliott *et al.*, 1999). Induces (Imajoh-Ohmi *et al.*, 1995; Bellas *et al.*, 1997; Cui *et al.*, 1997; Soldatenkov and Dritschilo, 1997; Delic *et al.*, 1998; He *et al.*, 1998; Boutillier *et al.*, 1999; Dou *et al.*, 1999; Glockzin *et al.*, 1999; Kitagawa *et al.*, 1999; Wu *et al.*, 1999), inhibits (Fujita *et al.*, 1996; Grimm *et al.*, 1996; Sadoul *et al.*, 1996; Chandra *et al.*, 1998; Hirsch *et al.*, 1998; Ivanov and Nikolic-Zugic, 1998; Knepper-Nicolai *et al.*, 1998; Lin *et al.*, 1998), or has no effect on apoptosis (Spinedi *et al.*, 1998; Squier *et al.*, 1999)	Aff, AB, BI, Cal, ICN
5-Methoxy-1-indanone-3-Ac-Leu-D-Leu-1-indanylamide	CVT-634	Inhibits ChT-L activity; does not inhibit calpain (Lum *et al.*, 1998a,b)	
Peptide aldehydes		Reversible; may also inhibit serine and cysteine proteases	
Ac-Leu-Leu-Arg-CHO	Leupeptin	Inhibits T-L activity; also inhibits calpains (Savory and Rivett, 1993; Figueiredo-Pereira *et al.*, 1995; Haire *et al.*, 1995; Mykles and Haire, 1995). Does not induce apoptosis (Drexler, 1997)	Cal, ICN, MP, Roc, Sig
Ac-Leu-Leu-nLeu-CHO	Calpain inhibitor I	Inhibits ChT-L, PGPH, and T-L activities; also inhibits calpains (Vinitsky *et al.*, 1992; Sherwood *et al.*, 1993; Figueiredo-Pereira *et al.*, 1994a; Rock *et al.*, 1994; Andró *et al.*, 1998; Geier *et al.*, 1999). Does not inhibit caspases (Juin *et al.*, 1998). Induces (Shinohara *et al.*, 1996; Drexler, 1997; Herrmann *et al.*, 1998; Mikulski *et al.*, 1998; Dou *et al.*, 1999; Kitagawa *et al.*, 1999; Wu *et al.*, 1999) or inhibits (Grimm *et al.*, 1996; He *et al.*, 1998; Knepper-Nicolai *et al.*, 1998; Spinedi *et al.*, 1998; Squier *et al.*, 1999; Stempien-Otero *et al.*, 1999) apoptosis	Aff, AB, BI, Cal, ICN, Roc
cyano-$(CH_2)_8CH$(cyclopentyl)CO-Arg(NO_2)-Leu-CHO	CEP1508	Inhibits ChT-L, PGPH, and BrAAP activities; may inhibit calpains (Harding *et al.*, 1995; Iqbal *et al.*, 1995; Mykles, 1997). Induces apoptosis (An *et al.*, 1998a)	

phthalamide-$(CH_2)_8CH$(cyclopentyl) CO-Arg(NO_2)-Leu-CHO	CEP1612	Inhibits ChT-activity; weaker inhibitor of calpain and cathepsin B (Harding *et al.*, 1995; Iqbal *et al.*, 1995). Induces apoptosis (An *et al.*, 1998a)	
Z-Gly-Leu-Ala-Leu-CHO	Z-GLAL-CHO	Inhibits ChT-L, BrAAP and SNAAP activities (Lee *et al.*, 1996; Cardozo and Kohanski, 1998)	
Z-Gly-Pro-Phe-Leu-CHO	Z-GPFL-CHO	Inhibits PGPH, SNAAP and BrAAP activities (Vinitsky *et al.*, 1994; Cardozo *et al.*, 1996, 1999; Mykles, 1997; Covi *et al.*, 1999)	
Z-Ile-Glu(O-t-Bu)-Ala-Leu-CHO	PSI	Inhibits ChT-L activity; also inhibits cathepsins and calpains (Figueiredo-Pereira *et al.*, 1994b, 1995; Mori *et al.*, 1995; Mykles, 1997; Ohtani-Kaneko *et al.*, 1998; El Khissiin and Leclercq, 1999). Induces apoptosis (Drexler, 1997; Lopes *et al.*, 1997; Tanimoto *et al.*, 1997; Wójcik *et al.*, 1997; Besançon *et al.*, 1998; Taglialatela *et al.*, 1998; McDade *et al.*, 1999; Stoklosa *et al.*, 1999)	Aff, AB, Cal, PI
Z-Leu-Leu-Glu-CHO	Z-LLE-CHO	Inhibits PGPH and BrAAP activities (Cardozo *et al.*, 1996; McCormack *et al.*, 1998)	
Z-Leu-Leu-Leu-CHO	MG132	Inhibits ChT-L, PGPH, T-L and BrAAP activities; also inhibits cathepsins and calpains (Tsubuki *et al.*, 1996; Bogyo *et al.*, 1997; Goldberg *et al.*, 1997b; McCormack *et al.*, 1998; Ohtani-Kaneko *et al.*, 1998). Inhibits apoptosis at 200 n*M*, but induces apotosis at 20 μM (Lin *et al.*, 1998). Induces (Fujita *et al.*, 1996; Shinohara *et al.*, 1996; Chandra *et al.*, 1998; He *et al.*, 1998; Meriin *et al.*, 1998; Boutillier *et al.*, 1999; Dou *et al.*, 1999; Glockzin *et al.*, 1999), inhibits (Grimm *et al.*, 1996; Meriin *et al.*, 1998; Hirsch *et al.*, 1998; Stefanelli *et al.*, 1998), or does not inhibit (Meriin *et al.*, 1998) apoptosis	Aff, AB, BI, Cal
Z-Leu-Leu-nVal-CHO	MG115	Inhibits ChT-L, PGPG, and T-L activities; also inhibits cathepsin B and calpain (Rock *et al.*, 1994; Mori *et al.*, 1995; Mellgren, 1997; McCormack *et al.*, 1998), but not caspases (Drexler, 1997). Induces (Fujita *et al.*, 1996; Drexler, 1997; Lopes *et al.*, 1997; Tanimoto *et al.*, 1997; Boutillier *et al.*, 1999; Dou *et al.*, 1999; Glockzin *et al.*, 1999) or inhibits (Stefanelli *et al.*, 1998) apoptosis	Aff, AB, Cal, PI
Z-Leu-Leu-Phe-CHO	Z-LLF-CHO	Inhibits ChT-L and BrAAP activities (Vinitsky *et al.*, 1992; McCormack *et al.*, 1998). Induces apoptosis (Orlowski *et al.*, 1998; Schauer *et al.*, 1998)	AB, Cal
Peptide chloromethyl ketones		Irreversible	
Z-Tyr-Ala-Glu-CH_2Cl	Z-YAE-CH_2Cl	Inhibits ChT-L activity (Savory *et al.*, 1993). Cell permeability not established	
Peptide diazomethyl ketones		Irreversible	
Z-Leu-Leu-Tyr-CHN_2	Z-LLY-CHN_2	Inhibits ChT-L activity; also inhibits cathepsins B and L, calpains (Angliker *et al.*, 1992; Savory *et al.*, 1993; Mellgren *et al.*, 1994; Mellgren, 1997)	ESP

continues

Table XI
continued

Inhibitor[a]	Other names	Comments/references	Commercial sources[b]
Z-Phe-Gly-Tyr-CHN_2	Z-FGY-CHN_2	Inhibits ChT-L activity (Savory *et al.*, 1993). Cell permeability not established	
Peptide α′β′-epoxy ketones		Irreversible	
Eponemycin		Inhibits ChT-L, T-L, and PGPH activities; weak inhibition of cathepsin B and chymotrypsin; no inhibition of calpain and trypsin (Kim *et al.*, 1999b; Meng *et al.*, 1999a). Induces apoptosis (Meng *et al.*, 1999a)	
Epoxomicin		Potent inhibitor of ChT-L activity; weaker inhibition of T-L and PGPH activities; no inhibition of trypsin, chymotrypsin, papain, calpain, and cathepsin B (Kim *et al.*, 1999b; Meng *et al.*, 1999b; Sin *et al.*, 1999)	Aff
Ac-homoPhe-Leu-Phe-Leu-epoxide	Ac-hFLFL- epoxide	Highly specific for ChT-L activity (Elofsson *et al.*, 1999)	
Peptidyl boronic acids		Reversible	
Pyz-Phe-boroLeu	PS-341, MG-341	Very stable in aqueous solutions; inhibits ChT-L and PGPH activities; weak inhibition of serine and cyseine proteases (Conner *et al.*, 1997; Adams *et al.*, 1998, 1999; Palombella *et al.*, 1998; Christie *et al.*, 1999; Elofsson *et al.*, 1999; Grisham *et al.*, 1999; Kalogeris *et al.*, 1999). Induces (Adams *et al.*, 1999) or inhibits (Brand *et al.*, 1999) apoptosis	
Z-Leu-Leu-Leu-$B(OH)_2$	PS-262	Inhibits ChT-L and PGPH activities (Adams *et al.*, 1998; Christie *et al.*, 1999; Grisham *et al.*, 1999)	Aff, BI
Peptide ethyl-ketoamides			
cyano-$(CH_2)_8$CH(cyclopentyl)CO-Arg(NO_2)-Leu-N-ethyl-ketoamide	Compound 8	Reversible; related to CEP1508; inhibits ChT-L activity; also inhibits calpains (An *et al.*, 1998a). A related compound with greater specificity for proteasome has been synthesized (Chatterjee *et al.*, 1999a).	
Peptidyl-α-keto aldehydes		Reversible	
Z-Leu-Leu-Tyr-COCHO	Z-LLY-COCHO	Inhibits ChT-L activity; also inhibits serine and cysteine proteases; cell-permeability assumed, based on related compounds (Lynas *et al.*, 1998)	Aff
Peptidyl-vinyl sulfones		Irreversible	
NIP-Leu-Leu-Leu vinyl sulphone	NLVS	Inhibits ChT-L, T-L, and PGPH activities (Bogyo *et al.*, 1997; Glas *et al.*, 1998; Elofsson *et al.*, 1999)	Cal
Z-Leu-Leu-Leu vinyl sulphone	ZL_3VS	Inhibits ChT-L, T-L, and PGPH activities (Bogyo *et al.*, 1997)	Cal

[a]Ac, acetyl; BrAAP, branched chain amino acid preferring; ChT-L, chymotrypsin-like; NIP, 4-hydroxy-5-iodo-3-nitrophenylacetyl; PGPH, peptidylglutamyl-peptide hydrolase; Pyz, 2,5-pyrazinecarboxylic acid; SNAAP, small neutral amino acid preferring; T-L, trypsin-like.
[b]See Table VII for abbreviations used for commercial sources.

Cell-permeable proteinase inhibitors have made it possible to define the role of proteinases and their substrates in intact cells. However, interpretation of experimental results depends on knowing with reasonable certainty that a compound has the intended effect on a particular enzyme. This is rarely achieved in practice. In selecting a compound, one must consider the following.

A. Inhibitors May Affect More Than One Proteinase

Proteinase inhibitors vary in specificity. Moreover, one cannot be absolutely confident in the specificity of a compound as there are few systematic studies comparing the effects of a compound on even a representative sample of the tens, or perhaps hundreds, of proteinases within cells. Generally, the confounding effects of inhibitors are not fully appreciated until a compound is available commercially and widely used, leading to the observation that an inhibitor becomes less specific with time. A good example is Ritonavir, a HIV-1 proteinase inhibitor that blocks viral replication. It was only discovered recently that Ritonavir also inhibits the chymotrypsin-like activity and stimulates the trypsin-like activity of the proteasome (André *et al.,* 1998). Fortunately, caspase inhibitors (Tables VIII and IX) are highly specific for caspases, except for the inhibition of granzyme B by Z-IETD-FMK (Thornberry *et al.,* 1997; Garcia-Calvo *et al.,* 1998) and inhibition of 26S proteasome (Beyette *et al.,* 1998) and calpain (Waterhouse *et al.,* 1998; Wolf *et al.,* 1999) by Z-VAD-FMK. Conversely, most calpain and proteasome inhibitors have no effect on caspases. The only exceptions are Z-VF-methyl ester and Z-LLY-CHN_2, which inhibit caspases-1 and -3 at higher concentrations (Sindram *et al.,* 1999).

Many peptide inhibitors of calpain and proteasome overlap in specificity to some degree and may also disrupt lysosomal function by inhibiting cathepsins. Compounds included in this category are leupeptin, calpain inhibitors I and II, CEP1508, CEP1612, PSI, MG115, MG132, Z-LLY-CHN2, Z-LLY-COCHO, compound 8 (a peptide ethylketoamide), aclacinomycin A, and Z-LLF-CH2Cl (Tables X and XI).

Deubiquitinases are cysteine proteinases, and the enzymes of the ubiquitin-conjugating system contain cysteine residues required for thiol-ester linkage with the C terminus of ubiquitin (Wilkinson, 1990, 1997). These enzymes are susceptible to sulfhydryl reageants, such as N-ethylmaleimide and *p*-hydroxymercuriphenyl sulfonyl acid (Wilkinson, 1990). However, they are relatively insensitive to peptide analogs that inhibit calpains and/or proteasome, such as leupeptin (Park *et al.,* 1997).

B. Inhibitors May Inhibit the Activities of Nonproteinase Enzymes

Protein–protein interactions are integral to the activities of many cellular enzymes, thus raising the possibility that peptide-based proteinase inhibitors may bind and disrupt the functions of other enzymes. The calpain inhibitor calpeptin inhibits a membrane-associated protein-tyrosine phosphatase, resulting in the enhanced phosphorylation of paxillin (Ariyoshi *et al.,* 1995; Schoenwaelder and Burridge, 1999). Because paxillin is also a calpain substrate (Yamaguchi *et al.,* 1994), the combined inhibition of both calpain and phosphatase stabilizes stress fibers in fibroblasts (Schoenwaelder and Burridge, 1999).

Aurintricarboxylic acid (ATA) inhibits apoptosis in a variety of cell types (Helgason *et al.*, 1995; Posner *et al.*, 1995; Rui *et al.*, 1998; Stefanis *et al.*, 1996; Efanova *et al.*, 1998; Andrew *et al.*, 1999; Ohguchi *et al.*, 1998; Aronica *et al.*, 1998; Rosenbaum *et al.*, 1998; Okada and Koizumi, 1997; Song *et al.*, 1998; Tan *et al.*, 1998). It is a reversible inhibitor of calpain (Posner *et al.*, 1995), but also affects a number of other cellular enzymes, many of which are involved in apoptosis. ATA inhibits topoisomerase II (Benchokroun *et al.*, 1995), DNA polymerase (Nakane *et al.*, 1988), phosphofructokinase (McCune *et al.*, 1989), calcium-dependent endonuclease (Urbano *et al.*, 1998; Hughes *et al.*, 1998), c-Jun N-terminal kinase (JNK) activation in serum-deprived PC12 cells (Park *et al.*, 1996), and ischemia-induced downregulation of GluR2 in hippocampal neurons (Aronica *et al.*, 1998). It also activates Jak2 (Rui *et al.*, 1998) and stimulates protein phosphorylation (Andrew *et al.*, 1999; Okada and Koizumi, 1997).

C. Other Compounds May Inhibit Proteinases

Various compounds have been found to inhibit calpain or proteasome. Calmodulin and protein kinase C antagonists, such as melittin, toluidinylnaphthalene sulfonate, felodipine, calmidazolium, W7, and trifuoperazine, inhibit calpain activity by binding to the calcium-binding domain (Hong *et al.*, 1990a,b; Johnson, 1990; Zhang and Johnson, 1988; Brumley and Wallace, 1989). Polyamines (spermine and spermidine), nitric oxide, and corticosteroids (dexamethasone, prednisolone, and methylprednisone) inhibit m-calpain (Johnson and Hammer, 1990; Michetti *et al.*, 1995; Banik *et al.*, 1997). Cyclosporin A, an inhibitor of calcineurin (protein phosphatase-2B), and lovastatin, an inhibitor of hydroxymethyl glutaryl-CoA reductase, inhibit the chymotrypsin-like activity of the proteasome (Marienfeld *et al.*, 1997; Meyer *et al.*, 1997; Rao *et al.*, 1999).

D. Exposure to Inhibitors May Induce Compensatory Responses by Cells

Short exposures to proteasome inhibitors induce the expression of heat-shock proteins. Hsp27, hsp72, and hsp90 are upregulated by lactacystin, calpain inhibitor I, MG115, or MG132 in yeast and mammalian cells in as little as 1–2 h (Bush *et al.*, 1997; Mimnaugh *et al.*, 1997; Lee and Goldberg, 1998a; Meriin *et al.*, 1998; Mathew *et al.*, 1998; Kim *et al.*, 1999a). These inhibitors increase the activity of heat-shock transcription factors (HSFs), possibly by stabilizing a short-lived kinase that phosphorylates HSFs (Kawazoe *et al.*, 1998; Mathew *et al.*, 1998; Kim *et al.*, 1999a). Increased levels of heat-shock proteins reduce c-Jun N-terminal kinase-dependent apoptosis induced by heat shock (Meriin *et al.*, 1998), as well as enhance thermotolerance (Bush *et al.*, 1997; Lee and Goldberg, 1998a).

Prolonged exposures may select for an inhibitor-resistant population of cells. Because proteasome, calpains, and other intracellular proteinases have overlapping functions, an extended culture with an inhibitor may produce a compensatory effect in which another proteinase substitutes for the inhibited enzyme. A good example is the apparent replacement of proteasome with tripeptidyl peptidase II in murine EL-4 lymphoma cells (Glas *et al.,* 1998; Geier *et al.,* 1999). Proteasome inhibitors cause most of the cells to die within 24–48 h. However, after 2–3 weeks of continuous culture with 10 μM NLVS, a small number of EL-4 cells (about 1 in 300) recovers and begins to grow (Glas *et al.,* 1998). Lactacystin (6 μM) has a similar effect (Geier *et al.,* 1999). In these adapted cells, proteasome activity is reduced, whereas tripeptidyl peptidase II activity is elevated up to 4.5-fold (Geier *et al.,* 1999).

V. General Guidelines for Using Proteinase Inhibitors

A. Use the Lowest Effective Concentration at the Shortest Effective Exposure

One should do a dose–response at several incubation intervals to determine optimal conditions and to minimize nonspecific effects. For cells in culture, concentrations of no more than 10 μM and exposures less than 1 h are usually sufficient for cell-permeable inhibitors. Not only is the response of a cell to an inhibitor concentration dependent, opposite effects are sometimes elicited by different inhibitor concentrations. Lower concentrations of proteasome inhibitors MG132 and PSI are more effective in stimulating neurite outgrowth in PC12 neurons than higher concentrations (Ohtani-Kaneko *et al.,* 1998). MG132 inhibits virus-induced apoptosis at 0.2 μM, but induces apoptosis at 20 μM in AT3 cells (Lin *et al.,* 1998).

Aside from membrane permeability, two factors that influence the effective concentration of an inhibitor are compound stability and potential binding to serum proteins often present in culture media. Unfortunately, there are very little published data, particularly for the newer reagents, to provide guidance in determining the appropriate concentration to use. In most cases, the reagents listed in Tables VIII–XI are stable in aqueous solutions for at least several hours. For long-term storage, stock solutions should be made in dimethyl sulfoxide, ethanol, or methanol and stored at -20°C. The choice of solvent is determined by the solubility properties of a compound (some compounds are not highly soluble in water and must first be dissolved in an organic solvent). Chemical properties are provided in Appendix III of Beynon and Bond (1989) and by suppliers (Table VII). The metabolic stabilities are largely unknown, but it is likely that most inhibitors currently in use are not metabolized rapidly in cells. For example, peptide inhibitors in which both termini are chemically blocked are resistant to endogenous aminopeptidases and carboxypeptidases.

B. Use at Least Two Chemically Distinct Inhibitors of Each Proteinase

There are at least three major proteolytic systems involved in apoptosis: caspases, calpains, and proteasome (for reviews, see Patel *et al.,* 1996; Grimm and Osborne, 1998; Lee and Goldberg, 1998b; Solary *et al.,* 1998; Chan and Mattson, 1999; Orlowski, 1999). Serine proteinases, metalloproteinases, and lysosomal cathepsins also have significant roles in some cells (Maeda *et al.,* 1996; Schnellmann and Williams, 1998; Isahara *et al.,*

1999; Gong *et al.,* 1999; Belaud-Rotureau *et al.,* 1999; Shimizu and Pommier, 1996; Tan *et al.,* 1998; Herren *et al.,* 1998). Thus, it is important to use a variety of compounds to determine which proteinase plays a role in apoptosis and other cellular processes. If different compounds that inhibit the same proteinase have the same effect, while others that inhibit other proteinases have no effect, then it is more likely that a particular proteinase is involved. For example, if spectrin cleavage is inhibited by E-64d or PD150606, but not by lactacystin, epoxomicin, Boc-D-FMK, or chloroquine, then calpains, but not proteasome, caspases, or lysosomal enzymes, probably mediate spectrin breakdown.

One should not put absolute faith in one compound. A good example is the disassembly of stress fibers in fibroblasts in response to forskolin or serum withdrawal (Schoenwaelder and Burridge, 1999). Calpeptin stabilizes the stress fibers and prevents disassembly, whereas other calpain inhibitors (E-64d, calpain inhibitor I, calpastatin, or EGTA) do not. This discrepancy led to the discovery that calpeptin inhibits a subset of protein-tyrosine phosphatases controlling cytoskeletal rearrangements (Schoenwaelder and Burridge, 1999).

C. Select Compounds with Narrow Specificities

There are several commercially available compounds that inhibit calpain while having no effect on caspases or proteasome at low concentrations. These include calpain inhibitor peptide, E-64c, E-64d, NCO-700, PD150606, calpeptin, MDL28170, and calpain inhibitors IV and V (Table X). Other selective inhibitors that are not yet available from commercial sources are PD151746, AK295, compound 4f, compound 69, and compound 8 (a quinoline carboamide). For those compounds that also inhibit cathepsins (MDL28170, compound 4f, calpain inhibitors IV and V, E-64c, E-64d, and perhaps NCO-700), controls using lysosomotropic reagents (ammonium chloride, 3-methyladenine, or chloroquine) and/or specific inhibitors of cathepsins, such as CA-074 for cathepsins B, K, and L (Okamura-Oho *et al.,* 1997; Yamashima *et al.,* 1998; Isahara *et al.,* 1999; Matsumoto *et al.,* 1999) and pepstatin A for cathepsins D and E (Isahara *et al.,* 1999; Kudo *et al.,* 1999), should be included.

Commercially available inhibitors specific for the proteasome are lactacystin, Z-LLF-CHO, epoxomicin, PS-262, PS-341, NLVS, ZL3VS, and gliotoxin (Table XI). Other compounds not available commercially are Z-GLAL-CHO, Z-GPFL- CHO, Z-LLE-CHO, Z-YAE-CH2Cl, Z-FGY-CHN2, eponemycin, AchFLFL-epoxide, and CVT- 634.

VI. Concluding Remarks

Intracellular proteinases are integral to processes underlying cellular growth and survival. The recognition that abnormal proteolysis causes, or at least contributes to, many human diseases and pathological conditions has led to the development of pharmacological reagents that inhibit specific proteinases. The research community has benefited from these efforts, as these reagents have become effective tools for the study of proteolytic systems in intact cells. New generations of compounds, some with greater potency and specificity, are being synthesized continually for eventual therapeutic applications (Chatterjee *et al.,* 1998, 1999a,b; Tripathy *et al.,* 1998; Roush *et al.,* 1998; Schirmeister, 1999; Loidl *et al.,* 1999; Adams *et al.,* 1999; Peet *et al.,* 1999). Although few compounds will make it through clinical trials, many could become valuable research reagents. Regardless, the three general guidelines still apply to any new compound. For any application using proteinase inhibitors, it is important for investigators to remember that absolute specificity is never achieved.

Acknowledgments

The author thanks Drs. Craig Crews, Marcus Groettrup, Mathias Kroll, and Simone Schoenwaelder for providing preprints.

References

Aberle, H., Bauer, A., Stappert, J., Kispert, A., and Kemler, R. (1997). β-Catenin is a target for the ubiquitin-proteasome pathway. *EMBO J.* **16,** 3797–3804.

Adachi, Y., Maki, M., Ishii, K., Hatanaka, M., and Murachi, T. (1990). Possible involvement of calpain in down-regulation of protein kinase C. *Adv. Second Messenger Phosphoprotein Res.* **24,** 478–484.

Adams, J., Behnke, M., Chen, S. W., Cruickshank, A. A., Dick, L. R., Grenier, L., Klunder, J. M., Ma, Y. T., Plamondon, L., and Stein, R. L. (1998). Potent and selective inhibitors of the proteasome: Dipeptidyl boronic acids. *Bioorg. Med. Chem. Lett.* **8,** 333–338.

Adams, J., Palombella, V. J., Sausville, E. A., Johnson, J., Destree, A., Lazarus, D. D., Maas, J., Pien, C. S., Prakash, S., and Elliott, P. J. (1999). Proteasome inhibitors: A novel class of potent and effective antitumor agents. *Cancer Res.* **59,** 2615–2622.

Alvarez, M. E., Houck, D. R., White, C. B., Brownell, J. E., Bobko, M. A., Rodger, C. A., Stawicki, M. B., Sun, H. H., Gillum, A. M., and Cooper, R. (1994). Isolation and structure elucidation of two new calpain inhibitors from *Streptomyces griseus*. *J. Antibiot. (Tokyo)* **47,** 1195–1201.

An, B., Goldfarb, R. H., Siman, R., and Dou, Q. P. (1998a). Novel dipeptidyl proteasome inhibitors overcome Bcl-2 protective function and selectively accumulate the cyclin-dependent kinase inhibitor p27 and induce apoptosis in transformed, but not normal, human fibroblasts. *Cell Death Differ.* **5,** 1062–1075.

An, W. G., Chuman, Y., Fojo, T., and Blagosklonny, M. V. (1998b). Inhibitors of transcription, proteasome inhibitors, and DNA-damaging drugs differentially affect feedback of p53 degradation. *Exp. Cell Res.* **244,** 54–60.

Anagli, J., Hagmann, J., and Shaw, E. (1993). Affinity labelling of the Ca^{2+}-activated neutral proteinase (calpain) in intact human platelets. *Biochem. J.* **289,** 93–99.

Andrew, D. J., Hay, A. W., and Evans, S. W. (1999). Aurintricarboxylic acid inhibits apoptosis and supports proliferation in a haemopoietic growth-factor dependent myeloid cell line. *Immunopharmacology* **41,** 1–10.

André, P., Groettrup, M., Klenerman, P., De Giuli, R., Booth, B. L., Jr., Cerundolo, V., Bonneville, M., Jotereau, F., Zinkernagel, R. M., and Lotteau, V. (1998). An inhibitor of HIV-1 protease modulates proteasome activity, antigen presentation, and T cell responses. *Proc. Natl. Acad. Sci. USA* **95,** 13120–13124.

Angliker, H., Anagli, J., and Shaw, E. (1992). Inactivation of calpain by peptidyl fluoromethyl ketones. *J. Med. Chem.* **35,** 216–220.

Ariyoshi, H., Oda, A., and Salzman, E. W. (1995). Participation of calpain in protein-tyrosine phosphorylation and dephosphorylation in human blood platelets. *Arterioscler. Thromb.* **15,** 511–514.

Aronica, E. M., Gorter, J. A., Grooms, S., Kessler, J. A., Bennett, M. V., Zukin, R. S., and Rosenbaum, D. M. (1998). Auritricarboxylic acid prevents GluR2 mRNA down-regulation and delayed neurodegeneration in hippocampal CA1 neurons of gerbil after global ischemia. *Proc. Natl. Acad. Sci. USA* **95,** 7115–7120.

Azanza, J.-L., Raymond, J., Robin, J.-M., Cottin, P., and Ducastaing, A. (1979). Purification and some physico-chemical and enzymic properties of a calcium ion-activated neutral proteinase from rabbit skeletal muscle. *Biochem. J.* **183,** 339–347.

Azuma, M., David, L. L., and Shearer, T. R. (1992). Superior prevention of calcium ionophore cataract by E64d. *Biochim. Biophys. Acta* **1180,** 215–220.

Baghdiguian, S., Martin, M., Richard, I., Pons, F., Astier, C., Bourg, N., Hay, R. T., Chemaly, R., Halaby, G., Loiselet, J., Anderson, L. V. B., De Munain, A. L., Fardeau, M., Mangeat, P., Beckmann, J. S., and Lefranc, G. (1999). Calpain 3 deficiency is associated with myonuclear apoptosis and profound perturbation of the IkappaBα/NF-kappaB pathway in limb-girdle muscular dystrophy type 2A. *Nature Med.* **5,** 503–511.

Banik, N. L., Matzelle, D., Terry, E., and Hogan, E. L. (1997). A new mechanism of methylprednisolone and other corticosteroids action demonstrated in vitro: Inhibition of a proteinase (calpain) prevents myelin and cytoskeletal protein degradation. *Brain Res.* **748,** 205–210.

Barnes, N. Y., Li, L., Yoshikawa, K., Schwartz, L. M., Oppenheim, R. W., and Milligan, C. E. (1998). Increased production of amyloid precursor protein provides a substrate for caspase-3 in dying motoneurons. *J. Neurosci.* **18,** 5869–5880.

Barnoy, S., Glaser, T., and Kosower, N. S. (1997). Calpain and calpastatin in myoblast differentiation and fusion: Effects of inhibitors. *Biochim. Biophys. Acta* **1358,** 181–188.

Barnoy, S., Glaser, T., and Kosower, N. S. (1998). The calpain-calpastatin system and protein degradation in fusing myoblasts. *Biochim. Biophys. Acta* **1402,** 52–60.

Barrett, A. J., Kembhavi, A. A., Brown, M. A., Kirschke, H., Knight, C. G., Tamai, M., and Hanada, K. (1982). L-*trans*-Epoxysuccinyl-leucylamido(4-guanidino)butane (E-64) and its analogues as inhibitors of cysteine proteinases including cathepsins B, H and L. *Biochem. J.* **201,** 189–198.

Barrett, A. J., Rawlings, N. D., and Woessner, J. F., eds. (1998). "Handbook of Proteolytic Enzymes." Academic Press, London.

Bartus, R. T., Baker, K. L., Heiser, A. D., Sawyer, S. D., Dean, R. L., Ellott, P. J., and Straub, J. A. (1994a). Postischemic administration of AK275, a calpain inhibitor, provides substantial protection against focal ischemic brain damage. *J. Cereb. Blood Flow Metab.* **14,** 537–544.

Bartus, R. T., Chen, E. Y., Lynch, G., and Kordower, J. H. (1999). Cortical ablation induces a spreading Calcium-dependent, secondary pathogenesis which can be reduced by inhibiting calpain. *Exp. Neurol.* **155,** 315–326.

Bartus, R. T., Hayward, N. J., Elliott, P. J., Sawyer, S. D., Baker, K. L., Dean, R. L., Akiyama, A., Straub, J. A., Harbeson, S. L., Li, Z., and Powers, J. (1994b). Calpain inhibitor AK295 protects neurons from focal brain ischemia: Effects of postocclusion intra-arterial administration. *Stroke* **25,** 2265–2270.

Behrens, M. M., Martínez, J. L., Moratilla, C., and Renart, J. (1995). Apoptosis induced by protein kinase C inhibition in a neuroblastoma cell line. *Cell Growth Differ.* **6,** 1375–1380.

Belaud-Rotureau, M. A., Lacombe, F., Durrieu, F., Vial, J. P., Lacoste, L., Bernard, P., and Belloc, F. (1999). Ceramide-induced apoptosis occurs independently of caspases and is decreased by leupeptin. *Cell Death Differ.* **6,** 788–795.

Bellas, R. E., FitzGerald, M. J., Fausto, N., and Sonenshein, G. E. (1997). Inhibition of NF_B activity induces apoptosis in murine hepatocytes. *Am. J. Pathol.* **151,** 891–896.

Benchokroun, Y., Couprie, J., and Larsen, A. K. (1995). Aurintricarboxylic acid, a putative inhibitor of apoptosis, is a potent inhibitor of DNA topoisomerase II in vitro and in Chinese hamster fibrosarcoma cells. *Biochem. Pharmacol.* **49,** 305–313.

Besançon, F., Afti, A., Gespach, C., Cayre, Y. E., and Bourgeade, M.-F. (1998). Evidence for a role of NF-κB in the survival of hematopoietic cells mediated by interleukin 3 and the oncogenic TEL/platelet-derived growth factor receptor β fusion protein. *Pro. Natl. Acad. Sci. USA* **95,** 8081–8086.

Beyette, J., Mason, G. G. F., Murray, R. Z., Cohen, G. M., and Rivett, A. J. (1998). Proteasome activities decrease during dexamethasone-induced apoptosis of thymocytes. *Biochem. J.* **332,** 315–320.

Beynon, R. J., and Bond, J. S. (eds.) (1989). "Proteolytic Enzymes: A Practical Approach." IRL Press, Oxford.

Bi, R. F., Bi, X. N., and Baudry, M. (1998). Phosphorylation regulates calpain-mediated truncation of glutamate ionotropic receptors. *Brain Res.* **797,** 154–158.

Bi, X., Chang, V., Molnar, E., McIlhinney, R. A. J., and Baudry, M. (1996). The C-terminal domain of glutamate receptor subunit 1 is a target for calpain-mediated proteolysis. *Neuroscience* **73,** 903–906.

Bi, X., Tocco, G., and Baudry, M. (1994). Calpain-mediated regulation of AMPA receptors in adult rat brain. *Neuroreport* **6,** 61–64.

Bi, X. N., Chen, J., Dang, S. D., Wenthold, R. J., Tocco, G., and Baudry, M. (1997). Characterization of calpain-mediated proteolysis of GluR1 subunits of α-amino-3-hydroxy-5-methylisoxazole-4-propionate receptors in rat brain. *J. Neurochem.* **68,** 1484–1494.

Blomgren, K., Hallin, U., Andersson, A. L., Puka-Sundvall, M., Bahr, B. A., McRae, A., Saido, T. C., Kawashima, S., and Hagberg, H. (1999). Calpastatin is up-regulated in response to hypoxia and is a suicide substrate to calpain after neonatal cerebral hypoxia-ischemia. *J. Biol. Chem.* **274,** 14046–14052.

Bochtler, M., Ditzel, L., Groll, M., Hartmann, C., and Huber, R. (1999). The proteasome. *Annu. Rev. Biophys. Biomol. Struct.* **28,** 295–317.

Bogyo, M., McMaster, J. S., Gaczynska, M., Tortorella, D., Goldberg, A. L., and Ploegh, H. (1997). Covalent modification of active site threonine of proteasomal β subunits and the *Escherichia coli* homolog Hs1V by a new class of inhibitors. *Proc. Natl. Acad. Sci. USA* **94,** 6629–6634.

Boutillier, A. L., Kienlen-Campard, P., and Loeffler, J. P. (1999). Depolarization regulates cyclin D1 degradation and neuronal apoptosis: A hypothesis about the role of the ubiquitin/proteasome signalling pathway. *Eur. J. Neurosci.* **11,** 441–448.

Brancolini, C., Lazarevic, D., Rodriguez, J., and Schneider, C. (1998). Dismantling cell-cell contacts during apoptosis is coupled to a caspase-dependent proteolytic cleavage of beta-catenin. *Cell Death Differ.* **5,** 1042–1050.

Brand, S. J., Morise, Z., Tagerud, S., Mazzola, L., Granger, D. N., and Grisham, M. B. (1999). Role of the proteasome in rat indomethacin-induced gastropathy. *Gastroenterology* **116,** 865–873.

Braun, C., Engel, M., Theisinger, B., Welter, C., and Seifert, M. (1999). CAPN 8: Isolation of a new mouse calpain-isoenzyme. *Biochem. Biophys. Res. Commun.* **260,** 671–675.

Brown, S. B., Bailey, K., and Savill, J. (1997). Actin is cleaved during constitutive apoptosis. *Biochem. J.* **323,** 233–237.

Brown, T. L., Patil, S., Cianci, C. D., Morrow, J. S., and Howe, P. H. (1999). Transforming growth factor β induces caspase 3-independent cleavage of αII-spectrin (α-fodrin) coincident with apoptosis. *J. Biol. Chem.* **274,** 23256–23262.

Brumley, L. M., and Wallace, R. W. (1989). Calmodulin and protein kinase C antagonists also inhibit the Ca^{2+}-dependent protease, calpain I. *Biochem. Biophys. Res. Commun.* **159,** 1297.

Budihardjo, I., Oliver, H., Lutter, M., Luo, X., and Wang, X. (1999). Biochemical pathways of caspase activation during apoptosis. *Annu. Rev. Cell Dev. Biol.* **15,** 269–290.

Buki, K. G., Bauer, P. I., and Kun, E. (1997). Isolation and identification of a proteinase from calf thymus that cleaves poly(ADP-ribose) polymerase and histone H1. *Biochim. Biophys. Acta* **1338,** 100–106.

Bush, K. T., Goldberg, A. L., and Nigam, S. K. (1997). Proteasome inhibition leads to a heat-shock response, induction of endoplasmic reticulum chaperones, and thermotolerance. *J. Biol. Chem.* **272,** 9086–9092.

Campbell, B., Adams, J., Shin, Y. K., and Lefer, A. M. (1999). Cardioprotective effects of a novel proteasome inhibitor following ischemia and reperfusion in the isolated perfused rat heart. *J. Mol. Cell. Cardiol.* **31,** 467–476.

Canu, N., Dus, L., Barbato, C., Ciotti, M. T., Brancolini, C., Rinaldi, A. W., Novak, M., Cattaneo, A., Bradbury, A., and Calissano, P. (1998). Tau cleavage and dephosphorylation in cerebellar granule neurons undergoing apoptosis. *J. Neurosci.* **18,** 7061–7074.

Cardozo, C., Chen, W. E., and Wilk, S. (1996). Cleavage of Pro-X and Glu-X bonds catalyzed by the branched chain amino acid preferring activity of the bovine pituitary multicatalytic proteinase complex (20S proteasome). *Arch. Biochem. Biophys.* **334,** 113–120.

Cardozo, C., and Kohanski, R. A. (1998). Altered properties of the branched chain amino acid-preferring activity contribute to increased cleavages after branched chain residues by the "immunoproteasome." *J. Biol. Chem.* **273,** 16764–16770.

Cardozo, C., Michaud, C., and Orlowski, M. (1999). Components of the bovine pituitary multicatalytic proteinase complex (proteasome) cleaving bonds after hydrophobic residues. *Biochemistry* **38,** 9768–9777.

Carrasco-Marín, E., Paz-Miguel, J. E., López-Mato, P., Alvarez-Domínguez, C., and Leyva-Cobián, F. (1998). Oxidation of defined antigens allows protein unfolding and increases both proteolytic processing and exposes peptide epitopes which are recognized by specific T cells. *Immunology* **95,** 314–321.

Castejon, M. S., Culver, D. G., and Glass, J. D. (1999). Generation of spectrin breakdown products in peripheral nerves by addition of m-calpain. *Muscle Nerve* **22,** 905–909.

Chan, S. L., Griffin, W. S. T., and Mattson, M. P. (1999). Evidence for caspase-mediated cleavage of AMPA receptor subunits in neuronal apoptosis and Alzheimer's disease. *J. Neurosci. Res.* **57,** 315–323.

Chan, S. L., and Mattson, M. P. (1999). Caspase and calpain substrates: Roles in synaptic plasticity and cell death. *J. Neurosci. Res.* **58,** 167–190.

Chandra, J., Niemer, I., Gilbreath, J., Kliche, K. O., Andreeff, M., Freireich, E. J., Keating, M., and McConkey, D. J. (1998). Proteasome inhibitors induce apoptosis in glucocorticoid-resistant chronic lymphocytic leukemic lymphocytes. *Blood* **92,** 4220–4229.

Chang, Y. C., Lee, Y. S., Tejima, T., Tanaka, K., Omura, S., Heintz, N. H., Mitsui, Y., and Magae, J. (1998). mdm2 and bax, downstream mediators of the p53 response, are degraded by the ubiquitin-proteasome pathway. *Cell Growth Differ.* **9,** 79–84.

Chard, P. S., Bleakman, D., Savidge, J. R., and Miller, R. J. (1995). Capsaicin-induced neurotoxicity in cultured dorsal root ganglion neurons: Involvement of calcium-activated proteases. *Neuroscience* **65,** 1099–1108.

Chatterjee, S., Ator, M. A., Bozyczko-Coyne, D., Josef, K., Wells, G., Tripathy, R., Iqbal, M., Bihovsky, R., Senadhi, S. E., Mallya, S., O'Kane, T., McKenna, B. A., Siman, R., and Mallamo, J. P. (1997). Synthesis and biological activity of a series of potent fluoromethyl ketone inhibitors of recombinant human calpain I. *J. Med. Chem.* **40,** 3820–3828.

Chatterjee, S., Dunn, D., Mallya, S., and Ator, M. A. (1999a). P'-extended α-ketoamide inhibitors of proteasome. *Bioorg. Med. Chem. Lett.* **9,** 2603–2606.

Chatterjee, S., Dunn, D., Tao, M., Wells, G., Gu, Z. Q., Bihovsky, R., Ator, M. A., Siman, R., and Mallamo, J. P. (1999b). P2-achiral, P'-extended α-ketoamide inhibitors of calpain I. *Bioorg. Med. Chem. Lett.* **9,** 2371–2374.

Chatterjee, S., Gu, Z. Q., Dunn, D., Tao, M., Josef, K., Tripathy, R., Bihovsky, R., Senadhi, S. E., O'Kane, T. M., McKenna, B. A., Mallya, S., Ator, M. A., Bozyczko-Coyne, D., Siman, R., and Mallamo, J. P. (1998). D-amino acid containing, high-affinity inhibitors of recombinant human calpain I. *J. Med. Chem.* **41,** 2663–2666.

Chatterjee, S., Josef, K., Wells, G., Iqbal, M., Bihovsky, R., Mallamo, J. P., Ator, M. A., Bozyczko-Coyne, D., Mallya, S., Senadhi, S. E., and Siman, R. (1996). Potent fluoromethyl ketone inhibitors of recombinant human calpain I. *Bioorg. Med. Chem. Lett.* **6,** 1237–1240.

Chen, F., Lu, Y. J., Kuhn, D. C., Maki, M., Shi, X. L., Sun, S. C., and Demers, L. M. (1997). Calpain contributes to silica-induced IkappaB-α degradation and nuclear factor-kappaB activation. *Arch. Biochem. Biophys.* **342,** 383–388.

Chen, S. J., Bradley, M. E., and Lee, T. C. (1998). Chemical hypoxia triggers apoptosis of cultured neonatal rat cardiac myocytes: Modulation by calcium-regulated proteases and protein kinases. *Mol. Cell. Biochem.* **178,** 141–149.

Christie, G., Markwell, R. E., Gray, C. W., Smith, L., Godfrey, F., Mansfield, F., Wadsworth, H., King, R., McLaughlin, M., Cooper, D. G., Ward, R. V., Howlett, D. R., Hartmann, T., Lichtenthaler, S. F., Beyreuther, K., Underwood, J., Gribble, S. K., Cappai, R., Masters, C. L., Tamaoka, A., Gardner, R. L., Rivett, A. J., Karran, E. H., and Allsop, D. (1999). Alzheimer's disease: Correlation of the suppression of β-amyloid peptide secretion from cultured cells with inhibition of the chymotrypsin-like activity of the proteasome. *J. Neurochem.* **73,** 195–204.

Cohen, G. M. (1997). Caspases: The executioners of apoptosis. *Biochem. J.* **326,** 1–16.

Conner, E. M., Brand, S., Davis, J. M., Laroux, F. S., Palombella, V. J., Fuseler, J. W., Kang, D. Y., Wolf, R. E., and Grisham, M. B. (1997). Proteasome inhibition attenuates nitric oxide synthase expression, VCAM-1 transcription and the development of chronic colitis. *J. Pharmacol. Exp. Ther.* **282,** 1615–1622.

Cooray, P., Yuan, Y. P., Schoenwaelder, S. M., Mitchell, C. A., Salem, H. H., and Jackson, S. P. (1996). Focal adhesion kinase ($pp125^{FAK}$) cleavage and regulation by calpain. *Biochem. J.* **318,** 41–47.

Corey, E. J., and Li, W. D. Z. (1999). Total synthesis and biological activity of lactacystin, omuralide and analogs. *Chem. Pharm. Bull. (Tokyo)* **47,** 1–10.

Corey, E. J., Li, W. D. Z., Nagamitsu, T., and Fenteany, G. (1999). The structural requirements for inhibition of proteasome function by the lactacystin-derived β-lactone and synthetic analogs. *Tetrahedron* **55,** 3305–3316.

Corsi, D., Galluzzi, L., Lecomte, M. C., and Magnani, M. (1997). Identification of α-spectrin domains susceptible to ubiquitination. *J. Biol. Chem.* **272,** 2977–2983.

Corsi, D., Paiardini, M., Crinelli, R., Bucchini, A., and Magnani, M. (1999). Alteration of α-spectrin ubiquitination due to age-dependent changes in the erythrocyte membrane. *Eur. J. Biochem.* **261,** 775–783.

Covi, J. A., Belote, J. M., and Mykles, D. L. (1999). Subunit compositions and catalytic properties of proteasomes from developmental temperature-sensitive mutants of *Drosophila melanogaster. Arch. Biochem. Biophys.* **368,** 85–97.

Craiu, A., Gaczynska, M., Akopian, T., Gramm, C. F., Fenteany, G., Goldberg, A. L., and Rock, K. L. (1997). Lactacystin and *clasto*-lactacystin β-lactone modify multiple proteasome β-subunits and inhibit intracellular protein degradation and major histocompatibility complex class I antigen presentation. *J. Biol. Chem.* **272,** 13437–13445.

Crawford, C. (1990). Protein and peptide inhibitors of calpains. *In* "Intracellular Calcium-Dependent Proteolysis" (R. L. Mellgren and T. Murachi, eds.), pp. 75–89, CRC Press, Boca Raton, FL.

Crawford, C., Mason, R. W., Wikstrom, P., and Shaw, E. (1988). The design of peptidyldiazomethane inhibitors to distinguish between the cysteine proteinases calpain II, cathepsin L and cathepsin B. *Biochem. J.* **253,** 751–758.

Cressman, C. M., Mohan, P. S., Nixon, R. A., and Shea, T. B. (1995). Proteolysis of protein kinase C: mM and μM calcium-requiring calpains have different abilities to generate, and degrade the free catalytic subunit, protein kinase M. *FEBS Lett.* **367,** 223–227.

Croall, D. E., and DeMartino, G. N. (1991). Calcium-activated neutral protease (calpain) system: Structure, function, and regulation. *Physiol. Rev.* **71,** 813–847.

Croall, D. E., Morrow, J. S., and DeMartino, G. N. (1986). Limited proteolysis of the erythrocyte membrane skeleton by calcium-dependent proteinases. *Biochim. Biophys. Acta* **882,** 287–296.

Cryns, V., and Yuan, J. Y. (1998). Proteases to die for. *Genes Dev.* **12,** 1551–1570.

Cui, H. L., Matsui, K., Omura, S., Schauer, S. L., Matulka, R. A., Sonenshein, G. E., and Ju, S. T. (1997). Proteasome regulation of activation-induced T cell death. *Proc. Natl. Acad. Sci. USA* **94,** 7515–7520.

Dear, N., Matena, K., Vingron, M., and Boehm, T. (1997). A new subfamily of vertebrate calpains lacking a calmodulin-like domain: Implications for calpain regulation and evolution. *Genomics* **45,** 175–184.

Dear, T. N., Möller, A., and Boehm, T. (1999). CAPN11: A calpain with high mRNA levels in testis and located on chromosome 6. *Genomics* **59,** 243–247.

Debiasi, R. L., Squier, M. K. T., Pike, B., Wynes, M., Dermody, T. S., Cohen, J. J., and Tyler, K. L. (1999). Reovirus-induced apoptosis is preceded by increased cellular calpain activity and is blocked by calpain inhibitors. *J. Virol.* **73,** 695–701.

Delic, J., Masdehors, P., Ömura, S., Cosset, J. M., Dumont, J., Binet, J. L., and Magdelénat, H. (1998). The proteasome inhibitor lactacystin induces apoptosis and sensitizes chemo- and radioresistant human chronic lymphocytic leukaemia lymphocytes to TNF-α-initiated apoptosis. *Br. J. Cancer* **77,** 1103–1107.

DeMartino, G. N., and Slaughter, C. A. (1999). The proteasome, a novel protease regulated by multiple mechanisms. *J. Biol. Chem.* **274,** 22123–22126.

De Moissac, D., Mustapha, S., Greenberg, A. H., and Kirshenbaum, L. A. (1998). Bcl-2 activates the transcription factor NFkappaB through the degradation of the cytoplasmic inhibitor IkappaBα. *J. Biol. Chem.* **273,** 23946–23951.

Denning, M. F., Wang, Y. H., Nickoloff, B. J., and Wrone-Smith, T. (1998). Protein kinase Cδ is activated by caspase-dependent proteolysis during ultraviolet radiation-induced apoptosis of human keratinocytes. *J. Biol. Chem.* **273,** 29995–30002.

Dou, Q. P., McQuire, T. F., Peng, Y. B., and An, B. (1999). Proteasome inhibition leads to significant reduction of Bcr-Abl expression and subsequent induction of apoptosis in K562 human chronic myelogenous leukemia cells. *J. Pharmacol. Exp. Ther.* **289,** 781–790.

Drexler, H. C. A. (1997). Activation of the cell death program by inhibition of proteasome function. *Proc. Natl. Acad. Sci. USA* **94,** 855–860.

Earnshaw, W. C., Martins, L. M., and Kaufmann, S. H. (1999). Mammalian caspases: Structure, activation, substrates, and functions during apoptosis. *Annu. Rev. Biochem.* **68,** 383–424.

Easwaran, V., Song, V., Polakis, P., and Byers, S. (1999). The ubiquitin-proteasome pathway and serine kinase activity modulate adenomatous polyposis coli protein-mediated regulation of β-catenin-lymphocyte enhancer-binding factor signaling. *J. Biol. Chem.* **274,** 16641–16645.

Edelstein, C. L., Yaqoob, M. M., Alkhunaizi, A. M., Gengaro, P. E., Nemenoff, R. A., Wang, K. K. W., and Schrier, R. W. (1996). Modulation of hypoxia-induced calpain activity in rat renal proximal tubules. *Kidney Int.* **50,** 1150–1157.

Efanova, I. B., Zaitsev, S. V., Zhivotovsky, B., Köhler, M., Efendic, S., Orrenius, S., and Berggren, P.-O. (1998). Glucose and tolbutamide induce apoptosis in pancreatic β-cells: A process dependent on intracellular Ca^{2+} concentration. *J. Biol. Chem.* **273,** 33501–33507.

El Khissiin, A., and Leclercq, G. (1999). Implication of proteasome in estrogen receptor degradation. *FEBS Lett.* **448,** 160–166.

Elliott, P. J., Pien, C. S., McCormack, T. A., Chapman, I. D., and Adams, J. (1999). Proteasome inhibition: A novel mechanism to combat asthma. *J. Allergy Clin. Immunol.* **104,** 294–300.

Elofsson, M., Splittgerber, U., Myung, J., Mohan, R., and Crews, C. M (1999). Towards subunit specific proteasome inhibitors: Synthesis and evaluation of peptide α',β'-epoxy ketones. *Chem. Biol.* **6,** 811–822.

Esser, R. E., Angelo, R. A., Murphey, M. D., Watts, L. M., Thornburg, L. P., Palmer, J. T., Talhouk, J. W., and Smith, R. E. (1994). Cysteine proteinase inhibitors decrease articular cartilage and bone destruction in chronic inflammatory arthritis. *Arthritis Rheum.* **37,** 236–247.

Eto, A., Akita, Y., Saido, T. C., Suzuki, K., and Kawashima, S. (1995). The role of the calpain-calpastatin system in thyrotropin-releasing hormone-induced selective down-regulation of a protein kinase C isozyme, nPKCϵ, in rat pituitary GH_4C_1 cells. *J. Biol. Chem.* **270,** 25115–25120.

Eymin, B., Sordet, O., Droin, N., Munsch, B., Haugg, M., Van de Craen, M., Vandenabeele, P., and Solary, E. (1999). Caspase-induced proteolysis of the cyclin-dependent kinase inhibitor $p27^{Kip1}$ mediates its anti-apoptotic activity. *Oncogene* **18,** 4839–4847.

Fenteany, G., Standaert, R. F., Lane, W. S., Choi, S., Corey, E. J., and Schreiber, S. L. (1995). Inhibition of proteasome activities and subunit-specific amino-terminal threonine modification by lactacystin. *Science* **268,** 726–731.

Fenteany, G., and Schreiber, S. L. (1998). Lactacystin, proteasome function, and cell fate. *J. Biol. Chem.* **273,** 8545–8548.

Figueiredo-Pereira, M. E., Banik, N., and Wilk, S. (1994a). Comparison of the effect of calpain inhibitors on two extralysosomal proteinases: The multicatalytic proteinase complex and m-calpain. *J. Neurochem.* **62,** 1989–1994.

Figueiredo-Pereira, M. E., Berg, K. A., and Wilk, S. (1994b). A new inhibitor of the chymotrypsin-like activity of the multicatalytic proteinase complex (20S proteasome) induces accumulation of ubiquitin-protein conjugates in a neuronal cell. *J. Neurochem.* **63,** 1578–1581.

Figueiredo-Pereira, M. E., Chen, W. E., Li, J. R., and Johdo, O. (1996). The antitumor drug aclacinomycin A, which inhibits the degradation of ubiquitinated proteins, shows selectivity for the chymotrypsin-like activity of the bovine pituitary 20 S proteasome. *J. Biol. Chem.* **271,** 16455–16459.

Figueiredo-Pereira, M. E., Chen, W.-E., Yuan, H.-M., and Wilk, S. (1995). A novel chymotrypsin-like component of the multicatalytic proteinase complex optimally active at acidic pH. *Arch. Biochem. Biophys.* **317,** 69–78.

Fraser, P. E., Levesque, G., Yu, G., Mills, L. R., Thirlwell, J., Frantseva, M., Gandy, S. E., Seeger, M., Carlen, P. L., and George-Hyslop, P. S. (1998). Presenilin 1 is actively degraded by the 26S proteasome. *Neurobiol. Aging* **19,** (Suppl.), S19–S21.

Fujita, E., Mukasa, T., Tsukahara, T., Arahata, K., Omura, S., and Momoi, T. (1996). Enhancement of CPP32-like activity in the TNF-treated U937 cells by the proteasome inhibitors. *Biochem. Biophys. Res. Commun.* **224,** 74–79.

Fukiage, C., Azuma, M., Nakamura, Y., Tamada, Y., Nakamura, M., and Shearer, T. R. (1997). SJA6017, a newly synthesized peptide aldehyde inhibitor of calpain: amelioration of cataract in cultured rat lenses. *Biochim. Biophys. Acta* **1361,** 304–312.

Fukiage, C., Azuma, M., Nakamura, Y., Tamada, Y., and Shearer, T. R. (1998). Nuclear cataract and light scattering in cultured lenses from guinea pig and rabbit. *Curr. Eye Res.* **17,** 623–635.

Fukuda, K. (1999). Apoptosis-associated cleavage of beta-catenin in human colon cancer and rat hepatoma cells. *Int. J. Biochem. Cell Biol.* **31,** 519–529.

Fukuda, S., Harada, K., Kunimatsu, M., Sakabe, T., and Yoshida, K. (1998). Postischemic reperfusion induces α-fodrin proteolysis by m-calpain in the synaptosome and nucleus in rat brain. *J. Neurochem.* **70,** 2526–2532.

Garcia-Calvo, M., Peterson, E. P., Leiting, B., Ruel, R., Nicholson, D. W., and Thornberry, N. A. (1998). Inhibition of human caspases by peptide-based and macromolecular inhibitors. *J. Biol. Chem.* **273,** 32608–32613.

Garcia-Calvo, M., Peterson, E. P., Rasper, D. M., Vaillancourt, J. P., Zamboni, R., Nicholson, D. W., and Thornberry, N. A. (1999). Purification and catalytic properties of human caspase family members. *Cell Death Differ.* **6,** 362–369.

Geier, E., Pfeifer, G., Wilm, M., Lucchiari-Hartz, M., Baumeister, W., Eichmann, K., and Niedermann, G. (1999). A giant protease with potential to substitute for some functions of the proteasome. *Science* **283,** 978–981.

Gellerman, D. M., Bi, X. N., and Baudry, M. (1997). NMDA receptor-mediated regulation of AMPA receptor properties in organotypic hippocampal slice cultures. *J. Neurochem.* **69,** 131–136.

Gervais, F. G., Xu, D. G., Robertson, G. S., Vaillancourt, J. P., Zhu, Y. X., Huang, J. Q., LeBlanc, A., Smith, D., Rigby, M., Shearman, M. S., Clarke, F. E., Zheng, H., Van Der Ploeg, L. H. T., Ruffolo, S. C., Thornberry, N. A., Xanthoudakis, S., Zamboni, R. J., Roy, S., and Nicholson, D. W. (1999). Involvement of caspases in proteolytic cleavage of Alzheimer's amyloid-β precursor protein and amyloidogenic Aβ peptide formation. *Cell* **97,** 395–406.

Gitler, D., and Spira, M. E. (1998). Real time imaging of calcium-induced localized proteolytic activity after axotomy and its relation to growth cone formation. *Neuron* **20,** 1123–1135.

Glas, R., Bogyo, M., McMaster, J. S., Gaczynska, M., and Ploegh, H. L. (1998). A proteolytic system that compensates for loss of proteasome function. *Nature* **392,** 618–622.

Glockzin, S., Von Knethen, A., Scheffner, M., and Brüne, B. (1999). Activation of the cell death program by nitric oxide involves inhibition of the proteasome. *J. Biol. Chem.* **274,** 19581–19586.

Goldberg, A. L., Akopian, T. N., Kisselev, A. F., and Lee, D. H. (1997a). Protein degradation by the proteasome and dissection of its *in vivo* importance with synthetic inhibitors. *Mol. Biol. Rep.* **24,** 69–75.

Goldberg, A. L., Akopian, T. N., Kisselev, A. F., Lee, D. H., and Rohrwild, M. (1997b). New insights into the mechanisms and importance of the proteasome in intracellular protein degradation. *Biol. Chem. Hoppe Seyler* **378,** 131–140.

Gong, B. D., Chen, Q., Endlich, B., Mazumder, S., and Almasan, A. (1999). Ionizing radiation-induced, Bax-mediated cell death is dependent on activation of cysteine and serine proteases. *Cell Growth Differ.* **10,** 491–502.

Graybill, T. L., Dolle, R. E., Osifo, I. K., Schmidt, S. J., Gregory, J. S., Harris, A. L., and Miller, M. S. (1995). Inhibition of human erythrocyte calpain I by novel quinolinecarboxamides. *Bioorg. Med. Chem. Lett.* **5,** 387–392.

Gressner, A. M., Lahme, B., and Roth, S. (1997). Attenuation of TGF-β-induced apoptosis in primary cultures of hepatocytes by calpain inhibitors. *Biochem. Biophys. Res. Commun.* **231,** 457–462.

Grimm, L. M., Goldberg, A. L., Poirier, G. G., Schwartz, L. M., and Osborne, B. A. (1996). Proteasomes play an essential role in thymocyte apoptosis. *EMBO J.* **15,** 3835–3844.

Grimm, L. M., and Osborne, B. A. (1998). Apoptosis and the proteasome. *In* "Apoptosis: Biology and Mechanisms" (S. Kumar, ed.), pp. 209–223, Springer Verlag, New York.

Griscavage, J. M., Wilk, S., and Ignarro, L. J. (1996). Inhibitors of the proteasome pathway interfere with induction of nitric oxide synthase in macrophages by blocking activation of transcription factor NF-kappaB. *Proc. Natl. Acad. Sci. USA* **93,** 3308–3312.

Grisham, M. B., Palombella, V. J., Elliott, P. J., Conner, E. M., Brand, S., Wong, H. L., Pien, C., Mazzola, L. M., Destree, A., Parent, L., and Adams, J. (1999). Inhibition of NF-kappaB activation *in vitro* and *in vivo*: Role of 26S proteasome. *Methods Enzymol.* **300,** 345–363.

Groettrup, M., and Schmidtke, G. (1999). Selective proteasome inhibitors: Modulators of antigen presentation? *Drug Discovery Today* **4,** 63–71.

Groll, M., Heinemeyer, W., Jäger, S., Ullrich, T., Bochtler, M., Wolf, D. H., and Huber, R. (1999). The catalytic sites of 20S proteasomes and their role in subunit maturation: A mutational and crystallographic study. *Proc. Natl. Acad. Sci. USA* **96,** 10976–10983.

Guttmann, R. P., and Johnson, G. V. W. (1998). Oxidative stress inhibits calpain activity *in situ*. *J. Biol. Chem.* **273,** 13331–13338.

Haass, C., Grünberg, J., Capell, A., Wild-Bode, C., Leimer, U., Walter, J., Yamazaki, T., Ihara, I., Zweckbronner, I., Jakubek, C., and Baumeister, R. (1998). Proteolytic processing of Alzheimer's disease associated proteins. *J. Neural Transm.* **105** (Suppl. 53), 159–167.

Haire, M. F., Clark, J. J., Jones, M. E. E., Hendil, K. B., Schwartz, L. M., and Mykles, D. L. (1995). The multicatalytic proteinase (proteasome) of the hawkmoth, *Manduca sexta:* Catalytic properties and immunological comparison with the lobster enzyme complex. *Arch. Biochem. Biophys.* **318,** 15–24.

Hajimohammadreza, I., Raser, K. J., Nath, R., Nadimpalli, R., Scott, M., and Wang, K. K. W. (1997). Neuronal nitric oxide synthase and calmodulin-dependent protein kinase IIα undergo neurotoxin-induced proteolysis. *J. Neurochem.* **69,** 1006–1013.

Han, Y. Q., Weinman, S., Boldogh, I., Walker, R. K., and Brasier, A. R. (1999). Tumor necrosis factor-α-inducible IkappaBα proteolysis mediated by cytosolic m-calpain: A mechanism parallel to the ubiquitin-proteasome pathway for nuclear factor-kappaB activation. *J. Biol. Chem.* **274,** 787–794.

Han, Z., Hendrickson, E. A., Bremner, T. A., and Wyche, J. H. (1972). A sequential two-step mechanism for the production of the mature p17 : p12 form of caspase-3 *in vitro*. *J. Biol. Chem.* **272,** 13432–13436.

Harbeson, S. L., Abelleira, S. M., Akiyama, A., Barrett, R., III, Carroll, R. M., Straub, J. A., Tkacz, J. N., Wu, C., and Musso, G. F. (1994). Stereospecific synthesis of peptidyl α-keto amides as inhibitors of calpain. *J. Med. Chem.* **37,** 2918–2929.

Harding, C. V., France, J., Song, R., Farah, J. M., Chatterjee, S., Iqbal, M., and Siman, R. (1995). Novel dipeptide aldehydes are proteasome inhibitors and block the MHC-I antigen-processing pathway. *J. Immunol.* **155,** 1767–1775.

Harris, A. L., Gregory, J. S., Maycock, A. L., Graybill, T. L., Osifo, I. K., Schmidt, S. J., and Dolle, R. E. (1995). Characterization of a continuous fluorogenic assay for calpain I: Kinetic evaluation of peptide aldehydes, halomethyl ketones and (acyloxy)methyl ketones as inhibitors of the enzyme. *Bioorg. Med. Chem. Lett.* **5,** 393–398.

Harvey, K. F., Harvey, N. L., Michael, J. M., Parasivam, G., Waterhouse, N., Alnemri, E. S., Watters, D., and Kumar, S. (1998). Caspase-mediated cleavage of the ubiquitin-protein ligase Nedd4 during apoptosis. *J. Biol. Chem.* **273,** 13524–13530.

Hauser, H. P., Bardroff, M., Pyrowolakis, G., and Jentsch, S. (1998). A giant ubiquitin-conjugating enzyme related to IAP apoptosis inhibitors. *J. Cell Biol.* **141,** 1415–1422.

Hayes, R. L., Wang, K. K. W., Kampfl, A., Posmantur, R. M., Newcomb, J. K., and Clifton, G. L. (1998). Potential contribution of proteases to neuronal damage. *Drug News Perspect,* **11,** 215–222.

He, H. L., Qi, X. M., Grossmann, J., and Distelhorst, C. W. (1998). c-Fos degradation by the proteasome: An early, Bcl-2-regulated step in apoptosis. *J. Biol. Chem.* **273,** 25015–25019.

Heissmeyer, V., Krappmann, D., Wulczyn, F. G., and Scheidereit, C. (1999). NF-kappaB p105 is a target of IkappaB kinases and controls signal induction of Bcl-3-p50 complexes. *EMBO J.* **18,** 4766–4778.

Helgason, C. D., Atkinson, E. A., Pinkoski, M. J., and Bleackley, R. C. (1995). Proteinases are involved in both DNA fragmentation and membrane damage during CTL-mediated target cell killing. *Exp. Cell Res.* **218,** 50–56.

Herren, B., Levkau, B., Raines, E. W., and Ross, R. (1998). Cleavage of beta-catenin and plakoglobin and shedding of VE-cadherin during endothelial apoptosis: Evidence for a role for caspases and metalloproteinases. *Mol. Biol. Cell* **9,** 1589–1601.

Herrmann, J. L., Briones, F., Jr., Brisbay, S., Logothetis, C. J., and McDonnell, T. J. (1998). Prostate carcinoma cell death resulting from inhibition of proteasome activity is independent of functional Bcl-2 and p53. *Oncogene* **17,** 2889–2899.

Hirata, H., Takahashi, A., Kobayashi, S., Yonehara, S., Sawai, H., Okazaki, T., Yamamoto, K., and Sasada, M. (1998). Caspases are activated in a branched protease cascade and control distinct downstream processes in Fas-induced apoptosis. *J. Exp. Med.* **187,** 587–600.

Hirsch, T., Dallaporta, B., Zamzami, N., Susin, S. A., Ravagnan, L., Marzo, I., Brenner, C., and Kroemer, G. (1998). Proteasome activation occurs at an early, premitochondrial step of thymocyte apoptosis. *J. Immunol.* **161,** 35–40.

Honda, T., Yasutake, K., Nihonmatsu, N., Mercken, M., Takahashi, H., Murayama, O., Murayama, M., Sato, K., Omori, A., Tsubuki, S., Saido, T. C., and Takashima, A. (1999). Dual roles of proteasome in the metabolism of presenilin 1. *J. Neurochem.* **72,** 255–261.

Hong, D., Huan, J., Ou, B., Yeh, J., Saido, T. C., Cheeke, P. R., and Forsberg, N. E. (1995). Protein kinase C isoforms in muscle cells and their regulation by phorbol ester and calpain. *Biochim. Biophys. Acta Mol. Cell Res.* **1267,** 45–54.

Hong, H., El-Saleh, S. C., and Johnson, P. (1990a). Fluorescence spectroscopic analysis of calpain II interactions with calcium and calmodulin antagonists. *Int. J. Biochem.* **22,** 399–404.

Hong, H., Johnson, P., and El-Saleh, S. C. (1990b). Effects of calcium and calmodulin antagonists on calpain II subunit conformations. *Int. J. Biol. Macromol.* **12,** 269–272.

Huang, Z., McGowan, E. B., and Detwiler, T. C. (1992). Ester and amide derivatives of E64c as inhibitors of platelet calpains. *J. Med. Chem.* **35,** 2048–2054.

Hughes, F. M., Jr., Evans-Storms, R. B., and Cidlowski, J. A. (1998). Evidence that non-caspase proteases are required for chromatin degradation during apoptosis. *Cell Death Differ.* **5,** 1017–1027.

Imajoh-Ohmi, S., Kawaguchi, T., Sugiyama, S., Tanaka, K., Omura, S., and Kikuchi, H. (1995). Lactacystin, a specific inhibitor of the proteasome, induces apoptosis in human monoblast U937 cells. *Biochem. Biophys. Res. Commun.* **217,** 1070–1077.

Iqbal, M., Chatterjee, S., Kauer, J. C., Das, M., Messina, P., Freed, B., Biazzo, W., and Siman, R. (1995). Potent inhibitors of proteasome. *J. Med. Chem.* **38,** 2276–2277.

Iqbal, M., Messina, P. A., Freed, B., Das, M., Chatterjee, S., Tripathy, R., Tao, M., Josef, K. A., Dembofsky, B., Dunn, D., Griffith, E., Siman, R., Senadhi, S. E., Biazzo, W., Bozyczko-Coyne, D., Meyer, S. L., Ator, M. A., and Bihovsky, R. (1997). Subsite requirements for peptide aldehyde inhibitors of human calpain I. *Bioorg. Med. Chem. Lett.* **7,** 539–544.

Isahara, K., Ohsawa, Y., Kanamori, S., Shibata, M., Waguri, S., Sato, N., Gotow, T., Watanabe, T., Momoi, T., Urase, K., Kominami, E., and Uchiyama, Y. (1999). Regulation of a novel pathway for cell death by lysosomal aspartic and cysteine proteinases. *Neuroscience* **91,** 233–249.

Ishii, H., Kuboki, M., Fujii, J., Hiraishi, S., and Kazama, M. (1990). Thiolprotease inhibitor, EST, can inhibit thrombin-induced platelet activation. *Thromb. Res.* **57,** 847–861.

Isoe, T., Naito, M., Shirai, A., Hirai, R., and Tsuruo, T. (1992). Inhibition of different steps of the ubiquitin system by cisplatin and aclarubicin. *Biochim. Biophys. Acta Gen. Subj.* **1117,** 131–135.

Ivanov, V. N., and Nikolic-Zugic, J. (1998). Biochemical and kinetic characterization of the glucocorticoid-induced apoptosis of immature $CD4^+$ $CD8^+$ thymocytes. *Int. Immunol.* **10,** 1807–1817.

Iwamoto, H., Miura, T., Okamura, T., Shirakawa, K., Iwatate, M., Kawamura, S., Tatsuno, H., Ikeda, Y., and Matsuzaki, M. (1999). Calpain inhibitor-1 reduces infarct size and DNA fragmentation of myocardium in ischemic/reperfused rat heart. *J. Cardiovasc. Pharmacol.* **33,** 580–586.

James, T., Matzelle, D., Bartus, R., Hogan, E. L., and Banik, N. L. (1998). New inhibitors of calpain prevent degradation of cytoskeletal and myelin proteins in spinal cord in vitro. *J. Neurosci. Res.* **51,** 218–222.

Johnson, P. (1990). Calpains (intracellular calcium-activated cysteine proteinases): Structure–activity relationships and involvement in normal and abnormal cellular metabolism. *Int. J. Biochem.* **22,** 811–822.

Johnson, P., and Hammer, J. L. (1990). Inhibitory effects of spermine and spermidine on muscle calpain II. *Experientia* **46,** 276–278.

Juin, P., Pelletier, M., Oliver, L., Tremblais, K., Grégoire, M., Meflah, K., and Vallette, F. M. (1998). Induction of a caspase-3-like activity by calcium in normal cytosolic extracts triggers nuclear apoptosis in a cell-free system. *J. Biol. Chem.* **273,** 17559–17564.

Kalogeris, T. J., Laroux, F. S., Cockrell, A., Ichikawa, H., Okayama, N., Phifer, T. J., Alexander, J. S., and Grisham, M. B. (1999). Effect of selective proteasome inhibitors on TNF-induced activation of primary and transformed endothelial cells. *Am. J. Physiol. Cell Physiol.* **276,** C856–C864.

Kass, G. E. N., and Orrenius, S. (1999). Calcium signaling and cytotoxicity. *Environ. Health Perspect.* **107** Suppl. 1, 25–35.

Kawazoe, Y., Nakai, A., Tanabe, M., and Nagata, K. (1998). Proteasome inhibition leads to the activation of all members of the heat-shock-factor family. *Eur. J. Biochem.* **255,** 356–362.

Kayalar, C., Ord, T., Testa, M. P., Zhong, L. T., and Bredesen, D. E. (1996). Cleavage of actin by interleukin 1 beta-converting enzyme to reverse DNase I inhibition. *Proc. Natl. Acad. Sci. USA* **93,** 2234–2238.

Khwaja, A., and Tatton, L. (1999). Caspase-mediated proteolysis and activation of protein kinase Cδ plays a central role in neutrophil apoptosis. *Blood* **94,** 291–301.

Kikuchi, H., and Imajoh-Ohmi, S. (1995). Antibodies specific for proteolyzed forms of protein kinase C α. *Biochim. Biophys. Acta Mol. Cell Res.* **1269,** 253–259.

Kilic, F., and Trevithick, J. R. (1998). Modelling cortical cataractogenesis XXIX: Calpain proteolysis of lens fodrin in cataract. *Biochem. Mol. Biol. Int.* **45,** 963–978.

Kim, D., Kim, S. H., and Li, G. C. (1999a). Proteasome inhibitors MG132 and lactacystin hyperphosphorylate HSF1 and induce hsp70 and hsp27 expression. *Biochem. Biophys. Res. Commun.* **254,** 264–268.

Kim, K. B., Myung, J., Sin, N., and Crews, C. M. (1999b). Proteasome inhibition by the natural products eponemycin and epoxomicin: Insights into specificity and potency. *Bioorg. Med. Chem. Lett.* **9,** 3335–3340.

Kim, T. W., Pettingell, W. H., Hallmark, O. G., Moir, R. D., Wasco, W., and Tanzi, R. E. (1997). Endoproteolytic cleavage and proteasomal degradation of presenilin 2 in transfected cells *J. Biol. Chem.* **272,** 11006–11010.

Kitagawa, H., Tani, E., Ikemoto, H., Ozaki, I., Nakano, A., and Omura, S. (1999). Proteasome inhibitors induce mitochondria-independent apoptosis in human glioma cells. *FEBS Lett.* **443,** 181–186.

Knepper-Nicolai, B., Savill, J., and Brown, S. B. (1998). Constitutive apoptosis in human neutrophils requires synergy between calpains and the proteasome downstream of caspases. *J. Biol. Chem.* **273,** 30530–30536.

Kohli, V., Madden, J. F., Bentley, R. C., and Clavien, P. A. (1999). Calpain mediates ischemic injury of the liver through modulation of apoptosis and necrosis. *Gastroenterology* **116,** 168–178.

Kothakota, S., Azuma, T., Rienhard, C., Klippel, A., Tang, J., Chu, K., McGary, T. J., Kirschner, M. W., Koths, K., Kwiatkowski, D. J., and Williams, L. T. (1997). Caspase-3-generated fragment of gelsolin: effector of morphological change in apoptosis. *Science* **278,** 294–298.

Kroll, M., Arenzana-Seisdedos, F., Bachelerie, F., Thomas, D., Friguet, B., and Conconi, M. (1999a). The secondary fungal metabolite gliotoxin targets proteolytic activities of the proteasome. *Chem. Biol.* **6,** 689–698.

Kroll, M., Margottin, F., Kohl, A., Renard, P., Durand, H., Concordet, J. P., Bachelerie, F., Arenzana-Seisdedos, F., and Benarous, R. (1999b). Inducible degradation of IkappaBα by the proteasome requires interaction with the F-box protein h-βTrCP. *J. Biol. Chem.* **274,** 7941–7945.

Kubbutat, M. H. G., Ludwig, R. L., Ashcroft, M., and Vousden, K. H. (1998). Regulation of Mdm2-directed degradation by the C terminus of p53. *Mol. Cell. Biol.* **18,** 5690–5698.

Kudo, S., Miyamoto, G., and Kawano, K. (1999). Proteases involved in the metabolic degradation of human interleukin-1β by rat kidney lysosomes. *J. Interferon Cytokine Res.* **19,** 361–367.

Kwak, K. B., Kambayashi, J., Sik Kang, M., Bong Ha, D., and Ha Chung, C. (1993). Cell-penetrating inhibitors of calpain block both membrane fusion and filamin cleavage in chick embryonic myoblasts. *FEBS Lett.* **323,** 151–154.

Lang, D., Beermann, M. L., Hauser, G., Cressman, C. M., and Shea, T. B. (1995). Phospholipids inhibit proteolysis of protein kinase Cα by mM calcium-requiring calpain. *Neurochem. Res.* **20,** 1361–1364.

LeBlanc, A., Liu, H., Goodyer, C., Bergeron, C., and Hammond, J. (1999). Caspase-6 role in apoptosis of human neurons, amyloidogenesis, and Alzheimer's disease. *J. Biol. Chem.* **274,** 23426–23436.

Lee, D. H., and Goldberg, A. L. (1998a). Proteasome inhibitors cause induction of heat shock proteins and trehalose, which together confer thermotolerance in *Saccharomyces cerevisiae. Mol. Cell. Biol.* **18,** 30–38.

Lee, D. H., and Goldberg, A. L. (1998b). Proteasome inhibitors: Valuable new tools for cell biologists. *Trends Cell Biol.* **8,** 397–403.

Lee, H. J., Sorimachi, H., Jeong, S. Y., Ishiura, S., and Suzuki, K. (1998). Molecular cloning and characterization of a novel tissue-specific calpain predominantly expressed in the digestive tract. *Biol. Chem. Hoppe Seyler* **379,** 175–183.

Lee, H. W., Smith, L., Pettit, G. R., Vinitsky, A., and Smith, J. B. (1996). Ubiquitination of protein kinase C-α and degradation by the proteasome. *J. Biol. Chem.* **271,** 20973–20976.

Lee, H. W., Smith, L., Pettit, G. R., and Smith, J. B. (1997). Bryostatin 1 and phorbol ester down-modulate protein kinase C-α and -ϵ via the ubiquitin/proteasome pathway in human fibroblasts. *Mol. Pharmacol.* **51,** 439–447.

Levkau, B., Herren, B., Koyama, H., Ross, R., and Raines, E. W. (1998). Caspase-mediated cleavage of focal adhesion kinase pp125FAK and disassembly of focal adhesions in human endothelial cell apoptosis. *J. Exp. Med.* **187,** 579–586.

Li, Q. X., Evin, G., Small, D. H., Multhaup, G., Beyreuther, K., and Masters, C. L. (1995). Proteolytic processing of Alzheimer's disease beta 4 amyloid precursor protein in human platelets. *J. Biol. Chem.* **270,** 14140–14147.

Li, Z., Patil, G. S., Golubski, Z. E., Hori, H., Tehrani, K., Foreman, J. E., Eveleth, D. D., Bartus, R. T., and Powers, J. C. (1993). Peptide α-keto ester, α-keto amide, and α-keto acid inhibitors of calpains and other cysteine proteases. *J. Med. Chem.* **36,** 3472–3480.

Li, Z. Z., Ortega Vilain, A. C., Patil, G. S., Chu, D. L., Foreman, J. E., Eveleth, D. D., and Powers, J. C. (1996). Novel peptidyl α-keto amide inhibitors of calpains and other cysteine proteases. *J. Med. Chem.* **39,** 4089–4098.

Lin, G. D., Chattopadhyay, D., Maki, M., Wang, K. K. W., Carson, M., Jin, L., Yuen, P., Takano, E., Hatanaka, M., DeLucas, L. J., and Narayana, S. V. L. (1997). Crystal structure of calcium bound domain VI of calpain at 1.9 Å resolution and its role in enzyme assembly, regulation, and inhibitor binding. *Nature Struct. Biol.* **4,** 539–547.

Lin, K. I., Baraban, J. M., and Ratan, R. R. (1998). Inhibition versus induction of apoptosis by proteasome inhibitors depends on concentration. *Cell Death Differ.* **5,** 577–583.

Liu, Z. Q., Kunimatsu, M., Yang, J. P., Ozaki, Y., Sasaki, M., and Okamoto, T. (1996). Proteolytic processing of nuclear factor kappaB by calpain in vitro. *FEBS Lett.* **385,** 109–113.

Loidl, G., Groll, M., Musiol, H. J., Huber, R., and Moroder, L. (1999). Bivalency as a principle for proteasome inhibition. *Proc. Natl. Acad. Sci. USA* **96,** 5418–5422.

Lopes, U. G., Erhardt, P., Yao, R. J., and Cooper, G. M. (1997). p53-dependent induction of apoptosis by proteasome inhibitors. *J. Biol. Chem.* **272,** 12893–12896.

Lu, Q., and Mellgren, R. L. (1996). Calpain inhibitors and serine protease inhibitors can produce apoptosis in HL-60 cells. *Arch. Biochem. Biophys.* **334,** 175–181.

Lu, Z. M., Liu, D., Hornia, A., Devonish, W., Pagano, M., and Foster, D. A. (1998). Activation of protein kinase C triggers its ubiquitination and degradation. *Mol. Cell. Biol.* **18,** 839–845.

Lum, R. T., Kerwar, S. S., Meyer, S. M., Nelson, M. G., Schow, S. R., Shiffman, D., Wick, M. M., and Joly, A. (1998a). A new structural class of proteasome inhibitors that prevent NF-kappaB activation. *Biochem. Pharmacol.* **55,** 1391–1397.

Lum, R. T., Nelson, M. G., Joly, A., Horsma, A. G., Lee, G., Meyer, S. M., Wick, M. M., and Schow, S. R. (1998b). Selective inhibition of the chymotrypsin-like activity of the 20S proteasome by 5-methoxy-1-indanone dipeptide benzamides. *Bioorg. Med. Chem. Lett.* **8,** 209–214.

Lynas, J. F., Harriott, P., Healy, A., McKervey, M. A., and Walker, B. (1998). Inhibitors of the chymotrypsin-like activity of proteasome based on di- and tri-peptidyl α-keto aldehydes (glyoxals). *Bioorg. Med. Chem. Lett.* **8,** 373–378.

Lynch, T., Vasilakos, J. P., Raser, K., Keane, K. M., and Shivers, B. D. (1997). Inhibition of the interleukin-1β converting enzyme family rescues neurons from apoptotic death. *Mol. Psychiatry* **2,** 227–238.

Ma, H., Fukiage, C., Azuma, M., and Shearer, T. R. (1998). Cloning and expression of mRNA for calpain Lp82 from rat lens: Splice variant of p94. *Invest. Ophthalmol. Vis. Sci.* **39,** 454–461.

Maeda, S., Lin, K. H., Inagaki, H., and Saito, T. (1996). Induction of apoptosis in primary culture of rat hepatocytes by protease inhibitors. *Biochem. Mol. Biol. Int.* **39,** 447–453.

Maki, M., Bagci, H., Hamaguchi, K., Ueda, M., Murachi, T., and Hatanaka, M. (1989). Inhibition of calpain by a synthetic oligopeptide corresponding to an exon of the human calpastatin gene. *J. Biol. Chem.* **264,** 18866–18869.

Marambaud, P., Da Costa, C. A., Ancolio, K., and Checler, F. (1998). Alzheimer's disease-linked mutation of presenilin 2 (N141I-PS2) drastically lowers APPα secretion: Control by the proteasome. *Biochem. Biophys. Res. Commun.* **252,** 134–138.

Margolin, N., Raybuck, S. A., Wilson, K. P., Chen, W., Fox, T., Gu, Y., and Livingston, D. J. (1997). Substrate and inhibitor specificity of interleukin-1β-converting enzyme and related caspases. *J. Biol. Chem.* **272,** 7223–7228.

Marienfeld, R., Neumann, M., Chuvpilo, S., Escher, C., Kneitz, B., Avots, A., Schimpl, A., and Serfling, E. (1997). Cyclosporin A interferes with the inducible degradation of NF-κB inhibitors, but not with the processing of p105/NF-κB1 in T cells. *Eur. J. Immunol.* **27,** 1601–1609.

Mashima, T., Naito, M., Noguchi, K., Miller, D. K., Nicholson, D. W., and Tsuruo, T. (1997). Actin cleavage by CPP-32/apopain during the development of apoptosis. *Oncogene* **14,** 1007–1012.

Mathew, A., Mathur, S. K., and Morimoto, R. I. (1998). Heat shock response and protein degradation: Regulation of HSF2 by the ubiquitin-proteasome pathway. *Mol. Cell. Biol.* **18,** 5091–5098.

Matsueda, R., Umeyama, H., Kominami, E., and Katunuma, N. (1990). Cysteine protease inhibitors with S-(3-nitro-2-pyridinesulfenyl)-cysteine residue in affinity analogs of peptide substrates. *Adv. Exp. Med. Biol.* **247B,** 265–270.

Matsumoto, K., Mizoue, K., Kitamura, K., Tse, W. C., Huber, C. P., and Ishida, T. (1999). Structural basis of inhibition of cysteine proteases by E-64 and its derivatives. *Biopolymers* **51,** 99–107.

McCormack, T. A., Cruikshank, A. A., Grenier, L., Melandri, F. D., Nunes, S. L., Plamondon, L., Stein, R. L., and Dick, L. R. (1998). Kinetic studies of the branched chain amino acid preferring peptidase activity of the 20S proteasome: Development of a continuous assay and inhibition by tripeptide aldehydes and *clasto*-lactacystin β-lactone. *Biochemistry* **37,** 7792–7800.

McCune, S. A., Foe, L. G., Kemp, R. G., and Jurin, R. R. (1989). Aurintricarboxylic acid is a potent inhibitor of phosphofructokinase. *Biochem. J.* **259,** 925–927.

McDade, T. P., Perugini, R. A., Vittimberga, F. J.,Jr., and Callery, M. P. (1999). Ubiquitin-proteasome inhibition enhances apoptosis of human pancreatic cancer cells. *Surgery* **126,** 371–377.

McGinnis, K. M., Gnegy, M. E., Park, Y. H., Mukerjee, N., and Wang, K. K. W. (1999a). Procaspase-3 and poly(ADP)ribose polymerase (PARP) are calpain substrates. *Biochem. Biophys. Res. Commun.* **263,** 94–99.

McGinnis, K. M., Wang, K. K. W., and Gnegy, M. E. (1999b). Alterations of extracellular calcium elicit selective modes of cell death and protease activation in SH-SY5Y human neuroblastoma cells. *J. Neurochem.* **72,** 1853–1863.

McGinnis, K. M., Whitton, M. M., Gnegy, M. E., and Wang, K. K. W. (1998). Calcium/calmodulin-dependent protein kinase IV is cleaved by caspase-3 and calpain in SH-SY5Y human neuroblastoma cells undergoing apoptosis. *J. Biol. Chem.* **273,** 19993–20000.

McGowan, E. B., Becker, E., and Detwiler, T. C. (1989). Inhibition of calpain in intact platelets by the thiol protease inhibitor E-64d. *Biochem. Biophys. Res. Commun.* **158,** 432–435.

Mehdi, S., Angelastro, M. R., Wiseman, J. S., and Bey, P. (1988). Inhibition of the proteolysis of rat erythrocyte membrane proteins by a synthetic inhibitor of calpain. *Biochem. Biophys. Res. Commun.* **157,** 1117–1123.

Mellgren, R. L. (1997). Specificities of cell permeant peptidyl inhibitors for the proteinase activities of μ-calpain and the 20 S proteasome. *J. Biol. Chem.* **272,** 29899–29903.

Mellgren, R. L., Lu, Q., Zhang, W. L., Lakkis, M., Shaw, E., and Mericle, M. T. (1996). Isolation of a Chinese hamster ovary cell clone possessing decreased μ-calpain content and a reduced proliferative growth rate. *J. Biol. Chem.* **271,** 15568–15574.

Mellgren, R. L., Shaw, E., and Mericle, M. T. (1994). Inhibition of growth of human TE2 and C-33A cells by the cell-permeant calpain inhibitor benzyloxycarbonyl-Leu-Leu-Tyr diazomethyl ketone. *Exp. Cell Res.* **215,** 164–171.

Meng, L. H., Kwok, B. H. B., Sin, N., and Crews, C. M. (1999a). Eponemycin exerts its autitumor effect through the inhibition of proteasome function. *Cancer Res.* **59,** 2798–2801.

Meng, L. H., Mohan, R., Kwok, B. H. B., Elofsson, M., Sin, N., and Crews, C. M. (1999b). Epoxomicin, a potent and selective proteasome inhibitor, exhibits *in vivo* antiinflammatory activity. *Proc. Natl. Acad. Sci. USA* **96,** 10403–10408.

Meredith, J., Mu, Z. M., Saido, T., and Du, X. P. (1998). Cleavage of the cytoplasmic domain of the integrin β_3 subunit during endothelial cell apoptosis. *J. Biol. Chem.* **273,** 19525–19531.

Meriin, A. B., Gabai, V. L., Yaglom, J., Shifrin, V. I., and Sherman, M. Y. (1998). Proteasome inhibitors activate stress kinases and induce Hsp72: Diverse effects on apoptosis. *J. Biol. Chem.* **273,** 6373–6379.

Meyer, S., Kohler, N. G., and Joly, A. (1997). Cyclosporine A is an uncompetitive inhibitor of proteasome activity and prevents NF-kappaB activation. *FEBS Lett.* **413,** 354–358.

Michetti, M., Salamino, F., Melloni, E., and Pontremoli, S. (1995). Reversible inactivation of calpain isoforms by nitric oxide. *Biochem. Biophys. Res. Commun.* **207,** 1009–1014.

Mikulski, S. M., Viera, A., Deptala, A., and Darzynkiewicz, Z. (1998). Enhanced *in vitro* cytotoxicity and cytostasis of the combination of onconase with a proteasome inhibitor. *Int. J. Oncol.* **13,** 633–644.

Milligan, S. A., Owens, M. W., and Grisham, M. B. (1996). Inhibition of IkappaB-α and IkappaB-β proteolysis by calpain inhibitor I blocks nitric oxide synthesis. *Arch. Biochem. Biophys.* **335,** 388–395.

Mimnaugh, E. G., Chen, H. Y., Davie, J. R., Celis, J. E., and Neckers, L. (1997). Rapid deubiquitination of nucleosomal histones in human tumor cells caused by proteasome inhibitors and stress response inducers: Effects on replication, transcription, translation, and the cellular stress response. *Biochemistry* **36,** 14418–14429.

Minger, S. L., Geddes, J. W., Holtz, M. L., Craddock, S. D., Whiteheart, S. W., Siman, R. G., and Pettigrew, L. C. (1998). Glutamate receptor antagonists inhibit calpain-mediated cytoskeletal proteolysis in focal cerebral ischemia. *Brain Res.* **810,** 181–199.

Miyamoto, S., Seufzer, B. J., and Shumway, S. D. (1998). Novel IkappaBα proteolytic pathway in WEHI231 immature B cells. *Mol. Cell. Biol.* **18,** 19–29.

Mizuno, K., Noda, K., Araki, T., Imaoka, T., Kobayashi, Y., Akita, Y., Shimonaka, M., Kishi, S., and Ohno, S. (1997). Proteolytic cleavage of protein kinase C isotypes, which generates kinase and regulatory fragments, correlates with Fas-mediated and 12-*O*-tetradecanoyl-phorbol-13-acetate-induced apoptosis. *Eur. J. Biochem.* **250,** 7–18.

Monti, B., Sparapani, M., and Contestabile, A. (1998). Differential toxicity of protease inhibitors in cultures of cerebellar granule neurons. *Exp. Neurol.* **153,** 335–341.

Mori, S., Tanaka, K., Omura, S., and Saito, Y. (1995). Degradation process of ligand-stimulated platelet-derived growth factor β-receptor involves ubiquitin-proteasome proteolytic pathway. *J. Biol. Chem.* **270,** 29447–29452.

Musleh, W., Bi, X. N., Tocco, G., Yaghoubi, S., and Baudry, M. (1997). Glycine-induced long-term potentiation is associated with structural and functional modifications of α-amino-3-hydroxyl-5-methyl-4-isoxazolepropionic acid receptors. *Proc. Natl. Acad. Sci. USA* **94,** 9451–9456.

Mykles, D. L. (1989). Purification and characterization of a multicatalytic proteinase from crustacean muscle: Comparison of latent and heat-activated forms. *Arch. Biochem. Biophys.* **274,** 216–228.

Mykles, D. L. (1990). Calcium-dependent proteolysis in crustacean claw closer muscle maintained in vitro. *J. Exp. Zool.* **256,** 16–30.

Mykles, D. L. (1997). Biochemical properties of insect and crustacean proteasomes. *Mol. Biol. Rep.* **24,** 133–138.

Mykles, D. L. (1998). Intracellular proteinases of invertebrates: Calcium-dependent and proteasome/ubiquitin-dependent systems. *Int. Rev. Cytol.* **184,** 157–289.

Mykles, D. L., and Haire, M. F. (1995). Branched-chain-amino-acid-preferring peptidase activity of the lobster multicatalytic proteinase (proteasome) and the degradation of myofibrillar proteins. *Biochem. J.* **306,** 285–291.

Mykles, D. L., and Skinner, D. M. (1986). Four Ca^{2+}-dependent proteinase activities isolated from crustacean muscle differ in size, net charge, and sensitivity to Ca^{2+} and inhibitors. *J. Biol. Chem.* **261,** 9865–9871.

Nakane, H., Balzarini, J., DeClercq, E., and Ono, K. (1988). Differential inhibition of various deoxyribonucleic-acid polymerases by Evans blue and aurintricarboxylic acid. *Eur. J. Biochem.* **177,** 91–96.

Nath, R., Raser, K. J., McGinnis, K., Nadimpalli, R., Stafford, D., and Wang, K. K. W. (1996a). Effects of ICE-like protease and calpain inhibitors on neuronal apoptosis. *Neuroreport* **8,** 249–255.

Nath, R., Raser, K. J., Stafford, D., Hajimohammadreza, I., Posner, A., Allen, H., Talanian, R. V., Yuen, P. W., Gilbertsen, R. B., and Wang, K. K. W. (1996b). Non-erythroid α-spectrin breakdown by calpain and interleukin 1β-converting-enzyme-like protease(s) in apoptotic cells: Contributory roles of both protease families in neuronal apoptosis. *Biochem. J.* **319,** 683–690.

Nguyen, H., Gitig, D. M., and Koff, A. (1999). Cell-free degradation of $p27^{kip1}$, a G_1 cyclin-dependent kinase inhibitor, is dependent on CDK2 activity and the proteasome. *Mol. Cell. Biol.* **19,** 1190–1201.

Ohguchi, M., Ishisaki, A., Okahashi, N., Koide, M., Koseki, T., Yamato, K., Noguchi, T., and Nishihara, T. (1998). *Actinobacillus actinomycetemcomitans* toxin induces both cell cycle arrest in the the G2/M phase and apoptosis. *Infect. Immun.* **66,** 5980–5987.

Ohtani-Kaneko, R., Takada, K., Iigo, M., Hara, M., Yokosawa, H., Kawashima, S., Ohkawa, K., and Hirata, K. (1998). Proteasome inhibitors which induce neurite outgrowth from PC12h cells cause different subcellular accumulations of multi-ubiquitin chains. *Neurochem. Res.* **23,** 1435–1443.

Ohtsu, M., Sakai, N., Fujita, H., Kashiwagi, M., Gasa, S., Shimizu, S., Eguchi, Y., Tsujimoto, Y., Sakiyama, Y., Kobayashi, K., and Kuzumaki, N. (1997). Inhibition of apoptosis by the actin-regulatory protein gelsolin. *EMBO J.* **16,** 4650–4656.

Okada, N., and Koizumi, S. (1997). Tyrosine phosphorylation of ErbB4 is stimulated by aurintricarboxylic acid in human neuroblastoma SH-SY5Y cells. *Biochem. Biophys. Res. Commun.* **230,** 266–269.

Okamura-Oho, Y., Zhang, S. Q., Callahan, J. W., Murata, M., Oshima, A., and Suzuki, Y. (1997). Maturation and degradation of β-galactosidase in the post-Golgi compartment are regulated by cathepsin B and a non-cysteine protease. *FEBS Lett.* **419,** 231–234.

Orford, K., Crockett, C., Jensen, J. P., Weissman, A. M., and Byers, S. W. (1997). Serine phosphorylation-regulated ubiquitination and degradation of β-catenin. *J. Biol. Chem.* **272,** 24735–24738.

Orian, A., Whiteside, S., Israël, A., Stancovski, I., Schwartz, A. L., and Ciechanover, A. (1995). Ubiquitin-mediated processing of NF-kB transcriptional activator precursor p105: Reconstitution of a cell-free system and identification of the ubiquitin-carrier protein, E2, and a novel ubiquitin-protein ligase, E3, involved in conjugation. *J. Biol. Chem.* **270,** 21707–21714.

Orlowski, R. Z., Eswara, J. R., Lafond-Walker, A., Grever, M. R., Orlowski, M., and Dang, C. V. (1998). Tumor growth inhibition induced in a murine model of human Burkitt's lymphoma by a proteasome inhibitor. *Cancer Res.* **58,** 4342–4348.

Orlowski, R. Z. (1999). The role of the ubiquitin-proteasome pathway in apoptosis. *Cell Death Differ.* **6,** 303–313.

Ostrowska, H., Wojcik, C., Omura, S., and Worowski, K. (1997). Lactacystin, a specific inhibitor of the proteasome, inhibits human platelet lysosomal cathepsin A-like enzyme. *Biochem. Biophys. Res. Commun.* **234,** 729–732.

Palombella, V. J., Conner, E. M., Fuseler, J. W., Destree, A., Davis, J. M., Laroux, F. S., Wolf, R. E., Huang, J. Q., Brand, S., Elliott, P. J., Lazarus, D., McCormack, T., Parent, L., Stein, R., Adams, J., and Grisham, R. B. (1998). Role of the proteasome and NF-kappaB in streptococcal cell wall-induced polyarthritis. *Proc. Natl. Acad. Sci. USA* **95,** 15671–15676.

Palombella, V. J., Rando, O. J., Goldberg, A. L., and Maniatis, T. (1994). The ubiquitin-proteasome pathway is required for processing the NF-kappaB1 precursor protein and the activation of NF-kappaB. *Cell* **78,** 773–785.

Pan, Y., and Haines, D. S. (1999). The pathway regulating MDM2 protein degradation can be altered in human leukemic cells. *Cancer Res.* **59,** 2064–2067.

Park, D. S., Stefanis, L., Yan, C. Y. I., Farinelli, S. E., and Greene, L. A. (1996). Ordering the cell death pathway: Differential effects of Bcl2, an interleukin-1-converting enzyme family protease inhibitor, and other survival agents on JNK activation in serum/nerge growth factor-deprived PC12 cells. *J. Biol. Chem.* **271,** 21898–21905.

Park, K. C., Woo, S. K., Yook, Y. J., Wyndham, A. M., Baker, R. T., and Chung, C. H. (1997). Purification and characterization of UBP6, a new ubiquitin-specific protease in *Saccharomyces cerevisiae. Arch. Biochem. Biophys.* **347,** 78–84.

Parkes, C., Kembhavi, A. A., and Barrett, A. J. (1985). Calpain inhibition by peptide epoxides. *Biochem. J.* **230,** 509–516.

Patel, T., Gores, G. J., and Kaufmann, S. H. (1996). The role of proteases during apoptosis. *FASEB J.* **10,** 587–597.

Peet, N. P., Kim, H. O., Marquart, A. L., Angelastro, M. R., Nieduzak, T. R., White, J. N., Friedrich, D., Flynn, G. A., Webster, M. E., Vaz, R. J., Linnik, M. D., Koehl, J. R., Mehdi, S., Bey, P., Emary, B., and Hwang, K. K. (1999). Hydroxyoxazolidines as α-aminoacetaldehyde equivalents: Novel inhibitors of calpain. *Bioorg. Med. Chem. Lett.* **9,** 2365–2370.

Peters, J.-M., Harris, J. R., and Finley, D. (eds.) (1998). "Ubiquitin and the Biology of the Cell." Plenum Press, New York.

Pike, B. R., Zhao, X. R., Newcomb, J. K., Posmantur, R. M., Wang, K. K. W., and Hayes, R. L. (1998a). Regional calpain and caspase-3 proteolysis of α-spectrin after traumatic brain injury. *Neuroreport* **9,** 2437–2442.

Pike, B. R., Zhao, X. R., Newcomb, J. K., Wang, K. K. W., Posmantur, R. M., and Hayes, R. L. (1998b). Temporal relationships between de novo protein synthesis, calpain and caspase 3-like protease activation, and DNA fragmentation during apoptosis in septo-hippocampal cultures. *J. Neurosci. Res.* **52,** 505–520.

Pinter, M., Stierandova, A., and Friedrich, P. (1992). Purification and characterization of a Ca^{2+}-activated thiol protease from *Drosophila melanogaster. Biochemistry* **31,** 8201–8206.

Pochampally, R., Fodera, B., Chen, L. H., Lu, W. G., and Chen, J. D. (1999). Activation of an MDM2-specific caspase by p53 in the absence of apoptosis. *J. Biol. Chem.* **274,** 15271–15277.

Pontremoli, S., Melloni, E., Viotti, P. L., Michetti, M., Salamino, F., and Horecker, B. L. (1991). Identification of two calpastatin forms in rat skeletal muscle and their susceptibility to digestion by homologous calpains. *Arch. Biochem. Biophys.* **288,** 646–652.

Posner, A., Raser, K. J., Hajimohammadreza, I., Yuen, P. W., and Wang, K. K. W. (1995). Aurintricarboxylic acid is an inhibitor of μ- and m-calpain. *Biochem. Mol. Biol. Int.* **36,** 291–299.

Potter, D. A., Tirnauer, J. S., Janssen, R., Croall, D. E., Hughes, C. N., Fiacco, K. A., Mier, J. W., Maki, M., and Herman, I. M. (1998). Calpain regulates actin remodeling during cell spreading. *J. Cell Biol.* **141,** 647–662.

Puri, R. N., Matsueda, R., Umeyama, H., Bradford, H. N., and Colman, R. W. (1993a). Modulation of thrombin-induced platelet aggregation by inhibition of calpain by a synthetic peptide derived from the thiol-protease inhibitory sequence of kininogens and *S*-(3-nitro-2-pyridinesulfenyl)-cysteine. *Eur. J. Biochem.* **214,** 233–241.

Puri, R. N., Matsueda, R., Umeyama, H., and Colman, R. W. (1993b). Specificity of the sequence in Phe-Gln-Val-Val-Cys(-3-nitro-2-pyridinesulfenyl)-Gly-NH_2: A selective inhibitor of thrombin-induced platelet aggregation. *Thromb. Res.* **72,** 183–191.

Pörn-Ares, M. I., Samali, A., and Orrenius, S. (1998). Cleavage of the calpain inhibitor, calpastatin, during apoptosis. *Cell Death Differ.* **5,** 1028–1033.

Rami, A., Ferger, D., and Krieglstein, J. (1997). Blockade of calpain proteolytic activity rescues neurons from glutamate excitotoxicity. *Neurosci. Res.* **27,** 93–97.

Rami, A., and Krieglstein, J. (1993). Protective effects of calpain inhibitors against neuronal damage caused by cytotoxic hypoxia in vitro and ischemia in vivo. *Brain Res.* **609,** 67–70.

Rano, T. A., Timkey, T., Peterson, E. P., Rotonda, J., Nicholson, D. W., Becker, J. W., Chapman, K. T., and Thornberry, N. A. (1997). A combinatorial approach for determining protease specificities: Application to interleukin-1 beta converting enzyme (ICE). *Chem. Biol.* **4,** 149–155.

Rao, S., Porter, D. C., Chen, X. M., Herliczek, T., Lowe, M., and Keyomarsi, K. (1999). Lovastatin-mediated G_1 arrest is through inhibition of the proteasome, independent of hydroxymethyl glutaryl-CoA reductase. *Proc. Natl. Acad. Sci. USA* **96,** 7797–7802.

Ravi, R., Bedi, A., and Fuchs, E. J. (1998). CD95 (Fas)-induced caspase-mediated proteolysis of NF-kappaB. *Cancer Res.* **58,** 882–886.

Reuther, J. Y., and Baldwin, A. S., Jr. (1999). Apoptosis promotes a caspase-induced aminoterminal truncation of IkappaBα that functions as a stable inhibitor of NF-kappaB. *J. Biol. Chem.* **274,** 20664–20670.

Rivett, A. J., Savory, P. J., and Djaballah, H. (1994). Multicatalytic endopeptidase complex: Proteasome. *Methods Enzymol.* **244,** 331–350.

Rock, K. L., Gramm, C., Rothstein, L., Clark, K., Stein, R., Dick, L., Hwang, D., and Goldberg, A. L. (1994). Inhibitors of the proteasome block the degradation of most cell proteins and the generation of peptides presented on MHC class I molecules. *Cell* **78,** 761–771.

Rosenbaum, D. M., D-Amore, J., Llena, J., Rybak, S., Balkany, A., and Kessler, J. A. (1998). Pretreatment with intraventricular aurintricarboxylic acid decreases infarct size by inhibiting apoptosis following transient global ischemia in gerbils. *Ann. Neurol.* **43,** 654–660.

Rosser, B. G., Powers, S. P., and Gores, G. J. (1993). Calpain activity increases in hepatocytes following addition of ATP: Demonstration by a novel fluorescent approach. *J. Biol. Chem.* **268,** 23593–23600.

Roush, W. R., Gwaltney, S. L., II, Cheng, J. M., Scheidt, K. A., McKerrow, J. H., and Hansell, E. (1998). Vinyl sulfonate esters and vinyl sulfonamides: Potent, irreversible inhibitors of cysteine proteases. *J. Am. Chem. Soc.* **120,** 10994–10995.

Rui, H., Xu, J., Mehta, S., Fang, H., Williams, J., Dong, F., and Grimley, P. M. (1998). Activation of the Jak2-Stat5 signaling pathway in Nb2 lymphoma cells by an anti-apoptotic agent, aurintricarboxylic acid. *J. Biol. Chem.* **273,** 28–32.

Saatman, K. E., Murai, H., Bartus, R. T., Smith, D. H., Hayward, N. J., Perri, B. R., and McIntosh, T. K. (1996). Calpain inhibitor AK295 attenuates motor and cognitive deficits following experimental brain injury in the rat. *Proc. Natl. Acad. Sci. USA* **93,** 3428–3433.

Sadoul, R., Fernandez, P. A., Quiquerez, A. L., Martinou, I., Maki, M., Schróter, M., Becherer, J. D., Irmler, M., Tschopp, J., and Martinou, J. C. (1996). Involvement of the proteasome in the programmed cell death of NGF-deprived sympathetic neurons. *EMBO J.* **15,** 3845–3852.

Saido, T. C., Kawashima, S., Tani, E., and Yokota, M. (1997). Up- and down-regulation of calpain inhibitor polypeptide, calpastatin, in postischemic hippocampus. *Neurosci. Lett.* **227,** 75–78.

Saitoh, Y., Kawahara, H., Miyamatsu, H., and Yokosawa, H. (1991). Comparative studies on proteasomes (multicatalytic proteinases) isolated from spermatozoa and eggs of sea urchins. *Comp. Biochem. Physiol. B* **99,** 71–76.

Salomon, D., Sacco, P. A., Roy, S. G., Simcha, I., Johnson, K. R., Wheelock, M. J., and Ben-Ze'ev, A. (1997). Regulation of β-catenin levels and localization by overexpression of plakoglobin and inhibition of the ubiquitin-proteasome system. *J. Cell Biol.* **139,** 1325–1335.

Sarin, A., Clerici, M., Blatt, S. P., Hendrix, C. W., Shearer, G. M., and Henkart, P. A. (1994). Inhibition of activation-induced programmed cell death and restoration of defective immune responses of HIV^+ donors by cysteine protease inhibitors. *J. Immunol.* **153,** 862–872.

Sasaki, T., Kikuchi, T., Fukui, I., and Murachi, T. (1986). Inactivation of calpain I and calpain II by specificity-oriented tripeptidyl chloromethyl ketones. *J. Biochem. (Tokyo)* **99,** 173–179.

Sasaki, T., Kishi, M., Saito, M., Tanaka, T., Higuchi, N., Kominami, E., Katunuma, N., and Murachi, T. (1990). Inhibitory effect of di- and tripeptidyl aldehydes on calpains and cathepsins. *J. Enz. Inhib.* **3,** 195–201.

Savory, P. J., Djaballah, H., Angliker, H., Shaw, E., and Rivett, A. J. (1993). Reaction of proteasomes with peptidylchloromethanes and peptidyldiazomethanes. *Biochem. J.* **296,** 601–605.

Savory, P. J., and Rivett, A. J. (1993). Leupeptin-binding site(s) in the mammalian multicatalytic proteinase complex. *Biochem. J.* **289,** 45–48.

Schauer, S. L., Bellas, R. E., and Sonenshein, G. E. (1998). Dominant signals leading to inhibitor kappaB protein degradation mediate CD40 ligand rescue of WEHI 231 immature B cells from receptor-mediated apoptosis. *J. Immunol.* **160,** 4398–4405.

Schirmeister, T. (1999). New peptidic cysteine protease inhibitors derived from the electrophilic α-amino acid aziridine-2,3-dicarboxylic acid. *J. Med. Chem.* **42,** 560–572.

Schnellmann, R. G., and Williams, S. W. (1998). Proteases in renal cell death: Calpains mediate cell death produced by diverse toxicants. *Ren. Fail.* **20,** 679–686.

Schoenwaelder, S. M., Kulkarni, S., Salem, H. H., Imajoh-Ohmi, S., Yamao-Harigaya, W., Saido, T. C., and Jackson, S. P. (1997). Distinct substrate specificities and functional roles for the 78- and 76-kDa forms of μ-calpain in human platelets. *J. Biol. Chem.* **272,** 24876–24884.

Schoenwaelder, S. M., and Burridge, K. (1999). Evidence for a calpeptin-sensitive protein-tyrosine phosphatase upstream of the small GTPase Rho: A novel role for the calpain inhibitor calpeptin in the inhibition of protein-tyrosine phosphatases. *J. Biol. Chem.* **274,** 14359–14367.

Shea, T. B., Spencer, M. J., Beermann, M. L., Cressman, C. M., and Nixon, R. A. (1996). Calcium influx into human neuroblastoma cells induces ALZ-50 immunoreactivity: Involvement of calpain-mediated hydrolysis of protein kinase C. *J. Neurochem.* **66,** 1539–1549.

Sherwood, S. W., Kung, A. L., Roitelman, J., Simoni, R. D., and Schimke, R. T. (1993). *In vivo* inhibition of cyclin B degradation and induction of cell-cycle arrest in mammalian cells by the neutral cysteine protease inhibitor *N*-acetylleucylleucylnorleucinal. *Proc. Natl. Acad. Sci. USA* **90,** 3353–3357.

Shimizu, T., and Pommier, Y. (1996). DNA fragmentation induced by protease activation in p53-null human leukemia HL60 cells undergoing apoptosis following treatment with topoisomerase I inhibitor camptothecin: Cell-free system studies. *Exp. Cell Res.* **226,** 292–301.

Shinohara, K., Tomioka, M., Nakano, H., Toné, S., Ito, H., and Kawashima, S. (1996). Apoptosis induction resulting from proteasome inhibition. *Biochem. J.* **317,** 385–388.

Siman, R., Card, J. P., and Davis, L. G. (1990). Proteolytic processing of β-amyloid precursor by calpain I. *J. Neurosci.* **10,** 2400–2411.

Sin, N., Kim, K. B., Elofsson, M., Meng, L. H., Auth, H., Kwok, B. H. B., and Crews, C. M. (1999). Total synthesis of the potent proteasome inhibitor epoxomicin: A useful tool for understanding proteasome biology. *Bioorg. Med. Chem. Lett.* **9,** 2283–2288.

Sindram, D., Kohli, V., Madden, J. F., and Clavien, P. A. (1999). Calpain inhibition prevents sinusoidal endothelial cell apoptosis in the cold ischemic rat liver. *Transplantation* **68,** 136–140.

Solary, E., Eymin, B., Droin, N., and Haugg, M. (1998). Proteases, proteolysis, and apoptosis. *Cell Biol. Toxicol.* **14,** 121–132.

Soldatenkov, V. A., and Dritschilo, A. (1997). Apoptosis of Ewing's sarcoma cells is accompanied by accumulation of ubiquitinated proteins. *Cancer Res.* **57,** 3881–3885.

Song, D.-K., Malmstrom, T., Kater, S. B., and Mykles, D. L. (1994). Calpain inhibitors block Ca^{2+}-induced suppression of neurite outgrowth in isolated hippocampal pyramidal neurons. *J. Neurosci. Res.* **39,** 474–481.

Song, Q., Mehler, M. F., and Kessler, J. A. (1998). Bone morphogenetic proteins induce apoptosis and growth factor dependence of cultured sympathoadrenal progenitor cells. *Dev. Biol.* **196,** 119–127.

Sorimachi, H., Ishiura, S., and Suzuki, K. (1993a). A novel tissue-specific calpain species expressed predominantly in the stomach comprises two alternative splicing products with and without Ca^{2+}-binding domain. *J. Biol. Chem.* **268,** 19476–19482.

Sorimachi, H., Toyama-Sorimachi, N., Saido, T. C., Kawasaki, H., Sugita, H., Miyasaka, M., Arahata, K., Ishiura, S., and Suzuki, K. (1993b). Muscle-specific calpain, p94, is degraded by autolysis immediately after translation, resulting in disappearance from muscle. *J. Biol. Chem.* **268,** 10593–10605.

Sorimachi, Y., Harada, K., Saido, T. C., Ono, T., Kawashima, S., and Yoshida, K. (1997). Downregulation of calpastatin in rat heart after brief ischemia and reperfusion. *J. Biochem. (Tokyo)* **122,** 743–748.

Spencer, M. J., Tidball, J. G., Anderson, L. V. B., Bushby, K. M. D., Harris, J. B., Passos-Bueno, M. R., Somer, H., Vainzof, M., and Zatz, M. (1997). Absence of calpain 3 in a form of limb-girdle muscular dystrophy (LGMD2A). *J. Neurol. Sci.* **146,** 173–178.

Spinedi, A., Oliverio, S., Di Sano, F., and Piacentini, M. (1998). Calpain involvement in calphostin C-induced apoptosis. *Biochem. Pharmacol.* **56,** 1489–1492.

Squier, M. K. T., Miller, A. C. K., Malkinson, A. M., and Cohen, J. J. (1994). Calpain activation in apoptosis. *J. Cell. Physiol.* **159,** 229–237.

Squier, M. K. T., and Cohen, J. J. (1997). Calpain, an upstream regulator of thymocyte apoptosis. *J. Immunol.* **158,** 3690–3697.

Squier, M. K. T., Sehnert, A. J., Sellins, K. S., Malkinson, A. M., Takano, E., and Cohen, J. J. (1999). Calpain and calpastatin regulate neutrophil apoptosis. *J. Cell. Physiol.* **178,** 311–319.

Stefanelli, C., Bonavita, F., Stanic, I., Pignatti, C., Farruggia, G., Masotti, L., Guarnieri, C., and Caldarera, C. M. (1998). Inhibition of etoposide-induced apoptosis with peptide aldehyde inhibitors of proteasome. *Biochem. J.* **332,** 661–665.

Stefanis, L., Park, D. S., Yan, C. Y. I., Farinelli, S. E., Troy, C. M., Shelanski, M. L., and Greene, L. A. (1996). Induction of CPP32-like activity in PC12 cells by withdrawal of trophic support: Dissociation from apoptosis. *J. Biol. Chem.* **271,** 30663–30671.

Steiner, H., Capell, A., Pesold, B., Citron, M., Kloetzel, P. M., Selkoe, D. J., Romig, H., Mendla, K., and Haass, C. (1998). Expression of Alzheimer's disease-associated presenilin-1 is controlled by proteolytic degradation and complex formation. *J. Biol. Chem.* **273,** 32322–32331.

Stempien Otero, A., Karson, A., Cornejo, C. J., Xiang, H., Eunson, T., Morrison, R. S., Kay, M., Winn, R., and Harlan, J. (1999). Mechanisms of hypoxia-induced endothelial cell death: Role of p53 in apoptosis. *J. Biol. Chem.* **274,** 8039–8045.

Stoklosa, T., Wójcik, C., Golab, J., Giermasz, A., and Wilk, S. (1999). Inhibition of proteasome, apoptosis and sensitization to tumour necrosis factor alpha: Do they always go together? *Br. J. Cancer* **79,** 375–376.

Sugita, H., Ishiura, S., Suzuki, K., and Imahori, K. (1980). Ca-activated neutral protease and its inhibitors: *In vitro* effect on intact myofibrils. *Muscle Nerve* **3,** 335–339.

Sun, X. M., MacFarlane, M., Zhuang, J. G., Wolf, B. B., Green, D. R., and Cohen, G. M. (1999). Distinct caspase cascades are initiated in receptor-mediated and chemical-induced apoptosis. *J. Biol. Chem.* **274,** 5053–5060.

Suzuki, K., Tsuji, S., and Ishiura, S. (1981). Effect of Ca^{2+} on the inhibition of calcium-activated neutral protease by leupeptin, antipain and epoxysuccinate derivatives. *FEBS Lett.* **136,** 119–122.

Suzuki, K. (1983). Reaction of calcium-activated neutral protease (CANP) with an epoxysuccinyl derivative (E64c) and iodoacetic acid. *J. Biochem. (Tokyo)* **93,** 1305–1312.

Taglialatela, G., Kaufman, J. A., Trevino, A., and Perez-Polo, J. R. (1998). Central nervous system DNA fragmentation induced by the inhibition of nuclear factor kappa B. *Neuroreport* **9,** 489–493.

Takata, T., Kudo, Y., Zhao, M., Ogawa, I., Miyauchi, M., Sato, S., Cheng, J., and Nikai, H. (1999). Reduced expression of $p27^{Kip1}$ protein in relation to salivary adenoid cystic carcinoma metastasis. *Cancer* **86,** 928–935.

Talanian, R. V., Quinlan, C., Trautz, S., Hackett, M. C., Mankovich, J. A., Banach, D., Ghayur, T., Brady, K. D., and Wong, W. W. (1997). Substrate specificity of caspase family proteases. *J. Biol. Chem.* **272,** 9677–9682.

Tan, S., Wood, M., and Maher, P. (1998). Oxidative stress induces a form of programmed cell death with characteristics of both apoptosis and necrosis in neuronal cells. *J. Neurochem.* **71,** 95–105.

Tanabe, F., Cui, S. H., and Ito, M. (1998). Ceramide promotes calpain-mediated proteolysis of protein kinase C β in murine polymorphonuclear leukocytes. *Biochem. Biophys. Res. Commun.* **242,** 129–133.

Tanaka, K., Ii, K., Ichihara, A., Waxman, L., and Goldberg, A. L. (1986). A high molecular weight protease in the cytosol of rat liver. I. Purification, enzymological properties, and tissue distribution. *J. Biol. Chem.* **261,** 15197–15203.

Tanimoto, Y., Onishi, Y., Hashimoto, S., and Kizaki, H. (1997). Peptidyl aldehyde inhibitors of proteasome induce apoptosis rapidly in mouse lymphoma RVC cells. *J. Biochem. (Tokyo)* **121,** 542–549.

Tao, M., Bihovsky, R., Wells, G. J., and Mallamo, J. P. (1998). Novel peptidyl phosphorus derivatives as inhibitors of human calpain I. *J. Med. Chem.* **41,** 3912–3916.

Thornberry, N. A., Rano, T. A., Peterson, E. P., Rasper, D. M., Timkey, T., Garcia-Calvo, M., Houtzager, V. M., Nordstrom, P. A., Roy, S., Vaillancourt, J. P., Chapman, K. T., and Nicholson, D. W. (1997). A combinatorial approach defines specificities of members of the caspase family and granzyme B. *J. Biol. Chem.* **272,** 17907–17911.

Tomoda, K., Kubota, Y., and Kato, J. (1999). Degradation of the cyclin-dependent-kinase inhibitor $p27^{Kip1}$ is instigated by Jab1. *Nature* **398,** 160–165.

Toyo-Oka, T., Kamashiro, T., Gotoh, Y., Fumino, H., Masaki, T., and Hosoda, S. (1986a). Temporary salvage of ischemic myocardium by the protease inhibitor bis[ethyl(2R,3R)-3-[(S)-methyl-1-[4-(2,3,4-trimethoxyphenyl-methyl) piperazin-1-ylcarbonyl]butyl-carbonyl] oxiran-2-carboxylate] sulfate. *Arzneimittel-Forschung* **36,** 671–675.

Toyo-Oka, T., Kamashiro, T., Hara, K., Nakamura, N., Kitahara, M., and Masaki, T. (1986b). Suppression of myocardial protein degradation by the protease inhibitor bis[ethyl(2R,3R)-3-[(S)-methyl-1-[4-(2,3,4-trimethoxyphenyl-methyl) piperazin-1-ylcarbonyl]butyl-carbonyl] oxiran-2-carboxylate] sulfate under hypoxia. *Arzneimittel-Forschung* **36,** 190–193.

Traenckner, E. B.-M., Wilk, S., and Baeuerle, P. A. (1994). A proteasome inhibitor prevents activation of NF-kappaB and stabilizes a newly phosphorylated form of IkappaB-α that is still bound to NF-kappaB. *EMBO J.* **13,** 5433–5441.

Traenckner, E. B.-M., Pahl, H. L., Henkel, T., Schmidt, K. N., Wilk, S., and Baeuerle, P. A. (1995). Phosphorylation of human IkappaB-α on serines 32 and 36 controls IkappaB-α proteolysis and NF-kappaB activation in response to diverse stimuli. *EMBO J.* **14,** 2876–2883.

Tripathy, R., Gu, Z. Q., Dunn, D., Senadhi, S. E., Ator, M. A., and Chatterjee, S. (1998). P_2-proline-derived inhibitors of calpain I. *Bioorg. Med. Chem. Lett.* **8,** 2647–2652.

Tsubuki, S., Saito, Y., Tomioka, M., Ito, H., and Kawashima, S. (1996). Differential inhibition of calpain and proteasome activities by peptidyl aldehydes of di-leucine and tri-leucine. *J. Biochem. (Tokyo)* **119,** 572–576.

Urbano, A., McCaffrey, R., and and Foss, F. (1998). Isolation and characterization of NUC70, a cytoplasmic, mematopoietic apoptotic endonuclease. *J. Biol. Chem.* **273,** 34820–34827.

Urthaler, F., Wolkowicz, P. E., Digerness, S. B., Harris, K. D., and Walker, A. A. (1997). MDL-28170, a membrane-permeant calpain inhibitor, attenuates stunning and PKC_ε proteolysis in reperfused ferret hearts. *Cardiovasc. Res.* **35,** 60–67.

Vanags, D. M., Pörn-Ares, M. I., Coppola, S., Burgess, D. H., and Orrenius, S. (1996). Protease involvement in fodrin cleavage and phosphatidylserine exposure in apoptosis. *J. Biol. Chem.* **271,** 31075–31085.

Verret, C., Poussard, S., Touyarot, K., Donger, C., Savart, M., Cottin, P., and Ducastaing, A. (1999). Degradation of protein kinase Mα by μ-calpain in a μ-calpain-protein kinase Cα complex. *Biochim. Biophys. Acta* **1430,** 141–148.

Villa, P., Kaufmann, S. H., and Earnshaw, W. C. (1997). Caspases and caspase inhibitors. *Trends Biochem. Sci.* **22,** 388–393.

Villa, P. G., Henzel, W. J., Sensenbrenner, M., Henderson, C. E., and Pettmann, B. (1998). Calpain inhibitors, but not caspase inhibitors, prevent actin proteolysis and DNA fragmentation during apoptosis. *J. Cell Sci.* **111,** 713–722.

Vinitsky, A., Michaud, C., Powers, J. C., and Orlowski, M. (1992). Inhibition of the chymotrypsin-like activity of the pituitary multicatalytic proteinase complex. *Biochemistry* **31,** 9421–9428.

Vinitsky, A., Cardozo, C., Sepp Lorenzino, L., Michaud, C., and Orlowski, M. (1994). Inhibition of the proteolytic activity of the multicatalytic proteinase complex (proteasome) by substrate-related peptidyl aldehydes. *J. Biol. Chem.* **269,** 29860–29866.

Vlach, J., Hennecke, S., and Amati, B. (1997). Phosphorylation-dependent degradation of the cyclin-dependent kinase inhibitor p27Kip1. *EMBO J.* **16,** 5334–5344.

Voges, D., Zwickl, P., and Baumeister, W. (1999). The 26S proteasome: A molecular machine designed for controlled proteolysis. *Annu. Rev. Biochem.* **68,** 1015–1068.

Wang, E. W., Kessler, B., Borodovsky, A., and Ploegh, H. L. (1999). Tight metabolic coupling of the proteasome with downstream oligopeptidases. *Mol. Biol. Cell* **10,** 90a.

Wang, K. K. W., Nath, R., Posner, A., Raser, K. J., Buroker-Kilgore, M., Hajimohammadreza, I., Probert, A. W., Jr., Marcoux, F. W., Ye, Q. H., Takano, E., Hatanaka, M., Maki, M., Caner, H., Collins, J. L., Fergus, A., Lee, K. S., Lunney, E. A., Hays, S. J., and Yuen, P. W. (1996a) An alpha-mercaptoacrylic acid derivative is a selective nonpeptide cell-permeable calpain inhibitor and is neuroprotective. *Proc. Natl. Acad. Sci. USA* **93,** 6687–6692.

Wang, K. K. W., Nath, R., Raser, K. J., and Hajimohammadreza, I. (1996b) Maitotoxin induces calpain activation in SH-SY5Y neuroblastoma cells and cerebrocortical cultures. *Arch. Biochem. Biophys.* **331,** 208–214.

Wang, K. K. W., Posner, A., Raser, K. J., Buroker-Kilgore, M., Nath, R., Hajimohammadreza, I., Probert, A. W., Marcoux, F. W., Lunney, E. A., Hays, S. J., and Yuen, P. W. (1996c) Alpha-mercaptoacrylic acid derivatives as novel selective calpain inhibitors. *Adv. Exp. Med. Biol.* **389,** 95–102.

Wang, K. K. W., Posmantur, R., Nadimpalli, R., Nath, R., Mohan, P., Nixon, R. A., Talanian, R. V., Keegan, M., Herzog, L., and Allen, H. (1998a) Caspase-mediated fragmentation of calpain inhibitor protein calpastatin during apoptosis. *Arch. Biochem. Biophys.* **356,** 187–196.

Wang, K. K. W., Posmantur, R., Nath, R., McGinnis, K., Whitton, M., Talanian, R. V., Glantz, S. B., and Morrow, J. S. (1998b) Simultaneous degradation of αII- and βII-spectrin by caspase 3 (CPP32) in apoptotic cells. *J. Biol. Chem.* **273,** 22490–22497.

Wang, K. K. W. (1999). Calpain substrates, assay methods, regulation, and its inhibitory agents. *In* "Calpain: Pharmacology and Toxicology of Calcium-Dependent Protease" (K. K. W. Wang and P.-W. Yuen, eds.), Taylor & Francis, Philadelphia. pp. 77–101.

Wang, K. K. W., and Yuen, P.-W. (eds.) (1999). "Calpain: Pharmacology and Toxicology of Calcium-Dependent Protease." Taylor & Francis, Philadelphia.

Wang, X., Luo, H. Y., Chen, H. F., Duguid, W., and Wu, J. P. (1998). Role of proteasomes in T cell activation and proliferation. *J. Immunol.* **160,** 788–801.

Ward, C., Chilvers, E. R., Lawson, M. F., Pryde, J. G., Fujihara, S., Farrow, S. N., Haslett, C., and Rossi, A. G. (1999). NF-κB activation is a critical regulator of human granulocyte apoptosis in *vitro*. *J. Biol. Chem.* **274,** 4309–4318.

Waterhouse, N. J., Finucane, D. N., Green, D. R., Elce, J. S., Kumar, S., Alnemri, E. S., Litwack, G., Khanna, K. K., Lavin, M. F., and Watters, D. J. (1998). Calpain activation is upstream of caspases in radiation-induced apoptosis. *Cell Death Differ.* **5,** 1051–1061.

Weideman, A., Paliga, K., Durrwang, U., Reinhard, F. B., Schuckert, O., Evin, G., and Masters, C. L. (1999). Proteolytic processing of the Alzheimer's disease amyloid precursor protein within its cytoplasmic domain by caspase-like proteases. *J. Biol. Chem.* **274,** 5823–5829.

Wells, G. J., and Bihovsky, R. (1998). Calpain inhibitors as potential treatment for stroke and other neurodegenerative diseases: Recent trends and developments. *Exp. Opin. Ther. Patents* **8,** 1707–1727.

Wen, L. P., Fahrni, J. A., Troie, S., Guan, J. L., Orth, K., and Rosen, G. D. (1997). Cleavage of focal adhesion kinase by caspases during apoptosis. *J. Biol. Chem.* **272,** 26056–26061.

Wilcox, D., and Mason, R. W. (1992). Inhibition of cysteine proteinases in lysosomes and whole cells. *Biochem. J.* **285,** 495–502.

Wilk, S., and Figueiredo-Pereira, M. E. (1993). Synthetic inhibitors of the multicatalytic proteinase complex (proteasome). *Enzyme Protein* **47,** 306–313.

Wilkinson, K. D. (1990). Detection and inhibition of ubiquitin-dependent proteolysis. *Methods Enzymol.* **185,** 387–397.

Wilkinson, K. D. (1997). Regulation of ubiquitin-dependent processes by deubiquitinating enzymes. *FASEB J.* **11,** 1245–1256.

Wójcik, C., Stoklosa, T., Giermasz, A., Golab, J., Zagozdzon, R., Kawiak, J., Wilk, S., Komar, A., Kaca, A., Malejczyk, J., and Jakóbisiak, M. (1997). Apoptosis induced in L1210 leukaemia cells by an inhibitor of the chymotrypsin-like activity of the proteasome. *Apoptosis* **2,** 455–462.

Wolf, B. B., Goldstein, J. C., Stennicke, H. R., Beere, H., Amarante-Mendes, G. P., Salvesen, G. S., and Green, D. R. (1999). Calpain functions in a caspase-independent manner to promote apoptosis-like events during platelet activation. *Blood* **94,** 1683–1692.

Wolf, B. B., and Green, D. R. (1999). Suicidal tendencies: Apoptotic cell death by caspase family proteinases. *J. Biol. Chem.* **274,** 20049–20052.

Wood, D. E., Thomas, A., Devi, L. A., Berman, Y., Beavis, R. C., Reed, J. C., and Newcomb, E. W. (1998). Bax cleavage is mediated by calpain during drug-induced apoptosis. *Oncogene* **17,** 1069–1078.

Wood, D. E., and Newcomb, E. W. (1999). Caspase-dependent activation of calpain during drug-induced apoptosis. *J. Biol. Chem.* **274,** 8309–8315.

Wu, L. W., Reid, S., Ritchie, A., Broxmeyer, H. E., and Donner, D. B. (1999). The proteasome regulates caspase-dependent and caspase-independent protease cascades during apoptosis of MO7e hematopoietic progenitor cells. *Blood Cells Mol. Dis.* **25,** 20–29.

Xie, H. Q., and Johnson, G. V. W. (1997). Ceramide selectively decreases tau levels in differentiated PC12 cells through modulation of calpain I. *J. Neurochem.* **69,** 1020–1030.

Xie, H. Q., and Johnson, G. V. W. (1998). Calcineurin inhibition prevents calpain-mediated proteolysis of tau in differentiated PC12 cells. *J. Neurosci. Res.* **53,** 153–164.

Yamaguchi, R., Maki, M., Hatanaka, M., and Sabe, H. (1994). Unphosphorylated and tyrosine-phosphorylated forms of a focal adhesion protein, paxillin, are substrates for calpain II in vitro: Implications for the possible involvement of calpain II in mitosis-specific degradation of paxillin. *FEBS Lett.* **356,** 114–116.

Yamashima, T., Kohda, Y., Tsuchiya, K., Ueno, T., Yamashita, J., Yoshioka, T., and Kominami, E. (1998). Inhibition of ischaemic hippocampal neuronal death in primates with cathepsin B inhibitor CA-074: A novel strategy for neuroprotection based on "calpain-cathepsin hypothesis." *Eur. J. Neurosci.* **10,** 1723–1733.

Yamazaki, T., Haass, C., Saido, T. C., Omura, S., and Ihara, Y. (1997). Specific increase in amyloid β-protein 42 secretion ratio by calpain inhibition. *Biochemistry* **36,** 8377–8383.

Yoshihara, Y., Ueda, H., Fujii, N., Shide, A., Yajima, H., and Satoh, M. (1990). Purification of a novel type of calcium-activated neutral protease from rat brain: Possible involvement in production of the neuropeptide kyotorphin from calpastatin fragments. *J. Biol. Chem.* **265,** 5809–5815.

Yuan, Y. P., Dopheide, S. M., Ivanidis, C., Salem, H. H., and Jackson, S. P. (1997). Calpain regulation of cytoskeletal signaling complexes in von Willebrand factor-stimulated platelets: Distinct roles for glycoprotein Ib-V-IX and glycoprotein IIb-IIIa (integrin $\alpha_{IIb}\beta_3$) in von Willebrand factor-induced signal transduction. *J. Biol. Chem.* **272,** 21847–21854.

Zhang, H., and Johnson, P. (1988). Inhibition of calpains by calmidazolium and calpastatin. *J. Enz. Inhib.* **2,** 163.

Zhang, J. L., Patel, J. M., and Block, E. R. (1998). Hypoxia-specific upregulation of calpain activity and gene expression in pulmonary artery endothelial cells. *Am. J. Physiol.* **275,** L461–L468.

Zhao, X., Pike, B. R., Newcomb, J. K., Wang, K. K. W., Posmantur, R. M., and Hayes, R. L. (1999). Maitotoxin induces calpain but not caspase-3 activation and necrotic cell death in primary septo-hippocampal cultures. *Neurochem. Res.* **24,** 371–382.

Zheng, J. Q., Felder, M., Connor, J. A., and Poo, M. (1994). Turning of nerve growth cones induced by neurotransmitters. *Nature* **368,** 140–144.

Zhu, W., Murtha, P. E., and Young, C. Y. F. (1995). Calpain inhibitor-induced apoptosis in human prostate adenocarcinoma cells. *Biochem. Biophys. Res. Commun.* **214,** 1130–1137.

CHAPTER 12

Identification and Analysis of Caspase Substrates: Proteolytic Cleavage of Poly(ADP-ribose)polymerase and DNA Fragmentation Factor 45

Claudia Boucher,* Stéphane Gobeil,*
Kumiko Samejima,† William C. Earnshaw,†
and Guy G. Poirier*

* Health and Environment Unit
Laval University Medical Research Center
CHUQ and Faculty of Medicine
Laval University, Quebec, Canada G1V 4G2

† Institute of Cell & Molecular Biology
University of Edinburgh
Scotland, United Kingdom EH9 3JR

METHODS IN CELL BIOLOGY, VOL. 66

0091-679X/01 $35.00

I. Introduction

During the last decade, apoptosis has emerged as an exciting area of research principally because of its pivotal involvement in many biological processes and systems. This evolutionary conserved form of cell suicide is characterized by a series of distinct morphological and biochemical changes (Kerr *et al.*, 1972; Wyllie, 1980; Wyllie *et al.*, 1980). These changes have been observed in almost all cell types, suggesting the presence of a common death pathway. Apoptosis is composed of three major steps: (1) decision to die, (2) execution, and (3) degradation (Mills *et al.*, 1999). This latter phase includes degradation of DNA and structural proteins.

Physiologic, biochemical, and genetic studies have suggested that a specialized proteolytic system is implicated in apoptosis and could play a critical and central role in the initiation and completion of this process. In 1986, Ellis and Horvitz (1986) demonstrated that the activity of specific genes was essential for initiating cell death in the nematode *Caenorhabditis elegans*. One of the essential genes was termed CED-3. In 1993, the CED-3 gene was cloned and sequence similarities were found with the mammalian protease interleukin-1β-converting enzyme (ICE) or caspase-1 (Yuan *et al.*, 1993). Caspases [cysteinyl aspartate-specific proteases (Alnemri *et al.*, 1996)] are a family of mammalian cysteine proteases that cleave their substrates after an aspartic acid residue (Earnshaw *et al.*, 1999). They are the central components of the proteolytic system implicated in apoptosis and are responsible for the apoptotic phenotype (Salvesen and Dixit, 1997; Wolf and Green, 1999). Currently, the caspase family is composed of at least 14 proteases (Medema *et al.*, 1999) that are implicated in the maturation of proinflammatory cytokines or in the process of apoptosis. Studies based on phylogenetic analysis have permitted the identification of three subfamilies of caspases: the ICE subfamily, comprising caspases-1, -4, -5, and -13 and the murine caspases-11 and -12; the CED-3 subfamily, which includes caspases -3, -6, -7, -8, -9, and -10; and a subfamily with only one member, caspase-2. Members of the ICE subfamily, especially caspases-1, -4, and -5, play a role in inflammation, whereas the other caspases of the CED-3 and caspase-2 subfamilies are involved in regulating apoptosis (Stroh and Schulze-Osthoff, 1998). During apoptosis, some caspases (i.e., caspases-8, -10) play a role in the initiation phase in response to proapoptotic signals, whereas others (i.e., caspases-3, -6, -7) function as apoptotic effectors and mediate cell disassembly.

Extensive work has been undertaken to identify caspase substrates and to elucidate the consequences of proteolytic cleavage on the process of apoptosis. As a result of caspase cleavage, caspase substrates may be functionally activated or inactivated. Moreover, for some substrates caspase cleavage can cause an alteration of their functions (Villa *et al.*, 1997). More than 60 proteins have been demonstrated to be caspase substrates, and this number is increasing rapidly. For an extensive list of caspase substrates, please refer to the review of Earnshaw and associates (1999).

This chapter details specific techniques that can be used to identify caspase substrates. Historically, random screening for *in vivo* and *in vitro* cleavage of proteins known or suspected to be implicated in apoptosis was routinely used. Sequencing of proteins processed during apoptosis was used to identify putative caspase cleavage sites. Recently, more targeted assays for caspase substrates have been employed, such as the yeast two- and three-hybrid system to screen for new caspase substrates (Kamada *et al.,* 1998; Van Criekinge *et al.,* 1998). The first section of this chapter focuses on protocols that are used to characterize the cleavage of poly(ADP-ribose)polymerase (PARP-1) during apoptosis. The second section describes the techniques used to determine the cleavage of DNA fragmentation factor 45 (DFF45/ICAD).

II. Poly(ADP-ribose)polymerase (PARP-1)

Poly(ADP-ribose)polymerase (PARP-1) is a nuclear enzyme that catalyzes the transfer of ADP-ribose polymers onto itself and other nuclear proteins in response to DNA strand breaks (D'Amours *et al.,* 1999). PARP-1 is a protein of 113 kDa and is cleaved to produce fragments of 89 and 24 kDa in a wide variety of cells undergoing apoptosis (Kaufmann, 1989; Kaufmann *et al.* 1993), and as a result this cleavage pattern has become a hallmark of apoptosis. In 1994, Lazebnik and associates found an enzymatic activity that cleaved bovine PARP-1 at the sequence $DEVD_{216}G$. This enzyme showed a similar but distinct substrate specificity from caspase-1 (Lazebnik *et al.,* 1994). Subsequently, that enzyme was identified as CPP32/Yama/apopain, now referred to as caspase-3 (Nicholson *et al.,* 1995; Tewari *et al.,* 1995). It was also demonstrated that PARP-1 could be a substrate for caspase-7 and that caspase-7 has a much higher affinity than caspase-3 for PARP-1 (Takahashi *et al.,* 1997). More recently, automodified PARP-1 appears to be preferentially cleaved by caspase-7 (Germain *et al.,* 1999).

A. Endogenous Cleavage during Apoptosis

1. Preparation of Cell and Tissue Extracts

a. Cell Extracts

1. After treatment, cells (approximately 10^6 cells) are harvested, centrifuged at 250*g*, 4°C, for 10 min. For example, 10^6 HL-60 human promyelocytic leukemia cells are treated with 68 μ*M* etoposide (Sigma) for 0 to 12 h and then harvested.

2. The pellet is rinsed in 1 ml of ice-cold phosphate-buffered saline (PBS) or Hepes buffer, pH 7.4.

3. An aliquot of 50 μl is kept for protein determination.

4. The pellet is lysed in a known volume of 1x reducing loading buffer [62.5 m*M* Tris–HCl, pH 6.8; 6 *M* deionized urea; 10% glycerol; 2% sodium dodecyl sulfate (SDS); 0.3% bromphenol blue; 5% β-mercaptoethanol (freshly added)] to have a recommended ratio of cells/lysis buffer of 5000 cells/μl.

5. Finally, the cell extract is sonicated on ice 2 × 20 s, microtips position 3 (Sonicator 550 Sonic Dismembrator, Fisher Scientific) to break the DNA.

6. Samples may be stored at −30°C or −80°C or directly incubated at 65°C for 15 min before analysis on a SDS–PAGE gel.

b. Tissue Extracts

1. Fresh tissue is homogenized 10–20 strokes on ice in an ice-cold extraction buffer [50 m*M* glucose; 25 m*M* Tris–HCl, pH 8.0; 10 m*M* EDTA; 1 m*M* phenylmethylsulfonyl

fluoride (PMSF) and 1x protease inhibitor cocktail tablets (Boehringer Mannheim)] with a glass homogenizer with a Teflon pestle.

2. An aliquot of 50 μl is kept for subsequent protein determination.

3. The sample is diluted in 0.5 volume of 2x reducing loading buffer and adjusted to have a concentration of 2.5–5 mg/ml.

4. Alternatively, frozen tissue can be pulverized in liquid nitrogen using a mortar and pestle and then resuspended in extraction buffer and diluted in 2x reducing loading buffer.

2. Western Blot Procedure

a. Electrophoresis and Transfer

1. Fifty thousand to 100,000 cells or 20 μg of protein is loaded on an 8 or 10% polyacrylamide gel.

2. The gel is run with an electrophoresis apparatus (Mini-PROTEAN II, BioRad) in running buffer (25 m*M* Tris; 192 m*M* glycine; 0.1% SDS) at 100 V (1 h and 15 min) or 200 V (45 min) until bromphenol blue runs off the gel.

3. Proteins are then transferred to a nitrocellulose membrane in a transblot cell apparatus (Bio-Rad) with precooled transfer buffer (25 m*M* Tris; 192 m*M* glycine; 20% methanol). Transfer is performed at 4°C with stirring at 100 V for 1 h or 35 V overnight. Alternatively, a semidry blotting apparatus can be used according to the manufacturer's recommendations.

4. The membrane can be stained for 1 min with Ponceau S [0.1% Ponceau S (w/v); 5% acetic acid (v/v)] and then washed extensively in deionized water. The position of the molecular weight standards can be identified and the protein profile of the samples visualized.

b. Immunoblotting

1. The nitrocellulose membrane is soaked in PBSMT [PBS 1X, pH 7.2; 5% nonfat powdered milk (w/v); 0.1% Tween 20] for 1 h.

2. The primary antibody is added. C-2-10 (Biomol) is a monoclonal antibody that recognizes the N-terminal region of PARP-1, which is not present in the other PARPs (Jacobson and Jacobson, 1999). It is diluted 1/10000 in PBSMT and is usually incubated with membrane overnight or for a minimum of 2 h. This antibody can be reused (up to five times) if 1 m*M* sodium azide is added and if the antibody solution is kept at 4°C. In mouse tissue where blood contamination is frequent, the use of an antipeptide antibody or rabbit polyclonal such as 419 or 422 (Biomol) is recommended (Fig. 1) (Duriez *et al.*, 1997).

3. The membrane is washed 3 × 10 min in PBSMT.

4. The secondary antibody, antimouse coupled to horseradish peroxidase (Jackson Immuno Research Laboratories) (at the initial concentration of 0.8 mg/ml), is added at a dilution of 1/2500 for 30 min.

5. The blot is then washed 3 × 10 min in PBSMT and 3 × 10 min in PBS 1X.

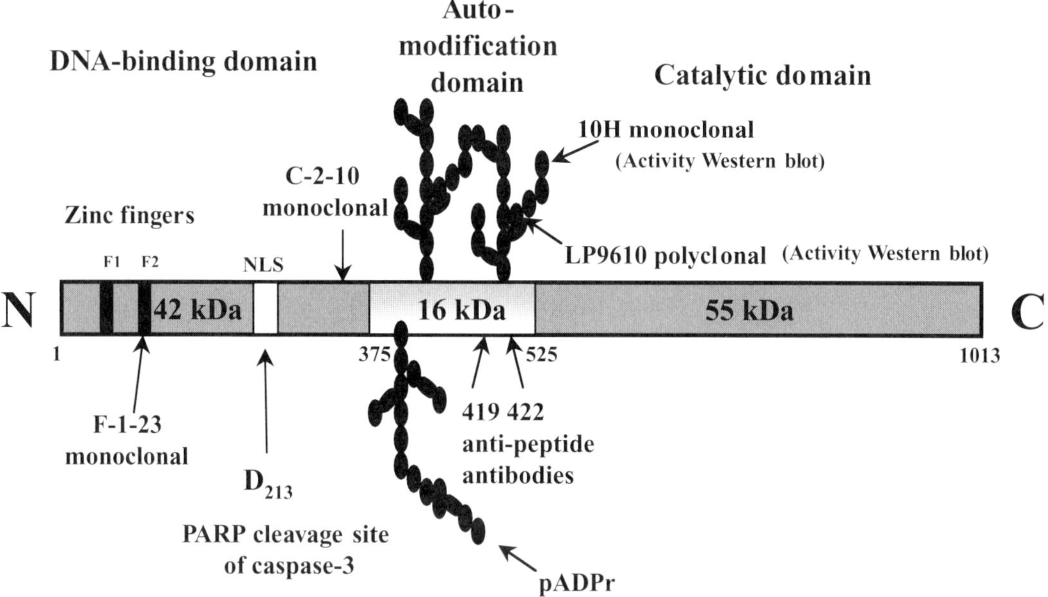

Fig. 1 Schematic representation of automodified human PARP-1 and antibodies mapping. PARP-1 is composed of three different domains: the DNA-binding domain (DBD) of 42 kDa, the automodification domain of 16 kDa, and the 55-kDa catalytic domain. These domains have been determined by papain and chymotrypsin digestion (Kameshita *et al.,* 1984; Shizuta *et al.,* 1986). The DBD is composed of two zinc fingers and a bipartite nuclear localization signal (NLS). Two monoclonal antibodies are available for the N-terminal part of PARP-1: the F-1-23 and the C-2-10 antibodies. For detection of the automodification domain, two antipeptide antibodies may be used: 419 and 422 (Duriez *et al.,* 1997). Identification of the intact catalytic domain can be done by the activity Western blot method using two different anti-pADPr antibodies: the 10H monoclonal and the LP9610 polyclonal antibodies. Cleavage of PARP-1 by caspase-3 occurs at position D_{213} in the NLS for human PARP-1 and at D_{216} for bovine PARP-1.

6. Proteins are detected by the chemiluminescence system from NEN Life Sciences. When more sensitivity is needed, a minimum of 2500–5000 cells can be used in conjunction with a kit for enhancing blocking and detection, such as SuperBlock blocking buffer (Pierce) and SuperSignal West Dura Extended Duration Substrate system (Pierce).

7. Other antibodies against PARP-1 are available such as LP96-72 (a polyclonal antibody raised against affinity-purified PARP; Biomol) or two new antibodies against the 89-kDa fragments: the anti-PARP p85 fragment pAB (Promega) and the cleaved PARP (89 kDa) antibody (New England BioLab). In all cases, however, it will be necessary to test the conditions of immunoblotting (e.g., the dilution of primary and secondary antisera and system of detection) to have optimal results.

c. Detection of Low Levels of PARP-1

When extracts from limited quantities of cells are used for Western blotting, we suggest the use of the SuperBlock blocking buffer (Pierce) for blocking. All other steps should be performed according to the company's recommendations. Attention must be paid to the dilution of primary and secondary antisera. For signal detection, we recommend using the SuperSignal West Dura Extended Duration Substrate system from Pierce.

d. Stripping and Reprobing the Blot

1. After detection of proteins by chemiluminescence, blots can be wrapped in Saran Wrap and stored at 4°C indefinitely or can be stripped and reblotted with the same or with another antibody.

2. Membranes are washed 4 × 10 min in PBST (PBS 1X; 0.1% Tween 20) and soaked for 30 min at 65°C in stripping buffer (62.5 m*M* Tris–HCl, pH 6.8; 2% SDS; 100 m*M* freshly added β-mercaptoethanol), followed by 6 × 5 min washes in PBST.

3. The membrane is now ready to be blocked for 1 h with PBSMT as for a regular Western blot. Complete stripping of the blot can be verified by incubating the membrane with the secondary antibody followed by the chemiluminescence assay.

3. Activity Western Blot

This is a very useful technique for the detection of poly(ADP-ribose) (pADPr) synthesized by PARP-1. This method enables the detection of PARP-1 in many species because pADPr remains the same no matter how the primary structure of PARP-1 varies. The activity Western blot also helps monitor the integrity of the catalytic domain of the polymerase. This technique is divided in four major steps: (1) reduction of the oxidized cysteines in the gel, (2) renaturation of proteins resolved on nitrocellulose membrane, (3) synthesis of polymer by PARP-1 present on the membrane, and (4) detection of pADPr by anti-pADPr antibodies (see Fig. 2 for description and comparison with Western blot).

a. Preparation of Cellular and Tissue Extracts

Cell extracts are processed as described for the Western blot (Section II,A,1).

b. Electrophoresis and Transfer

1. Electrophoresis is carried out as for normal Western blot (as described earlier).

2. After electrophoresis, the gel is soaked at 37°C for 1 h in running buffer containing 0.7 *M* β-mercaptoethanol freshly added. This treatment is necessary to avoid cross-linking between proteins and reduction of the cysteines present in PARP-1.

3. Proteins are transferred to a nitrocellulose membrane as for a classic Western blot (described earlier).

4. Prestained or biotinylated molecular weight standards are used to avoid staining under acidic conditions, which may destroy the polymerase activity.

c. Renaturation and Polymer Synthesis

1. The membrane is incubated for 1 h at room temperature in the renaturation buffer [50 m*M* Tris–HCl, pH 8.0; 100 m*M* NaCl; 1 m*M* dithiothreitol (DTT); 0.3% Tween 20] plus 2 m*M* $MgCl_2$. This step is important because it allows the restoration of the active conformation of PARP-1.

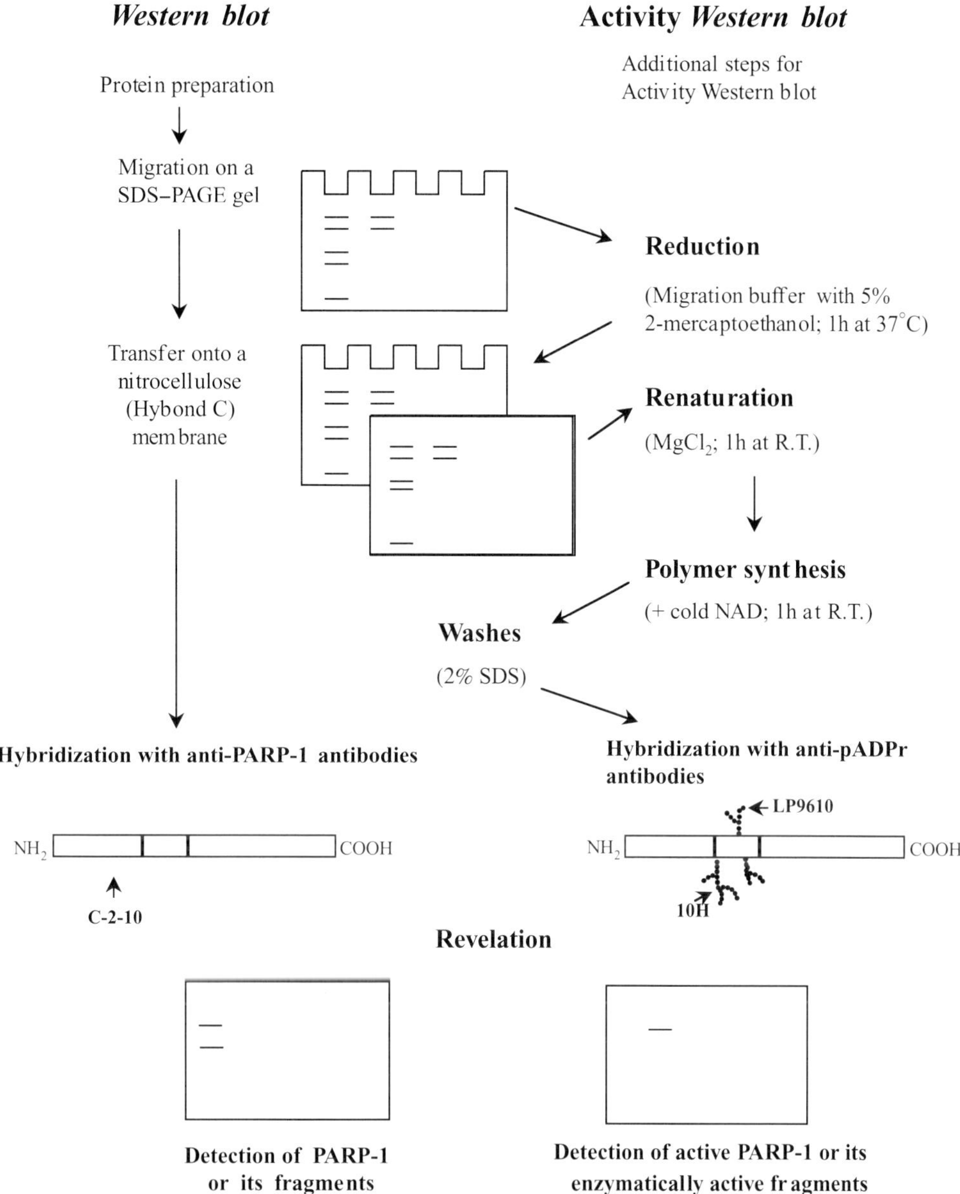

Fig. 2 Experimental differences between Western blot and activity Western blot. Schematic representation of the main steps describing the classical procedure for Western blot and the additional steps required for the activity Western blot of PARP-1. In brief, after resolving the proteins on a SDS–PAGE, the proteins in the gel are reduced in running buffer containing 5% β-mercaptoethanol. The gel transfer is performed as described in Section II,A,2, but the membrane for the activity Western blot is processed for the renaturation and polymer synthesis steps. Following washes in 2% SDS for the activity Western blot membrane, both membranes are submitted to the regular Western blotting protocol, differing only in the choice of antibody used. The classical Western blot is done using anti-PARP-1 antibodies such as C-2-10, F-1-23, 419, 422, or LP9672, whereas the membrane for the activity Western blot is incubated with 10H or LP9610.

2. The membrane is soaked for 1 h in the renaturation buffer containing 2 m*M* $MgCl_2$ and 100 μ*M* nicotinamide adenine dinucleotide (NAD). This is the step where pADPr is synthesized.

3. NAD is washed out by soaking the membrane in renaturation buffer 4 × 15 min.

4. The membrane is washed 4 × 15 min in SDS buffer (50 m*M* Tris–HCl, pH 8.0; 100 m*M* NaCl; 1 m*M* DTT; 2% SDS) to remove noncovalently bound pADPr.

d. Western Blotting

Western blotting using anti-pADPr antibodies is carried out as for a standard Western blot. Two anti-pADPr antibodies are currently available: mouse monoclonal 10H antibody (Serotec) (Kawamitsu *et al.,* 1984) and rabbit polyclonal LP9610 antibody (Biomol) (Affar *et al.,* 1999). Figure 3 shows results of the cleavage of PARP-1 during apoptosis with a Western blot and an activity Western blot.

B. *In Vitro* Cleavage of PARP-1

Once PARP-1 cleavage during apoptosis has been demonstrated in the system being analyzed, it is possible to identify the specific caspases that mediate this cleavage using purified substrate and caspases in an *in vitro* system. Cell-free extracts containing the caspase activity that cleaves the protein could also be used. Use of purified caspases or

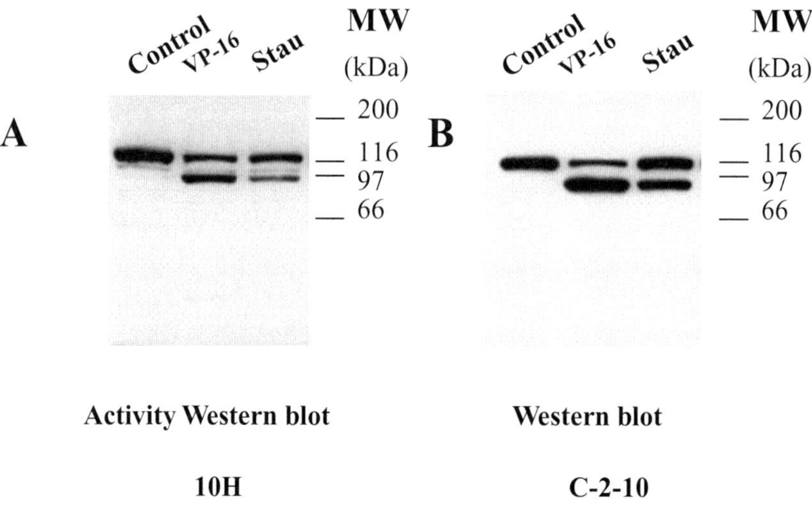

Fig. 3 Western blot and activity Western blot of PARP-1 cleavage during apoptosis. (A) Activity Western blot of PARP-1 cleavage in human T Jurkat cells treated with 150 μ*M* etoposide and 150 n*M* staurosporine for 6 h. The activity Western blot was performed with 20 μg of protein extract and incubated with the anti-pADPr monoclonal 10H antibody. Jurkat cells were grown according to ATCC instructions. (B) Western blot of the same membrane after the stripping protocol described earlier. Incubation of the Western blot was performed with the anti-PARP-1 monoclonal C-2-10 antibody. VP-16, etoposide; Stau, staurosporine.

subfamily-specific inhibitors facilitates the identification of the relevant protease. The caspase activity in the cell extracts can be analyzed by measuring the hydrolysis of commercially available fluorometric or colorimetric synthetic substrates (Enari *et al.,* 1996; Gurtu *et al.,* 1997; Martins *et al.,* 1997).

To study PARP-1 cleavage *in vitro,* we will take advantage of the ability of PARP-1 to synthesize poly(ADP-ribose) from NAD in response to DNA strand breaks on itself to produce ^{32}P-automodified PARP-1. Alternatively, we can label PARP-1 with [^{35}S]methionine.

1. Automodification Reaction of PARP-1

Bovine thymus PARP-1 (Biomol) is purified to homogeneity as described by Zahradka and Ebisuzaki (1984) with the modifications reported by Huletsky *et al.* (1989). We use 6 ng of PARP-1 μl of automodification reaction.

a. Automodification Buffer

50 m*M* Hepes, pH 7.4; 100 m*M* NaCl; 10% sucrose; 10 m*M* DTT; 6 μg/ml activated calf thymus DNA (Sigma); 200 μ*M* NAD; 1 n*M* [^{32}P]NAD (30 Ci/mmol) (NEN Life Sciences).

b. Automodification Reaction

1. Prepare the automodification buffer without the PARP-1 (e.g., 25 μl).
2. Preincubate for 5 min at 25°C.
3. Add the PARP-1 and incubate for 1 min at 25°C (6 ng of PARP-1μl of automodification buffer).
4. Stop the reaction by adding 1,5-dihydroxyisoquinoline (DHQ) (Sigma-Aldrich), a PARP-1 inhibitor, to reach a final concentration of 100 μM. Alternatively, 3-aminobenzamide can be used (1 m*M* final concentration).

An alternative method for generating labeled PARP-1 is to perform a coupled *in vitro* transcription/translation of PARP-1 cDNA using the TNT-coupled reticulocyte lysate systems (Promega) in the presence of [^{35}S]methionine. This approach avoids the use of radioactive [^{32}P]NAD, which can result in increased background compared to [^{35}S]methionine in the assay described later.

2. Preparation of Apoptotic Cell Extracts

1. After apoptotic treatments, 10^8 cells are harvested and centrifuged at 250*g,* 4°C, for 10 min.
2. The pellet is rinsed with ice-cold PBS or Hepes buffer, pH 7.4.
3. Cells are lysed in 1 ml of ice-cold hypotonic buffer [25 m*M* Hepes, pH 7.5; 1 m*M* EGTA; 5 m*M* $MgCl_2$; 0.1% Triton X-100; 250 μ*M* PMSF; 2 m*M* DTT; and 1x protease inhibitor cocktail tablets (Boehringer Mannheim)].
4. The cells are homogenized by 10–20 strokes in the lysis buffer with a glass homogenizer with a Teflon pestle (set aside an aliquot for subsequent protein determination).

5. The homogenate is centrifuged at 15,000*g*, 4°C, for 10 min.
6. Aliquots of 50 μl of the supernatant are frozen at −30°C or −80°C.

3. *In Vitro* Cleavage of PARP-1 with Purified Caspases or Apoptotic Cell Extracts

1. With automodified PARP-1, add 0.1% CHAPS (3-[3-cholamidopropyl) dimethylammonio]-1-propanesulfonate) to the automodification reaction and 1 unit of the recombinant caspase (−3 or −7) or the apoptotic cell extracts (30 ng/μl). For the cell extracts, add at least five times the concentration of PARP-1 to the assay.
2. Incubate at 37°C for different periods of time, typically ranging from 0 to 6 h. For incubation times of more than 3 h, 2 m*M* DTT must be readded every 3 h.
3. To stop the reaction, add 1x reducing loading buffer and heat the sample for 15 min at 65°C. The sample can be frozen at −30°C or analyzed immediately.
4. Analyze potential cleavage products by SDS–PAGE (as described earlier) and detect via autoradiography. Adjustments concerning the quantity of the caspase or the cell extract may be necessary depending on their activity. Different periods of incubation must also be tested to achieve optimal cleavage.
5. When using [^{35}S]PARP-1, add 4 μl (see translation procedure of Promega) of the translation reaction to 12.5 μl of reaction buffer (50 m*M* Hepes, pH 7.4; 100 m*M* NaCl; 10% sucrose; 10 m*M* EDTA; 2 m*M* DTT; and 0.1% CHAPS) containing the recombinant caspase or the apoptotic cell extract. Incubation and analysis of the digestion is done as for automodified PARP-1.

4. *In Vitro* Cleavage Inhibition Assay

Various caspase inhibitors are available to inhibit the cleavage of PARP-1 *in vitro*. The most commonly used inhibitors are the tetrapeptide inhibitor DEVD-CHO, Z-VAD, or YVAD-CHO. DEVD-CHO (acetyl-Asp-Glu-Val-Asp-aldehyde) is a caspase-3-like protease inhibitor and must be used at a concentration of 100 n*M* (Nicholson *et al.,* 1995). Z-VAD-fmk (*N*-benzyloxycarbonyl-Val-Ala-Asp(*O*-methyl)-fluoromethyl-ketone) is a broad-spectrum caspase inhibitor and abrogates etoposide-induced apoptosis in HL-60 cells at 50 μ*M* (Shao *et al.,* 1997). YVAD-CHO [*N*-(*N*-acetyl-tyrosinyl-valinyl-alaninyl)-3-amino-4-oxobutanoic acid] is a caspase-1 inhibitor, but studies have shown that it can also inhibit PARP-1 cleavage but at relatively high concentrations (1 μ*M*) (Nicholson *et al.,* 1995).

Typically, caspase inhibitors at the appropriate concentration are added to the cell extracts or to the recombinant caspases and preincubated for 30 min at room temperature. Then, it is added to the reaction mixture of *in vitro* cleavage of PARP-1 and processed as described earlier.

5. Detection of DEVDase Activity in Apoptotic Cell Extract

The measurement of DEVDase activity can be used to verify the presence of activated caspases in the cell extracts described earlier. This assay is based on the consensus cleavage site DEVD conjugated to a fluorescent molecule, 7-amino-4-trifluoromethyl coumarin (AFC), or a colorimetric molecule, *p*-nitroanilide (pNA) (Gurtu *et al.*, 1997). We routinely use the DEVD–pNA substrate because of its greater convenience.

1. Prepare a standard curve of pNA for 0 to 10,000 pmol in 25 m*M* Hepes, pH 7.5; 1 m*M* EGTA; 5 m*M* $MgCl_2$; 0.1% Triton X-100; 250 μ*M* PMSF; 2 m*M* DTT; and 1x protease inhibitor cocktail tablets (Boehringer Mannheim).
2. To confirm the specificity of the caspase activity being measured, add the DEVD-CHO inhibitor to a final concentration of 100 n*M* to one sample of the apoptotic cell extract (50 μg) and preincubate it for 30 min at room temperature.
3. Incubate each of the known standards prepared in step 1 and 50 μg of the apoptotic cell extract in 25 m*M* Hepes, pH 7.5; 1 m*M* EGTA; 5 m*M* $MgCl_2$; 0.1% Triton X-100; 250 μ*M* PMSF; 2 m*M* DTT; and 1x protease inhibitor cocktail tablets (Boehringer Mannheim) in a 96-well plate.
4. Start the reaction by adding the DEVD–pNA substrate at a final concentration of 50 μ*M*.
5. Incubate at 37°C and read at 405 nm on a spectrophotometer every 30 min.
6. The percentage increase of OD values in the treated sample relative to the control will indicate the percentage increase of the DEVDase activity following apoptotic treatment. Alternatively, the slope of the curve represents the specific activity of the activated caspase for substrate DEVD–pNa.

C. Characterization of Cleavage

The first step in identifying cleavage sites could be the use of Western blot and activity Western blot (antibody mapping), but the requirement for good antibodies directed against different parts of the protein must be considered. In the case of PARP-1, different antibodies are available to do that mapping (see Fig. 1). Moreover, necrotic cleavage of PARP-1 with antipeptide antibodies against specific regions of the PARP-1 sequence was performed in our laboratory (Sallmann *et al.*, 1997). For this analysis, cells are treated with apoptotic agents and analyzed by Western and activity Western blots. By comparing the results obtained with the different antibodies and with the analysis of the apparent molecular weights of the major fragments, it was possible to locate the cleavage site.

1. Protein Sequencing of the Proteolysis Fragment of PARP-1

While the different antibodies described in Section II,A,1 and Fig. 1 can be used to map general regions of PARP-1 cleavage, a more precise approach is to sequence the fragments generated by *in vitro* cleavage with recombinant caspases. Here, the availability of a sequencing service must be considered and the experimental conditions that result in a complete hydrolysis of the protein with the purified caspase must be determined. A minimum of 10–50 pmol of each fragment is necessary for a good sequencing reaction. The protocol used for the N-terminal sequencing by automatic Edman degradation performed on an Applied Biosystems Model 473A pulsed liquid protein sequencer is outlined.

1. After the *in vitro* cleavage reaction (see Section II,B,3), add 1x reducing loading buffer containing 5% β-mercaptoethanol freshly added and heat the sample for 15 min at 65°C.

2. Resolve the proteins on a 8 to 10% (depending on the size of the fragment to isolate) acrylamide SDS–PAGE gel and transfer them to a ProBlott membrane (Applied Biosystems) in 10 m*M* CAPS (3-[cyclohexylamino]-1-propanesulfonic acid) buffer, pH 11, containing 10% methanol.

3. To identify the fragment of interest, stain the membrane with Coomassie blue (0.25% Coomassie brilliant blue R-250; 50% methanol; 10% acetic acid) as described by Harlow and Lane (1999). Cut the band out and process it for sequencing by automatic Edman degradation.

2. Analysis of the Cleavage Site by Site-Directed Mutagenesis

Another approach to identify cleavage sites is the use of site-directed mutagenesis. Using the *in vitro* cleavage assay described earlier, the family of caspases responsible for the hydrolysis of the putative substrate should be known. Because the distinct cleavage specificities of the various caspases are known (Rano *et al.,* 1997; Thornberry *et al.,* 1997; Earnshaw *et al.,* 1999), the identity of the unknown caspase activity in cell extracts can be inferred. By using the protein sequencing data base, it will be possible to find the consensus amino acid sequence region that corresponds to these sequences and will fit with the apparent molecular weight fragment pattern obtained by the Western blot. We found in our laboratory that the QuikChange kit of Stratagene gives good and reliable results for site-directed mutagenesis [which should change the critical D (aspartate) residue adjacent to the cleavage site to either A (alanine) or E (glutamic acid)]. It should be noted that the cDNA of the protein must be available and that all the mutations done must be confirmed by DNA sequencing. Once the mutated cDNAs are translated with the TNT reticulocyte lysate from Promega, an *in vitro* cleavage assay with the specific caspase will confirm which are the caspase cleavage sites in the protein.

III. DNA Fragmentation Factor II (DFF45)

Cleavage of chromatin at internucleosomal sites was first reported in dying cells by the Czech radiobiologist Miroslav Skalka (Skalka *et al.,* 1976). Soon after, a "chromatin ladder" pattern of regulatory sized DNA fragments was recognized as characteristic of apoptosis (Wyllie, 1980). Despite intensive effort, the identities of the DNase involved in apoptosis were unknown for many years. Cytosolic fractions from apoptotic cells contain a factor(s) that induces DNA degradation in isolated nuclei (Lazebnik *et al.,* 1993; Newmeyer *et al.,* 1994; Enari *et al.,* 1995), suggesting that a nuclease or its activating molecule(s) is generated in apoptotic cells. Several candidates for the nuclease were reported, but their properties were diverse enough to preclude identity with one another. The enzyme variously called caspase-activated DNase (CAD)/CPAN (caspase-activated nuclease)/DNA fragmentation factor 40 (DFF40) is the first confirmed apoptotic nuclease (Enari *et al.*, 1998; Halenbeck *et al.*, 1998).

Cytosolic extracts from growing cells contain an inhibitory factor that blocks DNA fragmentation. However, treatment of extracts with caspase-3 or with dATP and cytochrome *c* converts them to apoptosis-inducing extracts, suggesting that living cells carry an inactive proform of the molecules(s) that induces DNA degradation (Liu *et al.*, 1996). A protein called inhibitor of CAD (ICAD)/DFF45 was purified from growing cells. ICAD has two forms, ICAD-L and ICAD-S (also called DFF45 and DFF35) (Gu *et al.*, 1999), arising by alternative splicing of a single transcript. Both isoforms have inhibitory activity; however, only ICAD-L/DFF45 seems to make a complex with CAD in living cells. ICAD-L also functions as a folding chaperone during the translation of CAD (Enari *et al.*, 1998). In apoptotic cells, ICAD/DFF45 is cleaved at two sites by caspases, liberating CAD, which then cleaves chromosomal DNA. Both caspase-3 and caspase-7 can cleave ICAD at the correct sites *in vitro* (Liu *et al.*, 1997; Sakahira *et al.*, 1998).

A. Endogenous Cleavage during Apoptosis

Cell extracts are prepared as described for PARP-1 in Section II,A,1, and Western blotting is performed as described for PARP-1 in Section II,A,2 except that the polyacrylamide gel used is either a 16 or a 10–20% gradient. This is required to resolve the smaller cleavage products of DFF45 (11 and 30 kDa). The precise conditions for immunoblotting must be adjusted for each antibody. These are examples of commercially available antibodies against DFF45: DFF45/C5 (Santa Cruz) and DFF45/NT (Millennum Biotechnology) used by Tang and Kidd (1998). A mouse monoclonal antibody is also available from MBL, Nagoya Japan (McIlroy *et al.*, 1999; Sakahira *et al.*, 1999). Polyclonal antibodies against ICAD can also be purchased from PharMingen and Oncogene (ICAD ab-1). Conditions for stripping and reprobing of the blot are the same as for PARP-1 (Section II,A,2).

B. *In Vitro* Cleavage of DFF45

To study ICAD cleavage *in vitro,* we have used either bacterially expressed ICAD or *in vitro*-translated ICAD (see Promega's instructions for the TNT *in vitro* transcription/translation kit).

1. Preparation of Apoptotic Cell Extracts

1. After apoptotic treatments (in case of chicken DU249 cells, treatment of 0.5 μM staurosporine for 12 h), cells (5×10^7 to 5×10^8) are harvested and centrifuged at 1400 rpm, 4°C, for 10 min.

2. The pellet is rinsed with ice-cold incomplete KPM buffer (50 mM PIPES–KOH, pH 7.0; 50 mM KCl; 10 mM EGTA; 2 mM $MgCl_2$) and then resuspended in a small

volume of complete KPM buffer (50 m*M* PIPES–KOH, pH 7.0; 50 m*M* KCl; 10 m*M* EGTA; 2 m*M* $MgCl_2$; 20 μ*M* cytochalasin B; 1 m*M* DTT; 0.1 m*M* PMSF; 1 μg/ml each chymostatin, leupeptin, antipain, pepstatin A). At this stage, cell pellets can be frozen quickly in liquid nitrogen and kept at −80°C until analysis.

3. Cells are lysed by several cycles of freezing and thawing and further disrupted by quick sonication on ice.

4. The resulting cell lysate is centrifuged at 139,000*g*, 4°C, for 2 h, yielding a clear cytosolic extract.

5. Remove an aliquot for protein determination.

6. Make aliquots of 10–20 μl and freeze them at −80°C.

2. Cleavage of ICAD by Purified Caspases or Cell Extracts

1. Mix recombinant caspase or apoptotic extract with ICAD protein and incubate at 37°C in complete KPM (e.g., final volume of 25 μl) for different periods of time.

2. To stop the reaction, add 1x reducing loading buffer and heat sample for 3 min at 95°C. Either freeze the reaction on liquid nitrogen and store at −80°C or analyze it immediately.

3. Analyze the product of the cleavage reaction by regular SDS–PAGE (as described earlier). Adjustments concerning the quantity of the caspase or the cell extract may be necessary depending on their activity. Different periods of incubation must also be tried to get optimal cleavage.

3. *In Vitro* Cleavage Inhibition Assay

See the protocols described for PARP-1 (Section II,B,4).

C. Characterization of Cleavage

See the experimental approaches described for PARP-1 (Section II,C).

D. Studies of the Biological Roles of Cleavage

The MCF-7 human breast carcinoma cell line has lost caspase-3 due to a 47- bp deletion within exon 3 of the caspase-3 gene. MCF-7 cells express caspases-2, -5, -7, -8, -9, and -10, but not detectable levels of caspase-1 (Janicke *et al.*, 1998). Whereas MCF-7 cells are still sensitive to tumor necrosis factor (TNF)- or staurosporine-induced apoptosis, no DNA fragmentation is observed. Introduction of the caspase-3 gene into MCF-7 cells resulted in DNA fragmentation (Janicke *et al.*, 1998). Despite the absence of caspase-3, intact DFF45 disappeared following TNF, cycloheximide, or staurosporine treatment of MCF-7 cells. The amino-terminal caspase site DFF45 (long/short) is cleaved in staurosporine-treated MCF-7 cells. However, it has been reported that carboxy-terminal caspase sites require caspase-3 for cleavage.

1. Phenotype of Mice Lacking ICAD/DFF45

The first three exons of the ICAD/DFF45 gene, which encode the first 142 amino acids, including the first caspase cleavage site, were deleted in a knockout mouse. Lack of ICAD/DFF45 protein and presence of CAD and caspase-3 in the mutant mice were confirmed by immunoblotting analysis. The DFF45 -/- mutant mice appeared healthy and did not show any obvious growth abnormalities. However, DNA fragmentation activity is abolished in cell extracts from DFF45 mutant tissues. In response to several apoptotic stimuli, splenocytes, thymocytes, and granulocytes from these mutant mice are resistant to DNA fragmentation and splenocytes and thymocytes are also resistant to chromatin condensation. Nevertheless, development of the immune system in the mutant mice is normal. These results demonstrate that although ICAD/DFF45 is critical for the induction of DNA fragmentation and chromatin condensation *in vivo,* it is not required for immune system development (Zhang *et al.,* 1998).

2. Overexpression of ICAD/DFF45

Human Jurkat T cells were transformed with plasmids expressing either wild-type ICAD-L or ICAD-S or mutants in which the caspase-3 cleavage sites had been eliminated. Transformed cells did not show DNA fragmentation after anti-Fas antibody or staurosporine treatment but were nonetheless killed by a caspase- dependent mechanism (Sakahira *et al.,* 1998). Contrasting results were obtained in CHO cells stably transfected with constructs expressing wild-type DFF45. In these cells, DNA fragmentation was not inhibited (Liu *et al.,* 1998) (data not shown).

References

Affar, E. B., Duriez, P. J., Shah, R. G., Winstall, E., Germain, M., Boucher, C., Bourassa, S., Kirkland, J. B., and Poirier, G. G. (1999). Immunological determination and size characterization of poly(ADP-ribose) synthesized in vitro and in vivo. *Biochim. Biophys. Acta* **1428,** 137–146.

Alnemri, E. S., Livingston, D. J., Nicholson, D. W., Salvesen, G., Thornberry, N. A., Wong, W. W., and Yuan, J. (1996). Human ICE/CED-3 protease nomenclature. *Cell* **87,** 171.

D'Amours, D., Desnoyers, S., D'Silva, I., and Poirier, G. G. (1999). Poly(ADP-ribosyl)ation reactions in the regulation of nuclear functions. *Biochem. J.* **342,** 249–268.

Duriez, P. J., Desnoyers, S., Hoflack, J. C., Shah, G. M., Morelle, B., Bourassa, S., Poirier, G. G., and Talbot, B. (1997). Characterization of anti-peptide antibodies directed towards the automodification domain and apoptotic fragment of poly(ADP-ribose)polymerase. *Biochim. Biophys. Acta* **1334,** 65–72.

Earnshaw, W. C., Martins, L. M., and Kaufmann, S. H. (1999). Mammalian caspases: Structure, activation, substrates and functions during apoptosis. *Annu. Rev. Biochem.* **68,** 383–424.

Ellis, H. M., and Horvitz, H. R. (1986). Genetic control of programmed cell death in the nematode *C elegans. Cell* **44,** 817–829.

Enari, M., Hase, A., and Nagata, S. (1995). Apoptosis by a cytosolic extract from Fas-activated cells. *EMBO J.* **14,** 5201–5208.

Enari, M., Sakahira, H., Yokoyama, H., Okawa, K., Iwamatsu, A., and Nagata, S. (1998). A caspase-activated DNase that degrades DNA during apoptosis, and its inhibitor ICAD. *Nature* **391,** 43–50.

Enari, M., Talanian, R. V., Wong, W. W., and Nagata, N. (1996). Sequential activation of ICE-like and CPP32-like proteases during Fas-mediated apoptosis. *Nature* **380,** 723–726.

Germain, M., Affar, E. B., D'Amours, D., Dixit, V. M., Salvesen, G. S., and Poirier, G. G. (1999). Cleavage of automodified poly(ADP-ribose)polymerase during apoptosis. *J. Biol. Chem.* **274,** 28379–28834.

Gurtu, V., Kain, S. R., and Zhang, G. (1997). Fluorometric and colorimetric detection of caspase activity associated with apoptosis. *Anal. Biochem.* **251,** 98–102.

Gu, J., Dong, R. P., Zhang, C., Mclaughlin, D. F., Wu, M. X., and Schlossman, S. F. (1999). Functional interaction of DFF35 and DFF45 with caspase-activated DNA fragmentation nuclease DFF40. *J. Biol. Chem.* **274,** 20759–20762.

Halenbeck, R., MacDonald, H., Roulston, A., Chen, T. T., Conroy, L., and Williams, L. T. (1998). CPAN, a human nuclease regulated by the caspase-sensitive inhibitor DFF45. *Curr. Biol.* **8,** 537–540.

Harlow, E., and Lane, D. (1999). "Using Antibodies: A Laboratory Manual." Cold Spring Harbor Laboratory Press, Cold Spring Harbor, New York.

Huletsky, A., de Murcia, G., Muller, S., Hengartner, M., Menard, L., Lamarre, D., and Poirier, G. G. (1989). The effect of poly(ADP-ribosyl)ation on native and H1-depleted chromatin: A role of poly(ADP-ribosyl)ation on core nucleosome structure. *J. Biol. Chem.* **264,** 8878–8886.

Jacobson, M. K., and Jacobson, E. L. (1999). Discovering new ADP-ribose polymer cycles: Protecting the genome and more. *Trends Biochem. Sci.* **24,** 415–417.

Janicke, R. U., Sprengart, M. L., Wati, M. R., and Porter, A. G. (1998). Caspase-3 is required for DNA fragmentation and morphological changes associated with apoptosis. *J. Biol. Chem.* **273,** 9357–9360.

Kamada, S., Kusano, H., Fujita, H., Ohtsu, M., Chirara Koya, R., Kuzumaki, N., and Tsujimoto, Y. (1998). A cloning method for caspase substrates that uses the yeast two-hybrid system: Cloning of the antiapoptotic gene gelsolin. *Proc. Natl. Acad. Sci. USA* **95,** 8532–8537.

Kameshita, I., Matsuda, Z., Taniguchi, T., and Shizuta, Y. (1984). Poly(ADP-ribose)synthase: Separation and identification of three proteolytic fragments as the substrate binding domain, the DNA binding domain and the automodification. *J. Biol. Chem* **259,** 4770–4776.

Kaufmann, S. H. (1989). Induction of endonucleolytic DNA cleavage in human acute myelogenous leukemia cells by etoposide, camptothecin, and other cytotoxic anticancer drugs: A cautionary note. *Cancer Res.* **49,** 5870–5878.

Kaufmann, S. H., Desnoyers, S., Ottaviano, Y., Davidson, N. E., and Poirier, G. G. (1993). Specific proteolytic cleavage of poly(ADP-ribose)polymerase: An early marker of chemotherapy-induced apoptosis. *Cancer Res.* **53,** 3976–3985.

Kawamitsu, H., Hoshino, H.-H., Miwa, M., Momoi, H., and Sugimura, T. (1984). Monoclonal antibodies to poly(adenosine diphosphate ribose) recognize different structures. *Biochemistry* **23,** 3771–3777.

Kerr, J. F. R., Wyllie, A. H., and Currie, A. R. (1972). Apoptosis: A basic biological phenomenon with wide-ranging implications in tissue kinetics. *Br. J. Cancer* **26,** 239–257.

Lazebnik, Y. A., Cole, S., Cooke, C. A., Nelson, W. G., and Earnshaw, W. C. (1993). Nuclear events of apoptosis in vitro in cell-free mitotic extracts: A model system for analysis of the active phase of apoptosis. *J. Cell Biol.* **123,** 7–22.

Lazebnik, Y. A., Kaufmann, S. H., Desnoyers, S., Poirier, G. G., and Earnshaw, W. C. (1994). Cleavage of poly(ADP-ribose)polymerase by a proteinase with properties like ICE. *Nature* **371,** 346– 347.

Liu, X., Aou, H., Slaughter, C., and Wang, X. (1997). DFF, a heterodimeric protein that functions downstream of caspase-3 to trigger DNA fragmentation during apoptosis. *Cell* **89,** 175–184.

Liu, X., Kim, C. N., Yang, J., Jemmerson, R., and Wang, X. (1996). Induction of apoptotic program in cell-free extracts: Requirement for dATP and cytochrome c. *Cell* **86,** 147–157.

Liu, X., Widlak, P., Zou, H., Luo, X., Garrard, W. T., and Wang, X. (1998). The 40-kDa subunit of DNA fragmentation factor induces DNA fragmentation and chromatin condensation during apoptosis. *Proc. Natl. Acad. Sci. USA* **95,** 8461–8466.

Martins, L. M., Kottke, T., Mesner, P. W., Basi, G. S., Sinha, S., Frigon, N. J., Tatar, E., Tung, J. S., Bryant, K., Takahashi, A., Svingen, P. A., Madden, B. J., McCormick, D. J., Earnshaw, W. C., and Kaufmann, S. H. (1997). Activation of multiple interleukin-1B converting enzyme homologues in cytosol and nuclei of HL-60 cells during etoposide-induced apoptosis. *J. Biol. Chem.* **272,** 7421–7430.

McIlroy, D., Sakahira, H., Talanian, R. V., and Nagata, S. (1999). Involvement of caspase 3-activated DNase in internucleosomal DNA cleavage induced by diverse apoptotic stimuli. *Oncogene* **18,** 4401–4408.

Medema, J. P., Hahne, M., Walczak, H., and DeLaurenzi, V. (1999). Rocking cell death. *Cell Death Differ.* **6,** 297–300.

Mills, J. C., Stone, N. L., and Pittman, R. N. (1999). Extranuclear apoptosis: The role of the cytoplasm in the execution phase. *J. Cell Biol.* **146,** 703–707.

Newmeyer, D. D., Farschon, D. M., and Reed, J. (1994). Cell-free apoptosis in *Xenopus* egg extracts: Inhibition by Bcl-2 and requirement for an organelle fraction enriched in mitochondria. *Cell* **79,** 353–364.

Nicholson, D. W., Ali, A., Thornberry, N. A., Vaillancourt, J. P., Ding, C. K., Gallant, M., Gareau, Y., Griffin, P. R., Labelle, M., Lazebnik, Y. A., Munday, N. A., Raju, S. M., Smulson, M. E., Yamin, T.-T., Yu, V. L., and Miller, D. K. (1995). Identification and inhibition of the ICE/CED-3 protease necessary for mammalian apoptosis. *Nature* **376,** 37–43.

Rano, T. A., Timkey, T., Peterson, E. P., Rotonda, J., Nicholson, D. W., Becker, J. W., Chapman, K. T., and Thornberry, N. A. (1997). A combinatorial approach for determining protease specificities: Application to interleukin-1beta converting enzyme (ICE). *Chem. Biol.* **4,** 149–155.

Sakahira, H., Enari, M., and Nagata, S. (1998). Cleavage of CAD inhibitor in CAD activation and DNA degradation during apoptosis. *Nature* **391,** 96–99.

Sakahira, H., Enari, M., and Nagata, S. (1999). Functional differences of two forms of the inhibitor of caspase-activated DNase, ICAD-L, and ICAD-S. *J. Biol. Chem.* **274,** 15740–15744.

Sallmann, F. R., Bourassa, S., Saint-Cyr, J., and Poirier, G. G. (1997). Characterization of antibodies specific for the caspase cleavage site on poly(ADP-ribose)polymerase: Specific detection of apoptotic fragments and mapping of the necrotic fragments of poly(ADP-ribose)polymerase. *Biochem. Cell Biol.* **75,** 451–456.

Salvesen, G. S., and Dixit, V. M. (1997). Caspases: Intracellular signaling by proteolysis. *Cell* **91,** 443–446.

Shao, R. G., Cao, C. X., and Pommier, Y. (1997). Activation of PKC alpha downstream from caspases during apoptosis induced by 7-hydroxystaurosporine or the topoisomerase inhibitors, camptothecin and etoposide, in human myeloid leukemia HL60 cells. *J. Biol. Chem.* **272,** 31321–31325.

Shizuta, Y., Kameshita, I., Ushiro, H., Matsuda, M., Suzuki, S., Mitsuuchi, Y., Yokoyama, Y., and Kurosaki, T. (1986). The domain structure and function of poly(ADP-ribose)synthetase. *Adv. Enzyme Regul.* **25,** 377–384.

Skalka, M., Matyasova, J., and Cejkova, M. (1976). DNA in chromatin of irradiated lymphoid tissues degrades in vivo into regular fragments. *FEBS Lett.* **72,** 271–273.

Stroh, C., and Schulze-Osthoff, K. (1998). Death by thousand cuts: An ever increasing list of caspase substrates. *Cell Death Differ.* **5,** 997–1000.

Takahashi, A., Hirata, H., Yonehara, S., Imai, Y., Lee, K. K., Moyer, R. W., Turner, P. C., Mesner, P. W., Okazaki, T., Sawai, H., Kishi, S., Yamamoto, K., Okuma, M., and Sasada, M. (1997). Affinity labeling displays the stepwise activation of ICE-related proteases by Fas, staurosporine, and CrmA-sensitive caspase-8. *Oncogene* **14,** 2741–2752.

Tang, D., and Kidd, V. J. (1998). Cleavage of DFF45/ICAD by multiple caspases is essential for its function during apoptosis. *J. Biol. Chem.* **273,** 28549–28552.

Tewari, M., Quan, L. T., O'Rourke, K., Desnoyers, S., Zeng, Z., Beidler, D. R., Poirier, G. G., Salvesen, G. S., and Dixit, V. M. (1995). Yama/CPP32 beta, a mammalian homolog of Ced-3, is a CrmA-inhibitable protease that cleaves the death substrate poly(ADP-ribose)polymerase. *Cell* **81,** 801–809.

Thornberry, N. A., Rano, T. A., Peterson, E. P., Rasper, D. M., Timkey, T., Garcia-Calvo, M., Houtzager, V. M., Nordstrom, P. A., Roy, S., Vaillancourt, J. P., Chapman, K. T., and Nicholson, D. W. (1997). A combinatorial approach defines specificities of members of the caspase family and granzyme B: Functional relationships established for key mediators of apoptosis. *J. Biol. Chem.* **272,** 17907–17911.

Van Criekinge, W., Van Gurp, M., Decoster, E., Schotte, P., Van de Craen, M., Fiers, W., Vandenabeele, P., and Beyaert, R. (1998). Use of the yeast three-hybrid system as a tool to study caspases. *Anal. Biochem.* **263,** 62–66.

Villa, P., Kaufmann, S. H., and Earnshaw, W. C. (1997). Caspases and caspase inhibitors. *Trends Biochem. Sci.* **22,** 388–393.

Wolf, B. B., and Green, D. R. (1999). Suicidal tendencies: Apoptotic cell death by caspase family proteinases. *J. Biol. Chem.* **274,** 20049–20052.

Wyllie, A. H. (1980). Glucocorticoid-induced thymocyte apoptosis is associated with endogenous endonuclease activation. *Nature* **284,** 555–556.

Wyllie, A. H., Kerr, J. F. R., and Currie, A. R. (1980). Cell death: The significance of apoptosis. *Int. Rev. Cytol.* **68,** 251–306.

Yuan, J. Y., Shaham, S., Ledoux, S., Ellis, H. M., and Horvitz, H. R. (1993). The *C. elegans* cell death gene ced 3 encodes a protein similar to mammalian interleukin 1 beta converting enzyme. *Cell* **75,** 641–652.

Zahradka, P., and Ebisuzaki, K. (1984). Poly(ADP-ribose)polymerase is a zinc metalloenzyme. *Eur. J. Biochem.* **142,** 503–509.

Zhang, J., Liu, X., Scherer, D. C., van Kaer, L., Wang, X., and Xu, M. (1998). Resistance to DNA fragmentation and chromatin condensation in mice lacking the DNA fragmentation factor 45. *Proc. Natl. Acad. Sci. USA* **13,** 12480–12485.

CHAPTER 13

Analysis of Reactive Oxygen Species in Cell Death

Ivan Stamenkovic

Department of Pathology
Harvard Medical School and Molecular Pathology Unit
Department of Pathology and MGH Cancer Center
Massachusetts General Hospital
Charlestown Navy Yard, Boston, Massachusetts 02129

I. Introduction

The notion that oxidants mediate apoptosis is supported by a number of observations. Treatment of cells with exogenous oxidizing species triggers apoptosis in a variety of cell types (Lennon *et al.*, 1991; Hampton and Orrenius, 1997). Increases in endogenous cellular oxidants have been detected with a range of stimuli that induce apoptosis (McGowan *et al.*, 1998), and antioxidants block apoptosis in these models (Verhaegen *et al.*, 1995). The ability of oxidants to trigger apoptosis depends, however, on the extent of the oxidative stress. Higher oxidant concentrations lead to necrosis, and the "window" for induction of apoptosis may be small, depending on the cell type and the oxidant being investigated (Vissers *et al.*, 1999). Under sufficient oxidative stress, ATP depletion will prevent cells from having an active role in their own demise, and necrosis will occur (Richter *et al.*, 1996; Leist *et al.*, 1997). It is also proposed that the inability of cells undergoing oxidative stress to initiate apoptosis

0091-679X/01 $35.00

results from the direct inhibition of caspase activity (Hampton and Orrenius, 1997; Samali *et al.,* 1999) and that oxidation regulates caspase function in various physiological and pathological conditions (Hampton and Orrenius, 1997; Mannick *et al.,* 1999; Fadeel *et al.,* 1998).

Caspases are a family of cysteine proteases that play a key role in the execution of apoptosis (Thornberry and Lazebnik, 1998). They are normally expressed as inactive zymogens until a specific stimulus triggers their proteolytic conversion to the active form. Caspases are broadly divided into two categories. Initiator/amplification caspases are the first to be activated by an apoptotic stimulus and include caspase-8, which mediates cell death as a result of engagement of the Fas receptor (Muzio *et al.,* 1996), and caspase-9, which, together with Apaf-1 and cytochrome *c*, forms the apoptosome necessary for induction of apoptosis by a broad range of receptor-independent stimuli (Li *et al.,* 1997). Effector/execution caspases are processed by initiator caspases and proceed to cleave an array of structural and regulatory proteins, leading to the irreversible dismantling of the cell. The best characterized member of this category is caspase-3 (Nicholson *et al.,* 1995).

Although several reports have shown that oxidation inhibits caspases (Hampton and Orrenius, 1997; Ueda *et al.,* 1998; Dimmeler *et al.,* 1997; Mohr *et al.,* 1997), caspase activity is required for apoptosis associated with oxidative stress. Several possible explanations may be proposed for these seemingly paradoxical observations. First, caspases may be relatively insensitive to oxidation at oxidant concentrations required to trigger apoptosis. Second, there may be a temporal dissociation between oxidant production and caspase activation, so that active caspases are not exposed to the oxidants. Third, cellular reductants may be potent enough to revert oxidized caspases to their active form.

Mammalian cells possess multiple sources of reactive oxygen species (ROS), including, among others, xanthine oxidase, mitochondrial NADH oxidoreductase, microsomal NADPH oxidoreductase, and plasma membrane-bound NADH/NADPH oxidases (Yu, 1994). Evidence suggests that plasma membrane-associated oxidases may provide one source of ROS associated with cell resistance to triggers of apoptosis. Transient resistance to apoptosis following stimulation of cells by phorbol ester has been shown to be associated with plasma membrane oxidase ROS overproduction triggered by phorbol ester-activated protein kinase C (PKC) (Watson *et al.,* 1991). Conversely, diphenylene iodinium (DPI), a flavoprotein inhibitor that supresses the plasma membrane oxidase-mediated generation of superoxide, increases cell sensitivity to apoptosis triggered by the anti-Fas antibody as well as by other stimuli. There are at least two flavin-containing oxidases in the plasma membrane: NADPH oxidase, responsible for the neutrophil oxidative burst, and NADH oxidase, both of which are inhibited by DPI (Griendling *et al.,* 1994; Thannickal and Fanburg, 1995).

Plasma membrane-bound NADPH oxidase was initially identified in phagocytic cells as a heterodimeric protein composed of $gp91^{phox}$ and $p22^{phox}$ subunits containing heme residues for ROS production and FAD for electron transfer from NADPH and requiring two cytosolic proteins, $p47^{phox}$ and $p67^{phox}$, for full activation (DeLeo and Quinn, 1996). Upon stimulation, NADPH oxidase produces an extracellular burst of ROS that provides a potent host defense mechanism against invading microorganisms, as well as a slower and more sustained intracellular increase in superoxide concentration. In addition to phagocytic cells, a variety of normal and malignant nonphagocytic cells display NADPH oxidase activity in their plasma membrane. However, although some cell types, including fibroblasts, vascular smooth muscle cells, renal mesangial cells, pulmonary neuroepithelial cells, and type I cells in the carotid body, display immunoreactivity with antibodies against phagocyte-associated NADPH oxidase components (Bunn and Poyton, 1996), the NADPH oxidase structure in these cells is still unknown. In addition to NADPH oxidase, the plasma membrane of most cells has another ROS-producing enzyme utilizing NADH as an electron donor. It is distinct from NADH oxidase of mitochondria because it is insensitive to inhibitors of mitochondrial oxidase, including cyanide and rotenone. It is also distinct from NADH oxidases of the endoplasmic reticulum and Golgi apparatus (Morre and Brightman, 1991). Because its activity is stimulated by hormones and growth factors and its inhibition arrests cell growth *in vitro* (Brightman *et al.,* 1992; Morre *et al.,* 1996), plasma membrane NADH oxidase may be implicated in the regulation of cell growth. Taken together, NADH and NADPH oxidases may provide candidate sources of superoxide anion production associated with cell resistance to apoptosis.

II. Methods

A. Role of NADH and NADPH Oxidases in the Regulation of Sensitivity to Apoptosis

1. Cell Culture

Cells are grown on culture dishes coated with collagen type I (Upstate Biotechnology Inc., Lake Placid, NY) for purification of plasma membranes and for NADH and NADPH oxidase assays in intact cells. After reaching confluence, cells are incubated in serum-free medium for 1 h prior to initiation of the experiments and throughout the experimental duration. Cell counts are performed immediately after every experiment in a hemocytometer.

2. Plasma Membrane Purification

1. Adherent cells are grown on collagen-coated dishes and detached by type I collagenase (1 mg/ml, in the presence of 1 mg/ml soybean trypsin inhibitor and 2 mg/ml bovine serum albumin).

2. Cells are collected by centrifugation at 200*g* for 10 min and the cell pellets are washed in isotonic Tris buffer (0.15 *M* NaCl, 1 m*M* EDTA, and 10 m*M* Tris–HCl, pH 7.2).

3. After centrifugation, the pellets are resuspended in hypotonic Tris buffer (1 m*M* EDTA and 10 m*M* Tris–HCl, pH 7.2) containing phenylmethylsulfonyl fluoride (0.2 m*M*), leupeptin (1 μg/ml), pepstatin (1 μg/ml), and chymostatin (1 μg/ml) for 15 min.

4. Cells are then lysed gently with a tight Dounce homogenizer until about 90% cell lysis without nuclear rupture is achieved as assessed by light microscopy. To prevent

rupture of intracellular organelles, a 1 *M* sucrose stock solution in Tris buffer is added to the homogenates to 0.25 *M* concentration immediately after homogenization.

5. The homogenates are centrifuged at 700*g* for 5 min to remove nonruptured cells and nuclei, and the supernatant is centrifuged at 2000*g* for 30 min to prepare a plasma membrane-enriched fraction.

6. Pellets are resuspended in 50% (w/w) sucrose solution in 10 m*M* Tris buffer (pH 7.2) and sequentially overlayed with 40/35/30/20/10% sucrose solutions. The sucrose gradient is then ultracentrifuged in a swing bucket rotor (SW28, Beckman) at 24,000 rpm for 15 h.

7. Purified plasma membranes are obtained from the interface between 35 and 30% sucrose layers. To remove excess sucrose, plasma membranes are washed by two rounds of centrifugation–resuspension with phosphate buffer containing 50 m*M* KH_2PO_4 (pH 7.4), 1 m*M* EDTA, and 10 μ*M* FAD to prevent removal of FAD from the oxidases.

8. All of the procedures are performed at 4°C, and protein concentrations are assayed by the BCA (Pierce, Rockford, IL) method.

9. The purity of the plasma membrane preparation is evaluated by assaying a plasma membrane-specific marker enzyme (ouabain-sensitive *p*-nitrophenylphosphatase) (Thannickal and Fanburg, 1995) and by electron microscopy.

10. In our hands, purified plasma membranes show a 50-fold increase in the activity of ouabain-sensitive *p*-nitrophenylphosphatase compared to crude homogenates, and yields of plasma membrane are about 25%.

11. Electron microscopy should reveal homogeneous membranes with little or no detectable contamination.

3. NADPH/NADH Consumption Assay

1. The activities of NADPH and NADH oxidases are assessed by NADPH/NADH consumption in purified plasma membranes (Morre *et al.,* 1995).
2. Purified plasma membranes are incubated with NADPH or NADH at 25°C, and the change of absorbance at 340 nm is monitored continuously.

4. Superoxide Anion Production Assay in Purified Plasma Membrane

1. NADPH/NADH oxidase activity is determined by measuring superoxide anion production by Lucigenin-based chemiluminescence (LC).

2. The reaction mixture contains 50 m*M* KH_2PO_4 (pH 7.4), 1 m*M* EDTA, 10 μ*M* FAD, 1 m*M* KCN to inhibit low levels of contaminant oxidases from internal membranes, and 10 μg protein of the plasma membrane.

3. After a 10-minute preincubation at 25°C, the enzyme reaction is started by the addition of NADPH/NADH and Lucigenin (Sigma). LC is monitored for 20 s in a Lumat LB 9501 luminometer (Wallac Inc., Gaithersburg, MD) and the reading is

expressed, after subtraction of the value corresponding to a cell blank, in relative light units (RLU).

4. The K_m and V_{max} value of NADPH oxidase and NADH oxidase are obtained using the double-reciprocal plot.

5. Superoxide Anion Production Assay in Intact Cells

1. Adherent cells are grown on collagen-coated dishes.
2. After reaching confluence, cells are harvested from the dish with the collagenase.
3. Cells are spun down at 200*g* at 4°C for 5 min and the pellet is resuspended in a cold balanced salt solution (130 m*M* NaCl, 2 m*M* KCl, 1 m*M* $MgCl_2$, 1.5 m*M* $CaCl_2$, 3 m*M* KH_2PO_4, and 20 m*M* Hepes, pH 7.4).
4. After an additional spin, cells are resuspended at 2×10^6 cells/ml in the just-described salt solution containing 10 m*M* glucose and 1 mg/ml bovine serum albumin and are stored on ice until use.
5. To measure oxidase activity, 6×10^5 cells are incubated in a cuvette with various concentrations of NADPH or NADH at 37°C for 5 min.
6. Lucigenin solution is injected into the cuvette automatically prior to reading and LC is monitored for 20 s. A cell blank is subtracted from each reading and data are expressed in RLU.

6. Cytotoxicity Assay

1. Cytotoxicity is assessed by using crystal violet staining of viable cells as described previously (Leist *et al.*, 1997).
2. Crystal violet staining is performed in 96-well microtiter plates. Adherent cells (2×10^4 per well) are plated in DMEM/5% FBS for 24 h.
3. Fresh serum-free medium is added 1 h prior to the assay.
4. After incubation with test reagents, medium is aspirated and replaced for 10 min by a staining solution containing 0.75% crystal violet in 50% ethanol, 0.25% NaCl, and 1.75% formaldehyde.
5. After washing with water and drying, the dye is eluted with PBS/1% SDS.
6. Cell viability is measured by dye absorbance at 595 nm on an ELISA reader.

7. PARP Cleavage

1. PARP cleavage is assessed as an index for apoptotic cell death and is detected by SDS–PAGE/Western blot analysis of cell lysates as described previously (Bunn and Poyton, 1996).
2. Nitrocellulose membranes containing the blotted PARP are incubated with 10,000-fold diluted anti-PARP mouse monoclonal antibody C-2-10 (Biomol Research Laboratories Inc., Plymouth Meeting, PA) for 2 h.

3. This is followed by an incubation with a 1:5000 dilution of a horseradish peroxidase-labeled goat antimouse immunoglobin antibody (Amersham, Arlington Heights, IL) for 2 h.

4. The signal is visualized by ECL (KPL, Gaithersburg, MD).

8. Comments

a. Optimization of the Lucigenin-Based Assay for Determining Superoxide Anion Concentration

Lucigenin-based chemiluminescence has been used for measuring intracellular superoxide anion production and concentration, but has proven to be an unreliable probe under some conditions. For example, in the presence of endothelial nitric oxide synthase and in *Escherichea coli*, lucigenin can induce superoxide production through redoxcycling, leading to an overestimation of superoxide production. To verify that lucigenin might be reduced directly by plasma membrane-associated NADPH/NADH oxidase, the rate of NADPH/NADH consumption can be measured in purified plasma membrane fractions in the presence of increasing concentrations of lucigenin. We have found that at high concentrations (100–500 μM), lucigenin increases the rate of NADH consumption but does not affect consumption at lower concentrations (25 μM). In contrast, lucigenin did not increase the rate of NADPH consumption at any of the concentrations tested. To determine the minimal concentration of lucigenin required for the detection of superoxide anion, we have measured LC in the presence of increasing concentrations of lucigenin using a defined concentration of xanthine–xanthine oxidase as a source of superoxide. LC increased with increasing concentrations of lucigenin in a dose-dependent manner, reaching a plateau at 5 μM of lucigenin. These results indicate that lucigenin can be used within the 5 to 25 μM range for assaying the superoxide anion derived from NADH oxidase and at higher concentrations for measuring superoxide anion produced by NADPH oxidase. These observations are also consistent with the notion that superoxide production by NADH oxidase is overestimated at high concentrations of lucigenin. Thus, LC values obtained in the presence of 20 μM lucigenin, in the NADH oxidase assay, were increased 10-fold when lucigenin was used at a 200 μM concentration. We therefore use 20 μM lucigenin to measure superoxide production by NADPH/NADH oxidase in most of our experiments.

To determine the specificity of LC for superoxide anion production and to estimate the proportion of extracellular to intracellular superoxide anion release, LC is measured following cell stimulation with NADH and NADPH in the presence of superoxide dismutase (SOD, 100 U/ml, Sigma) or tiron (Sigma) (1 mM), which respectively provide a membrane-impermeable and a membrane-permeable superoxide scavenger. In both cell types, NADH- and NADPH-induced LC is reduced significantly by SOD (a 60 and a 75% reduction for NADH and NADPH, respectively) while being nearly completely abolished by tiron (an 82 and a 96% decrease for NADH and NADPH, respectively). We interpret that the SOD-sensitive and the SOD-resistant, tiron-sensitive LC to respectively reflect extracellular and intracellular superoxide anion production. Based on this assumption, the ratios of extracellular to intracellular superoxide production by NADH and NADPH oxidase are about 2.5 : 1 and 3.5 : 1, respectively, in a variety of tumor cell lines tested, irrespective of their sensitivity or resistance to apoptosis. Exogenous NADPH/NADH therefore appear to increase both the intra- and the extracellular concentration of superoxide anion.

b. Exogenous NADH Has a Protective Effect against DPI-Induced Cell Apoptosis

Diphenylene iodinium (20 μM) inhibits both oxidases in purified plasma membrane, and the degree of inhibition of both oxidases increases with increasing incubation time. In this cell-free system, NADPH oxidase appears to be more sensitive to DPI than NADH oxidase. To compare the sensitivity of M14 (apoptosis-sensitive melanoma) and T24 (apoptosis-resistant bladder carcinoma) cells to DPI-induced cytotoxicity, cell viability is observed in the presence of culture medium supplemented with 20 μM DPI at a series of time points. M14 cells display a sharp drop in viability after 3 h of exposure to DPI. In contrast, T24 cells display greater resistance to DPI, with a significant reduction in viability being observed only after an 8-h incubation period.

To test the effects of exogeneous NADPH/NADH and antioxidants on DPI-induced cytotoxicity, we chose incubation conditions under which both cell types display roughly 50% viability (3-h incubation for human melanoma M14 cells and 8-h incubation for human bladder carcinoma T24 cells in culture medium containing 20 μM DPI). T24 cell viability was observed to decrease to $42.6 \pm 19.3\%$ (mean $\pm$ SD), whereas M14 cell viability was quantitated at $59.2 \pm 10.5\%$ by the crystal violet exclusion assay. The addition of NADH (40 μM) protected both T24 and M14 cells from DPI-induced cytotoxicity (T24 cell viability: $89.2 \pm 13.2\%$, $p < 0.001$ vs DPI-treated cells; M14 cell viability: $98.6 \pm 18.8\%$, $p < 0.002$ vs DPI-treated cells). In contrast, NADPH (40 μM) had little or no protective effect (53.1 ± 14.0 and $50.6 \pm 14.0\%$ M14 and T24 cell viability, respectively). To rule out possible effects of extracellular ROS released by the addition of NADH, cells were incubated with NADH in the presence of SOD (100 U/ml) and catalase (500 U/ml), neither of which can penetrate the plasma membrane, to remove any extracellular ROS. Neither SOD nor catalase could abolish the protective effect of NADH on DPI-induced cytotoxicity. In contrast, tiron (1 mM) abolished the antiapoptotic effect of NADH almost completely ($47.8 \pm 13.5\%$ T24 cell viability relative to DPI/NADH-treated cells, $p < 0.001$; $66.0 \pm 11.9\%$ M14 cell viability relative to DPI/NADH-treated M14 cells, $p < 0.005$).

B. Assessment of Caspase Activity in Response to Oxidants

1. Cell Culture and Preparation of Lysates

1. Cells are cultured in medium with 10% fetal bovine serum.
2. Cells are stimulated at 1×10^6/ml with 250 ng/ml anti-Fas antibody. After 2 h the cells are harvested, washed, and pelleted.
3. Because DTT is not included in the caspase assay buffer, the cells are either assayed immediately or frozen and assayed within 6 h to prevent detectable spontaneous caspase oxidation.

2. Caspase Assay

1. Pellets (10^6 cells) are resuspended in 100 μl of caspase buffer [100 mM Hepes, 10% sucrose, 0.1% CHAPS, 10^{-4}% NP-40, and 50 μM DEVD-AMC (Peptide Institute Inc. Osaka, Japan) at pH 7.25]. Samples are placed in a 100-μl cuvette from Hellma Cells, Inc. (New York), and substrate cleavage is monitored continuously at an excitation of 370 nm (band width 5 nm) and an emission of 445 nm (band width 10 nm) on a 37°C heat-jacketed fluorescent spectrophotometer (F-4500 Hitachi). Fluorescent units are converted to picomoles of free AMC using a standard curve generated with reagent AMC.

2. Recombinant caspase-3 (Merck-Frosst Canada Inc.) is supplied at a stock concentration of 40 U/μl in caspase buffer with 5 mM DTT (concentrations will vary according to the supplier).

3. The enzyme is typically recovered as separate p17 and p12 subunits from the inclusion bodies of *E. coli*, where it usually constitutes >99% of the total protein.

4. The stock is assayed at a 1 in 1000 dilution into 100 μl of DTT-free caspase buffer (4 U final). one unit is defined as the release of 1 pmol AMC min at 25°C.

5. Stock solutions of hydrogen peroxide, whose concentrations are determined by their $A240$ ($\epsilon = 43.6$), are diluted in water immediately before use.

6. Hydrogen peroxide is then added to cell lysates or recombinant caspase while the substrate is present to allow comparison and quantification of caspase activity before and after the addition of hydrogen peroxide.

7. Cells are suspended in fresh medium (includes 2 mg/ml glucose) and treated with varying concentrations of hydrogen peroxide or glucose oxidase.

8. After 1–3 h, the glucose oxidase-treated cells are pelleted, washed with PBS, and resuspended in fresh medium with or without glucose oxidase.

9. After 4–12 h, cells are harvested and caspase activity is measured. Cell viability is assessed by the crystal violet assay.

10. Hydrogen peroxide concentration is assessed at these times by the ferrous oxidation of the xylenol orange method (Watson *et al.*, 1991).

3. Comments

The rate equation for the reaction of recombinant caspase with hydrogen peroxide can be represented as

$$-d[\text{caspase}]/dt = k[\text{H}_2\text{O}_2]\,[\text{caspase}]$$

where k is the second-order rate constant.

Provided that hydrogen peroxide is in substantial excess, the pseudo first-order rate constant with respect to caspase activity ($k' = k\,[\text{H}_2\text{O}_2]$) can be obtained by fitting an exponential curve to the data measuring caspase activity versus time after treatment with hydrogen peroxide. The reaction is performed over a range of hydrogen peroxide (2–20 μM), and k is determined from the slope of plot of k' versus hydrogen peroxide concentration. Altering the initial concentration of caspase has no apparent effect on the value of the rate constant.

a. Example of a Typical Experiment

To induce apoptotic cell death, Jurkat T lymphocytes were cultured in the presence of anti-Fas antibody and harvested 2 h later. At this time >50% of the cells were observed to be undergoing apoptosis as determined by cytoplasmic membrane blebbing. The cells were then pelleted and lysed in caspase buffer containing the tetrapeptide DEVD-AMC, and caspase activity was monitored by the generation of fluorescent AMC. Consistent with the observations that most of the DEVDase activity is attributable to caspase-3, lysates of apoptotic cells typically contained 20 times more activity than those of control cells. The addition of hydrogen peroxide to the lysate/substrate mixture resulted in a rapid inhibition of activity. Comparison of caspase activity before and after the addition of hydrogen peroxide, at a range of hydrogen peroxide concentrations, revealed an IC_{50} of 7 μM. Concentrations in excess of 200 μM decreased the DEVDase activity to levels present in nonapoptotic cells, but did not completely eliminate it. This oxidant-insensitive activity was not saturable with increased substrate and had a much higher apparent K_m than the majority of DEVDase activity in apoptotic lysates, suggesting that it represents nonspecific proteolysis.

It is conceivable that hydrogen peroxide does not target caspases directly in the cell lysates, but rather a potential oxidant-sensitive cofactor necessary for caspase activity. To address this possibility, we tested the effect of hydrogen peroxide on recombinant caspase-3. The recombinant enzyme was inhibited rapidly by low concentrations of hydrogen peroxide. In contrast to lysates, the inhibition of the purified enzyme was complete at higher concentrations of hydrogen peroxide. Following the addition of catalase to remove the hydrogen peroxide, dithiothreitol was able to regenerate caspase activity, indicating that reversible thiol oxidation is responsible for the inactivation of caspase-3. However, the rate of reduction is slow, with only 50% recovery after 5 min. Similar slow kinetics are seen in apoptotic cell lysates treated with hydrogen peroxide, which is in direct contrast to the immediate reversal of caspase oxidation that occurs during the storage of cell lysates.

C. Assessing the Redox State of Caspases in Cells

Although caspases are sensitive to oxidation, the significance of alterations in the caspase redox state *in vivo* remains to be elucidated. Caspase-3 activity is mediated by cysteine-163, whose thiol group has to be in the reduced form for the enzyme to function, and dithiothreitol is included routinely in caspase activity assays. Reactive nitrogen species have been shown to inhibit caspase-3 activity via thiol oxidation *in vitro*, and caspase-3 is known to be highly sensitive to inactivation by hydrogen peroxide. Because of the inherent sensitivity of caspases to oxidation, it is possible that procaspases may be maintained in an oxidized state, thereby preventing aberrant activation. In support of this, the active site of caspase-3 has been reported to be S-nitrosylated in unstimulated cells and then denitrosylated on activation of apoptosis.

Reduced thiols can be detected by their ability to reduce dithionitrobenzoic acid to its strongly absorbent monomer, and this method has been used successfully with purified caspases. However, this approach cannot distinguish between different proteins in cell lysates. To provide selectivity, radiolabeled iodine reagents such as iodoacetic acid or iodoacetyltyrosine have been used in combination with immunoprecipitation and/or electrophoretic techniques. These compounds only label reduced thiols, and decreased labeling is indicative of protein oxidation. Here, we propose a method for addressing the redox state of caspases by combining these principles with the isoelectric focusing (IEF) techniques pioneered by Thomas and colleagues (1986, 1995). The reaction of reduced thiols with iodoacetic acid (IAA) lowers the isoelectric point (p*I*) of a protein, whereas oxidized protein does not react with IAA and maintains the same p*I*. Western blotting is used to monitor the chosen pro and active caspase in cellular lysates. This technique can be used to monitor the redox state of any caspase in the presence of various oxidants and in cells before and during apoptosis.

1. Resting cells, or those treated with a trigger of apoptosis (e.g., 250 ng/ml anti-Fas antibody for 2 h), are pelleted and resuspended at 10^6/100 μl in 100 m*M* Hepes, 10% sucrose, 0.1% CHAPS, and 0.5% NP-40 at pH 7.25.
2. When required, caspase assays with DEVD-AMC are performed as described earlier. Samples of 10 μl are taken for IEF and are incubated with or without 5 m*M* IAA (pH 7.25) for 15 min before an equal volume of sample buffer (9 *M* urea, 5% ampholytes, pH 3.5–10, 80 m*M* *n*-dodecyl-D-maltoside, and 100 m*M* *N*-ethylmaleimide) is added.
3. Samples are spun at 10,000*g* for 5 min to remove cellular debris and are kept on ice until ready to load.
4. Exogenous oxidants are added, as required, to the lysates 10 min before the IAA.
5. The IEF technique used is based on that described by Thomas *et al.* (1986, 1995).
6. The gels are made of 4% polyacrylamide (National Diagnostics, Atlanta, GA) containing 5% ampholytes, pH 3.5–10 (Pharmacia, Uppsala, Sweden), 7 *M* urea (American Bioanalytical, Natick, MA), and 2% Triton X-100 (Fisher, Fair Lawn, NJ).
7. The hydrophobic side of a GelBond PAG film sheet (Pharmacia), is placed down on a glass plate (26 × 12.5 cm), and a sheet of NetFix (Crescent Chemical Co, Hauppauge, NY) is layered on the GelBond to stabilize the gel during the immunoblotting transfer.
8. Spacers (2 × 12.5 cm) are placed across both ends of the NetFix, and 20 ml of gel mixture is poured directly onto the NetFix sheet. The gel is covered with a glass plate siliconized with SigmaCote (Sigma, St. Louis, MO) and allowed to polymerize for 2 h.
9. The gel is either used immediately or stored at 4°C for up to 1 week.
10. IEF gels are run on a flat-bed apparatus (Pharmacia), maintained at 10°C.
11. Gels typically have a length of 12.5 cm, but vary in width according to the number of samples being run.
12. Anodic and cathodic wicks are soaked with electrode buffers (Crescent Chemical Co, Hauppauge, NY) and blotted with adsorbent paper before application to the gel 1 cm from the top and bottom.

13. The samples are loaded in sample applicator strips (Pharmacia) located 3 cm from the anodic end of the gel.

14. The gels are run at a constant power of 1.1 W/cm gel width, a maximum of 2.7 mA/cm gel width, and a maximum of 1500 V for 90 min.

15. The IEF gel is then soaked in 0.7% (v/v) acetic acid for 10 min, separated from the GelBond backing, and placed in a transfer sandwich with the transfer membrane located on the cathodic side of the electrofocusing gel (opposite of a conventional transfer).

16. The transfer is performed overnight in 0.7% (v/v) acetic acid at 4°C.

17. A standard Western blot is performed with the appropriate anticaspase monoclonal antibody and the corresponding secondary HRP-conjugated antibody and visualized by ECL.

1. Comments

The approach just described provides a method to measure the redox state of the active site cysteine of caspases. The method is based on the ability of IAA to react with reduced but not oxidized thiols and the visualization of IAA incorporation by its ability to shift the mobility of the target protein on an IEF gel. Although several of the thiols in caspases may be labeled with IAA, resulting in multiple band shifts, it is possible to measure the active site thiol by lowering the pH at which labeling occurred. A conventional cysteine thiol has a pK_a of 8.5, but positively charged residues, as in the active site of caspase-3, for example, particularly His-121, stabilize the deprotonated thiolate anion and allow it to undertake a nucleophilic attack on the target substrate at physiological pH.

A variety of other methods have been used for visualizing IAA incorporation into thiol proteins. One approach consisted of immunoprecipitating the protein of interest following treatment with radiolabeled IAA and assessing the decrease in radiolabel incorporation in the pellet as a measure of oxidation. A drawback of this technique is that IAA incorporation into possible coimmunoprecipitating proteins within the pellet is included in the analysis. This potential flaw was overcome by Wu *et al.* (1998) where fluorescein IAA was used as the label. The lysates were subjected to SDS–PAGE, thereby separating the target protein by molecular weight. The extent of labeling was determined by Western blot analysis using an antifluorescein antibody. In a similar strategy, SDS–PAGE was used to separate proteins labeled with iodinated iodoacetyltyrosine. These approaches seem particularly well suited for screening the redox state of the major reactive cellular thiol proteins, but to focus on a specific target they still require an immunoprecipitation step. In preliminary experiments using the iodoacetyltyrosine technique combined with immunoprecipitation, we were able to measure labeled caspase-3, but large amounts of lysate were required to get a significant signal. In addition, the necessity of an immunoprecipitation step adds further complexity and is poorly suited for assaying large numbers of samples.

The IEF technique outlined here could be adapted for monitoring the redox state of any cellular thiol protein. It is highly sensitive and is only limited by the primary antibody used for Western blotting. In our hands, strong bands were detected in lysates from 10^5

cells and an ECL exposure time of less than 1 min. Another advantage is that the number of thiols labeled with IAA can be monitored, and the labeling conditions were adjusted accordingly to allow only the most reactive thiols to be assayed. A disadvantage of the technique is that quantitation by Western blot analysis is difficult. Therefore, the assay is primarily a qualitative measure for determining whether the protein of interest is the reduced or oxidized form. It would be difficult, for example, to detect a small fraction of oxidized protein within a predominantly reduced pool. This shortcoming could be overcome using the adaptation of Gitler *et al.* (1997), where NEM is applied to block reduced thiols and the oxidized thiols are then reduced with DTT and labeled. In the IEF assay, only the oxidized thiol proteins would shift, facilitating their detection.

A critical requirement for the IEF assay is a specific antibody to the target protein as cross-reactive bands, which cannot be excluded on the basis of their size by IEF, may interfere in the analysis. This problem could be overcome by using two-dimensional electrophoresis, but this technique is less amenable to comparing multiple samples.

References

Brightman, A. O., Wang, J., Miu, R. K., Sun, I. L., Barr, R., Crane, F. L., and Morre', D. J. (1992). *Biochim. Biophy. Acta* **1105,** 109–117.

Bunn, H. F., and Poyton, R. O. (1996). *Physiol. Rev.* **76,** 839–885.

DeLeo, F. R., and Quinn, M. T. (1996). *J. Leukocyte. Biol.* **60,** 677–691.

Dimmeler, S., Haendeler, J., Nehls, M., and Zeiher, A. M. (1997). *J. Exp. Med.* **185,** 601–607.

Fadeel, B.,Åhlin, A., Henter, J., Orrenius, S., and Hampton, M. B. (1998). *Blood* **92,** 4808–4818.

Gitler, C., Zarmi, B., and Kalef, E. (1997). *Anal.Biochem.* **252,** 48–55.

Griendling, K. K., Minieri, C. A., Ollerenshaw, J. D., and Alexander, R. W. (1994). *Circ. Res.* **74,** 1141–1148.

Hampton, M. B., and Orrenius, S. (1997). *FEBS Lett.* **414,** 552–556.

Leist, M., Single, B., Castoldi, A. F., and Nicotera, P. (1997). *J. Exp. Med.* **185,** 1481–1486.

Lennon, S. V., Martin, S. J., and Cotter, T. G. (1991). *Cell Prolif.* **24,** 203–214.

Li, P., Nijhawan, D., Budihardjo, I., Srinivasula, S. M., Ahmed, M., Alnemri, E. S., and Wang, X. (1997). *Cell* **91,** 479–489.

Mannick, J. B., Hausladen, A., Liu, L., Hess, D. T., Zeng, M., Miao, Q. X., Kane, L. S., Gow, A. J., and Stamler, J. S. (1999). *Science* **284,** 651–654.

McGowan, A. J., Bowie, A. G., O'Neill, L. A. J., and Cotter, T. G. (1998). *Exp.Cell Res.* **238,** 248–256.

Mohr, S., Zech, B., Lapetina, E. G., and Brune, B. (1997). *Biochem. Biophys. Res. Commun.* **238,** 387–391.

Morre, D. J., and Brightman, A. O. (1991). *J. Bioenerg. Biomembr.* **23,** 469–489.

Morre, D. J., Chueh, P.-J., and Morre', D. M. (1995). *Proc. Natl. Acad. Sci. USA* **92,** 1831–1835.

Morre, D. J., Sun, E., Geilen, C., Wu, L. Y., de Cabo, R., Krasagakis, K., Orfanos, C. E., and Morre, D. M. (1996). *Eur. J. Cancer* **32A,** 1995–2003.

Muzio, M., Chinnaiyan, A. M., Kischkel, F. C., O'Rourke, K., Shevchenko, A., Ni, J., Scaffidi, C., Bretz, J. D., Zhang, M., Gentz, R., Mann, M., Krammer, P. H., Peter, M. E., and Dixit, V. M. (1996). *Cell* **85,** 817–827.

Nicholson, D. W., Ali, A., Thornberry, N. A., Vaillancourt, J. P., Ding, C. K., Gallant, M., Gareau, Y., Griffin, P. R., Labelle, M., Lazebnik, Y.A., Munday, N. A., Raju, S. M., Smulson, M. E., Yamin, T.-T., Yu, V. L., and Miller, D. K. (1995). *Nature* **376,** 37–43.

Richter, C., Schweizer, M., Cossarizza, A., and Franceschi, C. (1996). *FEBS Lett.* **378,** 107–110.

Samali, A., Nordgren, H., Zhivotovsky, B., Petersen, E., and Orrenius, S. (1999). *Biochem. Biophys. Res.Commun.* **255,** 6–11.

Thannickal, V. J., and Fanburg, B. L. (1995). *J. Biol. Chem.* **270,** 30334–30338.

Thomas, J. A., and Beidler, D. (1986). *Anal. Biochem.* **157,** 32–38.

Thomas, J. A., Zhao, W., Hendrich, S., and Haddock, P. (1995). *Meth.Enzymol.* 423–429.

Thornberry, N. A., and Lazebnik, Y. A. (1998). *Science* **281,** 1312–1316.
Ueda, S., Nakamura, H., Masutani, H., Sasada, T., Yonehara, S., Takabayashi, A., Yamaoka, Y., and Yodoi, J. (1998). *J.Immunol.* **161,** 6689–6695.
Verhaegen, S., McGowan, A. J., Brophy, A. R., Fernandes, R. S., and Cotter, T. G. (1995). *Biochem.Pharm.* **50,** 1021–1029.
Vissers, M. C. M., Pullar, J. M., and Hampton, M. B. (1999). *Biochem. J.* **344,** 443–449.
Watson, F., Robinson, J., and Edwards, S. W. (1991). *J. Biol. Chem.* **266,** 7432–7439.
Wu, Y., Kwon, K., and Rhee, S. G. (1998). *FEBS Lett.* **440,** 111–115.
Yu, B. P. (1994). *Physiol. Rev.* **74,** 139–162.

CHAPTER 14

Methods for Studying Apoptosis and Phagocytosis of Apoptotic Cells in *Drosophila* Tissues and Cell Lines

Kristin White, Simonetta Lisi, Phani Kurada, Nathalie Franc, and Peter Bangs

Cutaneous Biology Research Center
Massachusetts General Hospital
Harvard Medical School
Charlestown, Massachusetts 02129

I. Introduction

Drosophila offers some unique advantages for the genetic dissection of cell death pathways. Similar to the nematode *Caenorhabditis elegans,* apoptosis can be visualized

0091-679X/01 $35.00

in the whole organism throughout development. This ability to assess death in many different tissues in a single preparation simplifies genetic screens and the evaluation of mutant phenotypes. In addition, the vast foundation of knowledge on the developmental genetics of this organism provides an important context to better understand the regulation and role of apoptosis in development.

Several techniques have been developed to label apoptotic cells in *Drosophila* tissues. Both acridine orange (AO) and the TUNEL technique can be used to assess apoptosis in the developing embryo and in the larval imaginal discs and nervous system (Abrams *et al.,* 1993; Bonini *et al.,* 1993; Robinow *et al.,* 1993; White *et al.,* 1994, 1996). The molecular basis of AO staining is controversial, but seems to involve specific retention of the dye in apoptotic cells. The TUNEL technique takes advantage of the DNA cleavage that is typically found in apoptotic cells (Gavrieli *et al.,* 1992). The free DNA ends are targeted by terminal deoxytransferase (TdT), incorporating digoxygenin-labeled nucleotides. This chapter provides protocols for AO and TUNEL labeling, as well as for ultrastructural studies.

Other assays can be used to label apoptotic cells in *Drosophila*. Activation of caspases is a stereotypic feature of apoptotic cells and can be detected *in situ* with affinity-labeling techniques (Kanuka *et al.,* 1999). The binding of Annexin V to exposed phosphatidylserine is another common technique used to label apoptotic cells. This has been used with limited success in flies (van den Eijnde *et al.,* 1998). In some cases, cell counting can also be used to quantify developmental apoptosis (Wolff and Ready, 1991; White *et al.,* 1994; Zhou *et al.,* 1997).

Often the stage at which apoptosis can be assayed is dictated by the tissue or gene being investigated. Apoptosis assays in the embryo provide an overall view of what is happening in all embryonic tissues at many time points. In addition, it may be critical to look at death in the embryo when assessing apoptosis mutants, as genes required to modulate or affect apoptosis may be essential for viability. Even large genomic deletions can be assayed for cell death defects. These deletions usually remove genes required for viability, but the large supply of maternal gene products allows development beyond the stage at which cell death begins (White *et al.,* 1994). Finally the newly developed RNA interference technique may be used in embryos to look at the effects of blocking gene activity on apoptosis (Kennerdell and Carthew, 1998). Assessment of apoptotic events at later stages may provide important information about the specific regulation of the death of particular cell types.

Not only can the regulation and execution of apoptosis be studied in the fly. Franc *et al.,* (1999) demonstrated that the genetic dissection of phagocytosis is also possible in *Drosophila*. Again, the investigation of this process in the embryo is advantageous for the study of genes that may have multiple functions during development. This chapter provides a protocol to assess embryonic phagocytosis of apoptotic cells. Phagocytosis also occurs later in development, a time when hemocytes are abundant and more accessible by bleeding (Rizki and Rizki, 1984).

One disadvantage of studying apoptosis in flies is the limited availability of cell lines. Schneider line 2 (S2) cells are used most commonly, as they are transfected easily. Vectors carrying both inducible and constitutive promoters are available to transfect these cells. Expression of pro- and antiapoptotic genes can be investigated for their ability to induce or protect against apoptosis. In addition, these cells are useful as a source of extracts to study protein interaction by coimmunoprecipitation. These techniques are described in detail in the following section.

II. Detection of Apoptosis in *Drosophila* Tissues

Two methods are particularly useful to detect apoptotic cells in *Drosophila* tissues or embryos. The first relies on the observation that the dye *acridine orange* is selectively retained in dying cells (Fig. 1). This allows one to observe the pattern of apoptosis in living embryos or freshly dissected larval tissues (Abrams *et al.*, 1993; Bonini *et al.*, 1993). The second method utilizes the TUNEL technique (Gavrieli *et al.*, 1992) (Fig. 2). This labels the DNA cleavage products that are generated in apoptotic cells. Each method has advantages. AO staining is very rapid and inexpensive, making it particularly appropriate for screening large numbers of strains. However, it is very transient and can be hard to photograph. TUNEL labeling is much more expensive and time-consuming, but the

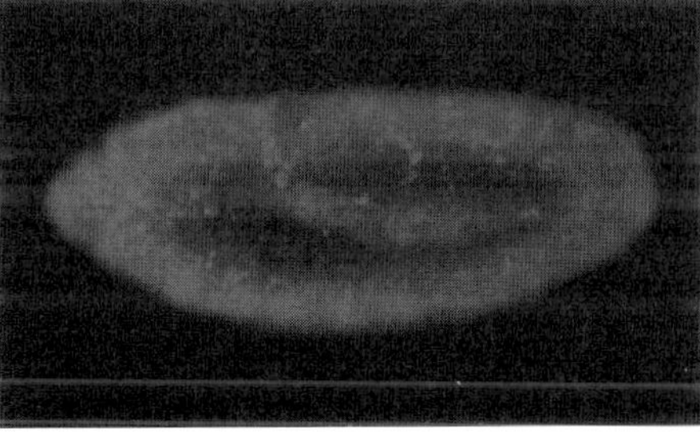

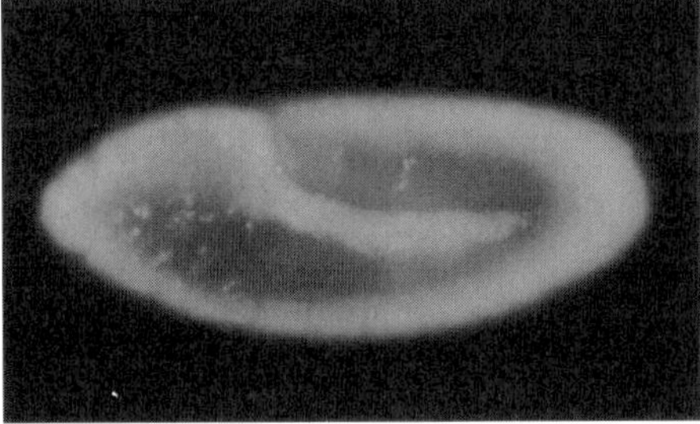

Fig. 1 Acridine orange staining of embryos. This embryo was stained with AO, and the staining was visualized using RITC (top) and FITC (bottom) filters. Each of the bright spots represents an apoptotic cell. At later stages, corpses can be seen clustered within macrophages. Note that more apoptosis is seen in the RITC channel, but that the general background is higher. (See Color Plate.)

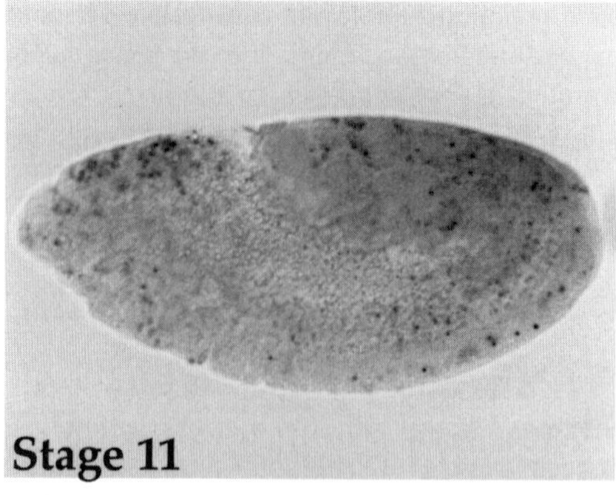

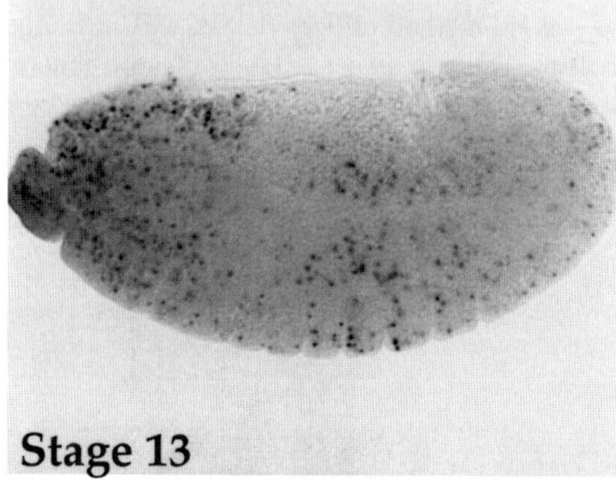

Fig. 2 TUNEL labeling of embryo. Two embryos of different stages are shown. Each dark dot represents the labeled nuclear material of an apoptotic cell.

resulting preparations are essentially permanent. In addition, this technique is performed on fixed tissues and can be combined with immunohistochemistry.

Sectioned material can also be used to visualize apoptotic cells. This allows for a more precise localization of the dead cells. Semithin plastic sections can be stained with methylene blue and toluidine blue and viewed by light microscopy. Transmission electron microscopy allows the most definitive view of apoptotic cells. Techniques for each of these methods are also presented.

In general, it should be noted that the pattern of apoptosis in embryos is very dynamic. Age-matched, wild-type controls are essential when looking for alterations in the pattern or amount of apoptosis. We find that collecting embryos for 1 h and then aging to the appropriate stage gives the most reliable results.

All procedures are at room temperature unless otherwise stated.

A. Acridine Orange (AO) Staining of *Drosophila* Embryos

Embryos are collected for 1 to 4 h at 25°C. For egg collection we use acrylic cages available from Harvard Biological Laboratories Machine Shop (Cambridge, MA). The smaller cages fit on a 60 × 15-mm petri dish. For media, mix 1250 ml water, 180 ml unsulfered molasses, and 47.5 g agar. Autoclave this mixture for 1 h and cool to 60°C. Add 1 ml 10% (w/v) Tegosept in 95% ethanol and 10 ml ethyl acetate. This mixture can then be poured into the petri dishes. A small amount of yeast paste should be put on the plates. Flies lay best after 24 h in the cages.

Eggs are aged until the embryos are 8 to 16 h after egg laying (AEL) Normal cell death starts at about 7.5 h. Embryos should not be too old, as AO staining does not work in embryos with a cuticle. Dechorionate the embryos by washing them into sieves (70-μm nylon cell strainers from Falcon) and soaking them in freshly prepared 50% bleach until the dorsal appendages are no longer visible (Ashburner, 1989). Wash extensively with distilled H_2O, as it is essential to remove all of the bleach.

After dechorionation, the embryos are still surrounded by a hydrophobic membrane. By shaking embryos hard at the interface of heptane and the aqueous dye, some of the dye is driven through this membrane and retained by the apoptotic cells. This same principle is used to fix embryos in later protocols (Zalokar and Erk, 1977). Using a brush, transfer the embryos into 1 ml of heptane in a 12 × 75-mm glass tube. Add 1 ml of 2.5, μg/ml AO (Molecular Probes) in 0.1 *M* phosphate buffer, pH 7.4. This solution should be kept in a foil-covered tube and is stable for months at room temperature. In fact, aged solutions seem to give lower background than fresh ones. A 10-mg/ml stock solution is kept at 4°C. Acridine orange is a mutagen and should be handled with gloves.

Cap the tube and shake hard by hand for 2 min. Place the tube in an Eppendorf shaker for 5 min or continue to shake hard by hand for an additional 5 min. Really hard shaking is necessary to keep the aqueous and organic phases mixed.

Using a glass Pasteur pipette that has be rinsed in heptane, remove embryos from the interface, being careful to take as little of the AO phase (lower) as possible.

Drop the embryos onto a glass slide and quickly remove as much liquid as you can with a Kimwipe. It is better to leave some liquid than to let the embryos dry out. Immediately cover the embryos with a few drops of halocarbon oil and a coverslip.

Observe the staining in either RITC or FITC channels. We use Zeiss filters Blue H 485 and Green H 546, respectively. Although there is less background in the RITC channel, the FITC channel photographs better. Background in the FITC channel fades after a couple of minutes. Staining lasts for about 30 to 45 min but is clearest for the first 15 min. There is strong autofluorescence of the yolk.

B. AO Staining of Imaginal Discs

Dissect discs in 0.1 *M* phosphate buffer, pH 7.4. Discs can be held briefly (10 min or less) in phosphate buffer on ice.

Incubate the discs with 1 μg/ml AO (Molecular Probes) in 0.1 *M* phosphate buffer, pH 7.4, for 5 min at room temperature. Rinse with phosphate buffer and mount in Vectashield

(Vector). Disc staining can be viewed with the same filters used for embryos. With discs, as with embryos, there is some variability in the wild-type patterns of apoptosis, so age-matched controls are essential.

C. TUNEL Staining of Embryos

1. Day 1

Collect the embryos at the desired stage, dechorionate them in 50% bleach, and wash extensively with distilled H_2O, as described earlier. Using a brush, transfer the embryos to 1 ml of heptane in a 12 × 75-mm glass tube. Add 1 ml of 4% paraformaldehyde/0.1 *M* phosphate buffer, pH 7.4. You can make your fix from commercially available 10% formaldehyde, as long as it is EM grade. Do not use formalin, as it has high levels of contaminants and gives inconsistent results. Cap the tubes and shake on an Eppendorf shaker for 30 min.

The embryos will be at the interface of the heptane and the fixative (aqueous) layer. Remove the fixative (lower phase), add 1 ml of methanol to the tube, and shake hard for 2 min to remove the vitelline membrane. Embryos that are devitellinized successfully will sink to the bottom of the tube. Remove almost all of the liquid (the embryos should never be completely dry) and rinse the embryos with methanol. Remove the methanol, add 1 ml of 100% ethanol, and transfer the embryos to a microfuge tube. At this point, the embryos can be stored at −20°C.

Rehydrate the embryos at room temperature through 75, 50, and 25% methanol/PBT [phosphate-buffered saline (PBS) with 0.1% Tween 20] for 5 min at each step. It is important not to have too many embryos in each tube for this assay. Less than 100 embryos per microfuge tube works best. If more embryos are needed, split them into several tubes.

Treat the embryos with 10 μg/ml proteinase K in PBS for 5 min. The timing of this step needs to be calibrated for each batch of enzyme. A 10-mg/ml stock solution of proteinase K should be stored as small aliquots at −20°C. Once thawed, a tube should be discarded after use. After proteinase K digestion, wash the embryos twice for 5 min in PBT.

Postfix the embryos in 4% paraformaldehyde in PBS for 20 min. Wash 5 × 5 min in PBT. The next steps require buffers provided in the Apoptag kit from Intergen.

Incubate the embryos for 1 h in equilibration buffer (100 or 200 μl per sample) at 37°C using a heating block placed on a platform shaker. Gentle agitation throughout the rest of the protocol is essential to obtain consistent staining.

Mix reaction buffer with terminal deoxytransferase (TdT) 2 : 1 and add Triton X-100 to 0.3%. As little as 60 μl of this buffer can be used per sample. Remove the equilibration buffer, add the freshly prepared reaction buffer, and incubate overnight at 37°C with continuous shaking.

2. Day 2

Stop the reaction with the stop/wash buffer diluted 1 : 34 in distilled H_2O. Incubate the embryos for 3–4 h at 37°C with shaking. Wash 3 × 5 min in PBT.

Detection

Block with 2 mg/ml bovine serum albumin (BSA)/5% goat serum in PBT for 1 h. Incubate at 4°C overnight with 1 : 2000 preabsorbed alkaline phophatase-conjugated sheep antidigoxygenin antibody from Boehringer. It is also possible to incubate for 2 h at room temperature and to complete the staining in 2 rather than 3 days. Alternatively, FITC or RITC-conjugated sheep antidigoxygenin antibodies are available. These allow one to visualize the TUNEL staining on the confocal scope and to perform double labelings, as described later.

3. Day 3

Wash 4 × 20 min in PBT. Rinse 2 × 20 min in NTMT buffer (0.1 *M* Tris, pH 9.5, 50 m*M* MgCl , 0.1 *M* NaCl, 0.1% Tween 20).

To 1 ml of NTMT, add 3.5 μl of 100 mg/ml nitroblue tetrazolium chloride (NBT) and 4.5 μl of 50 mg/ml BCIP (5-bromo-4-chloro-3- indolylphosphate) from Boehringer. Add the staining solution to the embryos and watch them under a microscope. The reaction can take up to 10 min. Stop the reaction by washing in PBT.

Mount the embryos in Vectashield (Vector Labs). Slides can be stored for several weeks at −20°C.

Inconsistent stainings are typically caused by (1) too many embryos in a tube, (2) insufficient agitation at all stages, or (3) too much or too little proteinase K.

D. TUNEL Labeling of Eye Discs

Dissect third instar larval eye discs in 0.1 *M* phosphate buffer, pH 7.4, leaving them attached to the mouth hooks. Discs can be held briefly (i.e., 10 min or less) in phosphate buffer on ice. All incubations are carried out in depression slides. Several pairs of discs can be placed in each well.

Fix the discs in 2% paraformaldehyde in 0.1 *M* phosphate buffer, pH 7.4, for 10 min. Wash in 0.1 *M* phosphate buffer, pH 7.4, for 5 min and then in BSS for 2 × 10 min. For 1 liter of BSS, use 2.21 g NaCl, 3.98 g KCl, 3.07 g $MgSO_4$, 0.74 g $CaCl_2 \cdot 1H_2O$, 1.76 g Tricine, 3.6 g glucose, 17.12 g sucrose, and 2 g of BSA, add distilled H_2O to 1 liter, pH to 6.95, and sterile filter. Store at 4°C.

Treat the tissue for 5 min with 10 μg/ml proteinase K in PBS. As described earlier, this step needs to be calibrated for each batch of enzyme. Excess proteinase treatment will cause the discs to fall apart, whereas too little lowers the sensitivity of staining. Wash 2 × 5 min in PBT (PBS + 0.1% Tween).

Postfix in 2% paraformaldehyde (in 0.1 *M* phosphate buffer) for 10 min. Wash 5 × 5 min in PBT.

1. TUNEL Reaction (Reagents from Intergen Apoptag Kit)

All of the following steps are done with gentle agitation on a platform shaker.

1. Incubate for 1 h at room temperature in equilibration buffer provided in the kit. As described earlier, mix the reaction buffer with TdT in a 2 : 1 ratio and add Triton X-100 to 0.3%. Remove the equilibration buffer and add enough of the mix to just immerse the sample. Cover the wells with coverslips and incubate overnight at 37°C in a humid chamber (a covered plastic box lined with a wet paper towel).

2. Make the stop reaction mix by diluting stop/wash buffer in distilled H_2O (1 : 34). Remove the reaction mix from the slide, add the stop solution, and incubate for 3–4 h at 37°C. Wash 3 × 5 min in PBT.

3. Block in BSN (5% goat serum/0.3% Triton X-100 in BSS) for 1 h.

4. To detect TUNEL positive cells, use RITC-conjugated sheep antidigoxygenin antibody (Boehringer) diluted 1 : 200 in BSN. Incubate overnight at 4°C. Wash 4 × 20 min in BSS.

5. Dissect off the mouth hooks and mount in Vectashield mounting media.

E. Double Labeling of Eye Discs for TUNEL and Antibody Staining

This protocol works well with some antigens that are resistant to digestion with proteinase K. We have had excellent experiences double labeling with TUNEL and an anti-ELAV antibody (Robinow and White, 1991). Alternatively, the proteinase K step can be left out, which results in less robust TUNEL labeling.

Proceed with the TUNEL protocol through the incubation with labeling mix and TdT, and the stop reaction.

Block in BSN for 1 h. Then incubate with the primary antibody. To label neurons we use mouse anti-ELAV (Developmental Studies Hybridoma Bank, University of Iowa) at a 1 : 200 dilution in BSN. Incubate overnight at 4°C. Wash 5 × 10 min in BSS.

Combine the secondary antibody with the antidigoxygenin antibody in BSN. For Elav staining, use FITC-conjugated goat antimouse at 1 : 200 and RITC-conjugated sheep antidigoxygenin antibody at 1 : 200. Incubate overnight at 4°C. Wash 4 × 20 min in BSS. Mount in Vectashield mount media. See Fig. 3 (Color Plate) for results.

F. Embedding Embryos in Plastic for Semithin Sections

Note: The fixative used in this protocol is particularly toxic and should be handled in a fume hood with gloves. Osmium should always be handled in the hood to prevent eye damage from the fumes.

Dechorionate the embryos in 50% bleach and wash extensively with distilled H_2O as described previously.

Fix the embryos at the interface of 1 ml fixative (2% EM grade glutaraldehyde/2% EM grade paraformaldehyde/0.1 *M* phosphate buffer, pH 7.2) and 1 ml heptane, shaking at 4°C for 30 min.

Remove the fixative and devitellinize as described in the TUNEL protocol, using 95% MeOH/5% 0.5 *M* EGTA (ME). Place the embryos in a fresh tube, and wash with ME and then with 0.1 *M* phosphate buffer, pH 7.2.

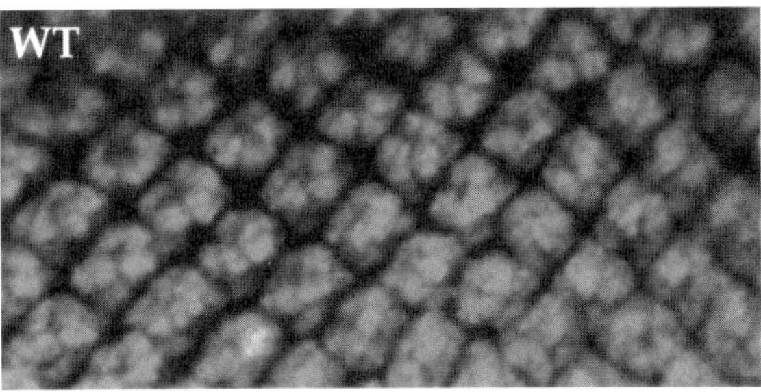

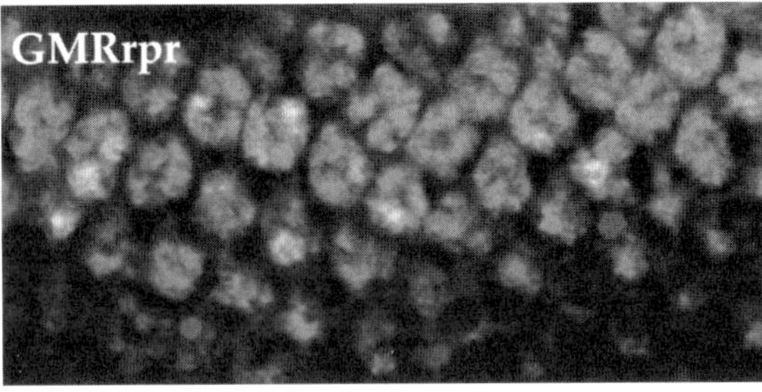

Fig. 3 Eye disc stained with TUNEL and anti-Elav. (Top) A wild-type disc. TUNEL staining for apoptosis is seen in red, while elav expression is seen in green. Very little apoptosis is seen in the differentiating region of the disc. However, both undifferentiated (red only) and differentiated (labeled with both, giving a yellow color) cells are undergoing apoptosis. (Bottom) A disc showing higher levels of apoptosis induced by overexpression of the proapoptotic gene *reaper* (*rpr*). Again both differentiated and undifferentiated cells are undergoing apoptosis. (See Color Plate.)

Postfix with the fixative for 30 min on ice. Wash 3 × 10 min in 0.1 *M* phosphate buffer, pH 7.2, on ice.

Fix in osmium tetroxide (2% in 0.1 *M* phosphate buffer, pH 7.2) for 15 min on ice and for 45 min at room temperature. The embryos will turn black. Wash 5 × 5 min in 0.1 *M* phosphate buffer.

Dehydrate the embryos through an ethanol series: 70% EtOH/30% 0.1 *M* phosphate buffer, 80% EtOH/20% 0.1 *M* phosphate buffer, 90% EtOH/10% 0.1 *M* phosphate buffer, and twice in 100% EtOH for 10 min each.

Incubate in propylene oxide for 2 × 10 min and then in a 1 : 1 mix of propylene oxide and Spurr' s mounting media for 30 min.

Put the embryos in 100% Spurr's resin (Polysciences "standard medium") for 2–5 h at room temperature and then embed in fresh Spurr's at 65–75°C overnight.

Section with a glass knife: 1- to 2.5 μm sections. Place the sections on drops of water on slides subbed with 1% gelatin/0.1% chrome alum (Ashburner, 1989), and dry the slides on a slide dryer. The embryos can be stained with 0.05% methylene blue/0.01% toluidine blue/0.05% di-Na-tetraborate (Ashburner, 1989). The slides are covered in stain and incubated at 55°C for 5 to 10 min. Wash slides in running cold tap water for a minute and then allow to air dry. When the slides are completely dry, mount with mounting medium and a coverslip. Using light microscopy, apoptotic cells can be seen as condensed, dark bodies.

G. Fixation of Embryos for Transmission EM

Note: See the warning on the fixative in the previous section. Uranyl acetate is radioactive and should be kept away from film.

Dechorionate embryos in 50% bleach and wash extensively with distilled H_2O.

Fix in EM fixative (2% glutaraldehyde/2% paraformaldehyde/2% acrolein/0.1 *M* phosphate buffer, pH 7.2), shaking at 4°C for 30 min. Remove the fixative phase, leaving the embryos in heptane. Devitellinize the embryos by hand. This is done most easily by sticking a piece of double-sided tape in the bottom of a petri dish. Drop a small drop of heptane containing some embryos onto the tape. Very quickly blow away the heptane and cover the embryos with 0.1 *M* phosphate buffer, pH 7.2. The embryos must not dry out. Place the dish under a dissecting microscope and, with a glass needle, push the embryos gently out of the vitelline membrane. Devitellinized embryos will float free of the tape. Collect the devitellinized embryos with a glass pipette and place in a glass tube containing 0.1 *M* phosphate buffer. The embryos are very sticky at this point so care must be taken to prevent them from attaching to the pipette.

Postfix the embryos for 30 min on ice. Wash 3 × 10 min in 0.1 *M* phosphate buffer on ice. Fix in 2% osmium tetroxide for 15 min on ice and for 45 min at room temperature. Embryos will turn black. Wash 5 × 5 min in distilled H_2O.

Stain in 1% uranyl acetate in distilled H_2O for 12–16 h at 45–55°C. Wash 5 × 5 min in distilled H_2O.

Dehydrate the embryos through an ethanol series: 70, 80, 90, 100, and 100% in water for 10 min each.

Incubate in propylene oxide for 2 × 10 min and then in a 1 : 1 mix of propylene oxide and Spurr's mounting media for 30 min.

Put the embryos in 100% Spurr's resin (Polysciences "standard medium") for 2–5 h at room temperature and then embed in fresh Spurr's at 65–75°C overnight. Embryos can then be sectioned for transmission EM, mounted on grids, and viewed without further staining (Fig. 4, see Color Plate).

III. Phagocytosis

To evaluate the efficiency of phagocytosis of apoptotic corpses in embryos, we developed an assay that utilizes antibody staining to mark the hemocytes and a DNA dye that allows the visualization of apoptotic corpses (Franc *et al.*, 1999). For hemocyte markers we have made use of the anti-peroxidasin antibody developed by L. Fessler (Nelson *et al.*, 1994) and the anti-croquemort antibody (Franc *et al.*, 1996); however, other markers should be usable. In particular, markers that are localized in the hemocyte cytoplasm or membrane allow one to score if a corpse has been engulfed by the hemocyte.

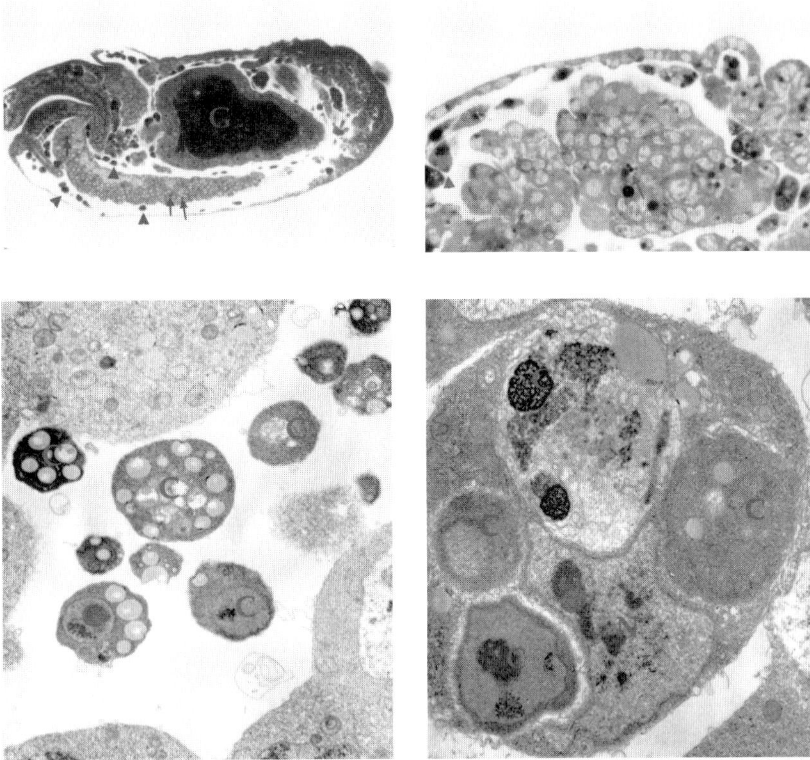

Fig. 4 Assessment of apoptosis by light microscopy of semithin sections and transmission electron microscopy. (Top) Embryos embedded in plastic, sectioned, and stained with methylene blue/toluidine blue/borax. (Right) A high magnification view of the head region. Many apoptotic cells are seen engulfed within macrophages (arrowheads). Unengulfed corpses can also be seen as intensely stained dots (arrows). The nucleolus in all cells can be seen as a dark spot within each nucleus, as can the lipid within the gut (G). (Bottom) Electron micrographs of unengulfed corpses on the left and a large macrophage on the right. Corpses are marked with a C and can be seen as electron-dense, condensed bodies. A large number of vesicles can also be seen within the unengulfed corpses. The macrophage on the right has engulfed at least four corpses. The electron-dense material in the macrophage nucleus (N) is the nucleolus. (See Color Plate.)

The DNA dye 7-amino actinomycin D (7AAD) labels all DNA. However, the signal in living cells is much weaker and more diffuse. Thus on the confocal microscope, it is easy to detect apoptotic corpses when the level of signal detection is decreased.

To quantify the number of corpses engulfed by an individual macrophage, optical sections are taken through an isolated macrophage. A fixed step size should be used to compensate for the variable sizes of macrophages. The number of 7AAD-bright bodies contained within the macrophage can be counted easily. To assess the overall efficiency of engulfment, macrophages in several regions of the embryo should be assayed.

All procedures are at room temperature unless otherwise stated.

A. Assay for Phagocytosis of Apoptotic Cells by Embryonic Macrophages

Collect embryos for 2 h, age to appropriate stages, and dechorionate in 50% bleach. Wash the embryos extensively with distilled H_2O, once quickly in a 0.1% Triton X-100 solution, and further with distilled H_2O. Transfer the embryos into the fixative (1 : 1 volume of 4% paraformaldehyde in PBS, pH 7.4, with heptane), fix for 15 min at room temperature, and devitellinize with methanol (as described earlier). Wash the devitellinized embryos twice quickly in 100% methanol to remove debris. Transfer the embryos to a new tube and wash with 100% EtOH. Embryos can be stored for several months at −20°C in 100% EtOH or processed immediately for the phagocytosis assay as follows.

All of the following steps are performed with moderate shaking on a platform shaker at room temperature unless otherwise specified. Rinse the devitellinized embryos twice in 100% ethanol and rehydrate through a series of ethanol/PBSS (PBS+ 0.0125% saponin) washes of 5 min each (70% EtOH, 50% EtOH, 30% EtOH, 100% PBSS). Block in PBSS/0.4% BSA/5% goat serum (PSN) for 45 min at room temperature. Incubate the embryos with mouse anti-peroxidasin and/or rabbit anti-croquemort (IgG purified) antibodies, each at a 1 : 1000 dilution in PSN, overnight at 4°C. Rinse the embryos three times in PBSS and wash three times for 15 min each in PBSS. Incubate the embryos with appropriate secondary antibodies at a 1 : 1000 dilution in PSN for 1 h. If fluorescent secondary antibodies are used, the tubes should be covered with foil. Rinse the embryos three times in PBSS and wash three times in PBS for 15 min. Stain the embryos with a solution of 5 μg/ml of 7AAD (Molecular Probes) for 30 min at room temperature, quickly wash three times in PBS, and mount immediately in mounting medium (Vectashield, Vector laboratories). Results are analyzed by confocal microscopy: individual macrophages can be observed using a 100x objective and a zoom of 4.0 (Fig. 5, see Color Plate).

IV. Testing for Apoptotic Activities in *Drosophila* S2 Cells

A variety of vectors are available for use in *Drosophila* Schneider line 2 (S2) cells. To test the killing and protective activities of pro- and antiapoptotic genes, we have found that the metal inducible pRmHA3 vector (Bunch *et al.*, 1988) is particularly useful. Because cultures can be split after transfection, the noninduced control is matched exactly for transfection efficiency with the test cells. In addition, the time from the induction of expression to the onset of morphological changes can be measured precisely. To mark transfected cells we use either lacZ or GFP reporter expressed from a constitutive promoter, using the pIE vector series (Novagen).

One drawback to using S2 cells is that they are phagocytic. Apoptotic bodies can be found as corpses within living, nontransfected cells, which presents problems for assessing the number of dead transfected cells. We have developed protocols to count dead cells by hand, but thus far have been unsuccessful in using flow cytometry to assay apoptosis in this cell line.

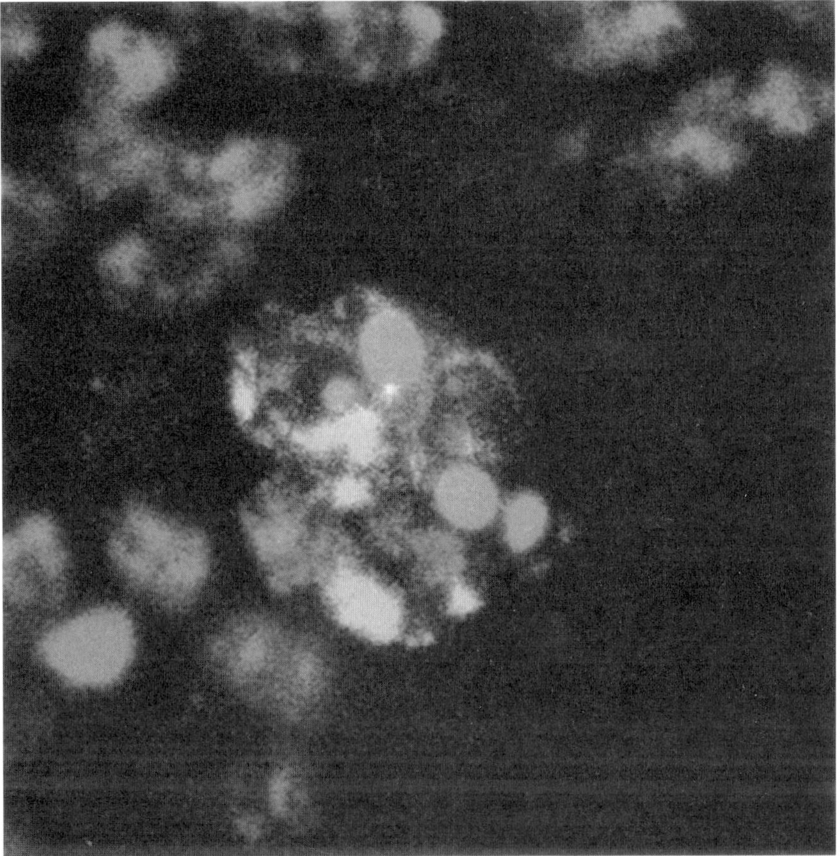

Fig. 5 Embryonic macrophage labeled with antiperoxidasin, anticroquemort, and 7AAD. The macrophage cytoplasm is labeled for peroxidasin (green), while the membranes surrounding the apoptotic corpses show croquemort (blue). Nuclei are labeled with 7AAD (red). Apoptotic nuclei can be seen as bright, condensed bodies. (See Color Plate.)

This section presents protocols for transfecting S2 cells, assaying apoptosis in these cells, and immunoprecitation to assay for physical interactions between transfected proteins.

All procedures are at room temperature unless otherwise stated.

A. Transfection of Cells

S2 cells can be obtained from the ATCC. It should be noted that strains of S2 cells show considerable variability in growth requirements. We have not used the ATCC strain.

Plate cells in complete medium (Cellgro DS2 medium supplemented with 2 m*M* L-glutathione and 10% fetal calf serum) at ~300,000 cells per square centimeter of plate surface (i.e., $\sim 3 \times 10^6$ cells for six-well plates or $\sim 7.5 \times 10^6$ cells for 25-cm^2 flasks). Allow the cells to settle for several hours so that they can adhere to the surface. Three hours is generally sufficient. Overnight is convenient, but if more than a single generation time passes (~12 h), the cells will become confluent and will no longer adhere to the plate.

Transfection of S2 cells works very well using the Cellfectin reagent from GIBCO/BRL and following the manufacturer's protocol. Briefly, two solutions are prepared separately and then combined prior to adding to the cells. The first solution consists of the Cellfectin reagent diluted into serum-free medium. The optimal amount of Cellfectin used per transfection may vary for specific applications, but 10 μl Cellfectin diluted into 100 μl of serum-free medium provides satisfactory results in six-well plates. The second solution consists of the DNA diluted into serum-free medium. Again, specific applications may require optimization, but 3–10 μg of DNA diluted into 100 μl of medium works well for six-well plates. Quantities of all reagents may be scaled up or down for plates of different sizes. DNA of sufficient quality can be obtained using Qiagen Midi- or Maxi-preps, as well as by a number of other methods. Circular plasmid works well, and we have not noted any effects of different plasmid backbones. Prepare the solutions separately in 12 × 70-mm polypropylene tubes and then combine to allow the DNA–lipid complexes to form (10–15 min). While the complexes are forming, aspirate the growth medium from the cells and rinse them once with serum-free medium. Be gentle with the cells to avoid dislodging them from the surface of the wells/plates. Add an appropriate amount of serum-free medium to the DNA–lipid complexes (800 μl for six-well plates) and overlay the cells with the solution. Incubate the cells at 26°C for 5–6 h, aspirate the transfection solution from the cells, and add 3–5 ml of complete medium (with 10% fetal calf serum). The amount of time that the cells are in the transfection solution can be varied for optimal transfection efficiencies. Toxicity becomes a problem with longer incubations. Incubate in complete medium for at least 24 h before assaying.

B. Assays for Cell Death

1. Using β-Galactosidase

Include a 1 : 10 molar ratio of a constitutively expressed β-galactosidase construct in the transfection. The pIE vector series available from Novagen use a baculovirus promoter and work very well. If using an inducible promoter for the expression of cell death genes, transfer equal numbers of transfected cells to two fresh plates and induce expression in one. We use the metallothionein promoter in the pRmHA3 vector and induce expression by adding $CuSO_4$ to a final concentration of 500–700 μ*M*. To inhibit caspases, the inhibitors Z-DEVD-FMK and Z-VAD-FMK (Calbiochem) can be added along with the $CuSO_4$ at a concentration of 50 μ*M* each.

After the appropriate period of expression, rinse the cells with PBS and fix with 2% formaldehyde in PBS for 5 min at room temperature. Rinse the cells again with PBS and stain for β-galactosidase expression by overlaying the cells with 0.1% X-Gal/5 m*M* $K_3Fe(CN)_6$/5 m*M* $K_4Fe(CN)_6$/2 m*M* $MgCl_2$/1X PBS.

Allow the cells to stain at room temperature (or 37°C if you are in a hurry) for 4 h and then rinse with PBS to avoid precipitation.

The extent of cell death can be assayed by determining the percentage of blue cells that remain attached to the plates when compared to uninduced controls. This provides a reasonable approximation, but bear in mind that dying cells do not always detach right away. It is also important to remember that S2 cells will engulf the corpses of their neighbors, so cells with large blue foci in their cytoplasm are probably not transfected with β-galactosidase but have instead engulfed a dead transfected cell.

An alternative assay for dying cells is to determine the percentage of blue-stained cells that have an apoptotic morphology. The advantage of this technique is that dying cells that have yet to detach from the plate are included in the nonviable pool. The disadvantage is that early stages of apoptosis are difficult to score, and the analysis can become overly subjective (Fig. 6, see Color Plate).

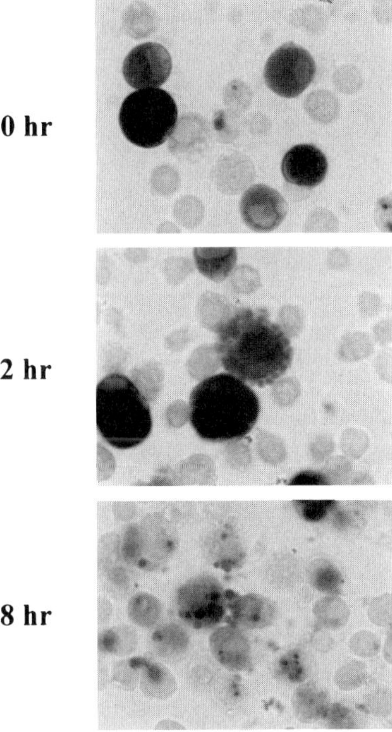

Fig. 6 β-Galactosidase assay. Cells were transfected with a construct containing the apoptosis-activating gene *reaper* (*rpr*) under the control of the metallothionein promoter and a 1 : 10 molar ratio of a constitutively expressed β-galactosidase construct. The transfected cells were transferred to fresh plates and Rpr expression was induced for the time periods shown before the cells were fixed and stained for β-galactosidase expression. The blue cells are those that have been transfected. Rpr expression leads to a characteristic apoptotic morphology within 2 h of induction, as shown by the blebbing cell. Engulfment of dying cells begins promptly, and by 8 h, many of the corpses are phagocytosed by neighboring cells. (See Color Plate.)

2. Using Green Fluorescence Protein

Include a 1 : 10 molar ratio of a constitutively expressed GFP construct in the transfection. Induce the expression of the test protein for the desired period of time, transfer the cells to a fresh tissue culture plate, and allow the cells to settle for 60–90 min. This technique takes advantage of the observation that dead and/or dying cells do not readhere to a fresh plate as well as viable cells do. After the cells have settled, rinse them with PBS and fix them with 2% formaldehyde in PBS for 5–10 min. Rinse them again with PBS and determine the percentage of adherent cells that are GFP positive. This technique is also imperfect as some nonviable cells will attach, but when compared with uninduced controls, reproducible numbers can be obtained. Engulfed corpses are fairly easy to distinguish from viable cells when they are viewed by fluorescence as they are considerably smaller and the engulfing cell is not itself fluorescent (Fig. 7, see Color Plate).

C. Immunoprecipitations

Approximately 48 h posttransfection, determine the number of cells in each well with a hemocytometer. Transfer equal numbers of cells (at least 1×10^6) to centrifuge tubes and pellet at $\sim$500*g* for 5 min at room temperature. Resuspend the cells in 1 ml of

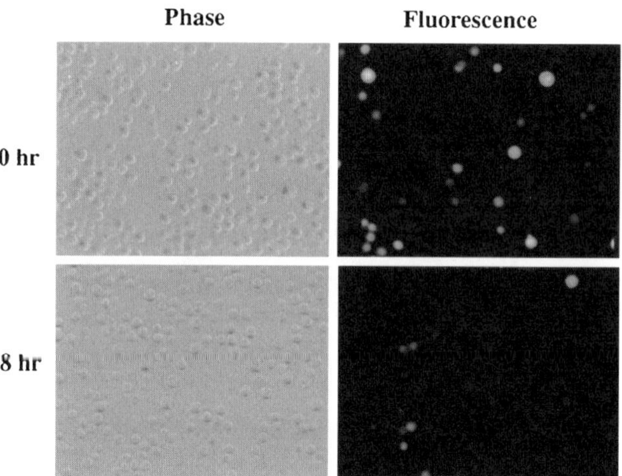

Fig. 7 GFP assay. Cells were transfected with a construct containing the apoptosis-activating gene *rpr* under the control of the metallothionein promoter and a 1 : 10 molar ratio of a constitutively expressed GFP construct. Transfected cells were induced to express Rpr for the indicated times and were then transferred to fresh plates and allowed to settle. The plates were rinsed and the remaining cells were fixed. Fields of cells were photographed in visible light to determine total cell number and then photographed with fluorescence activation to observe which of the cells were expressing GFP. The percentage of viable transfected cells (as determined by their ability to readhere to the plate) decreases dramatically by 8 h postinduction. (See Color Plate.)

complete medium supplemented with 50 μM each of the caspase inhibitors Z-DEVD-FMK and Z-VAD-FMK (Calbiochem). Induce the expression of transfected constructs by the addition of $CuSO_4$ to 500–700 μM and return the cells to a 35- or 60-mm tissue culture plate. We find that 4–5 h of expression is sufficient for easy detection of the proteins of interest, although this may vary with different proteins.

Recover the cells by transferring the medium to a centrifuge tube and rinsing the plate twice with 1 ml PBS. Add the PBS used for the rinses to the medium in the centrifuge tube. Pellet the cells in the centrifuge tube for 5 min at 1000*g* at room temperature, wash with 5 ml cold PBS, and repellet at 4°C. In the meantime, lyse adherent cells on the plate with 1 ml of ice-cold Nonidet P-40 (NP-40) lysis buffer (50 m*M* Tris–HCl (pH 8.0)/150 m*M* NaCl/1% NP-40/1 m*M* dithiothreitol/1 m*M* phenylmethyl sulfonyl fluoride). Recover the lysate from the plate and use it to resuspend the cell pellet from the rinses. This technique is more quantitative than trying to scrape all of the cells from the plate before lysing. Transfer the lysate to a 1.5-ml microfuge tube and tumble for 10–15 min at 4°C to ensure complete lysis of the cells. Pellet the cell debris at 4°C at top speed in a microfuge for 30 min. Transfer the supernatant to a fresh 1.5-ml microfuge tube.

Preclear the lysate by adding 30–50 μl of a 50% slurry of whichever beads will be used for the immunoprecipitation. We use Gammabind beads from Pharmacia, but protein A- or protein G-Sepharose will work as well depending on the antibody used. Tumble the lysate and beads for 60–90 min at 4°C and pellet the beads with a brief spin. Transfer the supernatant to a fresh tube, but keep the precleared beads for analysis later. Add 1–2 μg of the precipitating antibody to the precleared cell lysate and tumble at 4°C for 4 or more hours. We typically go overnight.

Precipitate the antibody–antigen complexes by adding 30–50 μl of beads and tumble for 90–120 min at 4°C. Recover the beads by pelleting briefly and wash by resuspending in 500 μl of cold lysis buffer. Wash the beads at least three times by pelleting and resuspending in fresh changes of lysis buffer. A convenient way to do this without losing the beads is to use a 25-gauge (or smaller) hypodermic needle to aspirate the buffer from the beads. The opening in the needle is too small to allow the passage of most beads. It is important to aspirate as much buffer as possible from the beads during each wash. Wash the preclear beads at this time as well.

Resuspend the beads in 40–50 μl of 1X Laemmli SDS–PAGE (Laemmli, 1970) loading buffer, denature by heating to 90°C for 3 to 5 min, pellet, and load the samples onto an appropriate gel. Try to avoid loading any beads into the wells of the gel. It is a good idea to include the resuspension from the preclear beads in the first few gels that are run to ensure that any observed bands are not the result of nonspecific interactions with the beads.

Acknowledgments

We thank Ilaria Mazzon and Christian Peterson who helped develop and refine these protocols. K. White's research is supported by a grant from the NIH and from the Shiseido Company of Japan to Massachusetts General Hospital. P. K. is supported by a fellowship from the American Cancer Society, and N. F. by a fellowship from the Irvington Institute.

References

Abrams, J. M., White, K., Fessler, L. I., and Steller, H. (1993). Programmed cell death during *Drosophila* embryogenesis. *Development* **117,** 29–43.

Ashburner, M. (1989). "*Drosophila*: A Laboratory Manual." Cold Spring Harbor Laboratory Press, Cold Spring Harbor, NY.

Bonini, N. M., Leiseron, W. M., and Benzer, S. (1993). *The eyes absent* gene: Genetic control of cell survival and differentiation in the developing *Drosophila* eye. *Cell* **72,** 379–395 .

Bunch, T. A., Grinblat, Y., and Goldstein, L. S. (1988). Characterization and use of the *Drosophila* metallothionein promoter in cultured *Drosophila melanogaster* cells. *Nucleic Acids Res.* **16,** 1043–1061.

Franc, N., Heitzler, P., Ezekowitz, R., and White, K. (1999). Requirement for croquemort in phagocytosis of apoptotic cells in *Drosophila*. *Science* **284,** 1994–1998.

Franc, N. C., Dimarcq, J. L., Lagueux, M., Hoffmann, J., and Ezekowitz, R. A. (1996). Croquemort, a novel *Drosophila* hemocyte/macrophage receptor that recognizes apoptotic cells. *Immunity* **4,** 431–43.

Gavrieli, Y., Sherman, Y., and Ben-Sasson, S. A. (1992). Identification of programmed cell death *in situ* via specific labeling of nuclear DNA fragmentation. *J. Cell Biol.* **119,** 493–501.

Kanuka, H., Hisahara, S., Sawamoto, K., Shoji, S., Okano, H., and Miura, M. (1999). Proapoptotic activity of *Caenorhabditis elegans* CED-4 protein in *Drosophila*: Implicated mechanisms for caspase activation. *Proc. Natl. Acad. Sci. USA* **96,** 145–150.

Kennerdell, J. R., and Carthew, R. W. (1998). Use of dsRNA–mediated genetic interference to demonstrate that frizzled and frizzled 2 act in the wingless pathway. *Cell* **95,** 1017–1026.

Kurada, P., and White, K. (1998). Ras promotes cell survival in *Drosophila* by downregulating hid expression. *Cell* **95,** 319–329.

Laemmli, E. K. (1970). Cleavage of structural proteins during the assembly of the head of bacteriophge T4. *Nature* **227,** 680–685.

Nelson, R., Fessler, L., Takagi, Y., Blumberg, B., Keene, D., Olson, P., Parker, C., and Fessler, J. (1994). Peroxidasin: A novel enzyme-matrix protein of *Drosophila* development. *EMBO J.* **13,** 3438–3447.

Rizki, T., and Rizki, R. (1984). The cellular defense system of *Drosophila* melanogaster. *In* "Insect Ultrastructure" (R. King and H. Akai, eds.), pp. 579–604, Plenum Press, New York.

Robinow, S., Talbot, W. S., S., Hogness, D. S., and Truman, J. W. (1993). Programmed cell death in the *Drosophila* CNS is ecdysone-regulated and coupled with a specific receptor isoform. *Development* **119,** 1251–1259.

Robinow, S., and White, K. (1991). Characterization and spatial distribution of the ELAV protein during *Drosophila melanogaster* development. *J. Neurobiol.* **22**(5), 443–461.

van den Eijnde, S., Boshart, L., Baehrecke, E., De Zeeuw, C., Reutelingsperger, C., and Vermeij-Keers, C. (1998). Cell surface exposure of phosphatidylserine during apoptosis is phylogenetically conserved. *Apoptosis* **3,** 9–16.

White, K., Grether, M. E., Abrams, J. M., Young, L., Farrell, K., and Steller, H. (1994). Genetic control of programmed cell death in *Drosophila*. *Science* **264,** 677–683.

White, K., Tahaoglu, E., and Steller, H. (1996). Cell killing by the *Drosophila* gene *reaper*. *Science* **271,** 805–807.

Wolff, T., and Ready, D. F. (1991). Cell death in normal and rough eye mutants of *Drosophila*. *Development* **113,** 825–839.

Zalokar, M., and Erk, I. (1977). Phase-partition fixation and staining of *Drosophila* eggs. *Stain Technol.* **52,** 89–95.

Zhou, L., Schnitzler, A., Agapite, J., Schwartz, L. M., Steller, H., and Nambu, J. R. (1997). Cooperative functions of the *reaper* and *head involution defective* genes in the programmed cell death of *Drosophila* central nervous system midline cells. *Proc. Natl. Acad. Sci. USA* **94,** 5131–5136.

CHAPTER 15

Phosphatidylserine Exposure and Phagocytosis of Apoptotic Cells

Patrick Williamson,* Stefan van den Eijnde,† and Robert A. Schlegel‡

* Department of Biology
Amherst College
Amherst, Massachusetts 01002

† Cardiovascular Research Institute Maastricht
Department of Biochemistry and Molecular Cell Biology and Genetics
University of Maastricht
6200 MD Maastricht, Netherlands

‡ Department of Biochemistry and Molecular Biology
Pennsylvania State University
University Park, Pennsylvania 16802

METHODS IN CELL BIOLOGY, VOL. 66

0091-679X/01 $35.00

I. Introduction

A functional component of apoptosis is a process of cell removal: presentation of a distinctive surface phenotype by the apoptotic cell results in its recognition and phagocytosis. The development of this phenotype is an early step in the apoptotic process (Verhoven *et al.,* 1995) and includes an alteration in the distribution of phospholipids across the plasma membrane, which brings the phospholipid phosphatidylserine (PS) to the cell surface (Fadok *et al.,* 1992; Martin *et al.,* 1995). Because animal cells generally restrict PS in the inner leaflet of their plasma membrane under normal conditions and display it when they become apoptotic (van den Eijnde *et al.,* 1997a, 1998), loss of lipid asymmetry and the consequent appearance of PS on the cell surface have become a widely used hallmark of apoptosis. This chapter describes methods for detecting apoptotic cells by loss of asymmetry and exposure of PS and methods for assessing phagocytosis of apoptotic cells.

A. Loss of Lipid Asymmetry as a Marker for Apoptotic Cells

The use of a single word, apoptosis, to connote the process of programmed cell death belies the variation and complexity of the process. Many different stimuli act through a variety of different pathways to trigger cell death (Kroemer *et al.*, 1998). Beyond these different induction pathways, however, the execution events are similarly not stereotyped: canonical events in the process, such as cell shrinkage, DNA degradation, and loss of lipid asymmetry, can occur at different times along the pathway, or even in different sequences, depending on the cell type and perhaps the stimulus (van den Eijnde *et al.*, 1999; Diaz *et al.*, 1999). Even more importantly, the initiation of the cell death program in most cases is not synchronous within a population (Verhoven *et al.*, 1995) so that the fraction of cells actually undergoing apoptosis at any time after stimulation may be quite small, and the fraction of cells engaging in a particular apoptotic event even smaller. Some assays, such as those for DNA degradation, loss of membrane integrity, or cell shrinkage and fragmentation, are insensitive to this reality because they are generally end-stage measures of the number of cells that have completed the apoptotic sequence at some time in the past.

Assays of lipid distribution depend on the integrity of the plasma membrane to define the difference between outer and inner leaflets because loss of membrane integrity allows probes access to the cell interior, negating the principle on which the assays are based. Therefore, by definition, assays of lipid asymmetry are not end-stage measures. At any time after the induction of apoptosis, within a population of intact normal cells with a normal asymmetric phospholipid distribution, and of lysed end-stage apoptotic cells, whose phospholipid distribution can no longer be reliably determined, there may exist relatively few intact apoptotic cells in which asymmetry has been lost. Bulk assays of populations provide information on loss of asymmetry averaged across the population and therefore may lack the sensitivity to detect these small numbers of apoptotic cells within the population. Thus, assessment of asymmetry is most sensitive and meaningful when carried out utilizing single-cell assays such as microscopy and flow cytometry, which examine large numbers of individual cells and can therefore identify and quantify subpopulations of apoptotic cells. The size of the subpopulation in which loss of asymmetry can be identified depends on how synchronously the cell populations enter apoptosis after stimulation, how long lipid symmetric cells persist before cell lysis, and how long the entire apoptotic process takes to reach completion, including phagocyte clearance and intraphagolysosomal degradation *in vivo*. For this reason, comparisons of the number of intact, lipid-symmetric cells between different cell types should be treated with caution (Fadeel *et al.*, 1999).

Several tools are available for assessing the distribution of phospholipids across the plasma membrane as well as the transbilayer lipid movements that underlie that distribution. The normal phospholipid distribution is one in which phosphatidylcholine (PC) and sphingomyelin are present mainly in the external leaflet, while most of the phosphatidylethanolamine (PE) and all of the PS are located in the internal leaflet (Devaux, 1991; Williamson and Schlegel, 1994). Loss of lipid asymmetry, through randomization of lipids across the bilayer, results in exposure of PS on the cell surface, where it can be detected by binding of the Ca^{2+}-dependent, PS-specific protein, Annexin V. Loss of asymmetry also alters the biophysical properties of the plasma membrane bilayer (Williamson and Schlegel, 1994) and can therefore also be detected in some circumstances by measuring the binding of the naturally fluorescent dye merocyanine 540 (MC540) (Choe *et al.*, 1985, 1986). Abolition of the normal lipid distribution requires inactivation of the aminophospholipid translocase, the enzyme responsible for maintaining lipid asymmetry, and activation of a membrane protein called the scramblase, which catalyzes transbilayer movement of all classes of phospholipids (Verhoven *et al.*, 1995). The activity of these enzymes that control the distribution of phospholipids can be assayed by measuring the transbilayer movements of phospholipid analogs.

B. Annexin Assays for Loss of Lipid Asymmetry in Apoptotic Cells: General Considerations

The most widespread assay for loss of lipid asymmetry in apoptotic cells is measurement of the level of binding of a fluorescent derivative of the Ca^{2+}-dependent, PS-binding protein Annexin V. These derivatives are available from commercial suppliers, a partial list of whom is given in Table I. Reagents are generally provided in units of "assays" (e.g., 100 assays) together with binding buffer and, in some cases, propridium iodide to make a kit. The "assay" unit is not standard, and there is no obvious relationship between the amount of functional Annexin V per "assay" for PS from different suppliers, or even for Annexin from the same supplier at different times. FITC–Annexin V is preferable for most flow cytometry assays because of the correspondence between the excitation maximum of fluorescein and the 488-nm excitation maximum of the argon laser. Of the other forms of Annexin available, the biotinylated one is potentially the most versatile. However, we have found that some commercially available forms of biotinylated Annexin contain little active Annexin, so some caution should be taken in designing experiments using these probes. In particular, the usual quality control data provided by most manufacturers consist of the difference in fluorescence between cells labeled with both biotinylated Annexin V and fluorescent streptavidin vs cells labeled with fluorescent streptavidin alone. Such assays may give satisfactory results *in vitro* even if the amount of active biotinylated Annexin is quite low, as large amounts of inactive Annexin will not prevent binding by a small amount of active Annexin in the preparation. However, it may take longer incubations to achieve normal signal/noise ratios, and the actual number of bound Annexin molecules on labeled cells may be small. In contrast, low amounts of active Annexin V may seriously hamper *in vivo* experimens, as these kind of experiments often do not permit increased incubation periods.

For those investigators who wish to prepare and characterize their own Annexin, either to have access to standardized preparatons or simply to avoid the high cost of commercial Annexin, it is possible to purify Annexin from bacteria containing an expressible Annexin V gene. A fairly simple purification scheme, which exploits the PS-specific binding properties of the protein, provides a reagent of sufficient purity and activity for most routine purposes. The following protocol is adapted from the method of Burger and co-workers (1993).

II. Annexin V Purification

A. Summary

Escherichia coli strain pRK6 from the American Type Culture Collection (Cat No. 67916; Manassas, VA; www.atcc.org) is grown in LB medium, and expression from

Table I

Company	Contact	Derivative	Catalog #
Boehringer (Roche)	biochem.roche.com/apoptosis	FITC	1828681
		Biotin	1828690
		Alexa 568	1985485
BioVision	biovision@aol.com	—	1005-100
		FITC	1001-1000
		Cy3	1002-1000
		EGFP	1004-1000
		Biotin	1003-100
Oncor	www.oncor.com	FITC	S7140-KIT
		Biotin	S7142-KIT
Nexins Research/ Molecular Probes	www.nexins.com/ www.probes.com	FITC	A700/A-13199
		Oregon Green	G650/A-13200
		Alexa 488	D650/A-13201
		Alexa 568	E650/A-13202
		Alexa 594	F650/A-13202
		Biotin	B700/A-13204
Calbiochem	www.calbiochem.com	FITC	PF032
		Biotin	PF036
R&D Systems	www.rndsystems.com	FITC	TA4638
		Biotin	TA4619
Clontech	www.clontech.com	—	8130-1
		FITC	8133-1
		Biotin	8131-1
		Cy3	8132-1
Trevigen	www.trevigen.com	FITC	4830-01-K
		Biotin	4835-01-K
Caltag	www.caltag.com	FITC	ANNEXINV01
		R-PE	ANNEXINV04
		APC	ANNEXINV05
		Biotin	ANNEXINV15
Bender MedSystems	www.bendermedsystems.com	—	BMS306a
		FITC	BMS306fi
		Biotin	BMS306bt
		PE	BMS306pe
		APC	BMS306apc
BD Pharmingen	www.pharmingen.com	—	65871A
		Biotin	65872X
		FITC	65874X
		PE	65875X

the lac promoter is induced with isopropylthiogalactoside (IPTG). A bacterial lysate is prepared and cleared, and vesicles containing PS added, in the presence of Ca^{2+}, to bind Annexin V. The vesicles are recovered by centrifugation, washed, and the bound Annexin released by the chelation of Ca^{2+}.

B. Preparation of Bacterial Spheroblasts

1. Inoculate 4 ml of LB medium containing 50 mg/ml Ampicillin (LB-amp) with a bacterial colony from a fresh plate or with an aliquot from a frozen stock; incubate overnight at 37°C with shaking.

2. Inoculate 500 ml of LB-amp in a 2-liter flask with the 4-ml bacteria culture; incubate overnight at 37°C with shaking.

3. Transfer 100 ml of this bacteria culture to each of five 2-liter flasks containing 500 ml of LB-amp; incubate at 37°C with shaking.

4. When the OD_{600} of the culture reaches 1.5 to 2.0, add IPTG to a concentration of 1 m*M*.

5. After 2–4 h of growth, harvest bacteria by centrifugation in six centrifuge bottles (250 ml each) at 5000*g* (5500 rpm in Sorvall GSA rotor) for 15 min at 4°C; pour off supernatant and refill the bottles with the balance of culture and centrifuge again so that all bacteria have been collected.

6. Resuspend the cells in each bottle in 10 ml of spheroblast buffer.

7. Add 1.1 ml of 10 mg/ml lysozyme in spheroblast buffer to each bottle (giving a final lysozyme concentration of approximately 1 mg/ml).

8. Immediately after addition of the lysozyme solution, add 70 ml of 0.5 × spheroblast buffer (spheroblast buffer diluted 1 : 1 with water) to each bottle.

9. Incubate the spheroblast suspension for 30 min on ice with gentle shaking.

10. Collect the spheroblasts by centrifugation for 30 min at 14,000*g* (9500 rpm in a Sorvall GSA rotor) at 4°C.

11. Resuspend spheroblasts in the largest volume of ultracentrifugation buffer that can be accepted by the available ultracentrifuge rotor (200 ml for a six-bucket Beckman SW28 rotor).

12. Centrifuge at 100,000*g* (27,000 rpm in SW28 rotor) at 4°C overnight. Collect supernatant.

C. Preparation of Liposomes

1. Dissolve 300 mg of crude phosphatidylserine (bovine brain extract type III, Sigma) in <3 ml of chloroform/methanol (2 : 1) in a wide-mouth glass tube (a 30-ml Corex tube works well); remove solvent under a stream of argon or dry nitrogen until lipids are deposited as a filmy residue in the bottom of the tube.

2. Resuspend the lipids in <3 ml of prewarmed (37°C) liposome buffer by vortexing vigorously until the film disappears; a cloudy suspension will result.

3. Sonicate using a probe sonicator in 30-sec bursts, chilling on ice for ~1 min between bursts; continue for ~15 min or until suspension is clear.

4. Spin at low speed (5000 rpm for 5 min in an SS-34 rotor) at 4°C. The supernatant is then used immediately.

D. Annexin-Liposome Binding

1. Combine the supernatant from ultracentrifugation (step 12, "Preparation of Bacterial Spheroblasts") with the sonicated liposome suspension.
2. Adjust the solution to >5 m*M* $CaCl_2$ by adding 0.12 g of $CaCl_2$ and stir on ice for 30 min.
3. Centrifuge mixture for 45 min at 40,000*g* (17,500 rpm in a Beckman SW28 rotor) at 4°C.
4. Remove supernatant and resuspend pellet in 35 ml/tube of liposome wash buffer.
5. Centrifuge mixture for 30 min at 50,000*g* (19,500 rpm in Beckman SW28 rotor) at 4°C.
6. Remove supernatant and resuspend pellet in 10 ml/tube of Annexin elution buffer.
7. Centrifuge liposomes for 1 h at 50,000*g* (19,500 rpm in a Beckman SW28 rotor) at 4°C.
8. Dialyze supernatant against 20 m*M* Bis–Tris, 0.02% (w/v) sodium azide, pH 6.0, using a membrane with a molecular mass cutoff of <25,000 kD.
9. Remove any precipitated protein by centrifugation for 30 min at 20,000*g* at 4°C.

E. Protein Storage

For extended storage, dialyze the protein into Annexin V storage buffer.

F. Reagents and Materials

E. coli ATCC 67916

Spheroblast buffer: 0.5 m*M* EDTA, 750 m*M* sucrose, 200 m*M* Tris, pH 8.0

Ultracentrifugation buffer: 2 m*M* EDTA, 5 m*M* $MgCl_2$, 100 m*M* NaCl, 0.1 mg/ml RNase, 0.1 mg/ml DNase I, 2 m*M* phenylmethylsulfonyl fluoride (PMSF), 0.5 mg/ml pepstatin A, 0.1% Triton X-100, 20 m*M* Tris, pH 8.0

Liposome buffer: 100 m*M* NaCl, 3 m*M* $MgCl_2$, 20 m*M* Tris, pH 8.0

Liposome wash buffer: 100 m*M* NaCl, 3 m*M* $MgCl_2$, 20 m*M* Tris, 5 m*M* $CaCl_2$, pH 8.0

Annexin V elution buffer: 100 m*M* NaCl, 3 m*M* $MgCl_2$, 20 m*M* Tris, 10 m*M* EDTA, pH 8.0

Annexin V storage buffer: 25 m*M* HEPES, 140 m*M* NaCl, 1 m*M* EDTA, pH 7.4

Crude phosphatidylserine: bovine brain extract type III, Folch fraction III, Sigma Cat No B1627, Sigma Chemical Co (St. Louis, MO; www.sigma-aldrich.com)

G. Tips and Hints

1. Liposomes should be prepared and used immediately, as they will aggregate upon storage, with a resulting loss in their capacity to bind protein.

2. A bath sonicator is not a substitute for the probe sonicator.

3. The final product is generally 80–90% pure, with low molecular weight protein impurities visible by sodium dodecyl sulfate (SDS)–polyacrylamide gel analysis. Removal of these impurities is possible using DEAE-Sepharose chromatography (Burger *et al.*, 1993), but this step often results in a substantial loss of Annexin.

4. The Annexin can be labeled with FITC or other fluorophores, or biotin. Substitution levels should be kept to about 1 : 1 protein : fluorophore mole ratio to prevent Annexin inactivation.

III. Annexin Staining for Flow Cytometric Analysis of Apoptotic Cells

Quantification of the frequency of cells exposing PS by measuring the number of cells that bind Annexin V is the one of the simplest of the assays available for assessing an early stage of apoptosis. Binding of Annexin V is Ca^{2+} dependent, requiring at least 2.5 m*M* Ca^{2+} levels for efficient binding. A simple buffer (0.15 *M* NaCl, 10 m*M* Hepes, pH 7.4, 2.5 m*M* $CaCl_2$) is sufficient for most analysis purposes. Binding is not sensitive to the presence of serum or to most of the components of growth medium, and thus assays of cells grown in suspension can usually be performed using samples taken directly from culture. Care should be taken not to remove or substantially dilute the Ca^{2+} in the binding buffer. Reduction in Ca^{2+} concentrations results in immediate dissociation of the Annexin from the cell surface, even in the cold.

As explained earlier, the binding of Annexin V is a significant marker only in cells that retain their membrane integrity; in the absence of membrane integrity, all eukaryote cells bind Annexin V, regardless of the mode of death. For this reason, monitoring membrane integrity, and eliminating from consideration cells that have lost it, is an important aspect of the measurement of PS exposure. Impermeant DNA stains, such as propidium iodide (PI), are used commonly for this purpose. PI can be added to samples immediately prior to analysis. When PI enters cells that have lost membrane integrity, its fluorescence [which is red (λ_{Ex} 540 nm, λ_{Em} 625 nm) and reasonably well separated from the green fluorescence of FITC) increases by over two orders of magnitude, even in cells in which apoptotic DNA degradation has resulted in subdiploid DNA levels. Because lipid asymmetry is no longer meaningful in cells that take up PI, these cells should be ignored in assessing loss of lipid asymmetry. Some authors refer to these cells as "secondary necrotic" cells, but this designation is somewhat misleading, as they have not undergone necrosis in any specific sense of the word.

For assessing apoptosis in cultured cells, size analysis is also important. Fragments of cells will bind Annexin V at a level per unit surface area comparable to (if intact) or greater than (if membrane integrity has been lost) that seen with the larger apoptotic cells that gave rise to them. Because the fragments may contain little or no DNA, and thus not stain with PI, quantitation of the frequency of apoptotic cells by simple determination of PI-negative, FITC-positive objects will include these cell fragments. Therefore, to obtain meaningful data, size analysis is required to remove small fragments from consideration.

PS-exposing cells bind Annexin V to a maximum level that is determined not only by Annexin–PS interactions, but also by protein–protein interactions between bound Annexin molecules. Under the conditions recommended by most commercial providers of Annexin V, these protein–protein interactions result in maximum binding levels, where fluorescence from apoptotic lymphoid cells can be up to 100-fold higher than fluorescence from normal cells in the same population. The rate at which this binding level is attained is dependent on the Annexin V concentration and thus differs for Annexin preparations from different suppliers, with times commonly ranging from 10 to 60 min at the manufacturer's recommended concentrations. If all that is needed is an estimate of the fraction of cells on which PS is exposed, much shorter incubations can be used because the Annexin-positive and -negative fractions can often be distinguished easily after 2–5 min of incubation with FITC7–Annexin and the number of cells in each subpopulation measured reliably without waiting for maximal binding to be completed. During most of the linear phase of the binding reactions, which usually last for about 10–30 min after addition of the fluorescent Annexin, binding to apoptotic cells will be 20- to 30-fold higher than binding to normal cells, and thus is often sufficient to distinguish these two subpopulations.

A. Simple Protocol for Rapid Determination of Phosphatidylserine (PS) Exposure by Cells in Suspension

1. Dilute 100 μl of cells suspended in medium at about 10^7 cell/ml into 500 μl of Annexin V-binding buffer (150 m*M* NaCl, 10 m*M* Hepes, pH 7.4, 2.5 m*M* $CaCl_2$) at room temperature.
2. Add 1 μl of FITC–Annexin V from manufacturer's stock. Mix briefly.
3. Add 0.4 μl of 1 mg/ml PI in water. Mix briefly.
4. Incubate for 2–5 min at room temperature.
5. Analyze by flow cytometry, collecting forward scatter, side scatter, log green fluorescence, and log red fluorescence data.
6. On a two-dimensional plot of forward versus side scatter, set amplifiers and photomultiplier voltages to give a pattern roughly as shown in Fig. 1a. Cells occupying the large area in the central/right region (R1) correspond to normal cells. Those in region 2 (R2) are smaller [lower forward scatter (FS)], are more convoluted [higher side scatter (SS)], and include apoptotic cells. Objects in the lower left-hand region (R3) include cell fragments, debris, and precipitates from the medium; this region should be excluded from cell counts in the determination of the fraction of cells that expose PS. Cells with a very high level of PI fluorescence have lost membrane integrity and should be excluded from determination of the fraction of cells that expose PS.

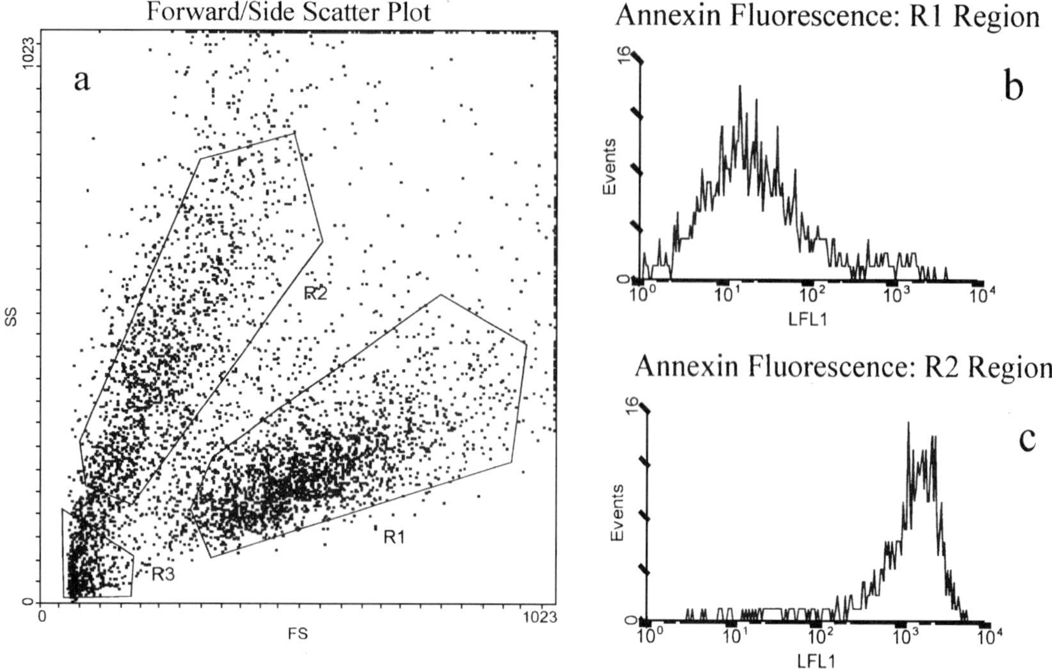

Fig. 1 EBV-transformed B cells containing an unusually high level of spontaneously apoptotic cells were stained with FITC–Annexin V and PI as described in the text and analyzed on a Coulter Epics cytometer. (a) Forward/scatter side scatter plots. (b) Annexin fluorescence (FL1 channel) of normal cells in region R1, after gating out PI-positive cells. (c) Annexin fluorescence (FL1 channel) of apoptotic cells in region R2, after gating out PI-positive cells.

B. Simple Protocol for Rapid Determination of PS Exposure by Adherent Cells

1. Carefully remove medium from cell monolayer and rinse gently with medium or saline lacking serum.
2. Add enough trypsin/EDTA to cover the surface (e.g., 0.5 ml/well in a six-well tissue culture plate). After 30 sec, gently remove trypsin/EDTA solution.
3. After 5 min, add a slight excess (e.g., 1.0 ml in a six-well tissue culture plate) of Hepes-buffered saline (150 m*M* NaCl, 10 m*M* Hepes, pH 7.4) and resuspend cells gently.
4. Proceed from step 1.

C. Tips and Hints

1. The fraction of apoptotic adherent cells is likely to be underestimated by the protocol given earlier because apoptotic cells are loosely adherent, and many are lost when the medium or trypsin is removed.

2. The frequency of objects in the small FS/SS class (region 3, Fig. 1) will be high in populations with substantial levels of apoptosis because of cell fragmentation.

3. Use of forward scatter, rather than log forward scatter, as the x axis for size determinations often makes it easier to distinguish apoptotic from normal cells.

4. Comparison of the Annexin V fluorescence of the morphologically normal subpopulation (Annexin fluorescence, R1 region, Fig. 1b) and of the morphologically apoptotic subpopulation (Annexin fluorescence, R2 region, Fig. 1c) gives a rapid estimate of the signal/noise ratio for the measurement. Occasionally, and particularly after cells have been disturbed, this ratio may be unexpectedly low because the Annexin V binding to the normal cell population is unusually high for unknown reasons. This elevated binding is often transient and is no longer observed when the same cells are resampled and analyzed again after 30 min. The number may also be unexpectedly low because the fluorescence of the apoptotic cell population is rather low. In some cell types (e.g., MCF-7 cells), low staining of apoptotic cells is reproducibly observed. However, where low staining levels are not reproducibly observed, they often result from inefficient Annexin V binding, which in turn is often a result of excess dilution of the Ca^{2+} in the Annexin V-binding buffer.

5. While all cytometers provide software for setting the regions (gates) indicated in the figures, it is often convenient to carry out these analyses off line. A very convenient free software package for this purpose is the WinMidi program, available for download from http://facs.scripps.edu/software.html.

IV. *In Vivo* Detection of Cell Surface-Exposed PS

A major challenge of apoptosis research is the understanding of (dis)organization of apoptosis and phagocytosis in tissues and organs, and the implications of these processes in disease. For such research, information about apoptosis of cells in their natural environment is often essential. In addition to detecting PS exposed on the surface of cells *in vitro,* human recombinant Annexin V can also be used to detect this plasma membrane alteration *in vivo* (van den Eijnde *et al.,* 1997c). *In vivo,* Annexin V binds almost exclusively to apoptotic cells, including sites of contact between the apoptotic cell and its phagocyte (van den Eijnde *et al.,* 1999) (Fig. 2, see Color Plate). The following method is used for detecting cell surface-exposed PS *in vivo* in whole mounts, and at the light microscopic and ultrastructural level using the developing mouse embryo as a model, according to protocols described previously (van den Eijnde *et al.,* 1997b) with minor alterations.

A. Injection of Annexin V into Mouse Embryos

1. To obtain embryos of known gestational age, mate female mice overnight and check for a vaginal plug the following morning. The presence of a vaginal plug is designated as embryonic day 0 (E0).

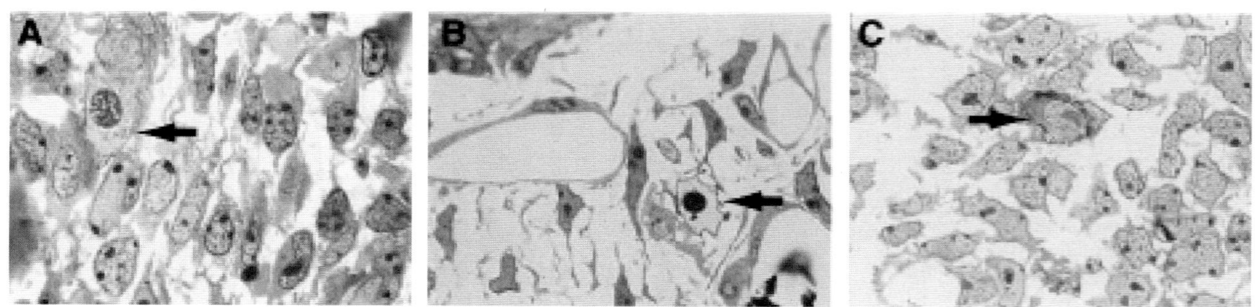

Fig. 2 *In vivo* labeling of cell surface-exposed phosphatidylserine. Annexin V-biotin labeling is shown (arrows) of early (A), late (B), and phagcytosed (C) apoptotic cells in mouse embryonic tissue (van den Eijnde *et al.*, 1998) (Fig. 1). Note in C that both the apoptotic cells and the surrounding membrane of the phagocyte are Annexin V positive (see also van den Eijnde *et al.*, 1999). (See Color Plate.)

2. At the appropriate gestational age, sacrifice the pregnant mouse via cervical dislocation after ether anesthesia. When the maternal circulation stops, embryos only survive for a limited period of time; an E13 embryo typically survives up to 30 min. Hence, it is essential to proceed through steps 4–7 rapidly.

3. Open the abdomen of the mouse with scissors, and dissect the uterus and place it in warm (37°C) Annexin V-binding buffer supplied with the Annexin V-kit or in Hepes buffer (20 m*M* Hepes, 132 m*M* NaCl, 2.5 m*M* $CaCl_2$, 6 m*M* KCl, 1 m*M* $MgSO_4$, 1.2 m*M* $K2HPO_4$, 5.5 m*M* glucose, 0.5% bovine serum albumin, pH 7.4).

4. In the same buffer at the same temperature, remove the embryos from the uterus using fine forceps; take care not to damage the embryonic membranes and leave the placenta attached.

5. Inject a dose of 30 mg/kg biotin-conjugated Annexin V (depending on embryo size, this corresponds to 2–3 μl of APOPTEST-BIOTIN, product B500) into the embryonic circulation using glass pipettes (tip diameter 15–25 μm) mounted on a Hamilton syringe. When Annexin V has been injected successfully into the heart or into the sigmoid vein of the head, a temporarily blanching of the bloodsteam can be observed.

6. Fix the embryos at a maximum of 30 min after injection with Annexin V or directly after the embryonic heart has stopped beating.

7. For Annexin V staining in whole mounts or in paraffin sections, embryos are fixed overnight at 4°C in a solution of 4% formaldehyde in Hepes buffer or 4% formaldehyde in Annexin V-binding buffer (pH 7.4). For light microscopy, embryos may be stored for several weeks in 70% ethanol at −20°C after rinsing with phosphate-buffered saline (PBS).

8. To detect Annexin V at the ultrastructural level, embryos are perfused intracardially with approximately 0.5 ml of an ice-cold solution of 2% glutaraldehyde and 2% paraformaldehyde in 0.1 *M* cacodylate buffer (pH 7.4). Subsequently, the tissue of interest is postfixed overnight in the same fixative and can be stored until further processing for 1–2 days in PBS at 4°C.

B. Detection of Cell Surface-Exposed PS in Whole Mounts

1. Wash fixed embryos with PBS at room temperature.
2. Wash the embryo with 0.3% Triton X-100 in PBS.
3. Digest the tissue with a solution of 0.01% proteinase K (Boehringer-Mannheim, Mannheim, Germany) in PBS for 10 min at room temperature.
4. Block endogenous peroxidase activity by incubating the samples in 1% H_2O_2 in Tris/EDTA for 60 min at room temperature.
5. Wash the specimens with PBS and, at the same time, prepare the avidin–peroxidase complex according to manufacturer's instructions (Vector ABC Elite kit, Vector Laboratories).

6. Incubate the specimens with the avidin–peroxidase complex overnight at room temperature.
7. Incubate the specimens with a solution of 3,3′-diaminobenzidine tetrahydrochloride (DAB; 0.05%) and 0.1% H_2O_2 in PBS at room temperature until the tissue stains brown.
8. Rinse the tissue repeatedly to remove unbound DAB.
9. Store specimens in Tris/EDTA until examination under a stereomicrosope.

C. Detection of Cell Surface-Exposed PS in Paraffin Sections

1. Wash fixed embryos with PBS at room temperature.
2. Dehydrate the embryos in a graded series of ethanol (50, 70, 80, 90, 98, 100%, each step several hours to overnight, depending on embryonic size, at room temperature).
3. Incubate the embryos with toluol at room temperature until clear.
4. Incubate embryos with toluol/paraffin (1 : 1, v/v) at 60°C overnight.
5. Embed embryos in paraffin (Paraplast, Sigma, St. Louis, MO) and section at 3–5 μm.
6. Mount sections on 4-aminopropyltriethoxysilane-coated slides.
7. Dewax the slides by heating for 60 min at 60°C in an incubator followed by rinsing with xylene (2 × 2 min) and 100% ethanol (2 × 1 min).
8. Block endogenous peroxidase activity by incubating the slides with 2% H_2O_2 in methanol for 20 min.
9. Rinse the slides with distilled water (1 min) and PBS (2 × 2 min), and at the same time prepare the avidin–peroxidase complex according to manufacturer's instructions (Vector ABC Elite kit, Vector Laboratories).
10. Incubate the specimens with the avidin–peroxidase complex for 30 min at room temperature.
11. Wash the slides with PBS (2 × 2 min).
12. Incubate slides for 7 min in the dark with a 0.05% DAB/PBS (w/v) solution containing 0.1% H_2O_2.
13. Rinse the slides with tap water for 6 min.
14. Stain the sections for approximately 10 s with undiluted Gills hematoxylin (Sigma Diagnostics, Deisenhofen, Germany).
15. Rinse the slides with tap water (10 s).
16. Wash slides with 0.25% HCl in 70% ethanol for 10 s.
17. Rinse slides with tap water for 6 min.
18. Dehydrate slides with a graded series of ethanol, ending with xylene (each step 1 min).
19. Mount the slides (Permount, Fisher Scientific, New Jersey).

D. Detection of Cell Surface-Exposed PS at the Ultrastructural Level

1. Cut the fixed tissue into 50-μm-thick sections on a vibratome (TPI Technical Products Int. Inc., St Louis, MO).
2. Block endogenous peroxidase activity by incubating the sections with 2% H_2O_2 in methanol for 20 min.
3. Rinse the sections with PBS (2 × 10 min), and at the same time prepare the avidin–peroxidase complex according to manufacturer's instructions (Vector ABC Elite kit).
4. Incubate the sections with the avidin–peroxidase complex overnight at room temperature.
5. Wash the sections with PBS (2 × 10 min).
6. Incubate sections with a 0.05% DAB/PBS (w/v) solution containing 0.1% H_2O_2 and 0.02% $CoCl_2$ until they stain dark brown.
7. Postfix the thick sections in 1.5% OsO_4 in a 8% glucose solution.
8. Rinse the sections with distilled water.
9. Stain *en block* in 3% uranyl acetate.
10. Dehydrate in dimethoxypropane.
11. Embed in Durcopan.
12. Cut semithin and ultrathin sections on an ultratome (Ultracut S, Reichert Jung, Rijswijk, The Netherlands).
13. Counterstain semithin sections with toluidine blue.
14. Counterstain ultrathin sections with uranyl acetate and lead citrate.
15. Examine semithin and ultrathin sections under a light microscope or an electron microscope (Philips CM100, Eindhoven, The Netherlands), respectively.

E. Tips and Hints

1. With the same technique, cell-surface-exposed PS has also been detected successfully in insect and various developing and adult vertebrates (Blankenberg *et al.,* 1998; van den Eijnde *et al.,* 1998; Watanabe *et al.,* 1998; van den Eijnde *et al.,* 1999).

2. Because there may be a large intra- and interstrain variation between gestational age and stage of development, staging the embryos after fixation but before processing for microscopy should be considered.

3. During embryogenesis, apoptosis takes place at specific spatiotemporal coordinates (Glucksmann, 1951), of which the E13 interdigital cell death is probably the best known developmentally regulated apoptosis pattern. Such apoptosis patterns can be used as a positive control for Annexin V staining. As a negative control, embryos can be used that were injected with Annexin V-binding buffer or Hepes buffer only or injected with heat-inactivated Annexin V-biotin (10 min at 56°C) with a destroyed PS-binding capacity (Reutelingsperger *et al.,* 1985).

4. Annexin V does not pass the intact plasma membrane; therefore, cells that were already located inside phagocytes before perfusion with Annexin V remain unlabeled with this marker.

V. MC540 Staining

An alternative method for identifying apoptotic cells in which lipid asymmetry has been lost is the use of the fluorescent dye MC540. The binding of this organic molecule to membranes is sensitive to lipid packing, and this physical property is changed upon loss of lipid asymmetry, with a consequent increase in MC540 binding (Schlegel *et al.*, 1993). This method has several advantages, not least of which it is very simple. Moreover, MC540 is a vital stain, and live cells stained with MC540 have been sorted and then grown successfully in culture (McEvoy *et al.*, 1988). Staining is reversible, and membrane-bound dye can be removed by washing with BSA-containing buffers. Because the emission spectrum of membrane-bound dye has a maximum near 600 nm, double staining with fluorescein-labeled antibodies is straightforward. Finally, the dye itself is very inexpensive; it can be obtained from Sigma for $32 for 100 mg, an amount sufficient to do thousands of assays. Several laboratories have shown that MC540 specifically stains apoptotic B and T lymphocytes (Fadok *et al.*, 1992; Schlegel *et al.*, 1993; Mower *et al.*, 1994). Whether the method is applicable to other types of cells has not been investigated.

Balanced against these advantages are several disadvantages. Because the difference in binding between normal and apoptotic cells is only about five fold, the signal/noise ratio is less advantageous for MC540 than for Annexin V. In addition, because the dye itself is phototoxic, normal cells exposed to it must be kept in the dark and excess dye removed carefully to prevent the induction of cell death.

A. Protocol

1. Suspend 3×10^6 cells in 1 ml of buffer containing 0.1% BSA.
2. Add MC540 to 10 μg/ml, from a stock solution at 1 mg/ml in 50% ethanol in water.
3. After 10 min at room temperature, dilute to 10 ml with buffer containing 0.1% BSA.
4. To stain and gate out nonviable cells, just prior to analysis, add 30 μl of a 1-mg/ml stock solution of PI in water to 1 ml of diluted, stained cells.
5. Analyze and/or sort.

B. Tips and Hints

1. When cells are double stained with both MC540 and antibody, antibody staining should be performed first, as usual, and then well-washed cells stained exactly as described earlier. As MC540 fluoresces in the red range, FITC or some other fluorophore that fluoresces in the green range should be used for antibody staining.

2. MC540 is light sensitive. Stock solution should be prepared in a brown bottle. Staining should be performed in tubes wrapped with foil.

3. Because MC540 permeates the membrane slowly, analyses should be performed within 5–10 min of staining.

C. Flow Cytometry

1. Analysis of cells stained with MC540 and PI: Excite at 514 nm, using a 514-nm laser-blocking filter. Monitor MC540 fluorescence through a 575-nm band-pass filter, reflect PI fluorescence from a 590-nm short-pass dichroic filter, and monitor through a 610-nm long-pass filter.

2. Analysis of cells stained with antibody, MC540, and PI: Excite at 488 nm, using a 457- to 504-nm laser-blocking filter. Monitor MC540 and PI fluorescence as in step 1. Reflect antibody fluorescence from a 550-nm long-pass filter and monitor through a 525-nm band-pass filter.

3. Sorting: To directly reanalyze sorted populations without the need for additional staining, collect sorted cells into buffer containing 0.1% BSA. To maintain the viability of sorted populations, collect sorted cells into culture medium containing 10% fetal calf serum to remove MC540 from the cells to avoid photolysis.

VI. Transbilayer Lipid Movement Assays

The appearance of PS on the surface of apoptotic cells depends on the regulation of two membrane activities that control transbilayer lipid movement and thereby the transbilayer equilibrium distribution of phospholipids. One of these activities is the aminophospholipid translocase, the membrane ATPase that catalyzes the transport of PS and PE from the outer to the inner leaflet of the plasma membrane, thus establishing and maintaining the normal asymmetric distribution of these phospholipids to the inner leaflet. The second of these activities is the lipid scramblase, which catalyzes the exchange of lipids between the two leaflets, thus randomizing lipid distribution between the two sides of the membrane. The exposure of PS results from the simultaneous downregulation of the translocase and upregulation of the scramblase. These activities can each be assayed independently, and measurement of the change in their activity is diagnostic and synonymous with PS exposure, thus providing an alternative to the measurement of PS exposure itself. In general, the signal/noise ratio of these measurements is not as large as the Annexin measurements, and the assays are less convenient to use. However, they have the advantage of being direct measurements of the mechanisms that regulate PS distribution. In addition, in some cells (such as MCF7 cells) the scramblase is not particularly active, and PS exposure in apoptotic cells is reduced correspondingly, even though the translocase is inactivated efficiently. In such cases, the measurement of translocase inactivation rather than Annexin binding may provide a useful level of sensitivity.

Assays for translocase and scramblase activity both depend on the use of fluorescent analogs of phospholipids, particularly the 7-nitro-2-1, 3-benzoadiazol-4-yl (NBD) derivatives of the major phospholipids, with the fluorescent NBD moiety introduced via a six-carbon linker to the second carbon of the glycerol backbone. These derivatives have several properties of importance to the assays in addition to their fluorescence (which is sufficiently similar to fluorescein to be detected with adequate sensitivity in a standard fluorescein channel of a flow cytometer). First, insertion of the fluorescent group on the fatty acid side chain preserves the structure of the phospholipid head group and hence the ability to distinguish translocase and scramblase activities based on their head group specificity. In addition, the short linker raises the critical micelle concentration of the probes to a level that permits their ready insertion into and removal from the plasma membrane, a critical requirement for the assays. Finally, the attachment at the second carbon avoids some (but not all) problems with hydrolytic cleavage of the probe.

Assays for translocase and scramblase are the same, differing only in the phospholipid analog whose movement is measured for routine purposes. Measurement of the internalization kinetics of NBD-labeled phospholipids differentiates between the two activities. The signature of the translocase is rapid internalization of PS and slow internalization of PC; its inactivation is signaled by a substantial drop in the rate of NBD-PS internalization, particularly at short (5 min) incubation times. The signature of the scramblase is internalization of all phospholipids, including PC, and its activation is signaled by a substantial increase in the uptake of NBD-PC. In both cases, the fluorescent probe is suspended in buffer (as micelles) and added to cells. Because of their hydrophobicity, the probes are incorporated rapidly (under 2 min or so) into the outer leaflet of the cell membrane. After incubation, the untransported probe remaining in the outer leaflet is removed by the addition of buffer containing fatty acid-free bovine serum albumin (BSA), and the sample is then introduced directly into the flow cytometer to determine fluorescence still associated with the cell, and thus internalized.

A. NBD-Phospholipid Transport Measurements

1. NBD probes (1-oleoyl-2-[6-[(7-nitro-2-1,3-benzoadiazol-4-yl) amino]caproyl]-sn-glycero-3-phosphocholine or -phosphoserine; Avanti Polar Lipids, Alabaster, AL) should be stored as a 10 m*M* stock in $CHCl_3$ under nitrogen at $-20°C$. Just prior to use, dry 5 μl of the stock on the bottom of glass test tube under a stream of nitrogen and then resuspend the dried probe in 50 μl Hepes/saline (150 m*M* NaCl, 10 m*M* Hepes, pH 7.4) by vigorous vortexing.

2. Collect cells by centrifugation at low speeds (1000 rpm in a clinical centrifuge) for 3 min, wash once in Hepes/saline, and resuspend in Hepes/saline at a concentration of about 10^7/ml.

3. For detailed kinetic analysis, initiate incubations by the addition of 0.4 μl of NBD-phospholipid to 0.5 ml of cell suspension. For single measurements of several samples, the volume of cells should be reduced and the NBD-labeled probe diluted to permit

incubations at similar dye : cell ratios. Take timed samples of 50 μl at predetermined intervals (up to about 15 min for detailed kinetics, usually at about 5 min for single measurements on multiple samples) into 0.5 ml of prechilled (0°) Hepes/saline with or without 0.1% fatty acid-free BSA, mixed, and incubate these on ice for at least 2 min before analysis. PI can be added to a final concentration of 1 μg/ml from a 1-mg/ml stock during this incubation.

4. Individual samples are analyzed by flow cytometry, with parameters similar to those for Annexin V measurements indicated earlier, and data analyzed as described in the next section.

B. Data Analysis

For each time point, set gates on the FS/SS plots (Fig. 1) to select cells with normal or apoptotic morphology. Use PI fluorescence to gate out cells in which membrane integrity has been lost. Cells with normal morphology are usually selected, and a histogram of NBD fluorescence is generated for each sample. The median fluorescence of cells sampled into buffer without BSA is a measure of the total amount of probe bound to the cells. The median fluorescence of cells sampled into buffer with BSA is a measure of the amount of the probe that has been transported to the cell interior. Cells in which the translocase has been inactivated are identified as a subpopulation with low fluorescence intensity in the histogram of cells labeled with NBD-PS and analyzed in the presence of BSA. The ideal time of transport for visualization of this fraction depends on the cell type, but is usually about 5 min for incubations at room temperature. Cells in which the scramblase has been activated are identified as a subpopulation with high fluorescence intensity when the cells are labeled with NBD-PC and analyzed in the presence of BSA. The ideal time for visualization of this subpopulation is more variable than the time required to identify translocase-inactive cells and may range from 2 to 15 min, depending on cell type.

C. Tips and Hints

1. In detailed kinetic measurements, because the most useful times are the shortest times (under 5 min), several samples at these times may be advantageous. These detailed measurements give the clearest indication of the transport properties of the cells.

2. In extended kinetic measurements, it is essential to make several measurements (in the absence of BSA) of the total amount of probe bound because the probe will redistribute between populations during incubations.

3. The amount of probe should be kept as low as possible, ideally to less than 2% of plasma membrane phospholipid. The latter depends on the size of the cell type being examined, but a useful rule of thumb is that 5×10^6 cells/ml corresponds to about 10 μM plasma membrane lipid. In general, it is the ratio of probe to cells, not the concentration of probe, that is the critical variable. Use of smaller numbers of cells at the probe concentrations given earlier may induce cell lysis or deformation.

4. Assays are carried out most conveniently at room temperature. In addition to convenience, probe degradation and background rates of probe movement into cells both disproportionately increase at 37°C.

5. Samples taken into cold buffer can be left on ice for up to 10 min before analysis. However, the probe is extracted slowly into BSA-containing buffers from cells even on ice so that samples left for different lengths of time before analysis may not be directly comparable.

VII. Phagocytosis of Apoptotic Cells

In vivo, the functional consequence of loss of lipid asymmetry is removal by phagocytosis. This process is far from understood, and assays for measuring it are still in their infancy. In *Caenorhabditis elegans,* microscopic observation of living animals using interference optics reveals that apoptotic cells are phagocytosed by neighboring cells that are not specialized for this purpose, i.e., are not professional phagocytes (Wu and Horvitz, 1998). In principle, microscopic observations of living cells have the advantage of being able to unambiguously distinguish between binding and phagocytosis, to determine the time course of the process for individual target/phagocyte pairs, and to observe the temporal relationship between phagocytosis and morphological aspects of apoptosis (cell blebbing, shrinkage, and lysis). However, the approach is laborious, generates only small numbers of observations, making establishing statistical significance difficult, and is not well adapted to the analysis of multiple assay conditions.

The most common method of studying the phagocytosis of apoptotic cells is adding a suspension of target cells to monolayers of adherent macrophages. This methodology has been used to identify most of the cell surface molecules that contribute to the phagocytosis of apoptotic cells by macrophages and is presented here. From studies employing this methodology, it is clear that macrophages differ in the mechanisms by which they recognize apoptotic cells, a difference that reflects in part differences in the way in which PS participates in the signal by which apoptotic cells are recognized. These differences are illustrated most readily by comparing recognition by two different classes of macrophages. The first class, referred to as "activated" macrophages, includes murine-elicited peritoneal macrophages or macrophages activated *in vitro* by one of several protocols; recognition by these cells is inhibited by PS vesicles. The second class are "unactivated" macrophages, such as murine bone marrow macrophages and human monocyte-derived macrophages; recognition by these cells is inhibited by erythrocytes, which express PS on their surface (lipid-symmetric erythrocytes) (Pradhan *et al.,* 1997). Recognition by both classes of macrophages is inhibited by pretreatment with Annexin V (Krahling *et al.,* 1999).

Several types of PS-expressing target cells with known characteristics are available to serve as positive controls when getting started. Phagocytosis of the lipid-symmetric erythrocytes described earlier, by both activated and unactivated macrophages, is inhibitable by both PS vesicles and pretreatment with Annexin V. Erythrocytes offer the advantage that phagocytosed vs adherent cells are easily distinguishable following a wash with NH_4Cl, which lyses uningested cells. Many target cells can be artificially induced to express PS on their surface within a matter of just a few minutes by treatment with the Ca^{2+} ionophore; most or all of the cells in the population express PS after such treatment. Therefore, these cells serve as a rapidly generated population of positive controls for PS-dependent (-inhibitable) phagocytosis. We have found the DO11.10 T-cell hybridoma line to be very useful. Cells are cultured easily, are readily inducible by ionophore treatment, and their phagocytosis is inhibitable by Annexin pretreatment and other appropriate inhibitors, just like apoptotic DO11.10 cells (Callahan *et al.*, 2000).

A. Preparation of Macrophages

We recommend the mouse J774A.1 macrophage cell line as the most convenient cells for studies where unactivated macrophages are required. Their phenotype with respect to phagocytosis of PS-presenting target cells is the same as primary unactivated macrophages, being inhibited by lipid-symmetric erythrocytes and by pretreating targets with Annexin V. However, in our hands, J744.1 macrophages activated in culture do not behave exactly as primary activated macrophages. Therefore, if activated macrophages are required, we recommend elicited mouse peritoneal macrophages.

1. J774A.1 Macrophages

1. Grow cells of the J774A.1 mouse monocyte-derived macrophage cell line (American Type Culture Collection) in 10% fetal bovine serum in DMEM at 37°C in 5% CO_2.
2. Twenty-four hours prior to phagocytosis assays, pipette 3×10^5 cells in 150 μl of culture medium onto 18-mm bicarbonate-treated glass coverslips kept in 60-mm petri dishes.

2. Elicited Mouse Peritoneal Macrophages

1. Inject into the peritoneal cavity of 6- to 8-week-old mice 1 ml of 3% Brewer's thioglycollate.
2. After 5–7 days harvest by peritoneal lavage using 10 ml of PBS containing 10 units/ml of heparin.
3. Centrifuge the exudate at 4°C for 10 min and wash in RPMI.
4. Suspend the cells in 5% calf serum in RPMI (4×10^6/ml) and pipette150 μl (6×10^5 cells) onto 18-mm bicarbonate-treated glass coverslips, pretreated for 2 h with calf serum, placed in 60-mm petri dishes.
5. Incubate for 2 h at 37°C, remove nonadherent cells by aspiration, and replace medium with 150 μl/coverslip of fresh RPMI containing 5% calf serum.
6. Incubate overnight at 37°C in 5% humidified CO_2 prior to use.

B. Preparation of Lipid-Symmetric Erythrocytes

1. Collect fresh human venous blood into ice-cold PBS containing 10 U/ml of heparin, centrifuge, and remove the supernatant and the top 10% of erythrocytes. Wash erythrocytes two more times.

2. Add an equal volume of PBS to the remaining pellet, remove the number of cells desired, centrifuge, and remove the supernatant.

3. To the pellet add 4 volumes of lysis buffer (0.1 × PBS containing 0.1 m*M* EGTA, 1 m*M* $MgCl_2$, and 0.1% BSA, plus 1 m*M* $CaCl_2$ for lipid-symmetric erythrocytes only), vortex, and place the lysing cells on ice.

4. After 2 min, reseal cells by the addition of 0.4 volumes (relative to the size of the original pellet) of 10 × PBS, invert the tube gently several times, and place at 37°C.

5. After 30 min, add 100 volumes of PBS, centrifuge the cells to pellet, wash three times with PBS, and resuspend in PBS for use.

C. Tips and Hints

1. Following step 2, the cell suspension loses its refractive properties and no longer scatters light; it appears as a clear, red solution.

2. Immediately upon addition of 10x buffer to the lysed cells, they should again scatter light, although not to the extent of the original unlysed cells.

3. Although fresh erythrocytes are preferable, washed erythrocytes can be stored at 4°C for a day or more and used for the preparation of lipid-symmetric erythrocytes, following washing to remove hemoglobin from the supernatant and removal of any lysed cells visible as a whitish layer at the top of the erythrocyte pellet.

4. It is best to use the lysed and resealed erythrocytes within hours of preparation. However, if stored overnight and washed as just described, they *may* still be useful, with the caveat that at least a portion of the control lipid-asymmetric cells may have lost their asymmetric distribution of PS.

D. Preparation of PS-Expressing DO11.10 Cells

1. Collect cells in late log phase ($<10^6$ cells/ml) and wash twice with medium (RPMI) containing penicillin/streptomycin, but without serum. Resuspend at 4×10^6 cells/ml in medium without serum.

2. Add 1/1000 volume 5 m*M* ionomycin (Calbichem) in ethanol.

3. Incubate for 10 min at room temperature.

4. Take an aliquot for Annexin/PI staining as described earlier to confirm that the cells have become positive for Annexin binding, and remain negative for PI staining.

E. Tips and Hints

1. A23187 can be substituted for ionomycin (at the same concentrations), but it has an intrinsic fluorescence that adds to background signals in the cytometer.

2. The washing step is required to remove serum, which will bind the ionophore and prevent activation of the cell scramblase.

3. Because the Ca^{2+} concentrations in RPMI do not allow efficient Annexin binding, cells should be diluted into Annexin-binding buffer before analysis.

4. While 10 min is sufficient for DO11.10 cells, other cells may require longer or shorter times of incubation before the membrane lipids are completely scrambled and, in some cells (such as fibroblasts), the process is very slow.

5. The lipid-scrambling process is complete when Annexin binding by the cells with normal morphology is the same as the binding by the spontaneously apoptotic cell fraction. However, prolonged incubation will result in the shrinkage of normal cells and transfer into the apoptotic cell region of the FS/SS plots.

6. Incubation at room temperature tends to preserve morphology, and for that reason is preferred to incubation at 37°C.

F. Preparation of PS Vesicles

1. Dry PS (or PC as a negative control) from a chloroform solution (Sigma) to a phospholipid film under a stream of nitrogen.

2. Place under vacuum for 4 h.

3. Hydrate the film to 150 μM by adding PBS and vortexing.

4. Sonicate in a bath-type sonicator to convert multilamellar to unilamellar vesicles.

G. Phagocytosis Assay

1. Remove media from macrophage coverslip cultures, prepared as described earlier, by aspiration and overlay onto the macrophage cultures 10^6 apoptotic cells or 15×10^6 erythrocytes (when used as targets) in 150 μl of DMEM, without serum.

2. Incubate at 37°C in 5% CO_2 for 30 min.

3. Wash coverslips three times with ice-cold PBS, fix either with 1.8% formaldehyde or with ice-cold acidic methanol, and stain with Diff-Quik (Baxter).

4. Enumerate phagcytosed cells by microscopy, selecting random fields and counting 400 macrophages per coverslip.

H. Inhibitors

1. All inhibitors are added to the target cells in a final volume of 150 μl with which the macrophage cultures are then overlaid. Phospholipid vesicles are added at 7.5 nmol/150 μl. When lipid-symmetric erythrocytes are used as inhibitors of phagocytosis, they are admixed with apoptotic target cells at 15×10^6 erythrocytes/10^6 target cells.

2. For pretreatment with Annexin, cells are incubated in 150 μl of various concentrations of Annexin V in buffer containing 2 mM Ca^{2+} for 15 min at room temperature and then washed twice with Ca^{2+}-containing buffer before targets in 150 μl are overlayed onto macrophages.

I. Tips and Hints

1. When erythrocytes are used as targets, treat coverslips with 17 m*M* Tris, 140 m*M* NH_4Cl, pH 7.2, prior to fixation to lyse uningested erythrocytes and stain erythrocytes for hemoglobin with benzidine prior to staining with Diff-Quik. Using this procedure, erythrocytes stain brown; thymocytes stain uniformly intensely blue; only the nuclei of macrophages stain intensely blue, with the cytoplasm staining less intensely, and the cell periphery is discerniable by its increased staining relative to the cytoplasm.

2. Although absolute numbers of cells phagocytosed in these assays increase beyond 30 min, the ratio of cells phagocytosed in apoptotic populations vs nonapoptotic populations does not increase beyond 30 min. Therefore, assays are conducted for only 30 min to avoid possible artifacts resulting from control cells spontaneously undergoing apoptosis, from reaching saturation levels of cells ingested by macrophages, or from ingested cells being destroyed.

3. Treating cultures with 0.25% trypsin for 5 min, after washing with ice-cold PBS, reduces adherent (nonphagocytosed) target cells about five-fold; however, not all adherent cells were removed by the treatment. However, the number of apoptotic cells counted as phagocytosed is no different than that of untreated identical coverslips. For this reason, a trypsin treatment does not remove the requirement for applying the following criteria for determining if apoptotic cells are adherent or phagocytosed.

4. Target cells are counted as phagocytosed if they are entirely within the outline and in the same focal plane as the macrophage. Target cells not entirely within the outline of the macrophage, but in the same focal plane and obviously being engulfed by a distended macrophage, are also counted as phagocytosed; they generally represent no more than 10% of cells counted as phagocytosed. Target cells not contained within the outline of the macrophage, but which appear to touch it, can be considered adherent, as can cells within or partially within the outline of the macrophage but in a different focal plane than the macrophage; the latter cells should be rare (e.g., less than 5% of thymocytes). Thus cells within or partially within the outline of the macrophage that are in the process of being phagocytosed are not counted as phagocytosed cells if they are not in the same focal plane as the macrophages. Although this underestimates the number of phagocytosed cells, the error is small because of the small number of cells that fall in this category.

References

Blankenberg, F. G., Katsikis, P. D., Tait, J. F., Davis, R. E., Naumovski, L., Ohtsuki, K., Kopiwoda, S., Abrams, M. J., Darkes, M., Robbins, R. C., Maecker, H. T., and Strauss, H. W. (1998). In vivo detection and imaging of phosphatidylserine expression during programmed cell death. *Proc. Natl. Acad. Sci. USA* **95,** 6349–6354.

Burger, A., Berendes, R., Voges, D., Huber, R., and Demange, P. (1993). A rapid and efficient purification method for recombinant Annexin V for biophysical studies. *FEBS. Lett.* **329,** 25–28.

Callahan, M. K., Williamson, P., and Schlegel, R. A. (2000). Surface expression of phosphatidylserine on macrophages is required for phagocytosis of apoptotic thymocytes. *Cell Death. Differ.* **7,** 645–653.

Choe, H. R., Schlegel, R. A., Rubin, E., Williamson, P., and Westerman, M. P. (1986). Alteration of red cell membrane organization in sickle cell anaemia. *Br. J. Haematol* **63,** 761–773.

Choe, H. R., Williamson, P., Rubin, E., and Schlegel, R. A. (1985). Disruption of phospholipid asymmetry in erythrocyte vesicles deficient in spectrin. *Cell. Biol. Int. Rep.* **9,** 597–606.

Devaux, P. F. (1991). Static and dynamic lipid asymmetry in cell membranes. *Biochemistry* **30,** 1163–1173.

Diaz, C., Lee, A. T., McConkey, D. J., and Schroit, A. J. (1999). Phosphatidylserine externalization during differentiation-triggered apoptosis of erythroleukemic cells. *Cell Death Differ.* **6,** 218–226.

Fadeel, B., Gleiss, B., Hogstrand, K., Chandra, J., Wiedmer, T., Sims, P. J., Henter, J. I., Orrenius, S., and Samali, A. (1999). Phosphatidylserine exposure during apoptosis is a cell-type-specific event and does not correlate with plasma membrane phospholipid scramblase expression. *Biochem. Biophys. Res. Commun.* **266,** 504–511.

Fadok, V. A., Voelker, D. R., Campbell, P. A., Cohen, J. J., Bratton, D. L., and Henson, P. M. (1992). Exposure of phosphatidylserine on the surface of apoptotic lymphocytes triggers specific recognition and removal by macrophages. *J. Immun.* **148,** 2207–2216.

Glucksmann, A. (1951). Cell deaths in normal vertebrate ontogeny. *Biol. Rev. Cam. Philos. Soc.* **26,** 59–86.

Krahling, S., Callahan, M. K., Williamson, P., and Schlegel, R. A. (1999). Exposure of phosphatidylserine is a general feature in the phagocytosis of apoptotic lymphocytes by macrophages. *Cell. Death. Differ* **6,** 183–189.

Kroemer, G., Dallaporta, B., and ResscheRigon, M. (1998). The mitochondrial death/life regulator in apoptosis and necrosis. *Annu. Rev. Physiol.* **60,** 619–642.

Martin, S. J., Reutelingsperger, C. P. M., Mcgahon, A. J., Rader, J. A., Vanschie, R. C. A. A., Laface, D. M., and Green, D. R. (1995). Early redistribution of plasma membrane phosphatidylserine is a general feature of apoptosis regardless of the initiating stimulus: Inhibition by overexpression of Bcl-2 and Abl. *J. Exp. Med.* **182,** 1545–1556.

McEvoy, L., Schlegel, R. A., Williamson, P., and Del Buono, B. J. (1988). Mercoyanine 540 as a flow cytometric probe of membrane lipid organization in leukocytes. *J. Leukocute Biol.* **44,** 337–344.

Mower, D. A., Peckham, D. W., Illera, V. A., Fishbaugh, J. K., Stunz, L. L., and Ashman, R. F. (1994). Decreased membrane phospholipid packing and decreased cell size precede DNA cleavage in mature mouse B cell apoptosis. *J. Immun.* **152,** 4832–4842.

Pradhan, D., Krahling, S., Williamson, P., and Schlegel, R. A. (1997). Multiple systems for recognition of apoptotic lymphocytes by macrophages. *Mol. Biol. Cell.* **8,** 767–778.

Reutelingsperger, C. P., Hornstra, G., and Hemker, H. C. (1985). Isolation and partial purification of a novel anticoagulant from arteries of human umbilical cord. *Eur. J. Bichem.* **151,** 625–629.

Schlegel, R. A., Reed, J. A., McEvoy, L., Algarin, L., and Williamson, P. (1987). Phospholipid asymmetry of loaded red cells. *Methods. Enzymol.* **149,** 281–293.

Schlegel, R. A., Stevens, M., Lumley-Sapanski, K., and Williamson, P. (1993). Altered lipid packing identifies apoptotic thymocytes. *Immunol. Lett.* **36,** 283–288.

van den Eijnde, S. M., Boshart, L., Baehrecke, E. H., DeZeeuw, C. I., Reutelingsperger, C. P. M., and VermeijKeers, C. (1998). Cell surface exposure of phosphatidylserine during apoptosis is phylogenetically conserved. *Apoptosis* **3,** 9–16.

van den Eijnde, S. M., Boshart, L., Reutelingsperger, C. P. M., DeZeeuw, C. I., and Vermeij-Keers, C. (1997a). Phosphatidylserine is exposed by apoptotic cells in mouse embryos in vivo. *Eur. J. Morphol.* **35,** 54–55.

van den Eijnde, S. M., Boshart, L., Reutelingsperger, C. P. M., de Zeeuw, C. I., and Vermeij-Keers, C. (1997b). Phosphatidylserine plasma membrane asymmetry in vivo: a pancellular phenomenon which alters during apoptosis. *Cell Death. Differ.* **4,** 311–316.

van den Eijnde, S. M., Lips, J., Boshart, L., Vermeij-Keers, C., Marani, E., Reutelingsperger, C. P., and De Zeeuw, C. I. (1999). Spatiotemporal distribution of dying neurons during early mouse development. *Eur. J. Neurosci.* **11,** 712–724.

van den Eijnde, S. M., Luijsterburg, A. J. M., Boshart, L., DeZeeuw, C. I., vanDierendonck, J. H., Reutelingsperger, C. P. M., and Vermeij-Keers, C. (1997c). In situ detection of apoptosis during embryogenesis with Annexin V: From whole mount to ultrastructure. *Cytometry* **29,** 313–320.

Verhoven, B., Schlegel, R. A., and Williamson, P. (1995). Mechanisms of phosphatidylserine exposure, a phagocyte recognition signal, on apoptotic T lymphocytes. *J. Exp. Med.* **182,** 1597–1601.

Watanabe, M., Choudhry, A., Berlan, M., Singal, A., Siwik, E., Mohr, S., and Fisher, S. A. (1998). Developmental remodeling and shortening of the cardiac outflow tract involves myocyte programmed cell death. *Development* **125,** 3809–3820.

Williamson, P., and Schlegel, R. A. (1994). Back and forth: The regulation and function of transbilayer phospholipid movement in eukaryotic cells. *Mol. Membr. Biol.* **11,** 199–216.

Wu, Y. C., and Horvitz, H. R. (1998). The *C. elegans* cell corpse engulfment gene ced-7 encodes a protein similar to ABC transporters. *Cell* **93,** 951–960.

CHAPTER 16

The (Holey) Study of Mitochondria in Apoptosis

Nigel J. Waterhouse, Joshua C. Goldstein, Ruth M. Kluck, Don D. Newmeyer, and Douglas R. Green

Division of Cellular Immunology
La Jolla Institute for Allergy and Immunology
La Jolla, California 92121

0091-679X/01 $35.00

I. Introduction

Mitochondria are dynamic organelles that are essential for the life of eukaryotic cells. They produce the majority of the cellular energy store in the form of ATP by oxidative phosphorylation and, as such, are often referred to as the powerhouse of the cell (reviewed in Saraste, 1999; Wallace, 1999; Willis, 1992). Accumulating evidence has shown that mitochondria also play an integral role in the death of a cell by apoptosis (reviewed in (Green and Reed, 1998; Waterhouse and Green, 1999).

Fractions enriched in mitochondria were shown to be required for apoptosis in extracts derived from *Xenopus laevis* eggs, and this mitochondrial proapoptotic activity was blocked by the oncogene Bcl-2 (Newmeyer *et al.,* 1994). Subsequently, Bcl-2 was shown to prevent apoptosis by blocking the translocation of cytochrome *c* from the mitochondrial intermembrane space (IMS) to the cytosol (Kluck *et al.,* 1997; Yang *et al.,* 1997). In contrast, proapoptotic proteins homologous to Bcl-2 (Bax and Bid) also act at the level of mitochondria to potentiate apoptosis by releasing cytochrome *c* from the IMS (Finucane *et al.,* 1999; Kuwana *et al.,* 1998). In the cytosol, cytochrome *c* forms part of a large proteinaceous complex, "the apoptosome," with apoptotic protease-activating factor (APAF-1), dATP/ATP, and caspase-9. This results in the activation of caspase-9 and the appearance of the morphological characteristics of apoptosis (Hu *et al.,* 1999; Saleh *et al.,* 1999; Zou *et al.,* 1999) (Fig. 1).

Caspases-2 and -9 (Susin *et al.,* 1999), apoptosis-inducing factor (AIF) (Susin *et al.,* 1999), adenylate kinase (Kluck *et al.,* 1999; Kohler *et al.,* 1999; Single *et al.,* 1998), sulfite oxidase (Kluck *et al.,* 1999), and HSP 10 and 60 (Samali *et al.,* 1999) are also released from mitochondria in various instances of apoptosis. Some of these proteins may possess proapoptotic activity (Susin *et al.,* 1999; Zamzami *et al.,* 1996).

Because mitochondrial cytochrome *c* release and its regulation by Bcl-2 family members appears to be a common mitochondrial change during apoptosis, we will focus on the methods used to study cytochrome *c* release. We will also discuss methods to examine other mitochondrial parameters that may be related to apoptosis, including the identification of proteins released from the mitochondria concomitantly with cytochrome *c*.

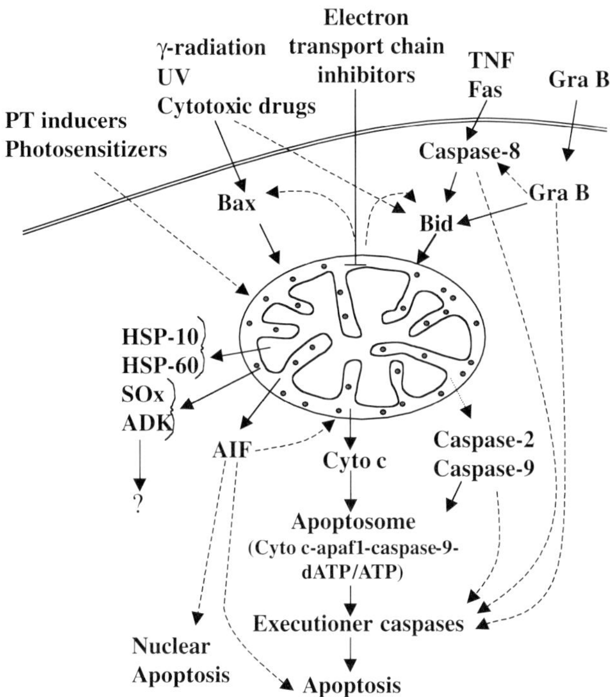

Fig. 1 Hypothetical pathways for mitochondrial-dependent and -independent apoptosis. Cytotoxic drugs and radiation appear to signal the release of several proteins, including cytochrome *c* from the mitochondrial IMS, via proapoptotic Bcl-2 family members (e.g., Bax, Bid). In the cytosol, the released cytochrome *c* forms part of the apoptosome (with apaf-1, caspase-9, and ATP/dATP). This results in the activation of caspase-9 and other downstream effector caspases. A second pathway is initiated by death receptors that activate caspase-8 directly. Active caspase-8 can either cleave Bid to activate the mitochondria/cytochrome *c*-dependent pathway to apoptosis or simply activate the executioner caspases directly. A third pathway involves granzyme B (Gra B), which can enter the cell and activate the executioner caspases directly, although it is possible that Gra B may activate the mitochondria/cytochrome *c*-dependent pathway via direct cleavage and activation of Bid. Fourth, inhibitors of the electron transport chain may induce mitochondrial cytochrome *c* release directly; however, it is also possible that some indirect signaling pathway dependent on proapoptotic Bcl-2 family members is utilized to signal the cytochrome *c*-dependent pathway to death. Fifth, photosensitizing agents and PT inducers appear to signal directly through the opening of the PT pore in the mitochondrial inner membrane, leading to mitochondrial swelling and the release of cytochrome *c*. Caspases-2 and -9, apoptosis-inducing factor (AIF), heat-shock protein (HSP) 60, HSP 10, sulfite oxidase (SOx), and adenylate kinase 2 (ADK) are also released from the mitochondria during apoptosis. The contribution of these proteins to apoptosis is not clear, however. AIF has been reported to be released prior to, and may signal for, cytochrome *c* release independently of the proapoptotic Bcl-2 family members. *Solid lines* indicate likely pathways; *dashed lines* indicate alternative pathways.

II. Induction of Cytochrome *c* Release from Mitochondria in Cultured Cells

A. Indirect Induction of Cytochrome *c* Release

Many inducers of apoptosis have been shown to induce the release of cytochrome *c* from the mitochondria of cultured cells. Radiation [e.g., ultraviolet (UV) and γ] (Bossy-Wetzel *et al.,* 1998; Chauhan *et al.,* 1997), topoisomerase inhibitors (e.g., etoposide) (Zhuang and Cohen, 1998), protein synthesis inhibitors (e.g., actinomycin D and cycloheximide) (Goldstein *et al.,* 2000; Tang *et al.,* 1999), and protein kinase inhibitors (e.g., staurosporine) (Finucane *et al.,* 1999; Heiskanen *et al.,* 1999; Wolf and Eastman, 1999) are all commonly used apoptotic stimuli that trigger cytochrome *c* release. None of these toxins appears to directly induce the release of cytochrome *c*; rather they use signaling pathways that may involve Bax, Bid, or other proapoptotic Bcl-2 family members (Fig. 1).

B. Induction of Cytochrome *c* Release via Direct Action on Mitochondria

Various toxins act directly on mitochondria to induce cytochrome *c* release (Fig. 1). Some of these toxins induce a phenomenon known as permeability transition (PT) in isolated mitochondria (e.g., calcium, *t*-butyl hydroperoxide, and atractyloside) (Cregan *et al.,* 1999; Lemasters *et al.,* 1998; Zamzami *et al.,* 1996) or in cells (e.g., lonidamine and MitoTracker orange) (Ravagnan *et al.,* 1999; Scorrano *et al.,* 1999). PT is believed to involve the opening of a pore in the mitochondrial inner membrane that allows molecules smaller than 1.5 kDa to pass and results in gross mitochondrial swelling and rupture of the outer mitochondrial membrane (reviewed in Bernardi *et al.,* 1998; Lemasters *et al.,* 1998). In this scenario, cytochrome *c* release is believed to occur only as a consequence of outer membrane rupture. Photodynamic therapy using chemicals that localize to mitochondria (e.g., Hypericin, Protoporforin IX, and MitoTracker red) may also directly induce PT and cytochrome *c* release in cells (Carthy *et al.,* 1999; Kubat *et al.,* 1997; Minamikawa *et al.,* 1999; Vantieghem *et al.,* 1998). Inhibitors of the electron transport chain have also been reported to induce apoptosis in various cell types (Marton *et al.,* 1997; Wolvetang *et al.,* 1994). Although it is not clear how these chemicals trigger an apoptotic response, in one case, apoptosis induced by complex I inhibitors was blocked by the PT inhibitor, cyclosporin A (Seaton *et al.,* 1998). Proapoptotic Bcl-2 family members also act directly on mitochondria to release cytochrome *c*; however, their mode of action does not necessarily appear to involve PT (Bossy-Wetzel *et al.,* 1998; Eskes *et al.,* 1998; Finucane *et al.,* 1999).

C. Inducers of Apoptosis That May Not Require Cytochrome *c* Release

Some pathways to apoptosis may not require mitochondria or cytochrome *c* release. These include apoptosis induced by death receptors [e.g., Fas, tumor necrosis factor (TNF), DR-3] (Adachi *et al.,* 1998; Chauhan *et al.,* 1997; Linsinger *et al.,* 1999) and granule-induced apoptosis (Andrade *et al.,* 1998; Darmon *et al.,* 1996; Namura *et al.,* 1998) (Fig. 1). During Fas-induced apoptosis, caspase-8 is activated at the receptor level (Muzio *et al.,* 1996) and may result in caspase-mediated apoptosis independently of mitochondria. Caspase-8 can also cleave and activate Bid, a proapoptotic Bcl-2 family member that can in turn directly induce the release of cytochrome *c,* thereby potentiating the apoptotic response (Kuwana *et al.,* 1998). TNF-induced death may also signal apoptosis through caspases recruited to the receptor complex (reviewed in Yuan, 1997), potentially bypassing the need for mitochondrial involvement. However, cytochrome *c* is also released during TNF-induced apoptosis (Goldstein *et al.,* 2000), and thus may actually activate or potentiate the activation of executioner caspases in this system. During granule-induced apoptosis, granzyme B can activate caspase-3 to initiate the death pathway, perhaps independently of cytochrome *c* release (Andrade *et al.,* 1998). Cytochrome *c* release has been seen during granzyme B-induced apoptosis (Heibein *et al.,* 1999), and this release amplifies the death signal in a caspase-dependent manner (MacDonald *et al.,* 1999).

Although the role of cytochrome *c* in the activation of caspases via the formation of an apoptosome (cytochrome *c,* apaf-1, dATP/ATP, and caspase-9) is well established in mammalian systems, it may be that the involvement of cytochrome *c* is a recent event in evolution. Genetic studies in *Caenorhabditis elegans* indicate that the Bcl-2 homologue (*ced-9*) and a short form of apaf-1 (*ced-4*) regulate the caspase homologue (*ced-3*) by a direct interaction that does not require cytochrome *c* (Chinnaiyan *et al.,* 1997). In *D. melanogaster,* a homologue of apaf-1 has been identified (Rodriguez *et al.,* 1999); however, cytochrome *c* release during apoptosis has not been shown formally in insect cells.

III. Assessment of Cytochrome *c* Release in Heterogeneous Populations of Cultured Cells

A. Cellular Fractionation

Assessment of a change in the intracellular distribution of cytochrome *c* may be simply assessed by cellular fractionation techniques. These include physical methods such as homogenization and nitrogen cavitation and chemical methods such as the use of detergents or pore-forming proteins. The specific question being addressed and the number of cells available will determine the most appropriate technique for the isolation of cellular fractions.

1. Homogenization

Homogenization is commonly used to separate the cytosolic and mitochondrial fractions from cells undergoing apoptosis (Bossy-Wetzel *et al.,* 1998; Finucane *et al.,* 1999; Granville *et al.,* 1998; Linsinger *et al.,* 1999). This involves physically disrupting the plasma membrane of the cells with a mortar and pestle while leaving the mitochondria intact. The cytosol, mitochondria, and nuclei/unbroken cells are then separated by differential centrifugation. Disadvantages to this method include the large numbers of cells required and the incidental damage to mitochondria that can result from excessive homogenization. Careful titration of the force used to isolate fractions should be employed to optimize this method for each cell type. This can be monitored by the measurement of cytochrome *c* in the cytosol isolated from cells in the absence of apoptotic stimuli, and both cytosolic and mitochondrial fractions should be analyzed during apoptosis studies.

The protocol reported here has been standardized for Jurkat cells. All steps are carried out at 4°C.

1. Harvest 2.5×10^7 Jurkat cells by centrifugation (800*g* for 10 min).
2. Transfer the pellet to a microfuge tube and wash the cells twice with ice-cold phosphate-buffered saline (PBS), pH 7.2.
3. Resuspend the cells in 500 μl of extraction buffer [220 m*M* mannitol, 68 m*M* sucrose, 50 m*M* PIPES–KOH, pH 7.4, 50 m*M* KCl, 5 m*M* EGTA, 2 m*M* $MgCl_2$, 1 m*M* EDTA, 1 m*M* dithiothreitol (DTT), and protease inhibitors].
4. Incubate the cells on ice for 30 min (with intermittent shaking) and then disrupt the cells by homogenization with 60 strokes using a glass dounce and a B-type pestle.
5. Centrifuge the lysate at 800*g* for 10 min to pellet the nuclei and unbroken cells.
6. Centrifuge the supernatant at 14,000*g* for 10 min.
7. The cytosol (supernatant) and pellet (mitochondria) can then be stored at −70°C for subsequent Western blot analysis.

2. Nitrogen Cavitation

Nitrogen cavitation is reported to be a more gentle method than homogenization for the isolation of mitochondria (Adachi *et al.*, 1998; Kjeldsen *et al.*, 1999). This technique uses specialized equipment, a "nitrogen bomb" in which pressurized nitrogen diffuses into the cells. When the pressure is reduced the nitrogen bubbles expand and rupture the cells from the inside. Although our laboratory has no experience with this technique, it has been used to isolate mitochondria from rat ascites cells (Shinohara *et al.*, 1997).

As reported, all steps were carried out at 4°C.

1. Rat ascites cells (6×10^9) are washed three times in medium containing 150 m*M* NaCl, 5 m*M* KCl, and 10 m*M* Tris–HCl (pH 7.4) and then once in 210 m*M* mannitol, 70 m*M* sucrose, 1 m*M* EGTA, 0.5% bovine serum albumin (BSA), 5 m*M* Hepes–KOH buffer (pH 7.2).
2. The pelleted cells are made into slurry and introduced into the nitrogen bomb.
3. Pressure (N_2 20 kg cm^{-2}) is applied for 30 min after which the lysate is recovered and fractionated by differential centrifugation as described in the homogenization protocol.

An additional advantage of this technique is that mitochondria are recovered with proteins attached to the outer membrane that are normally removed by homogenization (Shinohara *et al.*, 1997). Disadvantages of this method shared with homogenization are that large numbers of cells are required and mitochondrial damage can occur, thus necessitating standardization of the protocol for your individual purpose. Using this technique, some reports show that ~30% of all cytochrome *c* was found in the cytosolic fraction of untreated cells and ~15% was found in the nuclear fraction (Adachi *et al.*, 1997, 1998). This suggests that ~15% of cells were not lysed and that the outer membrane of ~30% of the mitochondria was damaged. This reiterates the point that great care should be used to standardize the chosen protocol with your cells and equipment.

3. Digitonin

Digitonin is a compound isolated from seeds of *Digitalis purpurea* (Bridges, 1977). It is capable of lysing both plasma and mitochondrial membranes. Using low concentrations of digitonin, however, plasma membranes can be selectively permeabilized without affecting mitochondrial permeability (Adam *et al.*, 1990; Boon and Zammit, 1988; Mannella and Wang, 1989). Selective permeabilization with low concentrations of digitonin has been used successfully to study mechanisms of nuclear import (Adam *et al.*, 1990) and is now becoming a common method for monitoring the release of cytochrome *c* release during apoptosis (Goldstein *et al.*, 2000; Heibein *et al.*, 1999; Vantieghem *et al.*, 1998). Advantages of this technique include that (1) only low numbers of cells are required, (2) it is rapid (~15 min), and (3) the majority of cells are lysed (~95%) so that a single centrifugation step can be used to separate mitochondrial and cytosolic cytochrome *c*.

The technique reported here has been optimized for Jurkat cells.

1. Harvest washed cells (1×10^6) in a microfuge tube by centrifugation (800*g* for 5 min) and resuspend in 100 μl of ice-cold lysis buffer (80 m*M* KCl, 250 m*M* sucrose, and 200 μg/ml digitonin in PBS).
2. After 5 min on ice, centrifuge the cells (10,000*g* for 5 min) and store the recovered pellet and supernatant at −70°C for Western blotting.
3. To follow the efficiency of lysis, incubate 5 μl of the lysing reaction with 5 μl of 0.1% trypan blue and monitor the percentage of blue permeabilized cells. The method should be standardized in control cells such that ~95% of cells are trypan blue positive with all cytochrome *c* remaining in mitochondria.

Note: Others have successfully used 35 μg digitonin/4×10^6 Jurkat cells diluted in 75 m*M* NaCl, 1 m*M* NaH_2PO_4, 8 m*M* Na_2HPO_4, 250 m*M* sucrose (Heibein *et al.*, 1999; Single *et al.*, 1998) and 200 μl of 0.02% digitonin in PBS/1.5×10^6 cells (Vantieghem *et al.*, 1998).

4. Streptolysin O

Streptolysin O is a compound isolated from *Streptococcus pyogenes* (Duncan, 1974) that can also selectively permeabilize the plasma membrane while leaving mitochondrial membranes intact (Bhakdi *et al.*, 1993; Wolf and Eastman, 1999). The main advantages of this technique are the low numbers of cells used and the efficiency of lysis.

This technique has been standardized for Jurkat cells.

1. Harvest cells (1×10^6) by centrifugation and resuspend the pellet in 100 μl of lysis buffer (20 m*M* Hepes–KOH, pH 7.5, 250 m*M* sucrose, 10 m*M* KCl, 1.5 m*M* $MgCl_2$,

1 m*M* Na-EDTA, 1 m*M* EGTA, 1 m*M* DTT, and proteinase inhibitors) containing 60 U of streptolysin O.

2. Incubate the cells for 20 min at 37°C and separate the cytosol and particulate fractions by centrifugation (10,000*g* for 5 min at 4°C).

3. The supernatants and pellets can be stored at −70°C for Western blot analysis. Again, the efficiency of lysis should be followed by trypan blue exclusion and the technique standardized for your particular application and cell type.

B. Detection of Cytochrome *c* in Cellular Fractions

Regardless of the technique used to isolate cellular fractions, Western blotting is the most common technique to detect cytochrome *c*, although ELISAs are now available commercially. The quantity of protein required for Western blotting will depend on the cell line and the cell fraction being analyzed.

1. Resuspend the particulate samples (mitochondria or cell pellets) in 100 μl of ice-cold universal immunoprecipitation buffer [50 m*M* Tris–HCl, pH 7.4, 150 m*M* NaCl, 2 m*M* EDTA, 2 m*M* EGTA, 0.2% Triton X-100, 0.3% Nonidet P-40 (NP-40)] and incubate with rocking at 4°C for 10 min.

2. Centrifuge the lysate (10,000*g* for 5 min) and store the supernatant for analysis.

3. For SDS–PAGE, boil the protein extracts (cytosol or proteins isolated from the particulate fractions) in 20% (v/v) SDS–PAGE loading dye (156 m*M* Tris–HCl, pH 7.4, 5% SDS, 10% glycerol, 12.5% β-mercaptoethanol, 0.025% bromphenol blue) for 5 min and resolve proteins on a 15% SDS–polyacrylamide gel.

4. Transfer proteins onto nitrocellulose membranes at 15 V overnight.

5. Incubate the membranes in blotto (5% skim milk powder in PBS) for 2 h followed by overnight incubation at 4°C in anticytochrome *c* primary antibody (7H8.2C12, Pharmingen, San Diego, CA) diluted in 1 : 1000 in blotto.

6. Following three washes in PBS (10 min), incubate the membranes for 1 h in any appropriate horseradish peroxidase-conjugated secondary anti-mouse antibody diluted 1 : 1000 in blotto.

7. After washing the membranes three times in PBS, the protein can be visualized using enhanced chemiluminescence (Pierce, Rockford, IL) according to the manufacturer's instructions.

IV. Assessment of Cytochrome *c* Release in Single Cells

The techniques just described are useful for analyzing cytochrome *c* release in homogeneous populations of cell; however, they have limited use in comparing the percentage of cells that have undergone cytochrome *c* release with other apoptotic events. Nor can they be used to determine the nature of cytochrome *c* release, for example, at a single cell level to determine if cytochrome *c* release is slow, fast, partial, or complete.

Techniques that assess cytochrome *c* release at a single cell level have been used to show that cytochrome *c* is released rapidly and completely from all mitochondria in a cell as it undergoes apoptosis (Goldstein *et al.,* 2000). Such techniques that analyze cytochrome *c* release in single cells are invaluable for the further study of this event.

A. Immunocytochemistry

Immunocytochemistry is a powerful technique for determining the cellular localization of proteins in individual cells. Using this technique, a punctate distribution of cytochrome *c* is observed in control cells (mitochondrially localized), whereas a diffuse distribution of cytochrome *c* is observed in apoptotic cells (cytosolic) (Bossy-Wetzel *et al.,* 1998; Goldstein *et al.,* 2000). Immunocytochemistry can be used to quantitate the percentage of cells that have undergone cytochrome *c* release by manually counting the number of cells with punctate versus diffuse staining. Because the cells are fixed, they can be stored for later analysis. By using antibodies to different proteins, this technique can be used to compare the localization of cytochrome *c* with the localization of other proteins or organelles in the cell. This technique, however, is quite laborious and expensive.

The following technique has been optimized for HeLa cells.

1. Grow cells to ~70% confluence, either in LabTek chamber slides or on sterile coverslips and treat the cells with the apoptotic stimuli being studied.
2. Freshly prepare paraformaldehyde (4%) by dissolving 8 g of paraformaldehyde in 100 ml of 2X PBS at 60°C (add a few drops of 1 *M* NaOH to aid solubilization) followed by the addition of 100 ml of water. Adjust the pH to 7.2 and store the paraformaldehyde solution at 4°C for up to 1 day.
3. Carefully remove the media from the cells and fix the cells in the paraformaldehyde for 15 min with gentle rocking at 4°C.
4. Remove the fixative and wash the cells six times in blocking buffer (0.05% saponin, 3% BSA in PBS, pH 7.2).
5. Incubate the cells overnight with anticytochrome *c* [clone 6H2.B4 (Pharmingen, San Diego, CA)] diluted 1 : 200 in blocking buffer.
6. Wash the cells six times in blocking buffer and incubate for 1 h at room temperature with mouse Ig Texas red diluted 1 : 200 in blocking buffer.
7. The fluorescence of the Texas red-conjugated antibody can be detected by confocal or fluorescence microscopy using 595 and 615 nm excitation and emission (Bossy-Wetzel *et al.,* 1998; Goldstein *et al.,* 2000) (Fig. 2A, see Color Plate). Secondary antibodies conjugated to other fluorescent probes (e.g., FITC) can also be used.

B. Cytochrome *c*–Green Fluorescent Protein

It has been shown that green fluorescent protein (GFP)-tagged cytochrome *c* expressed in cells behaves in a manner similar to that of endogenous cytochrome *c* during apoptosis (Goldstein *et al.,* 2000; Heiskanen *et al.,* 1999). That is, it leaves the mitochondria at

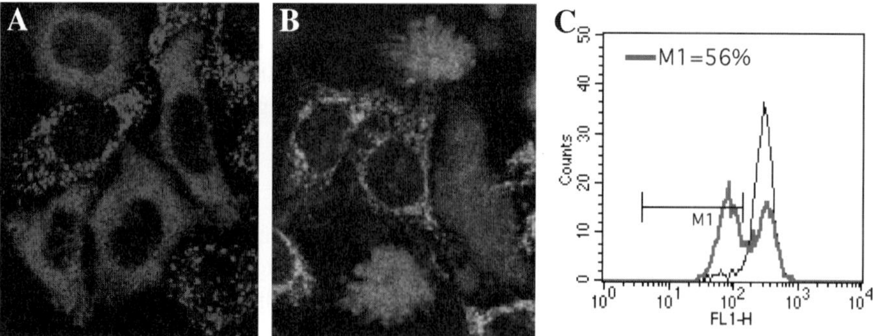

Fig. 2 Detection of cytochrome *c* release by fluorescent labeling. (A) A confocal image of a heterogeneous population of HeLa cells treated with UVC (180 mJ/cm^2) in the presence of zVAD-fmk (100 μM) and stained by immunocytochemistry after 6 h. (B) A confocal image of cells stably transfected with cytochrome *c*–GFP, 6 h after UVC. In A and B, diffuse staining indicates that cytochrome *c* has been released from mitochondria. Mitochondria in fixed cells have a more punctate distribution than live cells. (See Color Plates.) (C) Cells stably transfected with cytochrome *c*–GFP and either not treated (*thin black line*) or treated as in B (*thick grey line*) were permeabilized with digitonin and analyzed by flow cytometry. Permeabilized cells that had previously undergone mitochondrial cytochrome *c* release do not retain cytochrome *c*–GFP. In this case, 56% of the treated cells have released their mitochondrial cytochrome *c* (*thick gray line*). Only 45% of treated cells have phosphatidylserine (PS) exposed on their surface, in support of the concept that cytochrome *c* is released before PS exposure. In untreated cells, approximately 6% of cells had both released cytochrome *c* and had exposed PS on the outside of the plasma membrane. (See Color Plate.)

the same time as endogenous cytochrome *c* and becomes diffuse throughout the cell (Fig. 2B, see Color Plate). This tagged molecule offers significant advantages over immunocytochemistry for examining the release of cytochrome *c* over time in individual cells. The percentage of cells that release cytochrome *c* can be estimated without laborious fixing and staining. The actual release of cytochrome *c* can be assessed in real time by time-lapse microscopy and can be directly compared with apoptotic events such as phosphatidylserine exposure and loss of plasma membrane integrity.

To create cells expressing cytochrome *c*–GFP, first amplify cytochrome *c* from human cDNA by polymerase chain reaction using 5′tttgaattcaatgggtgatgttgag3′ and 5′tttggaatccttactcattagtagtagc3′ primers. Insert the resulting fragment between the *Eco*RI and the *Bam*HI sites of the plasmid pEGFP-C1. Transfect cells with the plasmid by any conventional technique and select a stable clone with cytochrome *c*–GFP expression in mitochondria. Mitochondrial dyes such as tetramethyl rhodamine ethyl ester (TMRE, discussed later) can be used to evaluate mitochondrial localization, and immunocytochemistry and Western blot can be used to monitor the concomitant release of endogenous and expressed cytochrome *c*. Cells can be treated with an apopototic stimulus and examined for the release of cytochrome *c*–GFP by confocal microscopy (Fig. 2B). The percentage of cells that have released cytochrome *c*–GFP from mitochondria can be assessed by comparing the number of cells with punctate versus those with diffuse green fluorescence.

The percentage of cells that have released cytochrome *c*–GFP from the mitochondria can also be quantitated by flow cytometry by permeabilizing the outer membrane and allowing the nonmitochondrial cytochrome *c* to exit the cell.

1. Harvest cells (2×10^4) in individual wells of a round-bottomed 96-well plate, aspirate the media, and resuspend the pellet in 100 μl of ice-cold cell lysis and mitochondria intact (CLAMI) buffer (250 m*M* sucrose, 80 m*M* KCl, 50 μg/ml digitonin) in PBS.
2. Transfer the cells to FACS inserts and incubate for 5 min on ice.
3. Dilute each sample with 200 μl ice-cold digitonin-free CLAMI buffer and immediately analyze the content of green fluorescent protein in the cells by flow cytometry (488 nm, FL-1).

As CLAMI buffer releases all cytosolic protein, cells in which cytochrome *c* has been released from the mitochondria will have lower fluorescence as compared to control cells (Fig. 2C).

V. Assessment of Mitochondrial Cytochrome *c* Release in Cell-Free Systems

Isolated mitochondria are commonly used in so-called cell-free assays to assess the effect of specific compounds, e.g., the mode of action of Bcl-2 family members and mitochondrial toxins. Techniques described in Section III (homogenization or nitrogen cavitation) can be used to isolate the mitochondrial fraction from cultured cells. However, further centrifugation steps that remove damaged mitochondria and remaining nuclei can increase mitochondrial purity for use in cell-free assays.

A. Isolation of Mouse Liver Mitochondria

1. Sacrifice a 2- to 6-month-old mouse by cervical dislocation, and rapidly excise the liver and place it in a petri dish on ice. *Note:* Perform the rest of the isolation procedure at 4°C.
2. Wash the liver in PBS, pH 7.4, and cut the liver into small pieces using a sterile scissors.
3. Separate the liver pieces into two parts and homogenize each half in 10 ml of mitochondrial isolation buffer (MIB: 220 m*M* mannitol, 68 m*M* sucrose, 10 m*M* Hepes–KOH, pH 7.4, 10 m*M* KCl, 1 m*M* EGTA, 1 m*M* EDTA, 0.1% BSA, and protease inhibitors) using a 15-ml Dounce with a tight-fitting Teflon pestle until all of the tissue has been dissociated (20–40 strokes).
4. Multiple steps of differential centrifugation in a Sorvall centrifuge with a swinging bucket rotor (HB-4) using round-bottom 15-ml polypropylene tubes can be used to isolate mitochondria. Centrifuge the cellular lysates (600*g* for 10 min) to pellet nuclei and unbroken cells and then centrifuge the supernatant (3500*g* for 15 min) to pellet mitochondria.

5. Resuspend the mitochondrial pellet in 15 ml of fresh MIB. Enrich this fraction further by centrifugation (1500*g* for 5 min) to remove any contaminating nuclei and centrifuge the supernatant from this spin (5500*g* for 10 min) to pellet mitochondria.

6. Rewash and enrich the mitochondrial fraction by repeating the last two centrifugation steps twice.

7. Resuspend mitochondria in 500 μl of ice-cold 200 m*M* mannitol, 50 m*M* sucrose, 10 m*M* succinate, 10 m*M* Hepes–KOH, pH 7.4, 0.1% BSA, and 5 m*M* potassium phosphate, pH 7.4. Mitochondria can be kept on ice for a maximum of 4 h.

B. Isolation of Mitochondria from Cultured Cells

To isolate mitochondria from human cells it is necessary to start with a large number of cells. The homogenization technique described for the isolation of mouse liver mitochondria can also be used to isolate mitochondria from human cells, but using a microfuge.

1. Harvest and wash cells (at least 5×10^8) by centrifugation (800*g* for 10 min) in ice-cold PBS (pH 7.4).

2. Resuspend the cells in 3 volumes of MIB (220 m*M* mannitol, 68 m*M* sucrose, 10 m*M* Hepes–KOH, pH 7.4, 10 m*M* KCl, 1 m*M* EGTA, 1 m*M* EDTA, 0.1% BSA, and protease inhibitors).

3. Disrupt the cells with 60–80 strokes in a 2-ml Dounce using a loose-fitting B-type pestle and centrifuge the lysate (600*g* for 10 min) to pellet nuclei and unbroken cells.

4. Centrifuge the supernatant (3500*g* for 15 min) to pellet mitochondria and resuspend the mitochondrial pellet in 1 ml of fresh MIB.

5. Recentrifuge the mitochondrial fraction (1500*g* for 5 min) to remove any contaminating nuclei and centrifuge the supernatant from this spin (5500*g* for 10 min) to repellet the mitochondria.

6. Rewash and enrich the mitochondrial fraction by repeating the last two steps twice.

7. Resuspend mitochondria in 500 μl of ice-cold mitochondria resuspension buffer (200 m*M* mannitol, 50 m*M* sucrose, 10 m*M* succinate, 10 m*M* Hepes–KOH, pH 7.4, 0.1% BSA, and 5 m*M* potassium phosphate, pH 7.4). Mitochondria can be maintained on ice for a maximum of 4 h.

C. Induction of Cytochrome *c* Release from Isolated Mitochondria

Recombinant truncated Bid and Bax directly induce cytochrome *c* release from isolated mitochondria (Finucane *et al.,* 1999; Kluck *et al.,* 1999). Because full-length Bax is relatively insoluble, we use a recombinant truncated form lacking 19 amino acids from the C terminus (encoded in pGEX-KG-BaxC19 from S. Korsmeyer) in cell-free assays (Finucane *et al.,* 1999). Because full-length Bid is less active (10X) prior to caspase cleavage, we use a truncated form of Bid (tBid, missing 7 kDa from the N terminus) produced by the cleavage of full-length Bid by caspase-8 (Kluck *et al.*, 1999; Kuwana *et al.*, 1998).

To induce cytochrome *c* release:

1. Incubate freshly isolated mitochondria (equivalent of 100 μg mitochondrial protein/ml) with tBid (10 μg/ml) or Bax (30–40 μg/ml) for 5–60 min at 37°C.
2. Cytochrome *c* release can be detected by pelleting mitochondria (10,000*g* for 5 min at 4°C) and assaying the pellet and supernatant for cytochrome *c* by Western blot.

Chemicals such as atractyloside (5 m*M*), $CaCl_2$ (100–500 μ*M*), and lonidamine (50 μ*M*) induce cytochrome *c* release from isolated mitochondria via the induction of PT (Finucane *et al.*, 1999; Ravagnan *et al.*, 1999).

To induce cytochrome *c* release with PT inducers:

1. Resuspend freshly isolated mitochondria (100 μg/ml) in PT buffer (125 m*M* KCl, 2.5 m*M* potassium phosphate, pH 7.4, 2.5 m*M* succinate, and 20 m*M* Hepes–KOH) containing the PT inducer.
2. Cytochrome *c* release can then be detected by pelleting mitochondria (10,000*g* for 5 min at 4°C) and assaying the pellet and buffer for cytochrome *c* by Western blot.

Note: Cyclosporin A (40 μ*M*) in the assay buffer can transiently inhibit PT and cytochrome *c* release induced by these chemicals.

VI. Detection of Proteins Released from Mitochondria Concomitant with Cytochrome *c*

A. Caspases

The release of certain caspases from the mitochondria can be detected by Western blotting. Caspases-2 and -9 have been shown to reside in the mitochondrial intermembrane space of certain cell types (Susin *et al.*, 1999). Because these proteins are more abundant in cytosol than in mitochondria, their translocation can be monitored by measuring mitochondrial rather than cytosolic content. Caspase zymogens vary in size but most can be resolved on 12–15% SDS–PAGE gels and probed by Western blotting with specific antibodies.

B. Apoptosis-Inducing Factor

Apoptosis-inducing factor was one of the first proapoptotic proteins shown to be released from mitochondria during apoptosis (Susin *et al.*, 1996). It is a 57-kDa flavoprotein with homology to bacterial oxidoreductases (Susin *et al.*, 1999). The recombinant protein induces chromatin condensation and large-scale DNA fragmentation in isolated nuclei and release of cytochrome *c* and caspases from isolated mitochondria. When microinjected into cells, it induces loss of membrane potential and phosphatidylserine externalization. While the caspase inhibitor zVAD-fmk does not block the effects of recombinant AIF, it does prevent internucleosomal DNA fragmentation, loss of mitochondrial membrane potential, and externalization of phosphatidyl serine in cells treated with apoptotic stimuli. It is therefore not clear what role, if any, AIF plays in apoptosis.

Until recently, AIF activity was determined by its action on isolated nuclei (Susin *et al.*, 1996, 1999; Zamzami *et al.*, 1996).

1. Incubate mitochondria (200 μg/ml) treated with PT inducers for 90 min at 37°C with isolated HeLa nuclei (10^3 nuclei/μl) in 220 n*M* mannitol, 68 m*M* sucrose, 2 m*M* NaCl, 2.5 m*M* PO_4H_2K, 0.5 m*M* EGTA, 2 m*M* $MgCl_2$, 5 m*M* Na pyruvate, 1 m*M* phenylmethylsulfonyl fluoride (PMSF), 1 m*M* DTT, and 10 m*M* Hepes–KOH, pH 7.4.
2. Stain the nuclei with 10 μ*M* 4′-6-diamidino-2-phenylindole dihydrochloride (DAPI) or similar DNA stain and examine nuclei by fluorescence microscopy for small condensed nuclei.

Note: The disadvantage of this assay for following the release of AIF from mitochondria is that it relies on the concept that AIF is the only protein capable of affecting these changes. Antibodies against peptides encompassing amino acids 150–170, 166–185, and 181–200 of AIF have been produced and used for Western blot analysis (Susin *et al.*, 1999), but are not yet available commercially.

C. Adenylate Kinase

Several isoforms of adenylate kinase (ADK) exist in cells: ADK1 is cytosolic, ADK2 is present in the mitochondrial IMS, and ADK3 is present in the mitochondrial matrix of untreated cells. In Jurkat cells, the majority of ADK activity is due to ADK2, which provides the basis for a functional assay to monitor ADK2 release from the mitochondrial IMS during apoptosis (Kluck *et al.*, 1999; Single *et al.*, 1998).

1. Lyse isolated mitochondria in small volumes of 1% Triton X-100, 210 m*M* mannitol, 60 m*M* sucrose, 10 m*M* Hepes–KOH, pH 7.4, 10 m*M* KCl, 5 m*M* EGTA, 10 m*M* succinate, and 0.5 m*M* DTT.
2. Make the assay buffer by adding 5 μl each of 100 m*M* ATP, 100 m*M* phosphoenol pyruvate, 1 m*M* rotenone, 1.5 m oligomycin, a mix of pyruvate kinase and lactate dehydrogenase (80 U/ml each), and 15 μl of 100 m*M* NADH to 1 ml of 130 m*M* KCl, 6 m*M* MgSO4, 100 m*M* Tris–HCl, pH 7.5.
3. Add 200 μl of this assay buffer to small (equal) volumes of mitochondrial lysates or cytosol in a microtiter plate and measure the consumption of NADH (decrease in $A_{366\,nm}$) over 10 min. Subtract the background values obtained in the presence of the adenylate kinase inhibitor (20 μ*M* diadenosine pentaphosphate) and calibrate activity against chicken muscle myokinase activity (Kluck *et al.*, 1999).

VII. Assessment of Mitochondrial Function

A. Mitochondrial Membrane Potential

The active transfer of electrons along the electron transport chain results in the transfer of protons out of the matrix, across the inner membrane to the mitochondrial IMS generating the mitochondrial transmembrane potential ($\Delta\Psi$m). This potential is used by ATP synthase to generate ATP from ADP and is required for the import of some proteins into mitochondria. A high $\Delta\Psi$m is therefore one measure of mitochondrial viability. $\Delta\Psi$m has been shown both to transiently increase (mitochondria become hyperpolarized) and to decrease entirely (mitochondria become depolarized or $\Delta\Psi$m is lost) during apoptosis. These observations became the driving force of some theories of how cytochrome *c* is released from mitochondria during apoptosis.

Measurement of $\Delta\Psi$m can be performed using a variety of commercially available cationic dyes that accumulate in the mitochondria in a $\Delta\Psi$m-dependent manner. Incubate cells for 10–20 min in growth media containing any of the dyes at the concentrations indicated in Table I. Stained cells can be analyzed by flow cytometry or microscopy using the excitation (Ex) and emission (Em) wavelengths noted in Table I. Cells should not be fixed prior to analysis (unless MitoTracker dyes have been used, as discussed later). FCCP (5–10 μ*M*) or CCCP (5–10 μ*M*) depolarizes $\Delta\Psi$m rapidly and should be used to estimate the fluorescence in cells that have lost $\Delta\Psi$m. Figure 3 shows fluorescence levels of the various dyes in cells with or without FCCP or CCCP treatment.

Many of the dyes also accumulate across nonmitochondrial membranes that possess a potential. It is therefore wise to compare the potential of stained cells in the presence and absence of 137 m*M* KCl in the growth medium. KCl neutralizes the plasma membrane potential and ensures that $\Delta\Psi$m is the predominant potential being measured. Some reports have used the dyes diluted in PBS to stain cells. However, because mitochondria

Table I
Fluorescent Dyes Used to Stain Mitochondria

Mitochondrial dye[a]	EX/EM	Staining conditions	Example of use in apoptosis
JC-1			
Green	510/527	5–10 μg/ml (20 min at 37°C)	Salvioli *et al.* (1997)
Red aggregate	485–585/590		
$DioC_6(3)$	484/500	40 n*M* (20 min at 37°C)	Finucane *et al.* (1999)
MitoTracker red	578/599	150 m*M* (20 min at 37°C)	Finucane *et al.* (1999)
MitoTracker orange	551/576	150 n*M* (20 min at 37°C)	Bossy-Wetzel *et al.* (1998)
MitoTracker green	490/516	150 n*M* (20 min at 37°C)	Metivier *et al.* (1998)
Rhodamine 123	507/529	2–5 μ*M* (20 min at 37°C)	Vander Heiden *et al.* (1997)
TMRE	550/573	50–100 n*M* (20 min at 37°C)	Fink *et al.* (1998)
Nonyl acridine orange	495/519	100 n*M* (30 min at 37°C)	Vander Heiden *et al.* (1997)

[a]All of the dyes listed can be obtained from Molecular Probes (Eugene, OR).

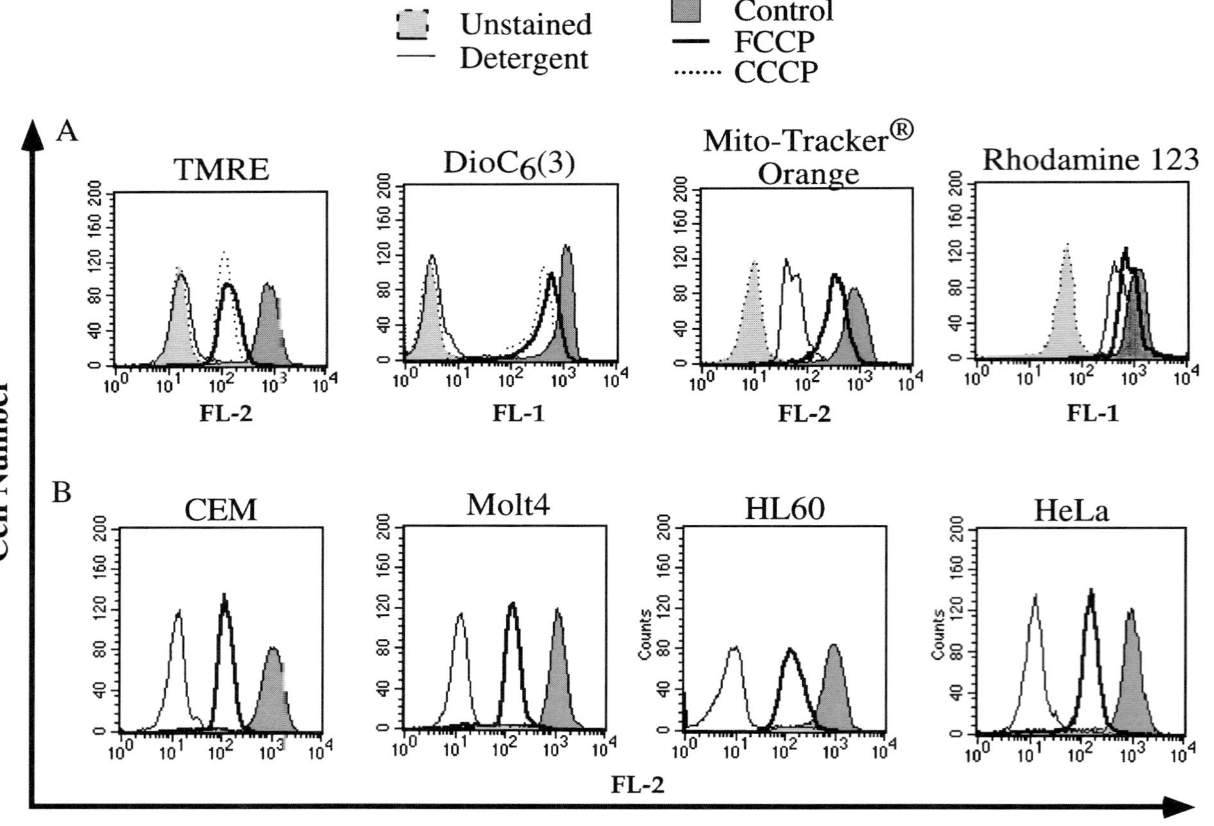

Fig. 3 A comparison of fluorescence levels of dyes used to measure $\Delta\Psi$m of mitochondria in cells. (A) Jurkat cells were stained with TMRE (50 n*M*), $DioC_6(3)$ (40 n*M*), MitoTracker orange (150 n*M*), or rhodamine 123 (2 μ*M*) (filled solid lines). FCCP (5 μ*M*; *thick solid line*) or CCCP (50 μ*M*; *dotted line*) was added during staining to depolarize mitochondria. Unstained cells (*filled dotted line*) and cells stained with the dyes followed by detergent (2% NP-40; *thin solid line*) served as controls. The dotted line for CCCP is masked in some cases by exact overlay of the FCCP trace, indicating that these dyes respond equally to the uncouplers. (B) Cells indicated were stained with TMRE (50 n*M*) in the presence or absence of FCCP (5 μ*M*). Detergent-treated cells show that some dye remains bound to the cells even in the presence of FCCP. *Note:* Much of the residual background staining can be eliminated by incubation with KCl (137 m*M*), indicating that much of this dye is retained by the low plasma membrane potential (not shown). This is also the case for the other dyes.

in cells react quickly to environmental changes, such as removal of glucose and the addition of mitochondrial toxins to the medium, the removal of cells from growth medium may alter the $\Delta\Psi$m in some instances. The growth medium does not affect flow cytometry.

The use of these dyes to measure $\Delta\Psi$m introduces several known complications (Metivier *et al.,* 1998; Salvioli *et al.,* 1997; Scaduto and Grotyohann, 1999). In untreated cells with high $\Delta\Psi$m, JC-1 becomes concentrated and forms J aggregates that fluoresces orange/red, while in depolarized cells, JC-1 is less concentrated and fluoresces green. These two states are easily distinguishable by flow cytometry or fluorescence microscopy (Salvioli *et al.,* 1997). Large changes in potential are required for JC-1 to change its fluorescence, and fluorescent imaging of untreated cells shows that J aggregates only form in a limited number of mitochondria within a cell. Rhodamine 123, also commonly used to measure $\Delta\Psi$m, has been shown to inhibit ATP synthase, meaning that incubation with this dye may itself alter $\Delta\Psi$m (Modica-Napolitano *et al.,* 1984). In addition, on accumulation in mitochondria, rhodamine also undergoes self-quenching (above a threshold, the signal decreases) (Emaus *et al.,* 1986). $DioC_6(3)$, a dye that fluoresces green, was first used to show mitochondrial depolarization during apoptosis (Zamzami *et al.,* 1995). This dye is known to stain nonmitochondrial membranes (it may be more sensitive to changes in plasma membrane potential than $\Delta\Psi$m) and to inhibit complex I of the electron transport chain (Anderson *et al.,* 1993; Jenssen *et al.,* 1986). MitoTracker dyes accumulate in the mitochondria and covalently bind to mitochondrial membranes. These dyes are therefore useful if the cells are to be fixed prior to analysis (Poot *et al.,* 1996), but they cannot be used to monitor the loss of membrane potential after staining. MitoTracker red has been shown to act as a photosensitizing agent that can induce permeability transition (Minamikawa *et al.,* 1999), and MitoTracker orange has been shown to induce PT and the release of cytochrome *c* from mitochondria but only at doses in excess of that used to monitor $\Delta\Psi$m (Scorrano *et al.,* 1999). MitoTracker green may not be a potential sensitive dye and may instead relate more to mitochondrial mass (Metivier *et al.,* 1998). TMRE is a red dye that has been shown to distribute across the mitochondrial membrane in a near-Nernstian fashion. TMRE is not toxic (Farkas *et al.,* 1989) and cells can be cultured and undergo cell division in the presence of low concentrations of TMRE (<50 n*M*). It does not become covalently attached to mitochondria, thereby allowing fluxes in $\Delta\Psi$m to be monitored. TMRE does not suffer from problems of self-quenching, although it does bind to nonmitochondrial membranes in the cell, which should be accounted for by the use of 137 m*M* KCl in the staining medium. Other problems associated with the use of TMRE [and a similar molecule tetramethylrhodamine methyl ester (TMRM)], such as high binding to polystyrene, have been reported (Scaduto and Grotyohann, 1999).

As can be seen, all membrane potential dyes have their positive and negative attributes. In our hands, most of the dyes yield similar results when quantifying the percentage of cells that have lost $\Delta\Psi$m; however, not all dyes appear suited for detecting subtle changes in $\Delta\Psi$m. In our experience, TMRE is the most robust and sensitive dye for measuring $\Delta\Psi$m in cells (Fig. 3) and isolated mitochondria.

B. Mitochondrial Mass

Mitochondrial mass can be analyzed simply by incubating cells in nonyl-acridine orange (NAO, 100 m*M*) or MitoTracker green (150 n*M*) in media and determining the fluorescence by flow cytometry using the specifications in Table I.

C. ATP Levels

Mitochondria are the main centers for energy production within a cell. Apoptosis is an energy-dependent process that can occur only if ATP is available, such that if intracellular ATP levels are depleted cells are forced to a necrotic death (Nicotera *et al.*, 1998). Two kits are available commercially for measuring ATP levels in cells [Sigma kit No. 366; ATP Bioluminescence assay kit CLS II (or HS II), Boehringer Mannheim]. A brief analysis of the method used in each kit is described.

If using Sigma kit No. 366:

1. Lyse 1×10^6 cells in 500 μl of solution A (6% TCA, 0.1 *M* sodium citrate, pH 7.0) at 4°C.
2. Centrifuge the lysate at 800*g* for 10 min at 4°C, add 400 μl of the supernatant to 1.25 ml of solution B (0.15 mg NADH, 500 μl of 18 m*M* 3-phosphoglyceric acid), and measure the A_{340}.
3. Add 20 μl of solution *c* (800 U/ml 3-phosphate dehydrogenase, 450 U/ml 3-phosphoglyceric phosphokinase) and measure the A_{340}. Levels of ATP are calculated based on the millimolar absorbtivity of NADH of 6.22.

If using ATP Bioluminescence Assay Kit CLS II:

1. Incubate 1×10^5 cells at 100°C in 200 μl of solution A (100 m*M* Tris–HCl, pH 7.75, 4 m*M* EDTA) for 10 min.
2. Centrifuge the lysates (10,000*g* for 1 min) and gently mix 50 μl of supernatant (supernatants can be stored at −70°C) with 50 μl of luciferase reagent.
3. Follow the luciferase activity for 10 s on a luminescence reader. Concentrations can be determined from a log log plot of an ATP standard curve performed at the same time.

D. Reactive Oxygen Species

Reactive oxygen species (ROS) are generated as a by-product of mitochondrial oxidative phosphorylation and have been implicated in various aspects of apoptosis (Hildeman *et al.*, 1999; Marchetti *et al.*, 1997; Ye *et al.*, 1999). Fluorogenic, chemiluminescent, and chromogenic compounds are available for detecting ROS in cells. One fluorescent probe commonly used to detect ROS generated during apoptosis is dihydroethidium (Heibein *et al.*, 1999). This dye is colorless in its chemically reduced form, but is easily oxidized to ethidium after which it intercalates with cellular DNA, staining the nucleus red. The degree of nuclear staining correlates with the amount of ROS present in the cell.

Incubate cells for 20 min at 37°C with 2 μ*M* dihydroethidium in medium. The oxidized dye in cells can then be detected by flow cytometry (488 nm, FL-2) or by fluorescence microscopy. Cells can be pretreated with ROS scavengers such as *N*-acetylcysteine, glutathione, and vitamin C for controls. At high concentrations, these compounds may be toxic and should be titrated in your system.

E. Permeability Transition

Permeability transition involves the opening of a pore in the mitochondrial membranes that allows passage of molecules less than 1.5 kDa and is followed by loss $\Delta\Psi m$, gross swelling, and outer membrane rupture. PT can therefore be detected by measuring $\Delta\Psi m$ as discussed earlier; however, a loss of $\Delta\Psi m$ can occur for reasons other than PT, e.g., in the presence of protonophores or loss of oxidative phosphorylation. PT can also be detected in isolated mitochondria by measuring mitochondrial swelling (Finucane *et al.,* 1999). This is seen as a reduction in absorbance at $A_{540\ nm}$. Finally, a change in fluorescence of mitochondria stained with calcein-acetoxymethyl ester (calcein-AM) can be used to detect PT. Calcein-AM is a small green fluorescent molecule that stains mitochondria. On induction of PT, calcein-AM can exit mitochondria, resulting in a detectable loss of fluorescence (Minamikawa *et al.,* 1999).

Mitochondrial swelling due to PT is best induced in hypotonic buffers (e.g., 125 m*M* KCL, 2.5 m*M* potassium phosphate, pH 7.4, 2.5 m*M* succinate, and 20 m*M* Hepes–KOH).

1. Place 300 μl PT buffer (containing the PT inducer, toxins, inhibitors, Bcl-2 family member, etc. being assayed) in individual wells of a microtiter plate.
2. Incubate the plate in a temperature-controlled microtiter plate reader until the buffer is 37°C.
3. Add mitochondria (10–50 μl, equivalent of 50–100 μg of isolated mitochondrial protein from the same preparation) to the wells, mix the samples, and follow A_{540}. Mitochondrial swelling is observed as a reduction in A_{540} (Fig. 4).

Swelling can also be detected during PT induced at room temperature using spectrophotometers without temperature control. In instruments with horizontal light beams, mitochondria may fall out of the light path over long periods of time due to gravity. To control for non-PT-induced swelling (due to buffer composition) or settling of mitochondria during the assay, cyclosporin A (an inhibitor of PT) can be added to the assay buffer. However, cyclosporin A only inhibits the onset of PT for short periods (20–60 min). A typical trace of $CaCl_2$-induced PT as assessed by mitochondrial swelling, together with transient inhibition afforded by cyclosporin A, is shown (Fig. 4).

To detect PT in isolated mitochondria using calcein-AM fluorescence:

1. Stain isolated mitochondria with 5 μ*M* calcein-AM in 200 m*M* mannitol, 50 m*M* sucrose, 10 m*M* succinate, 10 m*M* Hepes–KOH, pH 7.4, 0.1 % BSA, and 5 m*M* potassium phosphate, pH 7.4, for 15 min at room temperature.

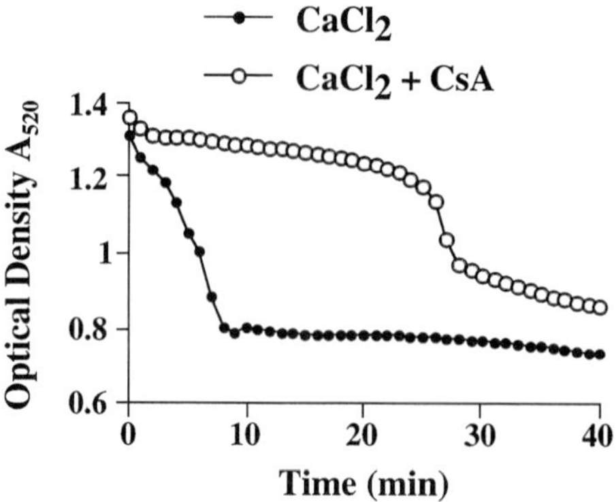

Fig. 4 Mitochondrial swelling due to permeability transition is transiently inhibited by cyclosporin A. Isolated mitochondria (~100 μg/ml) were incubated at 37°C with 100 μ*M* $CaCl_2$ in the presence or absence of cyclosporin A (40 n*M* CsA). A_{520} as a measure of light scatter indicative of swelling was recorded at 1-min intervals.

2. Wash mitochondria in calcein-free buffer and treat cells with a PT inducer. Wash mitochondria at the end of the assay period and determine the calcein fluorescence in mitochondria using a fluorescence spectrophotometer (Ex/Em 488/520 nm). PT is seen as a loss of fluorescence.

To detect PT in cells using calcein-AM fluorescence:

1. Stain cells with calcein-AM (1 μ*M* for 10 min at 37°C) in media and wash to remove extracellular stain.
2. Incubate the cells with $CoCl_2$ (1 m*M*) for 1 h at 37°C. Co^{2+} chelates the calcein localized in the nucleus and cytosol, resulting in disappearance of fluorescence in these areas.
3. Because mitochondria exclude Co^{2+}, mitochondrial calcein-AM fluorescence is maintained. Opening of the PT pore results in a fall in fluorescence of the mitochondrially localized calcein due to the increased permeability of the mitochondria to both calcein and Co^{2+}. This change in distribution can be followed by fluorescence microscopy (Ex 488 nm/Em 520 nm).

F. Outer Membrane Permeability

The permeability of the outer membrane to exogenous cytochrome *c* serves to measure the degree of outer mitochondrial membrane damage incurred during preparation or following treatment. The accessibility of exogenous cytochrome *c* to the inner membrane can be measured by the rate at which cytochrome *c* is either reduced by complex III or oxidized by complex IV. Intact mitochondria will exhibit minimal oxidation or reduction of exogenously added cytochrome *c*.

To measure complex III accessibility:

1. Add 25 μl mitochondrial preparation (1 mg/ml) to 300 μl buffer (125 m*M* sucrose, 60 m*M* KCl, 20 m*M* Tris–HCl, pH 7.4) containing 2 m*M* potassium cyanide (complex IV inhibitor).
2. Add 10 μl of 2.5 m*M* ferricytochrome *c* (horse heart).
3. Cytochrome *c* reduction is initiated by adding the complex III substrate decyl benzoquinol (DBH_2; 5 μl of a 3.5 m*M* stock in ethanol) and mixing.
4. The rate of cytochrome *c* reduction (increase in A_{550} over 30 s) is a measure of outer membrane permeability.

Complex IV accessibility is a similar but more convenient measure of outer membrane permeability, as the exogenously added reduced cytochrome *c* itself is the substrate.

1. The assay buffer [125 m*M* sucrose, 60 m*M* KCl, 20 m*M* Tris–HCl, pH 7.4, 100 μ*M* cytochrome *c* (horse heart), 1 μ*M* CCCP, 1 μ*M* rotenone, 1 μ*M* antimycin] contains inhibitors of complex I and III.
2. Sodium dithionate crystals are added until A_{550}/A_{565} is between 6 and 10 so that the majority of the cytochrome *c* is reduced.

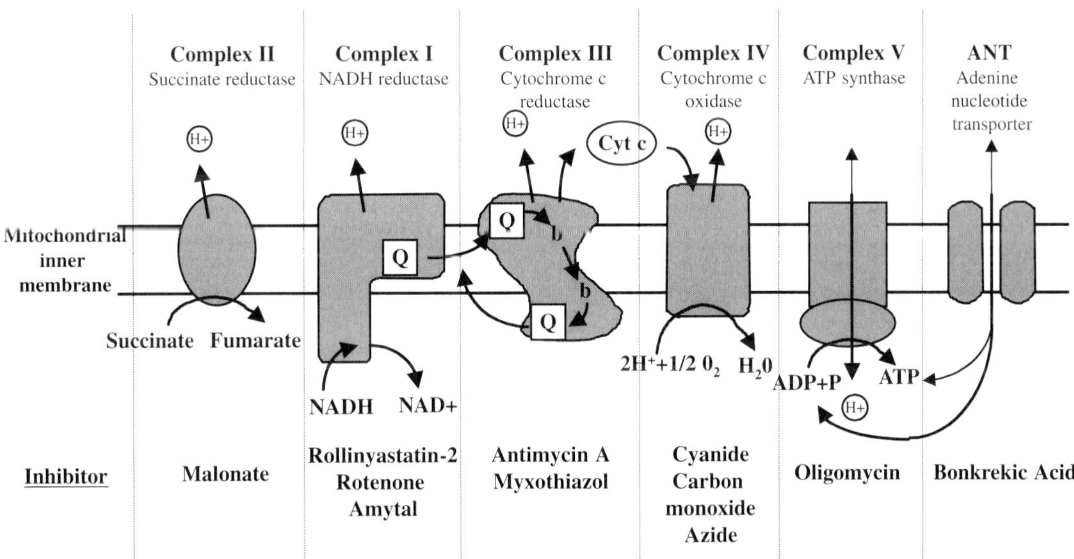

Fig. 5 Schematic view of the electron transport chain and some specific inhibitors. Electrons are passed via ubiquinone (Q) from complex I or II to complex III. Cytochrome *c* (cyt *c*) accepts the electron from complex III and transports it to complex IV. The flow of electrons is coupled to the transport of H^+ from the matrix to the mitochondrial intermembrane space. The generated electrical potential is used by complex V to generate ATP from ADP and free phosphate. This ATP is exported from the matrix via the ANT.

3. To start the assay, add 2 μl mitochondrial preparation (1 mg/ml) to 450 μl of assay buffer and mix.

4. The rate of cytochrome *c* oxidation (decrease in A_{550} over 30 s) is a measure of outer membrane permeability. Rates of complex III and IV accessibility can be compared to 100% outer membrane permeability produced by the pretreatment of mitochondria with 1% Triton X-100 (final assay concentration <0.1%).

G. Mitochondrial Toxins

As cytochrome *c* is a member of the electron transport chain (it transfers electrons from complex III to complex IV), as well as an important player in apoptosis, it is wise to be acquainted with a range of inhibitors of the electron transport chain. Some of the inhibitors of specific complexes are depicted in Fig. 5.

Acknowledgments

We thank Dr. Mauro Degli-Esposti for information regarding mitochondrial membrane potential and Dr. Helen Beere for critical comments.

References

Adachi, S., Cross, A. R., Babior, B. M., and Gottlieb, R. A. (1997). Bcl-2 and the outer mitochondrial membrane in the inactivation of cytochrome c during Fas-mediated apoptosis. *J. Biol. Chem.* **272,** 21878–21882.

Adachi, S., Gottlieb, R. A., and Babior, B. M. (1998). Lack of release of cytochrome C from mitochondria into cytosol early in the course of Fas-mediated apoptosis of Jurkat cells. *J. Biol. Chem.* **273,** 19892–19894.

Adam, S. A., Marr, R. S., and Gerace, L. (1990). Nuclear protein import in permeabilized mammalian cells requires soluble cytoplasmic factors. *J. Cell Biol.* **111,** 807–816.

Anderson, W. M., Wood, J. M., and Anderson, A. C. (1993). Inhibition of mitochondrial and Paracoccus denitrificans NADH-ubiquinone reductase by oxacarbocyanine dyes: A structure-activity study. *Biochem. Pharmacol.* **45,** 2115–2122.

Andrade, F., Roy, S., Nicholson, D., Thornberry, N., Rosen, A., and Casciola-Rosen, L. (1998). Granzyme B directly and efficiently cleaves several downstream caspase substrates: Implications for CTL-induced apoptosis. *Immunity* **8,** 451–460.

Bernardi, P., Colonna, R., Costantini, P., Eriksson, O., Fontaine, E., Ichas, F., Massari, S., Nicolli, A., Petronilli, V., and Scorrano, L. (1998). The mitochondrial permeability transition. *Biofactors* **8,** 273–281.

Bhakdi, S., Weller, U., Walev, I., Martin, E., Jonas, D., and Palmer, M. (1993). A guide to the use of pore-forming toxins for controlled permeabilization of cell membranes. *Med. Microbiol. Immunol. (Berl)* **182,** 167–175.

Boon, M. R., and Zammit, V. A. (1988). Use of a selectively permeabilized isolated rat hepatocyte preparation to study changes in the properties of overt carnitine palmitoyltransferase activity in situ. *Biochem. J.* **249,** 645–652.

Bossy-Wetzel, E., Newmeyer, D. D., and Green, D. R. (1998). Mitochondrial cytochrome c release in apoptosis occurs upstream of DEVD-specific caspase activation and independently of mitochondrial transmembrane depolarization. *EMBO. J.* **17,** 37–49.

Bridges, C. D. (1977). A method for preparing stable digitonin solutions for visual pigment extraction. *Vision Res.* **17,** 301–302.

Carthy, C. M., Granville, D. J., Jiang, H., Levy, J. G., Rudin, C. M., Thompson, C. B., McManus, B. M., and Hunt, D. W. (1999). Early release of mitochondrial cytochrome c and expression of mitochondrial epitope 7A6 with a porphyrin-derived photosensitizer: Bcl-2 and Bcl-xL overexpression do not prevent early mitochondrial events but still depress caspase activity. *Lab. Invest.* **79,** 953–965.

Chauhan, D., Pandey, P., Ogata, A., Teoh, G., Krett, N., Halgren, R., Rosen, S., Kufe, D., Kharbanda, S., and Anderson, K. (1997). Cytochrome *c*-dependent and -independent induction of apoptosis in multiple myeloma cells. *J. Biol. Chem.* **272,** 29995–29997.

Chinnaiyan, A. M., O'Rourke, K., Lane, B. R., and Dixit, V. M. (1997). Interaction of CED-4 with CED-3 and CED-9: A molecular framework for cell death. *Science* **275,** 1122–1126.

Cregan, S. P., Brown, D. L., and Mitchel, R. E. (1999). Apoptosis and the adaptive response in human lymphocytes. *Int. J. Radiat. Biol.* **75,** 1087–1094.

Darmon, A. J., Ley, T. J., Nicholson, D. W., and Bleackley, R. C. (1996). Cleavage of CPP32 by granzyme B represents a critical role for granzyme B in the induction of target cell DNA fragmentation. *J. Biol. Chem.* **271,** 21709–21712.

Duncan, J. L. (1974). Characteristics of streptolysin O hemolysis: Kinetics of hemoglobin and 86rubidium release. *Infect. Immun.* **9,** 1022–1027.

Emaus, R. K., Grunwald, R., and Lemasters, J. J. (1986). Rhodamine 123 as a probe of transmembrane potential in isolated rat-liver mitochondria: spectral and metabolic properties. *Biochim. Biophys. Acta* **850,** 436–448.

Eskes, R., Antonsson, B., Osen-Sand, A., Montessuit, S., Richter, C., Sadoul, R., Mazzei, G., Nichols, A., and Martinou, J. C. (1998). Bax-induced cytochrome c release from mitochondria is independent of the permeability transition pore but highly dependent on Mg2+ ions. *J. Cell Biol.* **143,** 217–224.

Farkas, D. L., Wei, M. D., Febbroriello, P., Carson, J. H., and Loew, L. M. (1989). Simultaneous imaging of cell and mitochondrial membrane potentials [published erratum appears in *Biophys. J.* 1990 Mar;57(3):following 684]. *Biophys. J.* **56,** 1053–1069.

Fink, C., Morgan, F., and Loew, L. M. (1998). Intracellular fluorescent probe concentrations by confocal microscopy. *Biophys. J.* **75,** 1648–1658.

Finucane, D. M., Bossy-Wetzel, E., Waterhouse, N. J., Cotter, T. G., and Green, D. R. (1999). Bax-induced caspase activation and apoptosis via cytochrome c release from mitochondria is inhibitable by Bcl-xL. *J. Biol. Chem.* **274,** 2225–2233.

Finucane, D. M., Waterhouse, N. J., Amarante-Mendes, G. P., Cotter, T. G., and Green, D. R. (1999). Collapse of the inner mitochondrial transmembrane potential is not required for apoptosis of HL60 cells. *Exp. Cell Res.* **251,** 166–174.

Goldstein, J. G., Waterhouse, N. J., Juin, P. I. E. G., and Green, D. R. (2000). The coordinate release of cytochrome c during apoptosis is rapid, complete and kinetically invariant. *Nature Cell Biol.* **2,** 156–162.

Granville, D. J., Carthy, C. M., Jiang, H., Shore, G. C., McManus, B. M., and Hunt, D. W. (1998). Rapid cytochrome c release, activation of caspases 3, 6, 7 and 8 followed by Bap31 cleavage in HeLa cells treated with photodynamic therapy. *FEBS Lett.* **437,** 5–10.

Green, D. R., and Reed, J. C. (1998). Mitochondria and apoptosis. *Science* **281,** 1309–1312.

Heibein, J. A., Barry, M., Motyka, B., and Bleackley, R. C. (1999). Granzyme B-induced loss of mitochondrial inner membrane potential (Delta Psi m) and cytochrome c release are caspase independent. *J. Immunol.* **163,** 4683–4693.

Heiskanen, K. M., Bhat, M. B., Wang, H. W., Ma, J., and Nieminen, A. L. (1999). Mitochondrial depolarization accompanies cytochrome c release during apoptosis in PC6 cells. *J. Biol. Chem.* **274,** 5654–5658.

Hildeman, D. A., Mitchell, T., Teague, T. K., Henson, P., Day, B. J., Kappler, J., and Marrack, P. C. (1999). Reactive oxygen species regulate activation-induced T cell apoptosis. *Immunity* **10,** 735–744.

Hu, Y., Benedict, M. A., Ding, L., and Nunez, G. (1999). Role of cytochrome c and dATP/ATP hydrolysis in Apaf-1-mediated caspase- 9 activation and apoptosis. *EMBO J.* **18,** 3586–3595.

Jenssen, H. L., Redmann, K., and Mix, E. (1986). Flow cytometric estimation of transmembrane potential of macrophages–a comparison with microelectrode measurements. *Cytometry* **7,** 339–346.

Kjeldsen, L., Sengelov, H., and Borregaard, N. (1999). Subcellular fractionation of human neutrophils on percoll density gradients. *J. Immunol. Methods* **232,** 131–143.

Kluck, R. M., Bossy-Wetzel, E., Green, D. R., and Newmeyer, D. D. (1997). The release of cytochrome c from mitochondria: a primary site for Bcl-2 regulation of apoptosis. Science **275,** 1132–1136.

Kluck, R. M., Esposti, M. D., Perkins, G., Renken, C., Kuwana, T., Bossy-Wetzel, E., Goldberg, M., Allen, T.,

Barber, M. J., Green, D. R., and Newmeyer, D. D. (1999). The pro-apoptotic proteins, Bid and Bax, cause a limited permeabilization of the mitochondrial outer membrane that is enhanced by cytosol. *J. Cell Biol.* **147,** 809–822.

Kohler, C., Gahm, A., Noma, T., Nakazawa, A., Orrenius, S., and Zhivotovsky, B. (1999). Release of adenylate kinase 2 from the mitochondrial intermembrane space during apoptosis. *FEBS Lett.* **447,** 10–12.

Kubat, P., Zelinger, Z., and Jirsa, M. (1997). The effect of the irradiation wavelength on the processes sensitized by protoporphyrin IX dimethyl ester. *Radiat. Res.* **148,** 382–385.

Kuwana, T., Smith, J. J., Muzio, M., Dixit, V., Newmeyer, D. D., and Kornbluth, S. (1998). Apoptosis induction by caspase-8 is amplified through the mitochondrial release of cytochrome c. *J. Biol. Chem.* **273,** 16589–16594.

Lemasters, J. J., Nieminen, A. L., Qian, T., Trost, L. C., Elmore, S. P., Nishimura, Y., Crowe, R. A., Cascio, W. E., Bradham, C. A., Brenner, D. A., and Herman, B. (1998). The mitochondrial permeability transition in cell death: A common mechanism in necrosis, apoptosis and autophagy. *Biochim. Biophys. Acta* **1366,** 177–196.

Linsinger, G., Wilhelm, S., Wagner, H., and Hacker, G. (1999). Uncouplers of oxidative phosphorylation can enhance a Fas death signal. *Mol. Cell. Biol.* **19,** 3299–3311.

MacDonald, G., Shi, L., Vande Velde, C., Lieberman, J., and Greenberg, A. H. (1999). Mitochondria-dependent and -independent regulation of Granzyme B-induced apoptosis. *J. Exp. Med.* **189,** 131–144.

Mannella, C. A., and Wang, Q. (1989). Permeability of the mitochondrial outer membrane to organic cations. *Biochim. Biophys. Acta* **981,** 363–366.

Marchetti, P., Decaudin, D., Macho, A., Zamzami, N., Hirsch, T., Susin, S. A., and Kroemer, G. (1997). Redox regulation of apoptosis: Impact of thiol oxidation status on mitochondrial function. *Eur. J. Immunol.* **27,** 289–296.

Marton, A., Mihalik, R., Bratincsak, A., Adleff, V., Petak, I., Vegh, M., Bauer, P. I., and Krajcsi, P. (1997). Apoptotic cell death induced by inhibitors of energy conservation: Bcl-2 inhibits apoptosis downstream of a fall of ATP level. *Eur. J. Biochem.* **250,** 467–475.

Metivier, D., Dallaporta, B., Zamzami, N., Larochette, N., Susin, S. A., Marzo, I., and Kroemer, G. (1998). Cytofluorometric detection of mitochondrial alterations in early CD95/Fas/APO-1-triggered apoptosis of Jurkat T lymphoma cells: Comparison of seven mitochondrion-specific fluorochromes. *Immunol. Lett.* **61,** 157–163.

Minamikawa, T., Sriratana, A., Williams, D. A., Bowser, D. N., Hill, J. S., and Nagley, P. (1999). Chloromethyl-X-rosamine (MitoTracker Red) photosensitises mitochondria and induces apoptosis in intact human cells. *J. Cell. Sci.* **112,** 2419–2430.

Minamikawa, T., Williams, D. A., Bowser, D. N., and Nagley, P. (1999). Mitochondrial permeability transition and swelling can occur reversibly without inducing cell death in intact human cells. *Exp. Cell Res.* **246,** 26–37.

Modica-Napolitano, J. S., Weiss, M. J., Chen, L. B., and Aprille, J. R. (1984). Rhodamine 123 inhibits bioenergetic function in isolated rat liver mitochondria. *Biochem. Biophys. Res. Commun.* **118,** 717–723.

Muzio, M., Chinnaiyan, A. M., Kischkel, F. C., O'Rourke, K., Shevchenko, A., Ni, J., Scaffidi, C., Bretz, J. D., Zhang, M., Gentz, R., Mann, M., Krammer, P. H., Peter, M. E., and Dixit, V. M. (1996). FLICE, a novel FADD-homologous ICE/CED-3-like protease, is recruited to the CD95 (Fas/APO-1) death–inducing signaling complex. *Cell* **85,** 817–827.

Namura, S., Zhu, J., Fink, K., Endres, M., Srinivasan, A., Tomaselli, K. J., Yuan, J., and Moskowitz, M. A. (1998). Activation and cleavage of caspase-3 in apoptosis induced by experimental cerebral ischemia. *J. Neurosci.* **18,** 3659–3668.

Newmeyer, D. D., Farschon, D. M., and Reed, J. C. (1994). Cell-free apoptosis in *Xenopus* egg extracts: Inhibition by Bcl-2 and requirement for an organelle fraction enriched in mitochondria. *Cell* **79,** 353–364.

Nicotera, P., Leist, M., and Ferrando-May, E. (1998). Intracellular ATP, a switch in the decision between apoptosis and necrosis. *Toxicol. Lett.* **102-103,** 139–142.

Poot, M., Zhang, Y. Z., Kramer, J. A., Wells, K. S., Jones, L. J., Hanzel, D. K., Lugade, A. G., Singer, V. L., and Haugland, R. P. (1996). Analysis of mitochondrial morphology and function with novel fixable fluorescent stains. *J. Histochem. Cytochem.* **44,** 1363–1372.

Ravagnan, L., Marzo, I., Costantini, P., Susin, S. A., Zamzami, N., Petit, P. X., Hirsch, F., Goulbern, M., Poupon, M. F., Miccoli, L., Xie, Z., Reed, J. C., and Kroemer, G. (1999). Lonidamine triggers apoptosis via a direct, Bcl-2-inhibited effect on the mitochondrial permeability transition pore. *Oncogene* **18,** 2537–2546.

Rodriguez, A., Oliver, H., Zou, H., Chen, P., Wang, X., and Abrams, J. M. (1999). Dark is a *Drosophila* homologue of Apaf-1/CED-4 and functions in an evolutionarily conserved death pathway. *Nature Cell Biol.* **1,** 272–279.

Saleh, A., Srinivasula, S. M., Acharya, S., Fishel, R., and Alnemri, E. S. (1999). Cytochrome *c* and dATP-mediated oligomerization of Apaf-1 is a prerequisite for procaspase-9 activation. *J. Biol. Chem.* **274,** 17941–17945.

Salvioli, S., Ardizzoni, A., Franceschi, C., and Cossarizza, A. (1997). JC-1, but not DiOC6(3) or rhodamine 123, is a reliable fluorescent probe to assess delta psi changes in intact cells: Implications for studies on mitochondrial functionality during apoptosis. *FEBS Lett.* **411,** 77–82.

Samali, A., Cai, J., Zhivotovsky, B., Jones, D. P., and Orrenius, S. (1999). Presence of a pre-apoptotic complex of pro-caspase-3, Hsp60 and Hsp10 in the mitochondrial fraction of jurkat cells. *EMBO J.* **18,** 2040–2048.

Saraste, M. (1999). Oxidative phosphorylation at the fin de siecle. *Science* **283,** 1488–1493.

Scaduto, R. C., Jr., and Grotyohann, L. W. (1999). Measurement of mitochondrial membrane potential using fluorescent rhodamine derivatives. *Biophys. J.* **76,** 469–477.

Scorrano, L., Petronilli, V., Colonna, R., Di Lisa, F., and Bernardi, P. (1999). Chloromethyltetramethylrosamine (Mitotracker Orange) induces the mitochondrial permeability transition and inhibits respiratory complex I: Implications for the mechanism of cytochrome c release. *J. Biol. Chem.* **274,** 24657–24663.

Seaton, T. A., Cooper, J. M., and Schapira, A. H. (1998). Cyclosporin inhibition of apoptosis induced by mitochondrial complex I toxins. *Brain Res.* **809,** 12–17.

Shinohara, Y., Sagawa, I., Ichihara, J., Yamamoto, K., Terao, K., and Terada, H. (1997). Source of ATP for hexokinase-catalyzed glucose phosphorylation in tumor cells: Dependence on the rate of oxidative phosphorylation relative to that of extramitochondrial ATP generation. *Biochim. Biophys. Acta* **1319,** 319–330.

Single, B., Leist, M., and Nicotera, P. (1998). Simultaneous release of adenylate kinase and cytochrome c in cell death. *Cell Death Differ.* **5,** 1001–1003.

Susin, S. A., Lorenzo, H. K., Zamzami, N., Marzo, I., Brenner, C., Larochette, N., Prevost, M. C., Alzari, P. M., and Kroemer, G. (1999a). Mitochondrial release of caspase-2 and -9 during the apoptotic process. *J. Exp. Med.* **189,** 381–394.

Susin, S. A., Lorenzo, H. K., Zamzami, N., Marzo, I., Snow, B. E., Brothers, G. M., Mangion, J., Jacotot, E., Costantini, P., Loeffler, M., Larochette, N., Goodlett, D. R., Aebersold, R., Siderovski, D. P., Penninger, J. M., and Kroemer, G. (1999b). Molecular characterization of mitochondrial apoptosis-inducing factor. *Nature* **397,** 441–446.

Susin, S. A., Zamzami, N., Castedo, M., Hirsch, T., Marchetti, P., Macho, A., Daugas, E., Geuskens, M., and Kroemer, G. (1996). Bcl-2 inhibits the mitochondrial release of an apoptogenic protease. *J. Exp. Med.* **184,** 1331–1341.

Tang, D., Lahti, J. M., Grenet, J., and Kidd, V. J. (1999). Cycloheximide-induced T-cell death is mediated by a Fas-associated death domain-dependent mechanism. *J. Biol. Chem.* **274,** 7245–7252.

Vander Heiden, M. G., Chandel, N. S., Williamson, E. K., Schumacker, P. T., and Thompson, C. B. (1997). Bcl-xL regulates the membrane potential and volume homeostasis of mitochondria. *Cell* **91,** 627–637.

Vantieghem, A., Assefa, Z., Vandenabeele, P., Declercq, W., Courtois, S., Vandenheede, J. R., Merlevede, W., de Witte, P., and Agostinis, P. (1998). Hypericin-induced photosensitization of HeLa cells leads to apoptosis or necrosis: Involvement of cytochrome c and procaspase-3 activation in the mechanism of apoptosis. *FEBS Lett.* **440,** 19–24.

Wallace, D. C. (1999). Mitochondrial diseases in man and mouse. *Science* **283,** 1482–1488.

Waterhouse, N. J., and Green, D. R. (1999). Mitochondria and apoptosis: HQ or high-security prison? *J. Clin. Immunol.* **19,** 378–387.

Willis, E. J. (1992). The powerhouse of the cell. *Ultrastruct. Pathol.* **16,** iii–vi.

Wolf, C. M., and Eastman, A. (1999). The temporal relationship between protein phosphatase, mitochondrial cytochrome c release, and caspase activation in apoptosis. *Exp. Cell Res.* **247,** 505–513.

Wolvetang, E. J., Johnson, K. L., Krauer, K., Ralph, S. J., and Linnane, A. W. (1994). Mitochondrial respiratory chain inhibitors induce apoptosis. *FEBS Lett.* **339,** 40–44.

Yang, J., Liu, X., Bhalla, K., Kim, C. N., Ibrado, A. M., Cai, J., Peng, T. I., Jones, D. P., and Wang, X. (1997). Prevention of apoptosis by Bcl-2: Release of cytochrome c from mitochondria blocked. *Science* **275,** 1129–1132.

Ye, J., Wang, S., Leonard, S. S., Sun, Y., Butterworth, L., Antonini, J., Ding, M., Rojanasakul, Y., Vallyathan, V., Castranova, V., and Shi, X. (1999). Role of reactive oxygen species and p53 in chromium(VI)-induced apoptosis. *J. Biol. Chem.* **274,** 34974–34980.

Yuan, J. (1997). Transducing signals of life and death. *Curr. Opin. Cell Biol.* **9,** 247–251.

Zamzami, N., Marchetti, P., Castedo, M., Zanin, C., Vayssiere, J. L., Petit, P. X., and Kroemer, G. (1995). Reduction in mitochondrial potential constitutes an early irreversible step of programmed lymphocyte death in vivo. *J. Exp. Med.* **181,** 1661–1672.

Zamzami, N., Susin, S. A., Marchetti, P., Hirsch, T., Gomez-Monterrey, I., Castedo, M., and Kroemer, G. (1996). Mitochondrial control of nuclear apoptosis. *J. Exp. Med.* **183,** 1533–1544.

Zhuang, J., and Cohen, G. M. (1998). Release of mitochondrial cytochrome c is upstream of caspase activation in chemical-induced apoptosis in human monocytic tumour cells. *Toxicol. Lett.* **102–103,** 121–129.

Zou, H., Li, Y., Liu, X., and Wang, X. (1999). An APAF-1.cytochrome c multimeric complex is a functional apoptosome that activates procaspase-9. *J. Biol. Chem.* **274,** 11549–11556.

Appendix

Materials

Tetramethylrhodamine ethyl ester (TMRE), CMTMRos (MitoTracker orange), MitoTracker green FM, CMXRos (MitoTracker red), CM-H_2Xros (Reduced MitoTracker red), 5,5′,6,6′-tetrachloro-1,1′,3,3′-tetraethylbenzimidazolylcarbocyanine iodide (JC-1) dihydroethidium (2-HE), 3,39-dihexyloxacarbocyanine iodide [DiOC$_6$(3)], rhodamine 123 (Rh123), bis(1,3-dibutylbarbituric acid) trimethyne oxonol [DiBAC(3)], bis(1,3-dibutylbarbituric acid) pentamethyne oxonol [DiBAC (5)], Nonyl acridine orange (NAO), 4′-6-diamidino-2-phenylindole dihydrochloride (DAPI), dihydroethidium, and calcein-AM can be obtained from Molecular Probes, Inc. (Eugene, OR); z-val-ala-asp-fluoromethylketone (zVAD-fmk) can be obtained from Enzyme Systems Products (Dublin, CA). LabTek Chamber slides can be obtained from Nalge Nunc (Napperville, IL). Bioluminescence assay kit CLS II and HS II and complete inhibitor can be obtained from Boehringer Mannheim. Cyclosporin A can be obtained from Novartis (San Diego, CA). The plasmid pEGFP-C1 can be obtained from Clontech (Palo Alto, CA). Anticytochrome *c* antibodies (7H8.2C12 for Western blotting and 6H2.B4 for immunocytochemistry) can be obtained from Pharmingen (San Diego, CA). Anti-HSP 60 and HSP-10 can be obtained from Stressgen (Canada). Texas red Ig can be obtained from Amersham Pharmacia Biotech (Piscataway, NJ). Dounce homogenizers, 2 ml with a B-type pestle glass Dounce and 15 ml with a tight Teflon pestle, can be obtained from Kimble-Kontes (Germany). Rotenone, malonate, antimycin, myxothiazol, sodium azide, oligomycin, *N*-acetyl-L-cysteine, vitamin C, glutathione, potassium cyanide, cobalt chloride, ferricytochrome *c*, 3-phosphoglyceric phosphokinase, 3-phosphate dehydrogenase,

3-phosphoglyceric acid, pyruvate, succinate, Sigma kit No. 366, staurosporine, actinomycin D, streptolysin O, digitonin, protoperforin IX, *t*-butyl hydroperoxide, etoposide, carbonyl cyanide *p*-(trifluoromethoxy)phenylhydrazone (FCCP), carbamoyl cyanide *n*-chlorophenylhydrazone (CCCP), hypericin, lonidamine, atractyloside, sodium dithionate crystals, sodium citrate, and all other general chemicals can be obtained from Sigma Chemicals (St. Louis, MO).

CHAPTER 17

In Situ Detection of Dying Cells in Normal and Pathological Tissues

Christos Valavanis,* Stephen Naber,† and Lawrence M. Schwartz*,‡

* Department of Biology
Morrill Science Center
University of Massachusetts
Amherst, Massachusetts 01003

† Department of Pathology
BayState Medical Center
Springfield, Massachusetts 01199

‡ Program in Molecular and Cellular Biology
Department of Biology
Morrill Science Center
University of Massachusetts
Amherst, Massachusetts 01003

METHODS IN CELL BIOLOGY, VOL. 66

0091-679X/01 $35.00

I. Introduction

Tissue homeostasis is essential for the proper development, maturation, and maintenance of organs in all multicellular organisms. Homeostasis is controlled through complex relationships among cell proliferation, differentiation, and programmed cell death (PCD). The last process has become a major focus of research during the past decade. Much of this interest has stemmed from the realization that PCD not only plays a critical role in normal physiological processes (Milligan and Schwartz, 1997), but can also be important for many of the pathological consequences of disease (Hetts, 1998). Dysregulation of PCD signaling cascades is one of the major pathogenetic mechanisms involved in a large number of human diseases, including viral infections (Gougeon and Montagnier, 1993; Famularo *et al.*, 1997; Young *et al.*, 1997); neurodegenerative diseases (Desjardins and Ledoux, 1998; Jellinger and Bancher, 1998; Mattson, 1999; Petersen *et al.*, 1999; Biros and Forrest, 1999); neurodevelopmental disorders (Margolis *et al.*, 1994; Nagy, 1999); autoimmune diseases (Mountz *et al.*, 1994); immunologic deficiencies (Peake *et al.*, 1999); cardiovascular diseases (Saraste *et al.*, 1997; Feuerstein, 1999); and tumors (Lyons and Clarke, 1997). In all of these maladies, the number of cells in affected tissues is highly correlated with the severity of the disease and the subsequent response to treatment. This is particularly true in oncology, where the rates of proliferation and cell death within a tumor provide clinically useful information regarding prognosis and can influence therapeutic decisions. This information is also invaluable in assessing the efficacy of chemotherapy and radiation therapy because these interventions act by inducing PCD (Debatin, 1999; Houghton, 1999; Dewey *et al.*, 1995). Consequently, the identification and quantification of dying cells in the context of heterogeneous cell populations within tissues are of fundamental relevance to the identification and treatment of disease. To identify these events on a cell-by-cell basis, appropriate methodologies must be employed. These methods should satisfy two basic requirements: (1) discrimination among the different morphological patterns of cell death (i.e., necrosis versus apoptosis) and (2) applicability at the tissue level with preservation of topological architecture.

Among the different morphological patterns of cell death that are observed in normal and pathological tissues, apoptosis is the best characterized. Two methods, terminal deoxynucleotidyl transferase (TdT)-mediated dUTP-biotin nick end labeling (TUNEL) and *in situ* end labeling (ISEL), provide easy, reproducible, and specific labeling of apoptotic cells within a heterogeneous cell population and are the most frequently employed. The principles and procedures for these techniques are described in the protocol section. Because these methods can also occasionally label necrotic cells, they should be coupled with both additional detection methods and morphological analysis to confirm the apoptotic nature of a cell. Morphological criteria for the differential diagnosis of apoptosis versus necrosis have been defined by many authors (Kerr *et al.,* 1972, 1995; Darzynkiewicz *et al.,* 1998; Raffray and Cohen, 1997; Majno and Joris, 1995) and are discussed in detail later.

This chapter describes basic methods for examining cell death in both normal and pathological tissue samples. We provide general methods that may require significant modification to obtain optimal results for the specific tissues in hand. Where possible, we will identify critical methodological variables and suggest ranges that can be tested.

II. Morphological Features of Cell Death: Apoptosis vs Necrosis

A. Apoptosis

In 1972, Kerr, Curie, and Wyllie coined the term apoptosis to describe a stereotypic pattern of cell death that was observed in a wide variety of tissues and species. Subsequent studies have determined that apoptosis is an endogenous cellular suicide program that is regulated by a phylogenetically conserved genetic cascade (Horvitz, 1999; Lorenzo *et al.,* 1999; Metzstein *et al.,* 1998; Hochman, 1997; Hengartner and Horvitz, 1994).

A number of morphological criteria are used to identify apoptotic cells (Fig. 1). Three specific phases can be discerned at the light microscopy level. During the first phase, chromatin condenses and forms crescent-shaped caps toward the inner nuclear membrane (pyknosis), nuclei disintegrate, the nucleus shrinks, and the total cellular volume is reduced. At the same time plasma membrane alterations occur and membrane blebbing is observed. During the second phase, which may overlap with the first, multiple, small, discrete, membrane-bound apoptotic bodies generated by the destruction of cytoplasm and nucleus (karyorrhexis) are formed. Later in this stage, apoptotic bodies may be phagocytosed by macrophages or neighboring cells. The average time during which an apoptotic body may remain visible at the light microscopy level within a tissue is between 2 and 18 h, depending on the tissue, e.g., 2–4 h in the thymus (Surh and Sprent, 1994) and 12–18 h in the adrenal gland (Wyllie *et al.,* 1980). Apoptotic bodies that become degraded following phagocytosis are difficult to identify. During the final phase, there is progressive disintegration of residual cell structures characterized with morphological changes that are also observed in necrosis.

B. Necrosis

In contrast to apoptosis, necrosis is a nongenetically based process that involves the random and chaotic death of cells. Morphologically, necrosis is characterized by irregular chromatin clumping followed in a later stage by degeneration and dissolution of chromatin (karyolysis), leaving the nucleus as a faintly stained ghost (Fig. 2). Early in the process, the cell swells, the cytoplasm becomes granular or finely vacuolated, and lysosomal rupture occurs. In late stages of necrosis, cellular outlines are poorly defined. Unlike apoptosis, which occurs discreetly in individual cells within a population, necrosis

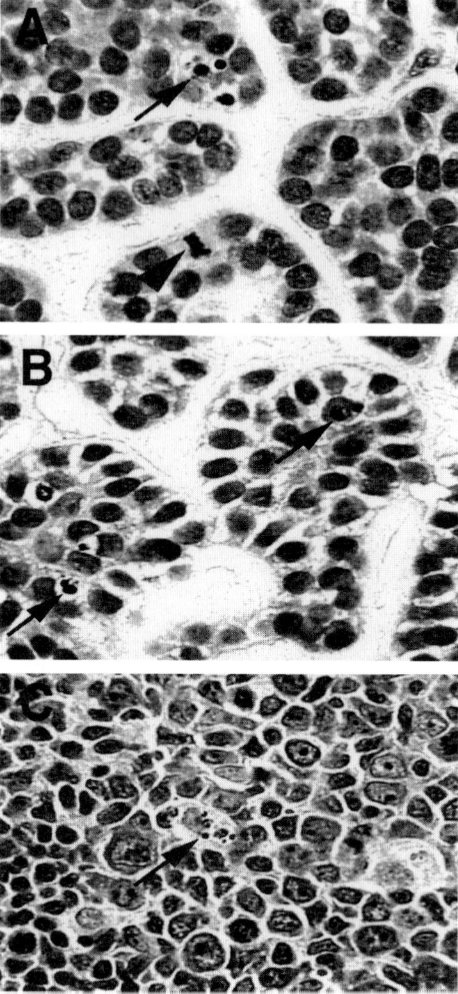

Fig. 1 Apoptosis. Light microscopic features of apoptotic bodies from paraffin-embedded human tissue sections (5 μm) stained with hematoxylin and eosin. (A and B) Lacrimal gland tumor from human. Compare the different appearance of apoptotic bodies (arrows) with that of a mitotic cell (arrowhead) in A. Magnification: 400×. (C) Apoptotic bodies phagocytosed by a macrophage (arrow) in a germinal center of human tonsil. Magnification: 500×.

usually involves a large group of contiguous cells. With time, there is an inflammatory response that involves infiltration by neutrophil leukocytes and mononuclear phagocytes (a process not observed as a consequence of apoptosis) (Fig. 2A). Late in the process, the necrotic area is characterized by the accumulation of nuclear debris and dead cell corpuscles that may become TUNEL or ISEL positive; however, the image of the dead tissue is so distinct that it cannot be mistaken for apoptosis.

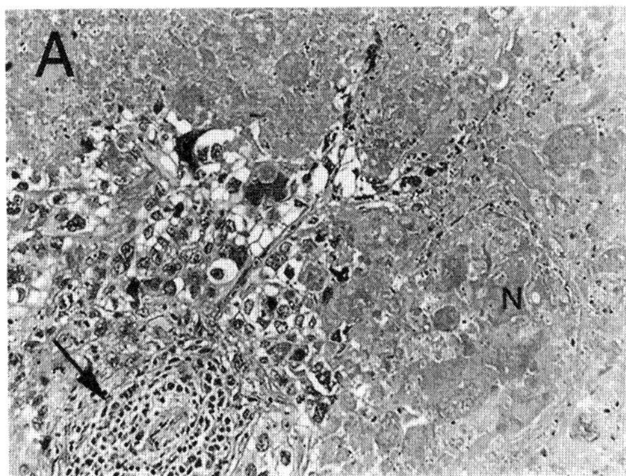

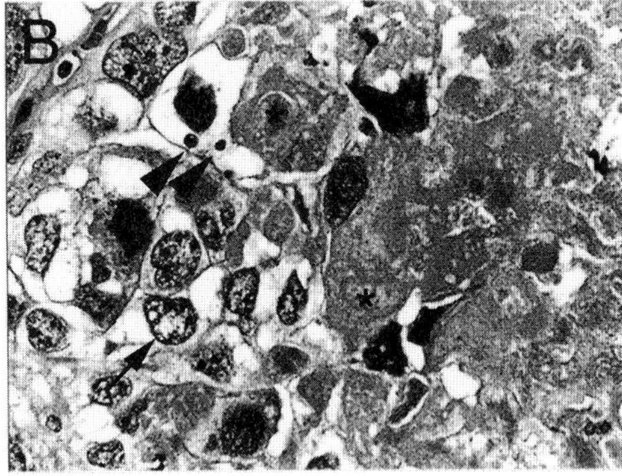

Fig. 2 Necrosis. (A) Necrotic area (N) from a human lung adenocarcinoma adjacent to a cluster of healthy cells and some degenerated tumor cells. Note the inflammatory infiltration characteristic of necrosis at the bottom left (arrow). Magnification: 200×. (B) Higher magnification of the same section as in A. Note the features of necrotic cells with ill-defined cellular margins and ghost nuclei (asterisk). Two apoptotic bodies (arrowheads) can be recognized within a cell with vacuolated cytoplasm and a pycnotic nucleus (arrow). Magnification: 400×.

III. Methods for Apoptotic Cell Identification

This volume details a variety of other methods that can be used to detect apoptosis in experimental materials, such as flow cytometry (see Chapters 3 and 4), cell-free systems (see Chapter 7), and protease assays (see Chapters 1 and 12). By and large, these methods are not applicable for fixed pathological tissues that are available in pathology departments. Instead, investigators are largely restricted to using histological methods. The following sections present some of the main methods used to identify apoptotic cells within fixed tissue. Depending on the method(s) employed, early or late apoptotic events will be detected (see Table I).

IV. TUNEL

This technique is based on the observation that the early stages of apoptosis typically involve the cleavage of chromatin. Initially, genomic DNA is cleaved to generate distinct high molecular weight fragments corresponding to >700 kb, followed by further degradation to 200- to 250- and 50-kb fragments, which are subsequently cleaved into low molecular weight fragments of 180- to 200-bp oligonucleosomal repeats (Filipiski *et al.*, 1990; Oberhammer, 1993). These oligonucleosomal-sized DNA fragments are responsible for the stereotypic DNA ladder that can be observed when DNA from apoptotic cells is fractionated by agarose gel electrophoresis (Wyllie, 1980).

The TUNEL method is used for *in situ* labeling of the DNA strand breaks that form in individual nuclei of apoptotic cells. Typically, frozen cells or fixed paraffin-embedded

Table I
Methods for Detecting Different Events in Apoptosis

	Apoptotic events during time				
Detection method	Very early	Early	Intermediate	Late	Very late
ICC, pAb	Caspase-9 activation				
ICC, IHC, pAb	Caspase-3 activation				
ICC, pAb	Caspase-7 activation				
IHC, mAb	Cytokeratin-18 cleavage				
IHC, pAb	Fractin generation				
ICC, pAb	Poly(ADP-ribose) polymerase cleavage				
IHC, biotin-annexin V		Phosphatidylserine externalization			
ICC, mAb APO 2.7		Mitochondrial membrane 7A6 antigen exposure			
TUNEL ISEL ISLG-LDF			DNA internucleosomal fragmentation		
Microscopy				Morphological changes	
Propidium iodide (for cell cultures and smears)					Plasma membrane leakage

tissue sections are employed (Gavrieli *et al.*, 1992; Ben-Sasson *et al.*, 1995). This method is based on the ability of the enzyme TdT to specifically bind at exposed 3′-OH termini of double- or single-stranded DNA fragments and synthesize a labeled polydeoxynucleotide by incorporation of modified deoxyuridine (X-dUTP, X=biotin, digoxigenin, or fluorescein) into the sites of DNA cleavage. The signal is then enhanced by (strept)avidin or antidigoxigenin antibody peroxidase and can be detected using conventional immunohistochemical methods and light microscopy (Fig. 3, see Color Plate). The use of fluorescein-dUTP allows the visualization of DNA strand breaks directly by fluorescence microscopy.

A. Materials

1. Reagents

Double-distilled water (DDW)
Paraffin
Xylene
Ethanol absolute, 95%, 70% diluted in distilled water
Ethanol : acetic acid 2 : 1 (v/v) for cryostat sections

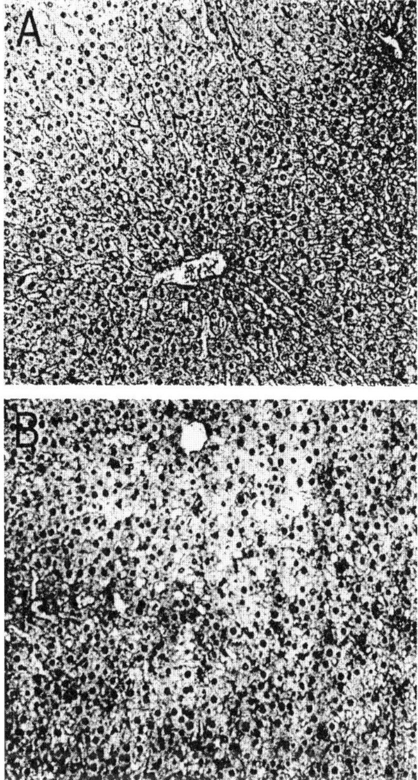

Fig. 3 TUNEL staining. (A) TUNEL-positive nuclei from a paraffin-embedded section of rat liver following 3 days of treatment with a high dose of a DNA-damaging agent. Counterstained with hematoxylin. Note that TUNEL-positive apoptotic nuclei are localized primarily in the centrilobular region. Magnification: 200×. (B) Positive control for the experiment outlined in A. Sections of normal liver were DNase I treated (200 mg/ml) as described in the text. Counterstained with hematoxylin. Note that almost all nuclei are TUNEL positive. Magnification: 200×. (See Color Plate.)

Permount mounting medium for glass support (Fischer)
Biotin-16-dUTP (Boehringer Mannheim)
Terminal deoxynucleotidyl transferase (GIBCO-BRL)
Extra avidin–peroxidase (Sigma)
3,3′-Diaminobenzidine tetrahydrochloride (DAB), storage at −20°C protected from light exposure (Sigma)
Sodium cacodylate (Sigma)
Cobalt chloride (Sigma)
Sodium citrate (Sigma)

Sodium phosphate (Sigma)
Trizma base (Sigma)
Sodium chloride (Sigma)
EDTA (Sigma)
Proteinase K (Sigma)
Bovine serum albumin (BSA; Sigma)

2. Solutions

10% (v/v) neutral-buffered formalin for tissue fixation prior of paraffin embedding
1% paraformaldehyde in phosphate-buffered saline (PBS), pH 7.4, for cryostat sections
PBS (50 m*M* sodium phosphate, 200 m*M* NaCl, pH 7.4)
Proteinase K [stock solution of 5 mg/ml in proteinase K buffer (50 m*M* Tris, 1 m*M* EDTA, pH 8.0), small aliquots at −20°C]
Hydrogen peroxide 30% solution, storage at 4°C (Sigma)
Mayer's or Harris's hematoxylin for counterstain (Sigma)

3. Equipment

Micropipettors
Glass coplin jars
Positively charged glass slides
Glass coverslips
Plastic coverslips
Microcentrifuge tubes
Tissue processor
Process/embedding cassettes
Flotation water bath at 37°C
Incubator(s) set at 37°C and 58°C
Microtome for paraffin blocks
Cryostat microtome for cryosections
Moisture chamber
Microwave oven
Light microscope

4. Working Solutions

Proteinase K (20 μg/ml) in proteinase K buffer (50 m*M* Tris, 1 m*M* EDTA, pH 8.0)
2% BSA in distilled H_2O

TdT buffer (30 m*M* Tris, 140 m*M* sodium cacodylate, 1 m*M* cobalt chloride, pH 7.2)

10× TB buffer (3 *M* NaCl, 0.3 sodium citrate) dilute to 1× before use

TdT/dUTP solution 1 : 50/1 : 10 in TdT buffer

3% hydrogen peroxide

Extravidin–peroxidase conjugate solution in PBS (1 : 10)

B. Protocol

1. Tissue Fixation

Tissues can be fixed by immersion in 10% neutral-buffered formalin or 4% paraformaldehyde in PBS, pH 7.4, for 1–7 days at 4°C prior to embedding in paraffin. Tissue cryosections can be fixed in 1% paraformaldehyde in a coplin jar for 10 min at room temperature or for up to 15 h at 4°C, followed by washing in PBS and fixation in precooled ethanol : acetic acid (2 : 1) for 5 min at −20°C. Sections 5–10 μm thick are spread on positively charged glass slides.

2. Processing of Tissue Sections

1. Heat the sections at 58°C for 30 min to remove the paraffin.
2. Wash the slides in three changes of fresh xylene for 5 min each.
3. Wash the slides in two changes of absolute ethanol for 5 min each.
4. Wash the slides once in 95% ethanol and once in 70% ethanol for 3 min each.
5. Put the slides in one change of PBS for 5 min.

3. Pretreatment of Tissue Sections

1. Incubate with freshly prepared proteinase K (20 μg/ml) for 30 min at room temperature in a coplin jar or applied directly on the section (60–80 μl/cm^2).
2. Wash the sections in four changes of DDW in a coplin jar for 2 min each.

4. Inactivation of Endogenous Peroxidase

1. Incubate tissues with freshly prepared 3% hydrogen peroxide in PBS for 5 min at room temperature (100–200 μl/section).
2. Rinse the sections twice with PBS or DDW for 5 min each time in a coplin jar.

5. Formation of Oligomers from 3′-OH Ends of Fragmented DNA

1. Cover the sections with 75 μl of TdT buffer for 2 min.
2. Remove excess liquid and carefully aspirate around the section.
3. Add 75 μl TdT/dUTP solution.

4. Incubate the sections covered with plastic coverslips in a moisture chamber at 37°C for 1 h.

6. Termination

1. Submerge the slides in TB buffer for 15 min at room temperature.
2. Rinse with DDW.
3. Block with 2% BSA for 10 min at room temperature.
4. Rinse with DDW.
5. Wash the sections in three changes of PBS for 2 min each.

7. Detection

1. Incubate the sections with extravidin–peroxidase conjugate in PBS for 30 min at 37°C using a moisture chamber.
2. Wash the slides in four changes of PBS in a coplin jar for 2 min each.
3. Blot off excess PBS.
4. Apply 100 μl of freshly prepared peroxidase substrate (DAB) on each section and incubate at room temperature for 3 to 6 min in the dark.
5. Wash the sections in three changes of DDW in a coplin jar for 1 min each.
6. Lightly counterstain with hematoxylin.
7. Wash with tap water.
8. Dehydrate the sections once in 70% ethanol for 3 min, once in 95% ethanol for 3 min, and twice in absolute ethanol for 5 min each time.
9. Clear twice in xylene for 5 min each time.
10. Drain out the slide from excess xylene and mount under a glass coverslip in a mounting medium such as Permount.

V. ISEL

ISEL, like TUNEL, relies on detecting the DNA strand breaks that accompany apoptosis. With ISEL, biotinylated nucleotides are incorporated at the sites of strand breaks using the enzymatic activity of DNA polymerase. *Escherichia coli* polymerase I, or its Klenow fragment, can be used in this technique to extend the DNA strand in the 5′ to 3′ direction by adding nucleotides to 3′-OH ends of the strand in the presence of a template. The incorporation of labeled (biotin-11-dUTP) and unlabeled nucleotides can occur in double-strand DNA breaks with recessed 3′ termini or in single-strand nicks. The signal is then enhanced by peroxidase-conjugated avidin and visualized by light microscopy using DAB as the chromogen (Wijsman *et al.*, 1993).

A. Materials

1. Reagents

Double distilled water
Xylene
Ethanol absolute, 95%, 70% diluted in distilled water
Ethanol : acetic acid 2 : 1 (v/v) for cryostat sections
Permount mounting medium for glass support
Biotin-11-dUTP (Boehringer Mannheim)
E. coli DNA polymerase I (Promega)
Klenow fragment of DNA polymerase I (Pharmacia)
Extra avidin–peroxidase (Sigma)
3,3′-DAB, storage at −20°C protected from light exposure (Sigma)
Bovine serum albumin (Sigma)
Triton X-100 or Tween 20
β-Mercaptoethanol (Sigma)

2. Solutions

10% (v/v) neutral-buffered formalin for tissue fixation prior to paraffin embedding
1% paraformaldehyde in PBS (50 m*M* sodium phosphate, pH 7.4, 200 m*M* NaCl) for cryostat sections
Concentrated PBS (CPBS) (74 g Na_2HPO_4, 9 g NaH_2PO_4, adjust volume to 1 liter, pH 7.5)
0.15 *M* PBS (8.77 g NaCl, 20 ml CPBS, adjust volume to 1 liter)
0.5 *M* PBS (29.22 g NaCl, 20 ml CPBS, adjust volume to 1 liter)
100 m*M* dNTP mix (dATP, dCTP, dGTP) (Boehringer Mannheim)
20× SSC (Boehringer Mannheim)
Hydrogen peroxide 30% solution, storage at 4°C (Sigma)
Mayer's or Harris's hematoxylin for counterstain (Sigma)

3. Equipment

Micropipettors
Glass coplin jars
Positively charged glass slides
Glass coverslips
Plastic coverslips
Microcentrifuge tubes
Water bath at 18°C

Incubator at 58°C

Microtome for paraffin blocks

Cryostat microtome for cryosections

Light microscope

4. Working Solutions

Pepsin 0.4% in 0.01 *M* HCl, pH 2.0

Trypsin 0.005% in 1× PBS

2× SSC (0.3 *M* NaCl, 30 m*M* Na-citrate, pH 7.0)

20 m*M* dNTP mix

3% hydrogen peroxide

Buffer A, pH 7.5: 50 m*M* Tris–HCl, 5 m*M* $MgCl_2$, 10 m*M* β-mercaptoethanol, and 0.005% BSA

Buffer B in 10 ml 0.5 *M* PBS: 1% BSA and 0.5% Tween 20

ISEL solution (for 20 slides at 50 μl per slide): 10 μl dNTPs mix (1 : 100), 2 μl biotin-11-dUTP (1 : 500), and 1 μl DNA Pol I (1 : 1000). Adjust volume to 1 ml with buffer A

ABC solution: 80 μl A + 80 μl B (from Vector elite kit, VECTOR) in 3840 μl buffer B prepared 30 min before use

DAB solution (Vector DAB kit, VECTOR or Sigma)

B. Protocol

1. Tissue Fixation

As in TUNEL protocol.

2. Processing of Tissue Sections

As in TUNEL protocol.

3. Inactivation of Endogenous Peroxidase

1. Incubate with freshly prepared 3% hydrogen peroxide in PBS for 30 min at room temperature (100–200 μl/section).
2. Rinse the sections twice with PBS or DDW for 5 min each time in a coplin jar.

4. Pretreatment of Tissue Sections

1. Preheat 2× SSC to 80°C.
2. Wash slides in 0.15 *M* PBS three times for 4 min each.

3. Incubate in preheated 2× SSC at 80°C for 20 min.
4. Rinse once in 0.15 *M* PBS and then wash in 0.15 *M* PBS three times for 4 min each at room temperature.
5. Incubate sections with pepsin or trypsin solution for 30–60 min.
6. Rinse once in 0.15 PBS and then wash in 0.15 PBS three times for 4 min each.
7. Equilibrate in buffer A three times for 4 min each.

5. Labeling

1. Prepare ISEL solution adding enzyme immediately before use.
2. Incubate the sections in ISEL solution at 18°C for 2 h.
3. Prepare ABC solution 20 min before the end of ISEL incubation.
4. Rinse the slides in buffer A and then wash in buffer A three times for 4 min each.
5. Wash in 0.5 *M* PBS three times for 4 min each.

6. Detection

1. Apply ABC solution on the sections and incubate at room temperature for 30 min.
2. Wash three times in 0.5 *M* PBS for 10 min each.
3. Prepare DAB solution with nickel for signal enhancement according to the manufacturer's instructions.
4. Apply DAB solution on the sections for 3–10 min at room temperature.
5. Check the intensity of the signal under the microscope; if it is the desired intensity, then rinse in running water for 2 min.
6. Counterstain with hematoxylin, dehydrate, and mount as in TUNEL protocol.

VI. Comments for Both TUNEL and ISEL

1. For negative controls, the required enzymes are omitted from the nucleotide mix (TdT for TUNEL and the DNA polymerase I or its Klenow fragment in ISEL). Positive controls are generated by treating sections with DNase I (Boehringer Mannheim) to introduce DNA breaks in all nuclei. For this purpose, adjacent sections are incubated with 200 ng/ml DNase I in buffer C (10 m*M* Tris–HCl, 10 m*M* NaCl, 5 m*M* $MgCl_2$, 0.1 m*M* $CaCl_2$, and 25 m*M* KCl, pH 7.4) for 30 min at 37°C before incubation with the nucleotide/enzyme mixture (Wijsman *et al.,* 1993).

2. Fixation times more than 3–5 weeks can decrease the sensitivity of the TUNEL and ISEL methods (Davidson *et al.,* 1995). Fixation in Carnoy's, Bouin's, or Histo-Choise fixatives can increase the background and are not recommended (Tornusciolo *et al.,* 1995). Fixation in ethanol or methanol may result in reduced sensitivity and it is better to avoid the use of these fixatives. Fixation-induced DNA–DNA and/or protein–DNA

cross-linkages can reduce the accesibility of 3′-OH termini for TdT or polymerase. This problem can be minimized by pretreating sections with proteinase K, pepsin, or trypsin, leading to increased sensitivity.

3. Microwave pretreatment may increase the sensitivity and eliminate the "false positive" TUNEL labeling seen in some paraffin-embedded tissues, such as the endometrium (Fertig, 1998).

Microwave pretreatment can be carried out after tissue dewaxing as follows:

a. Place the slides in a plastic jar containing 200 ml 0.1 *M* citrate buffer, pH 6.0.
b. Put the jar in a microwave oven and irradiate at 750 W for 1 min.
c. Cool immediately by adding 80 ml DDW at room temperature.
d. Transfer the slides into PBS at room temperature and omit proteolytic pretreatment.

4. When alkaline phosphatase (AP) is used for detection in place of peroxidase, it is recommended that endogenous AP activity should be blocked by adding 1 m*M* levamisole to the AP substrate solution.

5. High background or nonspecific labeling problems can often be overcome by diluting the labeling solutions 1 : 2 with the appropriate buffer.

6. Low sensitivity can occur with inappropriate fixation, a prolonged fixation time, or an insufficient access of reagents to the nuclei. Optimization of proteolytic pretreatment conditions (concentration, time, and temperature of incubation) for each type of tissue is recommended.

7. To optimize the conditions for ISEL it is recommended that one perform the labeling with and without proteolytic digestion using a range of dilutions of biotin-dUTP (1 : 100 to 1 : 500) and DNA Pol I (1 : 200 to 1 : 500).

VII. Other Methods

A. ISLG-LDF

ISLG-LDF stands for *in situ* ligation of labeled DNA fragments and is based on the generation of labeled oligonucleotide probes by polymerase chain reaction (PCR), followed by their *in situ* ligation to fragmented DNA in apoptotic nuclei. This technique relies on the observation that apoptotic nuclei contain DNA double-strand breaks with single-base 3′ overhangs at a much higher molar concentration than occurs in nuclei with DNA damage resulting from nonapoptotic processes (i.e., necrosis, *in vitro* autolysis, heating, and peroxide damage) (Didenko and Hornsby, 1996).

For the detection of single-base 3′ overhangs, 60- to 450-bp double-stranded DNA fragments are synthesized by *Taq* polymerase. These PCR-synthesized fragments have a single 3′ base (deoxyadenosine or deoxycytidine) extension beyond the template sequence (Hu, 1993) and can be ligated to the recessed 5′ base of the single-base 3′ overhangs of the apoptotic DNA (Didenko and Hornsby, 1996).

The primers used in the PCR reaction can be complementary to any common plasmid, such as pBluescript (Stratagene), which then serves as a template. The PCR products are then ligated to fragmented DNA in tissue sections using T4 DNA ligase. By incorporating labeled nucleotides during the PCR reaction, the DNA probes can be detected *in vivo.* Texas red-12-dUTP (Molecular Probes) can be used for fluorescence probes or digoxigenin-dUTP or biotin-16-dUTP (Boehringer Mannheim) for detection using a peroxidase-conjugated antidigoxigenin antibody or avidin and a common chromogen.

The advantage of this method over the TUNEL or the ISEL is that ISLG-LDF can exclusively identify DNA fragments with single-base 3′ overhangs, a feature characteristic of apoptotic nuclei, which is not seen in nuclei with necrotic or nonspecific damage. In contrast, nuclei with any kind of DNA damage may contain a high molar concentration of 3′-OH termini, which are substrates for terminal deoxynucleotidyl transferase and as such accessible to labeling with TUNEL or ISEL.

The major disadvantage of this technique that limits its utility is the high nonspecific background staining that occurs with the PCR-generated probes. To overcome this problem, synthetic biotin-labeled, double-stranded, hairpin-shaped oligonucleotides with a single-base 3′ overhang can be used as probes. A single biotin is positioned at the end of the hairpin opposite the 3′ overhang end used for ligation. (Didenko *et al.,* 1998, 1999).

1. Protocol

Note: Major material requirements are detailed in the protocols described earlier.

1. Sections are deparaffinized with heat and xylene as described earlier, rehydrated in a graded alcohol series, washed in DDW, and treated with proteinase K (50 μg/ml) in 0.1 *M* PBS for 15 min at room temperature.

2. Sections are rinsed with DDW and 80 μl of ligation buffer [66 m*M* Tris–HCl, pH 7.5, 5 m*M* $MgCl_2$, 0.1 m*M* dithiothreitol, 1 m*M* ATP, 15% polyethylene glycol (8000 MW, Sigma)] is applied on each section for equilibration and the sections are incubated for 15 min at room temperature.

3. After aspiration of the ligation buffer, 80 μl of ligation mixture [hairpin probe (35 μg/ml) and T4 DNA ligase (250 U/ml, Roche Molecular Biochemicals) in ligation buffer] is applied to the section. (As a control, some sections should only receive an equal volume of 50% glycerol in double-distilled water.) Sections are incubated in a moisture chamber for 16 h at 23°C.

4. The sections are washed in DDW three times for 10 min each at room temperature.

5. Avidin–fluorescein conjugate (4 μg/ml, Vector Laboratories) in 50 m*M* sodium bicarbonate, 15 m*M* NaCl, pH 8.2, is added onto the sections followed by incubation for 45 min at room temperature.

6. The sections are washed with the same buffer three times for 10 min each and are mounted in Vectashield (Vector Laboratoties) ready for fluorescence microscopy.

B. Detection of Activated Caspase-3, -7, and -9

The key initiation step of apoptosis is the activation of specific proteases known as caspases. This activation results either by triggering of death receptors, such as CD95 (Fas/Apo-1) or TNFR1, or by undefined intracellular signals due to noxious treatment of the cells that alter the mitochondrial membrane permeability and cause the release of cytochrome c (Budihardjo *et al.,* 1999). This event leads to the activation of caspase-9 and the subsequent activation of the effector caspases-3, -6, and -7 by proteolytic cleavage (Kuida, 2000). Immunocytochemical detection of activated caspases with antibodies raised against the cleaved fragments of these enzymes is thus a useful method to identify apoptotic cells.

Polyclonal or monoclonal antibodies that recognize the large active fragment (37 kDa) of caspase-9 (New England Biolabs), the large active subunit (17 kDa) of caspase-3 (Dai and Krantz, 1999; Fujita and Tsuruo, 1998), and the cleaved caspase-7 (20 kDa) active fragment (New England Biolabs) are used for this detection.

C. Cytokeratin-18 Neoepitope Detection

This immunocytochemical method is based on the observation that activated caspases can cleave specific substrate proteins at defined aspartate residues (Budihardjo *et al.,* 1999). As a consequence of these proteolytic events, new peptide epitopes are exposed in apoptotic cells (Caulin *et al.,* 1997; Yang *et al.,* 1998). The generation of these neoepitopes can be detected with specific antibodies against them, thus allowing apoptotic cells to be detected in tissue sections. One neoepitope that has been useful in this regard is mapped on a cytokeratin-18 (CK18) cleavage product generated by caspase cleavage at position Asp-396. This neoepitope is 10 residues long [387–396 (EDFNLGDALD)] and can be detected by the monoclonal antibody M30, which is specific for this site. M30 recognizes a 40-kDa CK18 fragment in the early phase of apoptosis, whereas subsequent cleavage in late apoptotic cells generates a 24-kDa fragment. M30 immunoreactivity precedes annexin V and TUNEL reactivity. It is visible in apoptotic epithelial cells, but not in viable or necrotic cells and is lost late in apoptosis. The cleavage and detection of this neoepitope are independent of the phosphorylation state of cytokeratin (Leers *et al.,* 1999). The method is applicable to both frozen and formalin-fixed material and cell culture.

D. β-Actin Neoepitope (Fractin) Detection

In this immunocytochemical method, a polyclonal antibody raised against a synthetic peptide is used for the detection of the C terminus of the 32-kDa actin fragment (fractin) produced by caspase cleavage at Asp-244. The antibody can recognize actin fragment residues 240–244 (YELPD) and requires the free C-terminal carboxyl group (Yang *et al.,* 1998; Suurmeijer *et al.,* 1999). The method is applicable to both frozen and formalin-fixed material and cultured cells. Best results in formalin-fixed tissues can be obtained using the antigen retrieval method of pressure cooking sections with 1 m*M* EDTA, pH 8.0 (Pileri *et al.,* 1997).

E. PARP Cleavage Detection

Poly(ADP-ribose)polymerase (PARP) is a zinc-finger DNA-binding protein involved in the DNA repair process, predominantly after DNA damage by genotoxic agents (Jeggo, 1998; see Chapter 12). During apoptosis, this enzyme is cleaved by caspases, mainly by caspase-3, between Asp-214 and Gly-215, generating two fragments of 24 and 89 kDa molecular mass (Lazebnik *et al.,* 1994; Tewari *et al.,* 1995). Immunocytochemical detection of the cleaved large fragment (89 kDa) with polyclonal antibodies can be used for the identification of dying cells in the early phase of apoptosis (New England Biolabs).

F. 7A6 Antigen Detection

During the apoptotic process, molecular alterations occur not only in the plasma membrane and nucleus, but also within mitochondria. It has been reported that an irreversible reduction of the mitochondrial transmembrane potential is observed prior to the classic morphological features of apoptosis (Zamzami *et al.,* 1995a,b). As a result of these alterations, the 7A6 antigen (38 kDa) is exposed on the surface of the mitochondrial membrane at an early stage of apoptosis and can be recognized by the specific monoclonal antibody APO 2.7 (Zhang *et al.,* 1996; Koester *et al.,* 1997). The mitochondrial membrane 7A6 antigen is a novel epitope and can be detected only in apoptotic cells. The APO 2.7 monoclonal antibody was generated from a hybridoma derived by the fusion of mouse myeloma P3-X63-Ag.8.653 cells with splenocytes from a mouse immunized with apoptotic Jurkat cells (Zhang *et al.,* 1996). Use of APO 2.7 for immunocytochemical detection facilitates the identification of apoptotic cells in the early phase of apoptosis.

G. Annexin V

One of the characteristics of apoptosis is a change in the asymmetric distribution of different phospholipids between inner and outer leaflets of the plasma membrane. The primary change is the translocation of an aminophospholipid, phosphatidylserine (PS), from the inner to the outer surface of the plasma membrane in intact cells (Van den Eijnde *et al.,* 1997; see also Chapter 15). PS can be detected readily using labeled Annexin V, a calcium-dependent PS-binding protein (Van Engeland *et al.,* 1998). Biotin-conjugated Annexin V-labeled cells can be fixed in cold methanol and visualized after fixation with FITC-labeled streptavidin (Van Engeland *et al.,* 1996). This procedure also enables the simultaneous detection of other intracellular antigens in different cell populations. The same technique can be used in experimental animals for the detection of apoptotic cells in formalin-fixed and paraffin-embedded normal adult tissues or during embryonic development (Van den Eijnde *et al.,* 1997, Van den 1999; detailed in Chapter 15).

In this procedure, biotin-labeled Annexin V (Molecular Probes) is injected into the animal and, after 30 min, the tissue of interest is removed, formalin fixed, paraffin embedded, and subsequently sectioned. Annexin V-labeled cells can be visualized using peroxidase-conjugated streptavidin and a common chromogen.

VIII. General Considerations

A number of variables influence the sensitivity, utility, and specificity of the methods outlined in this chapter. Two major issues are tissue quality and fixation. Unfortunately, these are hard variables to control in a clinical setting and mainly depend on the procedure by which tissue is acquired.

Obtaining human tissue for research purposes invariably requires the cooperation of pathologists. Pathologists are the physicians charged with the primary responsibility for making the histopathologic diagnosis that may determine the course of a patient's treatment. This requires strict adherence to standards of practice, and pathologists have the responsibility of acquiring, sampling, processing, and maintaining the integrity of the tissue specimens so that they are suitable for diagnostic testing. Histology laboratories use well-standardized protocols with strict quality control measures to perform these functions. Virtually all of the tissues that will be available in a pathology department archive are in the form of formalin-fixed, paraffin-embedded material. These are used to generate tissue sections stained with hematoxylin and eosin for routine diagnosis. Paraffin-embedded tissue is also suitable for use in immunohistochemical and *in situ* hybridization protocols. Despite standardization in tissue processing, variables that may influence the success of studies using this tissue remain.

A major potential pitfall in the use of these specimens is that the tissue being employed is not representative of the clinical problem under investigation. It is essential that a pathologist confirm the diagnosis for each block of tissue to be used because there are subtle, specific histopathologic criteria to consider when making a particular tissue diagnosis. Tumors, for example, are often heterogeneous in their growth pattern and cellular composition. Therefore, the detection of certain morphologic characteristics is crucial because it may be associated with a predictable biologic behavior. An accurate diagnosis is essential if a study is to draw valid conclusions about a particular biologic mechanism or disease entity and then correlate those observations with subsequent clinical behavior and response to therapy.

In addition to confirming the diagnosis, pathologists can provide an assessment of the state of tissue preservation. It is essential to know the fixation conditions of the tissue samples selected and to assure that the sections represent areas of viable tissue without necrosis and hemorrhage. Tissue integrity is particularly important to assess when using material obtained at autopsy. In some cases, there may be a long postmortem interval prior to sampling and many tissues may have undergone autolysis and may not be suitable for study (Fig. 4). Tissues obtained at autopsy or from large surgical specimens often have been fixed in formalin for a much longer period than is typical with tissues obtained by biopsy. Prolonged exposure to aldehyde fixatives, such as formalin, results in more extensive tissue protein cross-linking than is seen with precipitating fixatives such as ethanol. Tissues that are overfixed with aldehydes may require extensive protease pretreatment to obtain optimal immunohistochemical results. The preservation of antigens in fixed tissue can also be influenced by variations in the temperatures and protocol used during tissue processing. Similarly, DNA can degrade into smaller fragments during prolonged

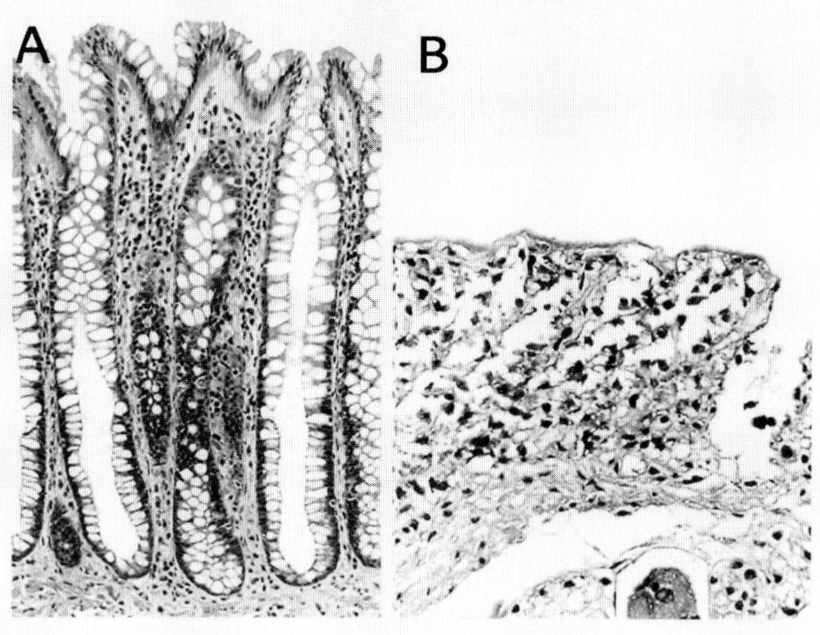

Fig. 4 Tissue preservation. Hematoxylin and eosin-stained, paraffin-embedded sections of human colon mucosa from well-fixed (A) or autolytic (B) samples. Magnification: 200× (A) and 400× (B).

formalin fixation and may be unsuitable for some analytic protocols such as TUNEL and ISEL. Preservation of RNA in formalin-fixed tissue is generally poor. Fortunately, DNA is a relatively durable molecule that is stable for decades in paraffin-embedded tissue and can be extracted readily for analysis (Diaz-Cano and Brady, 1997; Coombs *et al.,* 1999; Serth *et al.,* 2000). In recent years, molecular diagnostic techniques have exploited this property and archival tissue specimens can be used in studies designed to detect genetic alterations. *In situ* hybridization was the first of these histological analytical techniques and it provides a means for detecting specific nucleic acid sequences while maintaining tissue morphology (Speel *et al.,* 1999). PCR-based analysis can be applied to DNA extracted from minute specific areas of interest on routine histologic sections using microdissection protocols (Emmert-Buck *et al.,* 1996; Zhuang and Vortmeyer, 1998; Simone *et al.,* 1998; Sirivatanauksorn *et al.,* 1999). In addition, *in situ* PCR can be used with tissue sections and combined with subsequent hybridization (Long, 1998; Thaker, 1999), thereby allowing the assessment of genetic changes to be correlated with specific morphologic entities. Another resource available in most large modern pathology departments is snap-frozen fresh tissue. Obtaining this material also requires the cooperation of a pathologist because the tissue is examined within minutes of excision, usually in the intraoperative frozen section laboratory, and sampling of the tissue for research purposes cannot compromise routine diagnostic evaluation. Detailed protocols for freezing tissue samples are available and this frozen material is an excellent source of intact nucleic acids and protein (Naber, 1996; Karlsson and Toner, 1996). Another source of human tissues for research is from central tissue repositories such as the National Cancer Institute Cooperative Human Tissue Network (http://www-chtn.ims.nci.nih.gov). These facilities make specimens available that are suitable for both basic and developmental studies, as well as for clinical correlative studies on uniformly treated cancer patients.

The applicability of nucleic acid analysis to minute tissue samples has turned the paraffin-embedded and frozen tissue archives of pathology departments into de facto DNA banks. Concern over misuse of genetic information has led to concerted efforts by medical institutions to preserve the privacy of patient information. Consequently, the use of tissue samples in research generally requires that the investigators present evidence that their protocols do not result in a breach of patient confidentiality. The laws governing the use of human tissue for research purposes are becoming more stringent. These regulations vary depending on the locality, and the prudent course of action is for the investigator to obtain approval for a study from the institution's human studies committee.

References

Ben-Sasson, S. A., Sherman, Y., and Gavrieli, Y. (1995). Identification of dying cells: *In situ* staining. *Methods Cell Biol.* **46,** 29–39.

Biros, I., and Forrest, S. (1999). Spinal muscular atrophy: Untangling the knot? *J. Med. Genet.* **36,** 1–8.

Budihardjo, I., Oliver, H., Lutter, M., Luo, X., and Wang, X. (1999). Biochemical pathways of caspase activation during apoptosis. *Annu. Rev. Cell Dev. Biol.* **15,** 269–290.

Caulin, C., Salvesen, G. S., and Oshima, R. G. (1997). Caspase cleavage of keratin 18 and reorganization of intermediate filaments during epithelial cell apoptosis. *J. Cell Biol.* **138,** 1379–1394.

Coombs, N. J., Gough, A. C., and Primrose, J. N. (1999). Optimisation of DNA and RNA extraction from archival formalin-fixed tissue. *Nucleic Acids Res.* **27,** e12.

Dai, C., and Krantz, S. B. (1999). Interferon gamma induces upregulation and activation of caspases 1, 3, and 8 to produce apoptosis in human erythroid progenitor cells. *Blood* **93,** 3309–3316.

Darzynkiewicz, Z., Bedner, E., Traganos, F., and Murakami, T. (1998). Critical aspects in the analysis of apoptosis and necrosis. *Hum. Cell* **11,** 3–12.

Davison, F. D., Groves, M., and Scaravilli, F. (1995). The effects of formalin fixation on the detection of apoptosis in human brain by in situ end-labelling of DNA. *Histochem. J.* **27,** 983–988.

Debatin, K. M. (1999). Activation of apoptosis pathways by anticancer drugs. *Adv. Exp. Med. Biol.* **457,** 237–244.

Desjardins, P., and Ledoux, S. (1998). The role of apoptosis in neurodegenerative diseases. *Metab. Brain Dis.* **13,** 79–96.

Dewey, W. C., Ling, C. C., and Meyn, R. E. (1995). Radiation-induced apoptosis: Relevance to radiotherapy. *Int. J. Radiat. Oncol. Biol. Phys.* **33,** 781–796.

Diaz-Cano, S. J., and Brady, S. P. (1997). DNA extraction from formalin-fixed, paraffin-embedded tissues: Protein digestion as a limiting step for retrieval of high-quality DNA. *Diagn. Mol. Pathol.* **6,** 342–346.

Didenko, V. V., Boudreaux, D. J., and Baskin, D. S. (1999). Substantial background reduction in ligase-based apoptosis detection using newly designed hairpin oligonucleotide probes. *Biotechniques* **27,** 1130–1132.

Didenko, V. V., Tunstead, J. R., and Hornsby, P. J. (1998). Biotin-labeled hairpin oligonucleotides: Probes to detect double-strand breaks in DNA in apoptotic cells. *Am. J. Pathol.* **152,** 897–902.

Didenko, V. V., and Hornsby, P. J. (1996). Presence of double-strand breaks with single-base 3′ overhangs in cells undergoing apoptosis but not necrosis. *J. Cell Biol.* **135,** 1369–1376.

Emmert-Buck, M. R., Bonner, R. F., Smith, P. D., *et al.* (1996). Laser capture microdissection. *Science* **274,** 998–1001.

Famularo, G., De Simone, C., and Marcellini, S. (1997). Apoptosis: Mechanisms and relation to AIDS. *Med. Hypoth.* **48,** 423–429.

Fertig, G. (1998). "Apoptosis and Cell Proliferation," pp. 105–106. Boerhinger Mannheim.

Feuerstein, G. Z. (1999). Apoptosis in cardiac diseases; new opportunities for novel therapeutics for heart diseases. *Cardiovasc. Drugs Ther.* **13**(4), 289–294.

Filipski, J., Leblanc, J., Youdale, T., Sikorska, M., and Walker, P. R. (1990). Periodicity of DNA folding in higher order chromatin structures. *EMBO J.* **9,** 1319–1327.

Fujita, N., and Tsuruo, T. (1998). Involvement of Bcl-2 cleavage in the acceleration of VP-16-induced U937 cell apoptosis. *Biochem. Biophys. Res. Commun.* **246,** 484–488.

Gavrieli, Y., Sherman, Y., and Ben-Sasson, S. A. (1992). Identification of programmed cell death in situ via specific labeling of nuclear DNA fragmentation. *J. Cell Biol.* **119**(3), 493–501.

Gougeon, M.-L., and Montagnier, L. (1993). Apoptosis in AIDS. *Science* **260,** 1269–1270.

Margolis, R. L., Chuang, D. M., and Post, R. M. (1994). Programmed cell death: Implications for neuropsychiatric disorders. *Biol. Psychiat.* **35,** 946–956.

Hengartner, M. O., and Horvitz, H. R. (1994). The ins and outs of programmed cell death during *C. elegans* development. *Phil. Trans. R. Soc. Lond. B Biol. Sci.* **345,** 243–246.

Hetts, S. W. (1998). To die or not to die: An overview of apoptosis and its role in disease. *JAMA* **279,** 300–307.

Hochman, A. (1997). Programmed cell death in prokaryotes. *Crit. Rev. Microbiol.* **23,** 207–214.

Horvitz, H. R. (1999). Genetic control of programmed cell death in the nematode *Caenorhabditis elegans. Cancer Res.* **59,** 1701s–1706s.

Houghton, J. A. (1999). Apoptosis and drug response. *Curr. Opin. Oncol.* **11,** 475–481.

Hu, G. (1993). DNA polymerase-catalyzed addition of nontemplated extra nucleotides to the 3′ end of a DNA fragment. *DNA Cell Biol.* **12,** 763–770.

Jeggo, P. A. (1998). DNA repair: PARP—another guardian angel? *Curr. Biol.* **8,** R49–R51.

Jellinger, K. A., and Bancher, C. (1998). Neuropathology of Alzheimer's disease: A critical update. *J. Neural Transm. Suppl.* **54,** 77–95.

Karlsson, J. O., and Toner, M. (1996). Long-term storage of tissues by cryopreservation: Critical issues. *Biomaterials* **17,** 243–256.

Kerr, J. F., Gobe, G. C., Winterford, C. M., and Harmon, B. V. (1995). Anatomical methods in cell death. *Methods Cell Biol.* **46,** 1–27.

Kerr, J. F. R., Wyllie, A. H., and Currie, A. R. (1972). Apoptosis: Basic biological phenomenon with wide-ranging implications in tissue kinetics. *Br. J. Cancer* **26,** 239–257.

Koester, S. K., Roth, P., Mikulka, W. R., Schlossman, S. F., Zhang, C., and Bolton, W. E. (1997). Monitoring early cellular responses in apoptosis is aided by the mitochondrial membrane protein-specific monoclonal antibody APO2.7. *Cytometry* **29,** 306–312.

Kuida, K. (2000). Caspase-9. *Int. J. Biochem. Cell Biol.* **32,** 121–124.

Lazebnik, Y. A., Kaufmann, S. H., Desnoyers, S., Poirier, G. G., and Earnshaw, W. C. (1994). Cleavage of poly(ADP-ribose) polymerase by a proteinase with properties like ICE. *Nature* **371,** 346–347.

Leers, M. P. G., Kolgen, W., Bjorklund, V., Bergman, T., Tribbick, G., Persson, B., Bjorklund, P., Ramaekers, F. C. S., Bjorklund, B., Nap, M., Jornvall, H., and Schutte, B. (1999). Immunocytochemical detection and mapping of a cytokeratin 18 neo-epitope exposed during early apoptosis. *J. Pathol.* **187,** 567–572.

Long, A. A. (1998). In-situ polymerase chain reaction: Foundation of the technology and today's options. *Eur. J. Histochem.* **42,** 101–109.

Lorenzo, H. K., Susin, S. A., Penninger, J., and Kroemer, G. (1999). Apoptosis inducing factor (AIF): A phylogenetically old, caspase-independent effector of cell death. *Cell Death Differ.* **6,** 516–524.

Lyons, S. K., and Clarke, A. R. (1997). Apoptosis and carcinogenesis. *Br. Med. Bull.* **52,** 554–569.

Majno, G., and Joris, I. (1995). Apoptosis, oncosis, and necrosis: An overview of cell death. *Am. J. Pathol.* **146,** 3–15.

Mattson, M. P., Pedersen, W. A., Duan, W., Culmsee, C., and Camandola, S. (1999). Cellular and molecular mechanisms underlying perturbed energy metabolism and neuronal degeneration in Alzheimer's and Parkinson's diseases. *Ann. N.Y. Acad. Sci.* **893,** 154–175.

Metzstein, M. M., Stanfield, G. M., and Horvitz, H. R. (1998). Genetics of programmed cell death in *C. elegans:* Past, present and future. *Trends Genet.* **14,** 410–416.

Milligan, C. E., and Schwartz, L. M. (1997). Programmed cell death during animal development. *Br. Med. Bull.* **52,** 570–590.

Mountz, J. D., Wu, J., Cheng, J., and Zhou, T. (1994). Autoimmune disease: A problem of defective apoptosis. *Arthritis Rheum.* **37,** 1415–1420.

Naber, S. P. (1996). Continuing role of a frozen-tissue bank in molecular pathology. *Diagn. Mol. Pathol.* **5,** 253–259.

Nagy, Z. (1999). Mechanisms of neuronal death in Down's syndrome. *J. Neural Transm. Suppl.* **57,** 233–245.

Nicholson, D. W., Ali, A., Thornberry, N. A., Vaillancourt, J. P., Ding, C. K., Gallant, M., Gareau, Y., Griffin, P. R., Labelle, M., Lazebnik, Y. A., *et al.* (1995). Identification and inhibition of the ICE/CED-3 protease necessary for mammalian apoptosis. *Nature* **376,** 37–43.

Oberhammer, F., Wilson, J. W., Dive, C., Morris, I. D., Hickman, J. A., Wakeling, A. E., Walker, P. R., and Sikorska, M. (1993). Apoptotic death in epithelial cells: Cleavage of DNA to 300 and/or 50 kb fragments prior to or in the absence of internucleosomal fragmentation. *EMBO J.* **12,** 3679–3684.

Peake, J., Waugh, A., Le Deist, F., Priestley, A., Rieux-Laucat, F., Foray, N., Capulas, E., Singleton, B. K., de Villartay, J. P., Cant, A., Malaise, E. P., Fischer, A., Hivroz, C., and Jeggo, P. A. (1999). Combined immunodeficiency associated with increased apoptosis of lymphocytes and radiosensitivity fibroblasts. *Cancer Res.* **59,** 3454–3460.

Petersen, A., Mani, K., and Brundin, P. (1999). Recent advances on the pathogenesis of Huntington's disease. *Exp. Neurol.* **157,** 1–18.

Pileri, S. A., Roncador, G., Ceccarelli, C., Piccioli, M., Briskomatis, A., Sabbatini, E., Ascari, S., Santani, D., Piccaluga, P. P., Leone, O., Damiani, S., Ercolessi, C., Sandri, F., Leocini, L., and Flini, B. (1997). Antigen retrieval techniques in immunohistochemistry: Comparison of different methods. *J. Pathol.* **183,** 116–123.

Raffray, M., and Cohen, G. M. (1997). Apoptosis and necrosis in toxicology: A continuum or distinct modes of cell death? *Pharmacol. Ther.* **75,** 153–177.

Saraste, A., Pulkki, K., Kallajoki, M., Henriksen, K., Parvinen, M., and Voipio-Pulkki, L. M. (1997). Apoptosis in human acute myocardial infarction. *Circulation* **95,** 320–323.

Serth, J., Kuczyk, M. A., Paeslack, U., Lichtinghagen, R., and Jonas, U. (2000). Quantitation of DNA extracted after micropreparation of cells from frozen and formalin-fixed tissue sections. *Am. J. Pathol.* **156,** 1189–1196.

Simone, N. L., Bonner, R. F., Gillespie, J. W., Emmert-Buck, M. R., and Liotta, L. A. (1998). Laser-capture microdissection: Opening the microscopic frontier to molecular analysis. *Trends Genet.* **14,** 272–276.

Sirivatanauksorn, Y., Drury, R., Crnogorac-Jurcevic, T., Sirivatanauksorn, V., and Lemoine, N. R. (1999). Laser-assisted microdissection: Applications in molecular pathology. *J. Pathol.* **189,** 150–154.

Speel, E. J., Hopman, A. H., and Komminoth, P. (1999). Amplification methods to increase the sensitivity of in situ hybridization: Play card(s). *J. Histochem. Cytochem.* **47,** 281–288.

Surh, C. D., and Sprent, J. (1994). T-cell apoptosis detected *in situ* during positive and negative selection in the thymus. *Nature* **372,** 100–103.

Suurmeijer, A. J. H., Van der Wijk, Van Veldhuisen, D. J., Yang, F., and Cole, G. M. (1999). Fractin immunostaining for the detection of apoptotic cells and apoptotic bodies in formalin-fixed and paraffin-embedded tissue. *Lab. Invest.* **79,** 619–620.

Tewari, M., Quan, L. T., O'Rourke, K., Desnoyers, S., Zeng, Z., Beidler, D. R., Poirier, G. G., Salvesen, G. S., and Dixit, V. M. (1995). Yama/CPP32 beta, a mammalian homolog of CED-3, is a CrmA-inhibitable protease that cleaves the death substrate poly(ADP-ribose) polymerase. *Cell* **81,** 801–809.

Thaker, V. (1999). In situ RT-PCR and hybridization techniques. *Methods Mol. Biol.* **115,** 379–402.

Tornusciolo, D. R., Schmidt, R. E., and Roth, K. A. (1995). Simultaneous detection of TDT-mediated dUTP-biotin nick end-labeling (TUNEL)-positive cells and multiple immunohistochemical markers in single tissue sections. *Biotechniques* **9,** 800–805.

Van den Eijnde, S. M., Boshart, L., Reutelingsperger, C. P. M., De Zeeuw, C. L., and Vermeij-Keers, C. (1997).

Phosphatidylserine plasma membrane asymmetry in vivo: A pancellular phenomenon which alters during apoptosis. *Cell Death Differ.* **4,** 311–316.

Van den Eijnde, S. M., Lips, J., Boschart, L., Vermeij-Keers, C., Marani, E., Reutelingsperger, C. P. M., and De Zeeuw, C. I. (1999). Spatiotemporal distribution of dying neurons during early mouse development. *Eur. J. Neurosci.* **11,** 712–724.

Van den Eijnde, S. M., Luijsterburg, A. J., Boschart, L., De Zeeuw, C. I., Van Dierendonck, J. H., Reutelingsperger, C. P. M., and Vermeij-Keers, C. (1997). In situ detection of apoptosis during embryogenesis with annexin V: From whole mount to ultrastucture. *Cytometry* **29,** 313–320.

Van Engeland, M., Nieland, L. J. W., Ramaekers, F. C. S., Schutte, B., and Reutelingsperger, C. P. M. (1998). Annexin V-affinity assay: A review on an apoptosis detection system based on phosphatidylserine exposure. *Cytometry* **31,** 1–9.

Van Engeland, M., Ramaekers, F. C. S., Schutte, B., and Reutelingsperger, C. P. M. (1996). A novel assay to measure loss of plasma membrane asymmetry during apoptosis of adherent cells in culture. *Cytometry* **24,** 131–139.

Wijsman, J. H., Jonker, R. R., Keijzer, R., Van De Delde, C. J., Cornelisse, C. J., and Van Dierendonck, J. H. (1993). A new method to detecy apoptosis in paraffin sections: In situ end-labeling of fragmented DNA. *J. Histochem. Cytochem.* **41,** 7–12.

Wyllie, A. H. (1980). Glucocorticoid-induced thymocyte apoptosis is associated with endogenous endonuclease activation. *Nature* **284,** 555–556.

Wyllie, A. H., Kerr, J. F., and Currie, A. R. (1980). Cell death: The significance of apoptosis. *Int. Rev. Cytol.* **68,** 251–306.

Yang, F., Sun, X., Beech, W., Teter, B., Wu, S., Sigel, J., Vinters, H. V., Frautschy, S. A., and Cole, G. M. (1998). Antibody to caspase-cleaved actin detects apoptosis in differentiated neuroblastoma and plaque-associated neurons and microglia in Alzheimer's disease. *Am. J. Pathol.* **152,** 379–389.

Young, L. S., Dawson, C. W., and Eliopoulos, A. G. (1997). Viruses and apoptosis. *Br. Med. Bull.* **53,** 509–521.

Zamzami, N., Marchetti, P., Castedo, M., Decaudin, D., Macho, A., Hirsch, T., Susin, S. A., Petit, P. X., Mignotte, B., and Kroemer, G. (1995a). Sequential reduction of mitochondrial transmembrane potential and generation of reactive oxygen species in early programmed cell death. *J. Exp. Med.* **182,** 367–377.

Zamzami, N., Marchetti, P., Castedo, M., Zanin, C., Vayssiere, J. L., Petit, P. X., and Kroemer, G. (1995b). Reduction in mitochondrial potential constitutes an early irreversible step of programmed lymphocyte death in vivo. *J. Exp. Med.* **181,** 1661–1672.

Zhang, C., Ao, Z., Seth, A., and Schlossman, S. F. (1996). A mitochondrial membrane protein defined by a novel monoclonal antibody is preferentially detected in apoptotic cells. *J. Immunol.* **157,** 3980–3987.

Zhuang, Z., and Vortmeyer, A. O. (1998). Applications of tissue microdissection in cancer genetics. *Cell Vis.* **5,** 43–48.

CHAPTER 18

Model Cell Lines for the Study of Apoptosis *in Vitro*

Christos Valavanis,[*] Yanhui Hu,[†] Yili Yang,[‡] Barbara A. Osborne,[†,§] Salem Chouaib,[||] Lloyd Greene,[¶] Jonathan D. Ashwell,[‡] and Lawrence M. Schwartz[*,†]

[*] Department of Biology
Morrill Science Center
University of Massachusetts
Amherst, Massachusetts 01003

[†] Program in Molecular and Cellular Biology
University of Massachusetts
Amherst, Massachusetts 01003

[‡] Laboratory of Immune Cell Biology
National Cancer Institute
National Institutes of Health
Bethesda, Maryland 20892

[§] Department of Veterinary and Animal Sciences
University of Massachusetts
Amherst, Massachusetts 01003

[||] Cytokines et Immunite Antitumorale
Institut Gustave Roussy, INSERM
94805 Villejuif Cedex, France

[¶] Department of Pathology, Taub Center for Alzheimer's Disease Research
College of Physicians and Surgeons, Columbia University
New York, New York 10003

0091-679X/01 $35.00

I. Introduction

During the past decade, the study of apoptosis has become one of the most intensively investigated areas in cell biology. With tens of thousands of citations, it is clear that investigators from diverse fields have focused on the process of cell death. In some cases, the decision to study apoptosis has been a conscious one because this process is clearly implicated as a component of the biological phenomenon under investigation. In many cases, however, researchers have "stumbled" into the field as a result of other ongoing studies. For example, one can imagine that conducting a yeast two hybrid screen to identify interacting partners for a protein of interest might result in the isolation of Bcl-2. This observation would have obvious implications for the mechanism(s) by which the bait protein mediates its biological effects. In this case, the investigator is likely to search the literature to determine what is known about the role of Bcl-2 in the biological system under investigation and proceed from there (see, e.g., Chapter 2).

The next step is less obvious, however, if the protein isolated in the screen is not a known cell death regulator. For example, if the newly discovered protein partner is a novel BH3-containing protein, it may well be a Bcl-2 family member, but one of unknown function. Such a protein could conceivably function as either an anti- or a proapoptotic regulator (Sato *et al.,* 1994; Yang *et al.,* 1995; Wang *et al.,* 1996). While the investigator will ultimately want to study the role of this gene in a specific lineage or model, the first question to address is whether it is a cell death regulator.

One straightforward means of addressing this question is to transfect the gene of interest into a cell and determine if the rate of cell death is altered. [A variety of methods for assaying cell death can be found both in this volume and in its predecessor (Schwartz and Osborne, 1995)]. The obvious outcomes are that ectopic expression of the gene (1) has no effect, (2) increases the percentage of dying cells, or (3) retards cell death induced by specific triggers. In the last case, it would be valuable to have multiple independent triggers for apoptosis to determine if the gene of interest acts globally or in a specific signal transduction cascade.

This chapter provides detailed information about the use of six different cell lines for the study of apoptosis. The first, Jurkat cells, were originally derived from a patient with acute lymphoblastic T-cell leukemia (Gillis and Watson, 1980). These cells grow in suspension and can be induced to undergo apoptosis in response to a wide variety of well-defined physiological, pharmacological, and pathological triggers. Jurkat cells have functioned as the *in vitro* "white lab rat" of apoptosis research and have been one of the best exploited models for studying the biochemical steps that regulate apoptosis (Pimentel-Muinos and Seed, 1999; Karas *et al.,* 1999).

Two additional lymphoid-derived cell lines, 2B4.11 and D011.10, have also been popular models for the study of apoptosis (Pratt *et al.,* 1999; Ashwell *et al.,* 1987; Mercep *et al.,* 1988). Both of these lines are T-cell hybridomas with well-characterized and functional T-cell receptors (TCR). In addition to the standard triggers for apoptosis, these cells can be induced to die in response to TCR stimulation, such as occurs with specific antigen or cross-linking with anti-TCR antibodies. Like other lymphoid cells, 2B4.11 and D011.10 are also sensitive to glucocorticoid-induced apoptosis.

Three additional cell lines, MCF-7, PC-12, and C_2C_{12}, are useful for lineage-specific studies. MCF-7 is the standard *in vitro* model for human breast cancer (Clarke *et al.*, 1994; Kern *et al.*, 1994). PC-12 cells originate from rat pheochromocytoma, an adrenal gland tumor, and acquire neuronal morphology in response to nerve growth factor (NGF) treatment (Greene and Tischler, 1976; Biocca *et al.*, 1983). C_2C_{12} cells are mouse muscle satellite cells that can be induced to differentiate into muscle fibers when incubated with a serum-deficient medium (Yaffe and Saxel, 1977; Blau *et al.*, 1985). While not as well characterized as the cell lines just described, MCF-7, PC-12, and C_2C_{12} have proven to be valuable models for apoptosis research (Welsh, 1994; Mills *et al.*, 1995; Hu *et al.*, 1999).

The characteristics of these selected model cell lines for the study of apoptosis *in vitro*, their growth conditions and properties, and the methods used for their transfection are summarized in Tables II, III, and IV, respectively, and are discussed in detail later.

Table I
Additional Cell Line Models for the Study of Apoptosis

Cell type/tissue of origin	Designation	Organism	Commercial source/catalog number
Fibroblast/embryo	WI-38	*Homo sapiens*	ATCC/CCL-75
Fibroblast/embryo	NIH3T3	*Mus musculus*	ATCC/CRL-1658
Fibroblast/embryo	Rat-2	*Rattus norvegicus*	ATCC/CRL-1764
Lymphoblast/leukemia	K-562	*Homo sapiens*	ATCC/CCL-243
Promyeloblast/leukemia	HL-60	*H. sapiens*	ATCC/CCL-240
Rhabdomyosarcoma	HS 729	*H. sapiens*	ATCC/HTB-153
Rhabdomyosarcoma	Rh-39 (RD)	*H. sapiens*	ATCC/CCL-136
Osteoblast	hFOB 1.19	*H. sapiens*	ATCC/CRL-11372
Liposarcoma	SW-872	*H. sapiens*	ATCC/HTB-92
Chondrosarcoma	SW-1353	*H. sapiens*	ATCC/HTB-94
Neuroblastoma	SK-N-AS	*H. sapiens*	ATCC/CRL-2137
Glioblastoma	A172	*H. sapiens*	ATCC/CRL-1620
Astrocytoma	CCF-STTG1	*H. sapiens*	ATCC/CRL-1718
Umbilical vein endothelial cells	HUVEC	*H. sapiens*	ATCC/CRL 1730
Keratinocyte	CCD 1102 KERTr	*H. sapiens*	ATCC/CRL-2310
Squamous cell carcinoma	SCC-15	*H. sapiens*	ATCC/CRL-1623
Melanoma	MALME-3M	*H. sapiens*	ATCC/HTB-64
Adenocarcinoma/non-small, lung	NCI-H522	*H. sapiens*	ATCC/CRL-5810
Adenocarcinoma/gastric	NCI-N87	*H. sapiens*	ATCC/CRL-5822
Adenocarcinoma/colon	LoVo	*H. sapiens*	ATCC/CCL-229
Adenocarcinoma/colon	COLO 205	*H. sapiens*	ATCC/CCL-222
Adenocarcinoma/pancreas	HPAC	*H. sapiens*	ATCC/CRL-2119
Adenocarcinoma/prostate	PC-3	*H. sapiens*	ATCC/CRL-1435
Adenocarcinoma/prostate	DU 145	*H. sapiens*	ATCC/HTB-81
Adenocarcinoma/prostate	LNCaP	*H. sapiens*	ATCC/CRL-1740
Adenocarcinoma/ovary	OVCAR-3	*H. sapiens*	ATCC/HTB-161
Adenocarcinoma/endometrium	HEC-1-A	*H. sapiens*	ATCC/HTB-112
Clear cell carcinoma/kidney	CAKI-1	*H. sapiens*	ATCC/HTB-46
Hepatocellular carcinoma	HepG2	*H. sapiens*	ATCC/HB-8065

Table II
Characteristics of Selected Cell Lines for the Study of Apoptosis *in Vitro*

Cell line	Organism	Tissue	Morphology	Tumorigenic	Products	Commercial source/ catalog number
Jurkat	*Homo sapiens*	Acute T lymphoblastic leukemia	Lymphoblast, round to racquet shaped	Yes	Interferon-γ and IL-2 after activation by anti-TCR antibodies or PMA (20 ng/ml) plus ionomycin (1 μg/ml)	ATCC TIB-152
2B4.11	*Mus musculus*	Thymoma x splenic T-cell hybridoma	Round to racquet shaped	Yes	IL-2 after activation by specific antigen (pigeon cytochrome *c* 81–104, anti-TCR antibodies), mitogens (concanavalin A, 2 μg/ml), or PMA (20 ng/ml) plus ionomycin (1 μg/ml)	
DO11.10	*M. musculus*	Thymic leukemic x splenic T-cell hybridoma	Round	Yes	Produces no known cytokines unless stimulated	
MCF-7	*H. sapiens*	Adenocarcinoma, mammary gland	Epithelial	Yes	Estrogen receptor expression Insulin-like growth factor-binding proteins: IGFBP-2, IGFBP-4, and IGFBP-5 Secretion of IGFBPs can be modulated by antiestrogen treatment Caspase-3-deficient cells	ATCC HTB-22
PC-12	*Rattus norvegicus*	Pheochromocytoma, adrenal gland	Round/polygonal Neuronal morhology with extented neurites after treatment with NGF	Yes	Catecholamines Norepinephrine Dopamine Nerve growth factor receptor expression IL-1β synthesis and release Does not synthesize epinephrine	ATCC CRL-1721
C_2C_{12}	*M. musculus*	Skeletal myoblast	Fibroblast like, spindle shaped	No	TGF-β 3 Insulin-like growth factor II Muscle-specific proteins	ATCC CRL-1772

Table III
Culture Conditions of Selected Cell Lines for the Study of Apoptosis *in Vitro*

Cell line	Jurkat	2B4.11	DO11.10	MCF-7	PC-12	C_2C_{12}
Growth properties	Nonadherent In suspension in flasks	Nonadherent In suspension in flasks	Nonadherent In suspension in flasks	Adherent In plates or flasks	Poorly adherent Small tight clusters Adhere better on collagen or polylysine coated plates	Adherent On plates or flasks
Seeding density	2×10^5 cells/ml	10^5 cells/ml	5×10^4 or 10^5 cells/ml	5×10^4 cells/ml	$>10^6$ cells/ml	5×10^4 cells/ml
Growth medium	RPMI 1640 with 10% FCS–250 μg/ml gentamicin, 100 U/ml penicillin, 4 m*M* glutamine, and 5 μ*M* 2-mercaptoethanol	RPMI 1640 with 10% FCS–250 μg/ml gentamicin, 100 U/ml penicillin, 4 m*M* glutamine, and 5 μ*M* 2-mercaptoethanol	DME with 10% FCS or HS–100 units/ml penicillin–streptomycin	RPMI 1640 with 5% FBS–100 units/ml penicillin–streptomycin	RPMI 1640 with 10% horse serum and 5% FBS–100 units/ml penicillin–streptomycin	DMEM with 10% FBS–100 units/ml penicillin–streptomycin
Medium renewal	Every 3 days	Every other day	Every 2–3 days	Every 2–3 days	Every 2–3 days	Every 2 days
Differentiation medium					RPMI 1640 with 1% HS and 50–100 ng/ml NGF–100 U/ml penicillin–streptomycin	DMEM with 2% HS or 0.1% FBS, 5 μg/ml insulin, 5 μg/ml transferrin–100 U/ml penicillin–streptomycin
Temperature	37°C	37°C	37°C	37°C	37°C	37°C
Humidity	100%	100%	100%	100%	100%	100%
CO_2 concentration	5%	5%	10%	5%	7.5%	5%–10%

(continues)

Table III (*continued*)

Cell line	Jurkat	2B4.11	DO11.10	MCF-7	PC-12	C_2C_{12}
Doubling time	12–18 h	12–14 h	12–16 h	25–48 h	2.5–4 days	16 h
Subculturing	Every 3–4 days	Every 3–4 days	Every 3–4 days	Every 6–8 days Fresh 0.25% trypsin plus 0.03% EDTA	Every 7–10 days (80–90% confluent cultures) Mechanical detachment by squirting culture medium on cells	Every 3–4 days Fresh 0.25% trypsin Do not allow culture to become confluent
Split ratio	1 : 10	1 : 10	1 : 5 to 1 : 10	1 : 3 to 1 : 6	1 : 3 to 1 : 4	1 : 20
Freezing medium	90% FCS 10% DMSO	90% FCS 10% DMSO	85% FCS 15% DMSO	80% FBS 20% DMSO	90% GM 10% DMSO	92% FBS 8% DMSO
Storage in liquid nitrogen	10^6 cells/ml in freezing medium	10^6 cells/ml in freezing medium	10^6 cells/ml in freezing medium	10^6 cells/ml in freezing medium	$>5 \times 10^6$ cells/ml in freezing medium	10^6 cells/ml in freezing medium
Biosafety level	BL-2	BL-1	BL-1	BL-1	BL-1	BL-1
Comments		2B4.11 cells are quite fragile and will die if cell density is too high ($>10^6$ cells/ml)	DO11.10 cells are hardy and will become resistant to apoptosis over time if not split regularly	MCF-7 cells present several features of differentiated mammary epithelium They can form domes Their growth is inhibited by TNF-α	They respond reversibly to NGF (better in high cell density) by induction of the neuronal phenotype (neurites extension) Adhere loosely to tissue culture plastic especially after exposure to NGF Long-term culture on plastic selects for more adherent subpopulation with poorer NGF response	C_2C_{12} differentiate rapidly, forming contractile myotubes Treatment with bone morphogenetic protein 2 (BMP-2) shifts the differentiation process from myoblastic to osteoblastic

Table IV

Transfection Methods and Selection Conditions Used in Selected Cell Lines for the Study of Apoptosis *in Vitro*[a]

Cell line	Jurkat	2B4.11	DO11.10	MCF-7	PC-12	C_2C_{12}
Electroporation	Yes Transient Stable	Yes Transient Stable	Marginally effective	Yes	Yes Transient Stable	Yes
Retrovirus mediated	Yes Transient Stable	Yes Transient Stable	Yes Transient Stable		Yes Stable	Yes Stable
Adenovirus mediated			No adenovirus receptor expression	Yes Transient Stable	Yes Tansient Stable	
Liposome mediated			Marginally effective	Yes Transient Stable	Yes Transient Stable	Yes Transient Stable
DEAE-dextran mediated	Yes Transient	Yes Transient	Marginally effective			
Calcium-phosphate mediated			Marginally effective	Yes	Yes Transient Stable	Yes
Activated polyamidoamine dendrimer mediated	Yes Transient		Marginally effective			
Optimal promoters						MMLV-LTR CMV2
Antibiotic concentration for selection	G 418 or hygromycin B, 200 μg/ml to 2 mg/ml Add to the medium 48 h after transfection	G 418, 200 μg/ml to 2 mg/ml Add to the medium 36 h after transfection	G 418, 400–800 μg/ml Add to the medium 48 h after transfection	G 418 (neomycin), 500 μg/ml Add to the medium 48 h after transfection	G418 (neomycin), 500 μg/ml Add to the medium 48 h after transfection	G418 (neomycin), 500 μg/ml Puromycin, 3 μg/ml Hygromycin, 100–800 μg/ml Add to the medium 48 h after transfection
Incubation time with antibiotic	3–4 days	3–4 days	4–8 days	Few weeks Mechanical selection under the microscope using pipettors and yellow tips	At least 1 week Mechanical selection under the microscope using pipettors and yellow tips	G 418 Few weeks Puromycin or hygromycin, few days Mechanical selection under the microscope using pipettors and yellow tips

[a]Blank spaces in table mean that working data are not available.

In addition to testing the apoptosis-modulating activity of a candidate gene, cell lines can be employed for other preliminary studies in apoptosis. For example, one or more of these lines could be used to determine if a new pharmacological agent modulates apoptosis. Alternatively, an investigator might want to test a new or improved method for quantifying cell death. By using one of these well-characterized lines, the utility of the new assay can be compared readily to prior results. Once investigators have determined that their favorite new gene, treatment, or assay is relevant to the study of apoptosis, they may well want to focus on specific cellular lineages that are more appropriate to the interest of the laboratory. To help in the shift to other *in vitro* models, we provide a comprehensive list of additional cell lines (Table I). These are all well-characterized lines that can be purchased from the ATCC (American Tissue Type Collection; URL: http://phage.atcc.org/searchengine/ce.html).

II. Jurkat

This cell line was originally derived from a patient with acute lymphoblastic T-cell leukemia, and the clone E6-1 that is described in this chapter originates from the Jurkat-FHCRC cell line (Gillis and Watson, 1980). Morphologically, the cells are round to racquet shaped and express the T-cell antigen receptor. They produce interleukin (IL)-2 after stimulation with phorbol esters and lectins (Li *et al.,* 1999; Roosnek *et al.,* 1985) or monoclonal antibodies against the TCR (Wacholtz and Lipsky, 1993; Weiss *et al.,* 1984). IL-2 production can also be induced by 20–50 ng/ml phorbol-12-myristate-B-acetate (PMA) and 1 μ/ml ionomycin (Butscher *et al.,* 1998). Jurkat cells also produce interferon-γ after activation by either PMA or phytohemagglutinin (PHA) or antibodies to the TCR (Wiskocil *et al.,* 1985). They are glucocorticoid receptor negative (Northrop *et al.,* 1992) and as such are resistant to glucocorticoid-induced apoptosis.

Jurkat cells are nonadherent and grow in suspension at 37°C, 100% humidity, and 5% CO_2 using RPMI 1640 with 10% fetal calf serum (FCS), 250 μg/ml gentamicin, 100 U/ml penicillin, 4 m*M* glutamine, and 5 μM 2-mercaptoethanol as a growth medium. The optimum seeding density for this cell line is 2×10^5 cells/ml and the doubling time is 12–18 h. The culture should be maintained between 10^5 and 10^6 cells/ml and not allowed to exceed the cell concentration of 10^6 cells/ml. The medium should be renewed every 3 days and the cells can be subcultured every 3–4 days in a split ratio of 1:10. For long-term storage, 10^6 cells/ml are suspended in freezing medium [90% FCS with 10% dimethyl sulfoxide (DMSO)] using cryovials, kept in a container of isopropanol at -80°C for 24 h and stored in liquid nitrogen tanks. Jurkat cells are tumorigenic and should be handled carefully in a biosafety level 2 facility.

Jurkat cells can be easily transfected using a number of different methods. Electroporation (Boonen *et al.,* 1999) and retrovirus-mediated transfection are both good for transient and stable transfections, whereas DEAE-dextran (Demirhan *et al.,* 1998), liposome (Demirhan *et al.,* 1998), and activated polyamidoamine-dendrimer mediated (Tetsuka *et al.,* 2000) have been applied in transient ones. The most common antibiotics used for selection of the transfected clones are G418 (neomycin) or hygromycin B at a

Table V
Conditions and Chemical/Physical Agents Used for the Induction of Cell Death in Selected Cell Lines for the Study of Apoptosis *in Vitro*[a]

Cell line	Jurkat	2B4.11	DO11.10	MCF-7	PC-12	C_2C_{12}
Growth factor deprivation	Serum-free medium for 36 h	Serum-free medium for 24 h		Cells in 0.25% serum for 36 h	Serum-free medium (remove NGF if present)	Serum free medium for two days
Fas ligand/ antibodies	Anti-Fas Ab CH-11 100 ng/ml	FasL expressed on activated effector cells (e.g., d11S) or soluble FasL TCR cross-linking with Abs	Fas/FasL negative; induction of Fas/FasL by TCR cross-linking with Abs or with 10 n*M* PMA plus 500 n*M* A23187 Ca^{2+} ionophore	150 ng/ml antihuman Fas mAb CH11 and 1 μg cycloheximide		
Dexamethasone		10 n*M*–0.1 μ*M*	62 μ*M*			
Staurosporine	0.5–1 μ*M*	1–10 μ*M*		1 μ*M*	1–3 μ*M*	0.5–10 μ*M*
H_2O_2	50–500 μ*M*		30–40 μ*M*		150–200 μ*M*	
Ceramide	10–50 μ*M*				100 μ*M*	
TNF-α	100 ng/ml			40 or 30 ng/ml plus cycloheximide 10 μg/ml	50 ng/ml (for differentiated PC12 cells)	
2-Chloroadenosine						100 μ*M*
Tamoxifen	10 μ*M*			10 μ*M*		
Doxorubicin	10 μ*M*					
Daunorubicin	1–10 μ*M*					
Etoposide		3–10 μ*M*				
Etoposide/ cyclophosphamide				50–100 μ*M*/ 1.0–12.0 μg/ml		
Cisplatin	5 μ*M*	10 μ*M*			>10 μg/ml	
Melphalan	3 μ*M*					
5-Fluorouracil	30 μ*M*					
Ara-C	1–10 μ*M*				10 μ*M*	
Vincristin	0.01–10 μ*M*	10 μ*M*				
Genistein	30 μ*M*			0.15 m*M*		
UV irradiation	300 J/m^2				500–800 J/m^2	
Gamma irradiation	100 Gy		500 Gy			

[a]Blank spaces in table mean that working data are not available.

concentration range of 200 μg/ml to 2 mg/ml for either of them. The antibiotic is added to the medium 48 h after transfection and the incubation time is 3–4 days.

These cells can undergo apoptosis in response to a wide variety of well-defined physiological, pharmacological, and pathological triggers as shown in Table V, such as deprivation of growth factors (Borgatti *et al.,* 1997), anti-Fas monoclonal antibodies (Dhein *et al.,* 1995; Johnson *et al.,* 1999), staurosporine (Johnson *et al.,* 1999; Samali *et al.,* 1999), hydrogen peroxide (Stridh *et al.,* 1998), ceramide (Manna *et al.,* 2000), tumor necrosis factor (TNF)-α (Porn-Ares *et al.,* 1998; Manna *et al.,* 2000), tamoxifen (Ferlini *et al.,* 1999), doxorubicin (da Silva *et al.,* 1996), daunorubicin (da Silva *et al.,* 1996), etoposide (Boesen-de Cock *et al.,* 1999), cisplatin (Guchelaar *et al.,* 1998), melphalan (Vashishtha *et al.,* 1998), 5-fluorouracil (Guchelaar *et al.,* 1998), Ara-C (Guchelaar *et al.,* 1998), vincristin (da Silva *et al.,* 1996), genistein (Spinozzi *et al.,* 1994), and UV and gamma irradiation (McIlroy *et al.,* 1999; Chen *et al.,* 1996).

III. 2B4.11

2B4 is a mouse lymphoid cell line model for the study of apoptosis *in vitro*. It was generated by fusing lymph node T cells from pigeon cytochrome *c* (PPC)-immunized B10.A mice with BW5147 thymoma cells (Hedrick *et al.,* 1982). It was chosen for further study because it secreted IL-2 in response to PCC presented by B10.A splenocytes. It is now known to recognize the C-terminal fragment of PCC (81–104) presented by the I-E^k MHC class II molecule (Kelner *et al.,* 1993). Like all T-cell hybridomas, 2B4 has a tendency to lose expression of the T-cell antigen receptor.

2B4.11 is a widely used subclone that maintains TCR expression relatively well, although it still tends to be lost over time. Although initially positive for CD4, 2B4.11 (and most subclones of 2B4) is CD4 negative. 2B4.11 cells have a round/raquet-shaped morphology and produce IL-2 after activation by PPC, anti-TCR antibodies, mitogens (concanavalin A, 2 μg/ml) (Agata *et al.,* 1966), or PMA (20 ng/ml) plus ionomycin (1 μg/ml).

2B4.11 cells are nonadherent cells and grow in suspension at 37°C, 100% humidity, and 5% CO_2 using RPMI 1640 with 10% FCS, 250 μg/ml gentamicin, 100 U/ml penicillin, 4 m*M* glutamine, and 5 μ*M* 2-mercaptoethanol as a growth medium. The optimum seeding density for this cell line is 10^5 cells/ml and the doubling time is 12–14 h. The medium should be renewed every other day and the cells can be subcultured every 3–4 days in a split ratio of 1 : 10. For long-term storage, 10^6 cells/ml are suspended in freezing medium (90% FCS with 10% DMSO) using cryovials, kept in a container of isopropanol at −80°C for 24 h, and stored in liquid nitrogen tanks. 2B4.11 cells are tumorigenic and should be handled in a biosafety level 1 facility. These cells are quite fragile and will die if cell density is too high (>10^6 cells/ml). They also are rather finicky about the fetal calf sera used for their propagation, and prospective lots should be screened for their ability to support both low- and high-density growth.

2B4.11 cells can be transiently or stably transfected by electroporation and retrovirus-mediated methods. The DEAE-dextran-mediated method can be applied in transient transfection experiments (Memon *et al.,* 1995). The most common antibiotic used for selection of the transfected clones are G418 (neomycin) at a concentration range of 200 μg/ml to 2 mg/ml. The antibiotic is added to the medium 36 h after transfection and the incubation time is 3–4 days.

Apoptosis can be induced in these cells in response to a wide variety of well-defined physiological and pharmacological triggers as shown in Table V. Activation-induced T-cell growth inhibition and apoptosis were first demonstrated with 2B4.11 cells (Ashwell *et al.,* 1987; Mercep *et al.,* 1988, 1989) and is due to the upregulation of Fas ligand; unlike some other murine T-cell hybridomas, 2B4 cells constitutively express Fas (Yang *et al.,* 1995). Therefore, stimulation via Fas readily initiates cell death in this cell line. Note that the commonly used Jo2 antimurine Fas antibody does not kill 2B4.11 cells, but actually blocks apoptosis induced by the physiologic ligand (FasL) present on the activated T-cell line d11S (Yang *et al.,* 1995). These cells are also very sensitive to glucocorticoid-induced apoptosis (Zacharchuk *et al.,* 1990) and therefore are excellent models for the study of dexamethasone-mediated cell death pathways. Chemotherapeutic agents such as etoposide (Sarin *et al.,* 1995), vincristin, and cisplatin have also been used for the induction of apoptosis in this cell line.

IV. DO11.10

DO11.10 is a mouse T-cell hybridoma derived from a fusion between BALB/c splenocytes and the thymic leukemic cell line BW51479 (White *et al.,* 1983). Prior to fusion with BW5147, BALC/c splenocytes were stimulated to divide with ovalbumin. As a result, some of the T-cell hybrids derived from this fusion have TCRs with defined antigen specificity. DO11.10 is specific for the 323–339 peptide from chicken ovalbumin. Subsequently, DO11.10 cells were used as immunogens to raise monoclonal antibodies specific for the DO11.10 TCR. One such monoclonal antibody (F23.1) is very useful in stimulating apoptosis in DO11.10 cells (Haskins *et al.,* 1983). These cells have a round morphology and produce no known cytokines unless stimulated. They are Fas or FasL negative unless stimulated by TCR cross-linking with antibodies against the TCR (Pratt *et al.,* 1999) or with 10 n*M* PMA and 500 n*M* A23187 Ca^{2+} ionophore or 1 μ*M* ionomycin.

DO11.10 are nonadherent cells and grow in suspension at 37°C, 100% humidity, and 5% CO_2 using DME with 10% FCS or horse serum and 100 U/ml penicillin–streptomycin as a growth medium. The optimum seeding density for this cell line is 5×10^4 to 10^5 cells/ml and the doubling time is 12–16 h. The medium should be renewed every 2–3 days and the cells can be subcultured every 3–4 days in a split ratio of 1 : 5 to 1 : 10. These cells grow well but are hardy and will become resistant to apoptosis over time if not split regularly. For long-term storage, 10^6 cells/ml are suspended in freezing medium (85% FCS with 15% DMSO) using cryovials, kept in a container of isopropanol at −80°C for 24 h, and stored in liquid nitrogen tanks. DO11.10 cells can be handled in a biosafety level 1 facility.

DO11.10 cells are poorly transfected, and most transfection methods are only marginally effective with these cells. Retrovirus-mediated transfection is the method of choice for this cell line and works for both transient and stable transfections. G418 (neomycin) is used for the selection of transfected clones at a concentration range of 400–800 μg/ml. The antibiotic is added to the medium 48 h after transfection and the incubation time is 4–8 days.

Induction of cell death in DO11.10 can be triggered easily via a variety of methods as shown in Table V, such as TCR cross-linking (Pratt *et al.,* 1999), anti-Fas monoclonal antibodies after induction of Fas with activation, glucocorticoids (Verhoven *et al.,* 1999), hydrogen peroxide, and gamma irradiation. The main problem with these cells is that they tend to become resistant to various cell death induction strategies. Because of this tendency, it is recommended that clones be selected that die readily in response to a variety of death inducers in order to ensure a constant supply of apoptosis-competent cells.

V. MCF-7

MCF-7 is a typical cell line of epithelial origin and a standard model for the study of breast cancer. It was derived from the pleural effusion of a Caucasian female patient with adenocarcinoma of the mammary gland (Soule *et al.,* 1973). These cells show epithelial morphology and express estrogen receptors. They also produce the insulin-like growth factor-binding proteins IGFBP-2, IGFBP-4, and IGFBP-5, the secretion of which can be modulated by antiestrogen treatment (Pratt and Pollak, 1993). The MCF7 line presents several features of differentiated mammary epithelium (Bacus *et al.,* 1990). It can process estradiol via cytoplasmic estrogen receptors and form domes. MCF-7 cells are deficient in caspase-3 expression due to a 47-bp deletion in exon 3 (Jänicke *et al.,* 1998).

MCF7 cells are adherent and grow well on plates or flasks at 37°C, 100% humidity, and 5% CO_2 using RPMI 1640 with 5% fetal bovine serum (FBS) and 100 U/ml penicillin–streptomycin as a growth medium. Their growth is inhibited by TNF-α (Sugarman *et al.,* 1985). The optimum seeding density for this cell line is 5×10^4 cells/ml and the doubling time is 25–48 h. The medium should be renewed every 2–3 days and the cells can be subcultured every 6–8 days in a split ratio of 1 : 3 to 1 : 6 using fresh 0.25% trypsin with 0.03% EDTA to facilitate the dissociation of the cells from the surface of the plate. For long-term storage, 10^6 cells/ml are suspended in freezing medium (80% FBS with 20% DMSO) using cryovials, kept in a container of isopropanol at −80°C for 24 h, and stored in liquid nitrogen tanks. MCF-7 cells are tumorigenic and should be handled in a biosafety level 1 facility.

Adenovirus-mediated transfection (Katayose *et al.,* 1995; Seth *et al.,* 1997) is the method of choice for both transient and stable transfections. Electroporation works efficiently (Sreerama and Sladek, 1995). Conventional methods such as calcium phosphate (Manni *et al.,* 1995)- and liposome-mediated transfection (Jain and Gewirtz, 1998) can also be applied. G418 (neomycin) is used for the selection of transfected clones at a concentration of 500 μg/ml. The antibiotic is added to the medium 48 h after transfection and the incubation time is at least 1 week. The transfected clones are picked up by mechanical selection under the microscope using pipettors and sterile yellow tips.

Cell death can be induced in these cells in response to different conditions and treatments as shown in Table V, such as serum deprivation (Xie *et al.,* 1999; Oberhammer *et al.,* 1993), anti-Fas monoclonal antibodies (Bertin *et al.,* 1997), staurosporine (Hirota *et al.,* 1999), TNF-α (Hirota *et al.,* 1999; Bertin *et al.,* 1997), tamoxifen (Bursch *et al.,* 2000; Zhang *et al.,* 1999), etoposide/cyclophosphamide (Gibson *et al.,* 1999), doxorubicin (Ruiz-Ruiz and Lopez-Rivas, 1999), genistein (Constantinou *et al.,* 1998), 5-fluorouracil, and cisplatin (Geier *et al.,* 1995).

VI. PC12

PC12 is a rat pheochromocytoma cell line of adrenal gland origin and is considered the model system for studies of neuronal differentiation, function, and death as an alternative to cultured neurons (Greene and Tischler, 1976). In serum-containing medium, PC12 cells exhibit a round to polygonal phase-bright morphology and proliferate to high density forming small tight clusters loosely adherent onto the dish. Under these conditions, these cells display many features of immature adrenal chromaffin cells and sympathicoblasts. PC12 cells respond reversibly to nerve growth factor (NGF) treatment by ceasing division and acquiring a mature sympathetic neuronal phenotype. NGF-induced responses are better at high cell density. In response to physiological levels of NGF (50–100 ng/ml), these cells extend long branching neurites, become electrically excitable, and adhere even more loosely on the plastic plate.

PC12 cells synthesize and release catecholamines, norepinephrine, dopamine (Shafer and Atchison, 1993), and IL-1β (Schwartz *et al.,* 1994). They express NGF receptors and do not synthesize epinephrine (Tischler and Greene, 1978).

PC12 cells grow well on plates at 37°C, 100% humidity, and 7.5% CO_2 using RPMI 1640 with 10% horse serum, 5% FBS, and 100 U/ml penicillin–streptomycin as a growth medium. The horse serum should be heat inactivated at 56°C for 30 min before use. Because PC12 are poorly adherent cells, they need air-dried rat tail collagen or polylysine as a substrate for better adherence. Long-term culture on plastic will select for more adherent subpopulation with a poorer NGF response. The optimum seeding density for this cell line is $>10^6$ cells/ml and the doubling time is 2.5 to 4 days. The medium should be renewed every 2–3 days and the cells can be subcultured every 7–10 days when cultures are 80–90% confluent in a split ratio of 1 : 3 to 1 : 4. The cells can be detached from the surface of the plate mechanically by squirting culture medium on them. For long-term storage, $>5 \times 10^6$ cells/ml are suspended in freezing medium (full serum growth medium with 10% DMSO) using cryovials, kept in a container of isopropanol at −80°C in a freezer for 24 h, and stored in liquid nitrogen tanks (Greene *et al.,* 1998).

For the induction of neuronal phenotype, use RPMI 1640 with 1% horse serum and 100 units/ml penicillin–streptomycin supplemented with 50–100 ng/ml NGF. Low serum is recommended to reduce cell clumping. Neurite-bearing cells can be observed within 1–3 days of NGF treatment, and by 7–10 days of treatment almost 90% of the cells should generate neurite extensions (Teng *et al.,* 1998). PC12 cells are tumorigenic and should be handled in a biosafety level 1 facility.

PC12 cells can be transfected efficiently using electroporation (Akamatsu *et al.,* 1999), retrovirus (Castellon and Mirkin, 2000), adenovirus (Millecamps *et al.,* 1999; Shinoura *et al.*, 2000), liposome- or calcium phosphate-mediated methods (Muller *et al.*, 1990). Any of them can be employed for either stable or transient transfections. Retrovirus- and adenovirus-mediated transfections are more efficient using vectors at the highest possible titers. G418 (neomycin) is used for the selection of transfected clones at a concentration of 500 μg/ml. The antibiotic is added to the medium 48 h after transfection and the incubation time is at least 1 week. The transfected clones are picked up by mechanical selection under the microscope using pipettors and sterile yellow tips.

Cell death can be induced in these cells in response to a variety of biochemical and pharmacological agents as described in Table V, including staurosporine (Ivins *et al.*, 1999), hydrogen peroxide (Ivins *et al.*, 1999), ceramide (Hartfield *et al.*, 1997), TNF-α (Mielke *et al.*, 2000), cisplatin (Lindenboim *et al.*, 1998), Ara-C (Park *et al.*, 1998), UV irradiation (Park *et al.*, 1998; Mielke *et al.*, 2000), and serum starvation (Maroney *et al.*, 1999; Le-Niculescu *et al.*, 1999). Typically, 50% of these cells die by 24 h of serum deprivation and 90% by 3–4 days if cultured in the absence of trophic support (Teng *et al.*, 1998). Cell death rates can be assessed by a number of methods, although the MTT assay (McGahon *et al.*, 1995) should be used with caution given that NGF enhances MTT signals independent of cell death.

VII. C_2C_{12}

The C_2C_{12} cell line was derived from the C2 mouse (C3H strain) muscle cell line as a fast-fusing subclone and has the characteristics of a very proliferative muscle satellite cell (Yaffe and Saxel, 1977). It is considered the standard model for studies on skeletal muscle differentiation, function, and death (Yaffe and Saxel, 1977; Blau *et al.*, 1985; Hu *et al.*, 1999). These cells maintain a fibroblast-like, spindle-shaped (myoblast) phenotype and express a wide variety of adult muscle-specific markers during the differentiation process. Proliferating C_2C_{12} cells express MyoD and synthesize and secrete both TGF-β 3 (Lafyatis *et al.*, 1991) and insulin-like growth factor II (which functions as an autocrine factor) (Stewart *et al.*, 1996).

C_2C_{12} cells are adherent and grow well in plates or flasks at 37°C, 100% humidity, and 5 to 10% CO_2. The growth medium is Dulbecco's modified Eagle's medium (DMEM) supplemented with 10% FBS and 100 U/ml penicillin–streptomycin. The optimum seeding density for this cell line is 5×10^4 cells/ml and the doubling time is about 16 h. The medium should be renewed every 2–3 days and the cells can be subcultured every 3–4 days when cultures are 50–60% confluent in a split ratio of 1 : 20 using fresh 0.25% trypsin to facilitate detachment. It is important not to allow the culture to become confluent as cell–cell contact induces myotubes. Passing confluent cells will result in the selection of subpopulations of slower-fusing myoblasts. Passage of any C_2C_{12} stock for more than 2–3 weeks may render the cells unable to properly differentiate or form myotubes. Therefore, the regular freezing of short-term cultures is strongly recommended. For long-term storage, 10^6 cells/ml are suspended in freezing medium (92% FBS with 8% DMSO) using cryovials, kept in a container of isopropanol at −80°C for 24 h and subsequently stored in liquid nitrogen tanks.

Differentiation can be induced when the culture is ~80% confluent by changing the medium to DMEM with 2% horse serum or 0.1% FBS, 5 μg/ml insulin, 5 μg/ml transferrin, and 100 U/ml penicillin–streptomycin (Hu *et al.,* 1999). Both media induce the rapid differentiation of C_2C_{12} cells, and multinucleated contractile myotubes can be observed by 48–72 h. Treatment of C_2C_{12} cells with bone morphogenetic protein 2 (BMP-2) shifts the differentiation process from myoblastic to osteoblastic (Katagiri *et al.,* 1994). C_2C_{12} cells are not tumorigenic and can be handled in a biosafety level 1 facility.

C_2C_{12} cells can be transiently or stably transfected by electroporation (Tognarini and Villa-Moruzzi, 1998), liposome (Hu *et al.,* 1999), and calcium phosphate-mediated (Wechsler-Reya *et al.,* 1998) methods. Retrovirus-mediated transfection also works well for stable transfection (Fan *et al.,* 1999). G418 (neomycin) (500 μg/ml), puromycin (3 mg/ml), or hygromycin (100–800 μg/ml) can be used for the selection of transfected clones. The antibiotic is added to the medium 48 h after transfection and the incubation time is a few weeks for G418 and a few days for puromycin or hygromycin. The transfected clones are picked up by mechanical selection under the microscope using pipettors and sterile yellow tips.

Cell death can be induced in these cells in response to a variety of biochemical and pharmacological stimuli (Table V), such as deprivation of growth factors (Dominov *et al.,* 1998), staurosporine (Dominov *et al.,* 1998), and 2-chloroadenosine (Rufini *et al.,* 1997). It should be noted that the MTT assay should be used only with proliferating myoblasts and not with differentiated myotube cultures because the latter are very metabolically active and hydrolyze a disproportional amount of MTT.

References

Agata, Y., Kawasaki, A., Nishimura, H., Ishida, Y., Tsubata, T., Yagita, H., and Honjo, T. (1996). Expression of the PD-1 antigen on the surface of stimulated mouse T and B lymphocytes. *Int. Immunol.* **8,** 765–772.

Akamatsu, W., Okano, H. J., Osumi, N., Inoue, T., Nakamura, S., Sakakibara, S., Miura, M., Matsuo, N., Darnell, R. B., and Okano, H. (1999). Mammalian ELAV-like neuronal RNA-binding proteins HuB and HuC promote neuronal development in both the central and the peripheral nervous systems. *Proc. Natl. Acad. Sci. USA* **96,** 9885–9890.

Ashwell, J. D., Cunningham, R. E., Noguchi, P. D., and Hernandez, D. (1987). Cell growth cycle block of T cell hybridomas upon activation with antigen. *J. Exp. Med.* **165,** 173–194.

Bacus, S. S., Kiguchi, K., Chin, D., King, C. R., and Huberman, E. (1990). Differentiation of cultured human breast cancer cells (AU-565 and MCF-7) associated with loss of cell surface HER-2/neu antigen. *Mol. Carcinogen.* **3,** 350–362.

Bertin, J., Armstrong, R. C., Ottilie, S., Martin, D. A., Wang, Y., Banks, S., Wang, G. H., Senkevich, T. G., Alnemri, E. S., Moss, B., Lenardo, M. J., Tomaselli, K. J., and Cohen, J. I. (1997). Death effector domain-containing herpesvirus and poxvirus proteins inhibit both Fas- and TNFR1-induced apoptosis. *Proc. Natl. Acad. Sci. USA* **94,** 1172–1176.

Biocca, S., Cattaneo, A., and Calissano, P. (1983). A macromolecular structure favouring microtubule assembly in NGF-differentiated pheochromocytoma cells (PC12). *EMBO J.* **2,** 643–648.

Blau, H. M., Pavlath, G. K., Hardeman, E. C., Chiu, C. P., Silberstein, L., Webster, S. G., Miller, S. C., and Webster, C. (1985). Plasticity of the differentiated state. *Science* **230,** 758–766.

Boesen-de Cock, J. G., Tepper, A. D., de Vries, E., van Blitterswijk, W. J., and Borst, J. (1999). Common regulation of apoptosis signaling induced by CD95 and the DNA-damaging stimuli etoposide and gamma-radiation downstream from caspase-8 activation. *J. Biol. Chem.* **274,** 14255–14261.

Boonen, G. J., van Oirschot, B. A., van Diepen, A., Mackus, W. J., Verdonck, L. F., Rijksen, G., and Medema, R. H. (1999). Cyclin D3 regulates proliferation and apoptosis of leukemic T cell lines. *J. Biol. Chem.* **274,** 34676–34682.

Borgatti, P., Zauli, G., Colamussi, M. L., Gibellini, D., Previati, M., Cantley, L. L., and Capitani, S. (1997). Extracellular HIV-1 Tat protein activates phosphatidylinositol 3- and Akt/PKB kinases in $CD4^+$ T lymphoblastoid Jurkat cells. *Eur. J. Immunol.* **27,** 2805–2811.

Bursch, W., Hochegger, K., Torok, L., Marian, B., Ellinger, A., and Hermann, R. S. (2000). Autophagic and apoptotic types of programmed cell death exhibit different fates of cytoskeletal filaments. *J. Cell Sci.* **113,** 1189–1198.

Butscher, W. G., Powers, C., Olive, M., Vinson, C., and Gardner, K. (1998). Coordinate transactivation of the interleukin-2 CD28 response element by c-Rel and ATF-1/CREB2. *J. Biol. Chem.* **273,** 552–560.

Castellon, R., and Mirkin, B. L. (2000). Retroviral transfer of the beta-nerve growth factor gene into murine neuroectodermal tumor cells modulates cell proliferation rate, neurite formation, and NGF binding site expression. *J. Neurosci. Res.* **59,** 265–275.

Chen, Y. R., Wang, X., Templeton, D., Davis, R. J., and Tan, T. H. (1996). The role of c-Jun N-terminal kinase (JNK) in apoptosis induced by ultraviolet C and gamma radiation: Duration of JNK activation may determine cell death and proliferation. *J. Biol. Chem.* **271,** 31929–31936.

Clarke, R., Skaar, T., Baumann, K., Leonessa, F., James, M., Lippman, J., Thompson, E. W., Freter, C., and Brunner, N. (1994). Hormonal carcinogenesis in breast cancer: Cellular and molecular studies of malignant progression. *Breast Cancer Res. Treat.* **31,** 237–248.

Constantinou, A. I., Kamath, N., and Murley, J. S. (1998). Genistein inactivates bcl-2, delays the G2/M phase of the cell cycle, and induces apoptosis of human breast adenocarcinoma MCF-7 cells. *Eur. J. Cancer* **34,** 1927–1934.

da Silva, C. P., de Oliveira, C. R., da Conceicao, M., and de Lima, P. (1996). Apoptosis as a mechanism of cell death induced by different chemotherapeutic drugs in human leukemic T-lymphocytes. *Biochem. Pharmacol.* **51,** 1331–1340.

Demirhan, I., Hasselmayer, O., Chandra, A., Ehemann, M., and Chandra, P. (1998). Histone-mediated transfer and expression of the HIV-1 tat gene in Jurkat cells. *J. Hum. Virol.* **1,** 430–440.

Dhein, J., Walczak, H., Baumler, C., Debatin, K. M., and Krammer, P. H. (1995). Autocrine T-cell suicide mediated by APO-1. *Nature* **373,** 438–441.

Dominov, J. A., Dunn, J. J., and Miller, J. B. (1998). Bcl-2 expression identifies an early stage of myogenesis and promotes clonal expansion of muscle cells. *J. Cell Biol.* **142,** 537–544.

Fan, L., Owen, J. S., and Dickson, G. (1999). Construction and characterization of polycistronic retrovirus vectors for sustained and high-level co-expression of apolipoprotein A-I and lecithin-cholesterol acyltransferase. *Atherosclerosis* **147,** 139–145.

Ferlini, C., Scambia, G., Marone, M., Distefano, M., Gaggini, C., Ferrandina, G., Fattorossi, A., Isola, G., Benedetti Panici, P., and Mancuso, S. (1999). Tamoxifen induces oxidative stress and apoptosis in oestrogen receptor-negative human cancer cell lines. *Br. J. Cancer* **79,** 257–263.

Geier, A., Beery, R., Haimsohn, M., and Karasik, A. (1995). Insulin-like growth factor-1 inhibits cell death induced by anticancer drugs in the MCF-7 cells: Involvement of growth factors in drug resistance. *Cancer Invest.* **13,** 480–486.

Gibson, L. F., Fortney, J., Magro, G., Ericson, S. G., Lynch, J. P., and Landreth, K. S. (1999). Regulation of BAX and BCL-2 expression in breast cancer cells by chemotherapy. *Breast Cancer Res. Treat.* **55,** 107–117.

Gillis, S., and Watson, J. (1980). Biochemical and biological characterization of lymphocyte regulatory molecules. V. Identification of an interleukin 2-producing human leukemia T cell line. *J. Exp. Med.* **152,** 1709–1719.

Greene, L. A., Farenelli, S. E., Cunningham, M. E., and Park, D. S. (1998). Culture and experimental use of the PC12 rat pheochromocytoma cell line. *In* "Culturing Nerve Cells" (G. Banker and K. Goslin, eds.), 2nd Ed., pp. 161–187. MIT Press, Cambridge, MA.

Greene, L. A., and Tischler, A. S. (1976). Establishment of a noradrenergic clonal line of rat adrenal pheochromocytoma cells which respond to nerve growth factor. *Proc. Natl. Acad. Sci. USA* **73,** 2424–2428.

Guchelaar, H. J., Vermes, I., Koopmans, R. P., Reutelingsperger, C. P., and Haanen, C. (1998). Apoptosis- and necrosis-inducing potential of cladribine, cytarabine, cisplatin, and 5-fluorouracil in vitro: A quantitative pharmacodynamic model. *Cancer Chemother. Pharmacol.* **42,** 77–83.

Hartfield, P. J., Mayne, G. C., and Murray, A.W. (1997). Ceramide induces apoptosis in PC12 cells. *FEBS Lett.* **401,** 148–152.

Haskins, K., Kubo, R., White, J., Pigeon, M., Kappler, J., and Marrack, P. (1983). The major histocompatibility complex-restricted antigen receptor on T cells. I. Isolation with a monoclonal antibody. *J. Exp. Med.* **157,** 1149–1169.

Hedrick, S. M., Matis, L. A., Hecht, T. T., Samelson, L. E., Longo, D. L., Heber-Katz, E., and Schwartz, R. H. (1982). The fine specificity of antigen and Ia determinant recognition by T cell hybridoma clones specific for pigeon cytochrome *c*. *Cell* **30,** 141–152.

Hirota, J., Furuichi, T., and Mikoshiba, K. (1999). Inositol 1,4,5-trisphosphate receptor type 1 is a substrate for caspase-3 and is cleaved during apoptosis in a caspase-3-dependent manner. *J. Biol. Chem.* **274,** 34433–34437.

Hu, Y., Cascone, P. J., Cheng, L., Sun, D., Nambu, J. R., and Schwartz, L. M. (1999). Lepidopteran DALP, and its mammalian ortholog HIC-5, function as negative regulators of muscle differentiation. *Proc. Natl. Acad. Sci. USA* **96,** 10218–10223.

Ivins, K. J., Ivins, J. K., Sharp, J. P., and Cotman, C. W. (1999). Multiple pathways of apoptosis in PC12 cells. CrmA inhibits apoptosis induced by beta-amyloid. *J. Biol. Chem.* **274,** 2107–2112.

Jain, P. T., and Gewirtz, D. A. (1998). Estradiol enhances gene delivery to human breast tumor cells. *J. Mol. Med.* **76,** 709–714.

Jänicke, R. U., Sprengart, M. L., Wati, M. R., and Porter, A. G. (1998). Caspase-3 is required for DNA fragmentation and morphological changes associated with apoptosis. *J. Biol. Chem.* **273,** 9357–9360.

Johnson, V. L., Cooper, I. R., Jenkins, J. R., and Chow, S. C. (1999). Effects of differential overexpression of Bcl-2 on apoptosis, proliferation, and telomerase activity in Jurkat T cells. *Exp. Cell Res.* **251,** 175–184.

Karas, M., Zaks, T. Z., Liu, J. L., and LeRoith, D. (1999). T cell receptor-induced activation and apoptosis in cycling human T cells occur throughout the cell cycle. *Mol. Biol. Cell* **10,** 4441–4450.

Katagiri, T., Yamaguchi, A., Komaki, M., Abe, E., Takahashi, N., Ikeda, T., Rosen, V., Wozney, J. M., Fujisawa-Sehara, A., and Suda, T. (1994). Bone morphogenetic protein-2 converts the differentiation pathway of C2C12 myoblasts into the osteoblast lineage. *J. Cell Biol.* **127,** 1755–1766.

Katayose, D., Gudas, J., Nguyen, H., Srivastava, S., Cowan, K. H., and Seth, P. (1995). Cytotoxic effects of adenovirus-mediated wild-type p53 protein expression in normal and tumor mammary epithelial cells. *Clin. Cancer Res.* **1**(8), 889–897.

Kelner, G. S., Jenkins, M. K., and Jemmerson, R. (1993). A single amino acid substitution in a cytochrome *c* T cell stimulatory peptide changes the MHC restriction element from one isotype (I-Ak) to another (I-Ek). *Mol. Immunol.* **30,** 569–575.

Kern, F. G., McLeskey, S. W., Zhang, L., Kurebayashi, J., Liu, Y., Ding, I. Y., Kharbanda, S., Chen, D., Miller, D., Cullen, K., and (1994). Transfected MCF-7 cells as a model for breast-cancer progression. *Breast Cancer Res. Treat.* **31,** 153–165.

Lafyatis, R., Lechleider, R., Roberts, A. B., and Sporn, M. B. (1991). Secretion and transcriptional regulation of transforming growth factor-beta 3 during myogenesis. *Mol. Cell. Biol.* **11,** 3795–3803.

Le-Niculescu, H., Bonfoco, E., Kasuya, Y., Claret, F. X., Green, D. R., and Karin, M. (1999). Withdrawal of survival factors results in activation of the JNK pathway in neuronal cells leading to Fas ligand induction and cell death. *Mol. Cell. Biol.* **19,** 751–763.

Li, Y. Q., Hii, C. S., Der, C. J., and Ferrante, A. (1999). Direct evidence that ERK regulates the production/secretion of interleukin-2 in PHA/PMA-stimulated T lymphocytes. *Immunology* **96,** 524–528.

Lindenboim, L., Haviv, R., and Stein, R. (1998). Bcl-xL inhibits different apoptotic pathways in rat PC12 cells. *Neurosci. Lett.* **253,** 37–40.

Manna, S. K., Sah, N. K., and Aggarwal, B. B. (2000). Protein tyrosine kinase p56lck is required for ceramide-induced but not tumor necrosis factor-induced activation of NF-kappa B, AP-1, JNK, and apoptosis. *J. Biol. Chem.* **275,** 13297–13306.

Manni, A., Wechter, R., Grove, R., Wei, L., Martel, J., and Demers, L. (1995). Polyamine profiles and growth properties of ornithine decarboxylase overexpressing MCF-7 breast cancer cells in culture. *Breast Cancer Res. Treat.* **34,** 45–53.

Maroney, A. C., Finn, J. P., Bozyczko-Coyne, D., O'Kane, T. M., Neff, N. T., Tolkovsky, A. M., Park, D. S., Yan, C. Y., Troy, C. M., and Greene, L. A. (1999). CEP-1347 (KT7515), an inhibitor of JNK activation, rescues sympathetic neurons and neuronally differentiated PC12 cells from death evoked by three distinct insults. *J. Neurochem.* **73**(5), 1901–1912.

McGahon, A. J., Martin, S. J., Bissonnette, R. P., Mahboubi, A., Shi, Y., Mogil, R. J., Nishioka, W. K., and Green, D. R. (1995). The end of the (cell) line: Methods for the study of apoptosis *in vitro*. *Methods Cell Biol.* **46,** 153–185.

McIlroy, D., Sakahira, H., Talanian, R. V., and Nagata, S. (1999). Involvement of caspase 3-activated DNase in internucleosomal DNA cleavage induced by diverse apoptotic stimuli. *Oncogene* **18,** 4401–4408.

Memon, S. A., Petrak, D., Moreno, M. B., and Zacharchuk, C. M. (1995). A simple assay for examining the effect of transiently expressed genes on programmed cell death. *J. Immunol. Methods* **180,** 15–24.

Mercep, M., Bluestone, J. A., Noguchi, P. D., and Ashwell, J. D. (1988). Inhibition of transformed T cell growth in vitro by monoclonal antibodies directed against distinct activating molecules. *J. Immunol.* **140,** 324–335.

Mercep, M., Noguchi, P. D., and Ashwell, J. D. (1989). The cell cycle block and lysis of an activated T cell hybridoma are distinct processes with different Ca^{2+} requirements and sensitivity to cyclosporine A. *J. Immunol.* **142,** 4085–4092.

Mielke, K., Damm, A., Yang, D. D., and Herdegen, T. (2000). Selective expression of JNK isoforms and stress-specific JNK activity in different neural cell lines. *Brain Res. Mol. Brain Res.* **75**(1), 128–137.

Millecamps, S., Kiefer, H., Navarro, V., Geoffroy, M. C., Robert, J. J., Finiels, F., Mallet, J., and Barkats, M. (1999). Neuron-restrictive silencer elements mediate neuron specificity of adenoviral gene expression. *Nat. Biotechnol.* **17,** 865–869.

Mills, J. C., Wang, S., Erecinska, M., and Pittman, R. N. (1995). Use of cultured neurons and neuronal cell lines to study morphological, biochemical, and molecular changes occurring in cell death. *Methods Cell Biol.* **46,** 217–242.

Muller, S. R., Sullivan, P. D., Clegg, D. O., and Feinstein, S. C. (1990). Efficient transfection and expression of heterologous genes in PC12 cells. *DNA Cell Biol.* **9,** 221–229.

Northrop, J. P., Crabtree, G. R., and Mattila, P. S. (1992). Negative regulation of interleukin 2 transcription by the glucocorticoid receptor. *J. Exp. Med.* **175,** 1235–1245.

Oberhammer, F., Wilson, J. W., Dive, C., Morris, I. D., Hickman, J. A., Wakeling, A. E., Walker, P. R., and Sikorska, M. (1993). Apoptotic death in epithelial cells: Cleavage of DNA to 300 and/or 50 kb fragments prior to or in the absence of internucleosomal fragmentation. *EMBO J.* **12,** 3679–3684.

Park, D. S., Morris, E. J., Stefanis, L., Troy, C. M., Shelanski, M. L., Geller, H. M., and Greene, L. A. (1998). Multiple pathways of neuronal death induced by DNA-damaging agents, NGF deprivation, and oxidative stress. *J. Neurosci.* **18,** 830–840.

Pimentel-Muinos, F. X., and Seed, B. (1999). Regulated commitment of TNF receptor signaling: A molecular switch for death or activation. *Immunity* **11,** 783–793.

Porn-Ares, M. I., Samali, A., and Orrenius, S. (1998). Cleavage of the calpain inhibitor, calpastatin, during apoptosis. *Cell Death Differ.* **5,** 1028–1033.

Pratt, J. C., van den Brink, M. R., Igras, V. E., Walk, S. F., Ravichandran, K. S., and Burakoff, S. J. (1999). Requirement for Shc in TCR-mediated activation of a T cell hybridoma. *J. Immunol.* **163**(5), 2586–2591.

Pratt, S. E., and Pollak, M. N. (1993). Estrogen and antiestrogen modulation of MCF7 human breast cancer cell proliferation is associated with specific alterations in accumulation of insulin-like growth factor-binding proteins in conditioned media. *Cancer Res.* **53,** 5193–5198.

Roosnek, E. E., Brouwer, M. C., and Aarden, L. A. (1985). T cell triggering by lectins. I. Requirements for interleukin 2 production; lectin concentration determines the accessory cell dependency. *Eur. J. Immunol.* **15,** 652–656.

Rufini, S., Rainaldi, G., Abbracchio, M. P., Fiorentini, C., Capri, M., Franceschi, C., and Malorni, W. (1997). Actin cytoskeleton as a target for 2-chloro adenosine: Evidence for induction of apoptosis in C2C12 myoblastic cells. *Biochem. Biophys. Res. Commun.* **238,** 361–366.

Ruiz-Ruiz, M. C., and Lopez-Rivas, A. (1999). p53-mediated up-regulation of CD95 is not involved in genotoxic drug-induced apoptosis of human breast tumor cells. *Cell Death Differ.* **6,** 271–280.

Samali, A., Cai, J., Zhivotovsky, B., Jones, D. P., and Orrenius, S. (1999). Presence of a pre-apoptotic complex of pro-caspase-3, Hsp60 and Hsp10 in the mitochondrial fraction of jurkat cells. *EMBO J.* **18,** 2040–2048.

Sarin, A., Nakajima, H., and Henkart, P. A. (1995). A protease-dependent TCR-induced death pathway in mature lymphocytes. *J. Immunol.* **154,** 5806–5812.

Sato, T., Hanada, M., Bodrug, S., Irie, S., Iwama, N., Boise, L. H., Thompson, C. B., Golemis, E., Fong, L., Wang, H. G., and (1994). Interactions among members of the Bcl-2 protein family analyzed with a yeast two-hybrid system. *Proc. Natl. Acad. Sci. USA* **91,** 9238–9242.

Schwartz, L. M., and Osborne, B. A. (1995). Cell death. *Methods Cell Biol.* **46,** xv–xviii.

Schwartz, M., Sivron, T., Eitan, S., Hirschberg, D. L., Lotan, M., and Elman-Faber, A. (1994). Cytokines and cytokine-related substances regulating glial cell response to injury of the central nervous system. *Prog. Brain Res.* **103,** 331–341.

Seth, P., Katayose, D., Li, Z., Kim, M., Wersto, R., Craig, C., Shanmugam, N., Ohri, E., Mudahar, B., Rakkar, A. N., Kodali, P., and Cowan, K. (1997). A recombinant adenovirus expressing wild type p53 induces apoptosis in drug-resistant human breast cancer cells: A gene therapy approach for drug-resistant cancers. *Cancer Gene Ther.* **4,** 383–390.

Shafer, T. J., and Atchison, W. D. (1991). Transmitter, ion channel and receptor properties of pheochromocytoma (PC12) cells: A model for neurotoxicological studies. *Neurotoxicology* **12,** 473–492.

Shinoura, N., Satou, R., Yoshida, Y., Asai, A., Kirino, T., and Hamada, H. (2000). Adenovirus-mediated transfer of Bcl-X(L) protects neuronal cells from Bax-induced apoptosis. *Exp. Cell Res.* **254,** 221–231.

Soule, H. D., Vazguez, J., Long, A., Albert, S., and Brennan, M. (1973). A human cell line from a pleural effusion derived from a breast carcinoma. *J. Natl. Cancer Inst.* **51,** 1409–1416.

Spinozzi, F., Pagliacci, M. C., Migliorati, G., Moraca, R., Grignani, F., Riccardi, C., and Nicoletti, I. (1994). The natural tyrosine kinase inhibitor genistein produces cell cycle arrest and apoptosis in Jurkat T-leukemia cells. *Leuk. Res.* **18,** 431–439.

Sreerama, L., and Sladek, N. E. (1995). Human breast adenocarcinoma MCF-7/0 cells electroporated with cytosolic class 3 aldehyde dehydrogenases obtained from tumor cells and a normal tissue exhibit differential sensitivity to mafosfamide. *Drug Metab. Dispos.* **23,** 1080–1084.

Stewart, C. E., and Rotwein, P. (1996). Insulin-like growth factor-II is an autocrine survival factor for differentiating myoblasts. *J. Biol. Chem.* **271,** 11330–11338.

Stridh, H., Kimland, M., Jones, D. P., Orrenius, S., and Hampton, M. B. (1998). Cytochrome *c* release and caspase activation in hydrogen peroxide- and tributyltin-induced apoptosis. *FEBS Lett.* **429,** 351–355.

Sugarman, B. J., Aggarwal, B. B., Hass, P. E., Figari, I. S., Palladino, M. A., Jr., and Shepard, H. M. (1985). Recombinant human tumor necrosis factor-alpha: Effects on proliferation of normal and transformed cells in vitro. *Science* **230,** 943–945.

Teng, K. K., Angelastro, J. M., Cunningham, M. E., Farinelli, S. E., and Greene, L. A. (1998). Cultured PC12 cells: A model for neuronal function, differentiation and survival. *In* "Cell Biology: A Laboratory Handbook" (J. E. Celis, ed.), pp. 244–250. Academic Press, Orlando, FL.

Tetsuka, T., Uranishi, H., Imai, H., Ono, T., Sonta, S., Takahashi, N., Asamitsu, K., and Okamoto, T. (2000). Inhibition of nuclear factor-kappaB-mediated transcription by association with the amino-terminal enhancer of split, a Groucho-related protein lacking WD40 repeats. *J. Biol. Chem.* **275,** 4383–4390.

Tischler, A. S., and Greene, L. A. (1978). Morphologic and cytochemical properties of a clonal line of rat adrenal pheochromocytoma cells which respond to nerve growth factor. *Lab. Invest.* **39,** 77–89.

Tognarini, M., and Villa-Moruzzi, E. (1998). Protein phosphatase 1 isoforms in differentiating C2C12 myocytes. *Eur. J. Cell Biol.* **76,** 212–219.

Vashishtha, S. C., Nazarali, A. J., and Dimmock, J. R. (1998). Application of fluorescence microscopy to measure apoptosis in Jurkat T cells after treatment with a new investigational anticancer agent (N. C.1213). *Cell Mol. Neurobiol.* **18,** 437–445.

Verhoven, B., Krahling, S., Schlegel, R. A., and Williamson, P. (1999). Regulation of phosphatidylserine exposure and phagocytosis of apoptotic T lymphocytes. *Cell Death Differ.* **6,** 262–270.

Wacholtz, M. C., and Lipsky, P. E. (1993). Anti-CD3-stimulated Ca^{2+} signal in individual human peripheral T cells: Activation correlates with a sustained increase in intracellular Ca^{2+} 1. *J. Immunol.* **150,** 5338–5349.

Wang, H. G., Takayama, S., Rapp, U. R., and Reed, J. C. (1996). Bcl-2 interacting protein, BAG-1, binds to and activates the kinase Raf-1. *Proc. Natl. Acad. Sci. USA* **93,** 7063–7068.

Wechsler-Reya, R. J., Elliott, K. J., and Prendergast, G. C. (1998). A role for the putative tumor suppressor Bin1 in muscle cell differentiation. *Mol. Cell. Biol.* **18,** 566–575.

Weiss, A., Wiskocil, R. L., and Stobo, J. D. (1984). The role of T3 surface molecules in the activation of human T cells: A two-stimulus requirement for IL-2 production reflects events occurring at a pre-translational level. *J. Immunol.* **133,** 123–128.

Welsh, J. (1994). Induction of apoptosis in breast cancer cells in response to vitamin D and antiestrogens. *Biochem. Cell. Biol.* **72,** 537–545.

White, J., Haskins, K. M., Marrack, P., and Kappler, J. (1983). Use of I region-restricted, antigen-specific T cell hybridomas to produce idiotypically specific anti-receptor antibodies. *J. Immunol.* **130,** 1033–1037.

Wiskocil, R., Weiss, A., Imboden, J., Kamin-Lewis, R., and Stobo, J. (1985). Activation of a human T cell line: A two-stimulus requirement in the pretranslational events involved in the coordinate expression of interleukin 2 and gamma-interferon genes. *J. Immunol.* **134,** 1599–1603.

Xie, S. P., Pirianov, G., and Colston, K. W. (1999). Vitamin D analogues suppress IGF-I signalling and promote apoptosis in breast cancer cells. *Eur. J. Cancer* **35,** 1717–1723.

Yaffe, D., and Saxel, O. (1977). Serial passaging and differentiation of myogenic cells isolated from dystrophic mouse muscle. *Nature* **270,** 725–727.

Yang, E., Zha, J., Jockel, J., Boise, L. H., Thompson, C. B., and Korsmeyer, S. J. (1995). Bad, a heterodimeric partner for Bcl-XL and Bcl-2, displaces Bax and promotes cell death. *Cell* **80,** 285–291.

Yang, Y., Mercep, M., Ware, C. F., and Ashwell, J. D. (1995). Fas and activation-induced Fas ligand mediate apoptosis of T cell hybridomas: Inhibition of Fas ligand expression by retinoic acid and glucocorticoids. *J. Exp. Med.* **181,** 1673–1682.

Zacharchuk, C. M., Mercep, M., Chakraborti, P., Simons, S. S., Jr., and Ashwell, J. D. (1990). Programmed T lymphocyte death: cell activation- and steroid-induced pathways are mutually antagonistic. *J. Immunol.* **145,** 4037–4045.

Zhang, G. J., Kimijima, I., Onda, M., Kanno, M., Sato, H., Watanabe, T., Tsuchiya, A., Abe, R., and Takenoshita, S. (1999). Tamoxifen-induced apoptosis in breast cancer cells relates to downregulation of bcl-2, but not bax and bcl-X(L), without alteration of p53 protein levels. *Clin. Cancer Res.* **5,** 2971–2977.

CHAPTER 19

Programmed Cell Death Assays for Plants

Alan M. Jones,[*] Silvia Coimbra,[†] Angelika Fath,[‡] Mariana Sottomayor,[†] and Howard Thomas[§]

[*] Department of Biology
The University of North Carolina at Chapel Hill
Chapel Hill, North Carolina 27599

[†] Instituto de Biologia Molecular e Celular
Universidade do Porto
4150-180 Porto, Portugal

[‡] Department of Plant and Microbial Biology
University of California
Berkeley, California 94720

[§] Cell Biology Department
Institute of Grassland and Environmental Research
Aberystwyth, Wales SY23 3EB

0091-679X/01 $35.00

I. Introduction

As in animals, programmed cell death (PCD) is a critical part of normal development at all stages of the plant life cycle (Jones and Dangl, 1996). The earliest signs of PCD occur when synergid and microspore cells die during gametogenesis (Bell, 1996). As the sporophyte develops, suspensor cell death occurs during normal embryogensis (Nagl, 1977) and aleurone cell death occurs as the seed germinates (Bethke *et al.,* 1999). Throughout development of the juvenile and adult phases of the plant, cell death occurs to sculpture plant morphology such as during lysigenous arenchyma formation of cortical root tissues (Kawai *et al.,* 1998) and to bring about normal histogenesis such as during terminal differentiation of vessel members and tracheids (Groover and Jones, 1999). Finally, senesence is the process by which nitogen, phosphate, and carbon are recycled from the leaves to the roots or developing fruit (Yen and Yang, 1998). This process culminates in cell death in the leaves and there is good evidence indicating that cell dismantling is a defined and orchestrated event.

The approach that plant scientists have taken to understand programmed cell death is very different from the approaches taken in the animal PCD field. Whereas *in vitro* culture has been adopted in animal PCD studies, plant scientists tend to study PCD *in vivo* using a wide range of species to investigate developmental or pathological cell death. Moreover, plant scientists are only now characterizing cell death during plant development or in a specific plant species. Consequently, while many examples of PCD in plants have been described, detailed biochemical or genetic pathways have yet to be defined. The basic protocols described here are meant to help newcomers enter the plant PCD field and are adapted easily by educators who teach cell biology.

II. DNA Fragmentation by Gel Electrophoresis

A. Isolation and Analysis of DNA for Critical Applications

DNA degradation is an important diagnostic factor in programmed cell death, and cleavage of DNA into internucleosomal fragments of 180 bp is considered a hallmark of apoptosis (Hale *et al.,* 1996). The ability to isolate undegraded high molecular weight DNA from viable tissue is a requirement for studying PCD-induced DNA degradation because it ensures that any DNA degradation observed is not an artifact of the isolation procedure.

The barley aleurone layer is a model system to study hormonally induced PCD in plants (Bethke *et al.,* 1999; Fath *et al.,* 1999). In this tissue, gibberellic acid (GA) induces PCD, whereas abscisic acid (ABA) prevents this program. A method for DNA isolation from barley aleurone cells is described here that minimizes DNA degradation during the

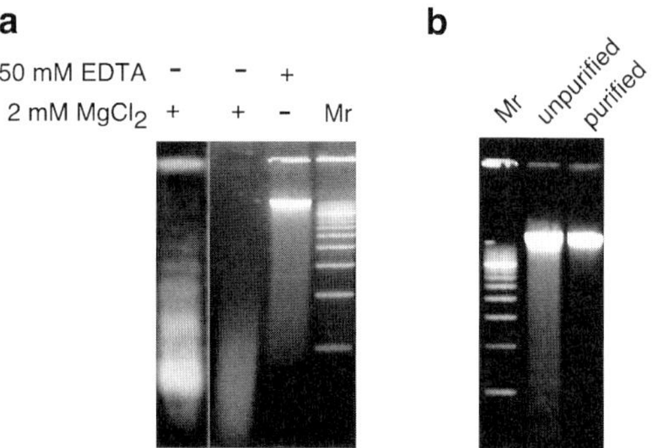

Fig. 1 Endogenous aleurone nucleases (a) and nucleases present in enzymes used for protoplast preparation (b) can cause extensive DNA degradation during isolation of DNA from aleurone cells. (a) DNA was isolated from aleurone layer homogenates that were incubated for 1 h at room temperature in buffer (0.1 *M* Tris–Cl, pH 7.5, and 0.5 *M* NaCl) containing 2 m*M* $MgCl_2$ and lacking 50 m*M* EDTA (independent samples shown in lanes 1 and 2) or lacking $MgCl_2$ and including 50 m*M* EDTA (lane 3). (b) DNA was isolated from freshly prepared protoplasts (unpurified) or from protoplasts purified over a Percoll density gradient (purified) that were frozen in liquid N_2 and thawed in lysis buffer A. DNA (2 μg/lane) was analyzed electrophoretically on a 1.5% agarose gel and then stained with ethidium bromide. DNA cleaved into fragments differing by 500 bp was used as a marker (Mr).

purification process. It has been shown to work well for high molecular weight DNA analysis in a variety of other tissue and cell preparations, e.g., *Arabidopsis thaliana* leaves.

In systems used to study cell death that require the preparation of protoplasts, extreme care must be followed because of the presence of contaminant nucleases in commercially available cell wall-degrading enzymes used for protoplast preparation (e.g., Onuzuka RS cellulase from Yakult Pharmaceutical Ind. Co., Tokyo, Japan). These enzymes hydrolyze aleurone DNA rapidly and can result in the formation of 180-bp DNA ladders, leading to the erroneous conclusion that barley aleurone cells die by apoptosis (Fig. 1b).

Nuclease activities are inhibited by the omission of $MgCl_2$ and inclusion of high amounts of EDTA (50 m*M*) in the homogenizing buffer. This is necessary to prevent endogenous nuclease activities from interfering with the isolation of high molecular weight DNA (Fig. 1a). In experiments where protoplasts are used to study PCD, the protoplasts must be purified from contaminating nucleases by centrifugation on a Percoll density gradient (Fig. 1b). The composition of the homogenizing buffer, the temperature at which cells are homogenized, and the incubation time in lysis buffer are critical parameters for the isolation of high molecular weight DNA (Fig. 2). Buffer composition is described later.

B. Protoplasts

1. Isolation of Protoplasts

Barley aleurone protoplasts are isolated from 500 quarter grains using the method of Lin *et al.* (1996). This method incorporates an extensive wash step that utilizes

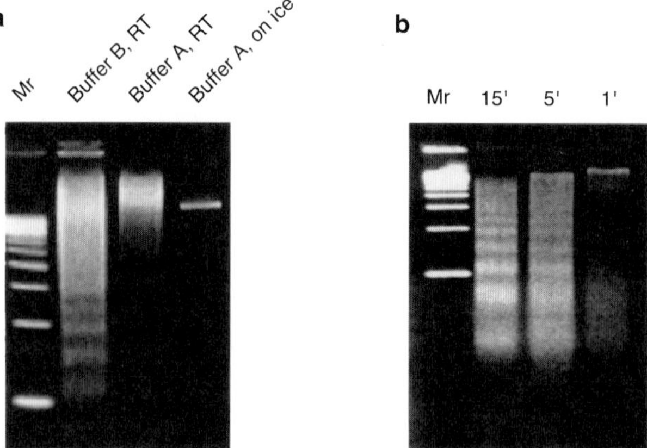

Fig. 2 The importance of lysis buffer composition, incubation time in lysis buffer, and incubation temperature is critical for the isolation of high molecular weight DNA. (a) The effect of buffer composition and temperature and (b) the effect of incubation time. DNA was isolated from freshly prepared, Percoll-purified protoplasts that were frozen in liquid N_2. (a) Frozen protoplasts were thawed in homogenizing buffer containing buffer B at room temperature, in buffer A at room temperature, or in buffer A on ice and incubated for an additional minute at the indicated temperature before DNA was extracted. (b) Frozen protoplasts were thawed in buffer A at room temperature and incubated for 15, 5, or 1 min at room temperature before DNA was extracted. DNA (2 μg/lane) was loaded on a 1.5% agarose gel. DNA cleaved into fragments differing by 500 bp was used as a marker (Mr).

centrifugation on the Percoll density gradient (step gradient of 5 ml 80%, 10 ml 60%, 7.5 ml 40%, and 7.5 ml 20% Percoll in Gamborg B5 medium) immediately after the release of protoplasts (10 ml suspension) from the testa/pericarp. The 40–60% layer of the Percoll density gradient containing living protoplasts is collected, diluted with 20 ml of Gamborg B5 medium (Gamborg *et al.,* 1968) added in steps of 5 ml every 2 min, mixed gently, and centrifuged for 2 min at 50*g*. The protoplast pellet is washed two more times with Gamborg B5 medium and resuspended in Gamborg B5 at 1×10^6 protoplasts/ml.

2. Isolation of DNA from Protoplasts

1. DNA is isolated from freshly prepared, Percoll-purified protoplasts that are frozen in liquid nitrogen. The frozen barley aleurone protoplasts are thawed on ice in 750 μl of buffer A, pH 7.5, and subsequently vortexed. Without further incubation in lysis buffer, NaOAc is added to a final concentration of 0.3 *M*.
2. The sample is mixed and then incubated for 20 min on ice.
3. After centrifugation for 1 min at 12,000*g* (4°C), the supernatant is transferred to a new tube.
4. An equal volume of ice-cold 100% isopropanol is added, mixed by inversion, incubated on ice for 1 min, and centrifuged for 5 min at 12,000*g* (4°C).

5. The supernatant is discarded and the pellet is resuspended in 250 μl TE buffer and 250 μl CTAB buffer and incubated for 15 min at 65°C.

6. DNA is extracted with an equal volume of chloroform and centrifuged for 5 min at 12,000*g* to separate the phases. The aqueous phase is transferred to a new tube and 2 volumes of ice-cold 100% EtOH are added.

7. DNA is allowed to precipitate for at least 20 min at −20°C and is centrifuged at 12,000*g* for 10 min. The DNA is washed in 70% EtOH, dried, and resuspended in 50 μl TE buffer supplemented with 1 mg/mL RNase A.

8. Undegraded or fragmented DNA is visualized on a 1.5% agarose gel.

C. Tissues

In general, DNA is isolated from barley aleurone layers using a protocol similar to that described for aleurone protoplasts except that 20 layers are first ground to a fine powder in liquid N_2. For the isolation of DNA from *A. thaliana* leaf tissue, 20 young leaves are frozen and ground to a fine powder in liquid N_2 and DNA is purified by carrying out the same DNA isolation procedure as described for barley aleurone protoplasts. DNA is quantified spectrophotometrically by standard techniques. The quality of the DNA is examined by loading the indicated amount of DNA on a 1.5% agarose gel followed by staining with 0.5 μg/ml (w/v, final concentration) ethidium bromide.

Using this method for DNA isolation, it has been shown that the DNA content of GA-treated barley aleurone cells is reduced before death, but DNA degradation does not result in the accumulation of low molecular weight internucleosomal DNA fragments (Bethke *et al.,* 1999; Fath *et al.,* 1999). The amount of DNA in ABA-treated aleurone cells does not change significantly over time (Bethke *et al.,* 1999; Fath *et al.,* 1999).

D. Tips and Hints

DNA should be isolated from freshly prepared, Percoll-purified protoplasts that are frozen in liquid N_2. DNA fragmentation is avoided when buffer A is used for DNA isolation. DNA degradation can be further diminished by thawing frozen protoplasts in buffer A on ice and incubating the mixture for an additional minute on ice rather than at room temperature before resuming the purification protocol. The typical buffer containing 5 m*M* Tris–Cl, pH 8.0, 20 m*M* EDTA, 0.5 *M* NaCl, and 0.5% Triton X-100 (buffer B) used to homogenize cells or tissue causes artifactual DNA fragmentation.

E. In-Gel Assay for Nuclease Activity

Nucleases are known to play an important role in DNA cleavage during programmed cell death. In barley aleurone cells, nucleases are induced by the plant hormone gibberellic acid (Fig. 3), and their appearance is correlated with a decrease in the amount of nuclear DNA in living aleurone cells prior to death (Fath *et al.,* 1999).

The in-gel assay for nuclease activities provides a quick and simple method for separating different nuclease activities and characterizing their action requirements and

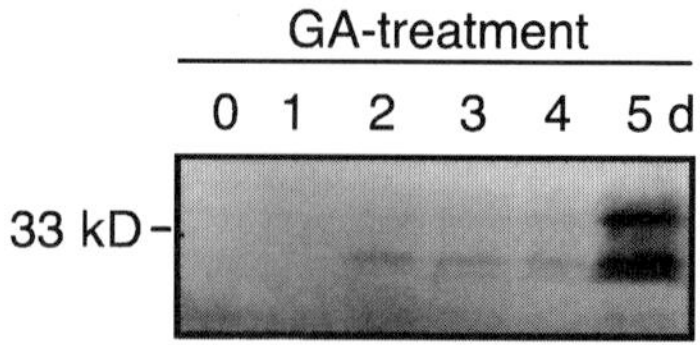

Fig. 3 GA treatment of barley aleurone protoplasts increases nuclease activities.The nuclease activity gel of protein extracts (5 μg protein/lane) from barley aleurone protoplasts was incubated in 25 μ*M* GA for the indicated times. Visualization of nuclease activity is as described in the text.

substrate specificities. The method used to determine nuclease activity in cells described here is essentially as described by Thelen and Northcote (1989) with modifications for protoplasts and aleurone tissue; however, the protocol can be used for a variety of other cell sources.

1. 12.5% SDS–PAGE gels are prepared containing single-stranded DNA (50 μg/ml boiled and snap-cooled, calf thymus DNA; Sigma, St. Louis, MO) and 50 μg/ml bovine fibrinogen (Sigma) in the resolving gel but not the stacking gel.
2. Aleurone layers (15–20) are ground to a fine powder in liquid N_2 and extracted in buffer C, and the homogenate is centrifuged for 5 min at 12,000*g* (4°C). Protoplasts are mixed with an equal volume of 4× SDS–PAGE loading buffer without reducing buffer and centrifuged. The supernatant is removed carefully and placed in clean tubes. The samples are not boiled prior to electrophoresis.
3. Samples of the supernatant are separated by SDS–PAGE at 70 V and 4°C using a Bio-Rad minigel system (Bio-Rad, Hercules, CA). Following electrophoresis, gels are washed twice for 30 min in buffer D, washed for 30 min in buffer D minus the isopropanol, and subsequently washed twice for 15 min in buffer D minus isopropanol and dithiothreitol (DTT). Gels are incubated overnight at 37°C in the same buffer with gentle agitation. Nuclease activity is detected by staining the gel with 1 μg/ml (w/v, final concentration) ethidium bromide for 15 min and destained for 30 min with 1 m*M* $MgCl_2$. Nuclease activities are visualized under ultraviolet light as clear, DNA-depleted bands in a fluorescent background.

F. Reagents

Buffer A: 0.1 *M* Tris–HCl, 50 m*M* EDTA, 0.5 *M* NaCl, and 1% SDS

Buffer B: 5 m*M* Tris–HCl, 20 m*M* EDTA, 0.5 M NaCl, and 0.5% Triton X-100, pH 7.5

Buffer C: 150 m*M* Tris–Cl, 0.5 m*M* phenylmethylsulfonyl fluoride (PMSF), and 20 μ*M* leupeptin, pH 6.8

Buffer D: 25% isopropanol, 10 m*M* MOPS, 1 m*M* $CaCl_2$, 1 m*M* $MgCl_2$, 1 μ*M* $ZnCl_2$, and 1 m*M* DTT, pH 6.0

TE buffer: 10 m*M* Tris–HCl and 1 m*M* EDTA, pH 8.0

CTAB buffer: 0.2 *M* Tris–Cl, 50 m*M* EDTA, 2 *M* NaCl, and 2% CTAB (ethyltrimethylammonium bromide; Aldrich, Milwaukee, WI), pH 7.5

III. TUNEL

A. Leaves, Roots, and Whole Seedlings

In leaves, one of the earliest responses to pathogens is the rapid accumulation of reactive oxygen intermediates, causing what is called an oxidative burst (Mehdy, 1994). Concomitantly with the oxidative burst, cells at the site of pathogen entry suffer a rapid death with collapse of the host tissue termed the hypersensitive response. This cell death seems to be carried out by an intrinsic genetic program that leads to the self-destruction of infected cells and should thus be viewed as an example of programmed cell death (Levine *et al.*, 1996; Pozo and Lam, 1999). In order to develop an easy protocol for the use of leaves for the TUNEL assay, *C. roseus* and *A. thaliana* leaves are treated with DNase to obtain labeled nuclei as described later. Leaves are cut in small pieces (approximately 2×5 mm) and fixed immediately. TUNEL nuclei of leaf positive controls are observed in Fig. 4 (see Color Plate).

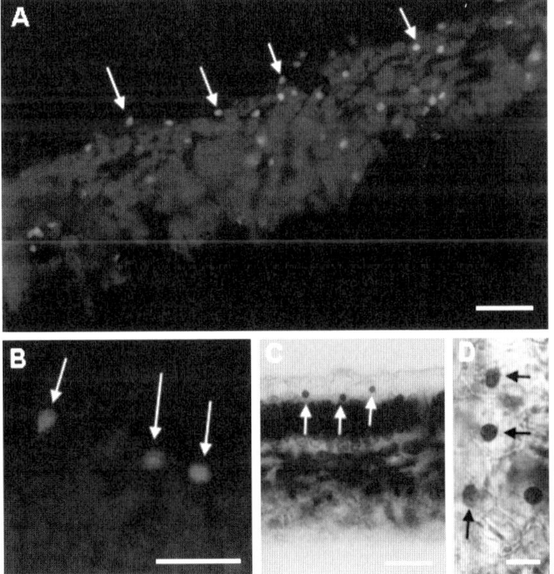

Fig. 4 TUNEL staining of *Catharanthus roseus* leaves treated as positive controls by incubation with DNase I. (A) TUNEL reaction detected by fluorescein labeling as observed by epifluorescent microscopy. Nuclei with cleaved DNA are stained bright green. Bar: 50 μm. (B) The same as A. Bar: 25 μm. (C) TUNEL reaction detected by peroxidase activity. Nuclei with cleaved DNA are stained dark brown. Bar: 50 μm. (D) The same as C. Bar: 12.5 μm. TUNEL-stained nuclei are indicated by arrows. (See Color Plate.)

Root cap cells are displaced continually to the root periphery by new cells. The peripheral cells die as a normal part of development and not as a consequence of abrasion during soil penetration. DNA staining of root cap cells shows that nuclei become condensed, and TUNEL experiments reveal the accumulation of 3′-OH groups in DNA, indicating that root cap cells die by PCD (Wang *et al.,* 1996). *Arabidopsis* seeds are germinated on filter paper in a petri dish and the whole plantlets are used for fixation and mounting on microscope slides. *Allium cepa* roots are allowed to grow from market onions and are used for TUNEL after 6 to 8 days of growth. The tips of the roots (4 mm) are fixed and used for the TUNEL assay.

The following TUNEL method has been optimized for roots and leaves using the *in situ* cell death detection kit manufactured by Roche Molecular Biochemical. Shown here are methods for fluorescent and nonfluorescent visualization of nuclei undergoing chromatin degradation. If detection is by fluorescence only, use kit catalog number 1-684-795 and, if by peroxidase, use kit catalog number 1 684 817. Additional buffer compositions are described.

1. Fix samples in 4% paraformaldehyde in phosphate-buffered saline (PBS) buffer, pH 7.4, in a small glass vial or microfuge tube for 30 min under gentle vacuum on ice. When tissue pieces sink, the fixative has infiltrated the tissue successfully. An alternative method is to fix under pressure by infiltrating tissue within a closed syringe barrel and placing pressure on the tissues by gently pushing the plunger until they sink.

2. Wash three times with PBS and, on the third wash, reinfiltrate under gentle vacuum as described in step 1.

3. Roots tips and other small tissues need not be sectioned. Hand section leaves into small pieces, approximately 2 × 5 mm, with a new sharp razor blade and put one leaf fragment on a microscope slide. Put one finger gently on top of the tissue, lean the blade on the finger next to one end of the tissue, and start cutting rapidly as you move the finger backwards.

4. Follow the instructions according to the *in situ* cell death kit from Roche Molecular Biochemical, starting with the permeabilization solution for 2 min on ice. Thin sections may not need any additional treatment to facilitate passage of the reagents through the cell wall, but thicker sections may require a 30-min treatment at room temperature in 2% β-glucanase (InterSpex Products Inc., Foster City, CA) in 0.05 *M* acetate buffer, pH 4.5, containing 0.1 *M* NaCl. Other cell wall hydrolases may also work, but it is very important that they are proven not to have significant DNase activity (see Section II). If a hydrolase treatment is applied, rinse sections in PBS by removing the drop of enzyme solution with a lint-free absorbent tissue and adding a drop of PBS.

5. Wash three times with PBS and dry area around tissue by absorption with a lint-free tissue.

6. Prepare a *positive control* slide by incubating the sample in 50 μl DNase I (250 U/ ml^{-1} in assay buffer) for 10 min on a glass slide at room temperature and wash three times with PBS. Keep slides in a humid chamber such as a sealed petri dish containing wet filter paper.

7. Wash all slides three times with TdT buffer and dry area around sample.

8. Prepare *negative controls* by omitting TdT (step 9).

9. Prepare TUNEL reaction mixture by adding the total volume (50 μl) of tube 1 (TdT) to the remainder of tube 2 (450 μl of label solution); 500 μl total for 10–15 microscope slides. This can be scaled down for fewer samples.

10. Apply 35–50 μl of reaction mixture to all samples except negative controls.

11. Incubate in a humid chamber for 1 h, in darkness, at 37°C.

12. Wash three times in PBS under low light conditions.

Note: For fluorescence visualization of TUNEL nuclei, observe tissue by fluorescence microscopy: excitation 470 nm and emission 510 nm. Keep slides in darkness.

13. Incubate with blocking solution, 3–5% H_2O_2 in methanol for 5 min at room temperature.

14. Wash twice with PBS for 5 min each.

15. Incubate with PBS buffer containing 2% bovine serum albumin (BSA), 0.3% Triton X-100 for 10 min at room temperature.

16. Wash three times with PBS and dry area around tissue.

17. Add 50 μl converter-POD (this is the antiserum to fluorescein conjugated to peroxidase supplied in the kit) and incubate in a humidified chamber for 30 min at 37°C.

18. Wash three times in PBS. Dry area around tissue.

19. Add 50 μl of DAB substrate solution supplied in the kit and incubate for 10 min at room temperature.

20. Wash three times in PBS.

21. Mount with mounting medium (Sigma Diagnostics, Cat. No. 1000-4) under a glass coverslip and analyze under a light microscope.

B. Tips and Hints

The sections may also be stained for DNA with 1 μg/ml 4′,6-diamidino-2-phenylindole (DAPI, Sigma) in 0.02 *M* sodium phosphate buffer, pH 7.2, and viewed immediately. Nuclei will appear blue when examined under UV epifluorescence microscopy. Be very careful to handle the sections as little as possible. Physical damage to the DNA, and subsequent artifactual TUNEL staining of nuclei, is very easy to induce, particularly in epidermal cells. For a negative control, omit the enzyme solution from the TUNEL reaction mixture (step 8). For a positive control, pretreat sections with micrococcal nuclease or DNase 1 as suggested in the Roche Molecular Biochemical kit instructions (step 6).

C. Cell Cultures

This method was adapted specifically for TUNEL analysis of tracheary elements (Groover and Jones, 1999) but has been shown to work equally well with tobacco BY2 cells. There are a number of precautions described that differ from the protocol for tissues.

The procedure has also been optimized to reduce high background signals and save on kit reagents. The protocol is based on the cell death detection kit (Cat No. 1 684 795, Roche Molecular Biochemicals). Reactions can be performed on cells in microfuge tubes or in multiwell plates.

1. Remove 0.35 to 1.0 ml suspension culture cells to microfuge tubes or multiwell plates. Gently spin cells in a clinical centrifuge at 1000*g* for 2 min at room temperature and draw off the supernatant carefully but completely. Cell loss can occur here.

2. To cells, add 100–200 μl fixative (see Section III,D) and mix by gentle shaking. Incubate for 30 min at room temperature, shaking occasionally. Samples then can be stored at 4°C for several days.

3. Add cell culture medium to cells, spin fixed cells down at 1000*g* for 2 to 3 min, and draw off supernatant.

4. Add 0.35 to 1 ml cell culture medium to wash the fixed cells again, spin cells at 1000*g* for 2–3 min, and draw off the supernatant.

5. Add 200 μl permeabilization solution and incubate for 5 min on ice (4°C).

6. Fill microfuge tube or the well with PBS, spin down the cells, and discard the supernatant carefully.

7. Repeat step 6. This time, remove as much of the supernatant as possible. [At this point, for a positive control, add DNase I (250 U/ml^{-1}) in assay buffer and incubate for 5 min at room tempratue, fill the tube or well with PBS, spin down the cells, and remove the supernatant.]

8. Remove 100 μl label solution for two negative controls, prepare TUNEL reaction mixture as described in the Roche Molecular Biochemical kit instructions, and mix well to equilibrate components.

Note: the TUNEL reaction mixture should be prepared immediately before use and kept on ice until use.

9. Add 37.5 to 50 μl reaction mixture to each sample and incubate for 1 h at 37°C.

10. In order to reduce the background, wash the cells once in PBS. Resuspend the cells in 50–100 μl PBS to continue the assay.

11. View cells under a fluorescence microscope. Zinnia or BY2 cells undergoing PCD contain fragmented DNA, which show green fluorescence (excitation: 470 nm, emission: 510 nm). Nuclei can be counterstained with 1 μg/ml DAPI or propidium iodide.

12. To calculate the percentage of TUNEL-stained nuclei, count the number of cells in light and then count the cells with green fluorescent nuclei under a fluorescence microscope in the same field.

D. Reagents

Cell culture media: for zinnia cells, make according to Roberts *et al.* (1992) and for BY2 cells, make according to Link and Cosgrove (1998)

DAB substrate solution: 0.1% DAB and 0.003% H_2O_2; dissolve DAB in HCl 0.1 *M*, dilute with PBS, and correct final pH to 6–7

DNase assay buffer: 40 m*M* Tris–HCl and 6 m*M* $MgCl_2$, pH to 7.5

Fixative for cells: 150 μl formaldehyde solution (37%) and 850 μl cell culture medium

PBS: 137 m*M* NaCl, 2.7 m*M* KCl, 1.5 m*M* KH_2PO_4, and 8.0 m*M* Na_2HPO_4

Permeabilization solution: 0.1% Triton X-100 and 0.1% sodium citrate

TdT buffer: 25 m*M* Tris–HCl, 200 m*M* sodium cacodylate, and 5 m*M* cobalt chloride

IV. Nondestructive Measurements of Plant Cell Senescence

The term "senescence" was the original term for the orchestrated molecular dismantling of leaf cells recycling nutrients to storage or reproductive sinks. It does not have the same meaning in animal biology, where it refers to aging. The most obvious sign of senescence, and a marker of its progression as well, is the change in leaf color due to pigment degradation.

A. Visual Scoring

The wavelength sensitivities of the human eye are attuned to the reflectance spectrum of foliage (Osorio and Bossomaier, 1992). This means that the subjective scoring of "greenness" is frequently comparable in resolution with imaging or pigment extraction methods, and correlations of better than 0.9 are common.

Objectivity may be introduced by the use of standard color references (Fig. 5, see Color Plate). A widely used index is that based on the Munsell system. Color notation takes the form of numerical values for hue, value, and chroma and is conventionally expressed in the form H V/C. A typical mature nonsenescent leaf of soybean may have the formula 7.5 GY (5/5), whereas a senescing leaf may be 5.0 Y (8/8). The general principles of applying Munsell analysis to plant tissues were established by Wilde and Voigt (1952). Charts are available commercially (GretagMacbeth LLC) to allow these estimates to be made with some confidence.

B. Colorimetry

From simple point-and-shoot devices to sophisticated detection and data-processing instruments, equipment is available commercially that can function essentially as an electronic Munsell color book, e.g., the Minolta Color Reader and Chroma Meter series. Minolta also makes a hand-held leaf color meter, the SPAD 502, which has been used extensively in monitoring crop performance (Finnan *et al.*, 1997). This device reads light transmittance at two wavelengths and displays a single numerical value, which may be related to pigment content by calibration. A green, presenescent leaf typically gives a meter reading in excess of 50, whereas tissue at advanced senescence often scores less than 10 (Fig. 5). Calibration is necessary to turn a SPAD value into chlorophyll content.

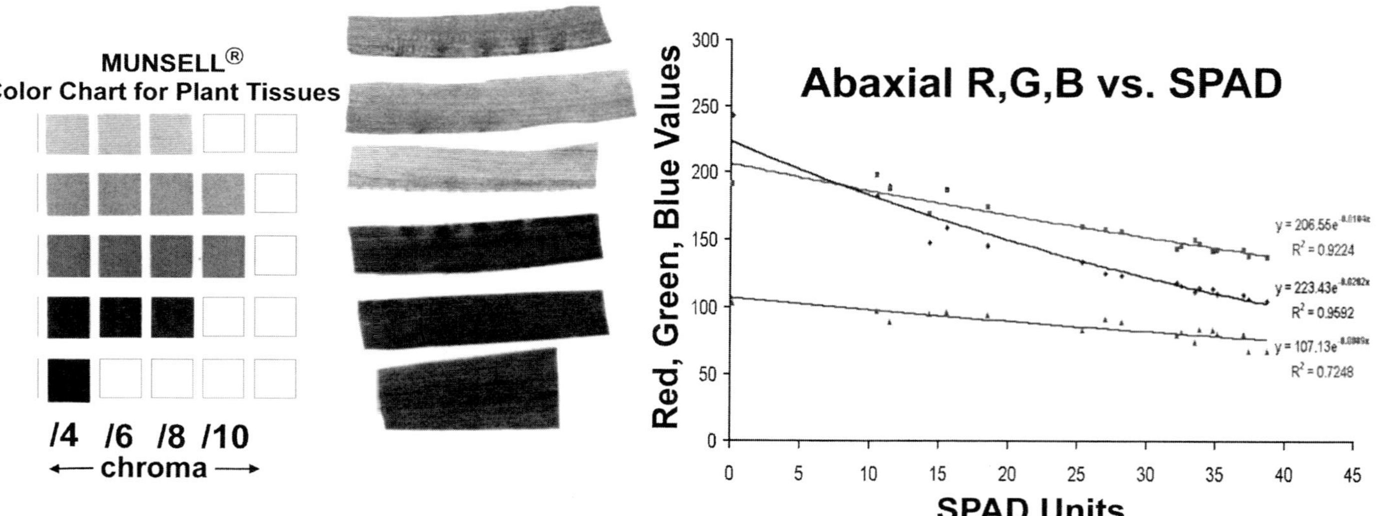

Fig. 5 Images of excised maize leaf tissues of different ages captured with a PC desk-top scanner. A Munsell color chart is included as a standard. The graph plots red, green, and blue values taken from scanned images of abaxial leaf surfaces against greenness as measured with a SPAD chlorophyll meter. (See Color Plate.)

C. Red, Green, and Blue Color Scanning

A simple, convenient, and high-resolution way of collecting and storing color information from leaves uses a flat-bed scanner attached to a PC. A suitable instrument is a UMAX PowerLook flat-bed scanner, operating in a reflective, true-color mode. Scan-to-scan variation can be corrected for by including an appropriate Munsell color chart (e.g., 5.0 GY). Scans taken at manufacturer's default gamma values and at a resolution of 400 dpi are saved as TIFF format images using the graphics package Adobe Photoshop 4.0 (Adobe, California). The Info palette in Photoshop can be used to determine red, green, and blue (RGB) values at 10 points selected randomly over each leaf or leaf segment, and the values are averaged. A large number of commercial and user-written software packages are available that automate and increase the accuracy of this manual method. Generally, the highest correlation is between R value and chlorophyll content (Fig. 5), although some studies have used statistical optimization to derive equations combining R, G, and B parameters (e.g., Kawashima and Nakatani, 1998). The user will need to carry out a calibration for each new species or growth condition.

D. Spectrometry and Hyperspectral Imaging

Spectrometers and digital cameras capable of capturing reflectance spectra or images are increasing in availability, affordability, and resolution and offer the possibility of determining subtle qualitative as well as quantitative changes in pigmentation. Hyperspectral imaging refers to the resolution of each pixel of an image into a large number of (commonly 256) color gradations, spanning the spectral range as determined by the digital capture device. For example, an IMSPECTOR imaging spectrograph (Spectral Imaging Ltd) runs from a PC and allows simultaneous determination of the reflectance spectrum in the range of 400–700 nm at each point along a selected area of tissue. For a discussion of the relationship between reflectance spectra and pigment changes during senescence, see Merzlyak *et al.* (1999). Difference spectra allow qualitative comparisons to be made, whereas statistical procedures such as principal component analysis (PCA) can resolve specific wavelength or wavelength combinations that relate directly to changes in pigment amounts. Hyperspectal imaging and PCA has allowed the nondestructive resolution of blockage points in the pathway of pigment degradation in senescence mutants (Ougham *et al.*, 1999). A digital imager attached to a microscope enables this approach to be applied to individual cells, protoplasts, and protophyta.

E. Chlorophyll Fluorescence

Pigment content is related in a general way to photosynthetic capacity and hence to the functional state during senescence and death of green cells. Nondestructive methods based on chlorophyll *a* fluorescence can give additional information on the functional state of the photosynthetic apparatus. The simplest and most widely used index is F_v/F_m, the ratio of maximal to variable fluorescence of chlorophyll *a* in leaves that have been previously dark adapted. F_v/F_m is also a direct measure of the maximum quantum efficiency of electron transport through photosystem II (PS II; Genty *et al.*, 1989). Tissue is generally held in the dark for at least 30 min to drain electrons from the acceptor side of PS II (all electron acceptors oxidized). Then, a time course of fluorescence emission following excitation first with a weak measuring light beam and immediately afterward with a pulse of intense white light, delivered by fiber optic, is measured by a photodetector. The light pulse must be of sufficient intensity (in the order of 15,000 μmol m^{-2} s^{-1}) to saturate all the PSII electron acceptors. F_m is the maximal level of chlorophyll *a* fluorescence after the saturating pulse (all acceptors reduced), F_o is the steady-state level of chlorophyll *a* fluorescence (in the presence of the measuring light), and $F_v = F_m - F_o$. Typically, presenescent tissue will give F_v/F_m values in the region of 0.8, whereas the figure for senescing or dying cells may be 0.2 or less. Suitable equipment for making these measurements includes the PAM Fluorometer series (Waltz, Effeltrich, Germany), the PK Morgan CF system (Andover, MA), and the Hansatech Plant Efficiency Analyser (Hansatech Ltd., Kings Lynn, UK).

Acknowledgments

These protocols were optimized in the laboratories of the authors using protocols that were published or developed in the author's laboratories. Thanks to Helen Ougham and Alison Kingston-Smith for their help with the pigment section. A. M. Jones thanks the National Science Foundation, Developmental Mechanisms Program for funding his work on PCD during tracheary element differentiation. A. Fath's work on PCD in aleurone degradation was supported by grants from the National Science Foundation and Novartis Agricultural Discovery Institute to R. L. Jones (University of California). The authors are very grateful for *in situ* cell death kits provided by Ms. Lisa Douglas of Roche Molecular Biochemicals (Chapel Hill, NC).

References

Bell, P. R. (1996). Megaspore abortion: A consequence of selective abortion? *Int. J. Plant Sci.* **157,** 1–7.

Bethke, P. C., Lonsdale, J. E., Fath, A., and Jones, R. L. (1999). Hormonally regulated programmed cell death in barley aleurone cells. *Plant Cell* **11,** 1033–1046.

Fath, A., Bethke, P. C., and Jones, R. L. (1999). Barley aleurone cell death is not apoptotic: Characterization of nuclease activities and DNA degradation. *Plant J.* **20,** 305–316.

Finnan, J. M., Burke, J. I., and Jones, M. B. (1997). A note on a non-destructive method of chlorophyll determination in wheat (*Triticum aestivum* L). *Irish J. Agric. Food Res.* **36,** 85–89.

Gamborg, O. L., Miller, R. A., and Ojima, K. (1968). Nutrient requirements of suspension cultures of soybean root cells. *Exp. Cell Res.* **50,** 151–158.

Genty, B., Briantais, J.-M., and Baker, N. R. (1989). The relationship between the quantum yield of photosynthetic electron transport and quenching of chlorophyll fluorescence. *Biochim. Biophys. Acta* **990,** 87–92.

Groover, A., and Jones, A. M. (1999). Programmed cell death in transdifferentiating tracheary elements is triggered by a secreted protease. *Plant Physiol.* **119,** 375–384.

Hale, A. J., Smith, C. A., Sutherland, L. C., Stoneman, V. E. A., Longthome, V. L., Culhane, A. C., and Williams, G. T. (1996). Apoptosis: Molecular regulation of cell death. *Eur. J. Biochem.* **236,** 1–26.

Jones, A. M., and Dangl, J. D. (1996). Logjam at the styx: The multiplicity of programmed cell death in plants. *Trends Plant Sci.* **1,** 114–119.

Kawai, M., Samaarajeewa, P. K., Barrero, R. A., Nishiguchi, M., and Uchimiya, H. (1998). Cellular dissection of the degradation pattern of cortical cell death during aerenchyma formation in rice roots. *Planta* **204,** 277–287.

Kawashima, S., and Nakatani, M. (1998). An algorithm for estimating chlorophyll content in leaves using a video camera. *Ann. Bot.* **81,** 49–54.

Levine, A., Pennell, R. I., Alvarez, M. E., Palmer, R., and Lamb, C. (1996). Calcium mediated apoptosis in a plant hypersensitive resistance response. *Curr. Biol.* **6,** 427–437.

Lin, W., Gopalakrishnan, B., and Muthukrishnan, S. (1996). Transcription efficiency and hormone-responsiveness of alpha-amylase gene promoters in an improved transient expression system from barley aleurone protoplasts. *Protoplasma* **192,** 93–108.

Link, B. M., and Cosgrove, D. J. (1998). Acid-growth response and a-expansins in suspension cultures of bright yellow 2 tobbacco. *Plant Physiol.* **118,** 907–916.

Mehdy, M. C. (1994). Active oxygen species in plant defence against pathogens. *Plant Physiol.* **105,** 467–472.

Merzlyak, M. N., Gitelson, A. A., Chivkunova, O. B., and Rakitin, V. Y. (1999). Non-destructive optical detection of pigment changes during leaf senesence and fruit ripening. *Physiol. Plant.* **106,** 135–141.

Nagl, W. (1977). Plastolysomes: Plastids involved in the autolysis of the embryo-suspensor in *Phaseolus*. *Z. Pflanzenphysiol.* **85,** 45–51.

Osorio, D., and Bossomaier, T. R. J. (1992). Human cone-pigment spectral sensitivities and the reflectances of natural surfaces. *Biol. Cyber.* **67,** 217–222.

Ougham, H., Alsberg, B., Rowland, J., Dickson, J., and Thomas, H. (1999). Determining plant pigment composition using imaging and multivariate analysis. Abstracts of XVI International Botanical congress, St. Louis, MO.

Pozo, O., and Lam, E. (1999). Caspases and programmed cell death in the hypersensitive response of plants to pathogens. *Curr. Biol.* **8,** 1129–1132.

Roberts, A. W., Koonce, L. T., and Haigler, C. H. (1992). A simplified medium for in vitro tracheary element differentiation in mesophyll suspension cultures from *Zinnia elegans* L. *Plant Cell Tissue Cult.* **28,** 27–35.

Schrauwen, J. A. M., Derks, M. W. M., Groot, P. F. M., Reijnen, W. H., Van Herpen, M. M. A., and Wullems, G. J. (1988). Differential gene-expression during microsporogenesis with *Nicotiana tabacum*. *In* "Sexual Reproduction in Higher Plants" (M. Cresti, P. Gori, and E. Pacini, eds.), pp. 3–7. Springer-Verlag, Berlin.

Thelen, M. P., and Northcote, D. H. (1989). Identification and purification of a nuclease from *Zinnia eleqans* L.: A potential molecular marker for xylogensis. *Planta* **179,** 181–195.

Wang, H., Li, J., Bostock, R. M., and Gilchrist, D. G. (1996). Apoptosis: A functional paradigm for programmed cell death induced by a host selective phytotoxin and invoked during development. *Plant Cell* **8,** 375–391.

Wilde, S. A., and Voigt, G. K. (1952). The determination of color of plant tissues by the use of standard charts. *Agron. J.* **44,** 499–500.

Yen, C. H., and Yang, C. H. (1998). Evidence for programmed cell death during leaf senescence in plants. *Plant Cell Physiol.* **39,** 922–927.

CHAPTER 20

Studies of Apoptosis Proteins in Yeast

Hong Zhang and John C. Reed

The Burnham Institute
La Jolla, California 92037

I. Introduction

Apoptosis is caused by the activation of caspase family cell death proteases (reviewed in Salvesen and Dixit, 1997; Thornberry and Lazebnik, 1998; Cryns and Yuan, 1999). Caspases are a family of cysteine proteases, which are synthesized as inactive proenzymes consisting of a N-terminal prodomain, followed by large and small catalytic subunits. These proenzymes undergo proteolytic processing at specific aspartic acid residues to generate active heterotetramic enzymes consisting of two large (~20 kDa) and two small (~10 kDa) catalytic subunits (Wallach *et al.*, 1997; Kang *et al.*, 1999; Hawkins *et al.*, 1999). Active caspases cleave substrates at aspartic acid residues within certain preferred sequence contexts. This specificity for aspartic acid thus allows caspases to cleave and activate themselves and each other, creating cascades of sequential protease activation (Thornberry and Lazebnik, 1998; Wallach *et al.*, 1997; Yuan, 1997; Ashkenazi and Dixit, 1998). Thus far, more than a dozen caspases have been identified in mammalian cells. Some of them, such as caspases-1, -2, -8, -9, and -10, contain long N-terminal prodomains that mediate interactions with other proteins that either facilitate or inhibit

METHODS IN CELL BIOLOGY, VOL. 66

0091-679X/01 $35.00

their activation. These caspases typically function at the apex of apoptotic pathways and are known as "initiator" caspases. In contrast, other caspases, such as caspases-3, -6, and -7, possess only short prodomains and are largely dependent on cleavage by upstream initiator caspases for their activation. These downstream "effector" caspases are thought to assume responsibility for most of the proteolytic cleavage of various substrate proteins, leading to apoptotic cell death (Salvesen and Dixit, 1997; Thornberry and Lazebnik, 1998; Cryns and Yuan, 1999).

The predominant mechanism for triggering the activation of upstream initiator caspases probably can be explained by the "induced proximity" model (Salvesen and Dixit, 1999). This model is predicated on the observation that the proenzyme forms of caspases actually are not completely inactive, possessing weak protease activity, at least in those cases where measured to date. Thus, when these zymogens are brought into close proximity through interacting adapter proteins, they can cleave each other, thereby generating the processed and fully active enzymes. The large N-terminal prodomains of initiator caspases thus participate in a variety of protein–protein interactions with other molecules, which form these protease-activation complexes ("apoptosomes"), accounting for their activation in response to specific cellular stimuli.

Several pathways for inducing caspase activation have been identified in mammalian cells, including (a) an "intrinsic" pathway in which mitochondria play a central role and (b) an "extrinsic" pathway in which cell surface receptors of the tumor necrosis factor (TNF) family transduce signals, resulting in activation of the cell death proteases. The mitochondrial pathway involves release from these organelles of cytochrome *c*. Once liberated into the cytosol, cytochrome *c* binds and activates a caspase-activating protein, Apaf-1. The Apaf-1 protein undergoes oligomerization in response to binding cytochrome *c* and exposes a protein interaction domain known as a caspase-associated recruitment domain (CARD) that binds specifically to a corresponding CARD found in the N-terminal prodomain of the initiator caspase, procaspase-9 (reviewed in Green and Kroemer, 1998). In the death receptor pathway for caspase activation, ligand binding to plasma membrane receptors of the TNF family induces a clustering of the corresponding cytosolic domains, which contain a protein interaction motif known as a death domain (DD). The aggregated DDs then recruit additional DD-containing adapter proteins, which contain additional domains, such as CARDs or death effector domains (DEDs), that allow them to recruit initiator caspases to the complex by interacting with similar domains located in their N-terminal prodomain region. For example, the death receptors TNFR1 and Fas bind the adapter protein Fadd directly or indirectly, which contains both a DD and a DED (reviewed in Wallach *et al.,* 1997; Yuan, 1997; Ashkenazi and Dixit, 1998). Fadd then binds via its DED to similar DEDs in the N-terminal prodomain of procaspases-8 and (possibly) procaspase-10, whereupon the closely approximated caspase zymogens are thought to trans-proteolyze each other within the death receptor complex at the membrane, subsequently releasing the active enzymes into the cytosol. Biochemical and genetic data indicate that the critical apical caspases in the extrinsic and intrinsic pathways are caspase-8 and caspase-9, respectively. Once activated, these proteases can then directly cleave and activate downstream members of the same family, particularly procaspase-3, thus bringing into action the downstream effector caspases that are ultimately responsible for the proteolytic cleavage of myriad protein substrates and generation of the apoptotic phenotype.

Each of these pathways for apoptosis induction is governed by multiple proteins that serve to either enhance or inhibit the caspase activation process. Not only do animal cells possess genes encoding various regulators of these pathways, but viruses and possibly other pathogens also harbor apoptosis-regulating genes within their genomes. Discovering the various apoptosis regulators in animal, viral, and possibly plant, fungal, and microbial genomes is currently a major goal of research in this field today.

Assays for function-based screens for genes (cDNAs) or drugs that modulate components of the apoptosis machinery can be created in yeast. Because homologues of the animal apoptosis system do not appear to exist in the budding yeast *Saccharomyces cerevisiae,* this organism provides an attractive system in which to interrogate the function of apoptosis proteins without interference from other family members. A variety of assays can be envisioned that monitor either the biochemical activity of apoptosis proteins, such as caspases, through various reporter methods or where ectopic expression of mammalian apoptosis proteins in yeast results in an easily scorable phenotype, such as cell death (Kang *et al.,* 1999). For example, overexpression of certain caspases in yeast results in either death or suppression of growth, probably through nonphysiological mechanisms, which nevertheless are exploitable as a phenotype for screens. Alternatively, cleavable reporter proteins can be engineered that permit the activity of caspases to be monitored in yeast in those cases where the particular caspase is nonlethal or nongrowth suppressive (Hawkins *et al.,* 1999).

In addition to caspases, some other types of regulators of the apoptosis machinery of animal cells exhibit screenable phenotypes when expressed ectopically in yeast. For example, Bcl-2 family proteins are central regulators of apoptosis in animal cells, which govern mitochondria-dependent steps in an evolutionarily conserved cell death pathway controlling the release of cytochrome *c* from these organelles (reviewed in Green and Reed, 1998; Adams and Cory, 1998; Reed, 1998; Kroemer and Reed, 2000). Bcl-2 family proteins regulate this pathway by either inducing (e.g., Bax) or repressing (e.g., Bcl-2) cytochrome *c* release, as well as affecting other aspects of mitochondrial physiology. These proteins share structural similarity with certain pore-forming proteins and thus possess an intrinsic biophysical function related to their capacity to create channels in membranes (reviewed in Schendel *et al.,* 1998). When ectopically expressed in yeast, the proapoptotic Bax protein confers a lethal phenotype. As in mammalian cells, the Bax protein targets mitochondria and induces cytochrome *c* release in yeast (Manon *et al.,* 1997). Furthermore, the antiapoptotic protein Bcl-2 negates the cytotoxic phenotype of Bax in both yeast and mammalian cells, without necessarily binding to Bax (Zha and Reed, 1997; Matsuyama *et al.,* 1998). The lethal phenotype of Bax in yeast has been exploited for structure–function analysis of Bax and its inhibitors, such as Bcl-2, as well as for discovery of novel genes that suppress Bax function in both yeast and animal cells (Xu and Reed, 1998; Matsuyama *et al.,* 1998; Zhang *et al.,* 2000).

This chapter describes methods developed and employed in our laboratory for the study of components of the apoptosis machinery, utilizing the budding yeast *S. cerevisiae*.

II. Methods

A. Plasmids for Expressing Apoptosis Proteins in Yeast

A variety of expression plasmids are available for expressing heterologous genes in *S. cerevisiae* in either a constitutive or a conditional manner. However, for proteins that confer lethal or growth-suppressive phenotypes when expressed in yeast, such as certain caspases (e.g., caspases-8 and -10), as well as certain proapoptotic Bcl-2 family proteins (e.g., Bax and Bak), it is preferable to employ inducible promoters for most experiments. Several conditional promoters have been employed successfully for expressing heterologous genes in *S. cerevisiae;* the most commonly used are the *GAL1* and *GAL10* promoters whose expression is repressed in the presence of glucose and derepressed in the presence of galactose. The particular plasmid we have employed for many experiments is YEp51, an episomal plasmid containing a yeast centromere and a *GAL10* promoter casette for driving galactose-inducible gene expression. We have generally not found it necessary to stably integrate plasmids for screens, including cDNA library screening, providing appropriate secondary screens are employed to exclude false positives (see later).

For some experiments, we have found it useful to simultaneously express in yeast an upstream initiator caspase, such as caspases-8 or -10, and a downstream effector caspase, such as capsase-3. In these cases, the cDNAs of the initiator caspases were cloned into YEp51, which drives expression from the *GAL10* promoter and contains a *LEU2* selectable marker, whereas the cDNA encoding a downstream effector caspase is introduced into the constitutively expressed yeast vector p423, which possesses the glyceraldehyde-3-phosphate dehydrogenase (GPD) promoter and the *HIS3* gene. Previous experiments have shown that the expression of caspases-8 or -10 from a constitutive vector, such as p423, results in few transformants due to cytotoxicity. In contrast, the constitutive expression of caspase-3 shows no effect on the number of transformants compared to control cells transformed with an "empty" vector.

Some proapoptotic molecules can be expressed as fusion proteins using standard yeast two-hybrid plasmids with satisfactory results (Matsuyama *et al.,* 1998; Sato *et al.,* 1994; Zha *et al.,* 1996). For example, the Bax protein can be expressed as a fusion with the DNA-binding domain (DBD) of LexA from two-hybrid plasmids such as pGilda, which is driven by the *GAL1* promoter, or pEG202, which uses a constitutive *ADH* promoter. The LexA DNA-binding domain produced via these plasmids lacks a nuclear targeting sequence and thus does not interfere with mitochondrial association of Bax. Indeed, the addition of LexA DBD to Bax seems to stabilize the protein, leading to higher levels of accumulation in yeast.

Several alternative yeast vectors containing various promoters and selectable markers are available from ATCC (American Type Culture Collection, Rockville, MD) or commerical sources.

B. Strains and Media

Laboratory *S. cerevisiae* strains are of diverse origins and thus strains from different sources can be distantly related with respect to genetic background. We have examined the cytotoxic effect of caspases-8 and -10, and Bax expression in a number of yeast strains. In all the strains tested, expression of these genes caused cell death. However, the efficiency of killing appeared to be strain dependent; some strains are killed more easily by these proteins, whereas some others exhibit some degree of resistance. Thus, it is advisable to test strains for sensitivity to the death-inducing genes prior to beginning experiments. Strains that produce acceptable results for caspase-mediated killing are BF264-15 Dau and EGY48. For Bax-induced killing, we prefer BF264-15 Dau, W303L, and EGY48.

Yeast cells containing auxotrophic mutations are used to select for transformed cells that have taken up successfully and that maintain episomally the expression plasmids described earlier. These cells are unable to grow on selective synthetic medium lacking certain nutrients, usually amino acids (leucine, histidine, tryptophan) or nucleotide precursors (uracil). Upon transformation with plasmids harboring "markers" that complement defects in biosynthetic pathways, successfully transformed cells grow and form colonies on semisolid medium. Four selectable markers are commonly employed in *S. cerevisiae* vectors: *LEU2, HIS3, TRP1,* and *URA3* genes.

We routinely maintain yeast cells on YPD medium. Once a caspase or Bax expression plasmid is transformed into yeast, cells should be cultured in synthetic dropout (SD) medium lacking the particular amino acid or nucleotide precursor necessary for plasmid selection. For example, yeast containing both YEp51-caspase-8 and YEp51-caspase-10 can be maintained in synthetic medium lacking leucine (SD-Leu), whereas those transformed with pGilda-Bax should be cultured in medium lacking histidine (SD-His), reflecting the presence in YEp51 and pGilda of a *LEU2* marker and a *HIS3* marker, respectively. In both instances, the SD medium is fortified with glucose to repress the *GAL1* promoter so that colonies of viable cells form on plates. Sometimes, the sensitivity of yeast to these proapoptotic factors can be increased by limiting the amount of nutrients in the medium, using, for example, Burkholder's minimal medium (Toh-e *et al.,* 1973).

The following media are used for assays for death-inducing activity of caspases and Bax in yeast. The preparation of yeast medium is described by Sherman *et al.* (1983).

YPD: 1% Bacto-yeast extract, 2% Bacto-peptone, 2% glucose (Dextrose)
SD Glu: 0.67% yeast nitrogen base without amino acid, 2% glucose
SD Gal: 0.67% yeast nitrogen base without amino acid, 2% galactose, 1% raffinose
SD Glu, -Leu: SD Glu lacking leucine
SD Gal, -Leu: SD Gal lacking leucine
SD Glu, -His: SD Glu lacking histidine
SD Glu, -Leu, His: SD Glu lacking leucine and histidine
SD Gal, -Leu, His: SD Gal lacking leucine and histidine
SD Glu, -Leu, His, Trp: SD Glu lacking leucine, histidine, and tryptophan
SD Gal, -Leu, His, Trp: SD Gal lacking leucine, histidine, and tryptophan

C. Introduction of Plasmids into Yeast

Plasmids are introduced into yeast by the LiOAc transformation procedure (Ito *et al.*, 1983). The transformation efficiency should be approximately 10^4 transformants/μg DNA.

1. Transformation

1. Inoculate a single colony into a 2-ml YPD culture in a round-bottom 15-ml polystyrene centrifuge tube with the cap loosened and grow overnight at 30°C with vigorous agitation.
2. Inoculate the 2-ml overnight culture into 200 ml YPD medium in a 1-liter Erylenmyer flask and continue to incubate at 30°C with vigorous agitation until OD_{600} reaches 0.5–1.0 (1 $OD_{600} = \sim 3 \times 10^7$).
3. Harvest cells by centrifugation in a swinging bucket rotor at 500 *g* for 5 min at room temperature.
4. Resuspend the cells in 50 ml distilled H_2O, mix thoroughly, and pellet by centrifugation at 500 *g* for 5 minutes at room temperature in a swinging bucket.
5. Resuspend the resulting cell pellet in 10 ml LiOAc solution (0.1 *M* LiOAc, 10 m*M* Tris–HCl, pH 8.0, and 1 m*M* EDTA), mix thoroughly, and pellet cells by centrifugation at 500 *g* for 5 minutes at room temperature.
6. Resuspend the cell pellet in about 2 ml LiOAc solution (0.1 ml LiOAc/10 ml culture).
7. Aliquot 0.1 ml into fresh polypropylene centrifuge tubes of any convenient size.
8. To each tube, add 10 μl freshly boiled carrier DNA (10 mg/ml sheared salmon sperm DNA, Sigma Chemical, St. Louis, MO) and 1 μg plasmid DNA [plasmid DNA should preferably be no more dilute than 0.1 μg/ml in TE buffer (10 m*M* Tris, pH 7.5, 1 m*M* EDTA)].
9. Add 0.7 ml PEG solution (40% PEG 4000, 0.1 *M* LiOAc, 10 m*M* Tris–HCl, pH 8.0, and 1 m*M* EDTA) to each tube. Vortex thoroughly.
10. Incubate the tubes at 30°C for 45 min with constant shaking.
11. Heat shock in a 42°C water bath for 10–15 min.
12. Pellet cells by centrifugation at 500 *g* for 5 min at room temperature.
13. Resuspend the cell pellet in 0.2 ml TE buffer and spread onto SD glucose plates lacking the specific amino acids or nucleotide precursors necessary to select for the introduced plasmids (e.g, SD-Glu, -Leu plates for YEp51-caspases-8 and YEp51-caspases-10; SD-Glu, -His plates for p423-caspase-3; and SD Glu, -Leu, His plates for YEp51-caspases-8 and YEp51-caspases-10 plus p423-caspase-3).
14. Incubate at 30°C. Transformants normally appear after 2–4 days.

D. Assessing Expression of Apoptosis Proteins in Yeast

The expression of caspases or Bcl-2 family proteins in yeast can be evaluated by immunoblotting. Alternatively, protease assays can be used to assess the activity of caspases

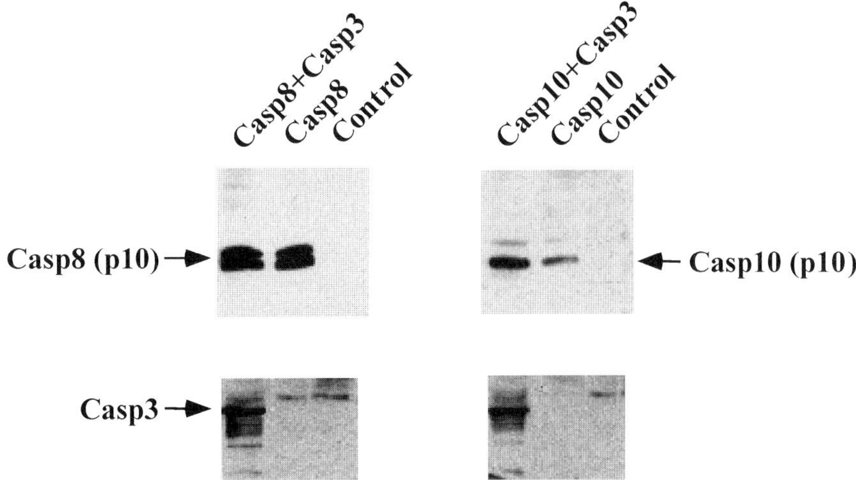

Fig. 1 Expression of caspases in transformed yeast. Either control vector or caspase encoding plasmids were transformed into yeast strain EGY48. Protein extracts were prepared from yeast transformants that contained control or caspase encoding plasmids. Cells were grown in glucose-containing medium and then transferred to galactose-containing medium for 12 h. Total protein extracts (20 μg) were subjected to SDS–PAGE and immunoblot analysis using corresponding anti-caspase antibodies.

in yeast cell lysates. Typically, transformed yeast are grown in liquid media (not plates) for expression studies. The medium in which cells are grown for expression studies depends, of course, on the specifics of the plasmids used, particularly whether the promoter is inducible versus constitutive. For most experiments we employ the *GAL1* promoter, which is galactose inducible, and thus the strategy entails first allowing the cells to grow in liquid culture in the presence of glucose (which represses the promoter) followed by switching to galactose-containing medium to induce expression. Cells are harvested for analysis before the cytotoxic effects of the induced proteins begin to kill cells. The optimal induction time depends on the yeast strain used and thus should be determined empirically. Typically, we induce the *GAL1* promoter for 12–24 h. Figure 1 shows an example of immunoblot analysis of Bax expression following the induction from a *GAL1* promoter in transformed EGY48 strain cells in galactose-containing medium for 12 h.

1. Preparation of Yeast Extracts

1. Inoculate a colony from each transformant plate into 1–2 ml SD glucose medium in a 15-ml polystyrene round-bottom centrifuge tube and grow at 30°C overnight with vigorous agitation with the cap loosened.

2. Expand the overnight culture into 10–20 ml SD glucose medium in a ~125-ml Erylenmyer flask and continue to grow with vigorous agitation at 30°C until $OD_{600} =$ 0.3–0.6.

3. Recover the yeast by centrifugation at 500 *g* for 10 min at room temperature using a swinging bucket rotor.

4. Aspirate the supernatant completely. Add 1 ml distilled H_2O to the pellet, resuspend the cell pellet thoroughly by pipetting up and down, and transfer to 1.5-ml polypropylene microcentrifuge tubes.

5. Pellet the cells by centrifugation in a microcentrifuge (Eppendorf-type with a fixed-angle rotor) at 10,000 *g* for 1 min at room temperature.

6. Repeat the wash with 1–1.5 ml of distilled H_2O two more times (the yeast must be completely free of glucose, as even a tiny amount of glucose will inhibit the galactose induction).

7. Resuspend the cell pellet in the same amount of galactose medium (10–20 ml).

8. Grow the yeast at 30°C for various times as desired (up to 12 h) in 125-ml Erylenmyer flasks with vigorous agitation.

9. Pipette 5–10 ml of the culture into centrifuge tubes and recover the yeast by centrifugation at 500 *g* in a swinging bucket rotor for 5 min at 4°C.

10. Resuspend the cell pellet in 1 ml distilled H_2O, mix well, and split into a 2-ml flat-bottom microcentrifuge tube. Recover cells by centrifugation at 10,000 *g* in a microcentrifuge for 1 min at 4°C. The size of the cell pellet should be approximately 50–100 μl.

11. Add an equal volume (50–100 μl typically) of yeast lysis buffer [100 m*M* NaCl, 50 m*M* Tris–HCl, pH 7.2, 1 m*M* EDTA, 1 m*M* dithiothreitol (DTT), 1% Nonidet P-40] and an equal volume of glass beads [acid washed, 425–600 μm diameter (Sigma)] to the cell pellet so that the portions are one-third cell pellet, one-third lysis buffer, and one-third glass beads.

12. Vortex for 1 min at 4°C, stop for 1 min. Repeat the vortexing 10 times.

13. Pellet debris and glass beads in a fixed-angle microcentrifuge at 10,000 *g* for 15–20 min at 4°C.

14. Aspirate supernatants carefully, leaving behind cell debris, and transfer the supernatants to fresh 0.5- or 1.5-ml microcentrifuge tubes.

15. Quantify the amount of protein in extracts using a protein assay, such as the bicinniconic acid method (Smith *et al.*, 1985), sacrificing 1 μl for the assay. (Kits are available from several manufacturers for protein assays, including Pierce, Inc. and Bio-Rad.)

16. Split the extracts into two portions. Add protease inhibitors (1 m*M* phenylmethylsulfonyl fluoride, 2 μg/ml aprotinin, 1 μg/ml pepstatin A, and 0.5 μg/ml leupeptin) to one portion and perform SDS–PAGE/immunoblot assays, using ~20 μg total protein per lane typically. The other portion is used directly for caspase assays.

2. Caspase Activity Measurements

Caspases all share the ability to cleave their substrates after aspartic acid residues, but can differ in terms of the sequence context in which aspartic acid cleavage sites are

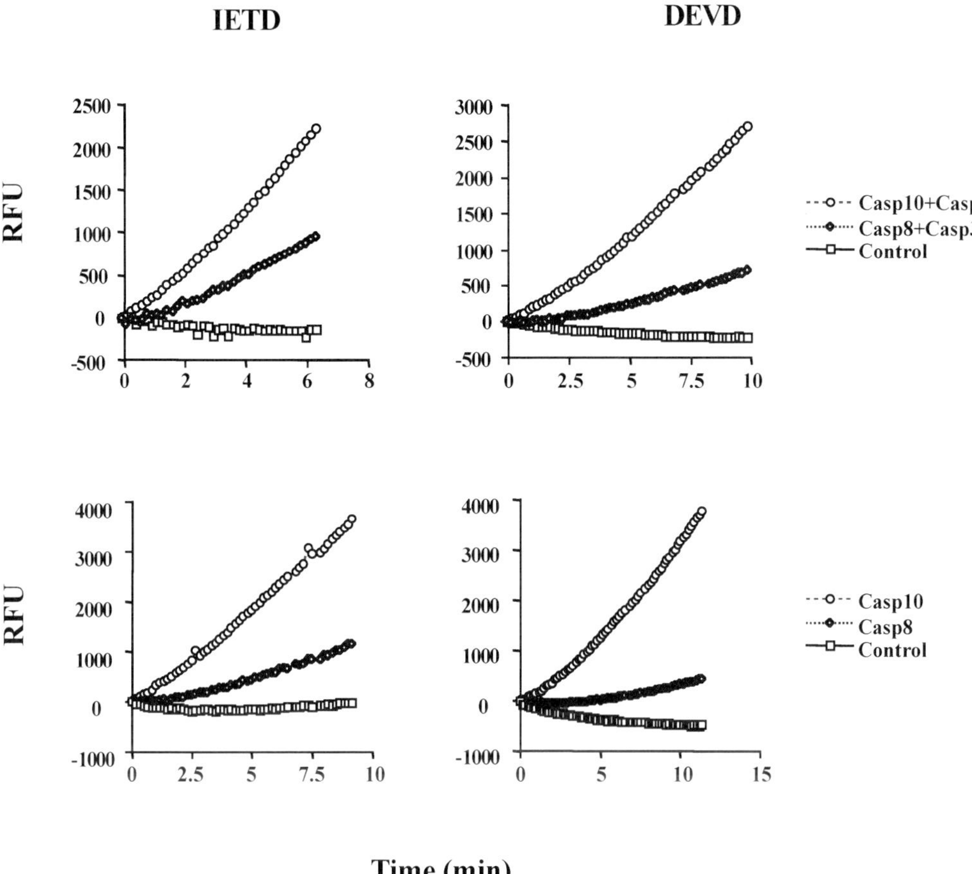

Fig. 2 Caspase activity in transformed yeast. Twenty micrograms of protein extracts from transformed yeast that contained control or caspase encoding plasmids was incubated with 100 μM IETD-AFC or DEVD-AFC. Enzyme activity was determined by the release of the AFC fluorophore (expressed as relative fluorescence units, RFU).

recognized. Structural studies of caspase family proteins suggest that they recognize a tetramer sequence, and combinatorial peptide library screens with human caspases have revealed the preferred tetrapeptide cleavage sequences (Rotonda *et al.*, 1996; Rano *et al.*, 1997; Thornberry *et al.*, 1997). For example, caspase-3 prefers the sequence DXXD, whereas caspases-8 and -10 are more effective at cleaving after the sequence IXXD. Therefore, different fluorigenic or colorimetric substrates can be employed for measuring distinct caspase activities. Typically, we employ IETD-AFC for caspases-8 and -10 and benzyloxycarbonyl DEVD-AFC for caspase-3. Upon cleavage, these substrates release fluorescent moiety, 7-amino-4-trifluoromethyl-coumarin (AFC). Figure 2 shows an example of a caspase activity assay using extracts from transformed yeast.

3. Measuring Caspase Activity

Reactions are performed in a final volume of 0.1 ml using a volume of extracts containing 20 μg of total protein in 1× caspase buffer with 100 μ*M* caspase substrate. However, because it is important to ensure that equivalent amounts of substrate are employed with each sample, we recommend premixing a master solution containing substrate and caspase buffer and then adding this to extracts.

1. Dissolve the synthetic substrates, such as z-IETD-AFC and z-DEVD-AFC (Calbiochem, La Jolla, CA) in dimethyl sulfoxide at a concentration of 10 m*M*. (These can be stored in small aliquots at −80°C.)
2. Prepare 2× caspase buffer (0.1 *M* Hepes, pH 7.2, 0.2 *M* NaCl, 2 m*M* EDTA pH 8.0, 0.2% CHAPS, 10 m*M* DTT, and 10% sucrose). This solution should be prepared fresh each time. DTT should be added from a 10–100X stock solution, which is stored in small aliquots at −20°C. Sucrose is prepared initially as a 20% (w:v) sucrose solution in water, passed through a 0.2-μm filter for sterilization, and stored at 4°C using sterile pipettes to avoid bacterial contamination.
3. Prepare a master mix sufficient for all samples (50 μl per sample) of 2× caspase buffer containing 1:50 (v:v) of substrate peptides (200 μ*M* peptide concentration). Keep on ice.
4. For caspase activity measurements, yeast extracts (prepared as described earlier) should ideally have a final total protein concentration of 5–10 μg/μl. Thaw samples on ice and adjust aliquots to 20 μg total protein per 50 μl total volume (0.4 mg/ml) by adding water immediately prior to performing reactions.
5. Pipette 50 μl of extract (at 0.4 mg/ml) into 0.5-ml microcentrifuge tubes on ice.
6. Add 50 μl of 2x caspase/200 μ*M* substrate master mix and mix thoroughly by gentle vortexing.
7. Simultaneously place all samples onto a fluorimeter plate reader (Perkin-Elmer, LS50B) preheated to 37°C and incubate for the desired time. Substrate cleavage (release of fluorigenic AFC) can be monitored continuously or end-point analysis can be performed by measuring AFC after a defined time (e.g., 20–30 min). For initial reactions, it is advisable to monitor continuously to ensure that the time chosen for end-point analysis is within the linear phase of the reactions. Release of AFC is measured in the kinetic mode using excitation and emission wavelengths of 405 and 510 nm, respectively (Smith *et al.,* 1985; Rotonda *et al.,* 1996).

E. Yeast Cell Death Assays

We have determined empirically that overexpressing certain caspases in yeast results in cell death. For example, when caspases-8 or -10 are expressed from the galactose-inducible YEp51 plasmid, few if any colonies grow on plates (Fig. 3). In contrast, some other caspases appear to be noncytotoxic when expressed in yeast, including caspase-3 and caspase-9. Monitoring of caspase activity using fluorigenic substrates indicates that expression of the proforms of caspases-8 and -10 generate active

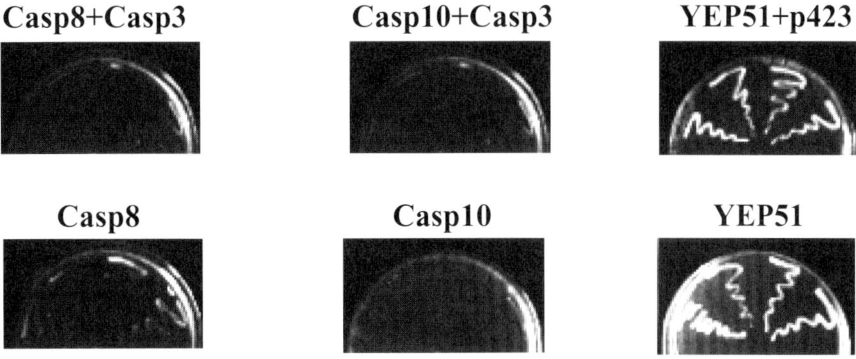

Fig. 3 Expression of caspases-8 or -10 inhibits yeast growth. Yeast transformants that contained control or caspase encoding plasmids were streaked onto plates containing galactose-containing synthetic medium lacking either leucine and histidine (top) or leucine only (bottom). Photograph was taken after a 4-day incubation at 30°C.

caspases in yeast, whereas expression of procaspases-3 or procaspase-9 is insufficient to result in caspase activity (Fig. 4). We presume that overexpression of procaspases-8 or -10 results in self association of these zymogens, via their N-terminal DED domains (which are known to be capable of binding), and results in caspase activation via

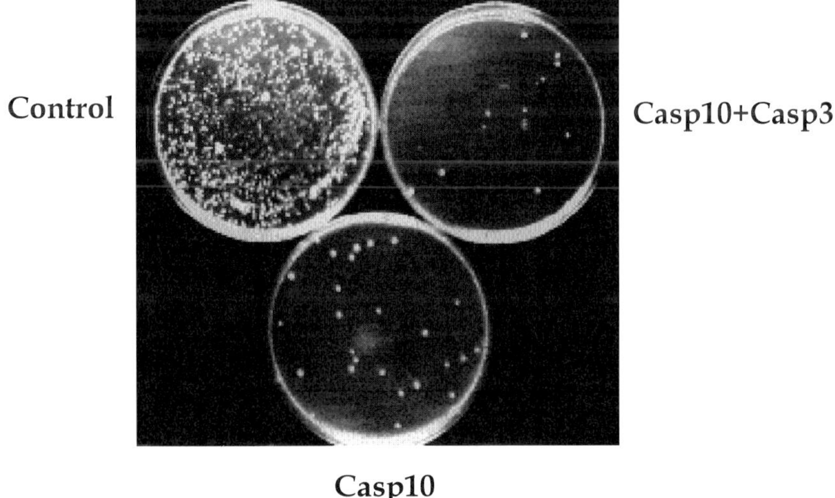

Fig. 4 Expression of caspase-10 in yeast-induced cell death. Yeast transformants that contained control, or caspase-10, or caspase-10 and caspase-3 were grown in glucose-containing medium to logarithm phase and then transferred to galactose-containing medium for 2 days. Cells were counted, and the same amount of cells from each sample was plated onto glucose-containing synthetic medium. Photograph was taken after a 4-day incubation at 30°C.

induced proximity. For caspases such as caspase-3 and caspase -9, where spontaneous activation does not occur, it is possible to use the technique of circular permutation to generate active proteases (Kang *et al.,* 1999; Hawkins *et al.,* 1999; C. Stehlik and J. Reed, unpublished observations). Circular permutation is based on empiric observation that in the three-dimensional structure of many proteins, N and C termini are located in close proximity. This suggests the possibility of circularizing the proteins by joining N and C termini. It has been observed that if a circularized protein is "nicked" at any random place, the protein generally still folds properly regardless of the site chosen to "linearize" the molecule. Thus, for expressing caspases in their active form, one can linearize a circularized molecule at the cleavage site located between the large and the small subunits, using recombinant DNA methodologies for engineering expression.

Alternatively, we have exploited the failure of these procaspases to spontaneously activate in yeast for devising systems where caspase activation is induced by other proteins, e.g., by coexpressing active Apaf-1 with procaspase-9 (C. Stehlik and J. Reed, unpublished observations). For caspases that do not confer a lethal phenotype when activated in yeast, it is possible to configure assays using cleavable reporter proteins. For example, a transcription factor can be expressed as a fusion protein with a membrane-anchoring domain separated from the transcription factor by a sequence that is recognized and cleaved by caspases. Cleavage of the membrane-bound protein releases the transcripition factor, allowing it to translocate into the nucleus and induce expression of reporter genes (Hawkins *et al.,* 1999; H. Hayashi, C. Stehlik, and J. Reed, unpublished observations).

For Bcl-2 family proteins, we have determined empirically that overexpression of Bax or Bak is lethal. In contrast, in our hands certain other proapoptotic Bcl-2 family proteins, including Bad and Bid, are not cytotoxic when expressed in yeast. The basis for these differences in phenotype in yeast is not entirely clear. Bid, for example, has been shown to possess a protein structure similar to pore-forming bacterial toxins and can create channels in synthetic membranes akin to Bax (Schendel *et al.,* 1999), but it lacks cytotoxic activity when expressed in yeast (B. Guo, S. Schendel, and J. Reed, unpublished observations).

Three types of assays for monitoring effects of proapoptotic proteins on yeast are described, including streaking on plates to monitor growth suppression, vital dye exclusion assays, and clonigenic survival assays. It is important to recognize that the growth inhibition observed upon expression of caspases and proapoptotic Bcl-2 family proteins in yeast can potentially be explained by two different possibilities: cytostatic effects versus cytotoxic effects. Thus, two assays are described that can differentiate between a mere cytostatic effect versus a cytotoxic phenotype: vital dye exclusion and clonigenic survival.

1. Growth Suppression Assay

1. Prepare transformants of the desired yeast strain by initially plating on glucose-containing medium that lacks the relevant amino acids required for plasmid marker selection. Incubate at 30°C until colonies form and then store the plates at 4°C for up to 2 months. Yeast containing YEp51-caspase-8 or -10, for example, are maintained on

SD glucose lacking leucine. Yeast containing pGilda-Bax are maintained on SD glucose lacking histidine. Control transformed cells contain YEp51 or pGilda without cDNA inserts.

2. Using a flamed loop, streak individual colonies onto SD galactose medium, again lacking the relevant amino acid (leucine for YEp51; histidine for pGilda, etc.).

3. Incubate at 30°C for 2–4 days and evaluate colony growth.

2. Vital Dye Exclusion Assays

1. Inoculate a single colony from each transformed plate into 2 ml SD glucose medium in a 15-ml polystyrene round-bottom centrifuge tube and grow at 30°C overnight with vigorous agitation to $OD_{600} = 0.3–0.4$. If cells are overgrown, dilute the culture with fresh medium and grow to this optimal density. To obtain efficient induction from the *GAL* promoter, it is important to keep the cell density at $OD_{600} < 0.6$ prior to transfer to galactose medium.

2. Collect cells by centrifugation at 1000 *g* for 5 min in a bench-top centrifuge and wash cells three times with distilled H_2O.

3. Resuspend cells in 2 μl SD galactose medium and continue culturing cells at 30°C with vigorous agitation.

4. Remove 20 μl culture at 6-h intervals and mix with an equal volume of 0.4% trypan blue in phosphate-buffered saline (Sigma). Count blue and white cells with a light microscope using a hemocytometer (Hausser Scientific, Horsham, PA). At least 400 cells are counted for each sample.

3. Clonigenic Survival Assays

1. Cell cultures are performed as described earlier for vital dye exclusion assays in steps 1, 2, and 3.

2. After switching cells to SD galactose, remove 100-μl aliquots at 6-h intervals for up to 48 h and dilute cells in distilled H_2O in a series of 1:10 serial dilutions.

3. Spread onto SD glucose plates.

4. Incubate the plates at 30°C for 3–4 days and count the number of colonies to obtain a ratio of clonigenic ability compared to control (e.g., YEp51- or pGilda-containing yeast).

III. Inhibition of Caspase- and Bax-Induced Cell Death in Yeast

The lethal phenotype in yeast associated with the expression of certain caspases and Bcl-2 family proteins can be suppressed by coexpression of antagonists of these proteins. For example, yeast can be rescued from the lethal phenotype of Bax or Bak by coexpression of antiapoptotic Bcl-2 family members, including Bcl-2, Bcl-X_L, and Mcl-1 (Sato *et al.*, 1994; Jurgensmeier *et al.*, 1997). With respect to caspases, the pox-virus protein

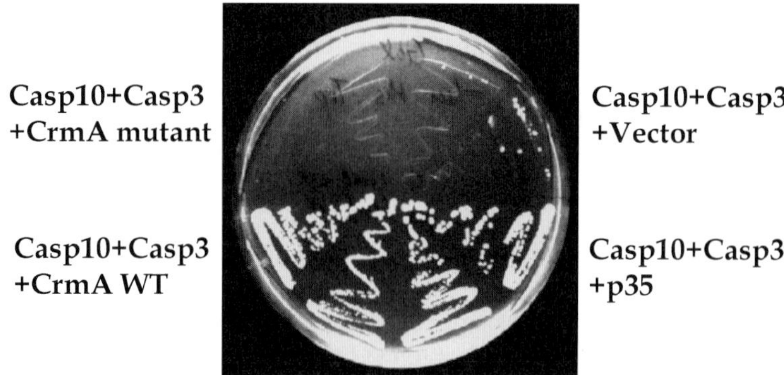

Fig. 5 CrmA and p35 suppress caspase-induced yeast cell death. Control vector, p35, wild-type CrmA, or mutant CrmA was transformed separately into a yeast strain containing caspase-10 and caspase-3. Transformants from each transformation were streaked on galactose-containing medium lacking leucine, histidine, and tryptophan. Photograph was taken after a 4-day incubation at 30°C.

CrmA is a direct and potent inhibitor of caspases-8 and -10. Other caspases–inhibitory proteins have similarly been employed, including baculovirus p35, which inhibits most caspases, and IAP family proteins, which inhibit downstream selected caspases in humans and *Drosophila* (reviewed in Deveraux *et al.*, 1999).

Figure 5 shows an example of an experiment where the lethality of caspase-10 was suppressed by coexpression of either CrmA or p35 in the EGY48 strain yeast. This experiment employed the plasmids YEp51, p423 and p424, which carry *LEU2*, *HIS3*, and *TRP1* markers, respectively. The YEp51 plasmid was used for galactose-inducible expression of procaspase-10, whereas p424 was used for constitutive expression from its *GPD* promoter of p35, wild-type CrmA, or a mutant CrmA lacking caspase inhibitory activity (Zhou *et al.*, 1997). In addition, pro-caspase-3 was co-expressed in cells (which we have found improves killing efficiency) from p423, which carries a *HIS3* marker. Empty plasmids lacking a cDNA insert were employed as additional controls. Transformants were selected and maintained on SD glucose medium lacking leucine, histidine, and tryptophan. Individual colonies were streaked onto galactose-containing medium (again lacking leucine, histidine, and tryptophan) and growth was monitored after 4 days of incubation at 30°C.

IV. Comments and Conclusions

The elucidation of the components of apoptotic pathways is critical for understanding normal development and the pathogenesis of many diseases. Caspases and Bcl-2 family proteins have been identified as central players in the regulation of apoptosis. Much remains unknown, however, about the structure–function relations that account for the ability of these proteins to regulate cell life or death decisions, and even less is known about the myriad upstream inputs into these proteins. Elucidating the functions of caspases and Bcl-2 family protein in animal cells may be hampered by functional redundancy and interference from endogenous proteins, particularly other family members. Yeast thus provide an unencumbered cellular context in which to interrogate the structure–function activities of caspases and Bcl-2 family proteins. These simple unicellular eukaryotes also permit the construction of screening assays that can be exploited for the analysis of cDNA libraries to identify novel genes that regulate caspases and Bcl-2 family proteins or for screening of chemical libraries in order to identify small molecule drugs that modify the activities of these apoptosis regulatory proteins. Finally, classical techniques of yeast genetics, including the production of mutant yeast strains that are resistant to the lethal effects of proteins such as Bax or caspases, can be applied in search of evolutionarily conserved components of the cell death machinery (Matsuyama *et al.,* 1998). The assays and methods described in this chapter represent only a small subset of the types of experimental approaches that can be applied using yeast as a tool for apoptosis research.

References

Adams, J., and Cory, S. (1998). The Bcl-2 protein family: Arbiters of cell survival. *Science* **281,** 1322–1326.

Ashkenazi, A., and Dixit, V. (1998). Death receptors: Signaling and modulation. *Science* **281,** 1305–1308.

Cryns, V., and Yuan, Y. (1999). Proteases to die for. *Genes. Dev.* **13,** 371.

Deveraux, Q., Stennicke, H., Salvesen, G., and Reed, J. (1999). Endogenous inhibitors of caspases. *J. Clin. Oncol.* **19,** 388–398.

Green, D., and Kroemer, G. (1998). The central executioners of apoptosis: Caspases or mitochondria? *Trends Cell Biol.* **8,** 267–271.

Green, D. R., and Reed, J. C. (1998). Mitochondria and apoptosis. *Science.* **281,** 1309–1312.

Hawkins, C., Wang, S., and Hay, B. (1999). A cloning method to identify caspases and their regulators in yeast: Identification of *Drosophila* LAPI as an inhibitor of the *Drosophila* caspase DCP-1. *Proc. Natl. Acad. Sci. USA* **96,** 2885–2890.

Ito, H., Fududa, Y., Murata, K., and Kimura, A. J. (1983). *Bacteriol.* **153,** 163–168.

Jürgensmeier, J. M., Krajewski, S., Armstrong, R., Wilson, G. M., Oltersdorf, T., Fritz, L. C., Reed, J. C., and Ottilie, S. (1997). Bax- and Bak-induced cell death in the fission yeast *Schizosaccharomyces pombe*. *Mol. Biol. Cell.* **8,** 325–229.

Kang, J., Schaber, M., Srinivasula, S., Alnemri, E., Litwack, G., Hall, D., and Bjornsli, M. (1999). Cascades of mammalian caspase activation in the yeast *Saccharomyces cerevisiae*. *J. Biol. Chem.* **274,** 3189–3198.

Kroemer, G., and Reed, J. C. (2000). Mitochondrial control of cell death. *Nature Med.* **6,** 513–519.

Manon, S., Chaudhuri, B., and Buérin, M. (1997). Release of cytochrome *c* and decrease of cytochrome *c* oxidase in Bax-expressing yeast cells, and prevention of these effects by coexpression of Bcl-XL. *FEBS Lett.* **415,** 29–32.

Matsuyama, S., Schendel, S., Xie, Z., and Reed, J. (1998). Cytoprotection by Bcl-2 requires the pore-forming $\alpha 5$ and $\alpha 6$ helices. *J. Biol. Chem.* **273,** 30995–31001.

Matsuyama, S., Xu, Q., Velours, J., and Reed, J. C. (1998). Mitochondrial F_0F_1-ATPase proton-pump is required for function of pro-apoptotic protein bax in yeast and mammalian cells. *Mol. Cell* **1,** 327–336.

Rano, T. A., Timkey, T., Peterson, E. P., Rotonda, J., Nicholson, D. W., Becker, J. W., Chapman, K. T., and Thornberry, N. A. (1997). A combinatorial approach for determining protease specificities: Application to interleukin-1beta converting enzyme (ICE). *Chem. Biol* **4,** 149–155.

Reed, J. (1998). Bcl-2 family proteins. *Oncogene* **17,** 3225–3236.

Rotonda, J., Nicholson, D. W., Fazil, K. M., Gallant, M., Gareau, Y., Labelle, M., Peterson, E. P., Rasper, D. M., Ruel, R., Vaillancourt, J. P., Thornberry, N. A., and Becker, J. W. (1996). The three-dimensional structure of apopain/CPP32, a key mediator of apoptosis. *Nat. Struct. Biol* **3,** 619–625.

Salvesen, G. S., and Dixit, V. M. (1997). Caspases: Intracellular signaling by proteolysis. *Cell* **91,** 443–446.

Salvesen, G. S., and Dixit, V. M. (1999). Caspase activation: The induced-proximity model. *Proc. Natl. Acad. Sci. USA* **96,** 10964–10967.

Sato, T., Hanada, M., Bodrug, S., Irie, S., Iwama, N., Boise, L. H., Thompson, C. B., Golemis, E., Fong, L., Wang, H.-G., and Reed, J. C. (1994). Interactions among members of the Bcl-2 protein family analyzed with a yeast two-hybrid system. *Proc. Natl. Acad. Sci. USA* **91,** 9238–9242.

Schendel, S., Azimov, R., Pawlowski, K., Godzik, A., Kagan, B., and Reed, J. (1999). Ion channel activity of the BH3 only Bcl-2 family member, BID. *JBC* **274,** 21932–21936.

Schendel, S., Montal, M., and Reed, J. C. (1998). Bcl-2 family proteins as ion-channels. *Cell Death Differ* **5,** 372–380.

Sherman, F., Fink, G. R., and Hicks, J. B. (1983). Methods in Yeast Genetics. Cold Spring Harbor Press, Cold Spring Harbor, NY.

Smith, P., Krohn, R., Hermanson, G., Mallia, K., Gartner, F., Prozenzano, M., Fujimoto, E., Goeke, N., Olson, B., and Klenk, D. (1985). Measurement of protein using bicinchoninic acid. *Anal. Biochem* **150,** 76–85.

Thornberry, N., and Lazebnik, Y. (1998). Caspases: Enemies within. *Science* **281,** 1312–1316.

Thornberry, N., Rano, T., Peterson, E., Rasper, D., Timkey, T., Garcia-Calvo, M., Houtzager, V., Nordstrom, P., Roy, S., Vaillancourt, J., Chapman, K., and Nicholson, D. (1997). A combinatorial approach defines specificities of members of the caspase family and granzyme B. *J. Biol. Chem* **272,** 17907–17911.

Toh-e, A., Ueda, Y., Kakimoto, S.-I., and Oshima, Y. J. (1973). *Bacteriol* **113,** 727–738.

Wallach, D., Boldin, M., Varfolomeev, E., Beyaert, R., Vandenabeele, P., and Fiers, W. (1997). Cell death induction by receptors of the TNF family: Towards a molecular understanding. *FEBS Lett* **410,** 96–106.

Xu, Q., and Reed, J. C. (1998). Bax inhibitor-1, a mammalian apoptosis suppressor identified by functional screening in yeast. *Mol. Cell* **1,** 337–346.

Yuan, J. (1997). Transducing signals of life and death. *Curr. Opin. Cell Biol* **9,** 247–251.

Zha, H., Fisk, H. A., Yaffe, M. P., Mahajan, N., Herman, B., and Reed, J. C. (1996). Structure-function comparisons of the proapoptotic protein Bax in yeast and mammalian cells. *Mol. Cell Biol* **16,** 6494–6508.

Zha, H., and Reed, J. C. (1997). Heterodimerization-independent functions of cell death regulatory proteins Bax and Bcl-2 in yeast and mammalian cells. *J. Biol. Chem* **272,** 31482–31488.

Zhang, H., Xu, Q., Krajewski, S., Krajewska, M., Xie, Z., Fuess, S., Kitada, S., Godzik, A., Pawlowski, K., Shabaik, A., Konenen, J., Bubendorf, L., and Reed, J. (2000). BAR: An apoptosis regulator at the intersection of caspase and bcl-2 family proteins. *Proc. Natl. Acad. Sci. USA* **97,** 2597–2602.

Zhou, Q., Snipas, S., Orth, K., Muzio, M., Dixit, V. M., and Salvesen, G. S. (1997). Target protease specificity of the viral serpin CrmA: Analysis of five caspases. *J. Biol. Chem* **272,** 7797–7800.

CHAPTER 21

Methods to Study Cell Death in *Dictyostelium discoideum*

Jean-Pierre Levraud, Myriam Adam, Sophie Cornillon, and Pierre Golstein

Centre d'Immunologie INSERM-CNRS de Marseille-Luminy
13288 Marseille Cedex 9, France

0091-679X/01 $35.00

I. Introduction

A. Why Study Cell Death in *Dictyostelium*?

Phylogenetically speaking, cell death seems to occur throughout evolution from primitive eukaryotic organisms to multicellular animals and plants (reviewed in Greenberg, 1996; Jacobson *et al.,* 1997; Vaux and Korsmeyer, 1999). From an anthropocentric point of view, studying cell death in evolutionarily ancient organisms may reveal core molecular mechanisms of cell death, either caspase dependent, but in a "simpler," more tractable version, or caspase independent, showing the details of such mechanisms in animals (and/or plants). In brief, important clues as to cell death in higher organisms may well be found in ancient organisms.

Among such evolutionarily ancient organisms, the choice of the protist *Dictyostelium discoideum,* a slime mold, to study programmed cell death (PCD) is appealing for several reasons. First, this organism can be considered as one of the probably several (Kaiser, 1986) evolutionary attempts at multicellularity. If cell death is a corollary to multicellular development, as suggested long ago (Weismann, 1890), it would seem of interest to characterize the type of death observed in the development of this "transitional" organism. Also, *Dictyostelium* is one of the most anciently diverged currently surviving eukaryotes (Christen *et al.,* 1991; Field *et al.,* 1988), stemming perhaps after divergence of the kingdom Plantae and before individualization of the kingdoms Animalia and Fungi (see, for instance, Baldauf and Doolittle, 1997). Demonstration of a common cell death mechanism between this organism and some of the higher eukaryotes would be a strong argument for a degree of generality of this mechanism. A second reason is ontogenetic: the relatively simple pattern of the development of *Dictyostelium* should facilitate the study of cell death occurring during this development. A third reason, which builds on the previous one, is that methods exist to trigger, *in vitro,* differentiation without morphogenesis (Town *et al.,* 1976; Kopachik *et al.,* 1983; Sobolewski *et al.,* 1983; Kay, 1987), and thus facilitate the isolation of dying *Dictyostelium* cells for study. Another reason *Dictyostelium* is a valuable model is due to the genetic tools available for this species. The genome of *Dictyostelium* is small ($\sim 3.4 \times 10^7$ bp), about 100-fold smaller than that of higher eukaryotes, and haploid. On the order of 300 genes may be involved in differentiation (Loomis, 1980) out of a total of about 10,000 genes (Loomis *et al.,* 1995). Importantly, this genome is being currently sequenced, both as cDNAs and genomic DNA (Kay and Williams, 1999). Genome haploidy makes it relatively easy to generate and select mutants of death-associated genes and to identify the latter, especially using recently developed genetic tools (Loomis, 1987; Kuspa and Loomis, 1992; Chang *et al.,* 1995; Kuspa *et al.,* 1995; Spann *et al.,* 1996). Because of the temporal separation between vegetative growth and development, developmental mutants (such as those related to cell death) can be propagated under vegetative conditions, thus behaving like conditional mutants (Loomis, 1987). Finally, and more trivially, *Dictyostelium* cells can be grown in large quantities on inexpensive media, are robust, and have been a popular model system for biochemical and physiological analysis of signal transduction for many years. Wild *Dictyostelium,* which are found in decaying leaf litter, feed on bacteria. While the initial

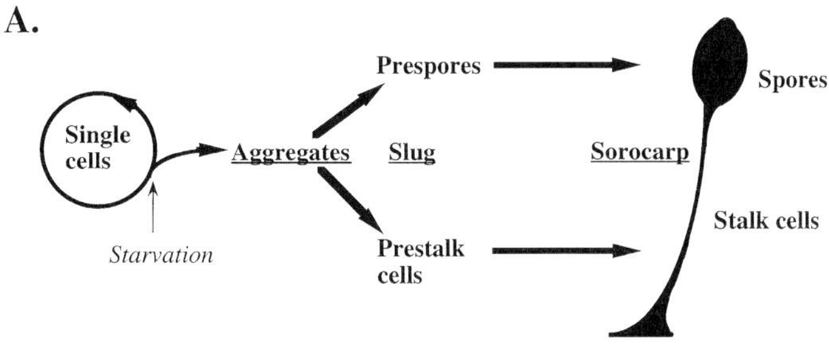

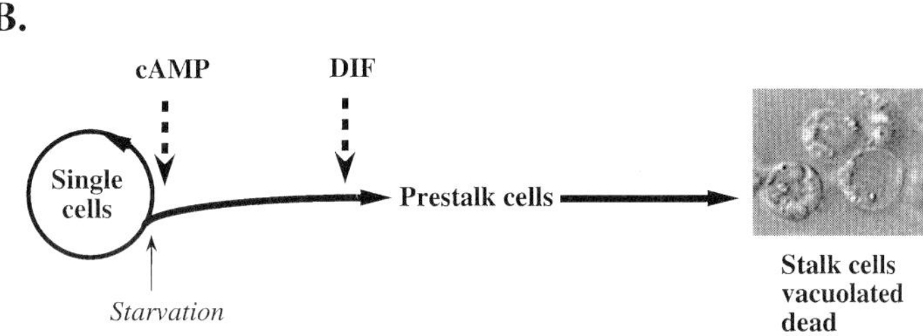

Fig. 1 Differentiation of *Dictyostelium.* (A) Normal development of *Dictyostelium* initiated upon starvation. (B) Differentiation of stalk cells in monolayer.

laboratory strains required bacteria to feed upon, axenic strains have been derived that facilitate *in vitro* work greatly.

B. Development of *Dictyostelium*

Dictyostelium multiplies as a unicellular protist under favorable conditions, but becomes multicellular by aggregation on starvation (Raper, 1935, 1984) (Fig. 1A). Aggregation is accompanied by cell differentiation and morphogenesis, leading to a migrating slug within 15 h and to a 1- to 2-mm-high, aesthetically pleasant fungus-like structure called sorocarp within 24 h. The sorocarp comprises two main populations of cells: viable spores in a bolus and vacuolated cells comprising the stalk. Differentiation to stalk cells (i.e., to vacuolated, dead cells) seems to result from the sequential action of at least two main factors under starvation conditions: (1) cyclic AMP (cAMP) and (2) the dichlorinated hexanone differentiation factor DIF-1 [which acts on starved cAMP subjected cells and promotes their differentiation into stalk cells (Town *et al.,* 1976; Town and Stanford, 1979; Sobolewski *et al.,* 1983; Morris *et al.,* 1987)]. The inducing effect of DIF-1 on stalk cell differentiation seems to involve a slow sustained increase in intracellular calcium (Ca^{2+}) levels (Schaap *et al.,* 1996) and activation of a signal transducer and activator of transcription (STAT) protein (Kawata *et al.,* 1997). Stalk cells are highly vacuolated (Raper and Fennell, 1952; Maeda and Takeuchi, 1969; George *et al.,* 1972; de Chastellier and Ryter, 1977; Quiviger *et al.,* 1980; Schaap *et al.,* 1981) and dead as demonstrated by their inability to regrow in culture medium (Whittingham and Raper, 1960). *Dictyostelium* stalk cells may thus be considered one the earliest known occurrences of developmental PCD in eukaryote evolution.

C. Morphotype of Programmed Cell Death (PCD) in *Dictyostelium*

Programmed cell death in *Dictyostelium* can be studied *in vitro* using conditions mimicking developmental circumstances that are more amenable to microscopic observation and to further genetic manipulations than *in vivo* experiments on stalk cell death (Whittingham and Raper, 1960). We took advantage of a protocol for differentiation in monolayers (Kay, 1987) and of a *Dictyostelium* mutant strain called HM44 (Kopachik *et al.,* 1983) derived from V12M2, adapted for axenic growth as HMX44 (J. G. Williams, now at the University of Dundee), and subcloned in our laboratory as HMX44A. HM44 produces very little DIF but is sensitive to exogenous DIF (Kopachik *et al.,* 1983). Upon starvation and addition of DIF, HM44 differentiates into stalk cells, however, without morphogenizing into a sorocarp. In brief, *Dictyostelium* HMX44A cells were subjected to the usual (Kay, 1987; Cornillon *et al.,* 1994) sequence of incubation in starvation medium in the presence of cAMP, which does not lead in itself to cell death, followed by another period of incubation together with DIF, which triggers cell death (Fig. 1B). In temporal succession after this triggering, programmed cell death began with an irreversible step leading to the inability to regrow at 8 h. At 12 h, massive vacuolization was visible by confocal microscopy, and prominent cytoplasmic condensation and focal chromatin condensation could be observed by electron microscopy (Fig. 2). Membrane permeabilization occurred only very late in the process (at 40–60 h), as judged by propidium iodide staining. Cell death could thus be described as a developmentally induced sequence of intracellular events, including an early irreversible step, subsequent massive vacuolization, cytoplasmic condensation, focal chromatin condensation, and very late membrane lesions. No early DNA fragmentation could be detected by standard or pulse field gel electrophoresis (Cornillon *et al.,* 1994) or by TUNEL assays (unpublished). In addition, no phosphatidylserine externalization could be detected (unpublished).

D. Approaches to Molecular Mechanisms of PCD in *Dictyostelium*

To identify genes involved in *Dictyostelium* cell death, one can examine genes whose homologues have been shown to be involved in higher eukaryotic cell death. Thus, several caspase inhibitors affected neither cell death per se nor spore formation. Strikingly, however, some caspase inhibitors that did not inhibit cell death impaired other developmental events (Olie *et al.,* 1998). The simplest interpretation of these results is that in *D. discoideum,* whether caspases exist or not (genome sequence will soon tell) and whether or not their activation is required for some developmental steps, caspase activation is not required for cell death itself (Olie *et al.,* 1998). An unbiased approach (Cornillon *et al.,* 1998) is to isolate cell death mutants generated by insertional mutagenesis. An improved insertional mutagenesis approach, called restriction enzyme-mediated integration (REMI), involves introducing a mix of a linearized plasmid and an appropriate

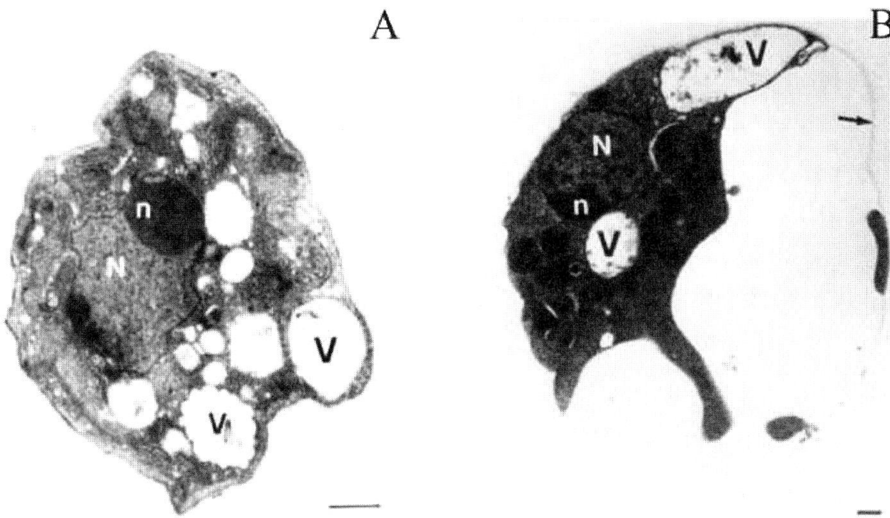

Fig. 2 Ultrastructure of live and dead *Dictyostelium* cells. Transmission electron micrographs of HMX44A cells. (A) Vegetative cell. (B) Stalk cell after 12 h in the presence of DIF. V; vacuolar system; N; nucleus; n; nucleolus. Arrow, cellulose wall. Note the huge phagocytic vacuole of the stalk cell. Bars: 1 μm in A and 0.5 μm in B. Reproduced from Cornillon *et al.* (1994), with permission.

restriction enzyme into cells (Kuspa and Loomis, 1992). As a result, the plasmid integrates into the genome with increased frequency at any potential site corresponding to the added enzyme (Kuspa and Loomis, 1992). The bsr-REMI approach makes use of a plasmid bearing a gene that confers resistance to the antibiotic blasticidin, thereby allowing easy selection of eukaryotic cells after plasmid integration (Adachi *et al.,* 1994). If the plasmid inserts into a gene required for cell death, gene disruption would prevent cell death in the corresponding cell. Selection of death-resistant cells followed by plasmid rescue may lead to the identification of genes required for *Dictyostelium* cell death.

This chapter makes no attempt to exhaustively list and describe all of the methods used for studying *Dictyostelium*. Instead, it focuses on those methods, quite often devised in other laboratories, that we have found useful for the identification, functional characterization, and molecular definition of *Dictyostelium* cell death. The methods detailed here should be easily applicable by readers unfamiliar with *Dictyostelium* but otherwise used to standard mammalian cell culture. Also, they require little specific equipment.

II. Cell Culture and Maintenance of Axenic Strains

Two major techniques are widely used to culture *D. discoideum* laboratory strains (Sussman, 1987). The first involves growth on bacteria (normal *Dictyostelium* food source), either on solid agar or in a simple salt buffer. The second, and the only one detailed here, is axenic growth (growth in the absence of bacteria) in a relatively well-defined rich medium, such as HL-5 (Sussman, 1987). Slight modifications to Sussman's protocol allow optimal growth of AX2 and HMX44A strains and allow the use of standard cell culture facilities (Cornillon *et al.,* 1994). AX2 is a developmentally competent strain derived from NC4 (Watts and Ashworth, 1970). HMX44A is a clonal isolate of V12M2 origin as detailed earlier, developmentally incompetent, that is used for monolayer experiments.

A. Media

1. Preparation of HL-5 Medium

14.3 g/liter bacteriological peptone (OXOID LTD, Basingstoke, Hampshire-England, ref. L37)

7.15 g/liter yeast extract (EZMix yeast extract, Sigma Chemical Co., St Louis, MO, ref. Y1626)

18 g/liter maltose (Sigma, ref. M5885)

0.93 g/liter Na_2HPO_4, $7H_20$ (3.6 m*M*)

0.49 g/liter KH_2PO_4 (3.6 m*M*)

Source water (we use Volvic) to reduce variations in the quality of water

Sterilize by autoclaving for 20 min. After cooling, filter through a 0.22-μm filter and store at 4°C. This medium is stable for several months at 4°C.

2. Important Notes

Use exclusively bacteriological peptone from OXOID; peptones from other sources cause a dramatic decrease of *Dictyostelium* growth and impair development after starvation (unpublished observations; see also Sussman, 1987).

While we routinely grow *Dictyostelium* in sterile conditions in the absence of antibiotics, accidental bacteriological contamination can usually be corrected by the addition of penicillin–streptomycin (100 units/ml and 100 μg/ml, respectively; GIBCO-BRL, Grand Island, NY) to the culture medium. Fungal contaminations are more problematic because *Dictyostelium* is killed by fungizone; nystatin or gentamycin may be of help.

B. Facilities and Routine Culture

Tissue culture plastic flasks kept horizontally (Falcon, Becton-Dickinson Labware, NJ) are suitable for growing *Dictyostelium*. Typically, a 175-cm^2 Falcon flask would receive 50 ml of HL-5 and yield up to 10^8 exponentially growing cells (see later). Cell cultures are maintained at 22–23°C in a water-satured normal atmosphere (no carbon dioxide is required). This optimum at "room temperature" means that an incubator regulated both by heating and by cooling is required (one of the very few specific "expensive" items in a *Dictyostelium* laboratory).

Dictyostelium cells die at temperatures in excess of 28°C. Thus, on hot days, bench work should be performed in an air-conditioned room. However, this allows use of the same culture hoods for both (axenic) *Dictyostelium* and mammalian cells because no growth of unwanted cells would occur following accidental cross-contamination.

Dictyostelium cells grow with a doubling time of 8 to 12 h in HL-5 medium at 22°C. Cultures may exceed concentrations of 4×10^6 cells/ml, but growth is only exponential up to 2×10^6 cells/ml. (Note that faster growth and higher cell concentrations may be obtained by culturing cells under agitation in a rotatory shaker instead of using stationary flasks; see Sussman, 1987). To avoid uncontrolled differentiation, it is important that cells be kept growing in the exponential phase; however, it is better not to dilute cultures to a point where cell density falls below 1×10^4 cells/ml. Therefore, some schedule for splitting the cultures has to be observed. We are inclined to culture *Dictyostelium* cells in 175-cm^2 Falcon flasks containing 50 ml of HL-5. These cultures are split (diluted about 50-fold) twice a week.

C. Storage

1. Short Term

For short-term storage, *Dictyostelium* amoebae can be maintained at 4°C in HL-5 medium in Falcon flasks or in the wells of microtiter plates for 10 to 15 days. Microtiter plates should be sealed with parafilm to prevent excessive evaporation of the medium.

2. Long Term

Dictyostelium cultures can be frozen and kept for years in liquid nitrogen and for many months at −80°C. This ensures the availability of permanent stocks of defined cells. It is important to thaw a vial of cells every 6 weeks or so to renew laboratory working stocks. This minimizes the variability that is a corollary of prolonged cell culture of any given strain. When cells are thawed, part of the resulting growing cultures should be refrozen soon (e.g., within less than a week) to maintain stocks.

a. Freezing **Dictyostelium** *Cells*

1. Pellet (700*g*, 5 min) about 5×10^6 cells (per vial) from an exponentially growing culture.
2. Discard the supernatant.
3. Keep cells on ice and add 1 ml (per vial) of cold HL-5 supplemented with 7% dimethylsulfoxide (DMSO; Merck, Darmstadt, Germany). Resuspend cells and transfer to vials suitable for storage in liquid nitrogen, kept on ice.
4. Place in −80°C freezer for 24 to 48 h. Then transfer vials to liquid nitrogen for long-term storage.

b. Thawing **Dictyostelium** *Cells*

1. Remove vials from liquid nitrogen and thaw quickly in a water bath at room temperature.

2. When there is still some ice in the vials, transfer contents of the vial to a Falcon tube containing 14 ml of HL-5. Pellet cells and discard supernatant. This removes DMSO, which is toxic for cells at room temperature.

3. Resuspend cells in 10 ml of HL-5 in a 25-cm^2 Falcon flask. Add appropriate antibiotic for selection if cells are transfectants.

D. Cloning

A major problem in cell culture is the preservation of a specific cell type and its specialized properties. A traditional approach to reduce culture heterogeneity is to isolate cell clones. *Dictyostelium* clones are often obtained by plating a small number of viable cells on a bacterial lawn where they form clonal plaques; however, this does not work for all strains (e.g., HMX44A does not produce sharp-edged plaques; Gérard Klein, Grenoble, personal communication). The easiest way to obtain clonal strains of *Dictyostelium* in axenic conditions is by limiting dilution cloning. Very small numbers of cells are distributed randomly into wells of microtiter plates, and using Poisson distributions, one can calculate the probability of positive wells to contain the progeny of a single cell.

1. Pellet exponentially growing cells (700*g*, 5 min).

2. Make serial dilutions in HL-5 (plus selection drug if required) in order to have the following cell concentrations: 10 cells/ml, 3 cells/ml, and 1 cell/ml. Prepare at least 12 ml of each suspension.

3. Inoculate three microtiter plates (one per cell concentration) with 100 μl of cell suspension per well. The plates will thus be seeded with, on average, 1 cell/well, 0.3 cell/well, and 0.1 cell/well, respectively. Seal each plate with parafilm to minimize evaporation.

4. Incubate plates for at least 5 or 6 days at 22–23°C.

5. Compare the percentage of positive wells (containing growing cells) for each plate with the expected percentage given by the Poisson law (63% of positive wells for an inoculum of 10 cells/ml, 26% for 3 cells inoculated/ml, and 10% for 1 cell/ml).

6. Use positive wells from the most dilute inoculum yielding a positive well to reduce the probability of wells containing the progeny of more than one cell. As a rule of thumb, do not use wells from a plate containing more than 26% positive wells.

7. Transfer cells from the selected positive wells to 25-cm^2 Falcon flasks containing 10 ml of HL-5, expand, and freeze as soon as possible.

III. Induction of Developmental Cell Death

Programmed cell death in *Dictyostelium* is the outcome of terminal differentiation of stalk cells. This can be obtained in two different ways: by inducing normal development at an air–wet solid interface (stalk cells then constitute the stalks and basal disks of the resulting fruiting bodies) or by *in vitro* stalk cell differentiation in monolayers under submerged conditions. The first method is the “natural” one, but is less convenient than the second one for most applications because (1) only ~15% of the cells end up as stalk cells; the remainder differentiate into viable spores. Regrowth of stalk cells exclusively is thus difficult to score (although not impossible; see Whittingham and Raper, 1960) and (2) Microscopic observation of cells in a stalk is not easy; not only are further manipulations required to place the stalk on a microscope slide, but cell morphology is more difficult to assess as cells are enclosed in the cellulose sheath tube (in addition to their own casing) and tightly packed.

A. Stalk Cell Differentiation in Monolayer

Note: This protocol is applicable with most of the usual strains. However, the percentage of cells differentiating into stalk cells is strongly strain dependent: cells of V12M2 origin (such as HMX44) differentiate more efficiently than cells of NC4 origin (such as AX-2). This is largely because of a difference in sensitivity of inhibition of the DIF-dependent step by cAMP (Berks and Kay, 1988), implying that an additional washing step is recommended for some strains. The protocol detailed here, derived from the one described by Kay (1987), has been optimized for HMX44A.

1. Reagents

cAMP (cyclic AMP; 3′:5′ adenosine cyclic monophosphate, sodium salt) (Sigma A6885). Make stock solution 60 m*M* in water, filter sterilize, and keep at −20°C in 4-ml aliquots.

DIF-1 (differentiation-inducing factor 1, 1-(3,5-dichloro-2,6-dihydroxy-4-methoxyphenyl)-hexan-1-one) (Molecular Probes, Eugene, OR, ref. D-3450). Make stock solution 10^{-2} *M* in absolute ethanol. Working stocks are diluted to 10^{-4} *M* (in absolute ethanol) in 1-ml aliquots. Store at −20°C.

Soerensen buffer (SB; 50x stock solution): 100 m*M* Na_2HPO_4, 735 m*M* KH_2PO_4, pH 6.0; filter sterilize and store at 4°C. SB 1× buffer is obtained by adding 10 ml of 50× solution to 490 ml of autoclaved source water.

2. Protocol

1. Collect vegetative cells growing in HL-5 medium in late log phase (i.e., no more than 2×10^6 cells/ml if cells are grown without stirring).
2. Wash twice with SB buffer and count.
3. In each of two 25-cm^2 flasks, add 10^6 cells in 2.5 ml of SB +3 m*M* cAMP (final). Gently swing the flasks so that the liquid covers the whole surface. Incubate the flasks horizontally.
4. Incubate for 8 h at 22°C. Most (>80%) cells should adhere firmly to the plastic. Because of the high concentration of cAMP, almost no aggregation should be seen at this stage, and the cells should be randomly scattered and isolated.
5. Carefully remove the liquid and wash once with 5 ml of SB.

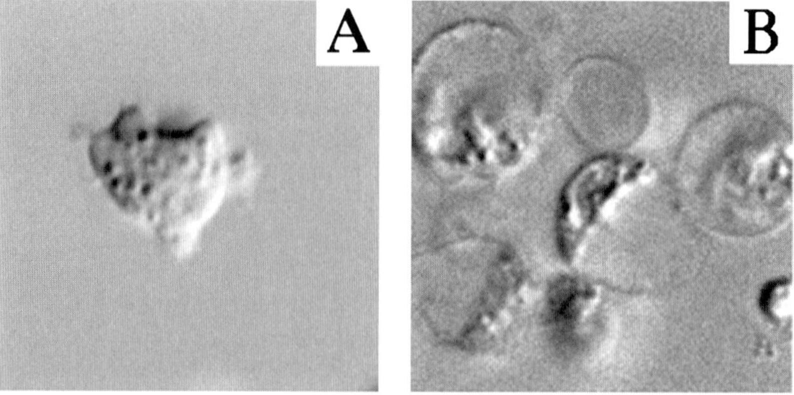

Fig. 3 Appearance of amoeboid and stalk cells using Nomarski optics. HMX44A starved in the absence (A) or presence (B) of DIF.

6. Replace with 2.5 ml of SB + DIF-1 $10^{-7}M$ (final) in one flask and 2.5 ml of SB with no DIF in the other flask.
7. Incubate for 24 h at 22°C.

At this stage, in the DIF-containing flask, most cells should be differentiated to stalk cells : highly vacuolated, cellulose-encased, nonrefringent cells as seen either by phase-contrast microscopy (see, e.g., Kay, 1987) or by using Nomarski optics (Fig. 3). Almost no stalk cells should be seen in the control flask if a cell line producing little endogenous DIF (e.g., HMX44) has been used.

Further incubation will not increase the number of dead cells, but vacuolization will become progressively more prominent as the cytoplasm of dying/dead cells continues to shrink.

B. Development of Wild-Type Strains

Note: The two protocols described here have been chosen because they are very simple to put into practice. Many *Dictyostelium* researchers prefer using another technique that involves filters on top of a buffer-saturated pad. This technique takes longer to set up than the ones described here and may be more problematic in terms of contamination. However, it has important advantages: (1) it allows modifying the medium conditions at any time point simply by moving the filter onto another pad and (2) cells can be easily harvested by introducing the filter into a tube and flushing with buffer. In case these features would be especially useful for your experiments, detailed protocols may be found in Sussman (1987) or at the web addresses mentioned in the Appendix.

1. Tilted Flask Protocol

1. Collect vegetative cells in late exponential growth phase.
2. Wash twice with SB.

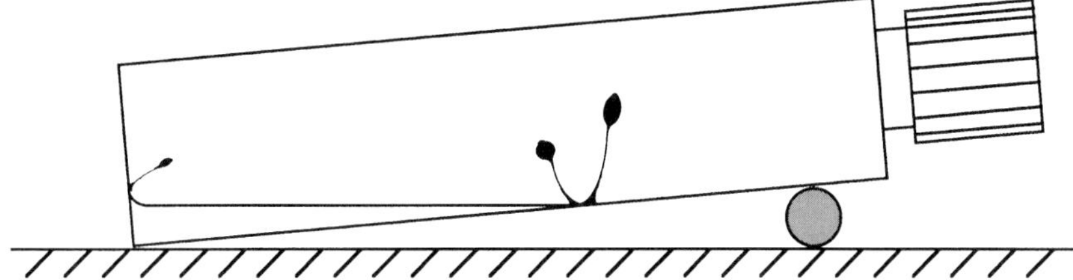

Fig. 4 Scheme of tilted flask for quick development assay (side view). Sorocarps are not to scale.

3. Resuspend in SB at a density of 6×10^6 cells/ml.
4. Dispense 1 ml of cell suspension in a 25-cm^2 culture-treated flask.
5. Gently swing the flask horizontally so that the liquid wets the whole lower surface of the flask
6. Incubate at 22°C in darkness in a slightly tilted position (about 15°, i.e., lean the flask on a pencil, as illustrated in Fig. 4).
7. Observe with a stereomicroscope at low magnification: aggregates and then sorocarps (after 18–24 h) will form, mostly at the air–liquid–plastic interface. Cells that remain fully submerged will aggregate, forming in this process nice spiral streaming patterns, but their development will not proceed further.

This technique is extremely simple and fast to set up, requires very few cells, allows for easy testing of medium conditions, and is very appropriate for newcomers eager to familiarize themselves with *Dictyostelium* development. However, it has some drawbacks: synchronicity of development is poor; flasks have to be moved with care to keep the correct pitch; and although organisms (slugs, fruiting bodies) are easy to observe, they are difficult to recover (this difficulty may be alleviated by using petri dishes, but tilting petri dishes is risky)

Synchronicity is much better with the following agar assay. Once the liquid is absorbed, plates may be moved with no special caution. One can also easily recover cells or organisms. It requires, however, more cells and a bit more time than the previous protocol and is not suitable for testing the effect of heat-labile drugs in the medium.

2. Agar Dish Protocol

1. Pour petri dishes with buffered 1% agar (1% agar in SB) (20 ml for a 90-mm-diameter dish).
2. Let the plates dry up a bit under a sterile hood.
3. Collect vegetative cells in late exponential growth phase.
4. Wash twice with SB and count.
5. Resuspend cells in SB at a density of 10^8 cells/ml.
6. Put 50-μl drops of this suspension onto the agar. For development to proceed

normally, the liquid within which the cells are added has to be absorbed by the agar.

7. Incubate at 22°C in the dark.

Using this protocol with AX-2 cells, the initially smooth cell lawn at the surface of the agar will display irregularities after about 3 h, loose aggregates will be formed by 5 h, tips will appear at around 10 h, slugs by 15–18 h, culmination begins at around 20 h, and mature sorocarps will be formed after 26 h. This is best viewed using a stereomicroscope at low magnification. For a side view, slices of agar may be cut with a scalpel, removed, and tilted onto another dish.

Cell density may be varied in this assay. Fewer cells mean smaller organisms that develop more slowly.

Development is affected by light condition. Not only do slugs move toward a light source, but overhead light stimulates culmination. This should be taken into account when taking pictures of the developing organism in an experiment where the timing of events is important; light exposure must remain as short as possible.

When development is complete, spores can be collected by inverting the dish and tapping it gently on the bench, which will make spores fall onto the dish cover.

Spores can be induced to germinate synchronously by heat shock (45°C for 30 min), which will also kill all nonspore cells, and then transferred to rich medium. Germination will then occur in a few hours, and small vegetative cells will emerge to begin a new cycle of vegetative growth.

IV. Analysis of Cell Death Characteristics

A. Regrowth Assay

This assay is a follow-up of the monolayer stalk cell differentiation assay described earlier. It provides a quantification of surviving cells.

1. Collect vegetative cells in late exponential growth phase.
2. Wash twice with SB buffer and count.
3. Plate two 25-cm^2 flasks each with 10^6 cells in 2.5 ml of SB containing 3 m*M* cAMP. Make sure that the liquid covers the whole surface.
4. Incubate for 8 h at 22°C.
5. Carefully remove the liquid, wash once with 5 ml of SB, and replace with 2.5 ml of SB + DIF-1 $10^{-7}M$ in the first flask and 2.5 ml of SB in the second flask.
6. Incubate for 24 h at 22°C. As described earlier, this leads to the death of most cells in DIF-containing flasks.
7. Add 2 volumes (5 ml) of HL-5 to each flask to initiate the regrowth of surviving cells.
8. Incubate the flasks at 22°C for 40 to 72 h.

9. Detach cells by shaking the flasks vigorously followed by flushes with a 5-ml pipette. Under an inverted microscope, check that all vegetative cells are detached; many stalk cells will still adhere, which is not a problem because they are not to be counted.

10. Count amoeboïd cells using hemocytometer and phase-contrast optics. The rare heavily vacuolated, nonrefringent stalk cells are easily distinguished and excluded. Trypan blue may be added to facilitate the identification of dead cells (see Section IV,B,5).

11. Calculate the ratio of the number of regrowing cells in the DIF flask to the number of cells in the control flasks. For HMX44, this should be around 0.15. This ratio expresses the percentage of cells surviving after DIF-induced cell death. About 15% is the usual background of surviving HMX44A cells. Cell death-resistant mutants give much higher values (see later).

Notes: To be rigorous, a similar amount of absolute ethanol (the solvent of DIF) should be added to the control flask. However, we have never seen a significant effect of the addition of 0.1% ethanol in this assay.

Because this test is sensitive to variations in initial density, cells should be counted carefully before plating into flasks.

If cells are incubated in starvation medium with DIF for longer than 24 h, vacuolization may seem more complete, but other phenomena may interfere with the results. In particular, and unexpectedly, differentiation into what appears to be macrocysts may occur with HMX44A cells maintained in SB without DIF (unpublished observation), lowering the frequency of regrowing cells.

Results of the test are collected after a period of exponential growth. Slight variations in culture conditions may thus significantly affect the results, although expression as a ratio prevents excessive departure from the usual values.

B. Staining Methods

1. Remarks Concerning Microscopy and Cytometry

Flow cytometry is widely used to study the characteristics of cell death in mammalian cells. It allows quantitative measurement of fluorescence, size, and granularity of cells that can be applied to statistically significant numbers of cells. However, flow cytometry can only be employed with isolated cells in suspension. When they differentiate into stalk cells, *Dictyostelium* cells often adhere strongly to their substrate and form very tight cell clumps that are bound together with cellulose. Thus they are impossible to analyze by flow cytometry. Vegetative cells or early differentiating cells, however, are of course amenable to such analyses.

The staining methods detailed here are meant for direct microscopic examination, mostly under a fluorescence microscope. A technical problem linked with such examination is the fact that when *Dictyostelium* cells differentiate in monolayers, some adhere very tightly to the substrate whereas others are found in suspension (most of them clustered). To get a representation of the total population, ideally both cell pools should be considered.

If an inverted fluorescence microscope is available, differentiation may be carried out conveniently in plastic flasks, and staining and washes may be performed directly in the flasks. Most cells adhere tightly under these conditions and nonadherent cells can be ignored.

To observe cells under a standard microscope, differentiation may be carried out directly on coverslips on the bottom of small petri dishes or on the wells of six-well plates. In these cases, a significant fraction of the cells do not adhere (even after centrifugation of the plates), and these cells have to be stained and observed separately. Using cell culture-treated coverslips may alleviate this problem (which we have not tested).

2. Propidium Iodide

Propidium iodide (PI) is a DNA-intercalating dye that cannot cross cell membranes freely: thus, cells will fluoresce only if membranes have become permeable, a late sign of cell death. This implies that cells must be stained fresh, e.g., they cannot be fixed for this test.

Propidium iodide (Sigma P4170) stock solution: 8×10^{-5} M (53 μg/ml) in sterile water, keep at 4°C protected from light. Gloves should be worn because PI is carcinogenic.

1. Add to cells at 4 μ*M* final in SB.
2. Incubate for 10 min at room temperature, away from light.
3. Wash twice carefully with SB.
4. View under a fluorescence microscope and observe red fluorescence.

3. Fluorescein Diacetate

In contrast to PI, fluorescein diacetate (FDA) stains living cells. The nonfluorescent, hydrophobic compound enters the cell freely where it is cleaved by cytoplasmic lipases of metabolically active cells into a green fluorophore unable to leave the cell if membrane integrity is not compromised. Again, cells must not have been fixed.

Fluorescein diacetate (Sigma F7378) stock solution: 10 mg/ml in acetone; store at 4°C away from light.

1. Wash cells once with SB.
2. Add to cells at 0.05 mg/ml in SB.
3. Incubate for 10 min at room temperature away from light.
4. Wash twice with SB.
5. View under a fluorescence microscope and observe green fluorescence.

FDA and PI can be used for double staining (directly mixing the two dyes), as shown in Fig. 5 (see Color Plate).

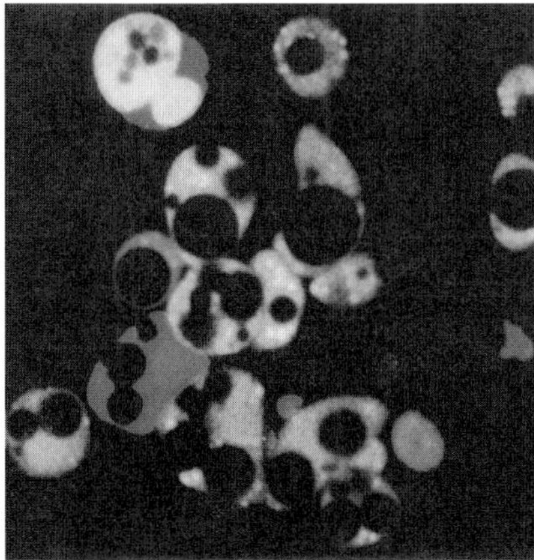

Fig. 5 Illustration of double fluorescein diacetate/propidium iodide staining of stalk cells. HMX44A cells viewed by confocal microscopy after a 24-h DIF incubation. Reproduced from Cornillon *et al.* (1994), with permission. (See Color Plate.)

4. Calcofluor

As they differentiate, stalk cells encase themselves in a cellulose shell that may be labeled with calcofluor. Positive staining does not constitute evidence of cell death, but is nevertheless a useful differentiation marker. One should be aware that other cell types (e.g., spores and macrocysts) also secrete cellulose coats, but cell sizes are very different. A thin cellulose trail is also left on the substrate before the cell finally stops migrating and fully differentiates (Blanton, 1993). Cells may be fixed before staining (0.5% glutaraldehyde, for at least 30 min at 4°C).

Calcofluor (Sigma F6259; now sold under Cat. No. F3397; also named as "fluorescent brightener 28"; "calcofluor white M2R"; "C.I. (color index) 40622"; or "tinopal LPW"): stock solution 1% (w/v) in H_2O, keep at 4°C protected from light.

1. Wash cells with SB.
2. Add calcofluor at 0.1% final in SB.
3. Incubate for 5 min at room temperature protected from light.
4. Wash twice with SB.
5. View under a fluorescence microscope and observe blue fluorescence (Fig. 6, see Color Plate).

5. Trypan Blue

Trypan blue is a commonly used dye to discriminate between live and dead mammalian cells because it stains only cells with a compromised plasma membrane. Staining with trypan blue or propidium iodide thus provides comparable information. Trypan

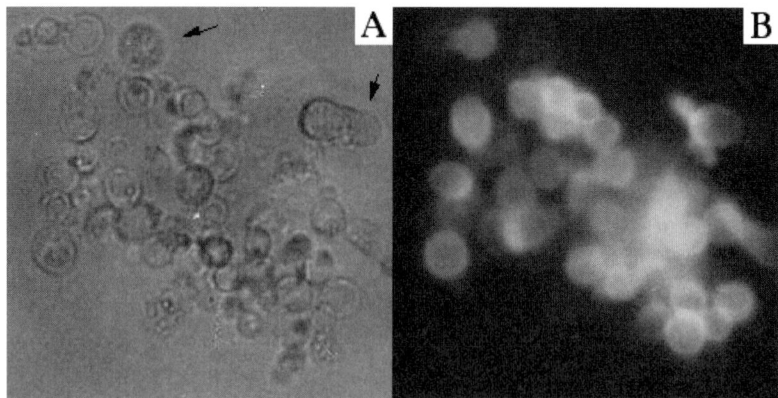

Fig. 6 Calcofluor staining of stalk cells. HMX44A cells incubated for 24 h in DIF on a glass coverslip. (A) Transmission image. (B) Blue fluorescence. Arrows point to cells that are not encased in cellulose (i.e., nonstalk cells). (See Color Plate.)

blue-stained cells are less obvious than PI-labeled cells, but because fluorescence is not required, trypan blue is more convenient for the routine quantification of live vs dead cells in a sample (Fig. 7, see Color Plate).

Trypan blue (Sigma T8154) is sold as a 0.4% stock solution. Dilute four times with SB and store at room temperature under sterile conditions (may be frozen; 0.1% sodium

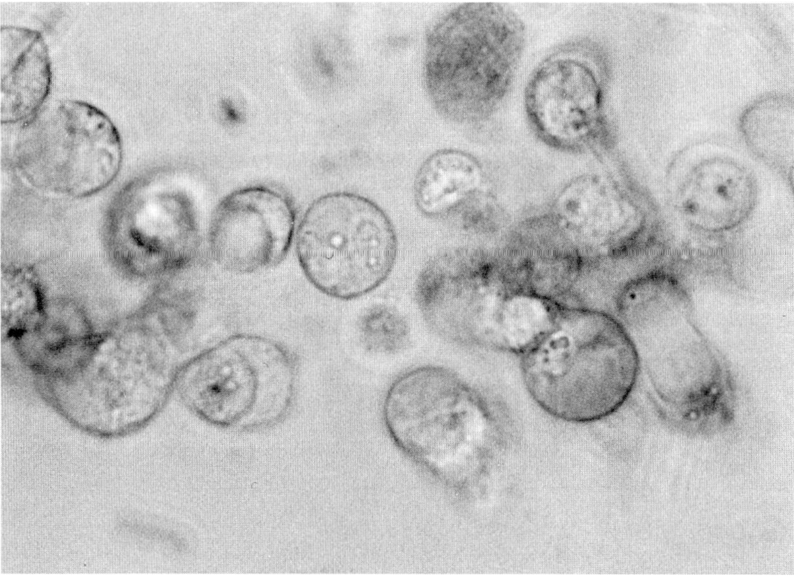

Fig. 7 Trypan blue staining of stalk cells. HMX44A cells incubated for 20 h in DIF on a glass coverslip. (See Color Plate.)

azide can be added to prevent contamination). Gloves should be worn when manipulating trypan blue, which is teratogenic and may also include carcinogenic compounds.

Trypan blue is simply added to cells at a final concentration of 0.1 to 0.03% for at least 10 min, and the sample is examined directly using standard microscopy. The liquid layer has to be thin; for cells in 25-cm^2 flasks, 1 ml of trypan blue solution is suitable. It is important not to allow cells to dry, as the liquid tends to accumulate as a meniscus at the corners of the flask. Results will be better optically if the cells are placed between a microscopic slide and a coverslip. Use of a blue filter enhances contrast.

V. Genetic Manipulation of *Dictyostelium*

A major advantage of *Dictyostelium* as a model organism is the ease with which it can be genetically manipulated (in particular because of the small size and the haploidy of its genome). Transfection is easy, homologous recombination is very efficient, and insertional mutagenesis is routinely performed in many laboratories. These three manipulations involve the same basic protocol to introduce DNA into *Dictyostelium,* which is given first, after which the specifics of each procedure are detailed. Convenient analysis through molecular genetics contrasts with the difficulty of generating diploid cells through the fusion of two haploid strains and to segregate again haploid progeny from the diploid (as detailed in Loomis, 1987). Such manipulations are time-consuming and two different selectable markers are required for the two partners of the fusion.

A. Selectable Markers

The first attempts to transform *Dictyostelium* with exogenous DNA in the 1980s used vectors containing a neomycin resistance gene to select transfected cells. Indeed, *Dictyostelium* is more sensitive to this class of protein synthesis inhibitors than many other eukaryotic cells: resistant cells can be selected in the presence of only 5 to 15 μg/ml of G418 (Nellen *et al.,* 1984). Several other selection procedures have been developed, including the ability to grow in the presence of hygromycin, phleomycin, bleomycin, or blasticidin (Egelhoff *et al.,* 1989; Leiting and Noegel, 1991; Chang *et al.,* 1991; Sutoh, 1993; Adachi *et al.,* 1994), as well as the selection of prototrophs from populations of auxotrophs (thymidine and uracil selection: Dynes and Firtel, 1989; Kalpaxis *et al.,* 1991). Several laboratories have used a selection system based on the complementation of the uracil pathway through transfection with a plasmid bearing the *Dictyostelium pyr5-6* gene (Kuspa and Loomis, 1992). However, strains to be transformed must carry a mutated *pyr5-6* gene in order not to grow in minimal medium in the absence of uracil (Kalpaxis *et al.,* 1991).

When possible we use blasticidin as a selection drug because we have found it to be simple and reliable. Cells are selected by adding blasticidin (from Invitrogen, Groningen, Netherlands) at a final concentration of 10 μg/ml. One copy of the blasticidin-resistance gene is sufficient to confer resistance, making this selection suitable for homologous recombination or insertional mutagenesis (Sutoh, 1993). When it is not possible to use blasticidin resistance (e.g., when transfecting an already blasticidin-resistant strain), we use neomycin selection, namely G418 sulfate (from GIBCO-BRL) at a final concentration of 10 to 15 μg/ml. It is wise to check the sensitivity of the different cell lines and the activity of different drug batches. G418-resistant cells appear to usually contain multiple copies of the plasmid (Knecht *et al.,* 1986), although homologous recombination has been achieved successfully using this marker (Manstein *et al.,* 1989). It is possible (in some cases at least) to select transfectants that have integrated a very large number of plasmid copies by gradually increasing the G418 concentration to 50 μg/ml (Luderus *et al.,* 1992).

B. General Protocol for Introducing DNA into *Dictyostelium* Cells

Initial attempts at transfecting *Dictyostelium* with exogenous DNA involved adapting the calcium phosphate procedure using G418 selection (Nellen *et al.,* 1984). Transfection frequency was low and results were poorly reproducible. Improvements came from using plasmids conferring a selectable advantage when only one copy is inserted in the genome (e.g., *thyA, pyr5-6,* blasticidin-resistance). Transfection frequency was also increased using DNA electroporation techniques (Howard *et al.,* 1988). The following protocol includes electroporation and a selection procedure based on blasticidin resistance and derives from the one described in Adachi *et al.* (1994).

1. Protocol

1. Collect exponentially growing vegetative cells (4×10^7 cells are needed for each transfection).
2. Chill the medium containing the cells on ice for 10 min; also cool the electroporation cuvettes (0.4-cm gap width; Bio-Rad, Hercules, CA) on ice.
3. Spin the cells down by centrifuging at 4°C, 700*g*, for 5 min.
4. Discard the growth medium and wash cells twice in ice-cold sterile electroporation buffer (10 m*M* $NaPO_4$, pH 6.1, 50 m*M* sucrose).
5. Count cells and resuspend at 5×10^7 cells/ml in electroporation buffer.
6. Per ice-cold cuvette, distribute 0.8-ml aliquots of cell suspension with 10 μg of transforming DNA and electroporate quickly in a Bio-Rad gene pulser (1 kV and 3 μF; expected pulse time: 0.6 to 1.1 ms).
7. After electroporation, cuvettes are left on ice for an additional 10 min.
8. Add 8 μl of a 1 : 1 mixture of 0.1 *M* $CaCl_2$ and 0.1 *M* $MgCl_2$ to the cell suspension in each cuvette and transfer into a 175-cm^2 Falcon flask (Becton-Dickinson Labware, NJ).
9. Incubate for 15 min at room temperature.
10. Add 50 ml of HL-5 medium to each flask and incubate at 22–23°C.
11. After ~24 h, add blasticidin to a final concentration of 10 μg/ml.
12. Five days later, almost all cells (presumably dead) have rounded up and do not adhere to the flask. Remove medium, pellet at 700*g* for 5 min, discard supernatant,

resuspend the pellet in 50 ml of fresh HL-5 containing 10 μg/ml of blasticidin, reintroduce into the flask, and let the selection proceed for an additional 5 days.

13. At this stage (10–12 days after the transfection), all growing cells are blasticidin resistant.

Notes: The plasmid DNA used for transfection can be purified on a CsCl gradient or prepared using Qiagen or Nucleobond plasmid extraction kits.

Several electroporations can be performed in succession on the same initial batch of cells distributed in several cuvettes, but it is important to proceed quickly to avoid a drop in transformation efficiency.

Blasticidin is diluted as a 4-mg/ml stock solution in source water, sterilized though a 0.22-μm filter, and stored at −20°C in 1-ml aliquots.

The population may be cloned directly in the presence of the selection drug 24 h after electroporation. The yield of stable transfectants depends on the cell line and DNA used, but one may expect a frequency in the order of 10^{-5}. In a first attempt it is advisable to plate three microtiter plates with 10^4, 3×10^4, and 10^5 cells/well, respectively. Positive wells should become evident after 10 days of selection.

For G418 selection, the protocol is similar; because nonresistant cells take a little more time to die than with blasticidin, two changes of medium instead of one (performed on days 3 and 7) are recommended.

Regarding expression of heterologous proteins, it should be stressed that proteins from other species are often poorly expressed in *Dictyostelium,* most likely because of the strong codon bias due to the extreme A/T richness of the genome, which reduces translation efficiency (Warrick and Spudich, 1988).

C. Insertional Mutagenesis and Gene Rescue

Even though electroporation significantly increases transfection efficiency, generally this was not sufficient for the isolation of genes based on the mutant phenotypes of the transformants. A further improvement in efficiency was realized by the use of the REMI approach. This was derived from experiments performed with yeast by Schiestl and Petes (1991), showing that adding the restriction enzyme *Bam*HI to the transfection mixture with a *Bam*HI-cut plasmid increased the number of transfectants. Moreover, the introduced plasmid was often integrated into *Bam*HI sites in the genome. Kuspa and Loomis (1992) adapted this technique to increase the efficiency of integration into the genome of *Dictyostelium* (Kuspa and Loomis, 1992). As a result, the plasmid integrates into the genome with increased frequency (4×10^{-5} or more) (Kuspa and Loomis, 1994) (in our hands, REMI performed on HMX44A yields stable blasticidin-resistant cells at a frequency of 3×10^{-5}). Plasmid integration leads to gene disruption, and the corresponding mutants often display developmental defects. An additional advantage of this technique is that the plasmid "tags" the disrupted gene, allowing for its subsequent identification through the analysis of the plasmid-flanking sequences. Haploidy of the *Dictyostelium* genome allows one to select recessive mutations (few organisms allow this). Selection of transfected cells in the initial procedure was based on uracil

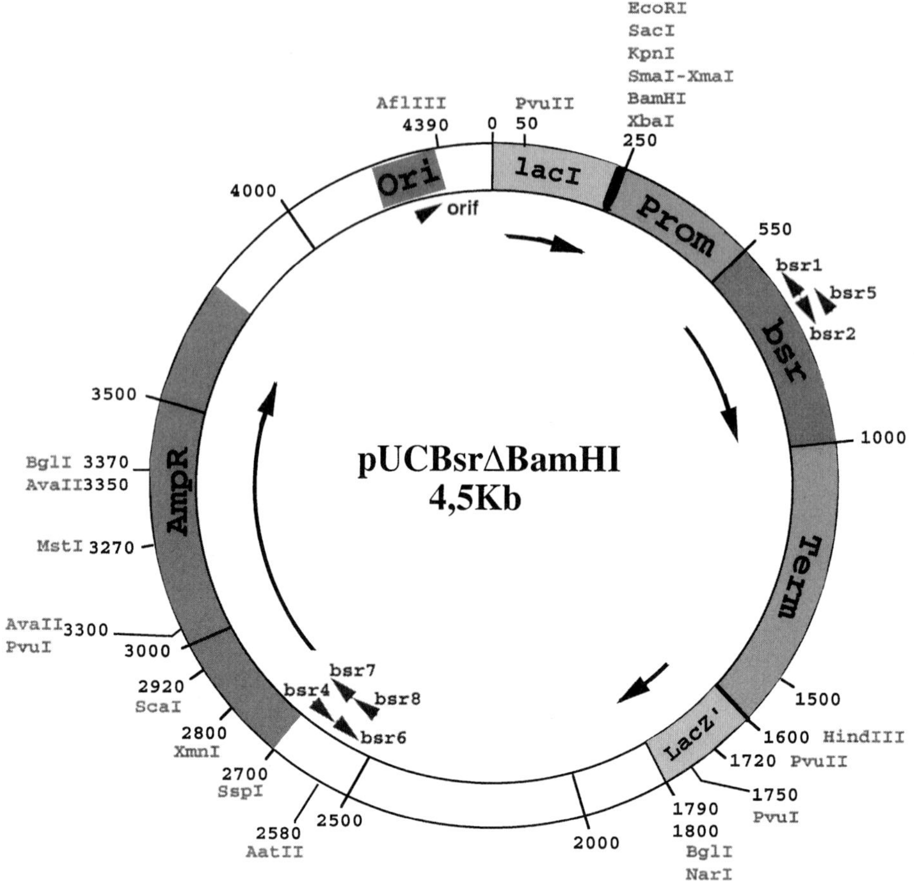

Fig. 8 Restriction map of pucBsrΔ*Bam*HI.

auxotrophy (Kuspa and Loomis, 1992) ; Adachi and colleagues (1994) later adapted it to blasticidin resistance selection using the pUCBsrΔ*Bam*HI plasmid (see map in Fig. 8).

The general strategy for generating insertional mutants in *Dictyostelium* and rescuing genomic sequences flanking the plasmid insertion is presented in Fig. 9.

1. Modifications to the Transfection Protocol of Section VB for REMI

a. The pucBsrΔ*Bam*HI plasmid bearing the blasticidin resistance cassette is linearized with *Bam*HI, purified by phenol–chloroform extraction, and ethanol precipitated.

b. At the sixth step, in addition to the 10 μg of linearized plasmid added to the cells, 10 units of the *Dpn*II restriction enzyme (New England Biolabs), which generates ends compatible with *Bam*HI, are added to the transfection mixture.

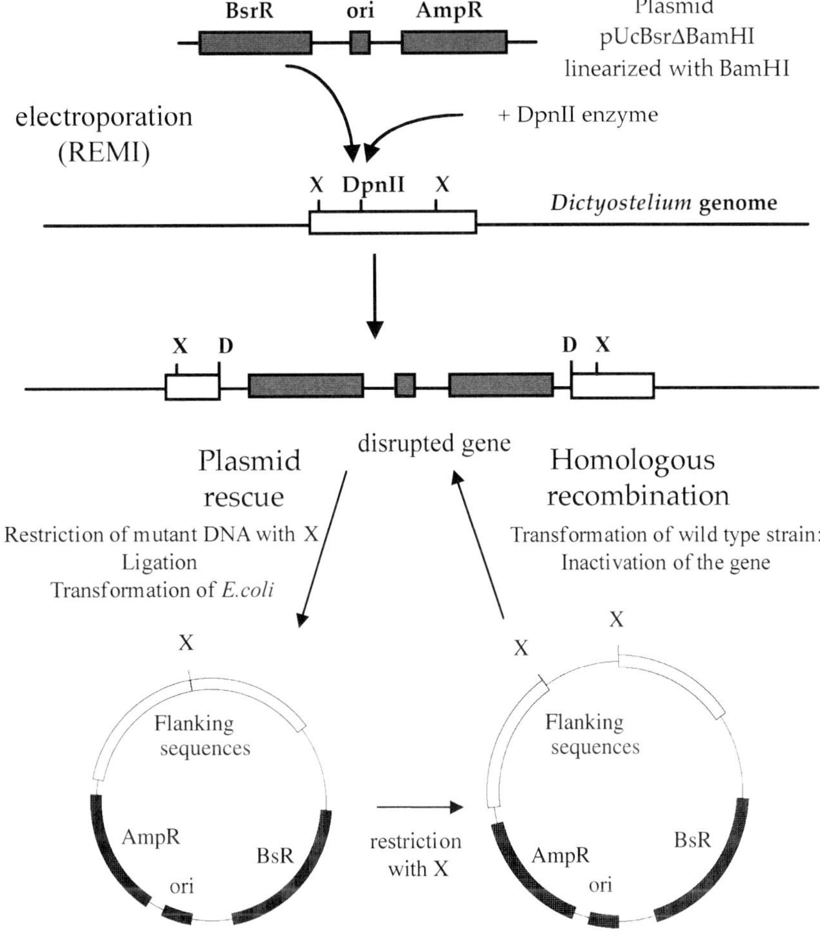

Fig. 9 Principle of insertional mutagenesis, gene rescue, and homologous recombination.

Blastidicin resistance means that the selected cells have incorporated a plasmid in their genome. Among these cells, one can further select for individuals in which plasmid insertion would have disrupted a gene involved in the function of interest (e.g., cell death, see Section VI).

Once a mutant of interest is obtained, it is in principle easy to identify the plasmid-disrupted gene by isolating the genomic sequences flanking the plasmid insertion site. The inserted plasmid and its flanking sequences are recovered from mutant *Dictyostelium* cells and cloned. This "plasmid rescue" is performed by restriction of the DNA from mutant cells, followed by recircularization of the plasmid and transformation in *Escherichia coli,* thanks to the presence of the origin of replication and ampicillin resistance gene on the recircularized plasmid.

2. Protocol to Clone the Inserted Plasmid Together with Its Flanking Insertion Sequences

1. Prepare genomic DNA from the mutant *Dictyostelium* cells. We had good results rescuing flanking sequences with DNA purified with the protocol of Chang *et al.* (1995) without CsCl gradient purification.

2. Make a restriction map of the genomic DNA of the mutant using enzymes that do not cut into the integrated plasmid (enzymes will therefore cut flanking sequences). Probe these restriction fragments on Southern blots with radiolabeled vector. Choose the restriction enzyme that gives a positive hybridization signal from fragments of at least 6 kb (i.e., including at least 1.5 kb of flanking sequences) and shorter than 10 kb to be sure that the recircularized plasmid can be stably maintained in *E. coli.* In addition, when using the pUCBsrΔ*Bam*HI plasmid, cutting genomic DNA with *Pvu*II (see map of the vector) should generate a 3-kb fragment that hybridizes with a probe containing the origin of replication and the ampicillin resistance gene from the vector. The presence and conserved length of this positive hybridization signal suggest that no modification impairing plasmid rescue occurred in this critical region of the plasmid.

3. Digest 3 μg of genomic DNA with the selected enzyme, precipitate it with ethanol, and then dissolve it in 100 μl of water.

4. Ligation is performed in a volume of 500 μl with 20 units of T4 DNA ligase (GIBCO-BRL) overnight at 16°C. This low DNA concentration favors plasmid recircularization rather than ligation of plasmid concatemers. The ligation product is ethanol precipitated (do not add salts) and resuspended in 80 μl of water.

5. Typically, 10 transformations of electrocompetent SURE *E. coli* cells (Stratagene) are performed using a Bio-Rad gene pulser (2.5 kV, 400 Ω, 25 μF), each with 2 μl ligated DNA. Using very small aliquots of ligated DNA prevents a decrease of transformation efficiency (DNA concentration above 50 ng is toxic to *E. coli*).

6. Transformed bacteria harboring the rescued plasmid are selected on ampicillin plates.

D. Homologous Recombination

Targeting genes by homologous recombination is very efficient and reliable in *Dictyostelium*. Theoretically, high frequency homologous recombination only requires about 1 kb of the targeted gene flanking each side of a dominant selectable gene (such as blasticidin resistance). In fact, shorter fragments are often sufficient; homologous recombination can be obtained with a construct containing 1.2 kb of *Dictyostelium* DNA on one side of the selectable gene and as little as 200 bp of flanking DNA on the other side (unpublished results).

A major application of homologous recombination in molecular genetics of *Dictyostelium* relates to insertional mutagenesis and plasmid rescue (Fig. 9). After obtaining a plasmid bearing the sequences flanking the insertion site (plasmid rescue), it is important to use it to verify that the original mutant phenotype is indeed a consequence of the insertion, i.e., of the disruption of this gene. To generate a similar insertion, the rescued plasmid has to be linearized (with the enzyme used to digest the mutant genomic DNA in the rescue process), purified by phenol–chloroform extraction, and then transfected into the initial cell line used for mutagenesis using the protocol of Section V,B. Homologous recombination allows the integration of the plasmid within the corresponding gene in a high proportion of transfectants: it is usually sufficient to screen (by Southern blotting) a dozen blasticidin-resistant clones stemming from the transfection to obtain several homologous recombinants. Identity of the phenotypes of this transformant and of the original mutant strongly suggests that disruption of this gene caused the mutant phenotype. Definitive proof would be the restoration of a wild-type phenotype through complementation of the mutant by ectopic expression of the corresponding gene.

The rescued plasmid can also be used to disrupt the gene in various genetic backgrounds: for example, in our study of PCD in *Dictyostelium,* insertional mutagenesis is performed on HMX44A cells to select death-resistant mutants (see Section VI) but it is of interest to check the possible developmental phenotype of the same mutation in a strain that can undergo morphogenesis (e.g., AX2).

VI. Selection and Analysis of Cell Death-Resistant Mutants

Programmed cell death can be studied in *Dictyostelium* cells using HMX44A strain in *in vitro* conditions involving differentiation without morphogenesis (see Section III,A). For insertional mutagenesis (see Section V), we electrotransfect the pUCBsrΔ*Bam*HI plasmid into vegetatively growing cells. Blasticidin-resistant cells are then tested for resistance to death when induced to differentiate under starvation conditions. Surviving cells (putative cell death mutants) are recovered by their ability to regrow in rich medium. As 10 to 20% of untransfected cells do not die (see Section IV,A), successive cycles of differentiation–regrowth are required to allow death-resistant mutants to emerge from this background (Cornillon *et al.,* 1994, 1998). The selection of a cell death-resistant mutant (if present) may require up to eight rounds of differentiation–regrowth, which takes at least 2 months after the initial electrotranfection. In our experience, use of the techniques described here results in a frequency of death-resistant mutants of about 10^{-9}.

The general strategy used to obtain cell death mutants is summarized in Fig. 10.

1. Protocol to Select Cell Death Mutants

1. Populations of blasticidin-resistant cells from independent electroporations are pooled two by two.

2. Logarithmically growing cells (3×10^6) from each pool and HMX44A cells (as control) are washed twice in SB.

3. Cell pellets are resuspended in 7.5 ml SB buffer to which 375 μl of 60 m*M* cAMP is added. Cell suspensions are poured into 75-cm^2 Falcon flasks maintained horizontally.

4. Cells are allowed to differentiate for 8 h at 22–23°C.

5. Trying to minimally disturb cells, medium is removed and a wash with 15 ml SB is performed to eliminate excess of cAMP.

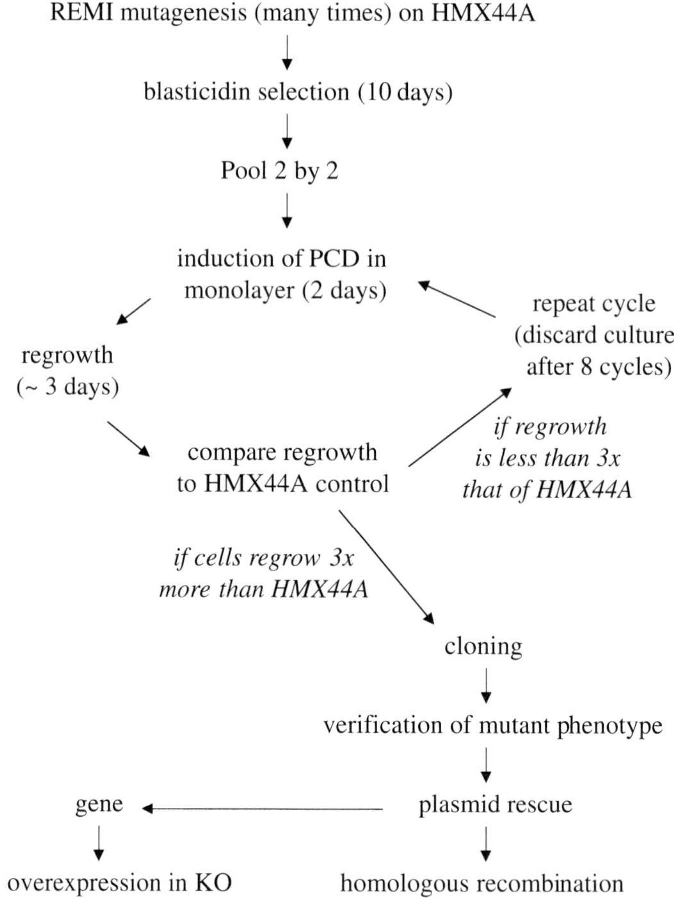

Fig. 10 Summary of the procedure to select cell death mutants in *Dictyostelium*.

6. SB with 100 n*M* DIF-1 is added to the cells (typically 7.5 ml SB buffer supplemented with 7.5 μl of 10^{-4} *M* DIF-1 in absolute ethanol). Cells are incubated for an additional 24 h at 22–23°C. This leads to the death of most cells.

7. The medium is then removed and HL-5 plus 10 μg/ml of blasticidin (or without blasticidin for HMX44A cells) is added to allow regrowth of surviving cells for 40 to 64 h.

8. Cells are then recovered, counted (for an evaluation of the surviving transfected cells/wild-type cells ratio, see Section IV,A), and subjected to another cycle of differentiation–regrowth beginning at step 2. Typically, up to eight cycles of differentiation–regrowth are performed with pools of blasticidin-resistant cells and HMX44A control cells.

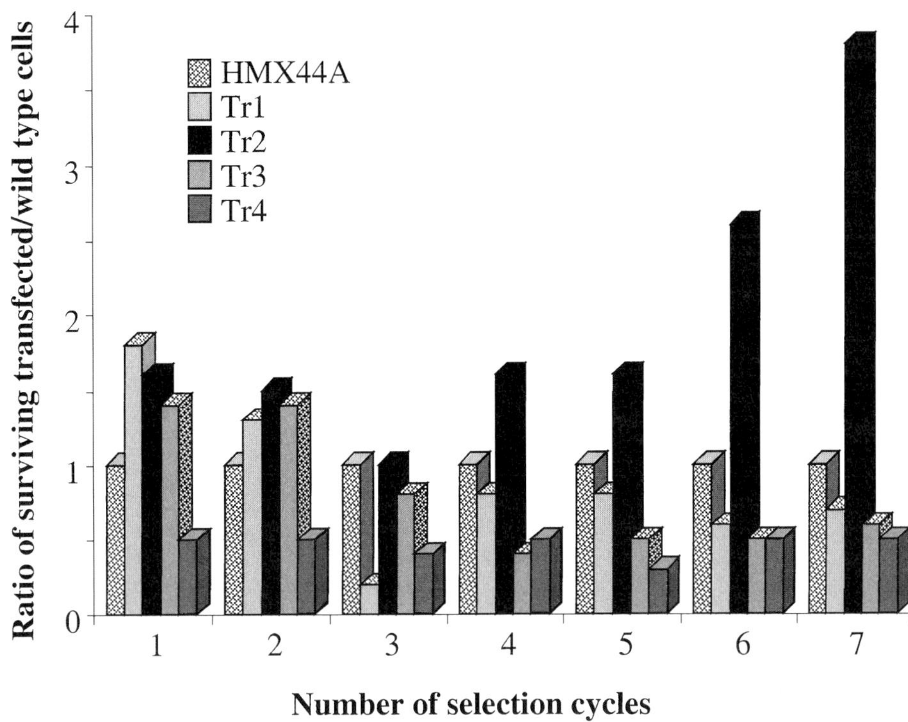

Fig. 11 Representative course of the emergence of a cell death mutant on repeated differentiation–regrowth cycles.

9. When cell death-resistant mutants within a population start to emerge from the background (see an example in Fig. 11; mutants emerged at cycle 6), subsequent cycles of differentiation–regrowth show an important increase of the ratio of surviving transfected cells/wild-type cells.

10. Emerging cell death-resistant cell populations are then cloned by limiting dilution in microplates (see Section II,D).

The cell death-resistant phenotype of the clones obtained is ascertained by the methods described in Section IV. Genuine death-resistant cells are then eligible for plasmid rescue, as described in Section V,C, in order to identify the disrupted gene. The causality of the disruption of the tagged gene for the cell death-resistant phenotype is then verified by disrupting this gene again in the original cell line (and possibly cell lines of other backgrounds) by homologous recombination using the rescued plasmid (see Section V,D). Complementation by the wild-type gene introduced in a mutant cell line formally demonstrates the involvement of this gene in cell death.

In addition to the investigative methods detailed in Section IV, an obvious step is to compare the expression of relevant genes (i.e., developmental markers) in wild-type and mutant cells by Northern blot analysis. For example, for each cell death-resistant mutant we usually study the expression level of two markers of prestalk and stalk cells, *ecmA* and *ecmB*, whose transcription is DIF-1 dependent (Williams *et al.,* 1987; Jermyn *et al.,* 1987). Because cells subjected to DIF adhere very tightly to the substrate, cells must be detached with a cell scraper for RNA extraction; otherwise the Northern blot analysis of *Dictyostelium* cells is fairly standard. We use Trizol (GIBCO-BRL) to extract RNA.

A. Appendix: Addresses of a Few Useful Dicty Web Sites

For a description of more techniques: http://dicty.cmb.nwu.edu/Chis_lab/Lab%20 Manual/dictyostelium_techniques.htm

The *Dictyostelium* genome information site: http://glamdring.ucsd.edu:80./others/ dsmith/dictydb.html

Much information (virtual library): http://dicty.cmb.nwu.edu/dicty/dicty.html

To perform BLAST search on the genome data gathered so far: http://genome.imb-jena.de/dictyostelium/BlastDicty.html

Access to the EST database: http://www.csm.biol.tsukuba.ac.jp/cDNAproject.html

Finally, a very useful mailing list for Dicty addicts: http://dicty.cmb.nwu.edu/dicty/ listserv.html

Acknowledgments

We thank the many members of the *Dictyostelium* community for advice, reagents, and discussions, particularly Richard A. Firtel (University of California, San Diego), Julian D. Gross (University of Oxford), Robert R. Kay (Medical Research Council, Cambridge), Michel Satre and Gérard Klein (Commissariat à l'Energie Atomique, Grenoble), Michel Véron (Institut Pasteur, Paris), and Jeffrey G. Williams (University of Dundee); for institutional support, INSERM (Institut National de la Santé et de la Recherche Médicale) and CNRS (Centre National de la Recherche Scientifique); and for additional support, the European Community, ARC (Association pour la Recherche contre le Cancer) and LNCC (Ligue Nationale Contre le Cancer). J.-P. Levraud is a fellow from the Pasteur Institute and M. Adam is the recipient of a TMR fellowship from the EEC.

References

Adachi, H., Hasebe, T., Yoshinaga, K., Ohta, T., and Sutoh, K. (1994). Isolation of *Dictyostelium discoideum* cytokinesis mutants by restriction enzyme-mediated integration of the blasticidin S resistance marker. *Biochem. Biophys. Res. Commun.* **205,** 1808–1814.

Baldauf, S. L., and Doolittle, W. F. (1997). Origin and evolution of the slime molds. *Proc. Natl. Acad. Sci. USA* **94,** 12007–12012.

Berks, M., and Kay, R. R. (1988). Cyclic AMP is an inhibitor of stalk cell differentiation in *Dictyostelium discoïdeum. Dev. Biol.* **126,** 108–114.

Blanton, R. L. (1993). Prestalk cells in monolayer cultures exhibit two distinct modes of cellulose synthesis during stalk cell differentiation in *Dictyostelium. Development* **119,** 703–710.

Chang, A. C., Hall, R. M., and Williams, K. L. (1991). Bleomycin resistance as a selectable marker for transformation of the eukaryote, *Dictyostelium discoideum. Gene* **107,** 165–170.

Chang, W. T., Gross, J. D., and Newell, P. C. (1995). Trapping developmental promoters in *Dictyostelium*. *Plasmid* **34,** 175–183.

Christen, R., Ratto, A., Baroin, A., Perasso, R., Grell, K. G., and Adoutte, A. (1991). An analysis of the origin of metazoans, using comparisons of partial sequences of the 28S RNA, reveals an early emergence of triploblasts. *EMBO J.* **10,** 499–503.

Cornillon, S., Foa, C., Davoust, J., Buonavista, N., Gross, J. D., and Golstein, P. (1994). Programmed cell death in *Dictyostelium*. *J. Cell Sci.* **107,** 2691–2704.

Cornillon, S., Olie, R. A., and Golstein, P. (1998). An insertional mutagenesis approach to *Dictyostelium* cell death. *Cell Death Differ.* **5,** 416–425.

de Chastellier, C., and Ryter, A. (1977). Changes of the cell surface and of the digestive apparatus of *Dictyostelium discoïdeum* during the starvation period triggering aggregation. *J. Cell Biol.* **75,** 218–236.

Dynes, J. L., and Firtel, R. A. (1989). Molecular complementation of a genetic marker in *Dictyostelium* using a genomic DNA library. *Proc. Natl. Acad. Sci. USA* **86,** 7966–7970.

Egelhoff, T. T., Brown, S. S., Manstein, D. J., and Spudich, J. A. (1989). Hygromycin resistance as a selectable marker in *Dictyostelium discoideum*. *Mol. Cell. Biol.* **9,** 1965–1968.

Field, K. G., Olsen, G. J., Lane, D. J., Giovannoni, S. J., Ghiselin, M. T., Raff, E. C., Pace, N. R., and Raff, R. A. (1988). Molecular phylogeny of the animal kingdom. *Science* **239,** 748–753.

George, R. P., Hohl, H. R., and Raper, K. B. (1972). Ultrastructural development of stalk-producing cells in *Dictyostelium discoïdeum,* a cellular slime mould. *J. Gen. Microbiol.* **70,** 477–489.

Greenberg, J. T. (1996). Programmed cell death: A way of life for plants. *Proc. Natl. Acad. Sci. USA* **93,** 12094–12097.

Howard, P. K., Ahern, K. G., and Firtel, R. A. (1988). Establishment of a transient expression system for *Dictyostelium discoideum*. *Nucleic Acid Res.* **16,** 2613–2623.

Jacobson, M. D., Weil, M., and Raff, M. C. (1997). Programmed cell death in animal development. *Cell* **88,** 347–354.

Jermyn, K. A., Berks, M., Kay, R. R., and Williams, J. G. (1987). Two distinct classes of prestalk-enriched mRNA sequences in *Dictyostelium discoideum*. *Development* **100,** 745–755.

Kaiser, D. (1986). Control of multicellular development: *Dictyostelium* and *Myxococcus*. *Annu. Rev. Genet.* **20,** 539–566.

Kalpaxis, D., Zundorf, I., Werner, H., Reindl, N., Boy-Marcotte, E., Jasquet, M., and Dingermann, T. (1991). Positive selection for *Dictyostelium discoideum* mutants lacking UMP synthase activity based on resistance to 5-fluoroorotic acid. *Mol. Gen. Genet.* **225,** 492–500.

Kawata, T., Shevchenko, A., Fukuzawa, M., Jermyn, K. A., Totty, N. F., Zhukovskaya, N. V., Sterling, A. E., Mann, M., and Williams, J. G. (1997). SH2 signaling in a lower eukaryote: a STAT protein that regulates stalk cell differentiation in Dictyostelium. *Cell* **89,** 909–916.

Kay, R. R. (1987). Cell differentiation in monolayers and the investigation of slime mold morphogens. *In* "Methods in Cell Biology" (J. A. Spudich, ed.), Vol. 28, pp. 433–448. Academic Press, New York.

Kay, R. R., and Williams, J. G. (1999). The *Dictyostelium* genome project: An invitation to species hopping. *Trends Genet.* **15,** 294–297.

Knecht, D. A., Cohen, S. M., Loomis, W. F., and Lodish, H. F. (1986). Developmental regulation of *Dictyostelium discoideum* actin gene fusions carried on low-copy and high-copy transformation vectors. *Mol. Cell. Biol.* **6,** 3973–3983.

Kopachik, W., Oohata, A., Dhokia, B., Brookman, J. J., and Kay, R. R. (1983). *Dictyostelium* mutants lacking DIF, a putative morphogen. *Cell* **33,** 397–403.

Kuspa, A., Dingermann, T., and Nellen, W. (1995). Analysis of gene function in *Dictyostelium*. *Experientia* **51,** 1116–1123.

Kuspa, A., and Loomis, W. F. (1992). Tagging developmental genes in *Dictyostelium* by restriction enzyme-mediated integration of plasmid DNA. *Proc. Natl. Acad. Sci. USA* **89,** 8803–8807.

Kuspa, A., and Loomis, W. F. (1994). Transformation of *Dictyostelium:* Gene disruptions, insertional mutagenesis, and promoter traps. *Methods Mol. Gen.* **3,** 3–21.

Leiting, B., and Noegel, A. A. (1991). The *ble* gene of *Streptoalloteichus hindustanus* as a new selectable marker for *Dictyostelium discoideum* confers resistance to phleomycin. *Biochem. Biophys. Res. Commun.* **180,** 1403–1407.

Loomis, W. F. (1980). Genetic analysis of development in *Dictyostelium. In* "The Molecular Genetics of Development" (T. Leighton and W. F. Loomis, eds.), pp. 179–212. Academic Press, New York.

Loomis, W. F. (1987). Genetic tools for *Dictyostelium discoideum. In* "Methods in Cell Biology" (J. A. Spudich, ed.), Vol. 28, pp. 31–65. Academic Press, New York.

Loomis, W. F., Welker, D., Hughes, J., Maghakian, D., and Kuspa, A. (1995). Integrated maps of the chromosomes in *Dictyostelium discoideum. Genetics* **141,** 147–157.

Luderus, M. E., Kesbeke, F., Knetsch, M. L., Van Driel, R., Reymond, C. D., and Snaar-Jagalska, B. E. (1992). Ligand-independent reduction of cAMP receptors in *Dictyostelium discoideum* cells over-expressing a mutated ras gene. *Eur. J. Biochem.* **208,** 235–240.

Maeda, Y., and Takeuchi, I. (1969). Cell differentiation and fine structures in the development of the cellular slime molds. *Dev. Growth Differ.* **11,** 232–245.

Manstein, D. J., Titus, M. A., De Lozanne, A., and Spudich, J. A. (1989). Gene replacement in *Dictyostelium:* Generation of myosin null mutants. *EMBO J.* **8,** 923–932.

Morris, H. R., Taylor, G. W., Masento, M. S., Jermyn, K. A., and Kay, R. R. (1987). Chemical structure of the morphogen differentiation inducing factor from *Dictyostelium discoideum. Nature* **328,** 811–814.

Nellen, W., Silan, C., and Firtel, R. A. (1984). DNA-mediated transformation in *Dictyostelium discoideum:* regulated expression of an actin gene fusion. *Mol. Cell. Biol.* **4,** 2890–2898.

Olie, R. A., Durrieu, F., Cornillon, S., Loughran, G., Gross, J., Earnshaw, W. C., and Golstein, P. (1998). Apparent caspase independence of programmed cell death in *Dictyostelium. Curr. Biol.* **8,** 955–958.

Quiviger, B., Benichou, J.-C., and Ryter, A. (1980). Comparative cytochemical localization of alkaline and acid phosphatases during starvation and differentiation of *Dictyostelium discoïdeum. Biol. Cell.* **37,** 241–250.

Raper, K. B. (1935). *Dictyostelium discoideum,* a new species of slime mold from decaying forest leaves. *J. Agric. Res.* **50,** 135–147.

Raper, K. B. (1984). "The Dictyostelids," Princeton Univ. Press, Princeton.

Raper, K. B., and Fennell, D. I. (1952). Stalk formation in *Dictyostelium. Bull. Torrey Bot. Club* **79,** 25–51.

Schaap, P., van der Molen, L., and Konijn, T. M. (1981). The vacuolar apparatus of the simple cellular slime mold *Dictyostelium minutum. Biol. Cell* **41,** 133–142.

Schaap, P., Nebl, T., and Fisher, P. R. (1996). A slow sustained increase in cytosolic Ca^{2+} levels mediates stalk gene induction by differentiation inducing factor in *Dictyostelium. EMBO J.* **15,** 5177–5183.

Schiestl, R. H., and Petes, T. D. (1991). Integration of DNA fragments by illegitimate recombination in *Saccharomyces cerevisiae. Proc. Natl. Acad. Sci. USA* **88,** 7585–7589.

Sobolewski, A., Neave, N., and Weeks, G. (1983). The induction of stalk cell differentiation in submerged monolayers of *Dictyostelium discoïdeum*: Characterization of the temporal sequence for the molecular requirements. *Differentiation* **25,** 93–100.

Spann, T. P., Brock, D. A., Lindsey, D. F., Wood, S. A., and Gomer, R. H. (1996). Mutagenesis and gene identification in *Dictyostelium* by shotgun antisense. *Proc. Natl. Acad. Sci. USA* **93,** 5003–5007.

Sussman, M. (1987). Cultivation and synchronous morphogenesis of *Dictyostelium* under controlled experimental conditions. *In* "Methods in Cell Biology" (J. A. Spudich, ed.), Vol. 28, pp. 9–29. Academic Press, New York.

Sutoh, K. (1993). A transformation vector for *Dictyostelium discoideum* with a new selectable marker *bsr. Plasmid* **30,** 150–154.

Town, C. D., Gross, J. D., and Kay, R. R. (1976). Cell differentiation without morphogenesis in *Dictyostelium discoïdeum. Nature* **262,** 717–719.

Town, C., and Stanford, E. (1979). An oligosaccharide-containing factor that induces cell differentiation in *Dictyostelium discoïdeum. Proc. Natl. Acad. Sci. USA* **76,** 308–312.

Vaux, D. L., and Korsmeyer, S. J. (1999). Cell death in development. *Cell* **96,** 245–254.

Warrick, H. M., and Spudich, J. A. (1988). Codon preference in *Dictyostelium discoideum. Nucleic Acid Res.* **16,** 6617–6635.

Watts, D. J., and Ashworth, J. M. (1970). Growth of myxamoebae of the cellular slime mould *Dictyostelium discoïdeum* in axenic culture. *Biochem. J.* **119,** 171–174.

Weismann, A. (1890). Prof. Weismann's theory of heredity. *Nature* **41,** 317–323.

Whittingham, W. F., and Raper, K. B. (1960). Non-viability of stalk cells in *Dictyostelium. Proc. Natl. Acad. Sci. USA* **46,** 642–649.

Williams, J. G., Ceccarelli, A., McRobbie, S., Mahbybani, S., Kay, R. R., Early, A., Berks, M., and Jermyn, K. A. (1987). Direct induction of *Dictyostelium* prestalk gene expression by DIF provides evidence that DIF is a morphogen. *Cell* **49,** 185–195.

CHAPTER 22

Methods of Study of Tumor Necrosis Factor-Related Ligands in Apoptosis

Isabelle A. Rooney, Chris A. Benedict, Paula S. Norris, and Carl F. Ware

La Jolla Institute for Allergy and Immunology
San Diego, California 92121

0091-679X/01 $35.00

I. Introduction

The tumor necrosis factor (TNF)-TNF Receptor(TNFR) superfamily currently contains more than 20 ligand-receptor systems that play essential roles in inflammatory and immune responses, and during the development of lymphoid, neurologic, and ectodermal tissues (reviewed in Smith *et al.,* 1994; Ware *et al.,* 1998; Wallach *et al.,* 1999) (Table I).

Table I
Members of TNF-Related Cytokine Receptor Superfamily[a]

Receptor	Ligand	Function
TNFR60(R1)	TNF/LTα3/LTα2β	Apoptosis/inflammation
TNFR80(R2)	TNF/LTα3/LTα2β	Apoptosis/proliferation
LTβR	LTα1β2/LIGHT	Apoptosis/lymph node development
HveA(HVEM)	LIGHT/LTα	Herpesvirus entry/costimulation
Fas/CD95	Fas L	Apoptosis/immune privilege
CD40	CD40 L	Cell survival/isotype switch
CD30	CD30 L	Apoptosis/negative selection
CD27	CD27 L	Costimulation
OX40	OX40 L	Costimulation
41BB	41BB L	Costimulation
p75NTR	Neurotrophins, NGF	Cell survival
TRAIL-R1(DR4)	TRAIL	Apoptosis/NF-κB
TRAIL-R2(DR5)	TRAIL	Apoptosis/NF-κB
TRAIL-R3(TRID/DcRI)	TRAIL	Decoy receptor for TRAIL
TRAIL-R4(DcR2)	TRAIL	Apoptosis/NF-κB
TRAMP (DR3/WSL-1/LARD/APO-3)	?	Apoptosis
GITR	GTIRL	Inhibits TCR-induced apoptosis
TRANCE-R (RANK)	TRANCE (RANKL)	Osteoclast differentiation
Osteoprotegerin	TRANCE/TRAIL	Soluble regulator of TRANCE
?	VEGI(TL1)	Endothelial cell growth inhibition
?	TWEAK	Inducer of weak apoptosis
TACI	APRIL/BAFF (Blys, TALL, THANK, zTNF4)	B-cell proliferation
BCMA	APRIL/BAFF	B-cell proliferation
DcR3	FasLigand/LIGHT	Soluble decoy receptor
EDAR (Downless)	EDA1 (tabby)	X-linked anhidrotic ectodermal dysplasia
Viral Homologues:		
UL144	?	Human cytomegalovirus
T2	TNF/LTα	poxvirus
Envelope gD	HveA	Herpes simplex virus
CAR1	TRAIL	Avian leukocytosis virus

Apo, apoptosis; APRIL, a proliferation inducing ligand; BAFF, B-cell-activating factor belonging to the TNF family; BCMA, B-cell maturation antigen; CAR-1, cytopathic ALV receptor; CD, cluster of differentiation; DR3, 4, 5, death receptor-3, -4, -5; EDA, ectodermal dysplasin; HveA (HVEM); herpes virus entry mediator; GITR, glucocorticoid-induced TNF receptor; LT, lymphotoxin; LIGHT, LT-like ligand competitive with gD-1 HSV for HVEM expressed on T cells; NTR, neurotrophin (nerve growth factor) receptor; TACI, transmembrane activator and CAML interactor; TNF, tumor necrosis factor, TRAIL, TNF-related apoptosis-inducing ligand; TRAMP, TNF receptor-associated membrane protein; TRANCE, TNF-related activation-induced cytokine; TWEAK, TNF-related ligand with weak apoptosis activity; TRAIL, receptor without an intracellular domain; DcR, decoy receptor-1, -2, or -3; VEGI, vascular endothelial growth inhibitor.

TNF-related ligands are type II transmembrane proteins that form trimers; an exception is lymphotoxin-α (LTα), which can also form heterotrimers with LTβ (Ware *et al.,* 1995). All of the ligands are active as membrane proteins, and some have soluble secreted forms that can act on surrounding tissue. These multivalent ligands initiate signal transduction by aggregation of their specific cell surface receptors, which are type I transmembrane glycoproteins. Several TNF receptor-like proteins lack a signaling domain but retain ligand-binding activity and may function as decoys regulating the bioavailability of the ligand for the signaling receptor counterpart (Ashkenazi and Dixit, 1999; Degli-Esposti *et al.,* 1997; Marsters *et al.,* 1997; Simonet *et al.,* 1997; Emery *et al.,* 1998; Pitti *et al.,* 1998; Yu *et al.,* 1999). Ligand binding to cell surface receptors leads to the activation of signaling pathways that activate transcription factors, such as Nuclear Factor (NF)-κB or AP1, or the cytosolic caspase cascade that induces apoptosis. Although several ligands and receptors within the family share one or more cognate receptors, for the most part each cytokine-receptor system has unique nonredundant roles in cellular responses (Fig. 1).

Based on structural features of their cytoplasmic domains, two major subgroups of the TNFR family can be defined. One group utilizes a protein interaction motif known as the death domain (DD), which promotes homotypic and heterotypic interactions with other death domain-containing proteins (Boldin *et al.,* 1995). The second group utilizes members of the TNF receptor-associated factor (TRAF) family of zinc ring-finger proteins to initiate signaling events (Arch *et al.,* 1998). Both DD and TRAF family act as adaptors that propagate signals to downstream pathways. DD-containing receptors, as examples Fas (Itoh and Nagata, 1993; Nagata, 1997), TNFR1 (Tartaglia *et al.,* 1993), and TRAIL receptors 1 and 2 (Schneider *et al.,* 1997b), induce rapid apoptosis in certain cell types. Their basic signaling pathway involves recruitment of FADD, a DD-containing adaptor that recruits and activates caspase-8, which in turn proteolytically activates executioner caspases, such as caspase-3 (Green and Reed, 1998). Fas induces apoptosis in normal lymphoid cells within 2 to 8 h, but the time course of death varies with cell type and may take longer for some cells (24 h).

In contrast, TRAF-dependent receptors, such as LTβR, CD30, or TNFR2, induce a slow apoptotic death that requires 3–4 days (Browning *et al.,* 1996). This slow apoptotic death is dependent on TRAF3, although how this is connected to the caspase pathway has not been determined (VanArsdale *et al.,* 1997; Force *et al.,* 1997). Most normal nontransformed cells use complex mechanisms to resist apoptosis induced by these receptors, which includes in part gene induction by NF-κB (Van Antwerp *et al.,* 1996). TRAF-dependent receptors are potent activators of NF-κB via TRAF2 or TRAF5, which can confer resistance to apoptotic signaling. Induction of cell death by TNFR2 and CD30 may involve induction of TNF, which in turn acts via TNFR1 (Grell *et al.,* 1999).

TNFR can induce both apoptotic and necrotic cell death, the latter being characterized by cell swelling. Interferon-γ (INF-γ) provides a potent enhancing effect on apoptosis induced by TNF ligands, and in some tumor lines the combination of the two cytokines is required (Sugarman *et al.,* 1985). The HT29.14S human colon carcinoma cell line requires IFN-γ for death induced by either DD- or TRAF-containing receptors (Browning *et al.,* 1996). Treatment of a typical tumor line with TNF, Fas ligand, or LT induces only a fraction of the population to die. A nuance of many tumor lines is that both sensitive and resistance clones reside within the population. Sensitive clones can be isolated readily

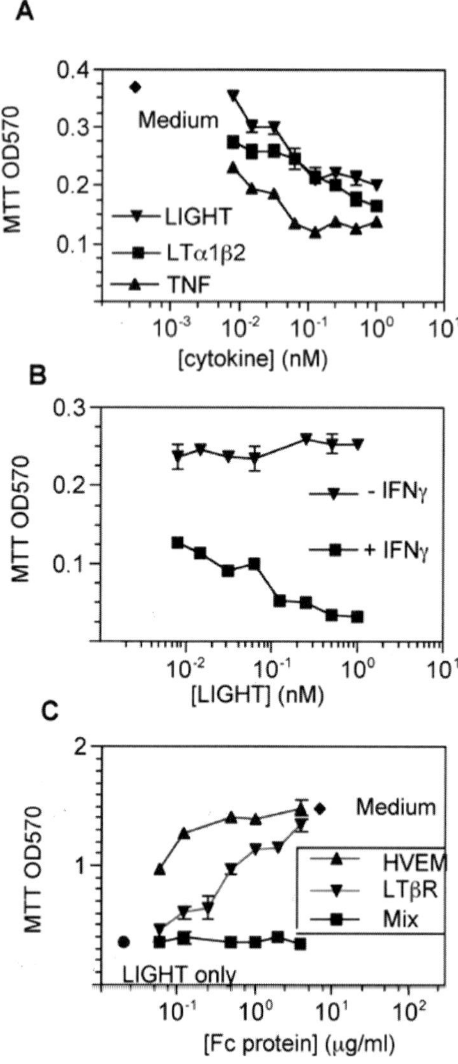

Fig. 1 LIGHT-mediated killing of HT29 cells. (A) HT29 cells were incubated with serial dilutions of LIGHT-FLAG, LTα1β2, or TNF in the presence of IFN-γ (80 U/ml.). After 72 h incubation at 37°C, cell viability was assessed by MTT dye reduction. (B) LIGHT-FLAG cytotoxicity is dependent on IFN-γ. HT29.14S cells were incubated with serial dilutions of LIGHT-FLAG in the presence or absence of IFN-γ (80 U/ml), and an MTT dye reduction assay was performed after 72 h incubation at 37°C. (C) LIGHT-FLAG cytotoxicity is blocked by coincubation with LTβR-Fc and HveA-Fc. LIGHT-FLAG (200 p*M*) was preincubated with varying dilutions of LTβR-Fc, HveA-Fc, or an equal mixture of Fas-Fc, TNFR1-FC, and TRAIL R-Fc for 30 min prior to its addition to HT29.14S cells in the presence of IFN-γ. Cell viability was assessed by MTT dye reduction after 72 h. Data represent the mean ± SD of triplicate wells and the results are representative of several different experiments. Reproduced with permission from Rooney *et al.* (2000).

by simple limit-dilution cloning and selecting for a sensitivity-based response to TNF (although the underlying molecular basis for this resistance is unknown). Typically the frequency of ligand-sensitive to ligand-insensitive clones is 1 : 10.

A wide variety of vertebrate viruses, including herpesvirus, adenovirus, poxvirus, and flavivirus, have evolved specific mechanisms that modify or mimic components of the TNF superfamily (Barry and McFadden, 1998). These mechanisms include secretion of decoy TNF receptors (poxvirus) (McFadden *et al.,* 1997), homologues of TNFR (human cytomegalovirus) (Benedict *et al.,* 1999), and a virokine (herpes simplex) that mimics the TNF-related cytokine LIGHT (Mauri *et al.,* 1998), among other examples (Zhu *et al.,* 1998; Mosialos *et al.,* 1995; Thome *et al.,* 1997).

This chapter describes production and characterization of standard reagents used in TNF research and discusses application of these tools to study the proapoptotic effects of TNF-related ligands on cells and their modulation by viruses.

II. Production and Characterization of Reagents

Direct measurement of apoptotic potential is performed by examining the effect of a ligand in cell culture models, either expressed on cell membranes or produced in soluble form. Therefore, an important preliminary step in the characterization of a new TNF family ligand is cloning and expression of the ligand in a form that allows accurate quantification and standardization of apoptotic responses. Key reagents include (1) soluble recombinant TNF family ligands; (2) cells expressing full-length, membrane-bound ligands; and (3) soluble, recombinant forms of TNF receptor proteins. This section presents methods for the production of these reagents.

A. Construction of Recombinant Soluble Forms of Tumor Necrosis Factor (TNF) Family Ligands

Lymphotoxin-α is secreted as a soluble homotrimer. Other TNF-related proteins described so far are type II membrane proteins with short N-terminal cytoplasmic domains. The ecto domain emerges from the membrane with a proline-rich stalk-like region of variable length, which precedes the bulk of the protein. To create soluble recombinant forms of the receptor-binding domain, the cytosolic and membrane domains are truncated at a position 10–20 residues prior to the start of the first β strand and replaced with a signal sequence to direct secretion. LTβ was truncated at Leu-69, whereas LIGHT was truncated at Gly-66. In some cases, an epitope tag is inserted to identify the molecule by immunochemical methods (e.g., LIGHT is produced in this laboratory with an N-terminal FLAG epitope tag and LTβ with an N-terminal c-myc tag) (Rooney *et al.,* 1999).

B. Expression of TNF Family Ligands in Insect Cells Using Baculovirus

The use of baculovirus as an expression vector to introduce eukaryotic cDNAs into insect cells results in high expression of recombinant protein (10 to 50 mg of protein/liter). This system allows correct folding and posttranslational modification of expressed protein and is used in this laboratory for the production of several recombinant, soluble TNF family ligands, including LTα and LT$\alpha 1\beta 2$. The system can also be used to introduce full-length TNF family ligands into insect cells, resulting in cell surface expression. Methods of producing TNF family ligands in soluble form by the baculovirus system (Rooney *et al.*, 1999) and for membrane-expressed ligands (Williams *et al.*, 1997) have been described in detail.

C. Expression of TNF Family Ligands in Mammalian Cells

We have obtained high production of some TNF family ligands, including soluble and membrane-bound LIGHT, in the human embryonic kidney (HEK 293) cell line (Rooney *et al.*, 2000). HEK293 cells (for production of stable transfectants) and 293T cells (for transient transfections) obtained from the American Type Culture Collection (ATCC, Rockville, MD) are cultured in Dulbecco's modified Eagle's medium (DMEM) containing 10% fetal calf serum (FCS) with glutamine (1 m*M*) and penicillin/streptomycin (100 μg/ml).

1. Full-Length LIGHT

Full-length LIGHT was cloned from activated II-23.D7 T-cell hybridoma cells by reverse transcription polymerase chain reaction (RT-PCR) (Mauri *et al.*, 1998). The LIGHT PCR product was subcloned into pCDNA3.1(+) (from Invitrogen) to create pCDNA3.1-LIGHT.

2. Soluble, FLAG-Tagged LIGHT

The extracellular domain of LIGHT (encoding Gly-66 to Val-240) was amplified from pCDNA3-LIGHT by PCR using the following primers: forward 5′-GTAGGAGAGATG-GTCACCCGCCT3′ and reverse 5′-GGAACGCGAATTCCCACGTGTCAGACCCAT-GTCCAAT-3′. The amplified LIGHT product was digested with *Eco*RI and ligated into the *Sna*BI and *Eco*RI sites of pCDNA3.1-VCAM-FLAG, which contains the VCAM1 signal sequence fused 5′ of the FLAG epitope.

3. Transfection of HEK 293 and HEK 293T Cells

Transfections are performed by the calcium phosphate method as follows:

a. Requirements

10-cm^2 sterile tissue culture dishes

DMEM containing 10% fetal bovine serum (FBS), glutamine, and penicillin/streptomycin

Sterile H_2O

DNA to be transfected, quantified

2 *M* $CaCl_2$

2× HBSS–16.4 g NaCl, 11.9 g Hepes acid, 0.21 g Na_2HPO_4, 800 ml H_2O, titrate to pH 7.05 with NaOH and add H_2O to 1 liter.

Sterile, 3-ml tubes

b. Procedure

1. Plate cells at 1.5×10^6/10-cm^2 dish with 10 ml DMEM/10% FCS with antibiotics and glutamine.
2. The next day, change medium 4 h prior to transfection.
3. Transfect 5 μg DNA/dish (circular DNA). All solutions should be room temperature and sterile. In one 3-ml tube, add DNA and sterile H_2O up to 0.5 ml. To a second 3-ml tube add 0.5 ml HBSS. Add 62 μl 2 *M* $CaCl_2$ to the DNA tube.
4. Add the DNA/$CaCl_2$ solution to the 2 × HBSS one drop at a time while swirling the tube. Let the mixture stand for a few minutes at room temperature. Add the mixture to the cells, drop by drop. Tilt the dish and rock back and forth; do not swirl.
5. Replace with normal medium the following morning.
6. For transient transfectants, medium containing secreted, transfected protein can be collected after 24–48 h. Cells expressing full-length protein can be used 24 h after transfection.

For selection of stable transfectants, HEK 293 cells are cultured in the presence of G418 (800 μg/ml) and subcloned to select for high secretors, assayed as described later.

D. Detection and Quantification of Expressed Ligands

Ligand concentration is measured in a number of ways, and during the investigation of a new ligand, the method of analysis that is chosen may depend on the available reagents. We recommend that when a new ligand or novel recombinant form of a ligand is examined, experiments addressing the activity of the ligand should be performed using unpurified material to eliminate potential damage to the protein during the purification process. Thereafter, purified material can be compared directly with unpurified material. Purified preparations of a new ligand may be used as a standard to quantify the amount of the ligand present in the tissue culture supernatant by immunochemical means.

1. ELISA

Two-site ELISA can be used to quantify TNF ligands for which two noncompeting antibodies are available or for which an antibody and a soluble form of the ligand's receptor are available. We routinely perform assays that use a TNF receptor : Fc construct (TNFR : Fc) as a capture molecule and a ligand-specific or epitope tag-specific antibody to detect bound ligand.

a. Requirements

96-well ELISA plates (F96 Maxisorp Nunc Immunoplates; Fisher, Pittsburgh, PA)

Binding buffer: 150 m*M* NaCl/20 m*M* Tris, pH 9.6

Wash buffer: phosphate-buffered saline (PBS)/Tween 20, 150 m*M* NaCl, 20 m*M* Na_2PO_4, pH 7.0, 0.5% Tween 20

Blocking buffer: PBS/3% bovine serum albumin/0.5% Tween 20

Capture molecule: Typically a fusion protein of the receptor's extracellular domain fused to the Fc portion of human IgG (e.g., TNFR : Fc)

Samples and standards to be tested

Antibody specific for the TNF ligand to be measured or for an epitope tag present on the ligand

Secondary antibody coupled to horseradish peroxidase (HRP)

Developing reagent: typically ABTS [2,2′azino-bis(3-ethylbenz-thiazoline-6 sulfonic acid; Sigma, St Louis, M)]

Developing buffer: 0.1 *M* sodium citrate, pH 4.2

30% hydrogen peroxide (H_2O_2)

Plate reader with the ability to read at 415 nm.

b. Procedure

The ELISA we use for measuring FLAG-tagged LIGHT is detailed next and can be adapted easily for measuring other ligands.

1. Coat ELISA plates with receptor protein (human HVEM : Fc, 3 μg/ml in binding buffer, 50 μl/well) overnight at 4°C. Wrap the plate in plastic wrap to avoid evaporation.
2. Remove unbound protein by inverting the plate sharply to decant the solution. Add 50 μl of blocking buffer for 30 min at room temperature.
3. Wash the wells five times in wash buffer.
4. Apply samples and standards, diluted in blocking buffer (50 μl/well). For most TNF ligand ELISAs, use a standard curve of doubling dilutions from 1 to 0.062 n*M*. Include wells with only blocking buffer as a negative control; incubate for 1 h at room temperature.
5. Wash five times with wash buffer.
6. Incubate wells with anti-FLAG monoclonal (M2; Sigma), 5 μg/ml in blocking buffer, 50 μl/well, for 1 h at room temperature.
7. Wash five times with wash buffer.
8. Incubate wells with goat antimouse-HRP (1 : 1000 in blocking buffer; 50 μl/well) for 1 h at room temperature; wash five times with wash buffer and once with distilled water.
9. Wash five times in wash buffer and once in H_2O.
10. Develop using ABTS, 400 μg/ml in 0.1 *M* sodium citrate buffer, pH 4.2, to which is added 1 μl (30%) H_2O_2/ml just before use. Add 50 μl of this solution per well. Yellow color develops over 5 to 30 min.
11. Read $OD_{415\ nm}$ in the plate reader.
12. Concentrations of LIGHT-FLAG in samples are then read from a standard curve of concentration vs $OD_{415\ nm}$.

2. Quantitative Western Blotting

Quantitative Western blotting is useful for determining the concentration of ligands for which an antibody is available, including epitope-tagged ligands. We have found this method useful for the preliminary investigation of ligands for which there is little information. Epitope-tagged ligands can be quantified by comparison with a different ligand bearing the same tag. Thus, the concentration of novel epitope-tagged ligands present in crude supernatants and cell extracts can be quantified in the absence of a ligand-specific receptor or antibody. In initial investigations of mutants of soluble LIGHT-FLAG, we used this method successfully to quantify the concentration of mutant proteins in tissue culture medium, thus facilitating experiments using known concentrations of the unpurified mutant ligand, and avoided experimental errors due to the potential damage of the mutants during the purification process (Rooney *et al.,* 2000).

a. Requirements

Samples and standards

Prestained electrophoresis markers

SDS–PAGE gels and electrophoresis equipment

Polyvinylidene difluoride (PVDF) membrane

Electrophoretic blotting apparatus

Blotting buffer: 10× stock containing 156 g SDS, 121 g Tris, 576 g glycine in 4 liters distilled H_2O (pH 8.3). Add 200 ml methanol to 800 ml × 1 buffer before use

Blocking buffer: PBS/10% dried milk protein/0.1% Tween 20

Primary antibody (ligand specific or epitope tag specific)

Appropriate secondary antibody coupled to HRP

Chemiluminescent detection reagents (e.g., Supersignal, Pierce, Rockford, IL)

X-ray film and cassette

b. Procedure

1. Fractionate samples and standards via 10% SDS–PAGE. For construction of an adequate standard curve, we recommend at least five threefold dilutions of each protein; in one lane run prestained marker proteins of an appropriate size range.

2. Soak the gel and PVDF membrane in blotting buffer and transfer proteins from the gel to the PVDF membrane (transfer times and procedures will depend on the equipment used).

3. Block unbound protein sites on the membrane in blocking buffer for at least 30 min.

4. Incubate the membrane in primary antibody at an appropriate dilution in blocking buffer for 2 h at room temperature.

5. Wash three times for 10 min each wash in blocking buffer.

6. Incubate with secondary antibody, diluted in blocking buffer according to the manufacturer's instructions, for 1 h at room temperature.

7. Wash three times in blocking buffer and twice in distilled water.

8. Detect the signals using X-ray film and cassette, making sure to do short exposure with ECL due to the limited linearity of this detection method.

9. Measure the intensity of ligand-specific bands using a densitometer.

10. Construct a standard curve of density versus concentration and use this to determine the concentration of ligand in samples.

3. Dot Blotting

This method is extremely useful for the rapid quantification of novel, soluble recombinant ligands that bear an epitope tag for which an antibody is available (e.g., FLAG or c-myc). An important advantage of this method is that it allows rapid quantification of multiple samples. However, these samples should also be examined by Western blot using the same antibodies as for dot blotting to ensure that the antibodies detect only the ligand under study. Aliquots (2 μl) of test ligand dilutions are "dotted" onto nitrocellulose, allowed to dry, and immunostained with antiepitope antibody followed by the HRP-linked secondary antibody; the intensity of HRP staining is measured by chemiluminescence or another detection method. The concentration of ligand is determined by comparing the intensity of dots with those of known concentrations of protein bearing the epitope tag (either a purified preparation of the protein under study or a different protein bearing the same epitope tag). A standard curve of intensity versus concentration can be prepared using a densitometer.

a. Requirements

Nitrocellulose membrane (not PVDF, which is hydrophobic)

Samples to be tested

Epitope-tagged protein standard of known concentration

Blocking buffer (PBS/5% dried milk containing 0.01% Tween 20)

Antiepitope antibody

HRP-linked secondary antibody

Developing method, e.g., Supersignal

X-ray film and cassette

b. Procedure

1. Prepare doubling dilutions of standards and samples. This can be performed conveniently using 96-well plates and a multichannel pipetter. For TNF ligands present in tissue culture supernatant obtained from transfected insect or mammalian cells, use standards from 0.1 to 10 μg/ml for an initial standard curve.

2. "Dot" 2 μl of samples and standards onto the nitrocellulose and allow membrane to dry.

3. Block unbound protein sites using blocking buffer for at least 30 min.

4. Incubate with primary antibody diluted in blocking buffer for 2 h at room temperature.

5. Wash three times in blocking buffer for 5 min each wash.

6. Incubate with HRP-labeled secondary antibody diluted in blocking buffer for 1 h at room temperature.

7. Wash three times in blocking buffer and twice in distilled water.

8. Develop using chemiluminescent reagent and film.

9. Measure intensity of dots using a densitometer.

E. Confirmation of Correct Assembly (Trimerization)

TNF ligands are trimeric, and TNF receptors bind ligands at sites formed between adjacent subunits, resulting in clustering of receptors and subsequent signaling. The cellular effects of TNF ligands are dependent on their proper trimerization. Therefore, when examining the effects of a new ligand, or a recombinant or mutated form of a known ligand, it is critical to determine if the ligand has trimerized correctly. We routinely use two methods to examine the subunit structure of ligands: cross-linking and gel-filtration analysis.

1. Cross-Linking of Membrane-Bound Ligands

a. Requirements

Cells expressing the ligand under study (e.g., transfected 293 cells)

Negative control cells (e.g., untransfected 293 cells)

Positive control cells expressing a ligand known to trimerize correctly

Cross-linking reagent: 3,3′-dithiobis (sulfosuccinimidyl propionate) (DTSSP; Pierce) is a water-soluble, homobifunctional *N*-hydroxysuccimide, thiol-cleavable ester. Water solubility ensures maximum cross-linking of surface proteins, with minimal cellular uptake of the ester

5 m*M* sodium citrate buffer, pH 5.0

1 *M* Tris, pH 7.4

14% SDS–PAGE gels and equipment for SDS–PAGE and Western blotting

Nonreducing SDS–PAGE sample buffer (omit the reducing agents 2-mercaptoethanol or dithiothreital)

PVDF membrane

Primary and secondary antibodies for detection of ligand after Western blotting

b. Procedure

1. Prepare cells for cross-linking. Harvest adherent cells using 5 m*M* EDTA in PBS to avoid cleavage of cell surface molecules. Suspend cells at 10^6/ml in PBS.

2. Prepare stock solution of DTSSP just prior to use. Dissolve DTSSP at 1–25 m*M* in sodium citrate buffer, pH 5.0.

3. Add DTSSP to cell suspension. We have obtained good results with a final concentration of 1 m*M* DTSSP. In initial experiments it may be useful to try a range

of concentrations between 0.25 and 5 m*M*. Include controls from which DTSSP is omitted.

4. Incubate cell suspension for 30 min at 4°C.

5. Stop reaction with 50 m*M* Tris, pH 7.4. Incubate for a further 15 min at 4°C.

6. Spin down cells at 10,000 rpm for 5 min and remove supernatant.

7. Extract cells in 100 μl nonreducing SDS–PAGE sample buffer.

8. Subject test and control samples to SDS–PAGE and Western blot. Stain with specific antibody. Trimeric ligands will appear as a "ladder" pattern of monomer, dimer, and trimer after cross-linking.

2. Cross-Linking of Soluble Ligands

It is advisable to use several reagents and concentrations. Insufficient cross-linker may fail to link the protein subunits, whereas excessive reagent concentration and/or reaction time may destroy epitopes. We routinely use bis [2-(sulfosuccinimidooxy-carbonyloxy)ethyl] sulfone (BSOCOES; 1 and 10 m*M*), ethylene glycolbis (succinimidylsuccinate) (EGS; 1 and 10 m*M*) (both from Pierce), and glutaraldehyde (0.1 and 1%).

a. Requirements

Sample of protein to be analyzed. Cross-linking may be performed on purified material, on crude tissue culture supernatant, or on detergent extracts of cells. We recommend initial experiments using crude supernatant to ensure that trimers have not been disassembled during the purification process

Glutaraldehyde (EM grade)

EGS (water-insoluble, homobifunctional *N*-hydroxysuccinimide ester) dissolved in dimethyl sulfoxide (DMSO) at 10–25 m*M* immediately prior to use

BSOCOES (water-insoluble, homobifunctional *N*-hydroxysuccinimide ester) dissolved in DMSO at 10–25 m*M* immediately prior to use

1 *M* Tris, pH 7.5 to stop reaction

14% SDS–PAGE gels and equipment for Western blotting

Specific primary and secondary antibodies for detection of protein after Western blotting

b. Procedure

1. Prepare protein sample (expressed protein present in crude tissue culture supernatant should first be dialyzed against PBS to remove primary amines and purified protein should be diluted in PBS).

2. Add cross-linkers in a range of concentrations, as listed earlier. Include a control in which cross-linker is omitted.

3. Incubate for 30 min to 2 h at 4°C.

4. Stop reaction by adding 50 m*M* Tris, pH 7.5, and incubate for a further 15 min at 4°C.

5. Analyze samples by SDS–PAGE (with reducing agents) and Western blotting.

The presence of dimers and trimers in cross-linked samples is evidence of trimerization.

3. Gel-Filtration Analysis

a. Requirements

Gel filtration column: For investigation of TNF family ligands, we routinely use a 30-ml column of Superose 12 (Pharmacia, Uppsala, Sweden)

Column loading apparatus: an automated system, e.g., FPLC apparatus, may be used; however, a simple peristaltic pump is adequate

Running buffer (PBS), filtered and degassed

Sterile H_2O, filtered and degassed

Sample of ligand to be investigated (detergent extract of cells expressing membrane-bound protein or tissue culture supernatant containing soluble protein) in >0.1 ml volume, 0.22 μm filtered

Calibration proteins: We routinely use a cocktail of blue dextran (2000 kDa), apoferritin (443 kDa), bovine serum albumin (66 kDa), carbonic anhydrase (29 kDa), and cytochrome *c* (12 kDa)

Fraction collector

Detection method for the TNF ligand under study (e.g., ELISA, dot blot, Western blot, or functional assay)

b. Procedure

1. Superose 12 is stored in 20% ethanol at 4°C. Prior to use the column should be washed with 2 column volumes sterile H_2O at 0.2 ml/min and then with 2 column volumes PBS at 0.5 ml/min.

2. Column calibration: Apply a mixture of calibration proteins (diluted to 1 mg/ml each) in a 0.2 ml volume to the top of the column. Elute proteins from the column using PBS at a flow rate of 0.5 ml/minute. Collect 0.5-ml fractions. Assay fractions for protein content (use UV detection or protein assay) and identify specific proteins by SDS–PAGE. Plot protein molecular weight versus elution volume.

3. Apply sample in a volume of 0.1–1 ml to the top of the column. Elute using PBS at a flow rate of 0.5 ml/min. Collect 0.5-ml samples.

4. Assay fractions for the ligand under study. Determine molecular weight from the calibration plot. Most TNF family ligands have a monomer size of 25–30 kDa and a trimer size of approximately three times the monomer.

F. Purification of Soluble TNF Family Ligands

While purification techniques must be evaluated for each new ligand, many TNF ligands can be purified successfully using a combination of ion exchange and affinity procedures. The following procedure is used for the purification of soluble FLAG-tagged LIGHT.

LIGHT-FLAG present in culture supernatant from stably transfected HEK 293 cells was purified by ion-exchange chromatography with an SP Hi-trap column (Pharmacia). Protein was loaded in 10 m*M* Tris/50 m*M* NaCl, pH 7.0, and after washing in loading buffer, bound protein was eluted from the column using 10 m*M* Tris/0.5 *M* NaCl, pH 7.0. LIGHT-containing fractions were identified by ELISA, pooled, dialyzed against PBS, and purified to homogeneity by affinity chromatography using a column of anti-FLAG (M2) coupled to Affigel (Bio-Rad, Hercules, CA). LIGHT-FLAG was eluted from the column using 20 m*M* glycine/150 m*M* NaCl, pH 3.0, and pH neutralized immediately by collection into 50 m*M* Tris, pH 7.4. The protein concentration was determined by amino acid analysis and absorbency at 280 nm.

Loading and elution conditions for ion-exchange chromatography vary for different ligands. Affinity chromatography may be performed using a ligand-specific antibody or ligand-specific receptor coupled to Affigel. For example, TNF receptor type 1 is used in the purification of LTα. At each stage of purification, the assembly (trimeric structure) of the new ligand should be confirmed by the methods detailed earlier and its activity tested by both receptor-binding and biological assay to determine whether the ligand is damaged by the conditions used.

G. Production of Soluble, Fc Fusion Constructs of TNF Receptor Proteins

Most TNFRs are transmembrane glycoproteins. However, engineered soluble forms of the ectodomains of TNFRs retain their ligand-binding characteristics and are extremely useful in the investigation of TNF ligands. Soluble, dimeric forms of TNFR proteins combining the ectodomain of the TNFR with the Fc portion of human IgG have become standard tools in this area of research (Schneider *et al.*, 1997a; Crowe *et al.*, 1994; Mauri *et al.*, 1998; Ettinger *et al.*, 1996). In this laboratory, TNFR : Fc fusion proteins are expressed in insect cells using the baculovirus system and are purified from the tissue culture supernatant in a one-step immunoaffinity procedure using a column of protein G-Sepharose (Amersham, Pharmacia, Uppsala, Sweden). Full details are provided in Rooney *et al.* (2000). Proteins are eluted for the column using 20 m*M* glycine/150 m*M* NaCl, pH 3.0, and fractions are collected into Tris, pH 8.0, so that the final concentration of Tris is 50 m*M*. As for TNF ligands, the binding characteristics of TNFR should be evaluated after purification to ensure that the protein is not damaged by these conditions. Proteins that are highly sensitive to acid elution may be removed from affinity columns using 50 m*M* diethylamine, pH 11.0.

III. Receptor Binding Characteristics of TNF Family Ligands

The ability of a TNF ligand to induce apoptosis is determined by the identity of the membrane receptor it binds. Furthermore, binding of a ligand to soluble or membrane-bound "decoy receptors" may reduce the amount of ligand available for binding to proapoptotic receptors. Therefore, examining the cytotoxic effects of a specific TNF ligand requires knowledge of both the receptor-binding characteristics of the ligand and the receptors that are expressed by the target cell chosen for experiments. A number of techniques are commonly used to investigate ligand-receptor binding.

A. Flow Cytometric Studies

1. Binding of Soluble Ligand to Cell Surface Receptors

a. Requirements

Cells expressing the TNFR, e.g., HEK 293 or 293T cells transfected with cDNA coding for the full-length receptor

Control cells, e.g., untransfected HEK 293 or 293T cells

FACS buffer (tissue culture medium containing 3% BSA)

Ligand to be tested

Positive control ligand or antibody known to bind the TNFR

Negative control ligand or antibody that does not bind the TNFR

Antiligand antibody

Secondary antibody coupled to R-phycoerythrin (RPE)

b. Procedure

1. If cells are adherent, harvest using 5 m*M* EDTA in PBS only (avoid proteases such as trypsin, which might digest the expressed receptor) and resuspend at 10^6/ml in FACS buffer.

2. Place 100 μl aliquots of cell suspension in wells of a round-bottomed microtiter plate. Incubate with dilutions of the ligand to be tested for 30 min at 4°C. Include positive and negative control ligands and an additional negative control to which no addition is made.

3. Wash the wells three times in FACS buffer. For each wash, centrifuge the plate at 1500 rpm for 5 min in a bench-top centrifuge to pellet the cells, aspirate the medium, and resuspend cells in 200 μl ice-cold FACS buffer.

4. Incubate the cells with antiligand antibody diluted in 200 μl FACS buffer/well for 30 min at 4°C.

5. Wash three times in FACS buffer.

6. Incubate the cells with RPE-labeled secondary antibody, diluted according to the manufacturer's instructions, for 30 min at 4°C.

7. Wash a further three times in FACS buffer, resuspend the cells in 400 μl FACS buffer, and analyze fluorescence intensity by flow cytometry using unstained cells to determine baseline fluorescence.

2. Binding of TNFR:Fc to an Expressed Ligand

If cells expressing membrane-bound ligand are available, these can conveniently be used to investigate the ability of multiple TNFR : Fcs to bind the ligand. The procedure used is the same as the one just described with the exception that soluble TNFR : Fc is incubated with ligand-expressing cells and bound TNFR : Fc is detected with antihuman IgG coupled to RPE.

3. Flow Cytometry-Binding Studies

Given the propensity of several of the TNF-related ligands to bind multiple receptors, it is prudent to analyze the direct binding of ligand to target cells. This can be accomplished by conducting direct binding studies using radiolabeled ligand and Scatchard analysis or, more conveniently, by using the epitope-tagged version of the ligand and detection of binding by flow cytometry (Schneider *et al.,* 1999). Titration of the ligand, typically with 1 n*M* as the midrange, is recommended. To determine whether a ligand binds to an additional receptor(s) on the cell surface, competition studies are performed using antibodies to known receptors, preincubating these antibodies with the cells of interest, and then, without washing, adding the ligand to the cells, followed by washing and detection of bound ligand with antiepitope tag. The antireceptor antibody, characterized for its ability to recognize a blocking epitope, will compete for (inhibit) ligand binding, providing that the antibody is used at saturation conditions. Alternatively, lack of blocking may indicate an alternative receptor.

B. Receptor-Mediated Ligand Precipitation

Soluble forms of TNF receptors can be used to precipitate their ligands from detergent extracts of radiolabeled cells. Precipitated proteins may then be identified by their electrophoretic mobility on SDS–PAGE, and their identity can be confirmed by Western blot or by preclearing the cell extracts with ligand-specific antibodies. Investigation of different cell types by this technique may reveal previously unidentified binding partners for TNF ligands and TNFR.

Mauri and colleagues (1998) used the II-23 T-cell hybridoma line, activated under different conditions, to identify the TNF-related ligand LIGHT. When activated with PMA alone, these cells express primarily LTα1β2, whereas when activated with PMA and ionomycin, the cells express LIGHT and expression of LTα1β2 is reduced. HVEM:Fc precipitated both LTα and LIGHT from activated cells, whereas LTβR-Fc precipitated LTα1β2 and LIGHT. Preclearing experiments using antibody to LTα and LTβ confirmed that LIGHT was distinct from these two ligands.

The procedure for precipitation of TNF ligands by a TNFR : Fc is called receptor-mediated ligand precipitation (RMLP) and is detailed next.

a. Requirements

Cells to be investigated

Radiolabel: For unknown ligands use Express-protein labeling mix (NEN, Boston, MA), which contains a mixture of [^{35}S]-methionine and [^{35}S]-cysteine. The ^{35}S has a half-life of 60 days and should not be used after 90 days

Appropriate tissue culture medium deficient in methionine and cysteine, containing 10% dialyzed fetal calf serum

Hepes buffer, 1 *M* stock sterile at pH 7.2–7.4. Should be less than 60 days old

TNFR : Fc

Human IgG

Protein G-Sepharose beads

Lysis buffer: 1% NP-40/150 m*M* NaCl/10 m*M* Tris, containing leupeptin (10 μg/ml), aprotinin (10 μg/ml), phenylmethylsulfonyl fluoride (PMSF) (1 m*M*), and iodoacetamide (20 m*M*)

Gels and equipment for SDS–PAGE;

Gel dryer

Detection equipment for radioactive gels (film and cassette, or phosphor imaging screen)

b. Procedure: Radiolabeling cells

1. Freshly prepare Met/Cys-deficient medium with 10% dialyzed FCS, 20 m*M* Hepes, and glutamine (0.5 m*M*) (10 ml is usually sufficient).

2. Thaw aliquots of ^{35}S label, open cap in fume hood, add radiolabel to medium at 250 μCi/ml, and warm to 37°C.

3. Use 10^6 cells for each RMLP test. Wash cells twice with 3 ml warm (ie., 37°C)PBS, aspirate, and immediately add radiolabeled medium to produce a cell suspension of density 5×10^6/ml. (Adherent cells can be labeled directly in the culture vessel and lysed thereafter). Incubate for 2–4 h at 37°C. Labeling too long will induce stress from amino acid starvation, and cold medium will shock the cells, reducing protein synthetic rate. For secreted proteins, label for 6–8 h.

4. For secreted proteins, save the supernatant. For cell-bound proteins, wash the cells by centrifugation and resuspension in ice-cold PBS, then centrifuge, aspirate the PBS, and resuspend cells by tapping and add cold lysis buffer (10^6 cells/ml). Incubate on ice for 20 min and mildly triturate the solution to ensure cell lysis. Centrifuge the lysate in a microfuge for 30 min to remove the cell pellet, transfer the supernatant to a clean tube, and centrifuge at 13,000*g* for 10 min to remove other insoluble debris. Transfer the supernatant to a clean tube.

1. Immunoprecipitation

1. Add 10 μg human IgG/ml of supernatant to preclear nonspecific binding components. Incubate for 30 min at 4°C. Add 50 μl protein G beads, incubate a further 10 min at 4°C, pellet the protein G by centrifugation, and transfer the supernatant to a clean tube.

2. Add TNFR : Fc (10 μg/ml) and 25 μl protein G beads to precipitate specific ligands. Include other TNFR : Fc that are to be compared and human IgG as a negative control.

Incubate for 60 min on ice, wash the protein G four times by repeated centrifugation and resuspension in 1 ml cold lysis buffer, wash once with PBS without detergent, aspirate the supernatant, add SDS–PAGE sample buffer, and heat at 100°C for 3 min or at 56°C for 20 min.

3. Samples are subjected to SDS–PAGE. Dry the gels and detect radiolabeled proteins by phosphor imaging or using film and a cassette after impregnation of the gel with sulfosalycilate [10% (w/v) solution in H_2O].

4. To confirm the identity of suspected ligands, preclear suspected ligands from the cell extract by incubation with 10 μg specific antibody followed by 50 μl protein G and remove the protein G before step 2.

C. ELISA Studies

Direct ELISA can be used for rapid screening of a ligand's ability to bind a panel of receptors. The basic technique is as detailed in Section II,D,a, and ELISA can give preliminary information regarding the relative affinity of ligands for different receptors. TNFR : Fcs are immobilized in wells of an ELISA plate and incubated with dilutions of the ligand under study. After washing, the bound ligand is detected by a specific antiligand antibody or antiepitope tag antibody. If using a monoclonal antiligand antibody, it is important to ensure that failure to detect ligand binding is not due to masking of the antibody's epitope by bound receptor. Negative results should be confirmed by detection with another antibody or by another assay, such as flow cytometry or competition ELISA.

ELISAs may also be performed in which soluble ligand (3 μg/ml) is bound to the ELISA plate and wells are then incubated with dilutions of soluble TNFR : Fc. Bound TNFR : Fc can be detected by a specific anti-TNFR antibody or by a monoclonal anti-human IgG antibody coupled to HRP (Sigma).

1. Competition ELISA

When lack of an antiligand antibody or other technical difficulties limit the usefulness of direct ELISA in detecting ligand–receptor interaction, the ligand's affinity for a specific receptor may be confirmed by its ability to inhibit interaction between the receptor and another known ligand. ELISA is performed according to the protocol detailed in Section D,II,a, with the following alterations.

1. After coating the plate with TNFR : Fc and blocking with 3% BSA, dilutions of the ligand under study (typically doubling dilutions from 10 to 0.1 μ*M*) are incubated in the wells (50 μl/well for 30 min at room temperature).

2. *Without removing the ligand,* a known ligand is added to each well to give a final, constant concentration previously determined to give 50% maximum binding. The known ligand should be added in as small a volume as possible (5–10 μl).

3. Incubate for 1 h at room temperature, wash, and proceed with detection as for direct ELISA. The ability to reduce binding of the known ligand with increasing concentrations of the ligand under study is strong evidence that this ligand binds the receptor.

D. Surface Plasmon Resonance

Surface plasmon resonance technology can provide valuable information about receptor–ligand interactions when soluble reactants are available. The technique depends on the availability of specialized equipment (Biacore, Uppsala, Sweden) and appropriate operator training. Detailed description of this technology is available from Biacore at www.biacore.com. We have used surface plasmon resonance to investigate the interaction of human LIGHT and LIGHT mutants with their receptors (Rooney *et al.,* 2000). Briefly, the TNFR : Fc (typically 50 μg/ml) is coupled to one channel of a CM5 "chip" coated with carboxymethylated dextran matrix (Biacore) by amine coupling in 0.1 *M* sodium acetate, pH 5.0. An inappropriate Fc-bearing protein (e.g., a TNFR : Fc that fails to bind the ligand under study or human IgG) is coupled to the chip's reference channel as a control. Thereafter, the soluble TNF ligand is injected over both channels (typically in a range of concentrations from 0.1 to 1 μ*M*) and the interaction of ligand with its receptor can be measured in real time due to a change in its refractive index. This method allows rapid calculations of on rate, off rate, and equilibrium constant. The surface of the chip must be regenerated between ligand pulses, typically by a pulse of 10 m*M* glycine, pH 3.0, although other methods may be chosen to dissociate receptor–ligand binding. It is clearly important to use dissociation methods that do not damage receptor integrity. Alternatively, protein A may be coupled to the chip and used as a matrix for binding of TNFR : Fc followed by the ligand under study.

IV. Induction of Apoptosis

A. Choice of Target Cell

The cell type chosen for study should obviously express receptors that engage the ligand under investigation, such as TNFR1 for LTα and LTβR for LTα1β2. Receptor expression by target cells can be determined by immunostaining using antibodies against either the receptor or the test ligand, followed by fluorescent secondary antibody staining and analysis by flow cytometry (see Section III,A). HT29.14S, a clone of the HT29 colon adenocarcinoma cell line, is sensitive to the proapoptotic activity of a wide variety of TNF-related ligands (Browning *et al.,* 1996) and is used routinely in our experiments. Cell viability is quantified by a dye reduction assay using (3-[4,5-dimethylthiazol-2-yl] 2,5 diphenyltetrazolium bromide, MTT). Live cells reduce the dye, which then appears as violet intracellular crystals. Dye reduction is proportional to the density of live cells, as only cells with active mitochondria reduce the dye. Figure 1A shows a typical experimental result, and Fig. 1B demonstrates the dependence of LIGHT cytotoxicity on IFN-γ.

B. Apoptosis Induced by Soluble Ligands

a. Requirements

Target cells (e.g., HT29.14S)

Sterile 96-well plates

Warmed tissue culture medium

IFN-γ

37°C incubator

MTT, made as a stock solution (5 mg/ml in sterile H_2O)

Acidified isopropanol (70% isopropanol to which is added 800 μl concentrated hydrochloric acid/100 ml)

Plate reader with a 570-nm detector

b. Procedure

1. Plate HT29.14S cells at 5000 cells/well in 50 μl DMEM. Wrap the plate in plastic wrap and incubate for 3 h or overnight at 37°C in a humidified incubator. Note that wrapping the plate well at each stage is important to prevent evaporation, which can distort results markedly.

2. Prepare dilutions of the ligand to be tested in complete DMEM in the presence and absence of IFN-γ. We recommend preparing a "master" 96-well plate with ligands and IFN-γ at twice the final concentration desired. IFN-γ should therefore be used at 160 U/ml for a final concentration of 80 U/ml. Ligands are usually tested over a range of dilutions from 1 to 0.02 n*M*. Therefore, prepare doubling dilutions starting at 20 n*M* in the master plate. Each dilution should be tested in a minimum of three replicates. Include a positive control known to be cytotoxic for the cells (e.g., TNF or LTα over the same concentration range) and negative controls (in the presence and absence of IFN-γ) from which TNF ligands are omitted.

3. Add 50 μl of the diluted ligands and controls from the master plate to the plate containing HT29.14S cells. Wrap the plate in plastic wrap and incubate at 37°C.

4. The cytotoxic effect of ligands that engage DD-containing receptors (e.g., TNF and LTα) will be apparent after 24 h and complete after 48 h. The cytotoxicity of ligands that engage non-DD-containing receptors (e.g., LTα1β2 and LIGHT) will not be apparent until after 48 h and will require 72 to 96 h for completion. Examine the plate microscopically after these time intervals to detect signs of cell death (HT29 cells become rounded and nonadherent) and decide when to perform a quantitative MTT assay. Do not allow cells in the negative control wells to overgrow.

5. Add 20 μl of a 5-mg/ml stock solution of MTT to each well. Wrap the plate in plastic wrap and incubate at 37°C for 4 h.

6. Aspirate the medium and MTT from the wells.

7. Add 200 μl acidified isopropanol to each well. Wrap the plate in plastic wrap and rock gently at room temperature for 4 h to achieve solubilization of crystals.

8. Read the color intensity at 570 nm in a microtiter plate reader.

9. Plot $OD_{570\ nm}$ vs ligand concentration.

10. To calculate the percentage viability using the MTT assay, subtract the background A_{570} of isopropanol alone (~0.040 OD units) from each well and divide the A_{570} of ligand-containing wells by the absorbance of the "no ligand" control.

C. Confirmation That Cytotoxicity Is Due to the Ligand Being Tested

1. Antagonism by Receptor

Preincubation of a cytotoxic ligand with its receptor should inhibit the activity of the ligand. A typical experiment is shown in Fig. 1C. To test this hypothesis, first construct a "master plate" in which the soluble ligand, at a constant final concentration previously determined to give half-maximal growth inhibition, is preincubated with varying dilutions of TNFR : Fc before addition to target cells.

The procedure follows the method described earlier with the following alterations. *Step 2.* In a "master plate," add 100 μl/well of ligand diluted in medium at twice the concentration required to produce half-maximal lysis. To the top row of wells add TNFR : Fc at 20 μg/ml so that the final concentration is 10 μg/ml. Prepare doubling dilutions of TNFR : Fc in a constant concentration of ligand by transferring 100 μl from the top row to the second row, from the second to the third, and so on. Perform tests in triplicate. The following should be employed: (1) an irrelevant TNFR : Fc and wells from which TNFR : Fc is omitted as negative controls; (2) a TNFR : Fc known to bind the ligand as a positive control; and (3) wells from which LIGHT is omitted to provide a measure of maximal growth. Incubate for 30 min at room temperature. Then add 50 μl from wells of the master plate to the plate containing HT 29.14S cells and proceed with steps 3 to 9.

2. Use of Agonistic and Antagonistic Antibodies

Preincubation of ligand with ligand-specific monoclonal and polyclonal antibodies, as described earlier for preincubation with TNFR : Fc, may reduce the cytotoxic action of the ligand by masking its receptor-binding site. Some soluble ligands, e.g., FasL, require aggregation with an antibody on the cell surface before the cytotoxic effect is seen (Peitsch and Tschopp, 1995).

D. Apoptosis Induced by Membrane-Bound Ligands

These assays can be performed on cells that express native forms of the ligands under study (e.g., lymphocytes activated to express LTα1β2 or LIGHT) or on cells transiently or stably transfected with full-length ligands. Irradiation of these cells prevents their proliferation; they are then used as *effector cells* when placed in contact with suitable *target cells* (e.g., HT29.14S). Engagement of receptors on the target cells by ligand on the irradiated effector cells can induce apoptosis of the targets.

a. Requirements

Effector cells (e.g., HEK 293 cells stably transfected with the ligand or 293T cells transiently transfected with the ligand). Cells should be harvested using 5 m*M* EDTA in PBS in order to prevent tryptic cleavage of surface proteins and then resuspended in complete DMEM with 10% FCS. Surface expression of the ligand should be confirmed by immunostaining and flow cytometry prior to performing the cytotoxicity assay

Negative control cells (e.g., untransfected 293T or HEK 293 cells)

Positive control cells (e.g., 293T or HEK 293 cells transfected with a ligand known to induce apoptosis in the target cells)

Target cells (HT29.14S cells)

IFN-γ

Sterile 96-well plates

Warmed medium (complete DMEM for HT29.14S)

MTT stock solution (5 mg/ml)

Acidified isopropanol (described earlier)

b. Procedure

1. Irradiate the effector cells with 2000 rad. Wash the cells once and resuspend them in warmed DMEM at a density of 10^6/ml.

2. Harvest HT29.14S cells using 5 m*M* EDTA in PBS and resuspend them in warmed DMEM at a density of 10^5 cells/ml.

3. Prepare a "master plate" with effector cells at varying densities, typically 10^6/ml down to 10^4/ml. This will allow preparation of cultures with effector : target ratios of from 10 : 1 to 0.1 : 1.

4. Mix 50-μl aliquots (5000) of HT29.14S cells with 50-μl aliquots of diluted effector cells in wells of a 96-well plate so that the final volume is 100 μl/well; perform the assay in the presence and absence of IFN-γ (final concentration 80 U/ml) and include additional negative controls from which effector cells are omitted. Seal the plate in plastic wrap and incubate at 37°C. We have found that after 4 to 5 days irradiated effector cells die and the MTT assay can be performed as described earlier.

V. Effect of Viral Infection on Cell Susceptibility to TNF Ligand-Mediated Cytotoxicity

A. Effect of Adenovirus

Infection of cells with virus can affect their sensitivity to cell killing induced by cytokines (Mahr and Gooding, 1999). Such is the case with adenovirus and some members of the TNF ligand superfamily. Infection of HT29.14S cells with wild-type adenovirus reduces the sensitivity of these cells to Fas-mediated killing (Shisler *et al.*, 1997). Because HT29.14S cells are readily infected with adenovirus and are sensitive to the effects of members of the TNF superfamily, these cells provide a useful system for testing how adenovirus and mutants thereof affect the susceptibility of cells to killing by cytokines. The following methods are similar to the assays described earlier using HT29.14S cells. However, modifications have been made to include a step for adenovirus infection, and concentrations have been adjusted to account for this step. Additionally, cytosine β-D-arabinofuranoside (Ara-C) is included as an inhibitor of viral replication. The presence of Ara-C is necessary to reduce cell death due to adenovirus infection, which would normally occur 24 to 30 h postinfection. Killing of HT29.14S by TNF superfamily members requires 24 to 72 h, depending on the ligand.

Note: Adenovirus is a human pathogen, and the appropriate safety precautions should be employed when doing experiments using this virus.

c. Requirements

HT29.14s adenocarcinoma cells (Browning *et al.,* 1996)

DMEM/10% FCS/100 m*M* L-glutamine and penicillin/streptomycin (100 μg/ml)

Adenovirus stock with a titer of at least 10^8 infectious particles/ml

Cytosine Ara-C (Sigma). A stock solution of 2 mg/ml should be made in PBS and sterile filtered. Ara-C is a selective inhibitor of DNA synthesis and does not inhibit RNA synthesis. It is used to inhibit viral replication and subsequent virus-induced cell death prior to ligand-induced death

TNF ligands are diluted in medium plus serum and sterile filtered. For ligands that are FLAG epitope tagged the addition of anti-FLAG antibody (M2, Sigma) is required to facilitate ligand aggregation and cell killing. Alternatively, antibodies to some of the TNF receptor family members can be used instead of the ligand to initiate death signaling from the receptor

Human IFN-γ

Sterile 96-well tissue culture dishes, flat and U bottom

Humidified incubator with 10% CO_2 set to 37°C

Reagents and equipment for MTT assay as described earlier

d. Procedure

1. Plate HT29.14S cells in 96-well dishes at a concentration of 10^4 cells per well in 150 μl DMEM plus 10% fetal calf serum 1 day prior to infection;

2. The following morning, remove medium from wells and replace with 50 μl of DMEM containing 1 μl adenovirus and Ara-C (20 μg/ml final concentration) or medium plus Ara-C for control wells. The final multiplicity of infection (MOI) should be at least 30. Allow infection to proceed at 37°C for 1–3 h.

3. Make cytokine dilutions in sterile 96-well U-bottom dishes in medium containing 20 μg/ml Ara-C and 160 units/ml IFN-γ. All dilutions should be twice the final concentrations desired and diluted 1 : 3 over 9 to 10 dilutions. The IFN-γ is also twice the final concentration of 80 units/ml, but the Ara-C is at the final concentration of 20 μg/ml

because it is already present in wells during virus infection. Each dilution should be tested in triplicate. Controls should include wells containing medium plus Ara-C and medium with Ara-C and IFN-γ.

4. Add 50 μl of diluted cytokine per well of cells. Wrap dish in plastic wrap to minimize evaporation and place in a humidified incubator at 37°C.

5. Check plates daily for cell death. The ideal time to measure viability is when most cells are dead in the wells containing the highest concentration of cytokine, but a considerable number of viable cells remain in the wells with the lowest concentration.

6. Proceed with the MTT assay.

7. Calculate the percentage of viable cells. When comparing the sensitivity of adenovirus-infected cells and noninfected cells to killing by cytokines, it is important to normalize data to that of untreated controls. Untreated controls should consist of samples exposed to identical concentrations of IFN-γ and Ara-C with or without infection. Cells treated with IFN-γ alone do not make good controls because the presence of Ara-C increases the sensitivity of HT29 cells to killing by IFN-γ. A_{570} values for cytokine-treated samples are divided by the average A_{570} value for control samples. The result is multiplied by 100 to determine percentage viability, and that value is plotted versus the cytokine concentration. An ideal plot of values from uninfected samples will range from 100% viability at the lowest concentration of cytokine to less than 10% at the highest concentration. It is not acceptable to simply graph A_{570} values versus the concentration of cytokine as the presence of virus can have a significant effect on cell viability. An example of a graph showing the decreased sensitivity of HT29.14S cells infected with wild-type adenovirus to killing by Fas is presented in Fig. 2.

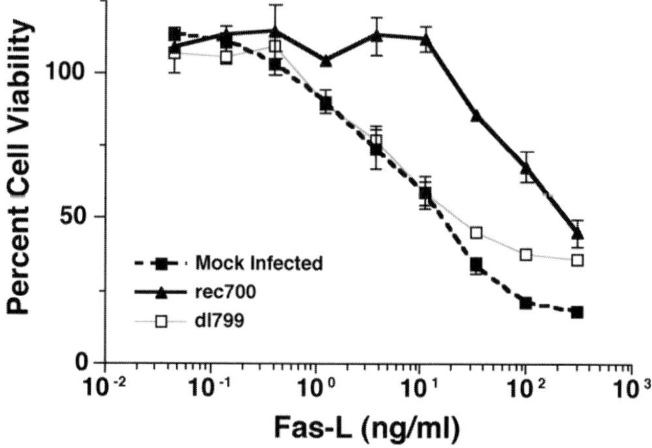

Fig. 2 Wild-type adenovirus desensitizes cells to killing by Fas-L. HT29.14S cells were infected with a wild-type adenovirus (rec700), an adenovirus deletion mutant lacking E3 genes 10.4K and 14.5K (dl799), or were mock infected. All cells were then exposed to increasing concentrations of Fas-L for x hours. Cells infected with a wild-type adenovirus were less sensitive to the killing effects of Fas-L, as determined by the MTT assay than those cells infected with the virus lacking 10.4K and 14.5K or the mock-infected cells.

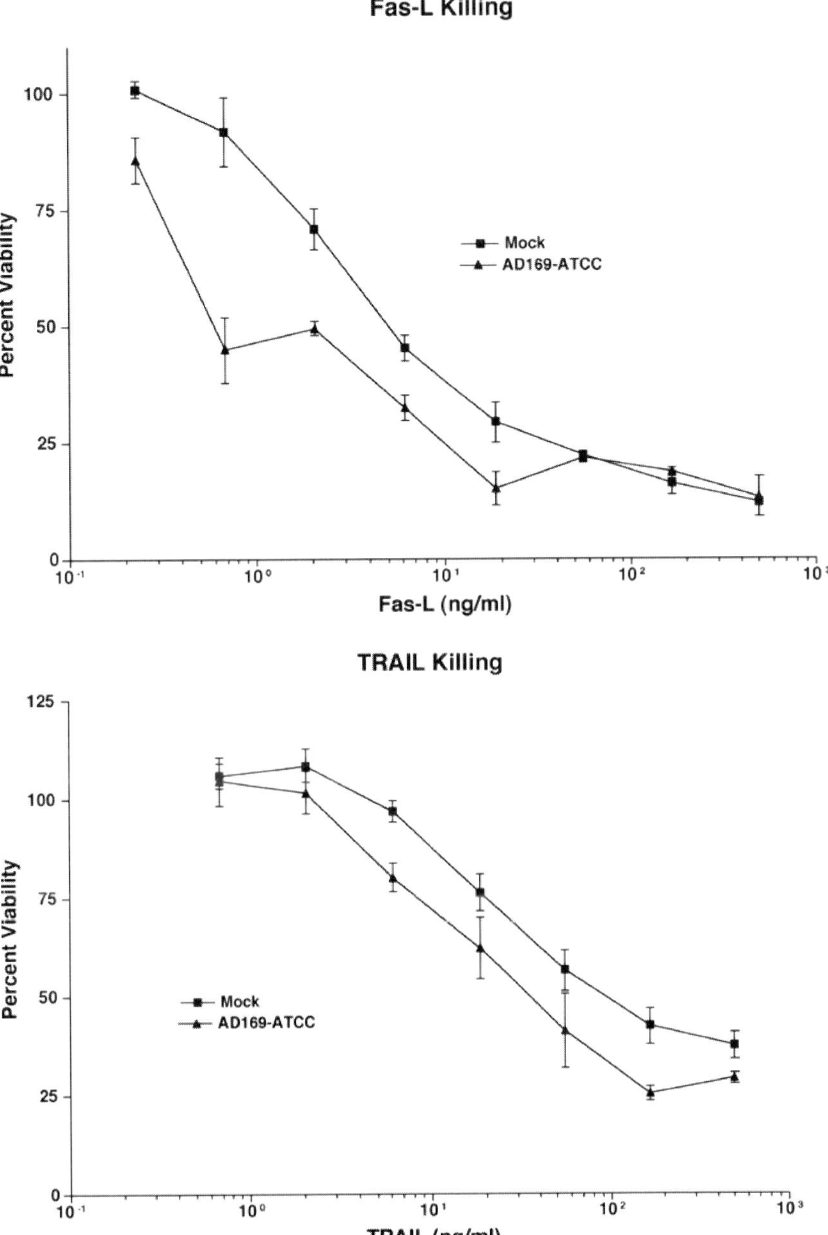

Fig. 3 Sensitivity of HCMV-infected cells to Fas-L and TRAIL. NHDF cells were infected with HCMV-AD169 (multiplicity of infection = 5) or mock infected. Twenty-four hours postinfection, medium containing either Fas-L or TRAIL plus 10 μg/ml cycloheximide was added for a 16-h incubation. MTT was then added for 4 h, and A_{570} was measured to allow for calculation of cell viability.

B. Effect of Human Cytomegalovirus

Because human cytomegalovirus (HCMV) infection and replication *in vitro* are largely restricted to primary normal human dermal fibroblasts (NHDF), the following protocol describes how to measure the TNF ligand-dependent death of HCMV-infected NHDF cells. This assay can be used easily to test the sensitivity of various HCMV isolates to TNF ligands. However, the assay only works well with cell-free virus due to the cell debris associated with cell sonicate viral preparations. Because HCMV is a human pathogen, take the appropriate safety precautions.

e. Procedure

1. Plate NHDF cells (passage 1–15) at a density of 1×10^4 cells per well into 96-well dishes the day before infection. NHDF are grown in DMEM containing 10% FCS, 5 μg/ml insulin, and 1 ng/ml basic fibroblast growth factor.
2. The next day, add HCMV in a 100-μl volume to each well at a MOI ~5.
3. After 24 h, aspirate medium and add "killing medium" containing limiting dilutions of TNF ligand(s) plus 10 μg/ml cyclohexamide (CHX). Perform triplicate measurements for each concentration of ligand. To conserve ligand, as little as 80 μl of supernatant can be added to each well. Include at least four wells of "no ligand" controls, which are incubated in medium containing only CHX.
4. Sixteen to 20 h postinfection perform an MTT assay.

We have found that NHDF cells infected with HCMV-AD169 show similar sensitivity to killing by Fas-L and TRAIL when compared to mock-infected cells (Fig. 3).

Acknowledgments

This work was supported in part by grants from U.S. Public Health Service National Institute of Heath Grants AI03368, P01CA69381, U19AI40038, American Cancer Society Grant RPG-92-033-06-CIM (to C.F.W.), and National Institutes of Health Training Grant T32AG00252 (to C.A.B.).

References

Arch, R., Gedrich, R., and Thompson, C. (1998). Tumor necrosis factor receptor-associated factors (TRAFs): A family of adapter proteins that regulates life and death. *Genes Dev.* **12,** 2821–2830.

Ashkenazi, A., and Dixit, V. M. (1999). Apoptosis control by death and decoy receptors. *Curr. Opin. Cell Biol.* **11,** 255–260.

Barry, M., and McFadden, G. (1998). Apoptosis regulators from DNA viruses. *Curr. Opin. Immunol.* **10,** 422–430.

Benedict, C., Butrovich, K., Lurain, N., Corbeil, J., Rooney, I., Schenider, P., Tschopp, J., and Ware, C. (1999). Cutting edge: A novel viral TNF receptor superfamily member in virulent strains of human cytomegalovirus. *J. Immunol.* **126.**

Boldin, M. P., Mett, I. L., Varfolomeev, E. E., Chumakov, I., Shemer-Avni, Y., Camonis, J. H., and Wallach, D. (1995). Self-association of the "death domains" of the p55 tumor necrosis factor (TNF) receptor and Fas/APO1 prompts signaling for TNF and Fas/APO1 effects. *J. Biol. Chem.* **270,** 387–391.

Browning, J. L., Miatkowski, K., Sizing, I., Griffiths, D. A., Zafari, M., Benjamin, C. D., Meier, W., and Mackay, F. (1996). Signalling through the lymphotoxin-β receptor induces the death of some adenocarcinoma tumor lines. *J. Exp. Med.* **183,** 867–878.

Crowe, P. D., VanArsdale, T. L., Walter, B. N., Ware, C. F., Hession, C., Ehrenfels, B., Browning, J. L., Din, W. S., Goodwin, R. G., and Smith, C. A. (1994). A lymphotoxin-beta-specific receptor. *Science* **264,** 707–710.

Degli-Esposti, M. A., Smolak, P. J., Walczak, H., Waugh, J., Huang, C.-P., DuBose, R. F., Goodwin, R. G., and Smith, C. A. (1997). Cloning and characterization of TRAIL-R3, a novel member of the emerging TRAIL receptor family. *J. Exp. Med.* **186,** 1165–1170.

Emery, J. G., McDonnell, P., Burke, M. B., Deen, K. C., Lyn, S., Silverman, C., Dul, E., Appelbaum, E. R., Eichman, C., DiPrinzio, R., Dodds, R. A., James, I. E., Rosenberg, M., Lee, J. C., and Young, P. R. (1998). Osteoprotegerin is a receptor for the cytotoxic ligand TRAIL. *J. Biol. Chem.* **273,** 14363–14367.

Ettinger, R., Browning, J. L., Michie, S. A., van Ewijk, W., and McDevitt, H. O. (1996). Disrupted splenic architecture, but normal lymph node development in mice expressing a soluble lymphotoxin-β receptor-IgG1 fusion protein. *Proc. Natl. Acad. Sci. USA* **93,** 13102–13107.

Force, W. R., Cheung, T. C., and Ware, C. F. (1997). Dominant negative mutants of TRAF3 reveal an important role for the coiled coil domains in cell death signaling by the lymphotoxin-β receptor (LTβR). *J. Biol. Chem.* **272,** 30835–30840.

Green, G. R., and Reed, J. C. (1998). Mitochondria and apoptosis. *Science* **281,** 1309–1312.

Grell, M., Zimmerman, G., Gottfried, E., Chen, C., Grunwald, U., Huang, D. C. S., Lee, Y. H. W., Durkop, H., Engelmann, H., Scheurich, P., Wajant, H., and Strasser, A. (1999). Induction of cell death by tumour necrosis factor (TNF) receptor 2, CD40 and CD30: A role for TNF-R1 activation by endogenous membrane-anchored TNF. *EMBO J.* 3034–3043.

Itoh, N., and Nagata, S. (1993). A novel protein domain required for apoptosis: Mutational analysis of human Fas antigen. *J. Biol. Chem.* **268,** 10932–10937.

Mahr, J. A., and Gooding, L. R. (1999). Immune evasion by adenovirus. *Immunol. Rev.* **1688,** 121–130.

Marsters, S. A., Sheridan, J. P., Pitti, R. M., Huang, A., Skubatch, M., Baldwin, D., Yuan, J., Gurney, A., Goddard, A. D., Godowski, P., and Ashkenazi, A. (1997). A novel receptor for Apo2L/TRAIL contains a truncated death domain. *Curr. Biol.* **7,** 1003–1006.

Mauri, D. N., Ebner, R., Montgomery, R. I., Kochel, K. D., Cheung, T. C., Yu, G.-L., Ruben, S., Murphy, M., Eisenbery, R. J., Cohen, G. H., Spear, P. G., and Ware, C. F. (1998). LIGHT, a new member of the TNF superfamily and lymphotoxin α are ligands for herpesvirus entry mediator. *Immunity* **8,** 21–30.

McFadden, G., Schreiber, M., and Sedger, L. (1997). Myxoma T2 protein as a model for poxvirus TNF receptor homologs. *J. Neuroimmunol.* **72,** 119–126.

Mosialos, G., Birkenbach, M., Yalamanchili, R., VanArsdale, T., Ware, C., and Kieff, E. (1995). The Epstein-Barr virus transforming protein LMP1 engages signaling proteins for the tumor necrosis factor receptor family. *Cell* **80,** 389–399.

Nagata, S. (1997). Apoptosis by death factor. *Cell* **88,** 355–365.

Peitsch, M. C., and Tschopp, J. (1995). Comparative molecular modelling of the Fas-ligand and other members of the TNF family. *Mol. Immunol.* **32,** 761–772.

Pitti, R. M., Marsters, S. A., Lawrence, D. A., Roy, M., Kischkel, F. C., Dowd, P., Huang, A., Donahue, C. J., Sherwood, S. W., Baldwin, D. T., Godowski, P. J., Wood, W. I., Gurney, A. L., Hillan, K. J., Cohen, R. L., Goddard, A. D., Botstein, D., and Ashkenazi, A. (1998). Genomic amplification of a decoy receptor for Fas ligand in lung and colon cancer. *Nature* **396,** 699–702.

Rooney, I., Butrovich, K., and Ware, C. F. (2000). Expression of lymphotoxins and their receptor-Fc fusion proteins by baculovirus. *Methods Enzymol.* **322,** 345–363.

Rooney, I. A., Butrovich, K. D., Glass, A. A., Borboroglu, S., Benedict, C. A., Whitbeck, J. C., Cohen, G. H., Eisenberg, R. J., and Ware, C. F. (2000). The lymphotoxin-beta receptor is necessary and sufficient for LIGHT-mediated apoptosis of tumor cells. *J. Biol. Chem.* **275,** 14307–14315.

Schneider, P., Bodmer, J. L., Holler, N., Mattman, C., Scuderi, P., Terskikh, A., Peitsch, M. C., and Tschopp, J. (1997a). Characterization of Fas (Apo-1, CD95)-Fas ligand interaction. *J. Biol. Chem.* **272,** 18827–18833.

Schneider, P., Bomer, J. L., Thome, M., Hofmann, K., Holler, N., and Tschopp, J. (1997b). Characterization of two receptors for TRAIL. *FEBS Lett.* **416,** 329–334.

Schneider, P., MacKay, F., Steiner, V., Hofmann, K., Bodmer, J. L., Holler, N., Ambrose, C., Lawton, P., Bixler, S., Acha-Orbea, H., Valmori, D., Romero, P., Werner-Favre, C., Zubler, R. H., Browning, J. L., and Tschopp, J. (1999). BAFF, a novel ligand of the tumor necrosis factor family, stimulates B cell growth. *J. Exp. Med.* **189,** 1747–1756.

Shisler, J., Yang, C., Walter, B., Ware, C., and Gooding, L. (1997). The adenovirus E3-10.4K/14.5K complex mediates loss of cell surface fas (CD95) and resistance to fas-induced apoptosis. *J. Virol.* **71,** 8299–8306.

Simonet, W. S., Lacey, D. L., Dunstan, C. R., Kelley, M., Chang, M.-S., Lüthy, R., Nguyen, H. Q., Wooden, S., Bennett, L., Boone, T., Shimamoto, G., DeRose, M., Elliott, R., Colombero, A., Tan, H.-L., Trail, G., Sullivan, J., Davy, E., Bucay, N., Renshaw-Gegg, L., Hughes, T. M., Hill, D., Pattison, W., Campbell, P., Sander, S., Van, G., Tarpley, J., Derby, P., Lee, R., and Boyle, W. J. (1997). Osteoprotegerin: A novel secreted protein involved in the regulation of bone density. *Cell* **89,** 309–319.

Smith, C. A., Farrah, T., and Goodwin, R. G. (1994). The TNF receptor superfamily of cellular and viral proteins: Activation, costimulation, and death. *Cell* **76,** 959–962.

Sugarman, B., Aggarwal, B. B., Hass, P. E., Figari, I. S., Palladino, M. A., and Shepard, H. (1985). Recombinant human tumor necrosis factor-alpha: Effects on proliferation of normal and transformed cells in vitro. *Science* **230,** 943–945.

Tartaglia, L. A., Rothe, M., Hu, Y. F., and Goeddel, D. V. (1993). Tumor necrosis factor's cytotoxic activity is signaled by the p55 TNF receptor. *Cell* **73,** 213–216.

Thome, M., Schneider, P., Hofmann, K., Fickenscher, H., Meinl, E., Neipel, F., Mattmann, C., Burns, K., Bodmer, J.-L., Schröter, M., Scaffidi, C., Krammer, P., Peter, M. E., and Tschopp, J. (1997). Viral FLICE-inhibitory proteins (FLIPs) prevent apoptosis induced by death receptors. *Nature* **386,** 517–521.

Van Antwerp, D. J., Martin, S. J., Kafri, T., Green, D. R., and Verma, I. M. (1996). Suppression of TNF-α-induced apoptosis by NF-κB. *Science* **274,** 787–789.

VanArsdale, T. L., VanArsdale, S. L., Force, W. R., Walter, B. N., Mosialos, G., Kieff, E., Reed, J. C., and Ware, C. F. (1997). Lymphotoxin-β receptor signaling complex: Role of tumor necrosis factor receptor-associated factor 3 recruitment in cell death and activation of nuclear factor κB. *Proc. Natl. Acad. Sci. USA* **94,** 2460–2465.

Wallach, D., Varfolomeev, E. E., Malinin, N. L., Goltsev, Y. V., Kovalenko, A. V., and Boldin, M. P. (1999). Tumor necrosis factor receptor and Fas signaling mechanisms. *Annul. Rev. Immunol.* **17,** 331–367.

Ware, C. F., Santee, S., and Glass, A. (1998). Tumor necrosis factor-related ligands and receptors. *In* "The Cytokine Handbook" (A. Thompson, ed.), pp. 549–592. Academic Press, San Diego.

Ware, C. F., VanArsdale, S. L., and VanArsdale, T. L. (1995). Cytokines and receptors of the lymphotoxin system. *In* "Pathways for Cytolysis" (G. Griffiths and J. Tschopp, eds.), pp. 175–218. Springer-Verlag, Berlin.

Williams-Abbot, L., Walter, B. N., Cheung, T. C., Goh, C. R., Porter, A. G., and Ware, C. F. (1997). The lymphotoxin-α (LTα) subunit is essential for the assembly, but not for the receptor specificity, of the membrane-anchored LTα1β2 heterotrimeric ligand. *J. Biol. Chem.* **272,** 19451–19456.

Yu, K. Y., Kwon, B., Ni, J., Zhai, Y., Ebner, R., and Kwon, B. S. (1999). A newly identified member of tumor necrosis factor receptor superfamily (TR6) suppresses LIGHT-mediated apoptosis. *J. Biol. Chem.* **274,** 13733–13736.

Zhu, N., Khoshman, A., Schneider, R., Matsumoto, M., Dennert, G., Ware, C., and Lai, M. M. (1998). Hepatitis C virus core protein binds to the cytoplasmic domain of tumor necrosis fector (TNF) receptor 1 and enhances TNF-induces apoptosis. *J. Virol.* **72,** 3691–3697.

INDEX

C

N

X

Y

Z

VOLUMES IN SERIES

Founding Series Editor
DAVID M. PRESCOTT

Volume 1 (1964)
Methods in Cell Physiology
Edited by David M. Prescott

Volume 2 (1966)
Methods in Cell Physiology
Edited by David M. Prescott

Volume 3 (1968)
Methods in Cell Physiology
Edited by David M. Prescott

Volume 4 (1970)
Methods in Cell Physiology
Edited by David M. Prescott

Volume 5 (1972)
Methods in Cell Physiology
Edited by David M. Prescott

Volume 6 (1973)
Methods in Cell Physiology
Edited by David M. Prescott

Volume 7 (1973)
Methods in Cell Biology
Edited by David M. Prescott

Volume 8 (1974)
Methods in Cell Biology
Edited by David M. Prescott

Volume 9 (1975)
Methods in Cell Biology
Edited by David M. Prescott

Volume 10 (1975)
Methods in Cell Biology
Edited by David M. Prescott

Volume 11 (1975)
Yeast Cells
Edited by David M. Prescott

Volume 12 (1975)
Yeast Cells
Edited by David M. Prescott

Volume 13 (1976)
Methods in Cell Biology
Edited by David M. Prescott

Volume 14 (1976)
Methods in Cell Biology
Edited by David M. Prescott

Volume 15 (1977)
Methods in Cell Biology
Edited by David M. Prescott

Volume 16 (1977)
Chromatin and Chromosomal Protein Research I
Edited by Gary Stein, Janet Stein, and Lewis J. Kleinsmith

Volume 17 (1978)
Chromatin and Chromosomal Protein Research II
Edited by Gary Stein, Janet Stein, and Lewis J. Kleinsmith

Volume 18 (1978)
Chromatin and Chromosomal Protein Research III
Edited by Gary Stein, Janet Stein, and Lewis J. Kleinsmith

Volume 19 (1978)
Chromatin and Chromosomal Protein Research IV
Edited by Gary Stein, Janet Stein, and Lewis J. Kleinsmith

Volume 20 (1978)
Methods in Cell Biology
Edited by David M. Prescott

Advisory Board Chairman
KEITH R. PORTER

Volume 21A (1980)
Normal Human Tissue and Cell Culture, Part A: Respiratory, Cardiovascular, and Integumentary Systems
Edited by Curtis C. Harris, Benjamin F. Trump, and Gary D. Stoner

Volume 21B (1980)
Normal Human Tissue and Cell Culture, Part B: Endocrine, Urogenital, and Gastrointestinal Systems
Edited by Curtis C. Harris, Benjamin F. Trump, and Gray D. Stoner

Volume 22 (1981)
Three-Dimensional Ultrastructure in Biology
Edited by James N. Turner

Volume 23 (1981)
Basic Mechanisms of Cellular Secretion
Edited by Arthur R. Hand and Constance Oliver

Volume 24 (1982)
The Cytoskeleton, Part A: Cytoskeletal Proteins, Isolation and Characterization
Edited by Leslie Wilson

Volume 25 (1982)
The Cytoskeleton, Part B: Biological Systems and *in Vitro* Models
Edited by Leslie Wilson

Volume 26 (1982)
Prenatal Diagnosis: Cell Biological Approaches
Edited by Samuel A. Latt and Gretchen J. Darlington

Series Editor
LESLIE WILSON

Volume 27 (1986)
Echinoderm Gametes and Embryos
Edited by Thomas E. Schroeder

Volume 28 (1987)
Dictyostelium discoideum: Molecular Approaches to Cell Biology
Edited by James A. Spudich

Volume 29 (1989)
Fluorescence Microscopy of Living Cells in Culture, Part A: Fluorescent Analogs, Labeling Cells, and Basic Microscopy
Edited by Yu-Li Wang and D. Lansing Taylor

Volume 30 (1989)
Fluorescence Microscopy of Living Cells in Culture, Part B: Quantitative Fluorescence Microscopy—Imaging and Spectroscopy
Edited by D. Lansing Taylor and Yu-Li Wang

Volume 31 (1989)
Vesicular Transport, Part A
Edited by Alan M. Tartakoff

Volume 32 (1989)
Vesicular Transport, Part B
Edited by Alan M. Tartakoff

Volume 33 (1990)
Flow Cytometry
Edited by Zbigniew Darzynkiewicz and Harry A. Crissman

Volume 34 (1991)
Vectorial Transport of Proteins into and across Membranes
Edited by Alan M. Tartakoff

Selected from Volumes 31, 32, and 34 (1991)
Laboratory Methods for Vesicular and Vectorial Transport
Edited by Alan M. Tartakoff

Volume 35 (1991)
Functional Organization of the Nucleus: A Laboratory Guide
Edited by Barbara A. Hamkalo and Sarah C. R. Elgin

Volume 36 (1991)
Xenopus laevis: Practical Uses in Cell and Molecular Biology
Edited by Brian K. Kay and H. Benjamin Peng

Series Editors
LESLIE WILSON AND PAUL MATSUDAIRA

Volume 37 (1993)
Antibodies in Cell Biology
Edited by David J. Asai

Volume 38 (1993)
Cell Biological Applications of Confocal Microscopy
Edited by Brian Matsumoto

Volume 39 (1993)
Motility Assays for Motor Proteins
Edited by Jonathan M. Scholey

Volume 40 (1994)
A Practical Guide to the Study of Calcium in Living Cells
Edited by Richard Nuccitelli

Volume 41 (1994)
Flow Cytometry, Second Edition, Part A
Edited by Zbigniew Darzynkiewicz, J. Paul Robinson, and Harry A. Crissman

Volume 42 (1994)
Flow Cytometry, Second Edition, Part B
Edited by Zbigniew Darzynkiewicz, J. Paul Robinson, and Harry A. Crissman

Volume 43 (1994)
Protein Expression in Animal Cells
Edited by Michael G. Roth

Volume 44 (1994)
Drosophila melanogaster: Practical Uses in Cell and Molecular Biology
Edited by Lawrence S. B. Goldstein and Eric A. Fyrberg

Volume 45 (1994)
Microbes as Tools for Cell Biology
Edited by David G. Russell

Volume 46 (1995) (in preparation)
Cell Death
Edited by Lawrence M. Schwartz and Barbara A. Osborne

Volume 47 (1995)
Cilia and Flagella
Edited by William Dentler and George Witman

Volume 48 (1995)
Caenorhabditis elegans: Modern Biological Analysis of an Organism
Edited by Henry F. Epstein and Diane C. Shakes

Volume 49 (1995)
Methods in Plant Cell Biology, Part A
Edited by David W. Galbraith, Hans J. Bohnert, and Don P. Bourque

Volume 50 (1995)
Methods in Plant Cell Biology, Part B
Edited by David W. Galbraith, Don P. Bourque, and Hans J. Bohnert

Volume 51 (1996)
Methods in Avian Embryology
Edited by Marianne Bronner-Fraser

Volume 52 (1997)
Methods in Muscle Biology
Edited by Charles P. Emerson, Jr. and H. Lee Sweeney

Volume 53 (1997)
Nuclear Structure and Function
Edited by Miguel Berrios

Volume 54 (1997)
Cumulative Index

Volume 55 (1997)
Laser Tweezers in Cell Biology
Edited by Michael P. Sheez

Volume 56 (1998)
Video Microscopy
Edited by Greenfield Sluder and David E. Wolf

Volume 57 (1998)
Animal Cell Culture Methods
Edited by Jennie P. Mather and David Barnes

Volume 58 (1998)
Green Fluorescent Protein
Edited by Kevin F. Sullivan and Steve A. Kay

Volume 59 (1998)
The Zebrafish: Biology
Edited by H. William Detrich III, Monte Westerfield, and Leonard I. Zon

Volume 60 (1998)
The Zebrafish: Genetics and Genomics
Edited by H. William Detrich III, Monte Westerfield, and Leonard I. Zon

Volume 61 (1998)
Mitosis and Meiosis
Edited by Conly L. Rieder

Volume 62 (1999)
Tetrahymena Thermophila
Edited by David J. Asai and James D. Forney

Volume 63 (2000)
Cytometry, Third Edition, Part A
Edited by Zbigniew Darzynkiewicz, J. Paul Robinson, and Harry Crissman

Volume 64 (2000)
Cytometry, Third Edition, Part B
Edited by Zbigniew Darzynkiewicz, J. Paul Robinson, and Harry Crissman

Volume 65 (2001)
Mitochondria
Edited by Liza A. Pon and Eric A. Schon

ISBN 0-12-544165-7

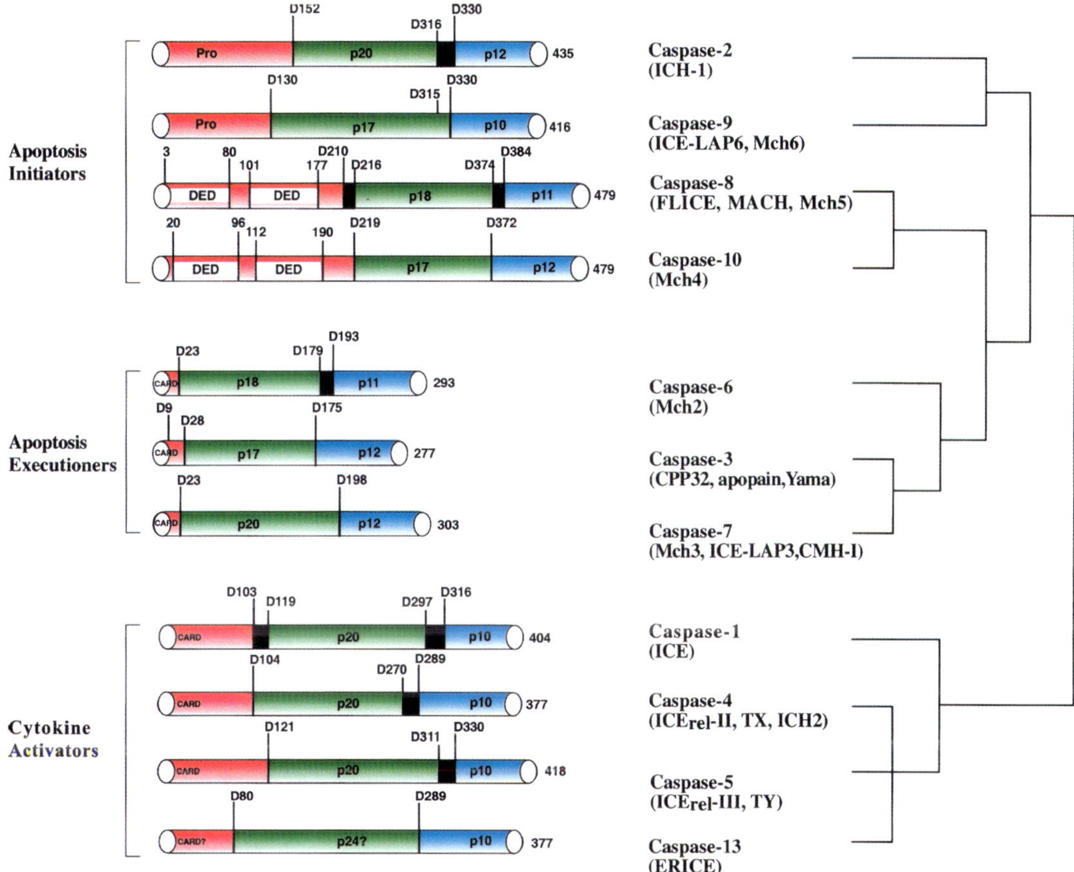

CHAPTER 1, FIGURE 1 Structure and organization of human caspases, as well as their functional and phylogenic relationships. Prodomains and large and small subunits are each shown in different colors with markings to indicate the position of the cleavage sites. The small interdomain link region between subnunits is also marked. (*Left*) Deduced biological functions of these caspases. (*Right*) The phylogenic relationship among caspases.

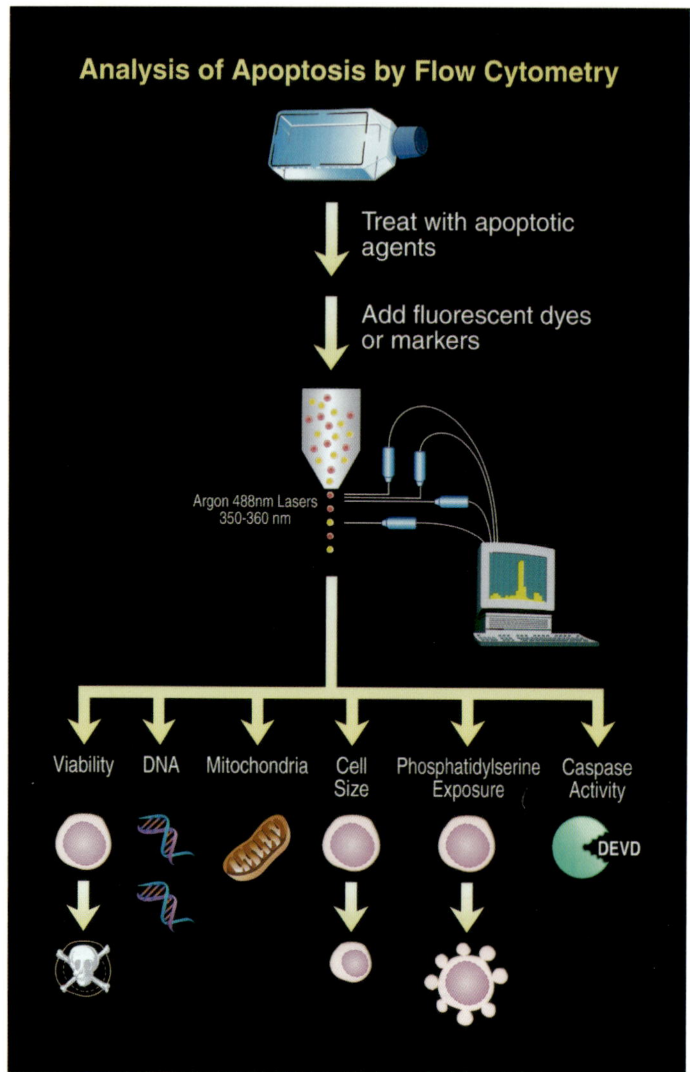

CHAPTER 3, FIGURE 1 Analysis of apoptosis by flow cytometry. Primary or cultured cell lines are treated with various apoptotic agents. After a period of time, numerous fluorescent dyes or markers can be added to the cells and the cells immediately or subsequently examined. During flow cytometry, the cells individually pass by a laser that excites the dye or marker. The fluorescent emission then passes through various filters and is detected by photomultiplier tubes. Several different cellular characteristics can be determined for a single cell and analyzed by various computer programs. In the study of apoptosis, characteristics such as cellular viability, cell size, DNA content, changes in the mitochondrial membrane potential, phosphatidylserine exposure, and caspase activity can be determined using flow cytometry.

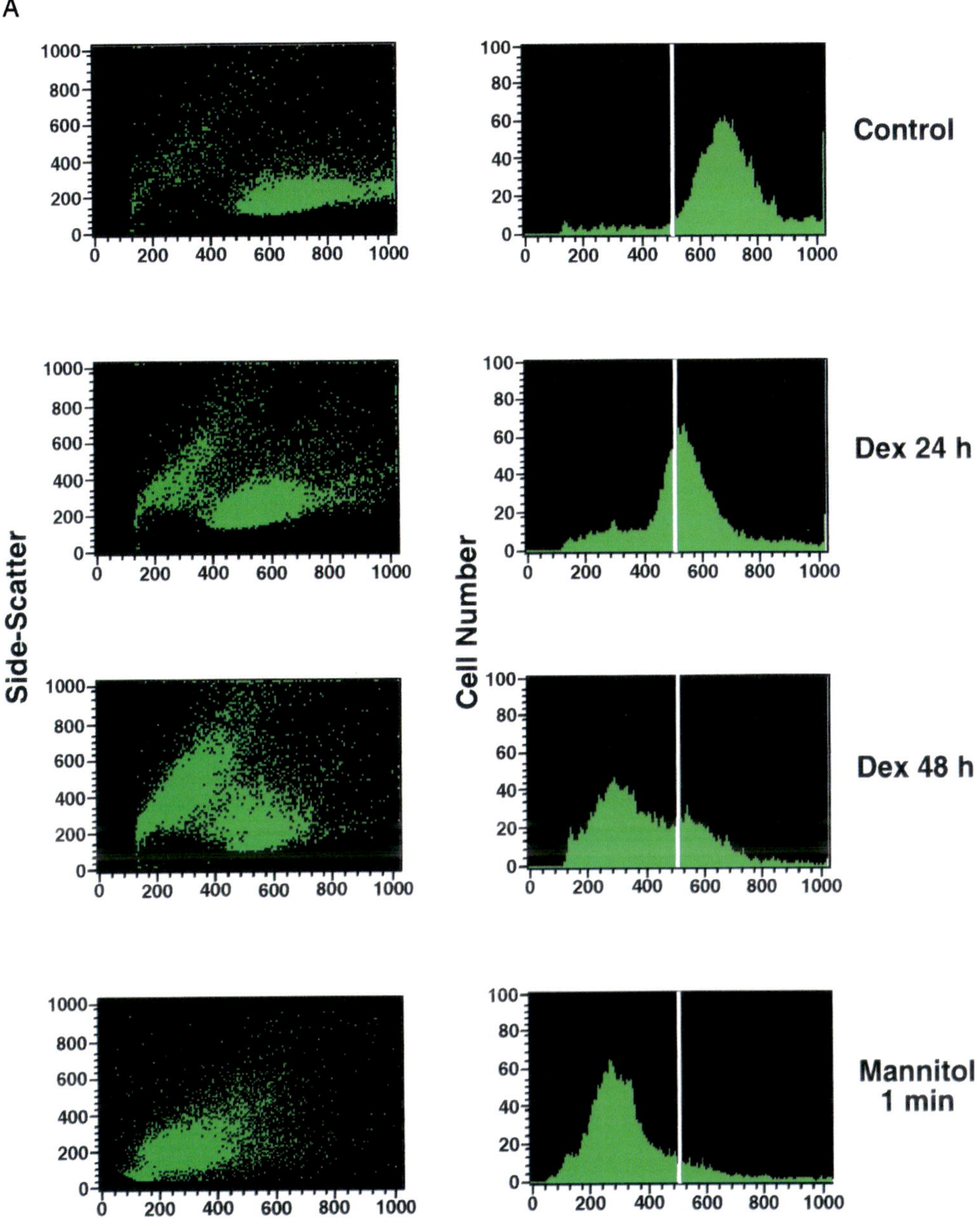

CHAPTER 3, FIGURE 2 Changes in cell size can be determined by flow cytometry. (A) S49 Neo cells treated with dexamethasone or mannitol were initially examined on a forward-scatter versus side-scatter dot plot. In this example, 10,000 cells were examined on a FACSort flow cytometer (Becton-Dickinson, San Jose, CA). Following 48 h of Dex treatment, we observed an increase in a population of cells that had a decrease in forward-scattered light, indicating a loss of cell size. Additionally, an increase in side scatter, indicating an increase in cellular granularity, was also observed during the course of apoptosis. In contrast, cells placed in a hypertonic environment using mannitol displayed an immediate loss of cell volume, without the increase in cell granularity. Forward-scatter histograms can be used to determine the position of the cells at any time during the cell death process.

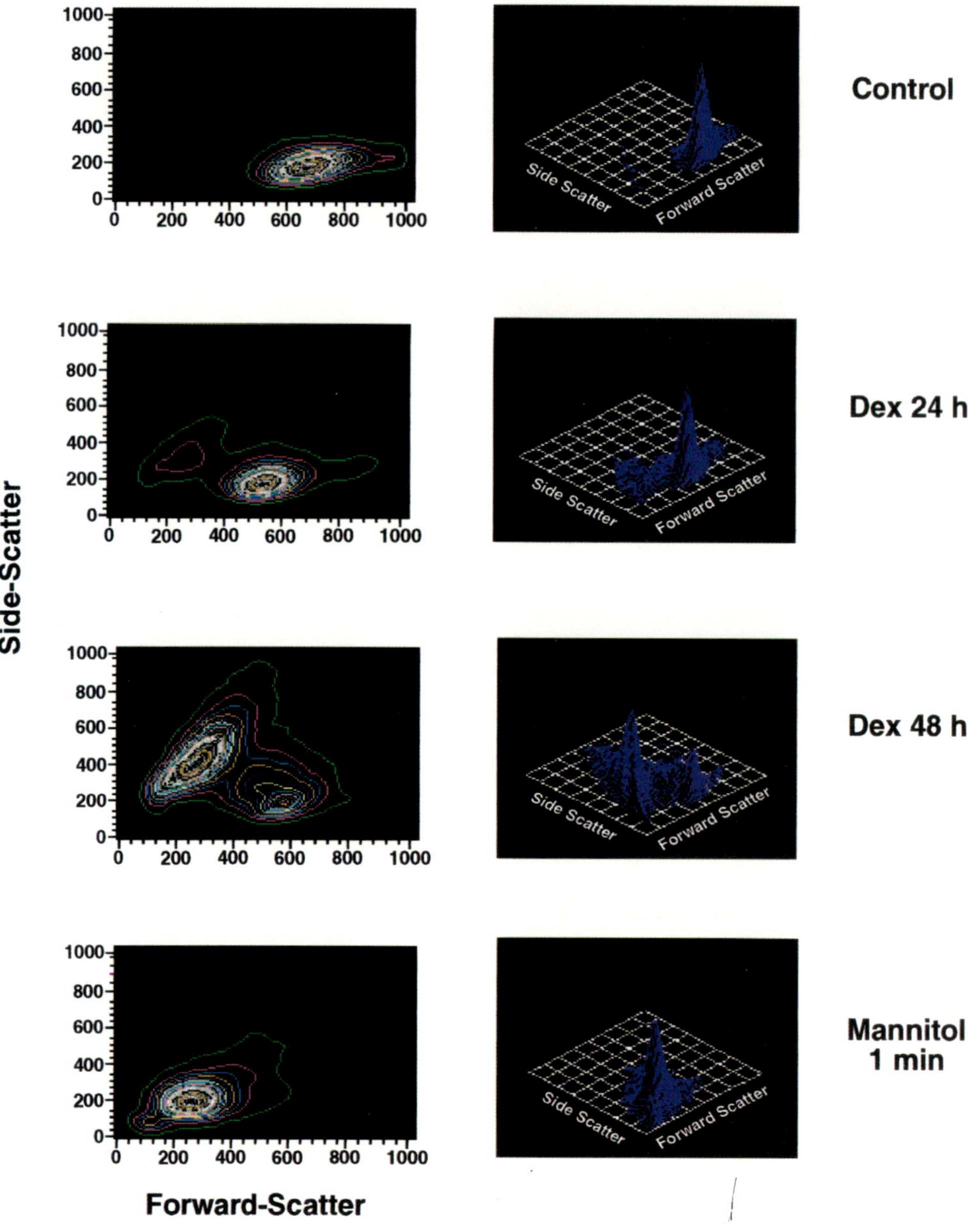

CHAPTER 3, FIGURE 2 (B) Numerous flow cytomeric plots can be employed to examine changes in cell size. Cells can be examined on a forward-scatter versus side-scatter contour plot to give a three-dimensional view of the cell population on a two-dimensional medium. Additionally, the cells can be examined on a three-dimensional plot to observed specific changes in cell number at various times during the apoptotic process.

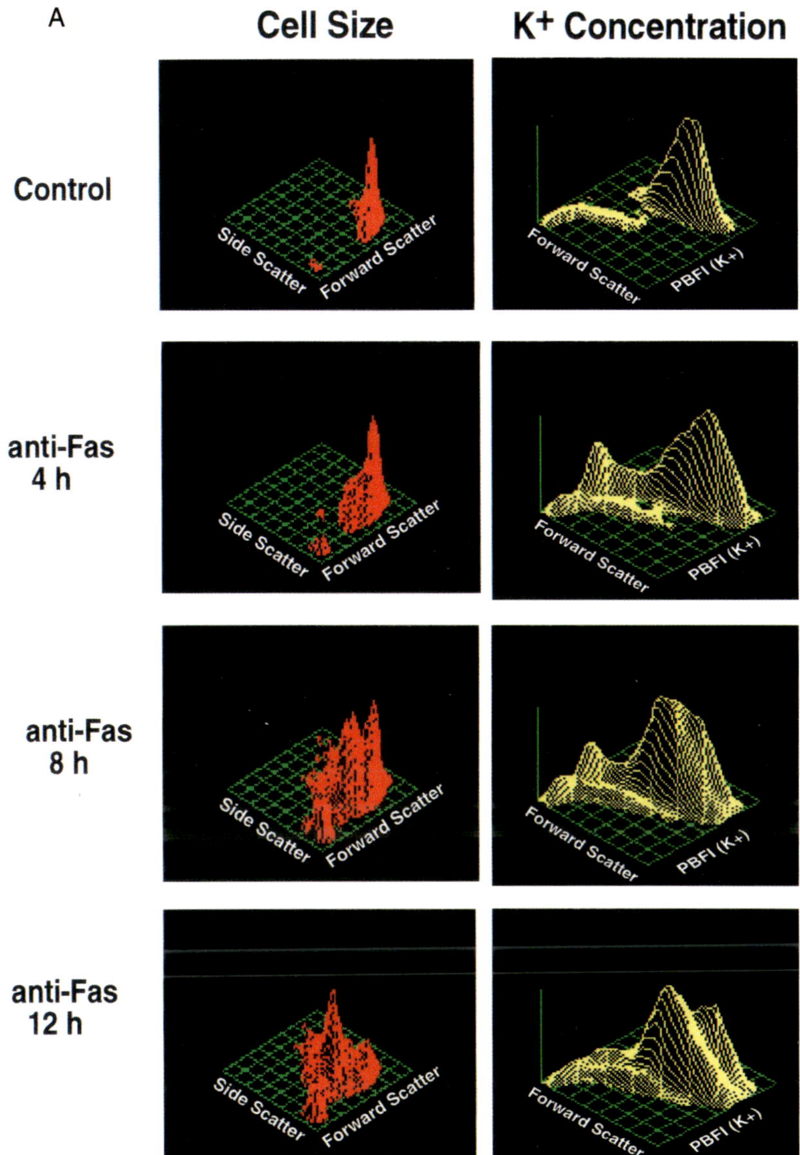

CHAPTER 3, FIGURE 3 Changes in cell size can be compared to those in intracellular potassium during apoptosis by flow cytometry. (A) Jurkat cells were treated with an anti-Fas antibody for 4, 8, and 12 h. Flow cytometry was used to assess the light-scattering properties of the cell and the loss of intracellular potassium. Ten thousand cells were examined on a FACSVantage flow cytometer (Becton-Dickinson, San Jose, CA). A loss in cell size, as indicated by a decrease in forward-scattered light, was observed in a time-dependent manner after the addition of anti-Fas. Additionally, a slight increase in side-scattered light, indicating an increase in cellular granularity, was also observed in a time-dependent manner. Gating on only viable or PI-negative anti-Fas-treated cells in the presence of the potassium indicator PBFI-AM showed that only the shrunken population of apoptotic cells had a decrease in intracellular potassium.

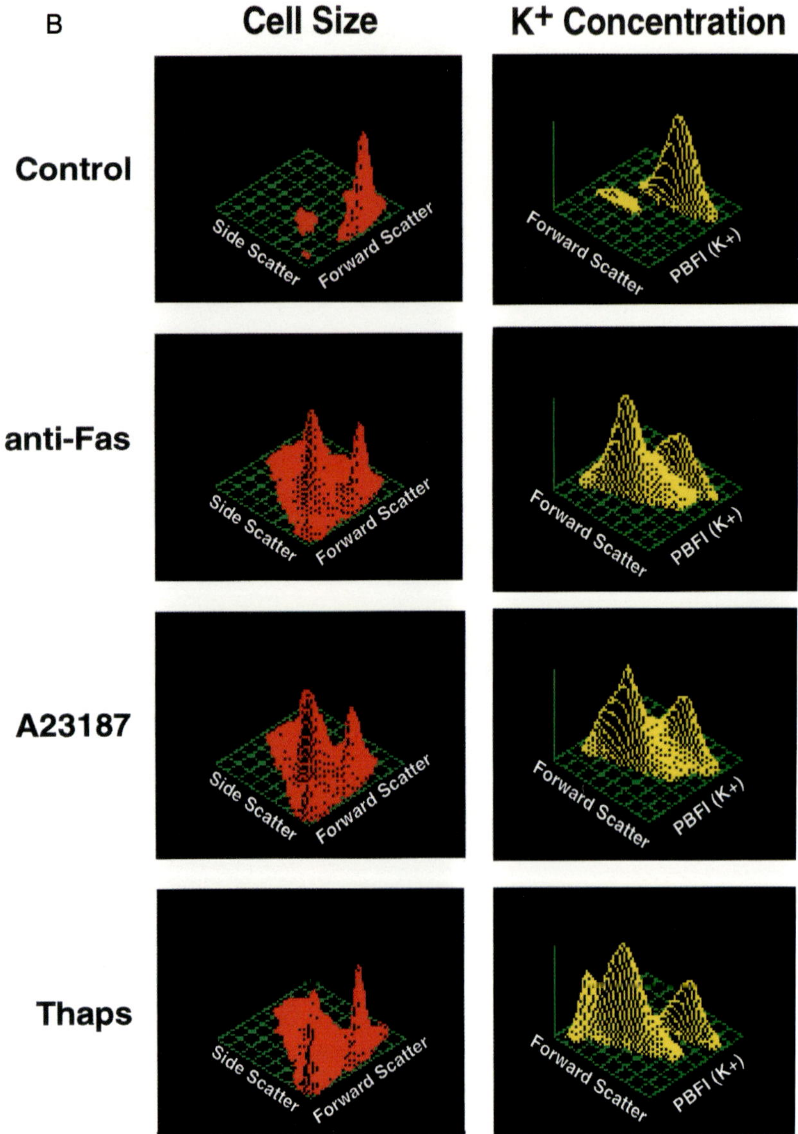

CHAPTER 3, FIGURE 3 (B) Jurkat cells treated with anti-Fas, A23187, or thapsigargin were examined for changes in cell size and intracellular potassium. A loss in cell size occurred with each apoptotic stimulus as a decrease in the ability of a population of cells to scatter light in the forward direction was observed. Comparing the loss of cell size to changes in intracellular potassium showed that only the shrunken population of cells had a decrease in intracellular potassium. Reprinted with permission from *J. Biol. Chem.* **274,** 21953–21962 (1999). Copyright (1999) The American Society for Biochemistry and Molecular Biology.

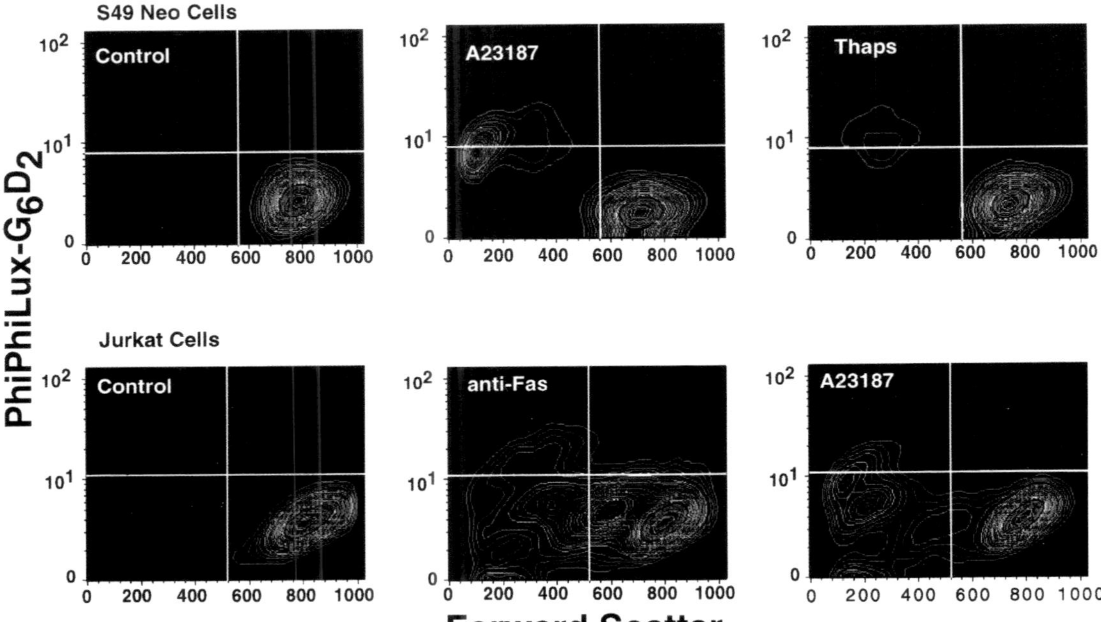

CHAPTER 3, FIGURE 4 Changes in cell size can be compared to changes in caspase-3-like activity during apoptosis by flow cytometry. S49 Neo cells were treated with either A23187 or thapsigargin (*top*) or Jurkat cells were treated with either anti-Fas or A23187 (*bottom*). Cells were exposed to the fluorescent caspase-3-like substrate (PhiPhiLux) for 1 h prior to flow cytomeric examination. Ten thousand cells were examined on a FACSort flow cytometer (Becton-Dickinson, San Jose, CA). When these cells were examined on a FSC vs PhiPhiLux fluorescence contour plot, only the shrunken population of cells had an increase in caspase-3-like activity. Interestingly, a loss of cell volume was clearly observed prior to detection of caspase-3-like activity in the Jurkat cell model system, supporting the conclusion that cell shrinkage must occur prior to effector caspase activation.

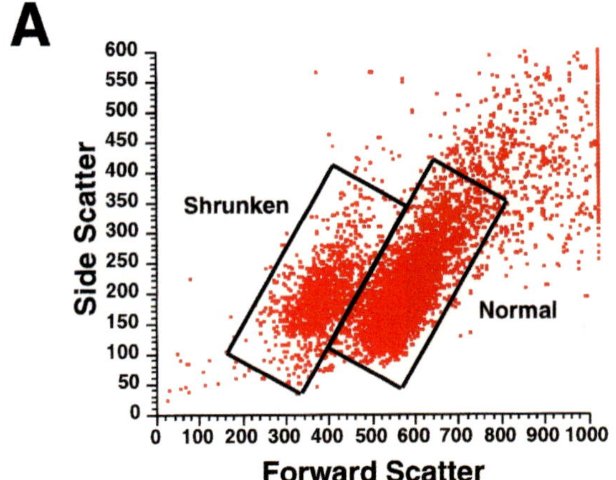

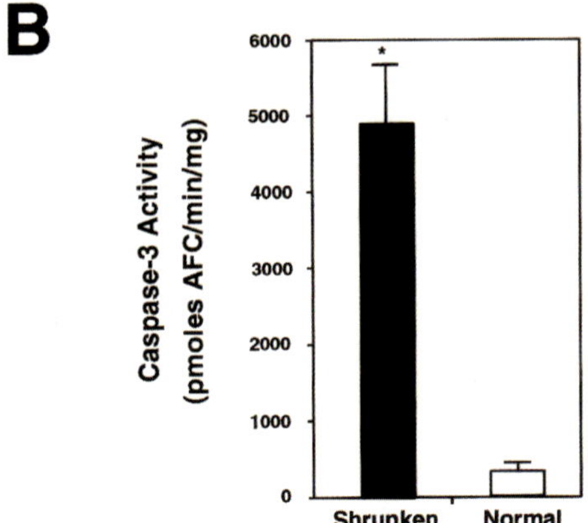

CHAPTER 3, FIGURE 5 Simultaneous sorting of normal and shrunken Dex-treated primary rat thymocytes for additional biochemical analysis. (A) Primary rat thymocytes were isolated and placed in culture in the presence of dexamethasone to induce apoptosis. These cells were initially examined by flow cytometry on a FSC vs SSC dot plot where two distinct populations of cells were observed. Sort gates were drawn around the normal and shrunken population of cells, which were sorted simultaneously into collection tubes. (B) Each individual sorted population of cells was examined for caspase-3-like activity using a fluoromeric caspase assay. Only the shrunken population of apoptotic cells had an increase in caspase-3-like activity. Reprinted with permission from *J. Biol. Chem.* **272,** 30567–30576 (1997). Copyright (1997) The American Society for Biochemistry and Molecular Biology.

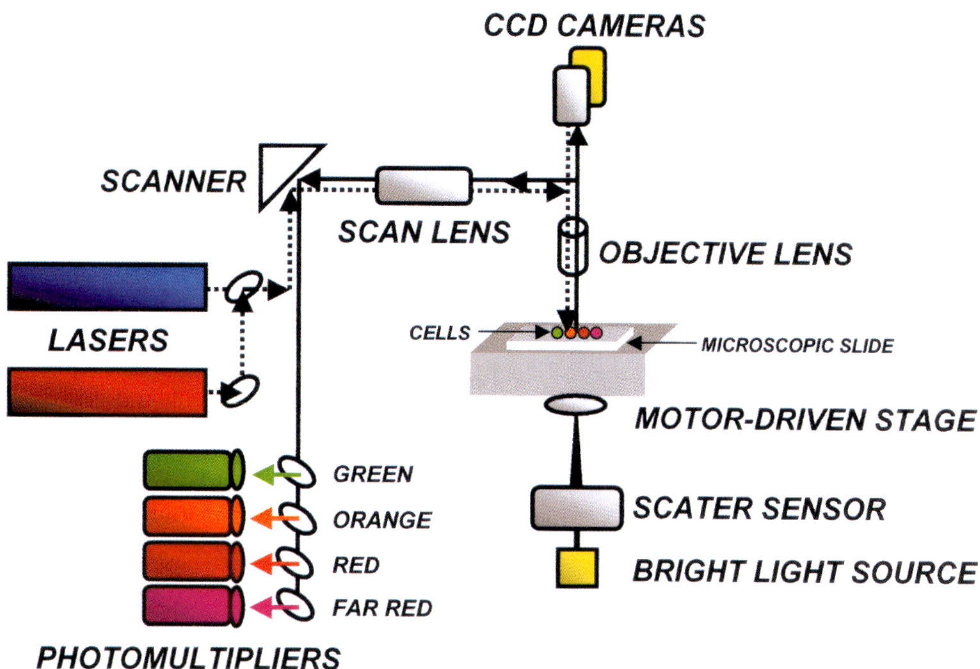

CHAPTER 4, FIGURE 1 Scheme representing major components of the LSC. See text for explanation.

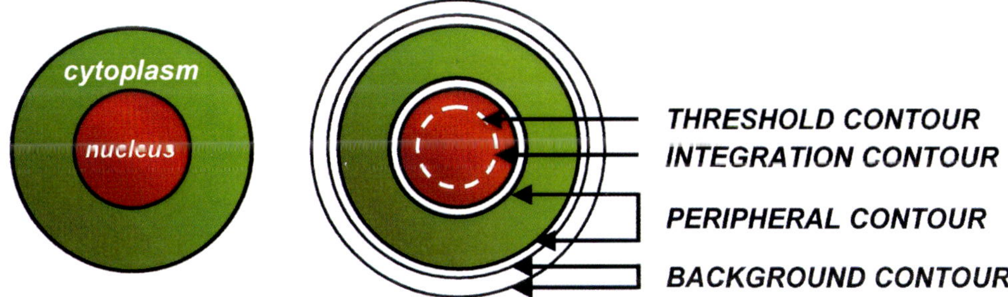

CHAPTER 4, FIGURE 2 The principle of analysis of nuclear and/or cytoplasmic fluorescence by LSC. When nuclear DNA is stained with the red fluorescing dye (e.g., 7-AAD) the threshold contour is set on red signal to detect the nucleus. To measure nuclear fluorescence the integration contour is set at a desired number of pixels outside of the threshold to ensure that all fluorescence emitted from the nucleus is measured and integrated. When fluorescence from the whole cell is measured the integration contour is set far from the threshold to ensure that fluorescence emitted from the cytoplasm is integrated as well. It is also possible to measure nuclear and cytoplasmic fluorescence separately. The peripheral contours are then set at the desired number of pixels outside of the nuclear integration contour, and fluorescence intensities emitted from the integration boundary (nuclear) and from the peripheral torus of the desired width (cytoplasmic) are measured and integrated separately. The background contour is automatically set outside the cell and background fluorescence is subtracted from nuclear, cytoplasmic, or total cell fluorescence.

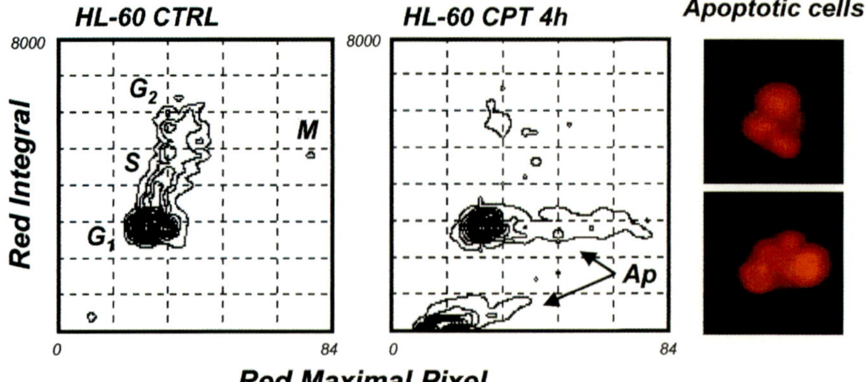

CHAPTER 4, FIGURE 3 Identification of apoptotic cells by LSC based on high values of maximal pixel detecting red fluorescence or fractional DNA content of propidium iodide (PI)-stained cells. Exponentially growing HL-60 cells untreated (CTRL) or induced to undergo apoptosis by treatment with $0.15\mu M$ camptothecin (CPT) for 4 h, were stained with PI in the presence of RNase A as described in the text. Contour maps represent bivariate distributions of cells with respect to their integrated red fluorescence (proportional to DNA content) vs maximal red fluorescence pixel value. Only mitotic cells (M) have a high maximal pixel value in the untreated culture. Apoptotic cells (Ap) that are present in CPT-treated cultures are characterized either by the increased intensity of maximal pixel of red fluorescence or by a low ("sub-G_1") DNA content. The relocation feature of LSC allows one to observe morphology of the cells selected from particular regions of the bivariate distributions. Upon relocation, cells with a high maximal pixel value or with a fractional DNA content show chromatin condensation and nuclear fragmentation typical of apoptosis (two panels on right).

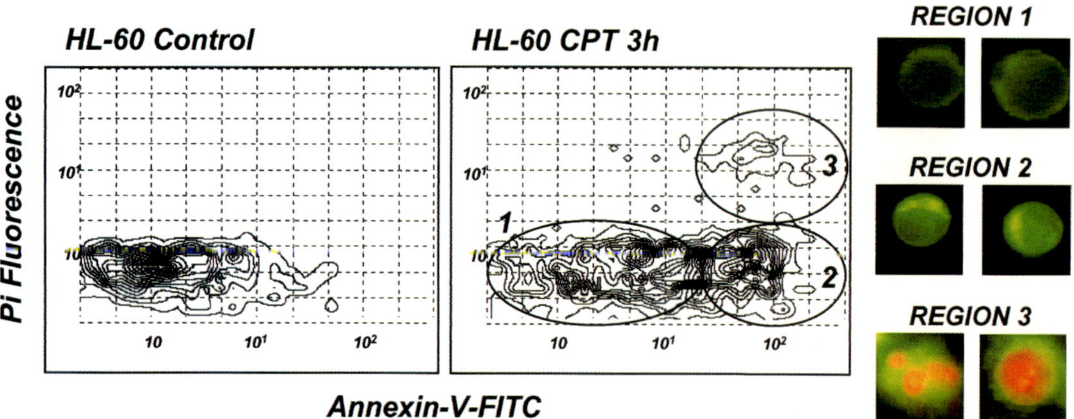

CHAPTER 4, FIGURE 5 Detection of early and late apoptotic cells by LSC after staining with annexin V–FITC conjugate and PI. To induce apoptosis, HL-60 cells were treated with CPT (see legend to Fig. 3), attached to a microscope slide, and subjected to the procedure of labeling with Annexin V–FITC and PI. Their fluorescence was measured by LSC as described in the text. Following the relocation for visual inspection, cells that bind neither Annexin V nor PI (region 1) appear to have normal morphology, cells that bind Annexin V but exclude PI (region 2, early apoptosis) display undulations of the plasma membrane and granularity in chromatin and cytoplasm, and cells that bind Annexin V and stain with PI (late apoptosis, "necrotic" stage of apoptosis) are shrunk and some show the formation and detachment of apoptotic bodies ("budding").

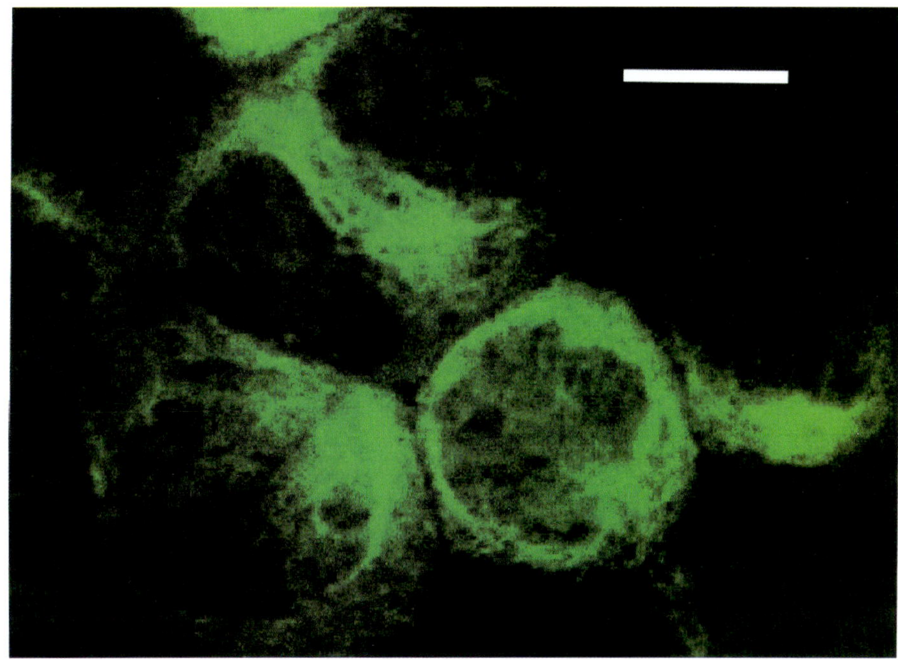

CHAPTER 5, FIGURE 1 Confocal image of apoptotic HL-60 cells. Intracellular glutamyl substrates were labeled by TGgase by the haptenized amine substrate DALP. The dinitrophenyl group of DALP was detected by immunofluorescence. Bar: 10 nm.

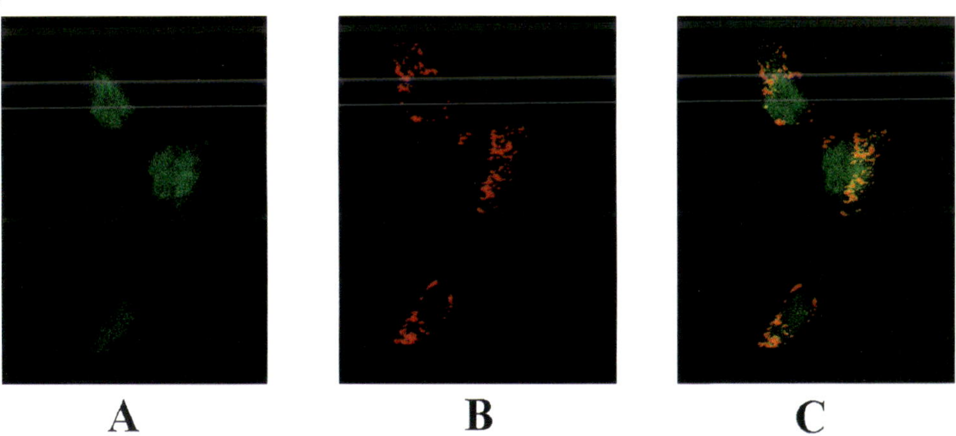

CHAPTER 10, FIGURE 3 Dual labeling of cells with a Ca^{2+} probe and a mitochondrial tracer. (A) Mitochondrial staining. PC-3 cells were labeled with the mitochondrial tracer rhodamine B ($1{:}10^6$ dilution) for 2 min. (B) Calcium dye fluorescence. The same cells presented in A were loaded with 5 μM fluo-3 AM for 30 min. (C) Images presented in A and B were overlaid. The intensity of orange/yellow fluorescence corresponds to levels of intramitochondrial Ca^{2+} concentrations.

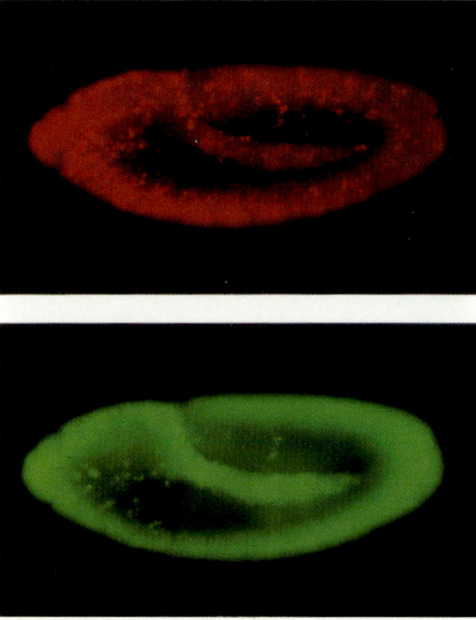

CHAPTER 14, FIGURE 1 Acridine orange staining of embryos. This embryo was stained with AO, and the staining was visualized using RITC (*top*) and FITC (*bottom*) filters. Each of the bright spots represents an apoptotic cell. At later stages, corpses can be seen clustered within macrophages. Note that more apoptosis is seen in the RITC channel, but that the general background is higher.

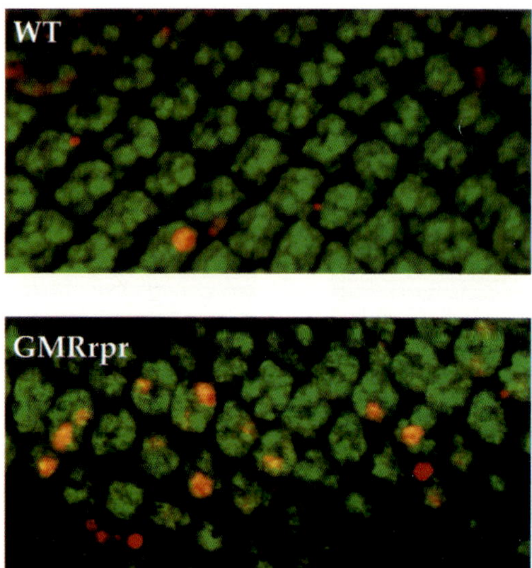

CHAPTER 14, FIGURE 3 Eye disc stained with TUNEL and anti-elav. (*Top*) A wild-type disc. TUNEL staining for apoptosis is seen in red, while elav expression is seen in green. Very little apoptosis is seen in the differentiating region of the disc. However, both undifferentiated (red only) and differentiated (labeled with both, giving a yellow color) cells are undergoing apoptosis. (*Bottom*) A disc showing higher levels of apoptosis induced by overexpression of the proapoptotic gene *reaper* (*rpr*). Again both differentiated and undifferentiated cells are undergoing apoptosis.

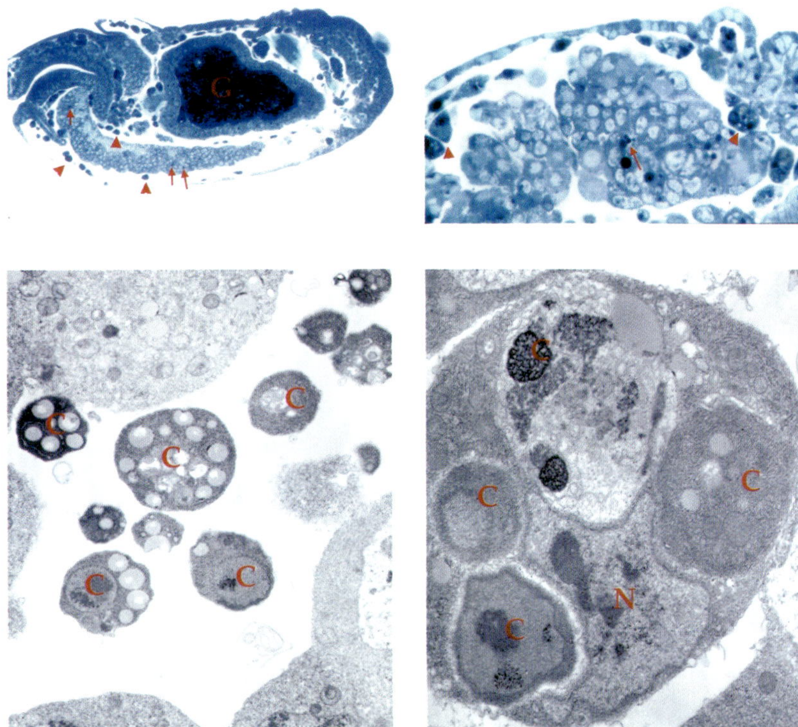

CHAPTER 14, FIGURE 4 Assessment of apoptosis by light microscopy of semithin sections and transmission electron microscopy. (*Top*) Embryos embedded in plastic, sectioned, and stained with methylene blue/toluidine blue/borax. (*Right*) A high magnification view of the head region. Many apoptotic cells are seen engulfed within macrophages (*arrowheads*). Unengulfed corpses can also be seen as intensely stained dots (*arrows*). The nucleolus in all cells can be seen as a dark spot within each nucleus, as can the lipid within the gut (G). (*Bottom*) Electron micrographs of unengulfed corpses on the left and a large macrophage on the right. Corpses are marked with a C and can be seen as electron-dense, condensed bodies. A large number of vesicles can also be seen within the unengulfed corpses. The macrophage on the right has engulfed at least four corpses. The electron-dense material in the macrophage nucleus (N) is the nucleolus.

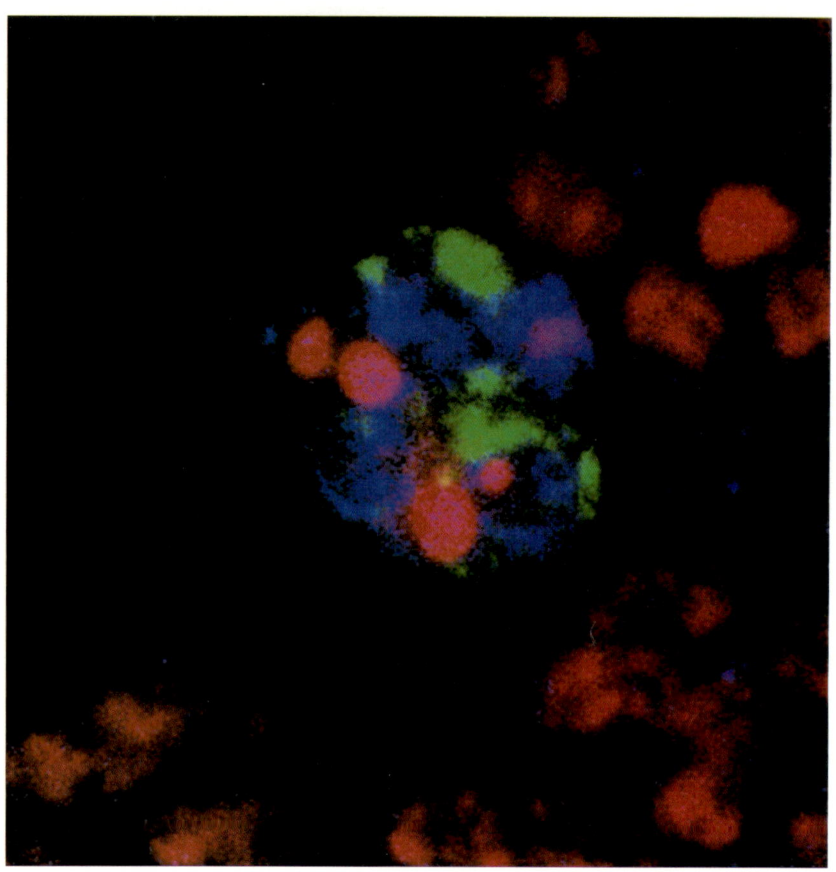

CHAPTER 14, FIGURE 5 Embryonic macrophage labeled with antiperoxidasin, anticroquemort, and 7 AAD. The macrophage cytoplasm is labeled for peroxidasin (green), while the membranes surrounding the macrophage show croquemort (blue). Nuclei are labeled with 7AAD (red). Apoptotic nuclei can be seen as bright, condensed bodies.

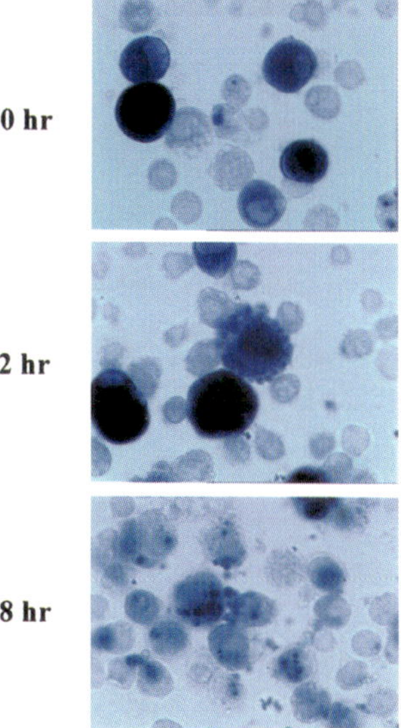

CHAPTER 14, FIGURE 6 β-Galactosidase assay. Cells were transfected with a construct containing the apoptosis-activating gene *reaper* (*rpr*) under the control of the metallothionein promoter and a 1:10 molar ratio of a constitutively expressed β-galactosidase construct. The transfected cells were transferred to fresh plates and RPR expression was induced for the time periods shown before the cells were fixed and stained for β-galactosidase expression. The blue cells are those that have been transfected. RPR expression leads to a characteristic apoptotic morphology within 2 h of induction, as shown by the blebbing cell. Engulfment of dying cells begins promptly, and by 8 h, many of the corpses are phagocytosed by neighboring cells.

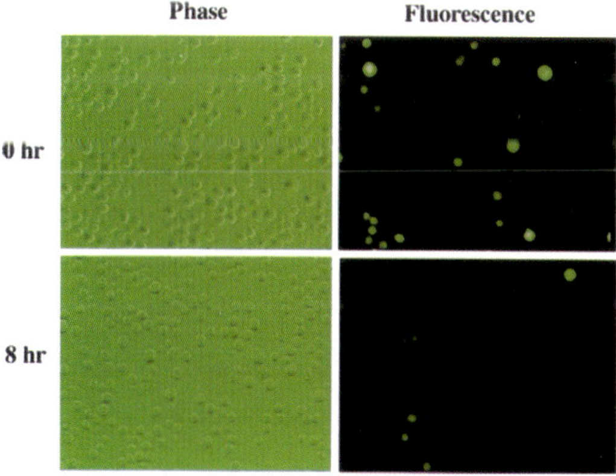

CHAPTER 14, FIGURE 7 GFP assay. Cells were transfected with a construct containing the apoptosis-activating gene *rpr* under the control of the metallothionein promoter and a 1:10 molar ratio of a constitutively expressed GFP construct. Transfected cells were induced to express RPR for the indicated times and were then transferred to fresh plates and allowed to settle. The plates were rinsed and the remaining cells were fixed. Fields of cells were photographed in visible light to determine total cell number and then photographed with fluorescence activation to observe which of the cells were expressing GFP. The percentage of viable transfected cells (as determined by their ability to readhere to the plate) decreases dramatically by 8 h postinduction.

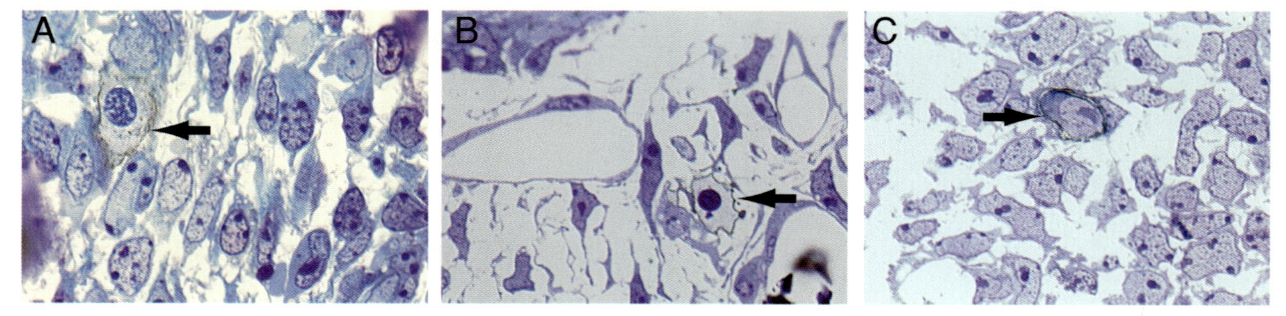

CHAPTER 15, FIGURE 2 *In vivo* labeling of cell surface-exposed phosphatidylserine. Annexin V-biotin labeling is shown (*arrows*) of early (A), late (B), and phagocytosed (C) apoptotic cells in mouse embryonic tissue (van den Eijnde *et al.,* 1998) (Fig. 1). Note in C that both the apoptotic cells and the surrounding membrane of the phagocyte are Annexin V positive (see also van den Eijnde *et al.,* 1999).

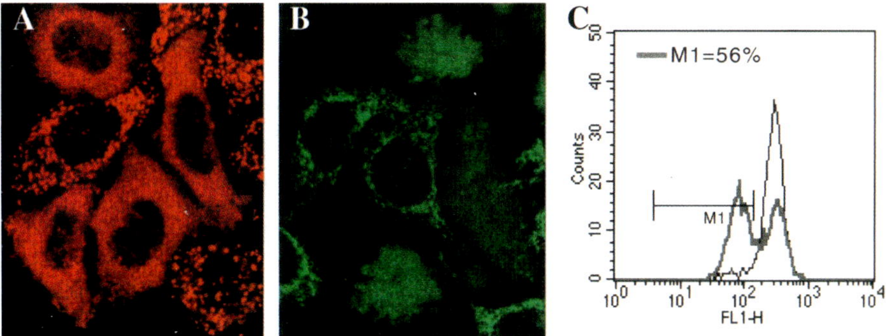

CHAPTER 16, FIGURE 2 Detection of cytochrome *c* release by fluorescent labeling. (A) A confocal image of a heterogeneous population of HeLa cells treated with UVC (180 mJ/cm^2) in the presence of zVAD-fmk (100 μM) and stained by immunocytochemistry after 6 h. (B) A confocal image of cells stably transfected with cytochrome *c*–GFP, 6 h after UVC. In A and B, diffuse staining indicates that cytochrome *c* has been released from mitochondria. Mitochondria in fixed cells have a more punctate distribution than live cells. (C) Cells stably transfected with cytochrome *c*–GFP and either not treated (*thin black line*) or treated as in B (*thick grey line*) were permeabilized with digitonin and analyzed by flow cytometry. Permeabilized cells that had previously undergone mitochondrial cytochrome *c* release do not retain cytochrome c–GFP. In this case, 56% of the treated cells have released their mitochondrial cytochrome c (*thick gray line*). Only 45% of treated cells have phosphatidylserine (PS) exposed on their surface, in support of the concept that cytochrome *c* is released before PS exposure. In untreated cells, approximately 6% of cells had both released cytochrome *c* and had exposed PS on the outside of the plasma membrane.

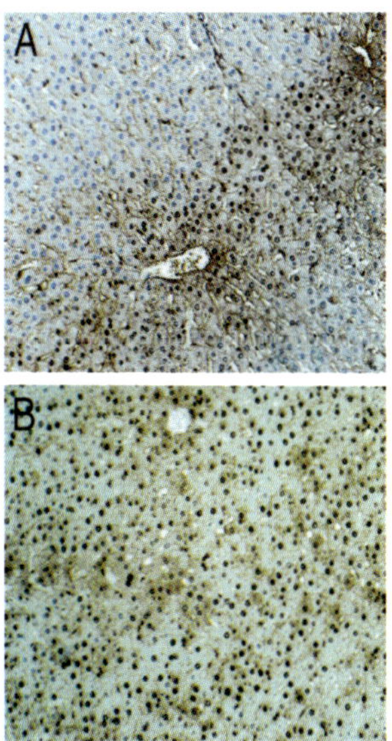

CHAPTER 17, FIGURE 3 TUNEL staining. (A) TUNEL-positive nuclei from a paraffin-embedded section of rat liver following 3 days of treatment with a high dose of a DNA-damaging agent. Counterstained with hematoxylin. Note that TUNEL-positive apoptotic nuclei are localized primarily in the centrilobular region. Magnification: 200×. (B) Positive control for the experiment outlined in A. Sections of normal liver were DNase I treated (100 mg/ml) as described in the text. Counterstained with hematoxylin. Note that almost all nuclei are TUNEL positive. Magnification: 200×.

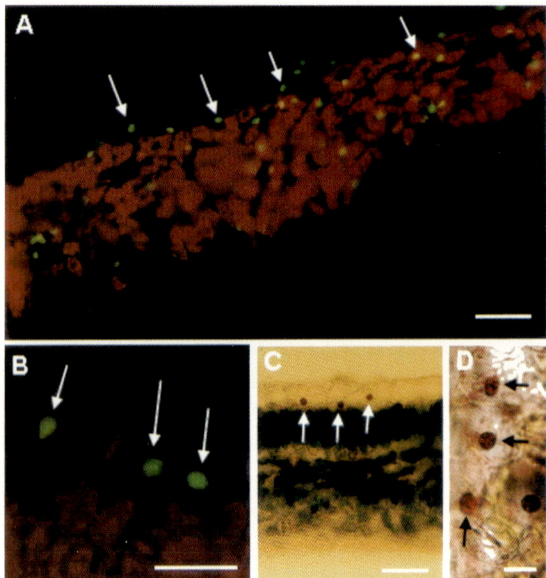

CHAPTER 19, FIGURE 4 TUNEL staining of *Catharanthus roseus* leaves treated as positive controls by incubation with DNase I. (A) TUNEL reaction detected by fluorescein labeling as observed by epifluorescent microscopy. Nuclei with cleaved DNA are stained bright green. Bar: 50 μm. (B) The same as A. 25 μm. (C) TUNEL reaction detected by peroxidase activity. Nuclei with cleaved DNA are stained dark brown. Bar: 50 μm. (D) The same as C. Bar: 12.5 μm. TUNEL-stained nuclei are indicated by arrows.

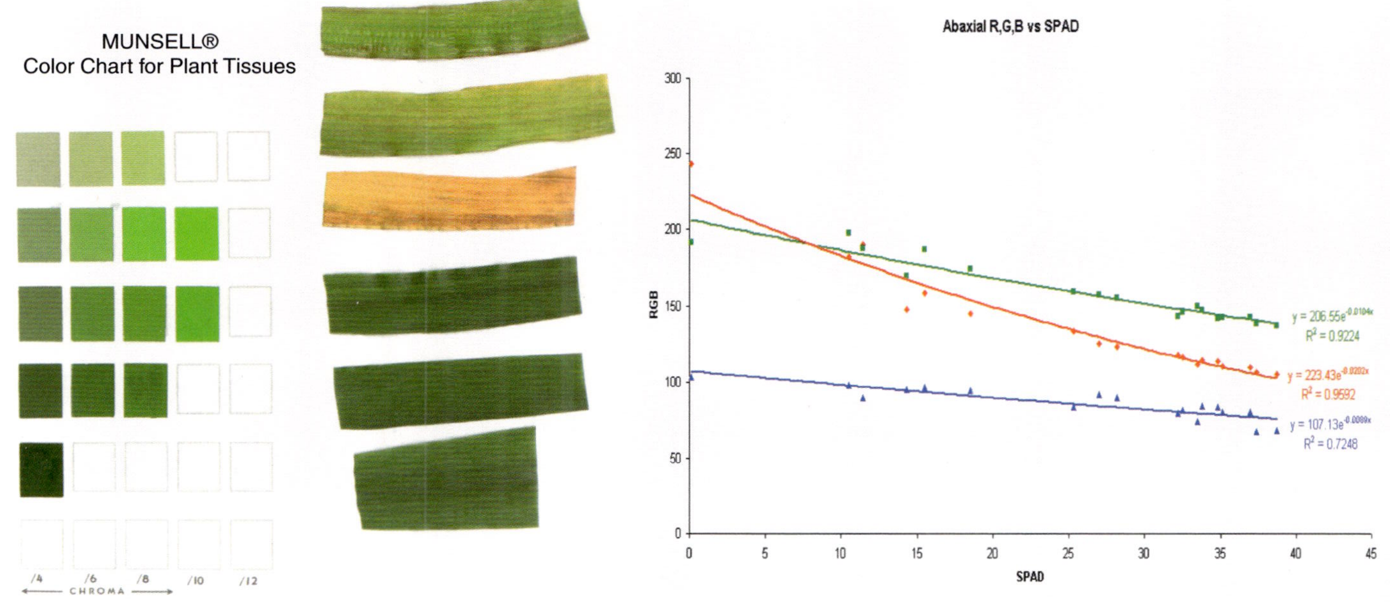

CHAPTER 19, FIGURE 5 Images of excised maize leaf tissues of different ages captured with a PC desk-top scanner. A Munsell color chart is included as a standard. The graph plots red, green, and blue values taken from scanned images of abaxial leaf surfaces against greenness as measured with a SPAD chlorophyll meter.

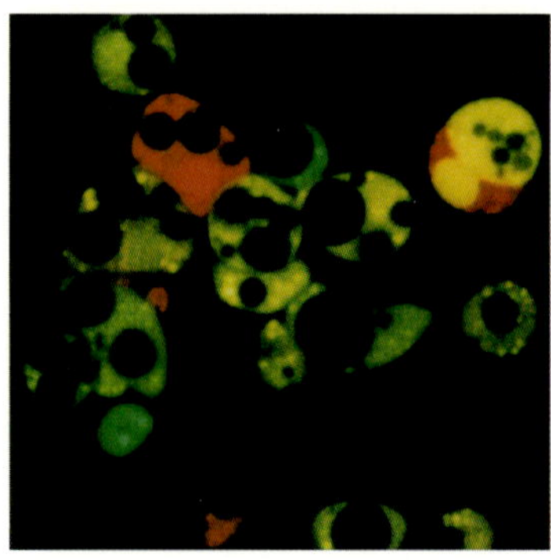

CHAPTER 21, FIGURE 5 Illustration of double fluorescein diacetate/propidium iodide staining of stalk cells. HMX44A cells viewed by confocal microscopy after a 24-h DIF incubation. Reproduced from Cornillon *et al*. (1994), with permission.

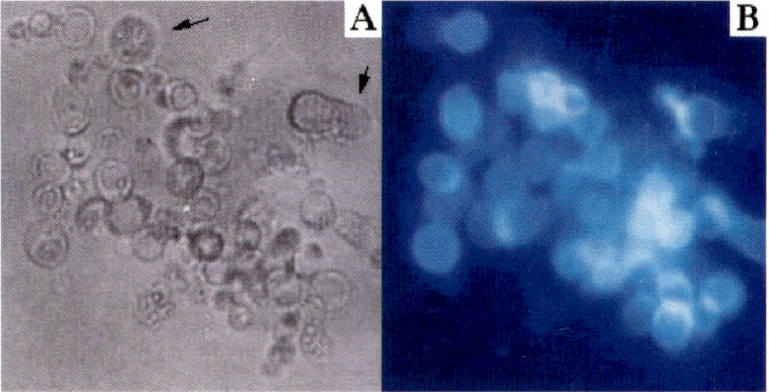

CHAPTER 21, FIGURE 6 Calcofluor staining of stalk cells. HMX44A cells incubated for 24 h in DIF on a glass coverslip. (A) Transmission image. (B) Blue fluorescence. Arrows point to cells that are not encased in cellulose (i.e., nonstalk cells).

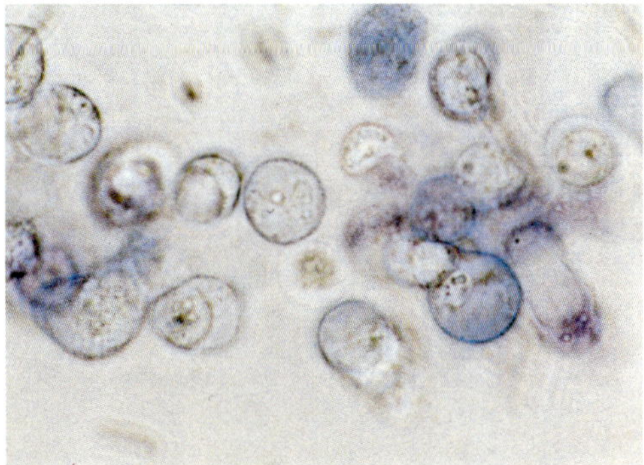

CHAPTER 21, FIGURE 7 Trypan blue staining of stalk cells. HMX44A cells incubated for 20 h in DIF on a glass coverslip.